BMW 5 Series (E28)

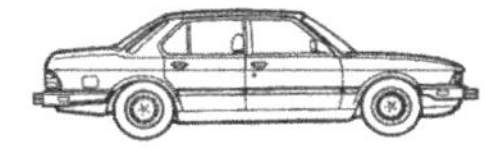

Service Manual

528e, 533i, 535i, 535is

1982, 1983, 1984, 1985, 1986, 1987, 1988

BentleyPublishers®
.com

Selected Books and Repair Information From Bentley Publishers

Motorsports

Alex Zanardi: My Sweetest Victory *Alex Zanardi and Gianluca Gasparini* ISBN 978-0-8376-1249-2

The Unfair Advantage *Mark Donohue and Paul van Valkenburgh* ISBN 978-0-8376-0069-7

Driving Forces: The Grand Prix Racing World Caught in the Malestrom of the Third Reich *Peter Stevenson* ISBN 978-0-8376-0217-2

Equations of Motion - Adventure, Risk and Innovation *William F. Milliken* ISBN 978-0-8376-1570-7

Engineering

Physics for Gearheads *Randy Beikmann* ISBN 978-0-8376-1615-5

Bosch Automotive Handbook *Robert Bosch GmbH* ISBN 978-0-8376-1732-9

Bosch Fuel Injection and Engine Management *Charles O. Probst, SAE* ISBN 978-0-8376-0300-1

Maximum Boost: Designing, Testing, and Installing Turbocharger Systems *Corky Bell* ISBN 978-0-8376-0160-1

Race Car Aerodynamics *Joseph Katz* ISBN 978-0-8376-0142-7

Scientific Design of Exhaust and Intake Systems *Phillip H. Smith & John C. Morrison* ISBN 978-0-8376-0309-4

Audi Repair Manuals

Audi A4 Service Manual: 2002-2008, 1.8L Turbo, 2.0L Turbo, 3.0L, 3.2L *Bentley Publishers* ISBN 978-0-8376-1574-5

Audi A4 Service Manual: 1996-2001, 1.8L Turbo, 2.8L *Bentley Publishers* ISBN 978-0-8376-1675-9

Audi TT Service Manual: 2000-2006, 1.8L turbo, 3.2 L *Bentley Publishers* ISBN 978-0-8376-1625-4

Audi A6 (C5 platform) Service Manual: 1998-2004 *Bentley Publishers* ISBN 978-0-8376-1670-4

BMW

Memoirs of a Hack Mechanic *Rob Siegel* ISBN 978-0-8376-1720-6

BMW X5 (E53) Service Manual: 2000-2006 *Bentley Publishers* ISBN 978-0-8376-1643-8

BMW X3 (E83) Service Manual: 2004-2010 *Bentley Publishers* ISBN 978-0-8376-1731-2

BMW 3 Series (E30) Service Manual: 1984-1990 *Bentley Publishers* ISBN 978-0-8376-1647-6

BMW 7 Series (E32) Service Manual: 1988-1994 *Bentley Publishers* ISBN 978-0-8376-1619-3

BMW 7 Series (E38) Service Manual: 1995-2001 *Bentley Publishers* ISBN 978-0-8376-1618-6

BMW Z3 (E36/7) Service Manual: 1996-2002 *Bentley Publishers* ISBN 978-0-8376-1617-9

Porsche

Porsche 911 (Type 996) Service Manual: 1999-2005 *Bentley Publishers* ISBN 978-0-8376-1710-7

Porsche 911 Carrera Service Manual: 1984-1989 *Bentley Publishers* ISBN 978-0-8376-1696-4

Porsche 911 SC Service Manual: 1978-1983 *Bentley Publishers* ISBN 978-0-8376-1705-3

Porsche 911 (964): Enthusiast's Companion *Bentley Publishers* ISBN 978-0-8376-0293-6

Porsche Boxster. Boxster S (986) Service Manual: 1997-2004 *Bentley Publishers* ISBN 978-0-8376-1645-2

Porsche: Excellence Was Expected *Karl Ludvigsen* ISBN 978-0-8376-0235-6

Ferdinand Porsche – Genesis of Genius *Karl Ludvigsen* ISBN 978-0-8376-1557-8

Porsche – The Man and His Cars *Richard von Frankenberg* ISBN 978-0-8376-0329-2

Porsche – Origin of the Species *Karl Ludvigsen* ISBN 978-0-8376-1331-4

Volkswagen

Volkswagen Jetta, Golf, GTI Service Manual: 1999-2005 *Bentley Publishers* ISBN 978-0-8376-1678-0

Volkswagen Jetta Service Manual: 2005-2010 *Bentley Publishers* ISBN 978-0-8376-1616-2

Volkswagen Passat Service Manual: 1998-2005 *Bentley Publishers* ISBN 978-0-8376-1669-8

Volkswagen Jetta, Golf, GTI: 1993-1999, Cabrio: 1995-2002 Service Manual *Bentley Publishers* ISBN 978-0-8376-1660-5

Volkswagen GTI, Golf, Jetta Service Manual: 1985-1992 *Bentley Publishers* ISBN 978-0-8376-1637-7

MINI Repair Manuals

MINI Cooper Service Manual: 2002-2006 *Bentley Publishers* ISBN 978-0-8376-1639-1

MINI Cooper Service Manual: 2007-2013 *Bentley Publishers* ISBN 978-0-8376-1730-5

Mercedes-Benz

Mercedes-Benz C-Class (W202) Service Manual 1994-2000 *Bentley Publishers* ISBN 978-0-8376-1692-6

Mercedes Benz E-Class (W124) Owner's Bible: 1986-1995 *Bentley Publishers* ISBN 978-0-8376-0230-1

Featured Repair Manuals

BMW 3 Series (E90, E91, E92, E93) Service Manual: 2006-2011 *Bentley Publishers* ISBN 978-0-8376-1723-7

BMW 5 Series (E60, E61) Service Manual: 2004-2010 *Bentley Publishers* ISBN 978-0-8376-1689-6

BMW 3 Series (E46) Service Manual: 1999-2005 *Bentley Publishers* ISBN 978-0-8376-1657-5

BMW 5 Series (E39) Service Manual: 1997-2003 *Bentley Publishers* ISBN 978-0-8376-1672-8

[B] BentleyPublishers®
Automotive Reference

Bentley Publishers has published service manuals and automobile books since 1950. For more information, please contact Bentley Publishers at 1734 Massachusetts Avenue, Cambridge, MA 02138 USA, or visit our web site at `BentleyPublishers.com`

BMW
5 Series
(E28) 

Service Manual
528e, 533i, 535i, 535is
1982, 1983, 1984, 1985, 1986, 1987, 1988

BentleyPublishers®
.com

[B] BENTLEY PUBLISHERS® | Automotive Reference™

Bentley Publishers, a division of Robert Bentley, Inc.
1734 Massachusetts Avenue
Cambridge, MA 02138 USA
800-423-4595 / 617-547-4170

Information that makes the difference®

BentleyPublishers®.com

Technical contact information

We welcome your feedback. Please submit corrections and additions to our technical discussion forum at:

`http://www.BentleyPublishers.com`

Updates and corrections

We will evaluate submissions and post appropriate editorial changes online as updates or tech discussion. Appropriate updates and corrections will be added to the book in future printings. Check for updates and corrections for this book before beginning work on your vehicle. See the following web address for additional information:

`http://www.BentleyPublishers.com/updates/`

WARNING—Important Safety Notice

Do not use this manual for repairs unless you are familiar with basic automotive repair procedures and safe workshop practices. This manual illustrates the workshop procedures for some maintenance and service work. It is not a substitute for full and up-to-date information from the vehicle manufacturer or for proper training as an automotive technician. Note that it is not possible for us to anticipate all of the ways or conditions under which vehicles may be serviced or to provide cautions as to all of the possible hazards that may result.

We have endeavored to ensure the accuracy of the information in this manual. Please note, however, that considering the vast quantity and the complexity of the service information involved, we cannot warrant the accuracy or completeness of the information contained in this manual.

FOR THESE REASONS, NEITHER THE PUBLISHER NOR THE AUTHOR MAKES ANY WARRANTIES, EXPRESS OR IMPLIED, THAT THE INFORMATION IN THIS MANUAL IS FREE OF ERRORS OR OMISSIONS, AND WE EXPRESSLY DISCLAIM THE IMPLIED WARRANTIES OF MERCHANTABILITY AND OF FITNESS FOR A PARTICULAR PURPOSE, EVEN IF THE PUBLISHER OR AUTHOR HAVE BEEN ADVISED OF A PARTICULAR PURPOSE, AND EVEN IF A PARTICULAR PURPOSE IS INDICATED IN THE MANUAL. THE PUBLISHER AND AUTHOR ALSO DISCLAIM ALL LIABILITY FOR DIRECT, INDIRECT, INCIDENTAL OR CONSEQUENTIAL DAMAGES THAT RESULT FROM ANY USE OF THE EXAMPLES, INSTRUCTIONS OR OTHER INFORMATION IN THIS MANUAL. IN NO EVENT SHALL OUR LIABILITY WHETHER IN TORT, CONTRACT OR OTHERWISE EXCEED THE COST OF THIS MANUAL.

Before attempting any work on your BMW, read the Warnings and Cautions on page ix and any warning or caution that accompanies a procedure in the manual. Review the warnings and cautions each time you prepare to work on your BMW.

Your common sense and good judgment are crucial to safe and successful service work. Read procedures through before starting them. Think about whether the condition of your car, your level of mechanical skill, or your level of reading comprehension might result in or contribute in some way to an occurrence which might cause you injury, damage your car, or result in an unsafe repair. If you have doubts for these or other reasons about your ability to perform safe repair work on your car, have the work done at an authorized BMW dealer or other qualified shop.

Part numbers listed in this manual are for identification purposes only, not for ordering. Always check with your authorized BMW dealer to verify part numbers and availability before beginning service work that may require new parts.

Special tools required to perform certain service operations are identified in the manual and are recommended for use. Use of improper tools may be detrimental to the car's safe operation as well as the safety of the person servicing the car.

The vehicle manufacturer will continue to issue service information updates and parts retrofits after the editorial closing of this manual. Some of these updates and retrofits will apply to procedures and specifications in this manual. We regret that we cannot supply updates to purchasers of this manual.

This manual is prepared, published and distributed by Bentley Publishers, 1734 Massachusetts Avenue, Cambridge, Massachusetts 02138 USA. All information contained in this manual is based on the information available to the publisher at the time of editorial closing. BMW has not reviewed and does not vouch for the accuracy or completeness of the technical specifications and work procedures described and given in this manual.

ISBN 978-0-8376-1694-0 **Bentley Stock No. B588** **Editorial Closing 07/1991** **Mfg. code: B588-15-D-1504**

Library of Congress Catalog Card No. 91-73492

The paper used in this publication is acid free and meets the requirements of the National Standard for Information Sciences-Permanence of Paper for Printed Library Materials. ∞

Manufactured in the United States of America.

202605R001

Contents:

Foreword

For the BMW owner with basic mechanical skills and for independent auto service professionals, this manual includes many of the specifications and procedures that were available in an authorized BMW dealer service department as this manual went to press. The BMW owner with no intention of working on his or her car will find that owning and referring to this manual will make it possible to be better informed and to more knowledgeably discuss repairs with a professional automotive technician.

The BMW owner intending to do maintenance and repair should have screwdrivers, a set of metric wrenches and sockets, and metric Allen and Torx wrenches, since these basic hand tools are needed for most of the work described in this manual. Many procedures will also require a torque wrench to ensure that fasteners are tightened properly and in accordance with specifications. Additional information on basic tools and other tips can be found in **1 Fundamentals**. In some cases, the text refers to special tools that are recommended or required to accomplish adjustments or repairs. These tools are usually identified by their BMW special tool number and illustrated.

Disclaimer

We have endeavored to ensure the accuracy of the information in this manual. When the vast array of data presented in the manual is taken into account, however, no claim to infallibility can be made. We therefore cannot be responsible for the result of any errors that may have crept into the text. Please also read the **Important Safety Notice** on the copyright page at the beginning of this book.

A thorough pre-reading of each procedure, and **WARNINGS and CAUTIONS** at the front of the book and those that accompany the procedure is essential. Reading a procedure before beginning work will help you determine in advance the need for specific skills, identify hazards, prepare for appropriate capture and handling of hazardous materials, and the need for particular tools and replacement parts such as gaskets.

Bentley Publishers encourages comments from the readers of this manual with regard to errors, and/or suggestions for improvement of our product. These communications have been and will be carefully considered in the preparation of this and other manuals. If you identify inconsistencies in the manual, you may have found an error. Please contact the publisher and we will endeavor to post applicable corrections on our website. Posted corrections (updates) should be reviewed before beginning work. Please see the following web address:

`http://www.BentleyPublishers.com/updates/`

BMW offers extensive warranties, especially on components of the fuel delivery and emission control systems. Therefore, before deciding to repair a BMW that may be covered wholly or in part by any warranties issued by BMW of North America, LLC, consult your authorized BMW dealer. You may find that the dealer can make the repair either free or at minimum cost. Regardless of its age, or whether it is under warranty, your BMW is both an easy car to service and an easy car to get serviced. So if at any time a repair is needed that you feel is too difficult to do yourself, a trained BMW technician is ready to do the job for you.

Bentley Publishers

Vehicle Identification and VIN Decoder

Vehicle Identification Number (VIN), decoding

Some of the information in this manual applies only to cars of a particular model year or range of years. For example, 1982 refers to the 1982 model year but does not necessarily match the calendar year in which the car was manufactured or sold. To be sure of the model year of a particular car, check the Vehicle Identification Number (VIN) on the car.

The VIN is a unique sequence of 17 characters assigned by BMW to identify each individual car. When decoded, the VIN tells the country and year of manufacture; make, model and serial number; assembly plant and even some equipment specifications.

The BMW VIN is on a plate mounted on the top of the dashboard, on the driver's side where the number can be seen through the windshield. The 10th character is the model year code. The letters I, O, Q and U are not used for model year designation for US cars. Examples: C for 1982, D for 1983, E for 1984, F for 1985, etc. The table below explains some of the codes in the VIN for 1982 through 1988 BMW E28 5 Series covered by this manual.

Sample VIN:	WBA	DC84	0	X	J	1	841989
position	1 2 3	4 5 6 7	8	9	10	11	12-17

VIN position	Description	Decoding information	
1	Country of Manufacture	W	Germany
2	Manufacturer	B	BMW AG
3	Manufacturing division	A S	BMW BMW Motorsport
4–7	Series, model	DB24 DB74 DB84 DC71 DC74 DC81 DC84 DC93 DK73 DK83	524tdA, 6-cylinder 2.4 liter (M21) 533i 6-cylinder 3.3 liter (M30) 533iA 6-cylinder 3.3 liter (M30) 535i Euro 6-cylinder 3.5 liter (M30) 535is 6-cylinder 3.5 liter (M30) 535iA Euro 6-cylinder 3.5 liter (M30) 535isA 6-cylinder 3.5 liter (M30) M5 6-cylinder 3.5 liter (S38) 528e 6-cylinder 2.7 liter (M20) 528eA 6-cylinder 2.7 liter (M20)
8	Restraint system	0 1 2	Manual belts Manual belts with supplemental restraint Manual belts with dual SRS airbags
9	Check digit		0 - 9 or X, calculated by NHTSA
10	Model year	C D E F G H J O	1982 1983 1984 1985 1986 1987 1988 European model
11	Assembly plant	0, 1, 2, 7, 9	Munich, Germany or Dingolfing, Germany or Regensburg, Germany
12-17	Serial number		Sequential production number for specific vehicle

Please read these warnings and cautions before proceeding with maintenance and repair work.

WARNINGS—
*See also **CAUTIONS** on next page.*

● Read the important safety notice on the copyright page at the beginning of the book.

● Some repairs may be beyond your capability. If you lack the skills, tools and equipment, or a suitable workplace for any procedure described in this manual, we suggest you leave such repairs to an authorized BMW dealer service department or other qualified shop.

● A thorough pre-reading of each procedure, and the **WARNINGS** and **CAUTIONS** that accompany the procedure is essential. Posted corrections (errata) should also be reviewed before beginning work. Please see `www.BentleyPublishers.com/updates/`

● BMW is constantly improving its cars. Sometimes these changes, both in parts and specifications, are made applicable to earlier models. Therefore, before starting any major jobs or repairs to components on which passenger safety may depend, consult your authorized BMW dealer about Technical Bulletins that may have been issued.

● Do not re-use any fasteners that are worn or deformed in normal use. Many fasteners are designed to be used only once and become unreliable and may fail when used a second time. This includes, but is not limited to, nuts, bolts, washers, self-locking nuts or bolts, circlips and cotter pins. Always replace these fasteners with new parts.

● Never work under a lifted car unless it is solidly supported on stands designed for the purpose. Do not support a car on cinder blocks, hollow tiles or other props that may crumble under continuous load. Never work under a car that is supported solely by a jack. Never work under the car while the engine is running.

● If you are going to work under a car on the ground, make sure that the ground is level. Block the wheels to keep the car from rolling. Disconnect the battery negative (–) terminal (ground strap) to prevent others from starting the car while you are under it.

● Never run the engine unless the work area is well ventilated. Carbon monoxide kills.

● Finger rings, bracelets and other jewelry should be removed so that they cannot cause electrical shorts, get caught in running machinery, or be crushed by heavy parts.

● Tie long hair behind your head. Do not wear a necktie, a scarf, loose clothing, or a necklace when you work near machine tools or running engines. If your hair, clothing, or jewelry were to get caught in the machinery, severe injury could result.

● Do not attempt to work on your car if you do not feel well. You increase the danger of injury to yourself and others if you are tired, upset or have taken medication or any other substance that may keep you from being fully alert.

● Illuminate your work area adequately but safely. Use a portable safety light for working inside or under the car. Make sure the bulb is enclosed by a wire cage. The hot filament of an accidentally broken bulb can ignite spilled fuel, vapors or oil.

● Catch draining fuel, oil, or brake fluid in suitable containers. Do not use food or beverage containers that might mislead someone into drinking from them. Store flammable fluids away from fire hazards. Wipe up spills at once, but do not store the oily rags, which can ignite and burn spontaneously.

● Always observe good workshop practices. Wear goggles when you operate machine tools or work with battery acid. Gloves or other protective clothing should be worn whenever the job requires working with harmful substances.

● Greases, lubricants and other automotive chemicals contain toxic substances, many of which are absorbed directly through the skin. Read the manufacturer's instructions and warnings carefully. Use hand and eye protection. Avoid direct skin contact.

● Disconnect the battery negative (–) terminal (ground strap) whenever you work on the fuel system or the electrical system. Do not smoke or work near heaters or other fire hazards. Keep an approved fire extinguisher handy.

● Friction materials (such as brake pads or shoes or clutch discs) contain asbestos fibers or other friction materials. Do not create dust by grinding, sanding, or by cleaning with compressed air. Avoid breathing dust. Breathing any friction material dust can lead to serious diseases and may result in death.

● Batteries give off explosive hydrogen gas during charging. Keep sparks, lighted matches and open flame away from the top of the battery. If hydrogen gas escaping from the cap vents is ignited, it will ignite gas trapped in the cells and cause the battery to explode.

● Battery acid (electrolyte) can cause severe burns. Flush contact area with water, seek medical attention.

● Connect and disconnect battery cables, jumper cables or a battery charger only with the ignition switched off. Do not disconnect the battery while the engine is running.

● Do not quick-charge the battery (for boost starting) for longer than one minute. Wait at least one minute before boosting the battery a second time.

● Do not allow battery charging voltage to exceed 16.5 volts. If the battery begins producing gas or boiling violently, reduce the charging rate. Boosting a sulfated battery at a high rate can cause an explosion.

● The air conditioning system is filled with chemical refrigerant, which is hazardous. The A/C system should be serviced only by trained technicians using approved refrigerant recovery/recycling equipment, trained in related safety precautions, and familiar with regulations governing the discharging and disposal of automotive chemical refrigerants.

continued on next page

Please read these warnings and cautions before proceeding with maintenance and repair work.

WARNINGS— (continued)

- Do not expose any part of the A/C system to high temperatures such as open flame. Excessive heat will increase system pressure and may cause the system to burst.

- Some aerosol tire inflators are highly flammable. Be extremely cautious when repairing a tire that may have been inflated using an aerosol tire inflator. Keep sparks, open flame or other sources of ignition away from the tire repair area. Inflate and deflate the tire at least four times before breaking the bead from the rim. Completely remove the tire from the rim before attempting any repair.

- Connect and disconnect a battery charger only with the battery charger switched off.

- Sealed or "maintenance free" batteries should be slow-charged only, at an amperage rate that is approximately 10% of the battery's ampere-hour (Ah) rating.

- Do not allow battery charging voltage to exceed 16.5 volts. If the battery begins producing gas or boiling violently, reduce the charging rate. Boosting a sulfated battery at a high charging rate can cause an explosion.

- The ignition system produces high voltages that can be fatal. Avoid contact with exposed terminals and use extreme care when working on a car with the engine running or the ignition switched on.

- Place jack stands only at locations specified by manufacturer. The vehicle lifting jack supplied with the vehicle is intended for tire changes only. A heavy duty floor jack should be used to lift vehicle before installing jack stands. See **2 Lubrication and Maintenance**.

- Aerosol cleaners and solvents may contain hazardous or deadly vapors and are highly flammable. Use only in a well ventilated area. Do not use on hot surfaces (engines, brakes, etc.).

- Do not remove coolant reservoir or radiator cap with the engine hot. Danger of burns and engine damage.

CAUTIONS—
*See also **WARNINGS** on previous page.*

- If you lack the skills, tools and equipment, or a suitable workshop for any procedure described in this manual, we suggest you leave such repairs to an authorized BMW dealer or other qualified shop.

- BMW is constantly improving its cars and sometimes these changes, both in parts and specifications, are made applicable to earlier models. Therefore, part numbers listed in this manual are for reference only. Always check with your authorized BMW dealer parts department for the latest information.

- Before starting a job, make certain that you have all the necessary tools and parts on hand. Read all the instructions thoroughly, and do not attempt shortcuts. Use tools appropriate to the work and use only replacement parts meeting BMW specifications. Makeshift tools, parts and procedures will not make good repairs

- Use pneumatic and electric tools only to loosen threaded parts and fasteners. Never use these tools to tighten fasteners, especially on light alloy parts. Always use a torque wrench to tighten fasteners to the tightening torque specification listed.

- Be mindful of the environment and ecology. Before you drain the crankcase, find out the proper way to dispose of the oil. Do not pour oil onto the ground, down a drain, or into a stream, pond or lake. Dispose of in accordance with Federal, State and Local laws.

- The control module for the anti-lock brake system (ABS) cannot withstand temperatures from a paint-drying booth or a heat lamp in excess of 203°F (95°C) and should not be subjected to temperatures in excess of 185°F (85°C) for more than two hours.

- Before doing any electrical welding on cars equipped with ABS, disconnect the battery negative (–) terminal (ground strap) and the ABS control module connector.

- Always make sure ignition is off before disconnecting battery.

- Label battery cables before disconnecting. On some models, battery cables are not color coded.

- Disconnecting the battery may erase fault code(s) stored in control module memory. Using special BMW diagnostic equipment, check for fault codes prior to disconnecting the battery cables.

- If a normal or rapid charger is used to charge battery, the battery must be disconnected and removed from the vehicle in order to avoid damaging paint and upholstery.

- Do not quick-charge the battery (for boost starting) for longer than one minute. Wait at least one minute before boosting the battery a second time.

- Connect and disconnect a battery charger only with the battery charger switched off.

- Sealed or "maintenance free" batteries should be slow-charged only, at an amperage rate that is approximately 10% of the battery's ampere-hour (Ah) rating.

- Do not allow battery charging voltage to exceed 16.5 volts. If the battery begins producing gas or boiling violently, reduce the charging rate. Boosting a sulfated battery at a high charging rate can cause an explosion.

FUNDAMENTALS

Contents

INTRODUCTION

This section of the manual is dedicated to helping the beginner get started smartly and safely with BMW maintenance and repair. The section begins with a **General Description** of the car, broken down into its individual systems, and a discussion on **How To Use This Manual**. It is a simple directory of the kind of information you can expect to find, and where to find it.

Safety and **General Advice For The Beginner** include tips on mechanic's skills and workshop techniques that can help the beginner do a faster and more thorough job. **Tools** describes the basic tools needed to do 90% of the procedures in this manual, and includes advice on how to buy tools wisely and use them effectively.

Finally, and once again of interest to any owner, this section ends with a quick reference guide to emergencies—what to do when the car won't start or when a warning light comes on, including basic troubleshooting and information on how to gauge the seriousness of a problem.

1. GENERAL DESCRIPTION

BMWs are sophisticated examples of today's automotive engineering, blending advanced design and manufacturing to provide an outstanding combination of performance, roadholding, and reliability. In spite of this sophistication, much of the necessary maintenance and repair can be accomplished by the average owner using this manual. While the complexity of the car may seem to make this a difficult challenge for the novice mechanic, it can be simplified and more easily understood by viewing the car as an assembly of simpler systems, each performing its own independent functions.

1.1 Body

The body is the basic building block. All of the BMW models covered in this manual feature unitized body construction, meaning that they do not have a separate frame.

A complex body shell is the main structural platform to which all the other systems are attached. See Fig. 1-1. Subassemblies attach engine, drivetrain, suspension, and steering systems to the basic body structure.

The doors, the instrument panel, the seats, and other interior trim pieces are also added to the body shell. Other parts of the body shell function as mounting points for the other major and minor subsystems. For more information, see **BODY AND INTERIOR**.

1.2 Engine

The engine produces the power to move the car. It burns a precise mixture of fuel and air, converting the fuel's stored en-

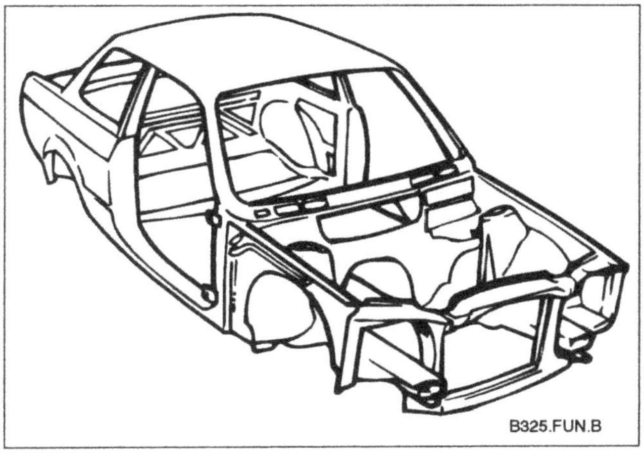

Fig. 1-1. BMW unitized body.

ergy into mechanical work, and delivering that mechanical work in a useful form.

All of the BMW engines covered in this manual are of reciprocating-piston design and operate on the four-stroke cycle. The combustion of the air-fuel mixture creates tremendous pressure in a closed space above a piston. This pressure forces the piston downward in its cylinder, translating the energy of combustion into mechanical force.

The crankshaft converts each piston's up and down motion into rotating motion, in much the same way that the up-and-down motion of a person's legs rotates the pedals of a bicycle. The power transmitted in this rotary form can then be used to move the car. The four-stroke cycle, the heart of how and why this all happens, is illustrated in Fig. 1-2.

Intake Stroke. The piston, traveling downward, creates low pressure inside the cylinder. With the intake valve open, this low pressure causes the fresh air-fuel mixture to rush in. When the piston is near the bottom of its travel, the intake valve closes, sealing the air and vaporized fuel in the cylinder.

Compression Stroke. As the piston begins its upward travel in the sealed cylinder, the air-fuel mixture is compressed to a small percentage of its original volume, creating a very flammable mixture in a very small space. This space is referred to as the combustion chamber. Just before the piston reaches the top of its travel, the air-fuel mixture is ignited by a precisely timed spark, and burns very rapidly.

Combustion or **Power Stroke.** As the confined air-fuel mixture burns, temperature and pressure rise very rapidly, forcing the piston downward, turning combustion energy into work. Generally, the faster an engine runs and the more often this combustion cycle happens, the more power is produced.

Exhaust Stroke. At the end of the power stroke, the piston is near the bottom of its travel and the cylinder is filled with the waste products of combustion. The exhaust valve opens and the piston, now traveling back upward, pushes the burned gasses out into the exhaust system. Near the top of the piston's travel, the intake valve opens, the exhaust valve closes, and the process begins anew with another intake stroke.

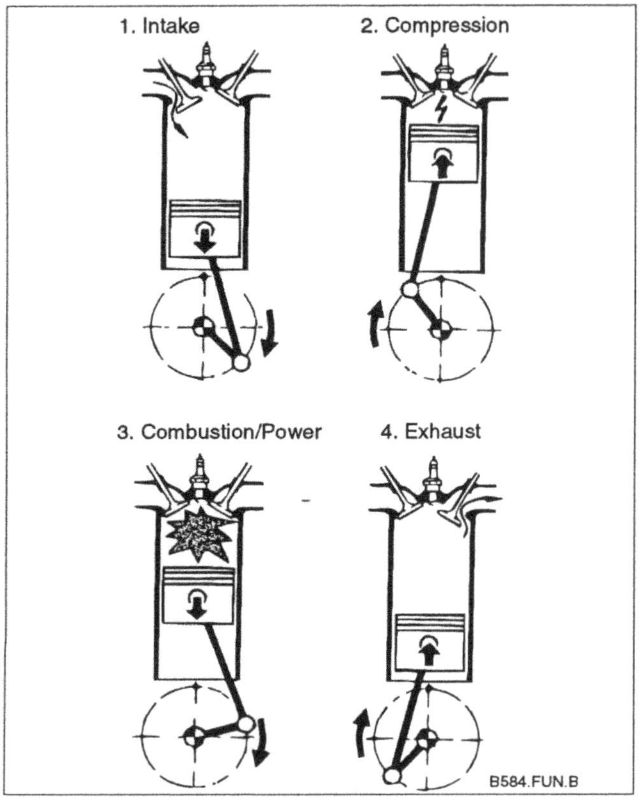

Fig. 1-2. The four-stroke cycle.

Engine Systems

The engine, which seems so complex, is a collection of simpler systems whose sole purpose is to efficiently convert energy into mechanical force and motion. All automobile engines are multi-cylinder designs, in which a number of individual pistons and cylinders are joined together in a common housing or cylinder block, transmitting their power to a common crankshaft.

A camshaft, driven by the crankshaft, opens and closes the intake and exhaust valves in a precisely timed sequence. Since each valve must cycle open and closed once for every two turns of the crankshaft, camshafts always turn at one-half crankshaft speed.

The rest of the engine assembly is made up of systems that supply the essential fuel, air, ignition, and lubrication to provide continuous operation of the pistons, crankshaft, and valves. For more information, see **ENGINE—SERVICE AND REPAIR** or **ENGINE—RECONDITIONING**.

The ignition system creates the high-voltage spark necessary to ignite the combustible air-fuel mixture in the cylinders. The ignition coil boosts the voltage so that the spark will be hot enough to ignite the air-fuel mixture. The Motronic control unit controls the timing of the spark.

The ignition distributor, synchronized to the rotation of the engine, delivers the spark to the right cylinder at precisely the right time. Since each cylinder has to have a spark once for every two revolutions of the crankshaft, the distributor always turns at one-half crankshaft speed. The basic system is shown schematically in Fig. 1-3. For more information, see **IGNITION**.

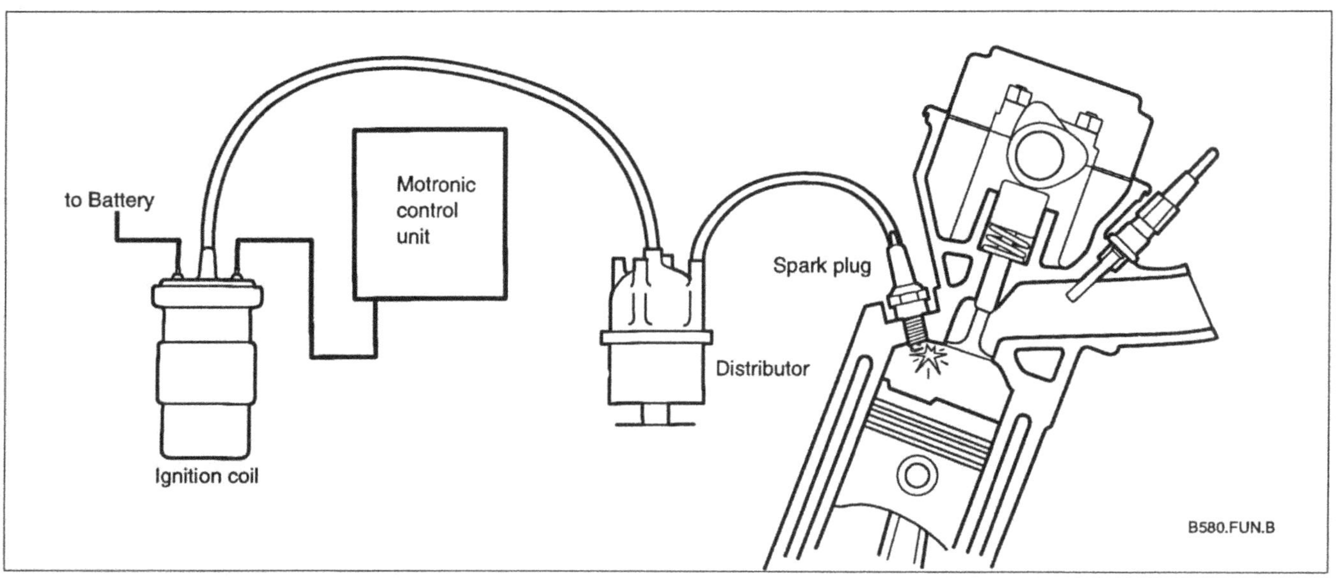

Fig. 1-3. Schematic representation of typical ignition system.

Fuel System

To run smoothly and produce power most efficiently, the engine requires the proper mixture of air and fuel. Depending on conditions, the optimum ratio for gasoline-fueled engines is about 14:1, fourteen parts of air for every one part of fuel. The throttle controls the amount of air entering the engine. The fuel system's job is to deliver and disperse fuel in the proper ratio to the incoming air.

Traditionally, the job of fuel delivery has been handled by a carburetor, a device carefully calibrated to dispense and atomize fuel in proportion to the amount of air passing through it. To meet the increasing demand for performance with economy and reduced exhaust emissions, many modern engines, including the BMW engines covered in this manual, use a more sophisticated fuel injection system.

The fuel injection system measures the incoming air more precisely than a carburetor, and in turn meters fuel more precisely for better control of the air-fuel ratio. This precise control means greater efficiency over a wider variety of operating conditions. In spite of all this sophistication, the fundamental task of the fuel injection system is to control the mixture of air and fuel entering the engine's combustion chambers.

The fuel system also includes a means of fuel storage. The fuel tank and a network of pump and lines transfer the fuel from the tank to the injection system. For more information on the fuel system, see **FUEL SUPPLY** and **FUEL INJECTION**.

Cooling System

Even the most advanced engines lose some of their combustion energy as heat, which must be dissipated to prevent damage to the engine parts. Some heat is carried away in the exhaust, but much of it is absorbed by the valves, the pistons, and the rest of the combustion chamber. Most modern automobile engines are liquid-cooled, using a network of passages around the cylinders and combustion chambers, filled with circulating water-based coolant, to carry away heat.

Coolant is circulated by an engine-driven pump, often called the water pump. The heat which the coolant absorbs from the hot engine is eventually dissipated to the surrounding atmosphere by the radiator at the front of the car.

A smaller radiator-like heater core, located near the interior of the car, radiates heat to warm the passenger compartment. A basic cooling system layout, similar to that used on cars covered by this manual, is shown in Fig. 1-4.

Since some heat is necessary for the engine to run most efficiently, a thermostat in the cooling system restricts the flow of coolant through the radiator until the engine has reached normal operating temperature. For more information on the entire engine cooling system, see **COOLING SYSTEM**.

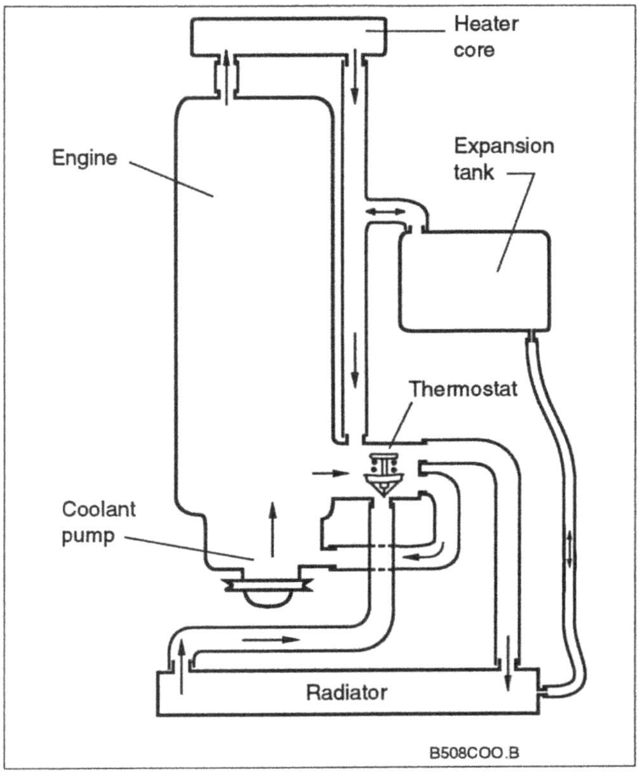

Fig. 1-4. Cooling system. Heat from engine is absorbed by coolant which is circulated by coolant pump and cooled by radiator.

Lubrication System

The crankshaft and camshaft rotate at speeds up to several thousand revolutions per minute (rpm). Valves and pistons accelerate at tremendous rates, abruptly changing direction between velocities of hundreds of feet per second.

In order to endure these harsh conditions, engine parts are manufactured to exact dimensions, assembled with precision clearances, and lubricated by a pressurized oiling system. The moving parts ride on a cushion of oil instead of each other. An engine-driven oil pump supplies oil under pressure to the engine where it is routed through a network of small passages that deliver it to each critical bearing surface.

The oil system includes a filter to clean the oil and a system to warn the driver of low oil level and pressure. A secondary function of the lubricating oil is to help carry away excess heat. More information on the lubrication system is found in **ENGINE—SERVICE AND REPAIR**. For information on oil and oil filter replacement intervals, see **LUBRICATION AND MAINTENANCE**.

Exhaust System

The exhaust system serves several functions, but the primary one is to carry spent combustion gasses from the engine and route them safely away from the passenger compartment. Modern exhaust systems include mufflers to reduce noise, chemically reactive components (catalytic converters) to reduce harmful emissions, and a sensor exposed to the exhaust gasses that provides feedback to the fuel injection system about engine efficiency. Fig. 1-5 shows a typical BMW exhaust system. For more information, see **EXHAUST SYSTEM**.

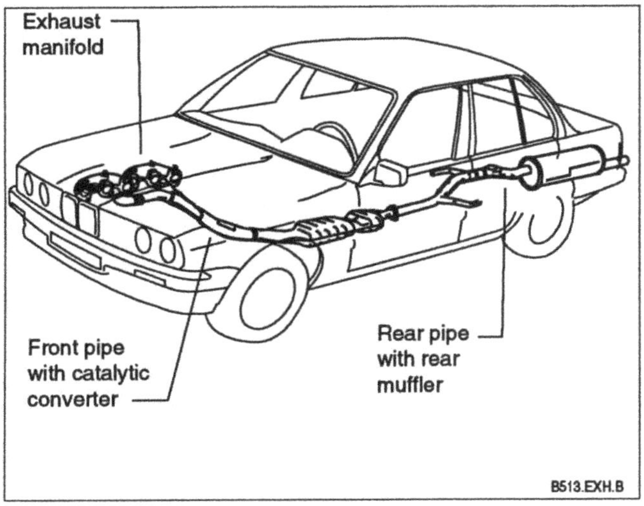

Fig. 1-5. Typical exhaust system.

1.3 Drivetrain

The drivetrain is a series of mechanisms that take the power developed by the engine and deliver it to the wheels in order to move the car. It consists of the clutch or torque converter, the transmission, the final drive, the drive axles, and the wheels and tires.

Clutch or Torque Converter

In a car with manual transmission, the clutch provides a way to connect and disconnect the engine from the drivetrain. The clutch assembly includes the spring-loaded pressure plate, the friction disc or clutch disc, and the engine flywheel.

The clutch disc, attached to the input shaft of the transmission, gets squeezed between the heavily spring-loaded pressure plate and the flywheel, both attached to the engine. The friction between pressure plate, clutch disc and flywheel, boosted by the heavy spring force, makes the transmission input shaft turn at the same speed as the engine.

For stopping, starting, and shifting gears, depressing the clutch pedal works against the spring force, relieving the friction bond and disconnecting the engine from the drivetrain. In normal use the clutch and flywheel wear, much like brakes, and need periodic maintenance or replacement. More information on the clutch is found in **MANUAL TRANSMISSION AND CLUTCH**.

In a car with automatic transmission, the friction clutch is replaced by a sophisticated fluid clutch called a torque converter. One part is attached to the engine and another to the transmission. As one part turns, power is transmitted to the other by a viscous fluid (automatic transmission fluid).

The design of the torque converter allows the engine to turn at idle speeds without transmitting much driving force to the transmission. A slight forward tug may be noticed when the car is in a forward gear, but at idle speed it is barely enough to drive the car. Above idle, the torque converter becomes increasingly resistant to slip, and transmits power through the fluid coupling to the transmission, delivering power through the rest of the drivetrain to the wheels. For more information, see **AUTOMATIC TRANSMISSION**.

Transmission

Although the engine develops a substantial amount of power, it does so best at relatively high revolutions per minute (rpm). To handle all driving conditions, it is necessary to use gearing to change the ratio of engine rpm to vehicle speed. A manual transmission arranges several sets of gears in a common housing.

A set of two gears determines a gear ratio, each suited to a particular range of driving speeds. A shifting mechanism allows the driver to change from one gear ratio to the next to match vehicle speed. For information on manual transmission maintenance and adjustments, see **MANUAL TRANSMISSION AND CLUTCH**.

In an automatic transmission, hydraulic fluid under pressure in a complex network of passages, valves and control mechanisms engage and disengage constantly meshed planetary gear sets. Hydraulic controls responding to vehicle speed, engine load, throttle position and gear shift position select the appropriate gear ratio. For information on automatic transmission maintenance and adjustments, see **AUTOMATIC TRANSMISSION**.

Driveshaft and Final Drive

The driveshaft transmits power from the engine and transmission to the final drive. The final drive is a gearset which transmits power to the drive axles.

When a car turns, the wheels on the outside of the turn have to turn slightly faster than those on the inside, since they have to travel a larger arc in an equal amount of time. The drivetrain must be able to transmit power to the wheels and still allow for these variations in wheel speed when cornering. The final drive includes a device called the differential, which allows wheels on opposite sides of the car to turn at different speeds. For more information see **DRIVESHAFT AND FINAL DRIVE**.

Drive Axles

The final step in the transfer of power from the engine to the wheels is the drive axles, which provide a connection between the differential and the wheel hubs. Information on the drive axles is found in **SUSPENSION—REAR**.

1.4 Suspension and Steering

Fig. 1-6 shows a typical front suspension and steering system, and their proximity to other systems in the car.

The suspension and steering systems are what allow the wheels to move and turn for a smooth ride, stability and directional control.

The suspension system is the combination of springs, shock absorbers, and other stabilizing devices that support the weight of the car and cushion the effects of bumps. For added control, the suspension system also includes dampers, or shock absorbers, which resist excessive movement of the springs. Stabilizer bars aid stability by transferring some of the cornering force acting on the suspension.

BMW's strut-type front suspension, like that of many modern cars, combines the spring and shock absorber into a single unit, performing the same jobs in less space and with fewer individual components.

The remainder of the suspension system are the parts that link it all together, designed with bushings, bearings, and joints which purposely allow or restrict movement.

The steering system is an assembly of gearbox mechanisms and linkages which translate the rotating motion of the steering wheel into the side-to-side motion of the front wheels. Cars covered by this manual use a recirculating ball type steering mechanism. The assembly is also referred to as the steering gearbox. Power-assisted steering uses hydraulic fluid under pressure to do some of the work normally done by the driver turning the steering wheel. For more information, see **STEERING AND WHEEL ALIGNMENT**.

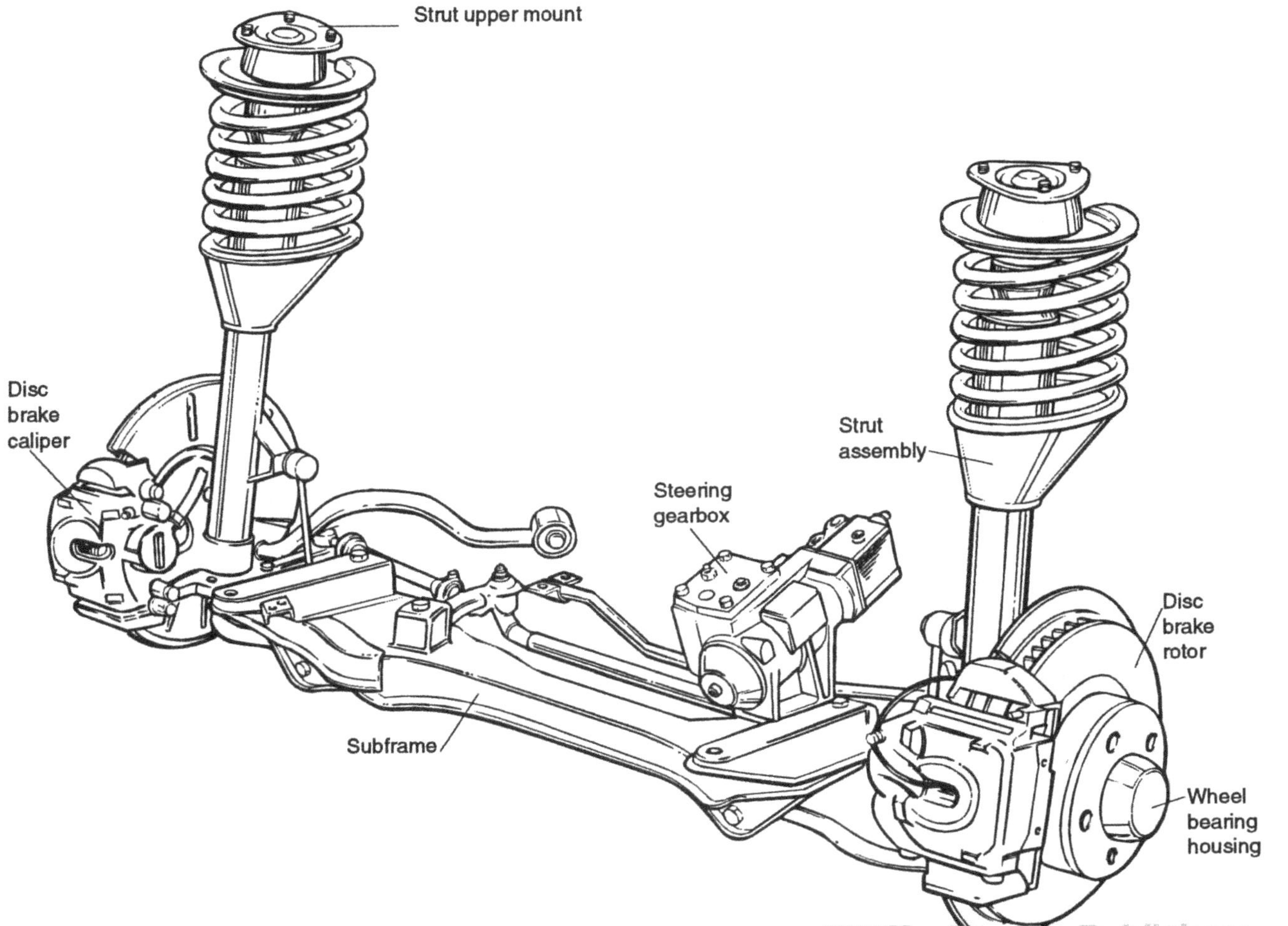

Fig. 1-6. Front suspension and steering system.

1.5 Brakes

The system for slowing and stopping the car is, not surprisingly, completely independent of the systems which make it go. Although the brakes are located at the wheels and mounted to parts of the suspension system, the brakes and suspension are completely separate systems.

The brakes act to slow or stop the car by causing friction. Since cars are relatively heavy, the friction required to stop safely and effectively is quite high, and generating this friction requires considerable force. The cars covered by this manual use either an engine vacuum boost system or hydraulic boost system to multiply the force applied to the brake pedal and to distribute it uniformly to the wheels.

The brake pedal is connected by a mechanical linkage to the first major hydraulic component, the master cylinder, mounted on the firewall at the back of the engine compartment. A piston in the master cylinder creates hydraulic pressure in the brake lines going to the wheels. Because the brakes are located at the wheels and move relative to the body, the final length of brake line at each wheel is flexible, so that it follows that movement.

At each wheel, the hydraulic pressure acts on the brake caliper to cause friction and slow the wheel. The sizes of the hydraulic components are such that the driver's force applied to the brake pedal is multiplied many times by the time it reaches the wheels.

All of the BMW models covered in this manual feature disc brakes at all four wheels. A disc brake squeezes pads lined with friction material against both sides of a flat, round brake disc, called a rotor. A disc brake assembly is shown below in Fig. 1-7.

Heat is a major enemy of brake efficiency. It affects the friction materials ability to grip. Under extreme conditions, excess heat from repeated heavy braking can cause the brake fluid to boil resulting in severely diminished braking performance, or brake fade. Disc brakes, because of their greater ability to dissipate heat, are more resistant to brake fade.

Anti-lock Braking System (ABS)

1985 535i models and all 1986 and later models covered in this manual are equipped with an anti-lock braking system (ABS). As the name implies, the purpose of this system is to prevent the wheels from locking during hard braking. Speed sensors at each wheel sense when the wheel is about to lock, and an electronic system modulates the braking force to that wheel.

Contrary to popular belief, research and testing have shown that the tires brake most effectively just before the point of locking up and skidding. Preventing the wheels from locking helps maintain directional control in emergencies, and is especially beneficial on slippery roads.

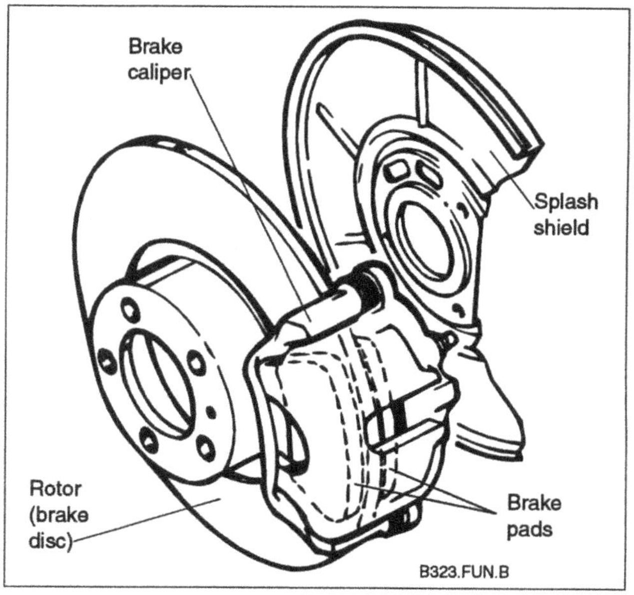

Fig. 1-7. Disc brake assembly showing disc, caliper, and splash shield. Caliper assembly holds pads with friction material.

1.6 Electrical System

Many components, and all electrical accessories, are powered by the car's electrical system. The electrical system uses a battery to store energy, an engine-driven alternator to generate electricity and recharge the battery, and various wiring harnesses and other circuits to distribute electric power to the rest of the car. The electrical system is represented in Fig. 1-8.

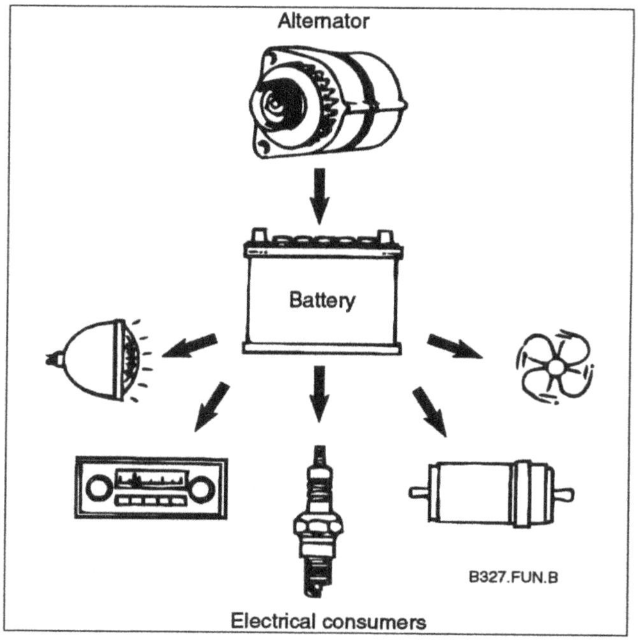

Fig. 1-8. The alternator generates electricity to recharge battery and power other electrical consumers.

Battery

Almost every electrical component in the car operates from 12-volt direct current (VDC). The battery converts electrical energy into chemical energy for storage, and converts its stored chemical energy back into electrical energy on demand.

Alternator

Left alone to meet all the electrical demands of an automobile, the battery would soon be completely discharged, so the electrical system includes a charging system.

The main component of the charging system is the alternator. Turned by the engine via a V-belt, the alternator takes over from the battery to supply electrical energy to the various electrical components. When the alternator generates more power than is needed, as it usually does at driving speeds, the extra energy recharges the battery.

Wiring Harness and Circuits

The flow of electricity depends upon a closed-loop path—a complete circuit. Electrical current flows through wires to the consumer, a light bulb for example, and back to the battery in a complete circuit. The electrical route back to the source, which completes the circuit, is called a path to ground. Every consumer of electrical power in the car must have a source of power and a path to ground in order to operate.

Commonly, the electrically conductive metal structure of the automobile is used as a ground path. The negative (–) terminal of the battery connects to the car body, and all of the electrical consumers in the car make a ground connection to the car body, thus eliminating the need for many feet of additional wire.

Electrical components near the engine are often grounded directly to the engine, which is then grounded to the body. Some components are grounded through their housings which are bolted to a ground. Electrically, the effect is the same.

2. How To Use This Manual

The manual is divided into 22 sections, **FUNDAMENTALS, LUBRICATION AND MAINTENANCE, ENGINE MANAGEMENT—DRIVEABILITY,** and 19 repair sections, each covering a particular system or portion of the car. Thumb tabs on the page margins help locate each section. A page listing section titles and showing their thumb-tab locations is near the front of the manual. An index is located at the back of the manual.

Each section has a Table of Contents listing the major subject headings within the section, and the pages on which they begin. Page numbers in the Table of Contents always refer to pages within that section. References to other numbered headings always refer to headings in the same section. Reference to a procedure in another section is by section title only, which will be in **BOLD TYPE.**

2.1 Fundamentals

This first section is **FUNDAMENTALS.** It contains basic information on equipment and safety which is important to any do-it-yourselfer, regardless of experience, as well as information on getting started and helpful suggestions for the novice. Anyone can use this manual. This section helps show how.

2.2 Lubrication and Maintenance

LUBRICATION AND MAINTENANCE is the section dedicated to taking care of the car and preventing future problems. BMW specifies certain periodic maintenance to prevent trouble and keep the car at its best. This section describes those maintenance tasks, shows how they are done, and tells what is needed to do them.

2.3 Engine Management—Driveability

In today's modern engines, the functions of the ignition system, the fuel delivery system, and the exhaust and emission control systems are closely related. On all of the cars covered by this manual, these subsystems are all controlled by one electronic control unit. It manages the related functions together to deliver the best possible combination of performance, fuel economy and clean exhaust, hence the term "engine management."

"Driveability" is a term used to describe the overall performance of the car, its ability to start quickly, run and accelerate smoothly, and deliver fuel economy and low exhaust emissions as well as power. Because engine management functions are so interrelated, it is often difficult to isolate the cause of a driveability problem.

The **ENGINE MANAGEMENT—DRIVEABILITY** section is intended to help the reader diagnose and remedy driveability problems using a logical, systematic approach. In other words, this section is a combined troubleshooting section for the engine and the subsystems responsible for driveability.

2.4 Repair Sections

The repair sections contain the more involved and more detailed information about system function, troubleshooting, and repair. For clarity and ease of use, each repair section begins with **1. General Description, 2. Maintenance,** and **3. Troubleshooting.**

General Description

The General Description is an overview of the system's technical features. It describes the general layout and function of the system, discusses unique aspects of different versions, and gives information on identifying each version and the repair information that applies to it.

Maintenance

Maintenance is a brief checklist of all routine maintenance specified by BMW for the system(s) being discussed. The listed maintenance items also include references to parts of the manual where particular maintenance procedures are described in detail.

Troubleshooting

A systematic approach to problem solving, based on carefully observing symptoms and isolating their causes, is called troubleshooting. Troubleshooting in each repair section begins with a discussion of the system's basic operating principles. Following that general discussion is a more specific list of symptoms—particular problems that may affect the car—and their probable causes. Suggested corrective actions include references to the numbered heading or section where the repair information can be found.

2.5 Index

A comprehensive index is found at the back of the manual. Each index entry is followed by a page reference giving the section and the section page number. For example, **4:16** refers to section four, **ENGINE—SERVICE AND REPAIR**, page 16.

2.6 Notes, Cautions, and Warnings

Throughout this manual are many passages with the headings **NOTE**, **CAUTION**, or **WARNING**. These very important headings have different meanings.

> ***WARNING —***
> *A warning is the most serious of the three. It warns of unsafe practices that are very likely to cause injury, either by direct threat to the person(s) doing the work or by increased risk of accident or mechanical failure while driving.*

> ***CAUTION —***
> *A caution calls attention to important precautions to be observed during the repair work that will help prevent accidentally damaging the car or its parts.*

> **NOTE —**
> A note contains helpful information, tips that will help in doing a better job and completing it more easily.

Please read every **NOTE**, **CAUTION**, and **WARNING** at the front of the manual and as they appear in repair procedures. They are very important. Read them before you begin any maintenance or repair job.

Some **CAUTION**s and **WARNING**s are repeated wherever they apply. Read them all. Do not skip any. These messages are important, even to the owner who never intends to work on the car.

3. GETTING STARTED

Most of the necessary maintenance and minor repair that a BMW will need can be done with ordinary tools, even by owners with little or no experience in car repair. Below is some important information on how to work safely, a discussion of what tools will be needed and how to use them, and a series of mechanic's tips on methods and workmanship.

3.1 Safety

Although an automobile presents many hazards, common sense and good equipment can ensure safety. Accidents happen because of carelessness. Pay attention and stick to these few important safety rules.

> ***WARNING —***
> - *Never run the engine in the work area unless it is well-ventilated. The exhaust should be vented to the outside. Carbon Monoxide (CO) in the exhaust kills.*
>
> - *Remove all neckties, scarfs, loose clothing, or jewelry when working near running engines or power tools. Tuck in shirts. Tie long hair and secure it under a cap. Severe injury can result from these things being caught in rotating parts.*
>
> - *Remove rings, watches, and bracelets. Aside from the dangers of moving parts, metallic jewelry conducts electricity and may cause shorts, sparks, burns, or damage to the electrical system when accidentally contacting the battery or other electrical terminals.*
>
> - *Disconnect the battery negative (–) terminal whenever working on the fuel system or anything that is electrically powered. Accidental electrical contact may damage the electrical system or cause fire.*

WARNING —
• Never work under a lifted car unless it is solidly supported on jack stands that are intended for that purpose. Do not support a car on cinder blocks, bricks, or other objects that may shift or crumble under continuous load. Never work under a car that is supported only by the lifting jack.

• The fuel system is designed to retain pressure even when the ignition is off. When working with the fuel system, loosen the fuel lines very slowly to allow the residual pressure to dissipate gradually. Avoid spraying fuel.

• Fuel is highly flammable. When working around fuel, do not smoke or work near heaters or other fire hazards. Keep an approved fire extinguisher handy.

• Illuminate the work area adequately and safely. Use a portable safety light for working inside or under the car. A fluorescent type is best because it gives off less heat. If using a light with a normal incandescent bulb, use rough service bulbs to avoid breakage. The hot filament of an accidentally broken bulb can ignite spilled fuel or oil.

• Keep sparks, lighted matches, and open flame away from the top of the battery. Hydrogen gas emitted by the battery is highly flammable. Any nearby source of ignition may cause the battery to explode.

• Never lay tools or parts in the engine compartment or on top of the battery. They may fall into confined spaces and be difficult to retrieve, become caught in belts or other rotating parts when the engine is started, or cause electrical shorts and damage to the electrical system.

• Some of the cars covered by this manual may be equipped with a Supplemental Restraint System (SRS) that automatically deploys an airbag. The airbag unit uses a pyrotechnic device to electrically ignite a powerful gas. On cars so equipped, any work involving the steering wheel should only be performed by an authorized BMW dealer. Performing repairs without disarming the SRS may cause serious personal injury.

Lifting The Car

For those repairs that require raising the car, the proper jacking points should be used to raise the car safely and avoid damage. To use the jack supplied with the car by BMW for changing wheels, there are four jacking points, two on each side of the car just behind the front wheel and just in front of the rear wheel. See Fig. 3-1. Use the same jacking points with a protective jack pad (wood, rubber. etc.) to lift the car with a floor jack or hydraulic lift.

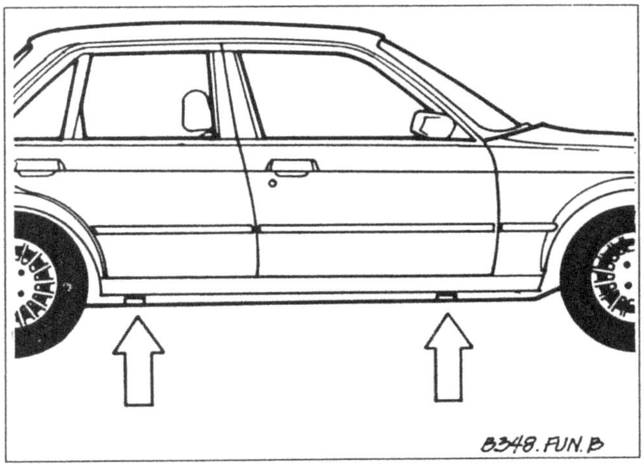

Fig. 3-1. Jacking points for use with BMW-supplied jack.

CAUTION —
• When raising the car at the rear jacking points using a floor jack or a hydraulic lift, carefully position the jack pad so that it does not contact the fuel tank. A suitable liner (wood, rubber. etc.) should be placed between the jack and the car so that the underbody will not be damaged

• Operating the car in gear while the rear wheels are suspended will cause damage to the axle shafts.

To raise the car safely:

1. Park the car on a flat, level surface.

2. Place the jack in position. Make sure the jack is resting on flat, solid ground. Use a board or other support to provide a firm surface for the jack, if necessary.

3. Raise the car slowly.

   WARNING —
   Watch the jack closely. Make sure that it stays stable and does not shift or tip. As the car is raised, the car will want to roll slightly and the jack will want to shift.

4. Once the car is raised, block the wheel that is opposite and farthest from the jack to prevent the car from unexpectedly rolling.

> ***WARNING —***
> *• Do not rely on the transmission or the emergency brake to keep the car from rolling. While they will help, they are not a substitute for positively blocking the opposite wheel.*
>
> *• Never work under a car that is supported only by a jack. Use jack stands that are properly designed to support the car. See **4. Tools**.*

To work safely under a car:

1. Disconnect the negative (–) battery cable so that no one else can start the car. Let others know what you will be doing.

> ***CAUTION —***
> *BMW anti-theft radios can be rendered useless by disconnecting the battery. If power to the radio is interrupted, a protection circuit engages and disables the radio. For the radio to operate, a code must be entered into the radio after power is restored. Make sure you know the correct code before disconnecting the battery. For more information, see the BMW owner's manual.*

2. Use at least two jack stands to support the car. A jack is a temporary lifting device and should not be used alone to support the car while you are under it. Use positively locking jack stands that are designed for the purpose of supporting a car. For more information on jack stands, see **4. Tools**.

> ***WARNING —***
> *Do not use wood, concrete blocks, or bricks to support a car. Wood may split. Blocks or bricks, while strong, are not designed for that kind of load, and may break or collapse.*

3. Place jack stands on a firm, solid surface, just like the jack. If necessary, use a flat board or similar solid object to provide a firm footing.

4. After placing the jack stands, lower the car slowly until its weight is fully supported by the jack stands. Watch to make sure that the jack stands do not tip or lean as the car settles on them, and that they are placed solidly and will not move.

5. Observe all jacking precautions again when raising the car to remove the jack stands.

3.2 General Advice For The Beginner

The tips in the paragraphs that follow are general advice to help any do-it-yourself BMW owner perform repairs and maintenance tasks more easily and more professionally.

Planning Ahead

Most of the repairs and maintenance tasks described in this manual can be successfully completed by anyone with basic tools and abilities. Some cannot. To prevent getting in too deep, know what the whole job requires before starting. Read the procedure thoroughly, from beginning to end, in order to know just what to expect and what parts will have to be replaced.

Cleanliness

Keeping things organized, neat, and clean is essential to doing a good job, and a more satisfying way to work. When working under the hood, fender covers will protect the finish from scratches and other damage. Make sure the car is relatively clean so that dirt under the cover does not scratch.

Avoid getting tools or clothing near the battery. Battery electrolyte is a corrosive acid.

Be careful with brake fluid, as it can cause permanent damage to the car's paint.

Finally, keep rubber parts such as hoses and belts free from oil or gasoline, as they will cause the material to soften and fail prematurely.

Non-Reusable Fasteners

Many fasteners used on the cars covered by this manual must be replaced with new ones when they are removed. These include but are not limited to: bolts, nuts (self-locking, nylock etc.), cotter pins, studs, brake fittings, roll pins, pins, clips and washers. Some bolts, for example, are designed to stretch during assembly and are permanently altered rendering them unusable again. Always replace fasteners where instructed to do so. Only genuine BMW parts should be used. See an authorized BMW dealer for applications and ordering information.

Tightening Fasteners

When tightening the bolts or nuts that attach a component, it is always good practice to tighten the bolts gradually and evenly to avoid misalignment or over stressing any one portion of the component. For components sealed with gaskets, this method helps to ensure that the gasket will seal properly and completely.

Where there are several fasteners, tighten them in a sequence alternating between opposite sides of the component. Fig. 3-2 shows such a sequence for tightening six bolts attaching a typical component. Repeat the sequence until all the bolts are evenly tightened to the proper specification.

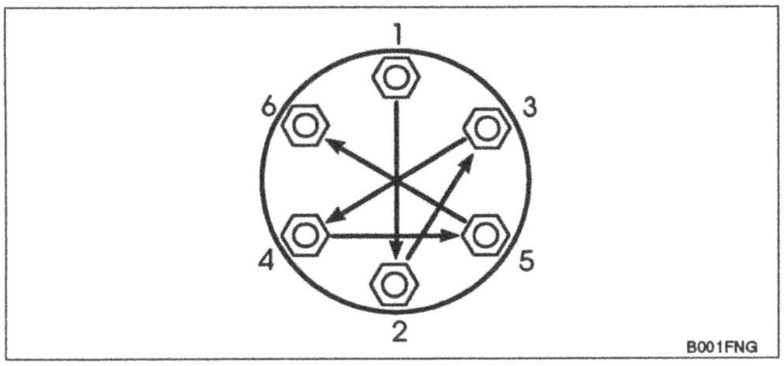

Fig. 3-2. Sequence for alternately tightening multiple fasteners.

For some repairs a specific tightening sequence is necessary, or a particular order of assembly is required. Such special conditions are noted in the text, and the necessary sequence is described or illustrated.

Bolt Torque

Tightening fasteners to a specified torque value using a torque wrench is a good way to ensure that bolts are correctly tightened. If a torque wrench is not used there is a danger of going too far and damaging the fastener or the threads in the mating part.

Too little torque on a fastener can also cause problems. Vibration of assembled parts can subject fasteners to stress alternating in opposite directions that will eventually cause them to loosen. To counter this loosening, fasteners are tightened more, and actually stretched, in order to prestress them. When tightened this way they are always stressed in one direction and are much less likely to work loose in spite of vibration.

The proper torque for a fastener is related to the amount of stretch necessary to prevent the fastener from working loose in normal use. Always use a torque wrench and follow BMW's torque specifications. See **4. Tools** for more information on torque wrenches.

Gaskets

The smoothest metal mating surfaces still have imperfections that can allow leakage. To prevent leakage at critical joints, gaskets of soft, form-fitting material are used to fill in the imperfections.

To be most effective, gaskets are designed to "crush," to become thinner as they are pressed together between mating parts. Once a gasket has been used and crushed, it is no longer capable of making as good a seal as when new, and is much more likely to leak. For this reason, gaskets should not be reused. Always plan to use new gaskets for any reassembly. Some gaskets, such as headgaskets, are directional. Make sure that these are installed correctly.

This same logic applies to any part used for sealing, including rubber O-rings and copper sealing washers.

Seals

In places where a shaft must pass through a housing, flexible lip seals are used to keep the lubricating oil or grease from leaking out past the rotating shaft.

Seals are designed to be installed in the housing only once and should never be reused. As long as they are not removed from the housing and not leaking, they need not be replaced. Seals, however, do age and deteriorate, and there is no easier time to replace them than when the car is already apart for some other repair.

When doing repairs that require removing a seal, be very careful not to scratch or otherwise damage the metal surfaces. Even minor damage to sealing surfaces can cause seal damage and leakage.

The key to seal installation is to get the seal in straight without damaging it. Use an object that is the same diameter as the seal housing to gently and evenly drive it into place. If a proper size seal driver is not available, a socket of the right size will do.

Coat the entire seal with a little grease or oil to help it go in more easily. Seals are directional. Make sure that it is being installed with the lip facing the correct way. Normally the lip faces the inside. Notice the installation direction of the old seal before removing it.

Wire Repairs

Repairs to a wiring harness to reconnect broken wires or correct shorts to ground deserve special care to make the repair permanent.

The wire ends must be clean. If frayed or otherwise damaged, cut off the end. If necessary, to maintain proper length, splice in a new piece of wire of the same size and make two connections.

Use connectors that are designed for the purpose. Crimped-on or soldered-on connectors are best. Crimp connectors and special crimping pliers are widely available. If soldering, use a needlenose pliers to hold the wire near the solder joint and create a "heat dam." This keeps the solder from "wicking" up the wire.

Always use a solder made specifically for electrical work, without the usual acid flux that will promote corrosion. Twisting wires together is a temporary repair at best, since corrosion and vibration will eventually spoil the connection.

Insulate the finished connection. Electronics stores can supply heat-shrinkable insulating tubing that can be placed onto the wire before connecting, slid over the finished joint, and shrunk to a tight fit with a heat gun or hair dryer. The next best alternative is electrical tape. Make sure the wire is clean and free of solder flux or other contamination. Wrap the joint tightly and completely to seal out moisture.

WARNING —
If the main ABS wiring harness is damaged in any way, it must be replaced. Do not attempt to repair the wiring harness. The ABS system is sensitive to very small changes in resistance. Repairing the wiring harness could alter resistance values and cause the system to malfunction.

Cleaning

Any repair job will be less troublesome if the parts are clean. For cleaning old parts, there are any number of solvents and parts cleaners available commercially.

For cleaning parts prior to assembly, commercially available aerosol cans of carburetor cleaner or brake cleaner are handy to use, and the cleaner will evaporate completely, leaving no residue.

WARNING —
Virtually all solvents used for cleaning parts are highly flammable, especially in aerosol form. Use with extreme care. Do not smoke. Do not use these products near any source of sparks or flame.

Let any solvent or cleaning product dry completely. Low-pressure, dry compressed air is helpful if available. Also, use only lint-free rags for cleaning and drying.

WARNING —
When drying roller or ball bearings with compressed air, do not allow them to spin. Unlubricated, they may fail and come apart, causing injury.

Electrical Testing

A great many electrical problems can be understood and solved with only a little fundamental knowledge of how electrical circuits function.

Electric current only flows in a complete circuit. To operate, every electrical device in the car requires a complete circuit including a voltage source and a path to ground. The positive (+) side of the battery is the original voltage source, and ground is any return path to the negative (–) side of the battery, whether through the wiring harness or the car body. Except for portions of the charging system, all electrical current in the car is direct current (DC) and flows from positive (+) to negative (–).

Switches are used to turn components on or off by completing or interrupting the circuit. A switch is "open" when the circuit is interrupted, and "closed" when the circuit is completed. Fig. 3-3 shows a complete circuit schematically.

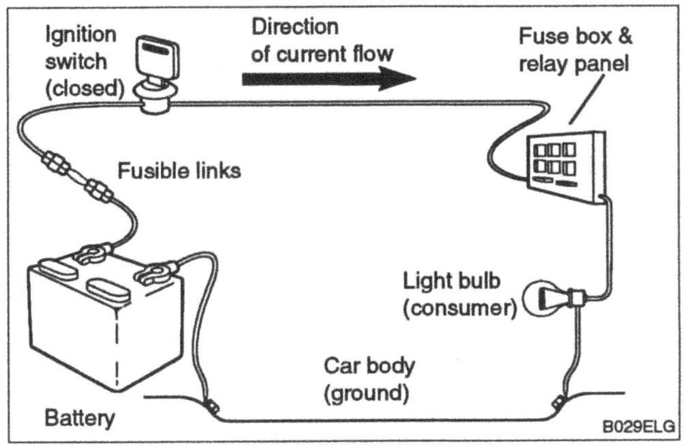

Fig. 3-3. Schematic representation of simple circuit for light bulb. Switch is shown closed, making circuit complete.

The first step in tracing an electrical system problem is to check with a test light, a voltmeter, or a multimeter (DC volts scale) to see that voltage is reaching the component. If so, then the circuit is sound as far as that point.

If voltage is not reaching the component, the circuit is open (interrupted), or shorted to ground, somewhere between the battery positive (+) terminal and the component. Look for a blown fuse, an open switch, a broken wire, or a failed component earlier in the same circuit.

Isolate the location of the problem by doing more voltage measurements at different points in the circuit. Voltage indicated by a test light or voltmeter at any point in the circuit means that the circuit is good at least up to that point. Look for problems after the last point in the circuit where voltage is indicated.

To test for voltage:

1. Start with an electrical connection closest to the component in question. If necessary, remove the terminal cover or disconnect the harness connector.

2. Connect the clip lead of a test light or the black (–) probe of a voltmeter or multimeter to ground (any clean, unpainted metal part of the engine or car).

3. Touch the probe of the test light or the red (+) probe of the meter to the terminal being tested. A meter reading or the test light lighting indicates that voltage is present in the circuit up to that point.

4. If no voltage is indicated, double check it by wiggling the probes and connections to make sure adequate contact is being made.

Once it is confirmed that voltage is reaching the component, check the remainder of the circuit by testing for continuity to ground. If the component is grounded through its mounting, make sure that the contact area is clean, dry, and free of corrosion.

> **NOTE —**
> Always make sure the circuit is turned off before making continuity checks using an ohmmeter. Voltage to the circuit may damage the ohmmeter.

To test for ground:

1. If the component is grounded through the wiring harness (usually a brown or brown striped wire), disconnect the harness connector or the ground wire. Connect the clip lead of a test light or the black (–) probe of a meter to the removed ground terminal.

2. If the component is grounded by its mounting to the car, connect one end of a test light or the black (–) probe of a meter to the clean metal surface of the component.

3. Briefly touch the remaining probe to a known source of battery voltage.

4. A meter reading or the test light lighting indicates current flow in a complete circuit and, therefore, a good connection to ground.

A continuity test, performed with an ohmmeter, is a universal test of any wire, connection, or component that will tell if current can flow through it. It can be used to check wires for breaks, to find out whether switches are open or closed, to find poor connections, and many other things. Continuity is a measure of resistance. A complete circuit has continuity—resistance is nearly zero. An open circuit due to a broken wire or an open switch has no continuity—there is infinite resistance. Do not make continuity tests on a live circuit.

To test the continuity of any conductor between any two points, connect one ohmmeter probe to each test point. If the circuit between points is uninterrupted, the ohmmeter should read nearly zero. If it reads significantly above zero, there may be a component in the circuit that is supposed to have some resistance. If not, there may be a poor connection or a damaged wire somewhere between the test points.

Making An LED Test Light

Some of the electrical tests in this manual require the use of a special LED test light, since the use of a more conventional test light with incandescent bulb can damage sensitive electronic circuits in the ignition, fuel injection, and emission control systems.

A low cost LED test light can be made using parts available from an electronics supply outlet. Assemble the components as shown in Fig. 3-4. Use a needlenose pliers to hold the parts and to act as a heat dam while soldering, as described above in **Wire Repairs**. Insulate all connections with heat-shrinkable tubing or electrical tape.

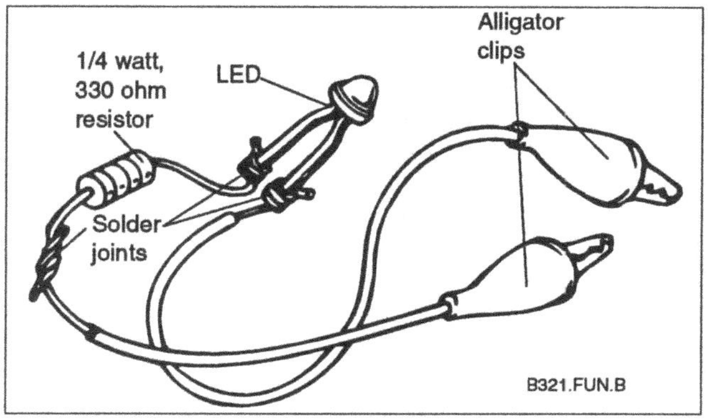

Fig. 3-4. Do-it-yourself LED test light for safe testing of ignition, fuel injection, and emission control circuits.

Parts:

1. (1) LED

2. (1) 1/4 watt, 330 ohm resistor

3. Wire and two alligator clips (Purchase a jumper wire with an alligator clip on each end, and cut it in half)

4. Solder and soldering iron

5. Heat-shrinkable tubing or electrical tape

Disconnecting Wiring Harness Connectors

BMW harness connectors used throughout the car are designed to positively lock into place to prevent them from coming loose. One common type of connector is equipped with an easy disconnect feature. To disconnect this type of connector, press on the wire clip to release the lock and carefully pull the connector loose. The other common type requires that a locking clip be pried from the connector before it can be removed. See Fig. 3-5.

> ***CAUTION —***
> *Always pull only on the connector body to disconnect it. Never pull on the wires themselves.*

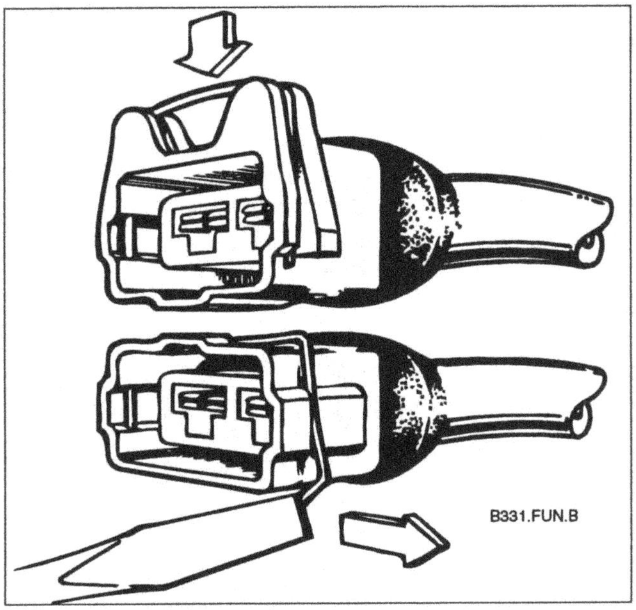

Fig. 3-5. Common versions of harness connectors. **Top:** Harness connector with quick disconnect feature. **Bottom:** Harness connector with retaining clip.

3.3 Buying Parts

Many of the maintenance and repair tasks in this manual call for the installation of new parts, or the use of new gaskets and other materials when reinstalling parts. Most often, the parts that will be needed should be on hand before beginning the job. Read the introductory text and the complete procedure to determine which parts will be needed.

> **NOTE —**
> For some bigger jobs, partial disassembly and inspection are required to determine a complete parts list. Read the procedure carefully and, if necessary, make other arrangements to get the necessary parts while your car is disassembled.

Genuine BMW Parts

Genuine BMW replacement parts from an authorized BMW dealer are designed and manufactured to the same high standards as the original parts. They will be the correct material, manufactured to the same specifications, and guaranteed to fit and work as intended by the engineers who designed the car. Some genuine BMW parts have a limited warranty.

Many independent repair shops make a point of using genuine BMW parts, even though they may be more expensive. They know the value of doing the job right with the right parts. Parts from other sources can be as good, particularly if manufactured by one of BMWs original equipment suppliers, but it is often difficult to know.

BMW is constantly updating and improving their cars, often making improvements during a given model year. BMW may recommend a newer, improved part as a replacement, and your authorized dealer's parts department will know about it and provide it. The BMW parts organization is best equipped to deal with any BMW parts needs.

Some caution is appropriate when buying parts. Parts that fit are not necessarily the same as parts that work. If someone else is buying parts for your BMW, make sure they are genuine BMW parts from an authorized BMW dealer or the equivalent from a quality supplier.

Non-returnable Parts

Some parts cannot be returned for credit, even if they are the wrong parts for the car. The best example is electrical parts, which are almost universally considered non-returnable because they are so easily damaged internally.

Buy electrical parts carefully, and be as sure as possible that a replacement is needed, especially for expensive parts such as control units. It may be wise to let an authorized BMW dealer or other qualified shop confirm your diagnosis before replacing an expensive part that cannot be returned.

Information You Need To Know

Model. When ordering parts it is important that you know the correct model designation for your car. Models covered in this manual are 528e, 533i, and 535i.

Model Year. This is not necessarily the same as date of manufacture or date of sale. A 1986 model may have been manufactured in late 1985, and perhaps not sold until early 1987. It is still a 1986 model. Model years covered by this manual are 1982 through 1988.

Date of Manufacture. This information is helpful when ordering replacement parts or determining if any of the warranty recalls are applicable to your car. The label on the driver's door below the door latch will specify the month and year that the car was built.

Vehicle Identification Number (VIN). This is a combination of letters and numbers that identify the particular car. The VIN appears on the state registration document, and on the car itself. One location, shown in Fig. 3-6, is on the dash near the driver's side of the windshield. It is most easily viewed from outside the car.

Copy down the VIN and date of manufacture and have it along whenever buying parts. If there was a mid-year change in specifications that affects replacement parts, the change will most often be defined in terms of the build date.

Beginning in 1987, a ruling by the National Highway and Traffic Safety Administration (NHTSA) requires passenger cars with a high theft rate to have the VIN marked on specific

Fig. 3-6. Location (arrow) of vehicle identification number (VIN).

parts of the car when manufactured. On BMW cars so affected, these parts are identified by an adhesive label.

Original parts installed during manufacture are identified by a label bearing the VIN and two BMW roundel logos. The replacement parts will have a similar label, bearing one BMW roundel logo and the letters DOT-R. See Fig. 3-7. Parts or assemblies bearing the label are the engine, transmission, front and rear bumpers, front fenders, rear quarter panels, hood, trunk lid and doors. These labels should not be removed as they will tear apart.

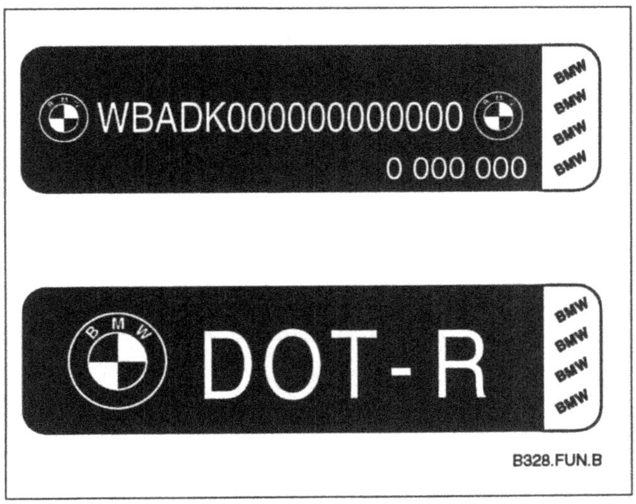

Fig. 3-7. Labels used to identify parts. Original equipment label with VIN number and roundel logos (top) and replacement part label with one roundel logo.

Engine. BMWs covered in this manual are powered by one of three different engines. All are six cylinder engines. 2.7 l (528e models), 3.2 l (533i models), and 3.5 l (535i models). For more information see **ENGINE—SERVICE AND REPAIR.**

Transmission Number. Although some internal repairs to the transmission are beyond the scope of this manual, the transmission number with its identifying code may be important when buying clutch parts, seals, gaskets, and other transmission-related parts for repairs that are covered.

Manual transmissions are identified by a manufacturer's stamp and code numbers and letters. The manufacturer's stamp is located on the case, near the clutch slave cylinder mounting as shown in Fig. 3-8. The code numbers and letters are located on top of the bellhousing as shown in Fig. 3-9.

Fig. 3-8. Manufacturer's stamp for Getrag manual transmissions.

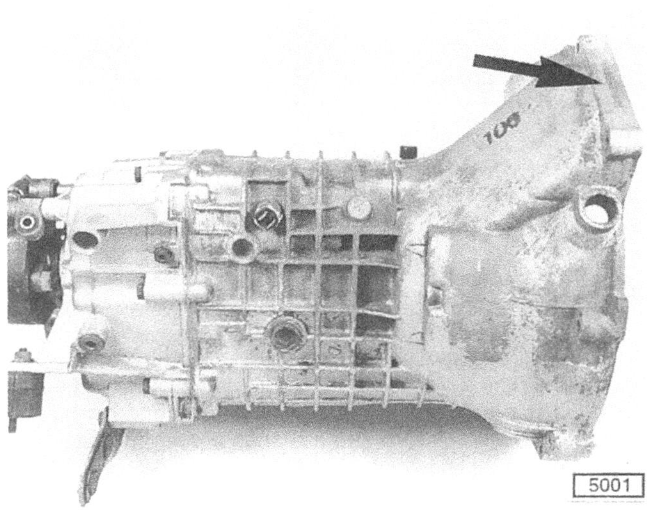

Fig. 3-9. Location (arrow) of transmission identification code for manual transmission.

More information on manual transmission codes and their meanings can be found in **MANUAL TRANSMISSION AND CLUTCH** and **MANUAL TRANSMISSION OVERHAUL**.

Automatic transmissions are identified by code letters and type numbers, located on a data plate behind the manual shift valve lever, as shown in Fig. 3-10. More information on automatic transmission codes and their meanings can be found in **AUTOMATIC TRANSMISSION**.

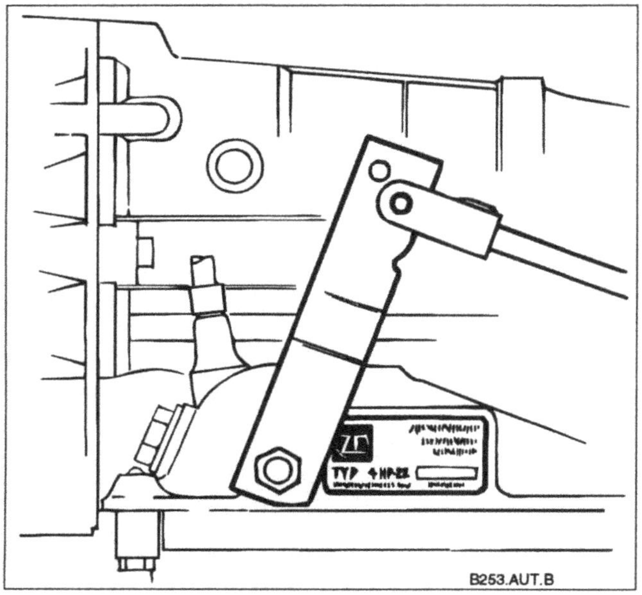

Fig. 3-10. Location of data plate on automatic transmission.

BMW dealers are uniquely qualified to provide service for BMW cars. Their relationship with the large BMW service organization means that they are constantly receiving new tools and equipment, together with the latest and most accurate repair information.

The BMW dealer's service technicians are highly trained and very capable. Unlike most independent repair shops, authorized BMW dealers are intensely committed to supporting the BMW product. They share every owner's interest in BMW value, performance, and reliability.

4. TOOLS

Most maintenance can be accomplished with a small selection of the right tools. Tools range in quality from inexpensive junk, which may break at first use, to very expensive and well-made tools which, to the professional, are worth every bit of their high cost. The best tools for most do-it-yourself BMW owners lie somewhere in between.

Cheap tools are not a bargain. They often do not hold up to even casual use, and they present a greater risk of personal injury. If they fit poorly, they can actually damage the fasteners they are intended to remove, making it that much harder to use a good tool the next time around.

Many reputable tool manufacturers offer good quality, moderately priced tools with a lifetime guarantee. A broken tool can be exchanged for a new one, for the life of the tool. These are your best buy. They cost a little more, but they are good quality tools that will do what is expected of them. Sears' Craftsman® line is one such source of good quality, reasonably priced, and guaranteed tools. Other sources of general and special tools are:

Baum Tools Unltd. Inc.
PO Box 5867
Sarasota, FL 34277-5867
1-800-848-6657

MAC Tools Inc.
Washington Court House, OH 43160
1-800-622-8665

Schley Products Inc.
5350 E. Hunter Ave.
Anaheim Hills, CA 92807
1-714-693-7666

Snap-On Tools Corporation
Kenosha, WI 53141-1410
local offices nationwide
1-414-656-5462

Basic Tool Requirements

NOTE —
BMWs are delivered with a tool kit mounted to the underside of the trunk lid. The kit contains a basic selection of tools that may fulfill some of the requirements listed in this section.

The basic hand tools described below can be used to accomplish most of the simple maintenance and repair tasks.

Screwdrivers. Two types, the common flat-blade type and the Phillips type, will handle 99% of all screws used on BMWs. Two or three different sizes of each type will be best, since a screwdriver of the wrong size will damage the screw head.

Screwdrivers are for screws. Do not use them for anything else, such as prying or chiseling. A complete set of screwdrivers can often be purchased for about the same money as the four or six individual ones that are really necessary. See Fig. 4-1.

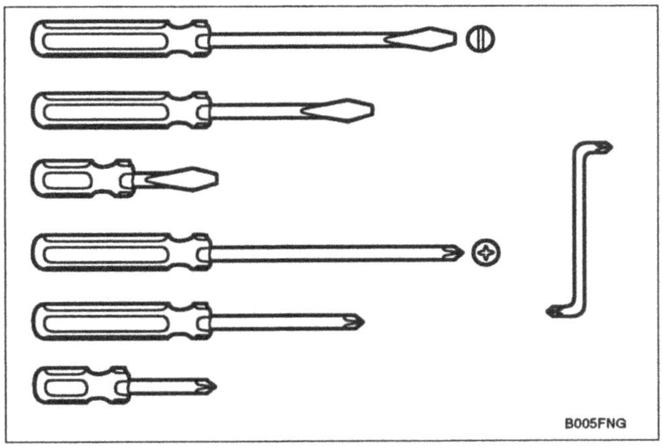

Fig. 4-1. Common flat-blade (top) and Phillips (bottom) screwdrivers. Offset screwdriver (right) is used for screws with limited access.

For a more complete tool box, include "stubby" screwdrivers or offset screwdrivers for use in tight spots where a normal length screwdriver will not easily fit and Torx®head screwdrivers.

Wrenches. Wrenches come in different styles for different uses. Fig. 4-2 shows several. The basic open-end wrench is the most widely used, but grips on only two sides. It can spread apart and slip off more easily. The box-end wrench has better grip, on all six sides of a nut or bolt, and is much less prone to slip.

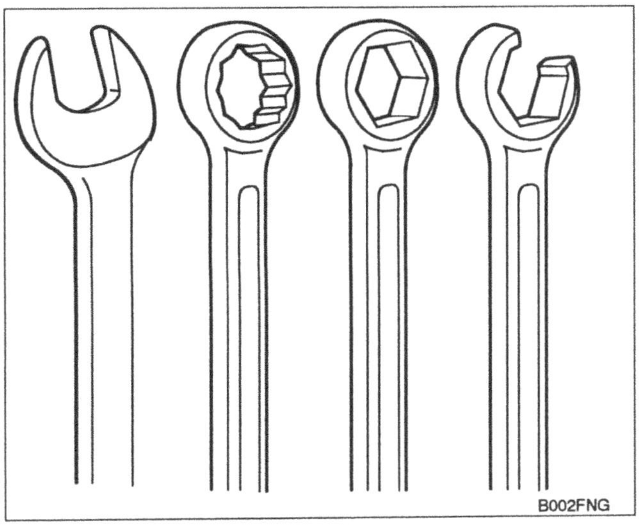

Fig. 4-2. Types of wrench heads. From left, open-end, 12-point box-end, 6-point box-end, flare nut.

A 12-point box-end can loosen a nut or bolt where there is less room for movement, while a 6-point box-end provides better grip. For hex fasteners on fluid lines, like brake lines and fuel lines, a flare-nut wrench offers the advantages of a box-end wrench with a slot that allows it to fit over the line.

The combination wrench, shown in Fig. 4-3, is the most universal. It has one open-end and one 12-point box-end. For BMWs, 10mm and 13mm wrenches are the most common sizes needed. A 17mm or 19mm wrench is needed to loosen and tighten the engine oil drain plug. A complete set should also include 6mm, 7mm, 8mm, 9mm, 11mm, 12mm, 14mm, and 15mm.

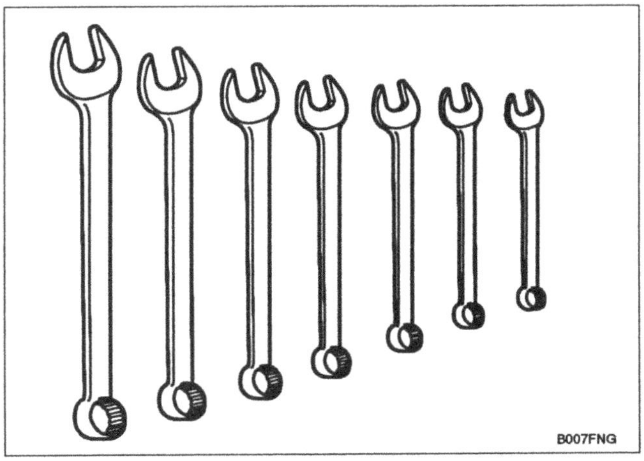

Fig. 4-3. Combination wrenches with one open-end and one 12-point box-end.

Sockets. Sockets perform the same job as box-end wrenches, but offer greater flexibility. They are normally used with a ratchet handle for speed and convenience, and can be combined with extensions to reach fasteners more easily. See Fig. 4-4.

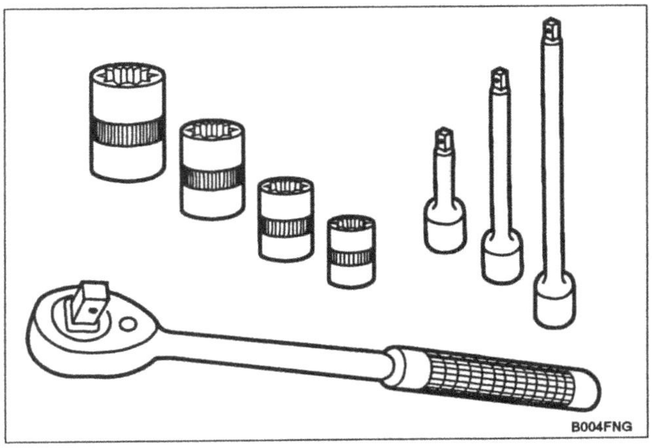

Fig. 4-4. Sockets, extensions, and a ratchet handle.

Standard sockets come in 6-point and 12-point styles. For use with a ratchet the 6-point offers a better grip on tight nuts and bolts. As with wrenches, 6mm to 15mm, 17mm, and 19mm are the most needed sizes.

Sockets come with different size connections to drive handles or extensions, called the drive size. The most common drive sizes are 1/4 in., 3/8 in., and 1/2 in.

As a start, 6-point sockets with a 3/8 in. square drive, two or three 3/8 in. extensions of different lengths, and a 3/8 in. drive ratchet handle will be suitable for most jobs.

For a more complete tool box, add deep sockets and a greater variety of handles and extensions. A universal joint extension can allow access from an angle where a straight extension will not quite fit.

Spark Plug Socket. A special socket for spark plugs is the correct size, is deep enough to accommodate a spark plug's length, and includes a rubber insert to both protect the spark plug from damage and grip it for easier removal. A typical spark plug socket is shown in Fig. 4-5.

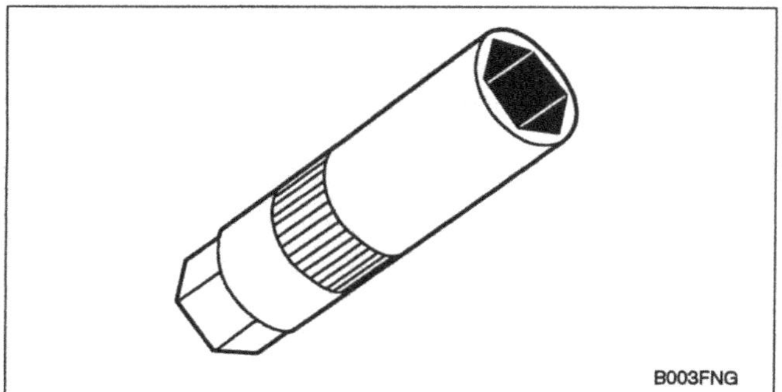

Fig. 4-5. Spark plug socket.

The spark plugs used in BMW engines require a 13/16 in. socket. Get one with the drive size to match your ratchet handle and extensions.

Pliers. A few of the many types of pliers are shown in Fig. 4-6. Most are used for holding irregular objects, bending, or crimping. Some have special applications.

A needlenose pliers is used for gripping small and poorly accessible objects, and is useful for wiring and other electrical work. A locking pliers such as the well-known Vise-Grip® is useful because of its tight grip.

Snap-ring and circlip pliers with special tipped jaws are used to remove and install snap-rings or circlips. A Channel-lock® or water pump pliers has adjustable jaws that can be quickly changed to match the size of the object being held to give greater leverage.

There are many different types and sizes of pliers. Start with a small selection of different types of medium size.

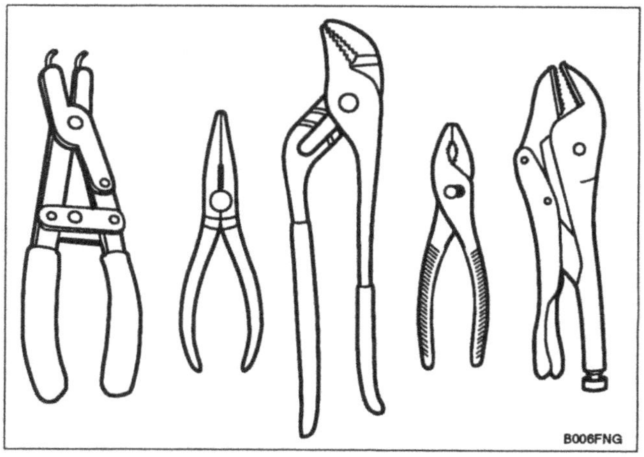

Fig. 4-6. Pliers. From left, snap-ring, needlenose, Channel-lock®, common, locking.

Adjustable wrench. An adjustable wrench, shown in Fig. 4-7, can be a useful addition to a small tool kit. It can substitute in a pinch, if two wrenches of the same size are needed to remove a nut and bolt. Use extra care with adjustable wrenches, as they especially tend to loosen, slip, and damage fasteners.

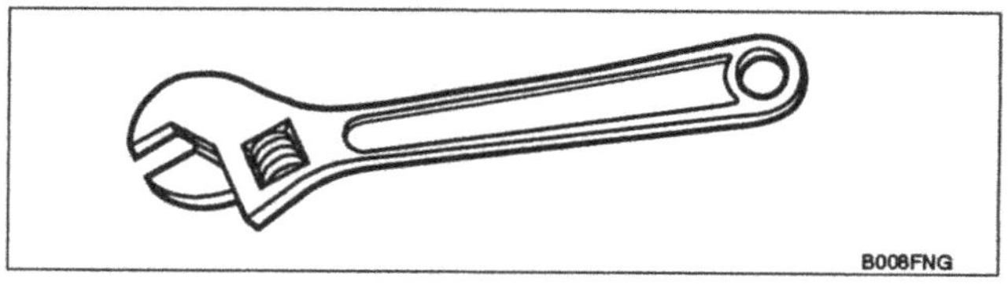

Fig. 4-7. Adjustable wrench.

Compared to a wrench of the correct size, an adjustable wrench is always second best. They should only be used when the correct size wrench is not available. Choose one of average size range, about 6 to 8 inches in length.

Jack Stands

Strong jack stands are extremely important for any work that is done under the car. Jacks are designed only for short term use and are not solid enough to support the car for a long period. A jack should never be used alone to support the car while working underneath.

Use only jack stands that are designed for the purpose. Blocks of wood, concrete, bricks, etc. are not safe or suitable substitutes.

Jack stands are available in several styles. A typical jack stand is shown in Fig. 4-8. The best ones are made of heavy material for strength, have a wide base for stability, and are equipped to positively lock in their raised positions. Get the best ones available.

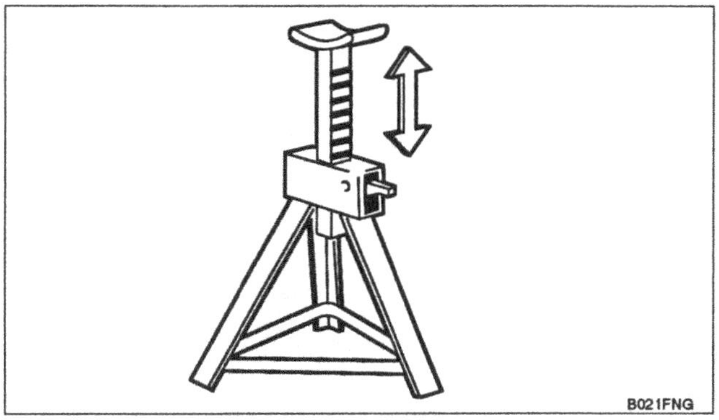

Fig. 4-8. Jack stand for safely supporting car to work underneath.

Oil Change Equipment

Changing engine oil requires a box-end wrench or socket to loosen and tighten the drain plug (17mm or 19mm), a drain pan (at least 7 qt. capacity), and an oil filter wrench. These items are shown in Fig. 4-9. A wide, low drain pan will fit more easily under the car. Use a funnel to pour the new oil into the engine.

An oil filter wrench is needed for some models to remove the oil filter. Be sure to get a filter wrench that will grip the BMW oil filter tightly.

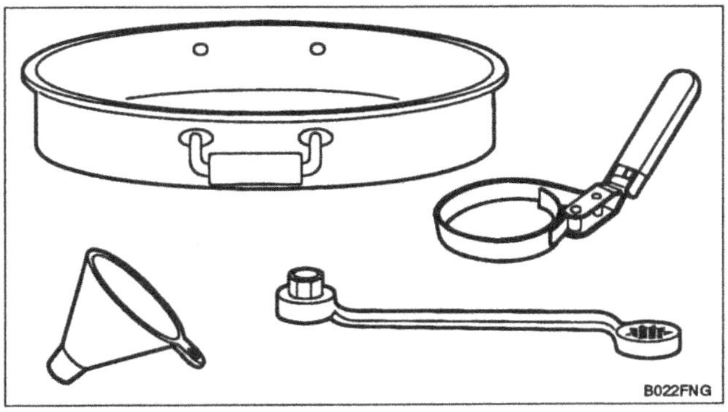

Fig. 4-9. Oil change equipment includes drain plug wrench (17mm or 19mm), 7 qt. drain pan, oil filter wrench, and funnel.

Torque Wrench

A torque wrench is used to precisely tighten threaded fasteners to a predetermined value. Nearly all of the repair procedures in this manual include BMW-specified torque values in Newton-meters (Nm) and the equivalent values in foot-pounds (ft-lb).

Several types of torque wrenches are widely available. They all do the same job, but offer different convenience features at different prices. Two typical torque wrenches are shown in Fig. 4-10. The most convenient ones have a built-in ratchet, and can be preset to indicate when a specific torque value has been reached. Follow the wrench manufacturer's directions for use to achieve the greatest accuracy.

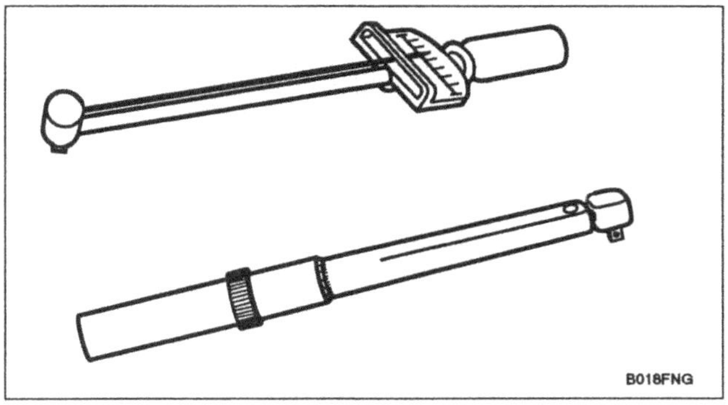

Fig. 4-10. Torque wrenches. **Top:** Inexpensive beam-type is adequate but must be read visually. **Bottom:** Ratchet-type can be preset to indicate when torque value has been reached.

A torque wrench with a range up to about 250 Nm (185 ft-lb) has adequate capacity for most of the repairs covered in this manual. For recommended torque values of 10 Nm or below, the English system equivalent is given in inch-pounds (in-lb). These small values may be most easily reached using a torque wrench calibrated in inch-pounds. To convert foot-pounds to inch-pounds, multiply by 12. To convert inch-pounds to foot-pounds, divide by 12.

Timing Light

A timing light is not needed to service the cars covered in this manual. The ignition timing cannot be adjusted. For more information see **IGNITION**.

Tachometer

An external tachometer is used to precisely measure engine speed (rpm) for various tests and adjustments. Most tachometers are powered by connection to the battery, and measure engine rpm through a connection to terminal 1 of the ignition coil. To locate terminal 1, see **IGNITION**.

Feeler Gauges

Feeler gauges are thin metal strips of precise thickness, used to measure small clearances. They are normally available as a set, covering a range of sizes. For BMWs, metric feeler gauges (in millimeters) are the best choice. Fig. 4-11 shows a set of feeler gauges.

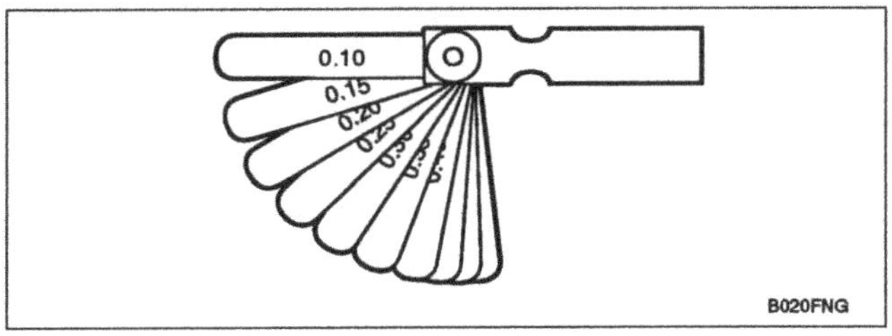

Fig. 4-11. Feeler gauge set, used for precise measurement of clearances between parts.

Micrometers

Precision measurements of internal engine parts and other critical dimensions are made with micrometers, some of which can accurately measure to within thousandths of a millimeter. These are expensive instruments, and are only recommended for those who plan to be repeatedly involved in engine overhauls or other similar work requiring detailed measurement. If such measurements are necessary on a one-time basis, a qualified machine shop can be called upon to make these measurements, particularly if they are also going to be doing the necessary machine work.

Test Light

A test light, shown in Fig. 4-12, is a simple tool used to check electrical circuits for voltage or continuity to ground when actual voltage values are unimportant. A bulb in the handle will light whenever current is flowing through the circuit. The use of a test light is described in **3.2 General Advice For The Beginner**.

> ***CAUTION —***
> *Ignition, fuel injection, emission controls and other electronic systems may be damaged by the high current draw of a test light with a normal incandescent bulb. For these applications, use a high impedance digital multimeter.*

Volt-Ohm Meter (VOM) or Multimeter

Many of the electrical tests in this manual call for the measurement of resistance (ohms) or voltage values. For safe and accurate tests of ignition, fuel injection, and emission control systems, the multimeter, shown in Fig. 4-13, should be digital, with high (at least 10,000 ohms) input impedance. Some meters have automotive functions such as dwell and pulse width that are useful for troubleshooting ignition and fuel injection problems.

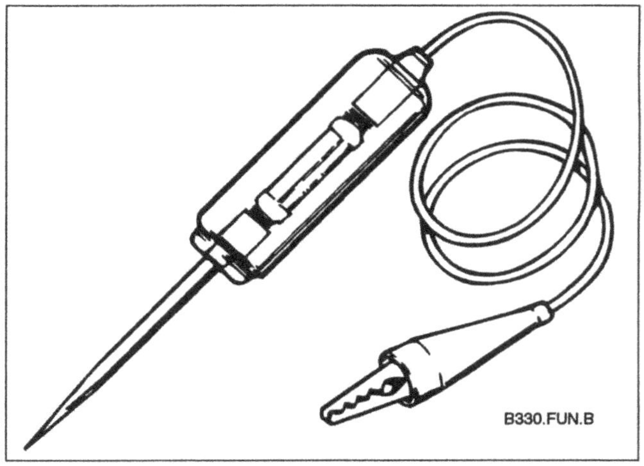

Fig. 4-12. Test light with alligator clip test lead.

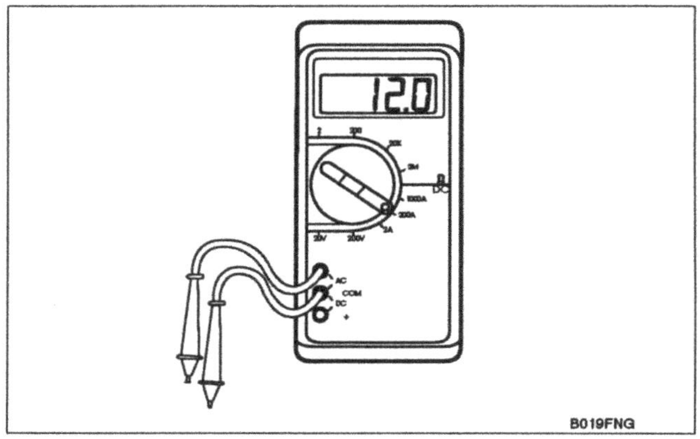

Fig. 4-13. Multimeter with test probes.

Jumper Wires

Some of the electrical tests in this manual require the use of extra jumper wires to bypass a component or a portion of the wiring harness. For most basic electrical tests, jumper wires with an in-line fuse and alligator clips at each end (made or purchased) are sufficient.

For tests involving harness connectors, hookup of jumper wires may damage the connector and cause inferior connections later on. To avoid this damage, jumper connections to harness connectors should be made using a small, flat-blade (spade) terminal that will mate properly with the connector. See Fig. 4-14.

BMW Special Tools

Some of the more challenging repairs covered in this manual call for the use of BMW special tools. This, however, does not automatically mean that the job is too complicated or out of reach of the novice.

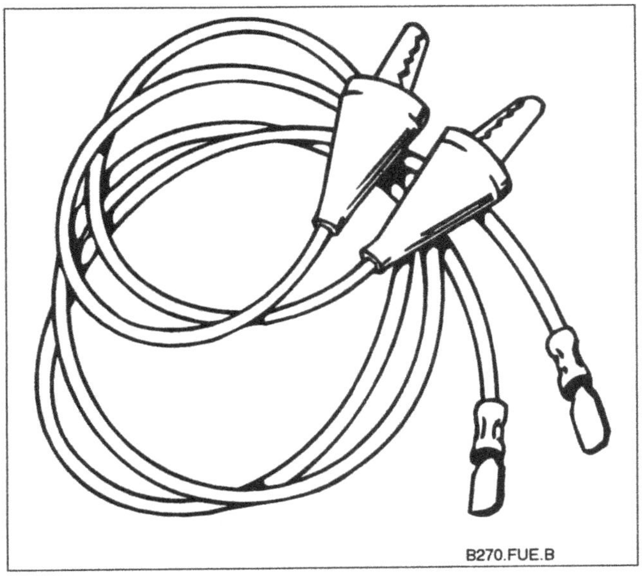

Fig. 4-14. Jumper wires with alligator clips. Flat connectors shown are for electrical testing at harness connectors.

Many of the BMW special tools mentioned in this manual are inexpensive and are simply the best thing to use to do the job correctly. In these cases, the tool is identified with a BMW part number. See your authorized BMW dealer parts department for information on how to order special tools.

There are some jobs for which expensive special tools are essential, and not a cost-effective purchase for one-time repair by the do-it-yourself owner. This manual includes such repairs for the benefit of those with the necessary experience and access to tools. For the do-it-yourselfer, the need for special tools is noted in the text, and whether or not BMW dealer service is recommended.

5. TROUBLESHOOTING FUNDAMENTALS

Troubleshooting is a systematic approach to identifying and solving a problem. The exact approach depends on the individual circumstances, but usually relies on carefully observing the symptoms. Paying attention to exactly what is happening, and under what conditions, is the most powerful tool available to get to the cause of a problem.

The basic rule for troubleshooting is to never overlook the obvious. Always start with the basics and work toward the more complex. Lots of time and money can be wasted on exotic testing only to find, eventually, that the problem is a loose wire or an empty fuel tank.

Following are some tips to initial analysis of common problems, and direction to other section or sections of this manual for more detailed troubleshooting information.

5.1 Starting

There are three main requirements for starting the engine:

1. The starting system (battery and starter) must provide adequate engine cranking speed
2. The ignition system must provide adequate spark at the proper time
3. The fuel system must deliver the proper amount of fuel at the proper time

Observing the symptoms of a starting problem will give clues to its cause. Slow cranking speed indicates problems with the electrical system, probably the battery or starter. Further troubleshooting should focus on the electrical system. Tests of the ignition or fuel systems at this point would be meaningless. See **BATTERY, STARTER, ALTERNATOR**.

An engine that cranks normally indicates that the battery and starter are fine, so other starting problems suggest an ignition or fuel system problem. If there is no sign whatsoever of starting, make sure there is adequate fuel in the tank. Check for loose wires around the coil and distributor. Check to see that the distributor cap and spark plug wires are dry.

Unless this basic inspection turns up a cause, the cause could be in either the ignition system or the fuel system. The ignition system is the more likely culprit and is also easier to evaluate. The condition and function of the ignition system should always be confirmed before suspecting the fuel system. See **ENGINE MANAGEMENT—DRIVEABILITY** for more information on troubleshooting starting problems.

5.2 Driveability

Problems with the way the engine runs, also known as driveability, may be caused by faults in either the ignition system or the fuel system. The ignition system should be investigated and its good condition confirmed before beginning any work on the fuel system.

The fuel injection system is far more likely to be influenced by temperature. Symptoms that are present only when the engine is cold, or only when it is warm, tend to suggest fuel system problems.

For an engine with high mileage, the general mechanical condition of the engine may also be a factor. Particularly in cases where driveability problems have developed slowly over time, troubleshooting should include evaluation of the engine's mechanical condition with a compression test. See **ENGINE—SERVICE AND REPAIR**.

When attempting to evaluate noise or vibrations that occur when the engine is running, try to eliminate other possible causes. With a manual transmission, symptoms that change depending on whether or not the clutch is engaged suggest that the problem may be in the clutch disc, the clutch release mechanism, or the transmission. See **MANUAL TRANSMISSION AND CLUTCH** or **MANUAL TRANSMISSION OVERHAUL**. With an automatic transmission, check the symptoms in different shifter positions. If the symptoms differ, the problem may be in the torque converter or the transmission. See **AUTOMATIC TRANSMISSION**.

5.3 Driving

To track down noise and vibrations that occur while driving, first try to learn more about the symptom. Does it occur all the time, or only at certain speeds? Does the symptom change depending on engine speed or vehicle speed?

Compare driving in different gears at the same approximate engine speed (rpm). A symptom that persists at a certain engine speed or speed range regardless of gear selection suggests an engine or exhaust system problem. See **ENGINE—SERVICE AND REPAIR** or **EXHAUST SYSTEM**.

For more analysis try driving the car at the speed where the symptom is most noticeable, then briefly shift into neutral and coast. If the symptom continues unchanged, then it is the car's speed that is a factor, and not engine speed. Symptoms that vary only with the car's speed suggest problems with running gear. See **SUSPENSION—FRONT, SUSPENSION—REAR** or **STEERING AND WHEEL ALIGNMENT** for more detailed troubleshooting information.

6. EMERGENCIES

6.1 Changing a Tire

If a tire goes flat while driving, pull well off the road. Changing a tire on a busy street or highway is very dangerous. If necessary, drive a short distance on the flat tire to get to a safe place. It is much better to ruin a tire or rim than to risk being hit.

Stop the car on as flat a surface as possible, in a place where you can be easily seen by other drivers. Avoid stopping just over the crest of a hill. Turn on the emergency flashers and set out flares or emergency markers well behind the car. Passengers should get out of the car and stand well away from the road. Take the jack, tools, and spare wheel from the trunk. Chock the wheel diagonally opposite to the one being changed.

Loosen the wheel bolts while the car is on the ground, but leave them a little snug. Place the jack under the lifting point nearest the wheel being changed (lifting points are described in **3.1 Safety**). Use a board to provide a firm footing for the jack if the ground is soft.

Raise the jack until it is just touching the lifting point, and adjust the jack so that its base is slightly under the car. Raise the car only far enough so that the wheel is off the ground, and then remove the wheel bolts and the wheel.

To install the spare wheel and tire, use the BMW-supplied centering pin. Install the pin into one of the holes, put the wheel on the pin, then install one wheel bolt. Remove the pin, install the remaining wheel bolts and tighten them by hand, then lower the car. With all wheels on the ground, fully tighten the bolts in a cross-wise pattern.

Torque the wheel bolts when installing the wheel.

Tightening torque	
• wheel lug bolts	100±10 Nm (74±7 ft-lb)

If torquing the wheel bolts is not possible, tighten them as much as possible, then loosen and retorque the bolts to the proper specification at the earliest opportunity. Check the inflation pressure of the spare tire. Inflation pressures are given in **LUBRICATION AND MAINTENANCE**.

6.2 Car Will Not Start

If the engine turns over slowly or not at all, especially on cold mornings, the battery may not be sufficiently charged. Jump-starting the battery from another car may help. Jump-starting is described below in **6.3 Jump-Starting**.

> **NOTE —**
> Be sure to read the cautions under **6.3 Jump Starting** prior to jump starting a low battery. Failure to follow the cautions may result in damage to the electronic control units for the on-board computer or the Anti-lock Braking System (ABS).

Push starting (or tow starting) a car with an insufficiently charged battery is another option. To push start the car, turn on the ignition, put the car in third gear and push in the clutch pedal. Push the car. When the car is moving at a fair speed, release the clutch pedal. After the engine has started, push the clutch pedal back in and allow the engine to idle.

> ***WARNING —***
> *Use extreme caution when push starting a car. Be aware of other traffic. Use the emergency flashers.*

> **NOTE —**
> On cars with automatic transmissions, the design of the transmission makes it impossible to start the engine by pushing the car.

If the starter seems to be operating but the engine does not turn over (indicated by a high-pitched whine or grinding when the ignition key is turned to START), then there is a problem with the starter. In this case jump starting will not help.

If the engine is turning over at normal speed, the battery and starter are fine. Check to make sure that there is fuel in the tank. Don't rely on the fuel gauge. It may be faulty. Instead, remove the gas filler cap and rock the car. If there is gas in the tank, you should hear a sloshing sound from the filler neck. If so, turn the ignition on and listen for the sound of the fuel pump. It should run for a few seconds, then stop. If it doesn't, fuel may not be reaching the engine.

The engine also may have difficulty starting because it has too much fuel, because the fuel system is vapor-locked on a hot day, or because the ignition system is wet on a very damp day. There will probably be a strong smell of gas if the engine has too much fuel (referred to as "flooded").

The common remedy of repeatedly cranking the engine with the gas pedal floored to clear a flooded engine may damage the catalytic converter. Instead, wait for a few minutes, and then try starting the engine again. If you suspect vapor-lock, raise the hood, let the engine cool, and then try to start the engine.

On damp days, check the distributor cap and spark plug wires for condensation. If they are wet, remove and replace the wires one at a time and dry them off with a clean dry cloth, then remove the distributor cap and wipe it dry inside and out.

6.3 Jump-Starting

Cars with partially discharged or completely dead batteries can be jump-started using the good battery from another car. When jump-starting the engine, always heed the following warnings and cautions.

WARNING —
- *Battery acid (electrolyte) can cause severe burns, and will damage the car and clothing. If electrolyte is spilled, wash the surface with large quantities of water. If it gets into eyes, flush them with water for several minutes and call a doctor.*

- *Batteries produce explosive and noxious gasses. Keep sparks and flames away. Do not smoke near batteries.*

- *Do not jump-start the engine if you suspect that the battery is frozen. Trapped gas may explode. Allow the battery to thaw first.*

CAUTION —
- *On models equipped with on-board computers, remove the computer fuses (fuse nos. 5, 6, 12) prior to quick-charging to prevent damaging the computer.*

- *Do not quick-charge the battery (for boost starting) for longer than one minute, and do not exceed 16.5 volts at the battery with the boosting cables attached. Wait at least one minute before boosting the battery a second time.*

To jump-start the engine, place the cars close together, but do not allow them to touch. Turn off the engine of the car with the good battery. Connect the jumper cables as shown in Fig. 6-1.

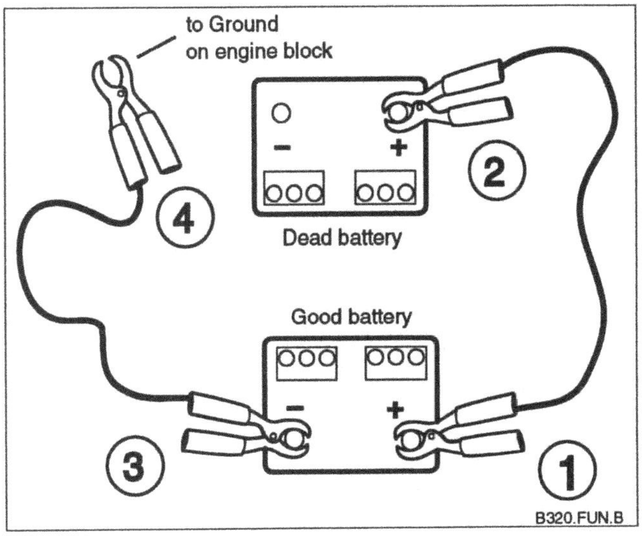

Fig. 6-1. Battery jumper cables connections. Numbers indicate correct sequence for cable attachment.

The battery is mounted in the engine compartment. Connect the end of one cable to the positive post of the good battery, and the other end of the same cable to the positive post of the dead battery. The positive post is marked with a plus (+) sign.

Connect one end of the other cable to the negative (–) post of the good battery, and connect the other end of the same cable to the engine block of the car with the dead battery. Make the connection as far away from the battery as possible, as there may be sparks.

Have a helper start the car with the good battery and race the engine slightly, then start the car with the dead battery. Leave the cars running and disconnect the cables in the reverse order in which they were installed. The car with the dead battery will need to run for at least 1/2 hour to recharge the battery.

6.4 Overheating

If the coolant temperature is too high, find a safe place to stop and turn the engine off. Open the hood and allow the engine to cool until the temperature gauge needle is at the lower third of the scale. Continuing to drive an overheated car can cause expensive engine damage.

> ***WARNING —***
> *Do not remove the coolant reservoir or radiator cap with the engine hot. Undoing either could spray hot coolant, and cause burns, or damage the engine.*

> **NOTE —**
> If the engine cannot be safely turned off, make sure the air conditioner is off and turn the heater to high. This will help cool the engine until a safe stopping place can be reached.

Overheating may be caused by the driving conditions, such as operating the air conditioner in slow traffic, or by low coolant level or a damaged V-belt. Visually check the coolant level and V-belts as described in **LUBRICATION AND MAINTENANCE**. If coolant is lost, check the filler cap, hoses, clamps and radiator for signs of leakage.

If no leaks are found, add coolant after the engine has cooled. The car can be driven, but have the cooling system thoroughly checked as soon as possible. If replacement coolant is not available, then plain water can be used, but the coolant should later be drained and refilled with the proper mixture of anti-freeze and water.

> ***CAUTION —***
> *Do not add cold water or coolant to a hot engine. Severe engine damage could result from the sudden temperature change.*

If steam is coming from the engine compartment then there is most likely a burst coolant hose or a large leak in the cooling system. To find the leak, look for signs of coolant leakage on hoses, at hose connections, or on the radiator. Let the engine cool thoroughly, then add coolant or water to fill the system and start the engine. If a great deal of water or coolant flows out of the hole, then the car should not be driven until repairs are made. If there is a slight seepage, then it may be possible to drive a short distance, adding coolant as needed.

6.5 Oil Pressure Warning Light

If the oil pressure warning light does not go out immediately after the engine starts, or if it comes on while driving the car, stop the engine immediately to prevent severe engine damage.

Check the oil level as described in **LUBRICATION AND MAINTENANCE**. If the level is low, add oil to the correct level and start the engine. If the light is still on, do not run the car at all. Have it towed.

6.6 Brake Fluid Level Warning Light

The red brake fluid level warning light is an indicator of brake fluid loss. Problems with the brake system should be checked and repaired immediately. See **BRAKES** for more information.

6.7 Anti-Lock Brake System Warning Indicator

If the anti-lock brake system warning indicator comes on at normal driving speeds, the anti-lock braking system is out of service. Under normal conditions, there will be no change in the effectiveness of the brakes. In an emergency situation, however, the normal anti-lock function is lost and the brakes could lock. Check the system as described in **BRAKES**.

6.8 Dim Lights

Headlights that are dim or gradually getting dimmer generally indicate a problem with the battery or charging system. The battery charge indicator light may come on as the lights are dimming. In either case, the engine and accessories are running off of the battery alone, and will soon discharge it altogether.

If possible, do not stop the engine unless you have the capability to jump start it. There may not be enough power in the starting system to restart the engine. Instead, turn off as many electrical consumers as possible. This will reduce the current drain and will allow the car to be driven farther before you lose all battery power.

With the engine and ignition off, check to see if the battery cables are firmly attached, or if there are any loose wires leading to the battery or to the alternator. Look for heavily corroded (covered by fluffy white deposits) wires and connectors.

Disconnecting, cleaning, and reinstalling corroded wires and connectors may solve the problem. Also check V-belt tension as described in **LUBRICATION AND MAINTENANCE**.

6.9 Towing

The cars covered by this manual can be towed either flat, on all four wheels, or by a tow truck using wheel lift or flat bed equipment.

> **NOTE —**
> Do not tow with sling-type equipment.

If flat-towing the car, use the towing eyes at the front of the car under the bumper. See Fig. 6-2. Set the transmission in neutral. BMW recommends using nylon tow ropes.

NOTE —
Installation of certain front spoilers may prevent access to the front towing eyes.

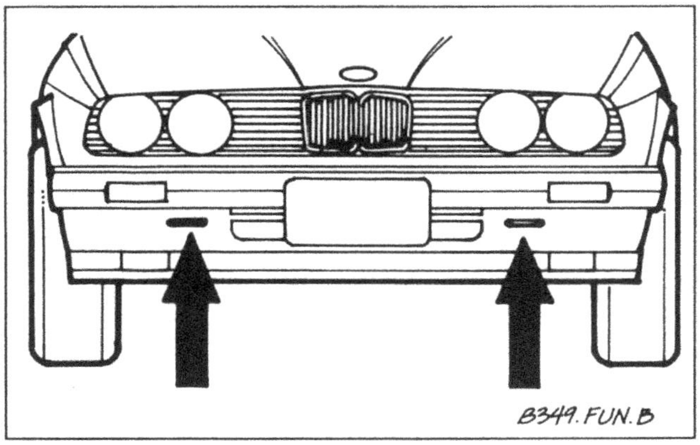

Fig. 6-2. Front towing eyes (arrows) used when flat-towing cars.

Towing a BMW with an automatic transmission while the rear wheels are on the ground can cause damage due to lack of lubrication. Always tow the car with the transmission lever in "N" (neutral) and the key in position "1." BMW recommends that cars with automatic transmission be towed with the rear wheels on the ground for no more than 30 miles (50 km), at no more than 30 mph (50 km/h). If the distance will be greater than 30 miles (50 km), either remove the driveshaft or add 1.05 quarts (1 liter) of ATF to the transmission. Reduce the fluid level to normal before driving the car.

NOTE —
There are no speed or distance restrictions when towing a car with a manual transmission, provided the transmission lubricant and final drive lubricant are filled to the correct levels.

6.10 Spare Parts Kit

Carrying a basic set of spare parts can prevent a minor breakdown from turning into a major annoyance. Many of the following items won't allow you to do major repair work on the car, but they will help in the event of the failure of something that can disable the car or compromise its safety.

Spare Parts Kit - Basic Contents:

1. V-belt for the alternator and water pump
2. one or two quarts of engine oil
3. a gallon container of engine coolant (premixed anti-freeze and water)
4. spare fuel pump relay, also spare main relay
5. a new, unopened bottle of brake fluid
6. 10 amp, 15 amp, and 20 amp fuses
7. upper and lower radiator hoses

Spare Parts Kit - Additional Contents:

1. replacement headlight (sealed beam or bulb)
2. brake light, turn signal light, and tail light bulbs
3. other relays such as main, headlight, turn signal, or load reduction
4. wiper blades
5. distributor cap and rotor

Section 2

LUBRICATION AND MAINTENANCE

Contents

TABLES

INTRODUCTION

The useful life of any car depends on the kind of maintenance it receives. The procedures described in this section of the manual include all of the routine checks and maintenance steps that are both required by BMW under the terms of their warranty protection and recommended by BMW to ensure long and reliable operation of your car. Also included are some instructions and recommendations for more basic car care.

BMW has taken a unique approach to establishing maintenance intervals for the cars covered by this manual. Most other manufacturers specify maintenance intervals strictly according to the number of miles driven or the number of months that have elapsed since the car's last service. It is well known, however, that mileage and time are not the only relevant factors that determine maintenance intervals. Aggressive driving, short trips, and frequent stops and starts are all harder on a car, and ideally call for more frequent maintenance. On the other hand, a more relaxed driving style, longer trips with the car fully warmed up, and mostly highway driving are easier on the car and can justify less frequent maintenance.

BMW's Service Indicator System computes maintenance intervals based not only on elapsed mileage, but also on such inputs as engine speed, engine temperature, number of starts, and lengths of trips. At the appropriate time, the system indicates that the next routine maintenance is due, based on the type of use experienced by that particular car.

BMW is constantly updating their recommended maintenance procedures and requirements. The information contained here is as accurate as possible at the time of publication. If there is any doubt about what procedures apply to a specific model or model year, or what intervals should be followed, remember that an authorized BMW dealer always has the latest information on factory-recommended maintenance.

1. GENERAL DESCRIPTION

Lubrication and maintenance refers to those routine procedures that are necessary to keep a car operating at its peak and to maintain the service requirements for full warranty coverage.

Maintenance Tables

These tables list all of the routine maintenance tasks for a particular model or model year that should be done at particular maintenance intervals. All of the applicable tables can be found under **2. Maintenance Tables.**

Fluid and Lubricant Specifications

The fluids and lubricants recommended for use in BMWs have been carefully chosen for their ability to perform under a wide range of conditions and to adequately protect your car. To maintain these high standards of performance, and to ensure that full warranty coverage remains in effect, use only the fluids and lubricants that meet the standards set forth by BMW and listed under **3. Fluid and Lubricant Specifications.**

Engine Oil Change

Regular changing of the engine lubricating oil and the engine oil filter is perhaps the single most important maintenance that a car can receive. It is also simple and easy.

The heading **4. Engine Oil Change** covers the basic details of checking and adding oil, as well as changing the oil and filter.

Tune-up

Much of what has traditionally been considered part of a tune-up has been rendered obsolete by sophisticated engine management technology. Therefore, tune-ups have become a less frequent and simpler part of maintenance.

The heading **5. Tune-up** covers those tasks that have traditionally been thought of as tune-up tasks, and that are still included by BMW as periodic routine maintenance.

Routine Maintenance — Engine Compartment

Many of the most important routine maintenance tasks are done under the hood within easy reach. They are grouped together so that more thorough maintenance can be planned and carried out most efficiently. See **6. Engine Compartment Maintenance** for more information.

Routine Maintenance — Chassis and Drivetrain

Thorough maintenance requires periodic inspection and servicing of parts that are only accessible by raising the car. Since this requires a suitable level workspace and the proper equipment to lift the car, a little more planning is required. As a convenient alternative, you may wish to leave these items to an authorized BMW dealer or other qualified and suitably equipped repair shop. See **7. Under-Car Maintenance** for more information. Please remember that the BMW body inspection for rust can only be carried out by an authorized BMW dealer. If any body work or repainting is performed, be certain to have an authorized BMW dealer inspect that work as well.

Routine Maintenance — Body and Interior

Periodic service and inspection of certain safety-related body and interior equipment is specified by BMW and covered under **8. Body and Interior Maintenance.**

Cleaning and Preserving

Aside from improving the car's appearance, cleaning and preserving can reduce the harmful effects of dirt and other contaminants which attack the finish. Information on recommended cleaning materials and methods can be found under **9. Cleaning and Preserving.**

2. MAINTENANCE TABLES

The tables under this heading list the routine maintenance tasks specified by BMW. As described in the introduction, the intervals for most of these tasks are determined by the BMW Service Indicator System. The intervals vary from car to car, depending on the way the car is used.

The maintenance intervals for a few additional items are based on either elapsed mileage or time. These intervals are clearly indicated in separate tables.

NOTE —
If the speedometer has been replaced be sure to take into account the different mileage readings so that your car is serviced when required.

2.1 BMW Service Indicator

The BMW Service Indicator notifies the driver when maintenance is required. The indicator, shown in Fig. 2-1, typically consists of nine light emitting diodes (LEDs)—five green, one yellow, and three red—as well as Oil Service and Inspection indicators.

When the ignition is turned on, the green LEDs come on. They go out when the engine is started. Immediately after a maintenance service, all five LEDs will be illuminated. As the car is driven in normal use, fewer and fewer green LEDs will be illuminated before start-up, indicating that the next maintenance interval is approaching.

When the car has accumulated sufficient use to require the next maintenance interval, the yellow LED will come on along with either the Oil Service indicator or the Inspection indicator. These will stay on after the engine is started. If maintenance service is delayed, the red LEDs will also illuminate, one by one, as a reminder that maintenance service is overdue.

Fig. 2-1. Service indicator display (arrow) in instrument panel.

An Oil Service interval will always be followed by an Inspection interval, which will then be followed by an Oil Service interval, and so on. Further explanations of these intervals are given below in **2.2 Oil Service** and **2.3 Inspection**.

Resetting Service Indicator

When the specified maintenance has been carried out, the service indicator memory needs to be reset. Resetting the service indicator turns out the instrument panel indicator lights.

The service indicator is reset using a special electronic tool. The tool is plugged into the diagnostic connector in the engine compartment, which is electrically connected to the service indicator circuit. See Fig. 2-2.

NOTE —
- Most of the models covered by this manual use a 15-pin diagnostic connector. Some late 528e models (built after March 1987) use a 20-pin connector. On models with a 20-pin connector, a special adaptor for the reset tool is usually required. Be sure to check with the tool manufacturer for the correct tool application.

- If the service indicator cannot be reset (lights stay on) after using the special reset tool, either the service indicator Ni-cad batteries or the service indicator board itself may be faulty. For more information on troubleshooting service indicator faults, see **ELECTRICAL SYSTEM.**

BMW specifies two individual tools, one for resetting the oil service memory (BMW tool no. 62 1 120) and one for resetting the inspection memory (BMW tool no. 62 1 100). On 528e

Fig. 2-2. 20-pin diagnostic connector (arrow) used on late 528e models. Other models use similar 15-pin connector.

models from built from March 1987 on, an additional adapter (BMW tool no. 62 1 140) is required. These tools are available from an authorized BMW dealer.

> ***CAUTION —***
> *Follow the manufacturer's directions when resetting the service indicator. If the reset procedures are done incorrectly, the reset tool or the electronic service indicator may be damaged.*

As an alternative, a single tool that resets both the inspection and oil service memory can be purchased from one of the following manufacturers:

Assenmacher Specialty Tools
6440 Odell Place
Boulder, CO 80301
(303) 530-2424

or

Peake Research, Automotive Products Division
P.O. Box 28776
San Jose, CA 95159-8776
(800) 231-6861

Table a, **Table b**, and **Table c** on the following pages list the maintenance tasks that need to be done at the intervals indicated by the BMW Service Indicator. Finally, additional maintenance tasks that are specified at particular time or mileage intervals are listed in **Table d**.

Except where noted, the maintenance items listed apply to all models covered by this manual. The boldface numbers after each listing are the headings in this section where the maintenance procedure is discussed. The columns on the right side of each table give quick-reference information about the job—whether tools are needed, whether the procedure requires new parts, whether the car should be warmed-up to normal operating temperature and, in some cases, a recommendation that the job be turned over to an authorized BMW dealer because of the need for special equipment or expertise.

> **NOTE —**
> For reference, BMW's inspection requirements are approximately equivalent to the maintenance requirements that other European manufacturers specify. Inspection I is normally due at intervals with a maximum of 15,000 miles or 12 months. Inspection II is normally due at intervals with a maximum of 30,000 miles or 24 months.

2.2 Oil Service

The Oil Service indicator signals the need for the most basic level of routine maintenance. BMW's required oil service specifies changing the engine lubricating oil and the engine oil filter after the engine has been warmed up. BMW-recommended additional maintenance for this same interval is also listed under oil service in **Table a** below.

> **NOTE —**
> • For reference, BMW's Oil Service requirements are approximately equivalent to the maintenance that other European manufacturers specify at intervals with a maximum of every 7,500 miles or 6 months.
>
> • Be sure to follow the instructions for resetting the Oil Service indicator light. The Inspection indicator can be accidentally reset rendering its function inaccurate.

2.3 Inspection

The Inspection indicator signals the need for more comprehensive maintenance and inspection. There are two sets of inspection requirements, Inspection I and Inspection II. These inspections alternate throughout a car's maintenance history. If the last inspection interval was Inspection I, the next inspection interval (following an oil service interval) will be Inspection II, the next after that will be Inspection I, and so on.

Inspection I tasks are listed in **Table b** below. Inspection II includes most of the tasks from Inspection I with additional Inspection II tasks. A complete listing of Inspection II tasks are listed in **Table c** below.

NOTE —
Aside from keeping your car in the best possible condition, proper maintenance plays a role in maintaining full protection under BMW's new-car warranty coverage. If in doubt about the terms and conditions of your car's warranty, an authorized BMW dealer should be able to explain them.

NOTE —
BMW specifies a one-time 1,200 mile inspection for all the cars covered by this manual. For more information on this inspection and on the BMW maintenance system, see your glove box information or an authorized BMW dealer.

Table a. Oil Service

Maintenance item	Tools required	New parts required	Warm engine required	Dealer service recommended
Engine compartment maintenance				
Change oil and oil filter **4.1**	*	*	*	
Check power steering fluid level **6.5**				
Check brake fluid level **7.2**				
Check clutch fluid level **7.4**				
Under car maintenance				
Check steering box and linkages. See **STEERING AND WHEEL ALIGNMENT**				*
Check brake calipers and rotors **7.2**				
Inspect brake system for damaged hoses and lines, leaks or damage **7.2**				
Check brake pad wear **7.2**				
Check hand brake, adjust cable if necessary **7.2**	*			
Check and adjust tire pressures including spare **7.1**				
Body and interior maintenance				
Check headlight and driving light aiming and adjust as necessary. See **BODY AND INTERIOR**	*			*
Check operation of headlights, parking lights, back-up lights, license plate lights, interior lights, glove box light, engine compartment light, trunk light, instrument panel lights, turn signals, emergency flashers, stop lights, horns, headlight flasher and dimmer switch				
Check active check control panel. See **ELECTRICAL SYSTEM**				
Check function of seat belts **8.3**				
Check windshield washer fluid level and concentration, add as necessary. Check operation of washer system. Check condition of wiper blades **8.1**				
Road Test				
Check braking performance, steering, heating and ventilation, manual or automatic transmission, and mirrors				

Table b. Inspection I

Maintenance item	Tools required	New parts required	Warm engine required	Dealer service recommended
Tune-up				
Check and adjust valve clearance **5.6**	*	*		
Engine compartment maintenance				
Change oil and oil filter **4.1**	*	*	*	
Check brake fluid level **7.2**				
Check clutch fluid level **7.4**				
Change automatic transmission fluid and strainer **7.5**	*	*	*	
Check battery acid level and correct as necessary **6.1**				
Check engine coolant level and anti-freeze protection and add as necessary. Inspect for cooling system leaks **6.4**	*			
Lubricate accelerator linkage and throttle linkage **6.2**				
Under car maintenance				
Check manual transmission oil level and add as necessary **7.4**	*			
Check final drive lubricant and add as necessary **7.7**	*			
Check fuel tank, fuel lines, and all connections for leaks **7.8**				
Inspect exhaust system **7.3**				
Check steering box and all linkages. See **STEERING AND WHEEL ALIGNMENT**				*
Check power steering system for leaks. Check fluid level **6.5**				
Check brake calipers and rotors. **7.2**				
Inspect brake system for damaged hoses and lines, leaks or damage **7.2**				
Check brake pad wear **7.2**				
Check hand brake, adjust cable if necessary **7.2**	*			
Inspect front suspension and steering for play **7.6**				*
Inspect wheels and tires, including spare, check tire pressure and condition **7.1**				
Body and Interior Maintenance				
Lubricate door hinges and hood latch **8.3**				
Check headlight and driving light aiming and adjust as necessary. See **BODY AND INTERIOR**	*			*
Check operation of headlights, parking lights, back-up lights, license plate lights, interior lights, glove box light, engine compartment light, trunk light, instrument panel lights, turn signals, emergency flashers, stop lights, horns, headlight flasher and dimmer switch				
Check active check control panel. See **ELECTRICAL SYSTEM**				

continued on next page

Table b. Inspection I (continued)

Maintenance item	Tools required	New parts required	Warm engine required	Dealer service recommended
Body and Interior Maintenance (cont'd)				
Check function of air conditioning and refrigerant charge. Tighten A/C compressor mounting bolts. See **BODY AND INTERIOR**	*			
Check function of seat belts **8.3**				
Check windshield washer fluid level and concentration, add as necessary. Check operation of washer system. Check condition of wiper blades **8.1**				
Tighten nuts and bolts for door locks and striker plates				
Check ground connection of SRS (airbag) front sensor screws to body, if applicable	*			*
Road Test				
Check braking performance, steering, heating and ventilation, manual or automatic transmission, and mirrors				

Table c. Inspection II

Maintenance item	Tools required	New parts required	Warm engine required	Dealer service recommended
Tune-up				
Check and adjust valve clearance **5.6**	*	*		
Replace spark plugs **5.2**	*	*		
Engine compartment maintenance				
Change oil and oil filter **4.1**	*	*	*	
Check engine timing belt tension.* See **ENGINE—SERVICE AND REPAIR**	*	*		*
Check brake fluid level **7.2**				
Check clutch fluid level **7.4**				
Check V-belt tension and condition **6.3**	*			
Check battery acid level and correct as necessary **6.1**				
Check engine coolant level and anti-freeze protection and add as necessary. Inspect for cooling system leaks **6.4**	*			
Lubricate accelerator linkage and throttle linkage **6.2**				
Change air filter **5.1**	*	*		

* Change timing belt every second Inspection II or 4 years

continued on next page

Table c. Inspection II (continued)

Maintenance item	Tools required	New parts required	Warm engine required	Dealer service recommended
Under car maintenance				
Check automatic transmission fluid **7.5**				
Change manual transmission oil **7.4**	*		*	
Check clutch plate for wear **7.4**	*			
Check drive axle boots for leaks **7.7**				
Check front wheel bearing play. See **SUSPENSION—FRONT**	*			
Change final drive fluid **7.7**	*		*	
Check fuel tank, fuel lines, and all connections for leaks **7.8**				
Change fuel filter.** See **FUEL SUPPLY**	*	*		
Inspect exhaust system **7.3**				
Check steering box and all linkages. See **STEERING AND WHEEL ALIGNMENT**				*
Check power steering system for leaks. Check fluid level **6.5**				
Check brake calipers and rotors. **7.2**				
Inspect brake system for damaged hoses and lines, leaks or damage **7.2**				
Check brake pad wear **7.2**				
Inspect parking brake cable, adjust as necessary **7.2**	*			
Inspect front suspension and steering for play **7.6**	*			*
Inspect wheels and tires, including spare, check tire pressure and condition **7.1**				
Body and Interior Maintenance				
Lubricate door hinges and hood latch **8.3**				
Check headlight and driving light aiming and adjust as necessary. See **BODY AND INTERIOR**	*			*
Check operation of headlights, parking lights, back-up lights, license plate lights, interior lights, glove box light, engine compartment light, trunk light, instrument panel lights, turn signals, emergency flashers, stop lights, horns, headlight flasher and dimmer switch				
Check active check control panel. See **ELECTRICAL SYSTEM**				
Check function of air conditioning and refrigerant charge. Tighten A/C compressor mounting bolts. See **BODY AND INTERIOR**	*			*
Check function of seat belts **8.3**				
Check windshield washer fluid level and concentration, add as necessary. Check operation of washer system. Check condition of wiper blades **8.1**				
Tighten nuts and bolts for door locks and striker plates				
Check ground connection of SRS (airbag) front sensor screws to body, if applicable	*			*
Road Test				
Check braking performance, steering, heating and ventilation, manual or automatic transmission, and mirrors				

** Recommended for California models.

Table d. Routine Maintenance — Time and Mileage Intervals

Maintenance item	every 12 months	every 24 months	every 30,000 miles (48,000 km)	every 50,000 miles (80,000 km)	every 60,000 miles (96,000 km)	Tools required	New parts required	Dealer service recommended
Replace oxygen sensor ***								
528e, 533i models without heated sensor			*		*	*	*	
528e models with heated sensor				*		*	*	
535i models				*		*	*	
Replace brake fluid	*					*		
Drain and flush cooling system and replace coolant		*				*		
Anti-corrosion warranty inspection ****	*							*

*** See **FUEL INJECTION** for information on oxygen sensors. A single wire oxygen sensor is unheated. Multi-wire (three or four wire) sensors are heated.
**** Must be performed and documented by an authorized BMW dealer every year.

> **NOTE —**
> The camshaft timing belt should be replaced every 60,000 miles (100,000 km), every 4 years (48 months), or every second inspection II, whichever comes first

3. FLUID AND LUBRICANT SPECIFICATIONS

The fluids and lubricants specified by BMW for use in the cars covered by this manual are listed below. **Table e** gives engine oil viscosity (SAE grade) vs. operating temperature range for the all BMW engine types covered in this manual. **Table f** lists fluid and lubricant specifications.

> ***CAUTION —***
> - *The use of fluids that do not meet BMW's specifications may impair performance and reliability, and may void warranty coverage.*
> - *Avoid high-speed long distance driving when using SAE 5W-20 or SAE 10W oil, especially if the outside temperature rises above the indicated limits. If maximum loads on the engine or continuous speeds above 60 mph (100 km/h) are expected, use an oil with the next higher viscosity rating.*

Table e. Oil Viscosity Requirement vs. Temperature

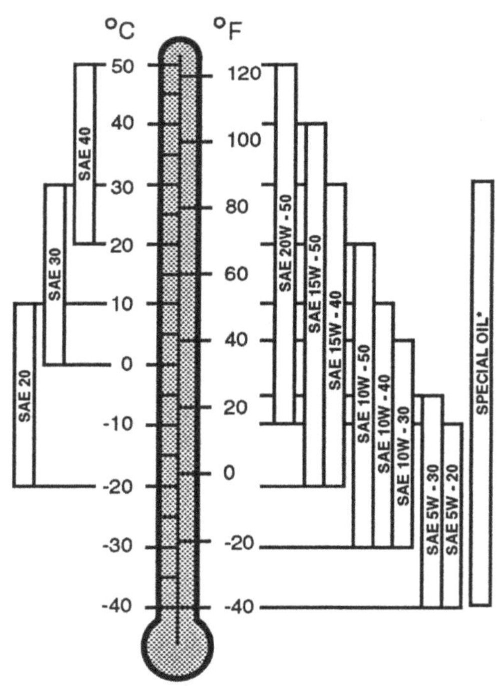

*Engine oils specially formulated and approved by BMW. See an authorized BMW dealer for more info.

Table f. Fluids and Lubricants

Fluid	Approximate capacity	Specification
Engine oil		
528e	4.0 l (4.2 US qt.)	API service SE or SF
with filter change	add 0.25 l (0.26 US qt.)	
533i	5.0 l (5.3 US qt.)	API service SE or SF
with filter change	add 0.75 L (0.8 US qt.)	
535i	5.0 l (5.3 US qt.)	API service SE or SF
with filter change	add 0.75 L (0.8 US qt.)	
Manual transmission oil		
1982 528e (Getrag 265)	1.6 l (1.7 US qt.)	
1983-1988 528e (Getrag 260)	1.25 l (1.3 qt.)	
1983-1984 533i (Getrag 260)	1.25 l (1.32 qt.)	
1985 535i (Getrag 265)	1.6 l (1.7 US qt.)	
1986-1988 535i (Getrag 260)	1.25 l (1.3 qt.)	SAE 80, API GL-4, MIL-L-2105 (non-hypoid type) For alternate oils see text
Automatic transmission fluid (ATF)(drain and fill)		
1982-1983 (528e, 533i)	2.0 l (2.1 US qt.) approx.*	Dexron® or Dexron II® ATF
1984-1988 (528e, 533i, 535i)	3.0 l (3.2 US qt.) approx.*	Dexron II® or Dexron® ATF
Final drive (drain and fill)		
528e, 533i	1.8 l (1.9** US qt.)	Hypoid gear oil SAE 90, GL-5 (see text)
535i	1.9 l (2.0** US qt.)	Hypoid gear oil SAE 90, GL-5 (see text)
Power steering fluid	permanently sealed, no drain plug	Dexron II® ATF
Brake fluid	as necessary	Dot 4
Engine coolant		50% phosphate-free ethylene glycol anti-freeze
528e	11.0 l (11.6 US qt.)	
533i, 535i	12.0 L (12.7 US qt.)	

* Additional fluid required when installing a dry torque converter. Add fluid as necessary.
** Additional fluid required when installing a rebuilt final drive.

Engine Oil

Engine oil provides a lubricating film between all moving parts, and also helps cool the engine. Maintaining an adequate supply of clean oil is one of the best ways of making an engine last. Some engine oil is consumed during normal operation, making it necessary to regularly check and "top up" the oil supply. Since oil becomes contaminated and breaks down over time, regular oil changes are necessary.

Engine oil requirements are defined by the oil's American Petroleum Institute (API) service rating and by the Society of Automotive Engineers (SAE) viscosity rating. This information can be found on the oil can or bottle, often on a standard label.

The API service rating designates the type of use suited for the oil. The rating is based on the additives that are used to resist break-down and carbon formation, inhibit corrosion, resist foaming, neutralize acids, and help remove deposits and keep contaminants suspended in the oil. Although SE rated oil can be used in the engines covered by this manual, BMW recommends using SF rated oils.

The SAE viscosity rating indicates resistance to flow. An oil designated SAE 40 has a higher viscosity (greater resistance to flow) than an oil designated SAE 30. While higher viscosity oils will generally offer greater engine protection, they may be too thick and resistant to flow and may inhibit starting in cold weather.

The correct engine oil viscosity depends on the operating temperature range. See the viscosity vs. temperature table above. Select a viscosity rating for the lowest anticipated temperature at which the engine must start.

Multi-viscosity oils have additives that make them suitable for use over a wider range of temperatures. For example, an oil rated SAE 10W-30 offers the flow characteristics of SAE 10 at low temperatures, but the protection capability of SAE 30 at engine operating temperature. The "W" in the SAE rating indicates that the oil is suitable for winter use.

Oils of different viscosity ratings can be mixed, but mixing oils of different API service ratings or brands is not recommended.

Manual Transmission Gear Oil

Gear oil requirements are also defined by API service rating and SAE viscosity rating, as described in **Engine Oil** above. For most driving conditions, a SAE 80, GL-4 rated non-hypoid gear oil is recommended by BMW for use in the manual transmission. In areas where cold temperatures are encountered, a BMW-approved straight weight engine oil or ATF can be used to help reduce the amount of effort required to shift the transmission.

Table g lists manual transmission oil viscosity (SAE grade) vs. operating temperature range. Using engine oil or ATF in the manual transmission will not affect the service life of the transmission, although BMW recommends that the heavier gear oil should be used during warmer temperatures.

Table g. Manual Transmission Oil vs. Temperature

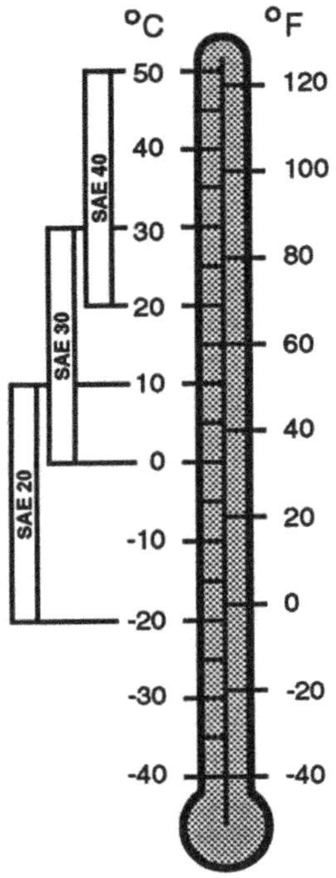

B415.LUB.B

CAUTION —
Multi-viscosity engine oils should not be used in the manual transmission. Use of such an oil could shorten the service life of the transmission.

NOTE —
- Changing the manual transmission gear oil to a less viscous engine oil (lower SAE viscosity rating) or ATF may increase the level of gear noise in the passenger compartment.

- On some 1984 528e models, the manual transmission was delivered from the factory filled with Mobil SHC 630 synthetic gear lubricant. These transmissions are identified with a green sticker on the case and can be refilled using the guidelines above. However, if refilling the transmission with synthetic lubricant, use only Mobil SHC 630.

- Synthetic gear lubricant may be compatible with other BMW transmissions. For information on BMW-approved synthetic lubricants and its use, contact an authorized BMW service department.

Final Drive Gear Oil

Owing to the demanding requirements of the final drive lubricant, BMW recommends using only a specially formulated gear oil that is available through an authorized BMW dealer parts department. For additional information on this lubricant and any other lubricant that may be compatible, contact an authorized BMW dealer service department.

Brake Fluid

Brake fluid deserves special consideration. It absorbs moisture easily, and moisture in the fluid affects brake performance and reliability. When replacing or adding brake fluid, use only new fluid from previously unopened containers. Do not use brake fluid that has been bled from the system, even if it is brand new. Use only DOT 4 brake fluid.

Engine Coolant
(Anti-freeze)

BMW recommends coolant that is a mixture of water and phosphate-free anti-freeze containing ethylene glycol. Anti-freeze raises the boiling point and lowers the freezing point of the coolant. It also contains additives that help prevent cooling system corrosion.

Power Steering Fluid

The power steering fluid is Dexron® or Dexron II® ATF. The system is permanently filled and does not have a drain. Routinely adding ATF is not required unless the system is leaking. On 533i and 535i models the power steering reservoir also serves as the hydraulic brake boost reservoir.

Gasoline Additive

Many gasolines do not contain the necessary additives to help prevent deposits on the fuel injectors and intake valves. For this reason, BMW recommends the periodic use of BMW Gasoline Additive (BMW Part No. 88 88 6 900 314). The additive is available from an authorized BMW dealer.

CAUTION —
Follow all label directions. Do not use a gasoline additive more than the manufacturer recommends. Fuel additives should not be used in conjunction with high-detergency fuel. Exceeding the recommended amount of fuel additive can lead to oil dilution and possible engine damage.

In extreme cases where clogged injectors and carbon deposits are severe, more comprehensive work may be required to completely solve the problem. For complete information on approved methods on injector cleaning and intake valve decarbonizing, consult an authorized BMW dealer.

Greases

Two different types of grease are used for lubrication of drive train and brake components. Multipurpose grease (lithium grease) has a wider temperature tolerance range than ordinary grease and should be used for most general lubrication purposes, including roller bearings.

Molybdenum grease is lithium grease with a friction-reducing molybdenum disulfide additive. This grease is recommended for certain applications including lubrication of the drive axle joints.

4. ENGINE OIL CHANGE

The engine oil level is checked with a dipstick located in the engine block behind the alternator. Check the level by pulling out the dipstick and wiping it clean. Reinsert it all the way and withdraw it again. The oil level is correct if it is between the two marks near the end of the stick. The location of the dipstick and the level marks are shown in Fig. 4-1. The upper (**MAX**) mark indicates full, the highest acceptable oil level. The lower mark (**MIN**) indicates the minimum acceptable level.

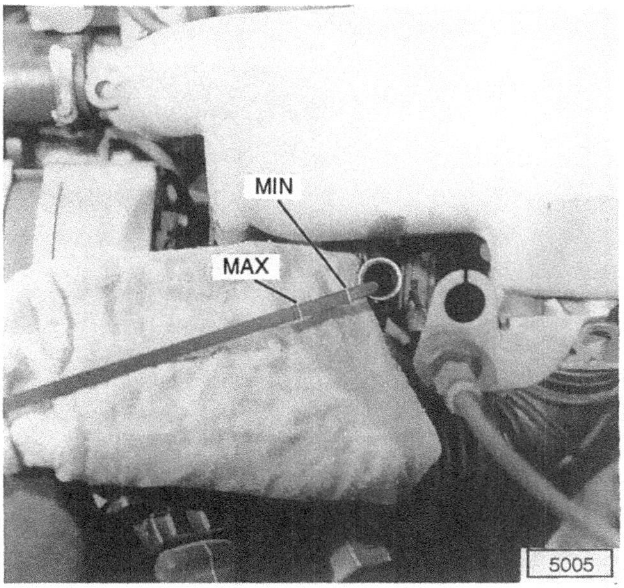

Fig. 4-1. Maximum (**MAX**) and minimum (**MIN**) oil level marks on dipstick.

Always check the oil with the car on a level surface, after the engine has been stopped for at least a few minutes. For the most accurate check, wait a few hours.

Add oil through the filler cap on the top of the cylinder head cover shown in Fig. 4-2. Add only the amount needed to bring the oil level to the **MAX** mark on the dipstick, using an oil of the correct viscosity and grade as described above in **3. Fluid and Lubricant Specifications.** Too much oil can be just as harmful as too little.

Fig. 4-2. Location of oil filler cap. Remove and install by turning (arrows).

The amount of oil that needs to be added between oil changes varies from one engine to another. Generally, a new engine or an engine operated routinely at high speeds will consume more oil. It is helpful to become familiar with the rate at which a particular engine requires oil. A sudden increase may be an early warning of engine mechanical problems.

4.1 Changing Engine Oil and Filter

The oil service light in the service interval indicator should be the basic guide to scheduling oil changes. Do not rely on the color of the oil on the dipstick to indicate when a change is needed. Because of the detergent additives in the oil, fresh oil can look dark after only a few hundred miles.

The oil service light should come on at or before 7,500 miles. If the car is used primarily for short trips in slow moving traffic, or routinely operated aggressively, the oil service light should come on earlier. In general, changing the oil at more frequent intervals will help better protect the engine and promote longer engine life.

A complete oil change requires approximately 5 qt. of new oil (see **3. Fluid and Lubricant Specifications**), a new oil filter, and a new drain plug sealing washer. Oil filter and sealing washer part numbers are listed in **Table h**. The tools needed—a drain plug socket or box wrench (17 or 19 mm), a drain pan of at least 6 US qt. (5.6 L) capacity, and an oil filter wrench—are described in **FUNDAMENTALS**.

There are two different types of oil filter mountings, depending on model. On 528e models, the filter is the common spin-on type. On 533i and 535i models, the filter is a cartridge type, contained in a metal housing. The procedure below covers both types of filter.

> **NOTE —**
> If using a "fast-lube" service facility for oil changes, make sure the technician hand-starts and torques the engine oil drain plug using only hand-tools. Using power tools can strip the threads of the plug and the oil pan.

Table h. Engine Oil Change Parts

Engine oil filter	
528e (spin-on)	BMW Part No.11 42 1 266 773
533i, 535i (cartridge)	BMW Part No.11 42 1 718 816
O-ring for 533i, 535i oil filter	BMW Part No.11 42 1 266 374
Oil pan drain plug sealing washer (all engines)	BMW Part No. 01 11 9 963 130

To change oil and filter:

1. Run the car for a few minutes to slightly warm the engine and the oil, then shut the engine off.

2. With the car on level ground, place a drain pan under the oil drain plug shown in Fig. 4-3.

Fig. 4-3. Engine oil drain plug (arrow) in oil pan underneath engine.

> **NOTE —**
> The car will not need to be raised if a shallow drain pan is used.

3. Using a socket or box wrench, loosen the drain plug. By hand, remove the plug and let the oil drain into the pan.

> ***CAUTION —***
> *Pull the loose plug away from the hole quickly to avoid being burned by the hot oil. It will run out quickly when the plug is removed.*

> **NOTE —**
> If possible, use gloves to protect your hands.

4. When the oil flow has diminished to an occasional drip, reinstall the drain plug with a new metal sealing washer and torque the plug.

Tightening torques	
• engine oil drain plug (17mm wrench size)	33±3 Nm (24±2 ft-lb)
• engine oil drain plug (19mm wrench size)	59 to 64 Nm (43 to 47 ft-lb)

5. Position the drain pan directly under the oil filter and loosen the filter by turning it clockwise. On 528e models use an oil filter wrench. On 533i and 535i models loosen the filter housing bolt. See Fig. 4-4.

Fig. 4-4. Typical engine oil filter (arrow) located in left-hand (driver's) side of engine. Air Filter housing removed from car.

6. After the oil stops dripping, wipe clean the oil filter gasket surface on the filter mounting flange.

7. On 528e models, lubricate the rubber gasket of the new oil filter with a light coating of clean engine oil. Install the filter by hand until the gasket contacts the mounting flange, then turn the filter another $^1/_2$ turn to tighten it.

8. On 533i and 535i models, remove the old O-ring. Lubricate and install the new O-ring in the groove. Install the housing with a new filter cartridge, then tighten the filter housing bolt.

Tightening torque	
• 533i, 535i engine oil filter housing bolt	30±3 Nm (22±2 ft-lb)

> ***CAUTION —***
> *Overtightening the oil filter will make the next change much more difficult, and may deform the gasket, causing leaks.*

9. Refill the crankcase with oil. Approximate oil capacity is listed in **Table f** above. Use the dipstick to check for the correct oil level. Oil specifications are found in **3. Fluid and Lubricant Specifications**.

10. Start the engine and check that the oil pressure warning light immediately goes out. Allow the engine to run for a few minutes to circulate the new oil, then check for leaks at the drain plug and around the oil filter. Stop the engine and recheck the oil level.

> ***CAUTION —***
> *If the oil pressure warning light does not immediately go out after the engine is started, quickly turn the engine off. Loosen the oil filter approximately 1/4 turn (90°) and restart the engine. As soon as oil begins to run out of the filter, turn the engine off and tighten the filter. This will release any trapped air that is blocking oil flow.*

> **NOTE —**
> Dispose of the used oil properly. Use tight-sealing containers and mark them clearly. Check with the place of purchase about disposal.

5. TUNE-UP

A tune-up is regular maintenance of the ignition and fuel systems to compensate for normal wear. Modern BMW electronic ignition and fuel injection systems have eliminated much of the work involved in a tune-up. For the BMWs covered by this manual, only limited tune-up maintenance is necessary to maintain peak performance and economy.

5.1 Air Filter

The specified maintenance intervals for the air filter are based on normal use. If the car is operated primarily in dusty conditions, the air filter should be serviced more frequently. A dirty air filter starves the engine for air, reducing power output and increasing fuel consumption. Fig. 5-1 shows the typical location of the air filter housing for the engines covered by this manual.

Fig. 5-1. Air filter housing in front left (driver's) side of engine compartment. Loosen nuts and unclip spring clips (arrows) to remove air filter.

The upper and lower parts of the air filter housing are fastened together with spring clips around the outside edge. To replace the air filter element, loosen the upper air filter housing mounting nuts and unfasten the clips. See Fig. 5-1 above. Separate the upper air filter housing from the lower housing just enough to remove the filter element. See Fig. 5-2. Take note of the filter's installed position. Wipe the inside of the air filter housing using a lint-free cloth and install the new filter. Reinstall the upper air filter housing, making sure that the two halves are mated correctly. Refasten the spring-clips and tighten the mounting nuts.

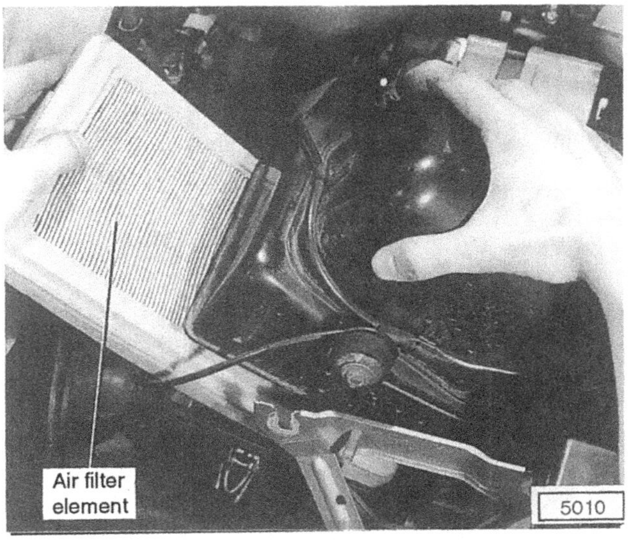

Fig. 5-2. Air filter being removed from air filter housing. Air filter housing will vary slightly between models.

CAUTION —
If a used air filter element is to be reinstalled, it is important to reinstall the filter in the same position that it was in before removal. Reversed installation will allow accumulated dirt to be drawn into the engine.

5.2 Spark Plugs

High temperature and high-voltage sparks eventually wear out the spark plug electrodes, and the spark plugs must be replaced. To replace a spark plug, gently remove the spark plug wire by pulling on the protective boot. Blow or brush away any dirt from around the base of the plug to prevent it from entering the engine when the plug is removed.

CAUTION —
Spark plugs should be replaced one at a time so that the spark plug wires do not get mixed up. If all of the wires need to be removed together, label each wire so that they can be reinstalled on the correct spark plug.

Use a 13/16 in. spark plug socket to remove spark plugs. The correct spark plugs for the different engines covered by this manual are listed in **Table i.** Use a spark plug gap gauge to check the gap. If necessary, bend the outer electrode slightly to adjust the gap to meet the specification. Do not bend or file the center electrode.

Table i. Spark Plug Specifications

Model	Spark plug	Electrode gap
	Bosch	
528e, 533i*, 535i	WR9LS	0.7±0.1 mm (0.027±0.004 in.)

* 1983 533i models use WR9DS

Lightly lubricate the new spark plug threads with a little oil. Thread the plugs into the cylinder head by hand to prevent cross-threading. Torque the spark plugs.

Tightening torque
• spark plugs . . . . 20 to 30 Nm (15 to 22 ft-lb)

Inspect the old plugs. Spark plug condition is a good indicator of combustion quality and can help diagnose engine faults. Fig. 5-3 shows examples of spark plug condition and what they mean.

NOTE —
Any of the abnormal spark plug conditions described below could also result from spark plugs of the wrong specification being installed. Check replacement plugs carefully and follow the spark plug manufacturer's recommendations.

Normal

Normal spark plug has gray or light tan color that indicates proper combustion

Oil-fouled

Oil-fouled spark plug has wet, oily black deposits caused by excess engine oil getting into combustion chamber, probably due to worn piston rings or valve guide oil seals.

Carbon-fouled

Carbon-fouled spark plug has dry, sooty black deposits caused by too much fuel that may indicate fuel injection or ignition problems.

Worn out

Worn out spark plug may have correct gray or light tan color, but shows physical deterioration (enlarged gap, eroded electrodes).

Fig. 5-3. Spark plug appearance that may indicate engine condition. For more information on interpreting spark plug condition, see **ENGINE—SERVICE AND REPAIR**. Photos courtesy of Champion Spark Plug Co.

Lightly-fouled spark plugs can be cleaned with a light wire brush. Remove all debris from around the electrode. Do not chip the ceramic insulator. Badly fouled spark plugs should be replaced and the cause of the fouling should be investigated and corrected.

5.3 Distributor Cap, Rotor, and Spark Plug Wires

The distributor cap, the rotor, and the spark plug wires deliver a high-voltage spark to the spark plugs. They are subject to insulation breakdown, corrosion fouling, and electrode wear and damage. The components should be check and replaced as necessary at the intervals listed under **2. Maintenance Tables** to ensure maximum ignition system efficiency. Guidelines for visual inspection and testing, and instructions for replacement are found in **IGNITION**.

Firing Order

Each spark plug wire leads from a specific terminal on the distributor cap to a specific spark plug. The ignition firing order is 1-5-3-6-2-4. Cylinder no. 1 is at the front of the engine and the rotor turns in a clockwise direction when viewed from the front of the engine.

When the engine is running, the rotor will contact terminal 1 in the cap, then terminals 5, 3, 6, 2, 4, and return to terminal 1. During this time, the crankshaft has made 2 complete rotations.

When removing the wires, label their positions so that they can be reinstalled in the proper places. If the wires get mixed up, see **IGNITION** for more information on the firing order.

> ***WARNING —***
> *Lethal voltages are present at the distributor, ignition coil, spark plug wires, and spark plugs when the key is on or the engine is running.*

5.4 Fuel Filter

Because of varying quality of gasoline, the fuel filter may become clogged enough to restrict fuel flow. To prevent any such problems, and to guarantee continued good performance, the filter should be replaced at the specified interval. The fuel filter is located beneath the right rear of the car, just in front of the gas tank. See Fig. 5-4.

Fig. 5-4. Fuel filter beneath car near right rear wheel. Direction of flow is indicated by arrow or markings on filter housing. Loosen clamping bracket at filter center (**A**) and two hose clamps (**B**). Replace hose clamps.

When replacing the fuel filter, disconnect the battery negative (–) cable and clamp the filter inlet and outlet hoses to lessen fuel spillage. Loosen the center mounting bracket and the two hose clamps on either end of the filter. Note the arrow or markings indicating direction of flow on the new filter. Install the filter and use new hose clamps.

> ***WARNING —***
> *Fuel will be expelled when the filter is removed. Do not smoke or work near heaters or other fire hazards. Keep a fire extinguisher handy.*

> ***CAUTION —***
> *Clean thoroughly around the filter connections before removing them, and make sure that no dirt gets into the fuel lines.*

5.5 Idle Speed

Engine idle speed can change due to a number of factors, including normal wear of engine and fuel injection components. The idle speed is electronically controlled and is non-adjustable. See **FUEL INJECTION** for detailed information.

5.6 Valve Adjustment

All of the engines covered by this manual may require periodic valve clearance adjustment. The complete valve adjustment procedure is covered in **ENGINE—SERVICE AND REPAIR**.

6. ENGINE COMPARTMENT MAINTENANCE

The information under this heading describes the routine maintenance—other than oil change and tune-up—that is done in the engine compartment. It is not necessary that the car be raised and supported off the ground. For information on oil change and tune-up, see **4. Engine Oil Change** and **5. Tune-up**.

6.1 Battery

Simple maintenance of the battery and its terminal connections will ensure maximum starting performance, especially in winter when colder temperatures reduce battery power. For a more detailed discussion of the battery and charging system, see **BATTERY, STARTER, ALTERNATOR**.

Checking and Cleaning Battery

Battery cable clamps should be tight. The terminals, the cable clamps, and the battery case should be clean and free of the white deposits that indicate corrosion and acid salts. Even a thin layer of dust containing conductive acid salts can cause the battery to discharge.

To remove battery corrosion, begin by disconnecting the cables. Disconnect the negative (–) cable first. Clean the terminal posts and the cable clamps with a wire brush. Clean the main chassis ground terminal next to the battery. Corrosion can be washed away with a baking soda and water solution that will neutralize the acid. Apply the solution carefully, though, since it will also neutralize the acid inside the battery. Avoid getting the solution into the battery cells through vent holes. Reconnect the cable clamps, positive (+) cable first. Lightly coat the outsides of the terminals, hold down screw, and clamps with petroleum jelly, grease, or a commercial battery terminal corrosion inhibitor.

WARNING —
- *Battery acid is extremely dangerous. Take care to keep it from contacting eyes, skin, or clothing. Wear eye protection. Extinguish all smoking materials and do not work near any open flames.*

- *Disconnecting the battery cables with the engine running, or reconnecting the cables to the incorrect posts will damage the electrical system.*

- *BMW Anti-theft radios can be rendered useless by disconnecting battery cables. See your owner's manual for more information.*

Battery electrolyte should be maintained at the correct level just above the battery plates and their separators. The correct level is approximately 5 mm (1/4 in.) above the top of battery plates or to the top of the indicator marks (if applicable). The battery plates and the indicator marks can be seen once the filler caps are removed. If the electrolyte level is low, replenish it by adding distilled water only.

For additional information on batteries, see **BATTERY, STARTER, ALTERNATOR**.

Replacing Battery

Batteries are rated by ampere hours (Ah), the number of hours they can sustain a specific current drain before complete discharge, or by cold cranking amps (CCA), the number of amps they produce to crank the engine in cold weather conditions. They may be rated according to European (DIN) standards, by Society of Automotive Engineers (SAE) standards, or both. In general, replacement batteries should always be rated equal or higher than the original battery.

The battery is held in place by a single hand screw and plate. A secure battery hold-down is important in order to prevent vibrations and road shock from damaging the plates.

NOTE —
Always disconnect the negative (–) cable first, and connect it last. While changing batteries, clean away any corrosion in or around the battery tray using a baking soda and water solution.

Charging Battery

A discharged battery is not necessarily faulty. It may be restored by recharging, using a battery charger. There are some limitations on the rate at which low-maintenance batteries may be charged. Frozen batteries should be recharged only after they have thawed. For complete information on battery charging and applicable cautions and warnings, see **BATTERY, STARTER, ALTERNATOR**.

CAUTION —
*Do not jump start the car without first removing the fuses for the on-board computer. If the fuses are not removed, the internal fuse in the on-board computer module may blow due to excessive voltage. See **BATTERY, STARTER, ALTERNATOR** or **FUNDAMENTALS** for more information.*

6.2 Accelerator and Throttle Linkage

The accelerator and throttle linkage should be lubricated at the intervals described under **2. Maintenance Tables.** Use a general purpose oil on the joints and bearings of the linkage. Use a multipurpose grease on the bearing points of the throttle plate.

6.3 V-Belts

V-belts and pulleys transfer power from the engine crankshaft to various accessories. Cars covered by this manual have at least one V-belt, and may have as many as three, depending on the accessories installed.

Inspecting and Adjusting V-Belts

Incorrect V-belt tension can decrease the life of the belt and the component it drives. Inspect belts with the engine off. Twist the belt to inspect its sidewalls and bottom. Belt structural damage, glazed or shiny sidewalls caused by a loose belt, or separation caused by oil contamination are all reasons to replace a belt. Some of these faults are illustrated in Fig. 6-1.

V-belt squealing is normally caused by incorrect belt tension (too loose) or by contamination between the belt and pulley. Extremely loud squealing may only be corrected by replacing the belt. Belt dressings should not be used to correct the problem. Many dressings contain oil-based compounds that can soften the rubber and reduce belt life.

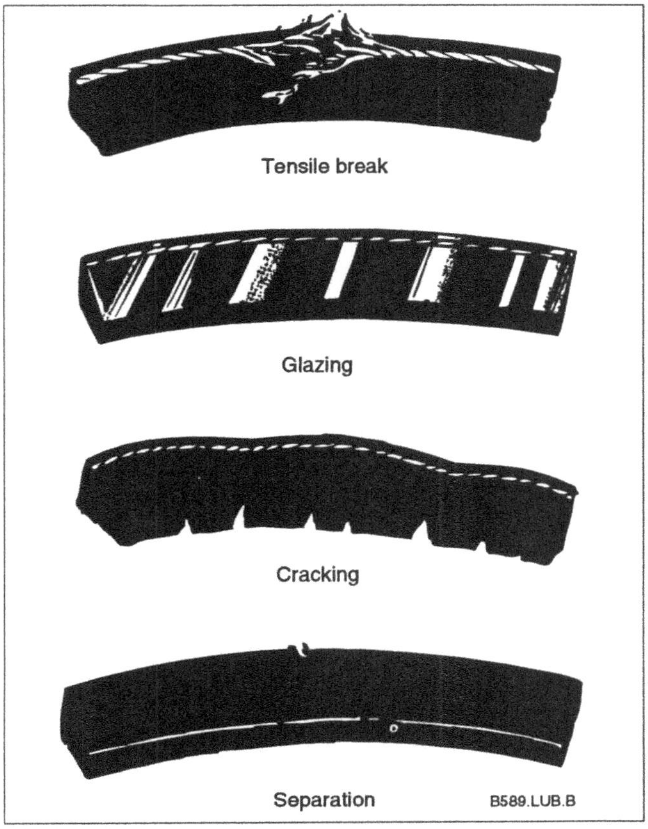

Fig. 6-1. Examples of belt failure. Courtesy of Gates Rubber Inc.

The drive belt tension for the alternator and power steering is adjusted through a toothed-rack mechanism. To accurately tension the drive belt for the air conditioning compressor, a special drive belt tensioning tool should be used.

Adjust the alternator or the power steering belt by first loosening all of the mounting nuts until the unit is able to move freely. Use a torque wrench and a crowfoot wrench to turn the tensioning gear bolt. See Fig. 6-2. Hold the wrench steady and tighten the locknut on the rear of the tensioning gear bolt. Tighten all other mounting nuts.

The air conditioning compressor V-belt is adjusted using a toothed-rack adjusting mechanism similar to that for the power steering pump and alternator. In general, the air conditioning compressor drive belt can be adjusted by loosening the compressor's mounting bolts and pivoting the compressor using the toothed-rack adjuster. Check the belt tension using a V-belt tension gauge.

NOTE —
When using a belt tension gauge on V-rib type belts, make sure the gauge's pulling hook is on the tip of a V-rib and not between ribs. Otherwise, inaccurate readings may result.

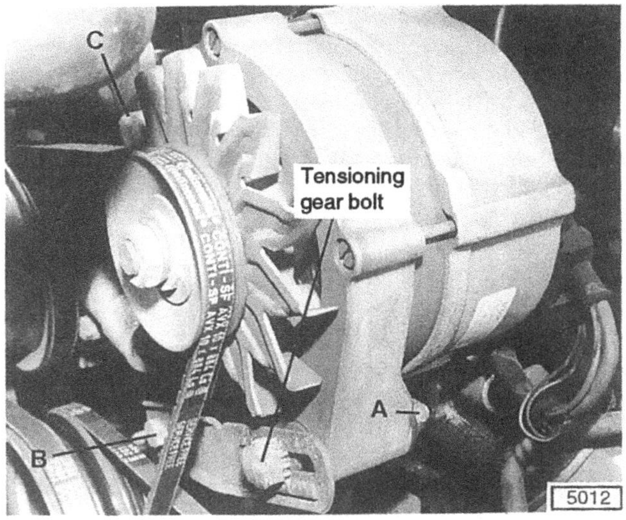

Fig. 6-2. Typical alternator mounting bolts (**A** and **B**), and pivot bolt (**C**). Power steering pump similar.

Tightening torques	
• tensioning gear (alternator belt tensioning)	7 Nm (62 in-lb)
• tensioning gear (power steering belt tensioning)	8 to 8.5 Nm (71 to 75 in-lb)
• tensioning gear locking nut	24 Nm (18 ft-lb)

Replacing V-Belts

To reduce the chance of V-belt failure while driving, replacement of the belts every four years is recommended. Loosen the mounting bolts and adjust until the belt tension is as loose as possible, then remove the belt by slipping it over the pulleys. In some cases it may be necessary to remove one V-belt to get to another. Cross section and length determine belt size. Use the old belt for comparison, or make sure that the new belt fits into the pulley groove as shown in Fig. 6-3.

NOTE —
When belts are replaced with new ones, keep the old set in the luggage compartment by the spare tire for emergency use.

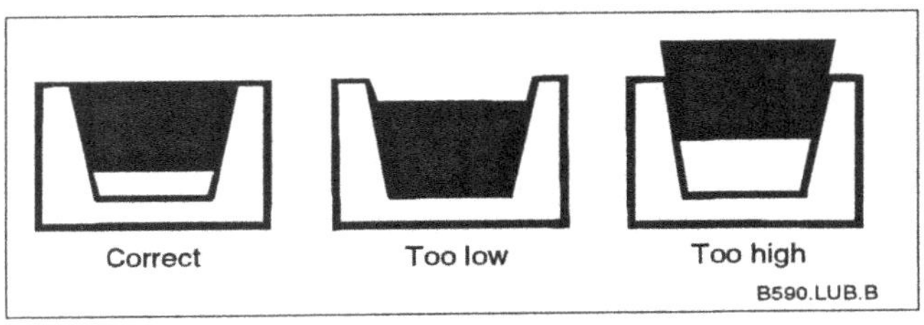

Fig. 6-3. Cross-section of correct V-belt position in pulley. Courtesy of Gates Rubber Inc.

With the belt off, clean the pulleys using a suitable solvent. Inspect the pulleys for wear or damage that may cause early failure of the new belt. A straight edge on either side of pulley is good way to check for wear. See Fig. 6-4. This is also a good opportunity to inspect the belt-driven accessory, checking for bearing wear and excess play, for example. When installing the new belt, gently pry it over the pulleys. Too much force may damage the belt or the accessory. Tension the belt(s), run the engine for a few minutes (at least 1500 rpm), then recheck the belt tension.

CAUTION —
Do not over tighten the V-belts. Overtightening will cause the bearings to fail prematurely.

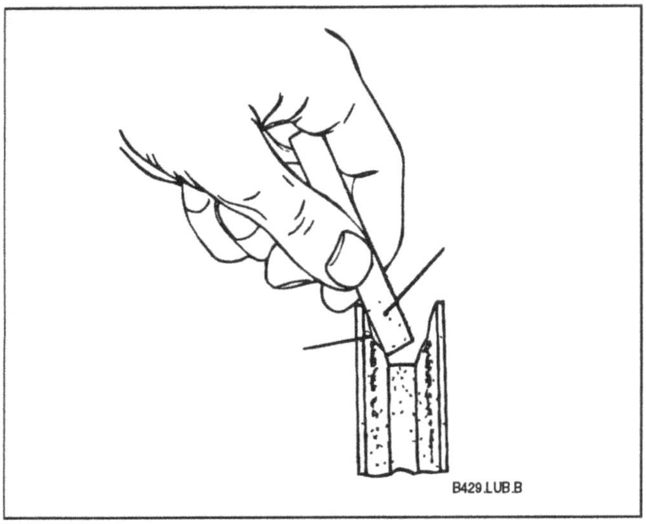

Fig. 6-4. V-belt pulley being checked for wear using a straight edge. A worn pulley may be the cause of a noisy belt, especially if the belt is new. Courtesy of Gates Rubber Inc.

6.4 Cooling System

Cooling system maintenance consists of maintaining the coolant level and inspecting the hoses. Because the coolant's anti-corrosion and anti-freeze additives gradually lose their effectiveness, replacement of the coolant every 2 years is recommended. As a preventive measure, replacement of the cooling system hoses every 4 years is also recommended.

CAUTION —
Use only BMW-approved phosphate-free anti-freeze when filling the cooling system. Use of anti-freeze containing phosphates is considered to be harmful to the cooling system and may void warranty coverage.

Checking Coolant Level

A translucent expansion tank, or overflow reservoir, provides easy monitoring of coolant level. Because the expansion tank is translucent, the coolant level can be checked visually without opening the system.

Always check the coolant level with the engine cold. The coolant level should be at the maximum mark on the expansion tank, as shown in Fig. 6-5.

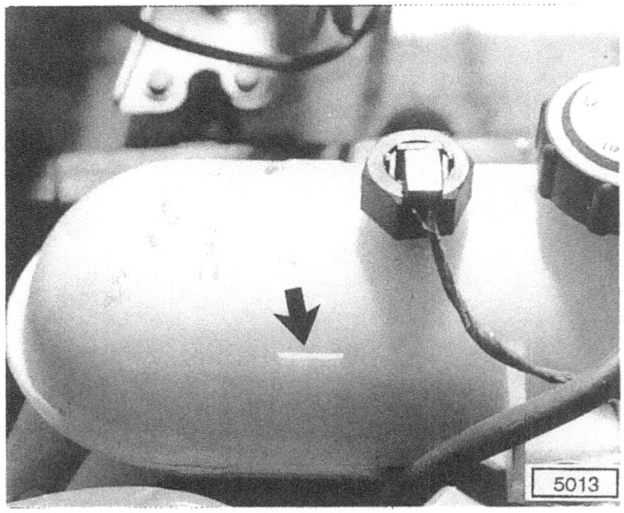

Fig. 6-5. Fill mark on coolant expansion tank. Coolant level should be at mark when engine is cold.

Inspecting Hoses

Connections should be tight and dry. Coolant seepage indicates either that the hose clamp is loose, that the hose is damaged, or that the connection is dirty or corroded. Dried coolant has a chalky appearance. Hoses should be firm and springy. Replace any hose that is cracked, that has become soft and limp, or has been contaminated by oil. See Fig. 6-6.

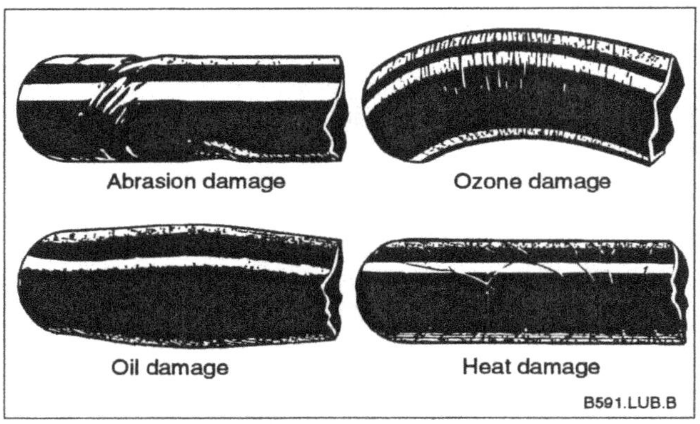

Fig. 6-6. Examples of damage to coolant hoses. Any of conditions shown is cause for replacement. Courtesy of Gates Rubber Inc.

6.5 Power Steering

Check the power steering fluid level in the fluid reservoir, just behind the fuse/relay panel. Park the car on level ground with the engine off.

On 528e models the fluid level is correct if it is between the **MAX** and **MIN** marks on the dipstick. See Fig. 6-7. If the level is below the **MIN** mark, start the engine and add fluid to the reservoir to bring the level up. Stop the engine and recheck the level. Hand-tighten the reservoir cap.

Specification	
• power steering fluid	Dexron II®

Fig. 6-7. Power steering fluid dipstick showing **MIN** and **MAX** marks on 528e models.

On 533i and 535i models pump the brakes at least 10 times or until the pedal is completely hard. Unscrew the center nut on the reservoir and check that the fluid level is no more than 10 mm below the rim of the reservoir. Add fluid as necessary. See Fig. 6-8.

> **NOTE —**
> If fluid is added to the correct level with the engine running and then rechecked after the engine is stopped, the level may be slightly above the **MAX** mark. This condition is normal.

Fig. 6-8. Power steering fluid/hydraulic brake boost reservoir on 533i and 535i models. Check level **A** (10mm) after pumping the brake pedal at least 10 times or until pedal is hard.

6.6 Oxygen Sensor

The oxygen sensor monitors engine combustion efficiency by measuring the oxygen content of the exhaust gasses. That information in turn is used to control the fuel injection system and reduce exhaust emissions. Any problems with the oxygen sensor will directly affect exhaust emissions and the way the engine runs.

Replacement of the oxygen sensor at the specified interval ensures that the engine and emission control system will continue to operate as designed. On some models, a lamp on the active check panel will come on to indicate its replacement is due. Extending the replacement interval may void the emission control warranty coverage.

The sensor is mounted either in the exhaust manifold and is accessible from inside the engine compartment or farther down in the exhaust pipe near the catalytic convertor. See Fig. 6-9. See **FUEL INJECTION** for more information on the oxygen sensor system.

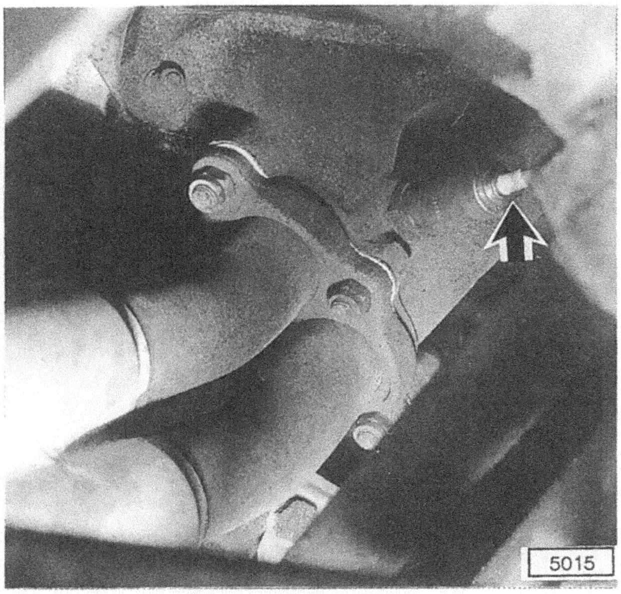

Fig. 6-9. Typical oxygen sensor location in exhaust manifold (arrow). Location of sensor varies between models.

Replacing Oxygen Sensor and Turning Off Oxygen Sensor Warning Lamp

The sensor is threaded into place and has a wire or wires extending from the back. Trace the sensor wiring back from the sensor and disconnect the electrical connector. When installing a new sensor, apply a light coat of anti-seize compound to the sensor threads. Torque the sensor and reconnect the wiring.

Tightening torque
• oxygen sensor. . . . . 55±5 Nm (40±4 ft-lb)

CAUTION —
Do not get any anti-seize compound on the sensor tip or in the sensor slits. The anti-seize compound will quickly foul the sensor element and render the sensor inoperative.

NOTE —
Special sockets for replacing the oxygen sensor are available from most automotive parts stores. The socket has a groove cut down one side to allow the sensor to be installed without damaging the wire harness. See **FUNDAMENTALS** for information on sources.

To turn off the oxygen sensor warning lamp on 528e and 533i models only, see **ELECTRICAL SYSTEM.**

7. UNDER-CAR MAINTENANCE

7.1 Tires and Wheels

For stability and car control, the wheels and tires must be of the correct size and in good condition. Tires must be inflated to the recommended air pressures and the wheels must be in proper alignment. For maximum safety and best all-around handling, always install replacement radial tires having the same specifications. When possible, all four tires should be replaced at once, or at least in pairs on the front or rear. New tires do not provide maximum traction, and should be broken in gently for the first 100 miles (160 kilometers) or so.

Tire Inflation Pressure

Correct tire inflation pressures are important to handling and stability, fuel economy, and tire wear. Tire pressures change with temperature. Pressures should be checked often during seasonal temperature changes. The correct inflation pressures for cars covered by this manual are listed in **Table j**. Notice that the pressures should be higher when the car is more heavily loaded.

All inflation pressures are for cold inflation. That is, when the car has not been driven for at least three hours, or for more than one mile after sitting for at least three hours.

WARNING —
Do not inflate any tire to a higher pressure than the tire's maximum inflation pressure listed on the sidewall. Use care when adding air to warm tires. Warm tire pressures can increase as much as 4 psi (0.3 bar) over their cold pressures.

Table j. Recommended Wheel and Tire Specifications

Model	Wheels	Tires	Tires pressures psi, cold			
			Max. 4 persons		Heavier load	
			front	rear	front	rear
528e	6 1/2J x 14 H2 6J x 14 H2	175 HR 14 88H	29	29	29	32
		175 SR 14 88Q M+S				
		195/70 R 14 90H	29	29	29	32
		195/70 HR 14				
		195/70 HR 14 89H				
		195/70 HR 14 M+S				
		195/70 R 14 90Q M+S				
		195/70 SR 14 89Q M+S				
	165 TR 390	200/60 HR 390 90H	29	29	32	36
		200/60 HR 390 90H TR				
		200/60 R 390 90H M+S TR				
		200/60 HR 390 90H M+S				
		200/60 HR 390 90H M+S TR				
533i	6 1/2J x 14 H2 6J x 14 H2	175 SR 14 88Q M+S	33	33	33	36
		195/70 VR 14	33	33	33	36
		195/70 SR 14 89Q M+S				
		195/70 HR 14 89T M+S				
	165 TR 390	200/60 VR 390	33	33	33	36
		200/60 HR 390 90H M+S				
535i	6J x 14 H2	175 SR 14 88Q M+S	33	33	35*	36
		195/70 HR 14	33	33	35*	36
		195/70 VR 14				
		195/70 R 14 90Q M+S				
	6J x 14 H2	195/70 SR 14 89Q M+S				
	165 TR 390	200/60 HR 390 TR	33	33	35	36
		200/60 VR 390 TR				
		200/60 R 390 90H M+S TR				
		200/60 HR 390 90H M+S TR				

*36 in 1985 only

Tire Rotation

BMW does not recommend tire rotation. Owing to the car's suspension design, the front tires begin to wear first at the outer shoulder and the rear tires begin to wear first at the middle of the tread or inner shoulder. Rotating the tires may adversely affect road handling and tire grip.

NOTE —
- The main purpose of tire rotation is to promote even wear and maximum tire life. Tire life may be decreased slightly if the tires are not rotated.
- Wheel lug bolts should be torqued to the correct value and rechecked after 600 miles (1000 km).

Tightening torque	
• wheel lug bolts	100±10 Nm (74±7 ft-lb)

Wheel Alignment

BMW recommends checking the front and rear alignment once a year and whenever new tires are installed. See **STEERING AND WHEEL ALIGNMENT** for a more detailed discussion of alignment requirements and specifications.

7.2 Brakes

Routine maintenance of the brake system includes maintaining an adequate level of brake fluid in the reservoir, checking brake pads for wear, checking hand brake function, and inspecting the system for fluid leaks or other damage.

WARNING —
- *Friction materials such as brake linings may contain asbestos fibers. Do not create dust by grinding, sanding, or cleaning the pads with compressed air. Avoid breathing asbestos fibers and asbestos dust, as it may result in serious diseases such as asbestosis and cancer, or in death.*
- *Brake fluid is poisonous. Do not siphon brake fluid by mouth. Wear gloves when working with brake fluid or brake pads to prevent contamination of cuts.*

Checking Brake Fluid Level

The level of the brake fluid will drop slightly as the brakes wear. Check the fluid level at the brake fluid reservoir, located on the driver's side, near the firewall. See Fig. 7-1. When filling the reservoir, use only new brake fluid from previously unopened containers. See **3. Fluid and Lubricant Specifications** for brake fluid specifications.

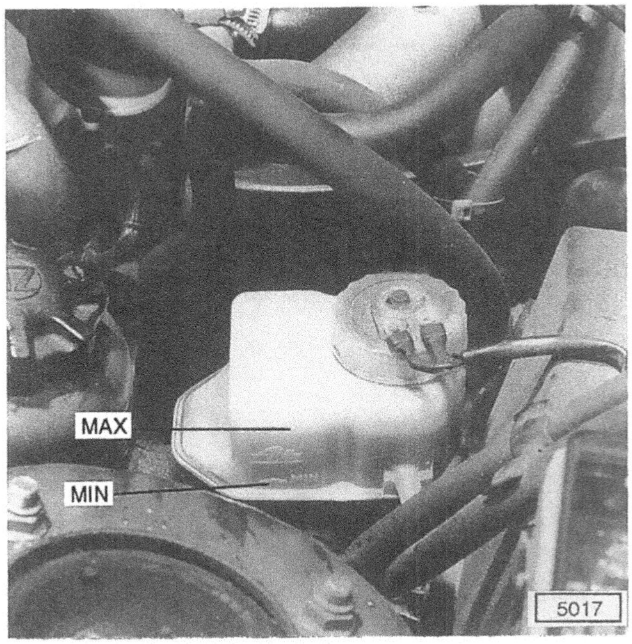

Fig. 7-1. Level indicators on brake fluid reservoir. Correct level is between **MIN** and **MAX** marks.

NOTE —
- The brake fluid also serves as the hydraulic fluid for the clutch. Checking brake fluid level automatically checks clutch fluid level.
- Low fluid level in the brake fluid reservoir may be due to leaks in the brake system or in the hydraulic clutch.

Inspecting Brake Hoses and Lines

Gently bend the hoses to check for cracks. Check that all hoses are correctly routed to avoid chafing or kinking. Inspect the unions and the brake calipers for signs of fluid leaks. Inspect the lines for corrosion, dents, or other damage. Replace faulty hoses or lines as described in **BRAKES**.

WARNING —
Incorrect installation or overtightening hoses, lines, and unions may cause chafing or leakage. This can lead to partial or complete brake system failure.

Checking Disc Brake Pad Wear

Although the cars covered by this manual are equipped with a brake pad warning system, the system only monitors one wheel per axle. There can be slight variations in brake pad wear at each wheel, so brake pad thickness should be checked whenever the wheels are off or brake work is being done.

Brake pad thickness can be inspected by looking through an opening in the caliper after removing the wheel. See Fig. 7-2. Specifications are given in **Table k**.

Fig. 7-2. Disc brake pad wear being checked through opening in caliper. Minimum brake pad thickness shown by dimension **A**.

Table k. Brake Pad Lining Minimum Thickness (not including brake pad backing plate)

Front and rear disc brake pads	2.0 mm (0.08 in.)

NOTE —
For parking brake lining wear, see **Parking Brake** below.

Replacing Brake Fluid

BMW strictly recommends replacing the brake fluid every year. This will help protect against corrosion and the effects of moisture in the fluid. The procedure is described in detail in **BRAKES**.

Parking Brake

The parking brake system is independent of the main braking system and may require periodic adjustment depending on use. Adjust the parking brake if the brake lever can be pulled up more than 8 clicks on the ratchet mechanism. Check that the cable moves freely in its housing. A complete description of the parking brake and parking brake adjustment can be found in **BRAKES**.

NOTE —
The parking brake may lose some of its effectiveness over time if it is not used frequently. This is due to corrosion build-up on the brake drum. To remove corrosion, apply the brake just until it begins to grip, then pull the lever up one more stop (click). Drive the car approximately 400 meters (1,300 ft.) and release the brake. Recheck the adjustment of the parking brake as described **BRAKES**.

7.3 Exhaust System

Exhaust system life varies widely according to driving habits and environmental conditions. If short-distance driving predominates, the moisture and condensation in the system will not fully dry out. This will lead to early corrosion damage and more frequent replacement.

Scheduled maintenance of the exhaust system is limited to inspection. Check for restrictions due to dents or kinks. Check for weakness or perforation due to rust. Check to see that all the hangers are in place and properly supporting the system and that the system does not strike the body. Alignment of the system and the location of the hangers are described in **EXHAUST SYSTEM**.

7.4 Manual Transmission Service

Manual transmission service consists of inspecting for leaks, checking and changing the gear oil, checking the clutch fluid level, and checking the clutch disc for wear.

Evidence of transmission leaks is most likely to be seen around the drive shaft mounting flange, and at the bottom of the bellhousing between the transmission and the engine. For more information on identifying oil leaks and their causes, see **MANUAL TRANSMISSION AND CLUTCH**.

Checking and Filling Manual Transmission Oil

BMW recommends checking the manual transmission oil level at every Inspection I interval and changing the oil at every Inspection II interval. Check and fill the transmission with the car on a level surface.

To check the oil level, remove the filler plug on the side of the transmission and carefully insert a finger into the hole. See Fig. 7-3. The oil level should be just flush with the bottom edge of the filler plug threads. In other words, if more oil is added, it would run out of the filler hole.

When changing the transmission oil, drive the car for few miles to warm up the oil. Place a drain pan under the oil drain plug on the bottom of the transmission. Remove both the filler plug and the drain plug. Reinstall and torque the drain plug when the oil is completely drained.

Fig. 7-3. Remove oil filler plug (**A**) to check transmission lubricant level. Remove oil drain plug (**B**) to replace lubricant. Use 17 mm hex wrench to remove plugs.

Add oil through the filler until it begins to run out of the hole. When no more oil can be added, reinstall and torque the oil filler plug.

Tightening torque
• manual transmission drain or fill plug. . . . . 40 to 60 Nm (29 to 44 ft-lb)

NOTE —
One method of filling the transmission is to use a length of clear plastic tubing (approximately 5 feet) inserted into the fill hole and extended up into the engine compartment. Attach a funnel to one end and slowly fill the transmission from above.

Checking Clutch Fluid Level

The hydraulic clutch and the brake system share the same reservoir and the same brake fluid. Clutch fluid level and brake fluid level are checked at the same time. See **7.2 Brakes**.

Checking Clutch Disc Wear

The clutch disc can be checked for wear using a special tool available from an authorized BMW dealer. The clutch wear tool is BMW tool no. 21 2 060.

Insert the tool in the opening between the transmission bellhousing and the clutch slave cylinder so that the tool touches the slave-cylinder pushrod. See Fig. 7-4. The slave cylinder is mounted to the driver's side of the transmission. If the tool stop (stepped portion) is more than 5 mm (0.2 in.) away from the slave cylinder, the clutch disc is sufficiently worn and will need to be replaced in the near future. See **MANUAL TRANSMISSION AND CLUTCH.**

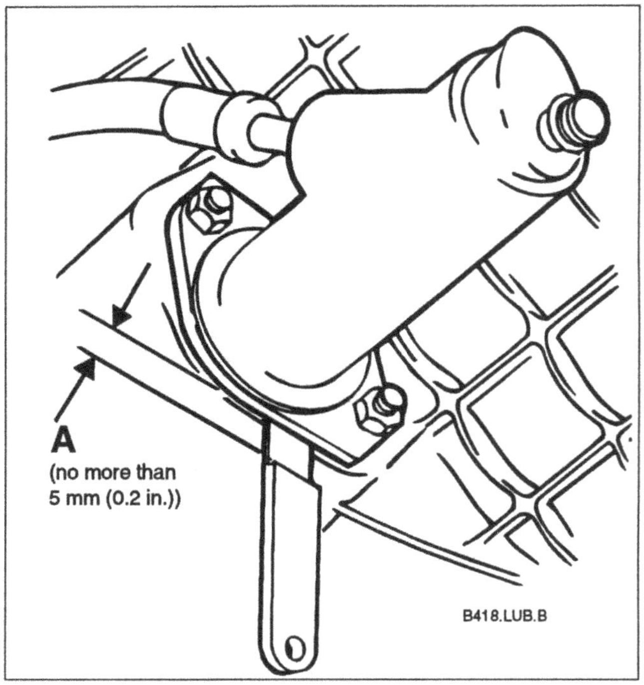

Fig. 7-4. Clutch disc being checked for wear using special tool. If gap **A** is more than 5 mm (0.2 in.), the clutch disc is worn and should be replaced soon.

7.5 Automatic Transmission Service

Smooth and efficient operation of the automatic transmission relies on the automatic transmission fluid (ATF). Many automatic transmission problems can be traced to an incorrect fluid level, incorrect type of fluid, a clogged ATF strainer, or contaminated fluid. With regular preventative maintenance, expensive and unnecessary automatic transmission repair may be avoided.

Before checking the ATF level, inspect for leaks. ATF leaks are most likely to be seen around the ATF oil sump gasket and at the bottom of the bellhousing, where the transmission joins the engine. All leaks should be corrected. If necessary, replace a leaky ATF oil sump gasket as described below. For more information, see **AUTOMATIC TRANSMISSION.**

CAUTION —
Extreme cleanliness is important when working on the automatic transmission. Use lint-free rags to check the level, and use a clean funnel when adding fluid.

Checking and Filling ATF

The location of the dipstick for checking the ATF is shown in Fig. 7-5. Two types of dipsticks are installed on cars covered by this manual, a short and a long version. See Fig. 7-6. Models produced before February 1985 are equipped with the short dipstick. All models produced after this date are equipped with the longer dipstick. The longer dipstick (BMW Part No. 24 11 1 207 767) can be fitted retroactively to all earlier models to ensure more accurate fluid checks.

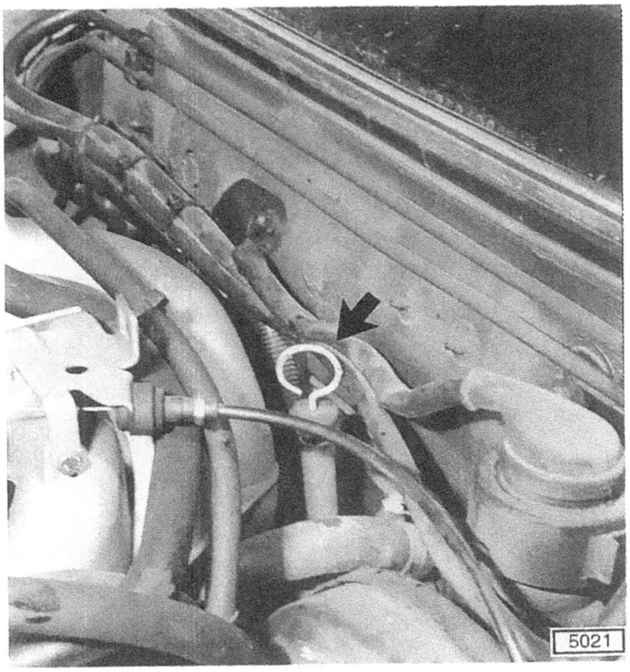

Fig. 7-5. ATF dipstick location (arrow).

CAUTION —
Use care when adding fluid to models equipped with the short dipstick. Unless the transmission fluid is fully warmed (ATF temperature above 175°F (80°C)), the ATF level may appear to be too low when checked. If too much ATF is added, it will overflow the transmission filler tube or the transmission vent under heavy engine load or cruising.

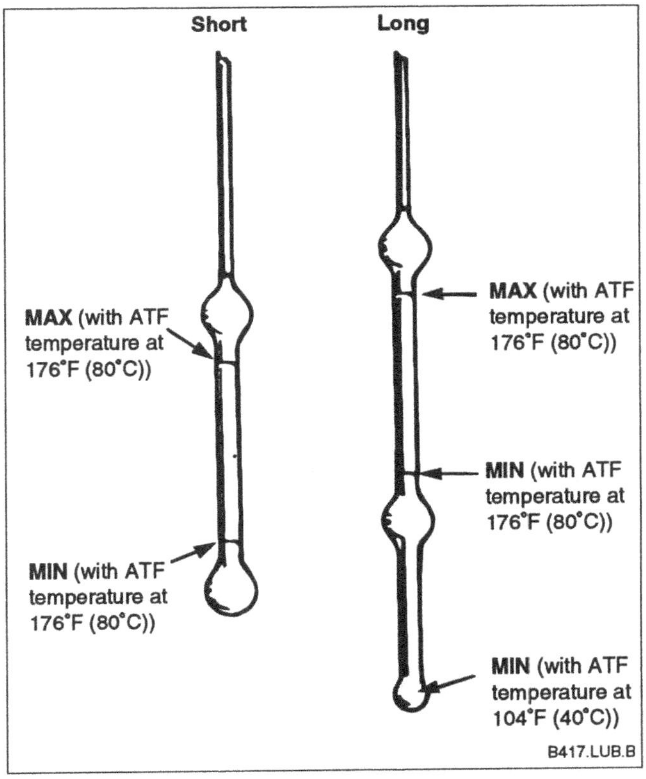

Fig. 7-6. Two versions of automatic transmission dipsticks installed on cars covered by this manual.

NOTE —
The area between the **MIN** mark and the **MAX** mark on the dipstick represents approximately 0.6 pints (0.3 liters) of ATF on models with 4-speed automatic transmissions, and approximately 0.8 pints (0.4 liters) of ATF on models with 3-speed automatic transmissions. See **AUTOMATIC TRANSMISSION** for information on transmission applications.

The ATF level should be checked with the car on a level surface with the engine idling and the transmission fully warmed. Firmly set the parking brake and place the transmission selector lever in park or neutral and remove the ATF dipstick. The ATF level is correct if it is between the **MIN** and **MAX** marks on the dipstick.

NOTE —
Driving the car for five to ten minutes around town, or approximately 12 miles (20 km) on the highway will ensure a fully warmed transmission.

If the level is too low, use a clean funnel to add ATF through the dipstick/filler tube as specified in **3. Fluid and Lubricant Specifications** until the fluid level is between the two marks on the dipstick.

Check the condition of the ATF by rubbing some between your fingers and sniffing it. The ATF should not be foamy, gritty, or have a burnt odor. Contaminated ATF should be drained and replaced to prevent further damage, but doing so will not repair any internal transmission damage that has already occurred.

> **NOTE —**
> Dexron II® ATF is a red/brown color that discolors to black/brown during normal use. ATF color may not be a good indicator of its condition.

Draining and Replacing ATF, and Cleaning ATF Strainer

With the car raised and supported securely on jack stands, place a drain pan of at least 6.5 l (7 US qt.) capacity under the transmission and remove the transmission drain bolt. See Fig. 7-7. Remove the dipstick/filler tube mounting nut from the front of the sump and remove the tube. See Fig. 7-8. Remove the sump mounting bolts and clamping brackets noting their lengths and locations. See Fig. 7-9. Remove the oil sump and the gasket.

Fig. 7-7. Automatic transmission fluid drain bolt (arrow).

> ***CAUTION —***
> *Towing the car or running the engine without ATF in the transmission will severely damage the transmission.*

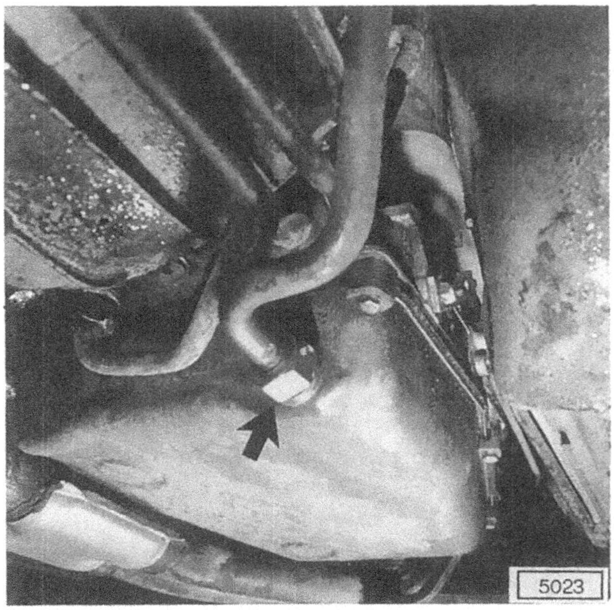

Fig. 7-8. Automatic transmission fluid filler dipstick/tube mounting nut (arrow).

Fig. 7-9. Automatic transmission sump mounting bolts and clamping brackets (arrows).

> **NOTE —**
> Two types of sump mounting brackets are used on the 4-speed automatic transmission (4 HP 22). Four rounded brackets are installed at each corner of the sump and two flat brackets are installed at the sump's side. Be sure the brackets are positioned correctly during installation.

Remove and clean the magnets in the oil sump. Note the position of the magnets. Remove and clean the transmission strainer. On models with 3-speed transmission, the strainer is held in place with twelve Torx® bolts of unequal length. On models with 4-speed transmission, the strainer is held in place with three Torx® bolts. See Fig. 7-10. On models with 4-speed transmission, remove the strainer O-ring.

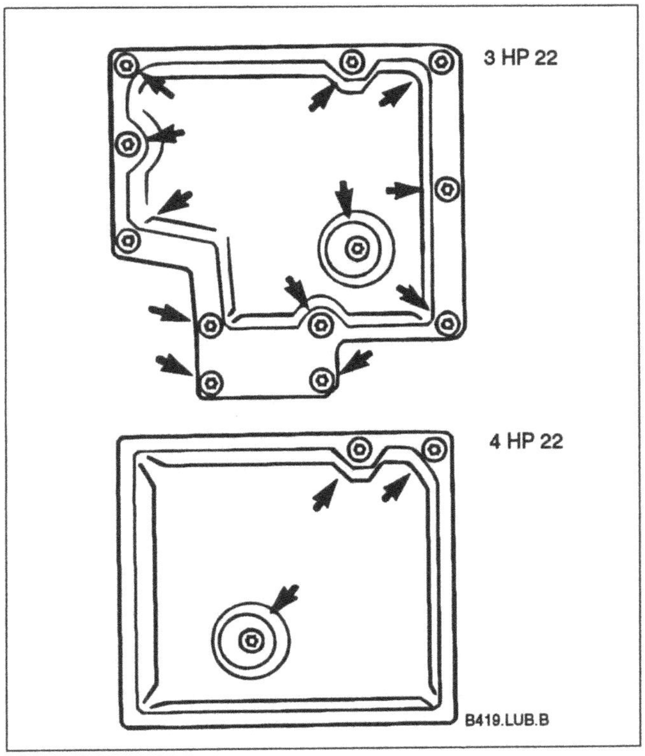

Fig. 7-10. Transmission fluid strainer mounting bolts (arrows). **Top**: 3-speed transmission (3 HP 22). **Bottom**: 4-speed transmission (4 HP 22).

CAUTION —
The strainer can be cleaned if it is dirty or clogged. The strainer should be replaced if a brown gummy residue has accumulated on the screen, if it cannot be thoroughly cleaned, or if it is in any way damaged. Using a damaged or partially plugged strainer may cause serious transmission damage.

NOTE —
See **AUTOMATIC TRANSMISSION** for more information on how to identify a particular transmission.

Clean the sump and install the sump magnets. Make sure the sump is completely dry. Remount the strainer using a new O-ring (if applicable) and tighten the mounting bolts to the torque listed below. Using a new sump gasket without any sealer, install the ATF sump. Install the clamping brackets so that the short leg of the bracket is contacting the sump. Tighten the bolts to the torque listed below. Install and tighten the oil filler tube. Refill the transmission with fluid according to the type and amount specified in **3. Fluid and Lubricant Specifications**. Firmly apply the handbrake, then start the engine and shift the transmission through all gears to circulate the fluid. Check the fluid level as described earlier. Check for leaks.

Tightening torques	
• ATF strainer to transmission	10 to 11 Nm (86 to 97 in-lb)
• ATF sump to transmission	
3 HP 22	8 to 9 Nm (70 to 80 in-lb)
4 HP 22	6 to 7 Nm (53 to 62 in-lb)
• ATF sump drain plug	15 to 17 Nm (11 to 12 ft-lb)
• ATF filler tube to oil sump	
3 HP 22	100 to 115 Nm (74 to 85 ft-lb)
4 HP 22	98 Nm (72 ft-lb)

7.6 Front Suspension and Steering

Inspection of the front suspension and steering includes a check of all moving parts for wear and excessive play. Also inspect the rubber seals and boots for cracks or tears that could allow the entry of dirt, water, and other contaminants. Complete front suspension inspection and troubleshooting information can be found in **SUSPENSION—FRONT** and **STEERING AND WHEEL ALIGNMENT**.

7.7 Final Drive and Rear Drive Axles

Final drive and drive axle service consists of checking and changing the gear oil, inspecting for leaks, and checking the rear drive axle rubber boots for damage.

The areas where leaks are most likely to occur are around the drive shaft and drive axle mounting flanges. For more information on identifying oil leaks and their causes, see **DRIVESHAFT AND FINAL DRIVE**.

Checking and Filling Final Drive Lubricant

Check the lubricant level with the car level. Remove the oil filler plug, shown in Fig. 7-11. The level is correct when the fluid just reaches the edge of the filler hole. See **3. Fluid and Lubricant Specifications**. Tighten the oil filler plug.

Tightening torque	
• final drive filler plug	55 Nm (40 ft-lb)

Fig. 7-11. Final drive oil filler plug (**A**) and oil drain plug (**B**).

To replace the final drive gear oil, drive the car for a few miles to warm up the lubricant. With the car on a level surface, remove the filler and drain plug as shown above. When the oil is fully drained, install and torque the drain plug. Add oil though the filler plug until it begins to run out of the filler hole. Install and torque the filler plug.

Tightening torque
• final drive drain or filler plug . . . . 55 Nm (40 ft-lb)

Drive Axle Joint Boots

The protective boots must be closely inspected for cracks and any other damage that will allow contaminants to get into the joint. If the rubber boots fail, the water and dirt that enter the joint will quickly damage it. Replacement of the drive axle joint boots and inspection of the joints are described in **SUSPENSION—REAR**.

7.8 Fuel Tank and Fuel Lines

Inspect the fuel tank, fuel lines, and fuel system for damage or leaks. If fuel odors are detected in the passenger compartment, the fuel tank sender cover O-ring (beneath the luggage compartment carpeting) may be faulty. See **FUEL SUPPLY** for replacement. Also check the expansion tank and hoses located above the right rear wheel in the luggage compartment.

Check for fuel leaks in the engine compartment. Check for faulty fuel lines by bending them. If any leaks are present, fuel should be expelled. Check for a damaged fuel tank. Incorrectly raising the car can damage the tank. Check for any evaporative emissions hoses that may have become disconnected, checking carefully at the charcoal canister and evaporative emissions purge valve (where applicable). See **FUEL SUPPLY** and **FUEL INJECTION** for component locations.

> ***WARNING —***
> *When checking for fuel leaks, the engine must be cold. A hot exhaust manifold or exhaust system could cause the fuel to ignite or explode causing serious personal injury. Ventilate the work area and clean up spilled fuel immediately.*

8. Body and Interior Maintenance

8.1 Windshield Wiper Blades

Common problems with the windshield wipers include streaking or sheeting, water drops after wiping, and blade chatter. Streaking is usually caused when wiper blades are coated with road film or car wash wax. Clean the blades using soapy water. If cleaning the blades does not cure the problem then they should be replaced. BMW recommends replacing the wipers blades twice a year, before and after the cold season. On older cars, check the tension spring that holds the wiper to the glass. Replace the springs if necessary.

Drops that remain behind after wiping are caused by oil, road film, or diesel exhaust coating the windshield. Use an alcohol or ammonia solution, or a non-abrasive cleanser to clean the windshield.

Wiper blade chatter may be caused by dirty or worn blades, by a dirty windshield, or by bent or twisted wiper arms. Clean the blades and windshield as described above. Adjust the wiper arm so that there is even pressure along the blade, and so that the blade is perpendicular to the windshield at rest. Lubricate the wiper linkage with a light oil. The linkage is located under the hood on the driver's side. If the problem persists, the blades are excessively aged or worn and should be replaced.

8.2 Body Lubrication

The door locks and lock cylinders can be lubricated with an oil that contains graphite. Such a lubricant is available from BMW, Part No. 81 22 9 407 421

The body and door hinges, the hood latch, and the door check rods should be lubricated with SAE 30 or SAE 40 engine oil. Lubricate the seat runners with multipurpose grease. Do not apply any oil to rubber parts. Lubricate the sunroof guide rails with silicone spray. If door weatherstrips are sticking, lubricate them with silicone spray or talcum powder. The hood release cable should be lubricated as well. The radio antenna mast should be cleaned and lubricated with a product such as WD-40®.

8.3 Seat Belts

Dirt and other abrasive particles will damage seat belt webbing. If it is necessary to clean seat belts, use a mild soap solution. Bleach and other strong cleaning agents may weaken the belt webbing.

> ***WARNING —***
> *Do not clean the seat belt webbing using dry cleaning or other chemicals. Allow wet belts to dry before allowing them to retract.*

The condition of the belt webbing and the function of the retractor mechanisms should be inspected. See **BODY AND INTERIOR** for seat belt inspection information.

9. CLEANING AND PRESERVING

9.1 Care of Exterior Finish

The longer dirt is left on the paint, the greater the risk of damaging the glossy finish, either by scratching or by the chemical effect dirt particles may have on the painted surface.

Washing

Do not wash the car in direct sunlight. If the engine hood is warm, allow it to cool. Beads of water not only leave spots when dried rapidly by the sun or heat from the engine, but also act as tiny magnifying glasses that can burn spots into the finish. Wash the car with a mixture of lukewarm water and a mild soap made expressly for washing cars. Using general detergents, even ones that claim to be gentle, can cause damage to the paint over time. Rinse using plenty of clear water under as little pressure as possible. Wipe the body dry with a soft terry-cloth towel or chamois to prevent water-spotting. BMW recommends washing the car once a week.

Waxing

For a long-lasting, protective, and glossy finish, apply a hard wax after the car has been washed and dried. Use carnauba or synthetic based products. Waxing is not needed after every washing. You can tell when waxing is required by looking at the finish when it is wet. If the water coats the paint in smooth sheets instead of forming beads that roll off, a new coat of wax is needed. Wax should not be applied to black trim pieces, rubber, or other plastic parts.

Polishing

Use paint polish only if the finish assumes a dull look after long service. Polish can be used to remove tar spots and tarnish, but afterwards a coat of wax should be applied to protect the clean finish. Do not use abrasive polish or cleaners on aluminum trim or accessories.

Washing Chassis

Periodic washing of the underside of the car, especially in winter, will help prevent accumulation of road salt and rust. The best time to wash the underside is just after the car has been driven in wet conditions. Spray the chassis with a powerful jet of water. Commercial or self-service car washes may not be best for this, as they may recycle the salt-contaminated water.

Special Cleaning

Tar spots can be removed with a bug and tar remover. Never use gasoline, kerosene, nail polish remover, or other unsuitable solvents. Insect spots also respond to tar remover. A bit of baking soda dissolved in the wash water will facilitate their removal. This method can also be used to remove spotting from tree sap.

9.2 Care of Interior

Dirt spots can usually be removed with lukewarm soapy water. Use spot remover for grease and oil spots. Do not pour the liquid directly on the carpet, but dampen a clean cloth and rub carefully, starting at the edge of the spot and working inward. Do not use gasoline, naptha, or other flammable substances to clean the carpeting.

Vinyl and Cloth Upholstery and Trim

Use a dry foam cleaner. Grease or paint spots can be removed by wiping with a cloth soaked with this cleaner. Use the same cleaner, applied with a soft cloth or brush, on the headliner and side trim panels. For cloth-covered seat areas, use the techniques described previously for cleaning the carpeting.

Leather Upholstery and Trim

Leather upholstery and trim should be periodically cleaned using a slightly damp cotton or wool cloth. The idea is to get rid of the dirt in the creases and pores that can cause brittleness and premature aging. On heavily soiled areas, use a mild detergent (such as Woolite®). Use two tablespoons to one quart of cold water. Dry the trim and upholstery completely using a soft cloth. Regular use of a good quality leather conditioner will reduce drying and cracking of the leather.

Section 3

ENGINE MANAGEMENT—DRIVEABILITY

Contents

INTRODUCTION

Driveability—the overall performance of the car, its ability to start quickly, run and accelerate smoothly, and deliver fuel economy as well as power—can only be achieved when all of the engine's major systems are working properly. This is even more true for today's cars with strict exhaust emission control.

One problem in troubleshooting an engine that performs poorly is knowing where to begin looking for the problem. There may be an ignition fault, a fuel delivery problem, a faulty emission control system, or a mechanical problem with the engine itself. A fault in any of these systems might cause poor driveability. Because top engine performance depends on the integrated functions of several systems, effective troubleshooting must include checking all these systems together.

This section offers advice for simple maintenance that can help prevent driveability problems. It also includes basic troubleshooting information that addresses symptoms of poor driveability and can help isolate problems to a specific system. It is a guide to problem solving, intended to be used in conjunction with the other sections in this manual.

1. GENERAL DESCRIPTION

Engine management is a term widely used to describe the control of all the functions that affect how the engine runs. While this obviously includes the fuel system and the ignition system, it may also include emission control systems and auxiliary functions such as idle speed control.

It is important to consider all of the engine management functions when troubleshooting. This section describes some likely causes of general driveability problems, and suggests ways of isolating other problems to faults in a particular system. Specific tests and repairs are covered in **FUEL SUPPLY**, **FUEL INJECTION**, **IGNITION**, **EXHAUST SYSTEM**, and **ENGINE—SERVICE AND REPAIR**.

1.1 Motronic Engine Management System

All 6-cylinder engines covered by this manual are equipped with a Bosch Motronic engine management system. Fuel injection, ignition, and other functions are combined under the control of one ECU. See Fig. 1-1.

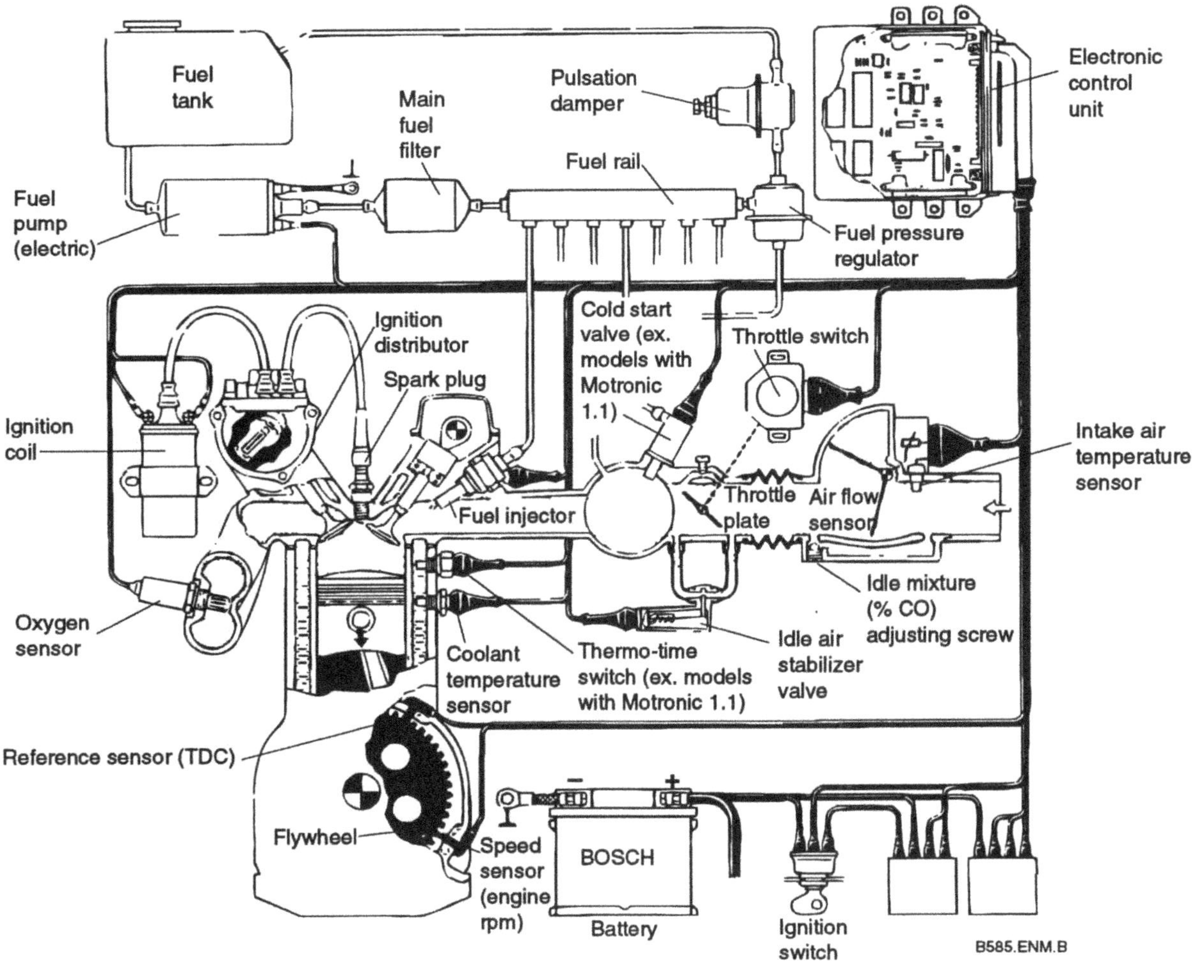

Fig. 1-1. Schematic of Motronic engine management system used on BMW 5-series models covered in this manual. Courtesy Robert Bosch Corp.

The Motronic engine management system uses the electrical signals from the air flow sensor, the air and coolant temperature sensors, the speed sensor, and an exhaust-mounted oxygen sensor as its primary inputs to electronically calculate fuel delivery and engine timing.

The ignition distributor is an integral part of the cylinder head, driven directly by the camshaft. It contains no mechanical timing-advance components, and basic ignition timing is non-adjustable. All ignition timing is determined by the Motronic ECU, interpreting signals from sensors that indicate engine crankshaft position and engine speed. Ignition timing is advanced and retarded electronically.

Fuel under pressure is injected via electronically controlled solenoid-type fuel injectors. The ECU electrically controls the opening and closing of the injectors by switching the ground side of each injector circuit. The exact amount of fuel injected is determined by the amount of time the injectors are open.

A second generation Motronic system, Motronic 1.1, was introduced in March of 1987 on the 528e models. It is a more sophisticated version of the "basic" Motronic system mentioned above. The differences between the two systems are mainly in the electronic circuitry. Both systems have adaptive abilities that actually adjust the system to compensate for things such as engine wear and vacuum leaks. On Motronic 1.1, the system has additional adaptive features that enable it to change the base settings in the control unit's memory over time. For a more detailed explanation of the Motronic 1.1 engine management system, see **FUEL INJECTION.**

> **NOTE —**
> All of the information in this section labeled "Motronic" applies to both Motronic systems, unless Motronic 1.1 is specifically mentioned as an exception.

2. MAINTENANCE

Carrying out maintenance at the specified interval is the key to preventing engine performance and driveability problems. The BMW service interval system has been precisely designed to provide optimum maintenance requirements based on personal driving habits.

Extending the specified service intervals will only lead to driveability problems later on. As an example, extending the replacement of the tune-up components can cause clogged fuel injectors, a damaged catalytic converter, or even severe engine wear. A worn out oxygen sensor will cause poor idle characteristics and higher than normal fuel consumption. Diluted engine oil or oil that has reached its "break down" point can lead to engine wear on the cylinder walls and other friction surfaces, such as bearings and bushings. Therefore, it is good insurance to service the car promptly when the maintenance reminder lights come on.

For information on BMW's recommended maintenance pertaining to engine management and driveability, see **LUBRICATION AND MAINTENANCE**, **FUEL SUPPLY**, **FUEL INJECTION**, **IGNITION**, and **EXHAUST SYSTEM**.

3

3. TROUBLESHOOTING— BASIC REQUIREMENTS

This heading covers general engine management principles and the basic requirements that allow an engine to run smoothly. Therefore, effective troubleshooting of specific running conditions can only take place after all of the common problem areas listed below have been eliminated as a source of trouble.

Most driveability problems are complex in nature. A logical method needs to be used to isolate the trouble area. Always begin with the simplest and most fundamental engine management basics. Jumping to conclusions or searching aimlessly for the problem can be time consuming and frustrating.

As with any troubleshooting, careful observation of symptoms is the key to identifying and isolating driveability problems. A test drive can help by demonstrating when the problem is most pronounced, such as a hesitation which occurs only when accelerating, or a steady miss at high speed.

How has the symptom developed? A symptom that develops quickly is probably caused by a problem that can be corrected by simple maintenance or repair. A symptom that has developed gradually over time, especially after sixty or seventy thousand miles is more likely an indication of general wear and the need for more comprehensive work.

Warnings and Cautions

For general safety, and to protect the sensitive electronic components, the following warnings and cautions should be adhered to during any troubleshooting, maintenance, or repair work. Always follow the proper repair and working procedures in the sections that are referenced.

> ***WARNING —***
> • *The ignition systems on the cars covered by this manual operate in a dangerous voltage range that could prove to be fatal if exposed terminals or live parts are contacted. Use extreme caution when working on a car with the ignition on or the engine running.*
>
> • *Do not touch or disconnect any of the high tension cables from the coil, distributor, or spark plugs while the engine is running or being cranked by the starter.*

WARNING —
• Connect and disconnect ignition system wires, multiple connectors, and ignition test equipment leads only while the ignition is switched off.

*• Before operating the starter without starting the engine (as when making a compression test), disable the ignition. See **IGNITION** for more information.*

• Do not disconnect terminal 4 (center terminal) from the coil or remove the distributor cap as a means of disabling the ignition.

• During any test where fuel is discharged, do not smoke or work near heaters or other fire hazards. Have a fire extinguisher handy.

CAUTION —
• Do not connect test instruments with a 12-volt power supply to terminal 15 (+) of the ignition coil. The current flow will damage the ignition control unit. In general, make test connections only as specified by BMW, as described in this manual, or as described by the test instrument's manufacturer.

• Do not disconnect the battery while the engine is running.

*• Do not exceed 16.5 volts at the battery with boosting cables attached, and do not quick-charge the battery (for boost starting) for longer than one minute. Wait at least one minute before boosting the battery a second time. On models equipped with on-board computers, the computer fuses should be removed prior to quick-charging to prevent damaging the computer. See **ELECTRICAL SYSTEM** for fuse identification.*

• Do not connect terminal 1 (–) of the ignition coil to ground as a means of preventing the engine from starting.

• Running the engine with a spark plug wire disconnected may damage the catalytic converter.

• Cleanliness is essential when working with fuel circuit components. Before disconnecting any fuel lines, thoroughly clean the unions. Use clean tools.

3.1 Engine

Before troubleshooting a poorly running engine or an engine that will not start, determine the general condition of the engine, especially if it has high mileage. If the engine is severely worn or has mechanical problems, the only remedy is overhaul or repair. If a tune-up or scheduled maintenance is due, it should be done before proceeding to other areas of this section.

Mechanical Condition

Only a few basic functions are required of the engine. The parts must fit together properly, operate smoothly, and seal well enough to create and maintain compression, and keep pistons, valve train, and ignition properly timed.

General engine condition can be easily assessed by performing a few simple diagnostic tests. Make sure that the valves are correctly adjusted before making these tests. If the engine cannot be started, perform a compression test. If the engine runs and can idle, perform a vacuum gauge test. These tests are covered in **ENGINE—SERVICE AND REPAIR**.

NOTE —
Compression and vacuum gauge tests require special test equipment. If the equipment is not available, most automotive repair shops can do these tests quickly and at a reasonable cost.

Carbon Deposits

Carbon deposits on the fuel injectors and the intake valves will affect the way the engine idles and runs. See Fig. 3-1. Even a ten percent decrease in the amount of fuel that the injectors deliver will cause driveability problems. These deposits normally form during the "hot soak" period immediately after the engine is turned off, when the engine temperature rises slightly for a short period.

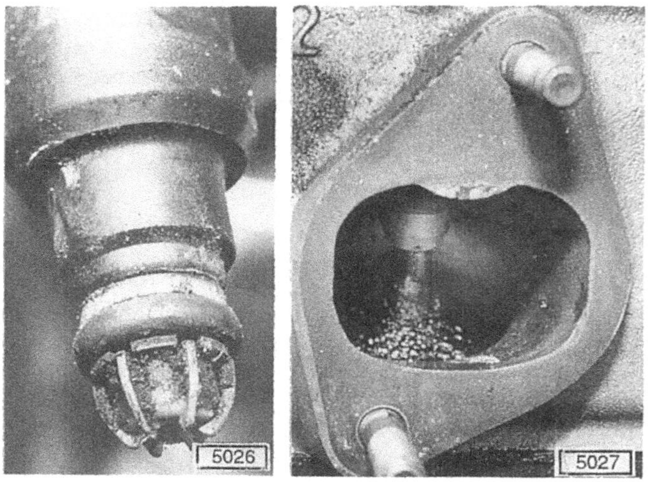

Fig. 3-1. Examples of carbon deposits on fuel injector (left) and intake valve (right). Carbon deposits can cause a rough idle, hard cold starting, and overall poor performance.

Driving style is thought to be the main contributor to the problem. A car that is predominantly driven on short trips around town or in city traffic seems to increase the likelihood of deposits forming.

Carbon deposits on the intake valves and injectors should be removed prior to troubleshooting driveability problems. Special fuel injector test equipment is required to accurately check for clogged injectors. If the injectors are severely clogged, they can be removed and visually inspected. Inspecting the intake valves is more difficult because the intake manifold needs to first be removed. Check with an authorized BMW dealer for the latest information on carbon deposits and the best methods used to remove them.

CAUTION —
Always follow the manufacturer's directions when using fuel additives designed to remove carbon deposits and clean injectors. It is recommended that high detergency fuels should not be used together with fuel additives. The excess amount of cleaner in the fuel can dilute engine oil and accelerate engine wear. Always check with an authorized BMW dealer before using fuel additives together with high detergency fuel.

Tune-up and Preventive Maintenance

The condition of the tune-up and emission control components can affect engine performance and driveability. Extending tune-up and maintenance schedules beyond those recommended by the manufacturer can adversely affect the way the engine runs. When experiencing driveability problems, a good starting point is to do a tune-up, especially if scheduled maintenance is overdue. Many driveability problems are eliminated simply by replacing these components.

A tune-up is regular maintenance of the ignition and fuel system components for normal wear and contamination. The ignition components all carry high voltage to deliver a precisely timed spark to ignite the air-fuel mixture. If any of these components are faulty or worn, the intensity and timing of the spark will be affected. Extending the replacement intervals of fuel system and emission control components can adversely affect fuel delivery, air-fuel mixture, and exhaust emissions.

Replacement schedules and procedures for the spark plugs, spark plug wires, distributor cap, ignition rotor, fuel filter, air filter, oxygen sensor, and oil and oil filter are given in **LUBRICATION AND MAINTENANCE**.

NOTE —
For information on inspecting ignition components, see **IGNITION**. For information on testing the oxygen sensor, see **FUEL INJECTION**.

Valve Adjustment

Correctly adjusted valves are necessary for efficient engine operation. If the valve clearances are too small, the valves may not close all the way, resulting in low compression and a loss in power. If the valve clearances are too large, the valves may not fully open causing a reduction in engine efficiency. Procedures for checking and adjusting valve clearances are described in **ENGINE—SERVICE AND REPAIR**.

3.2 Basic Adjustments

In addition to tune-up component replacement, it is important that all of the basic adjustments that can be made are correctly set. Check idle speed and idle mixture (%CO) to be sure they are all within the specified limits. All of the models covered by this manual are equipped with a non-adjustable or self-correcting, electronically controlled idle speed. All models have non-adjustable ignition timing. On 528e models with Motronic 1.1, the idle mixture is not adjustable. See **Table a.**

NOTE —
All of the basic adjustments require the use of specialized test equipment. If any of the test equipment is not available, it is recommended that the adjustments be done by an authorized BMW dealer or other qualified repair shop. These adjustments can be made quickly and at a reasonable cost.

Table a. Basic Adjustment Information

Model	Ignition timing	Idle speed	Idle mixture (%CO)
1982–1987 (up to March 1987) 528e, 533i	non-adjustable	non-adjustable	adjustable
1987–1988 (from March 1987) 528e	non-adjustable	non-adjustable	non-adjustable
1985–1988 535i	non-adjustable	non-adjustable	adjustable

The systems that adapt idle mixture, idle speed, and ignition timing can only correct engine operation within a limited range. Once these limits are exceeded, driveability problems will become noticeable. For example, the oxygen sensor can adapt idle mixture for things such as a small vacuum leak or minor engine wear. A large vacuum leak or a severely worn engine may exceed the operating range of the sensor, causing the engine to run lean. The same conditions are true for the electronic idle stabilization system. Keep in mind that if large adjustments are necessary, the faults that are causing these incorrect settings should be corrected prior to making any adjustments.

To check how much the oxygen sensor is adapting the fuel mixture, make sure the engine is fully warmed up. Then check the voltage at the oxygen sensor connector with the connector connected (closed-loop) and then with connector disconnected (open-loop). There should not be a big difference between the two readings. If there is a big difference, either a fault exists such as a vacuum leak (lean or low open-loop voltage), or a faulty fuel injector (rich or high open-loop voltage), or the idle mixture screw is out of adjustment (if applicable) due to normal engine wear. See **FUEL INJECTION** for more information on setting idle mixture.

Oxygen Sensor

The oxygen sensor adapts the air-fuel mixture by sending a varying voltage signal to the fuel injection control unit. The sensor is positioned in the exhaust stream and actually measures the amount of oxygen in the exhaust gas so that the fuel injection system can correctly adjust the air-fuel mixture. A high concentration of oxygen in the exhaust gas indicates a lean mixture and a low content indicates a rich mixture. Thus, the signal from the oxygen sensor influences engine performance and driveability.

NOTE —
The signal from the oxygen sensor is ignored by the fuel injection control unit until the engine reaches a specified temperature. Therefore, when troubleshooting cold engine driveability problems, the oxygen sensor can be ruled out as a possible cause.

As the sensor ages, it loses its ability to react quickly to changing conditions and it may eventually cease to produce any signal at all. When this happens, fluctuations in idle speed and increased fuel consumption may be noticed. The oxygen sensor should be replaced at the specified mileage interval as described in **LUBRICATION AND MAINTENANCE**.

Fig. 3-2. Intake air duct (arrow) can crack or become loose. Be sure to check bottom side of duct. Air bypass hoses and hose fittings are also suspect. Bend air duct and hoses to detect cracks.

3.3 Air Flow Measurement and Vacuum Leaks

To control fuel injection the engine management system uses an air flow sensor to precisely measure incoming air. The sensor sends an electrical signal proportional to the measured air flow to the control unit, which uses this signal to determine the amount of fuel the engine needs.

Because proper fuel metering depends on accurately measuring the intake air, any unmeasured air entering the system will cause a lean fuel mixture and poor running. To see how air leaks can affect engine running, remove the oil filler cap while the engine is running.

There are many possible places for unmeasured air to enter the engine. Carefully inspect all hoses, fittings, duct work, and seals and gaskets. Check the oil filler cap seal and the seal on the oil dipstick. Fig. 3-2 through Fig. 3-4 show some of the common areas where air leaks can develop. For a thorough inspection, it may be necessary to remove hoses and ducts that cannot be completely checked in their installed positions.

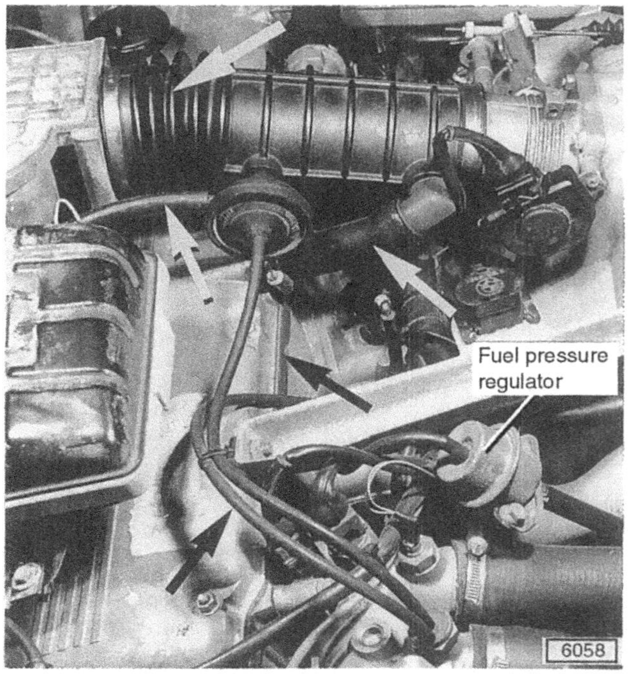

Fig. 3-3. A leaking cylinder head cover gasket or cracked or brittle hoses should be replaced (arrows). Also check the small vacuum hose to the fuel pressure regulator.

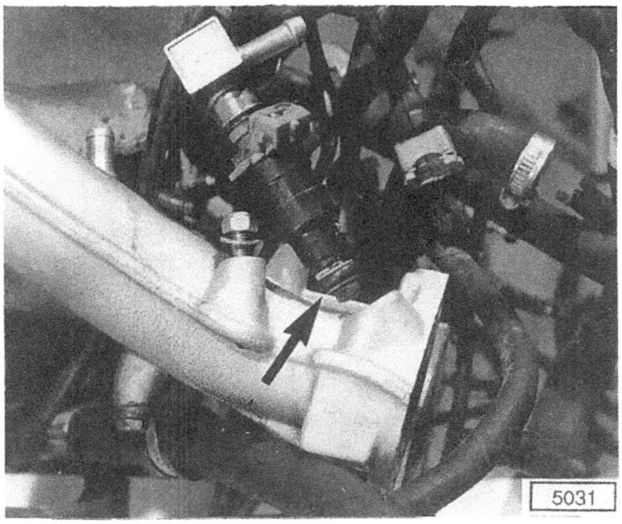

Fig. 3-4. Injector seals or O-rings (arrow) are a common source of air leaks, especially on high mileage engines. Injectors must first be removed to replace seals.

3.4 Electrical System

All the cars covered by this manual use engine management systems that rely on precise electrical signals for proper operation. If any of these signals are distorted, incorrect, or missing, the car can develop major driveability problems.

Battery Voltage

One of the most fundamental requirements in troubleshooting engine performance problems is to make sure the battery is fully charged and in good condition. Many of the sophisticated electronics used on the cars covered by this manual require a specified operating voltage to function correctly.

Battery voltage can be measured across its terminals with all cables attached. Do not eliminate the battery as a possible source of trouble until a load test has been performed, especially if starting problems are encountered. See **BATTERY, STARTER, ALTERNATOR**, and **ELECTRICAL SYSTEM** for battery testing information.

> **NOTE —**
> A digital voltmeter should be used to accurately measure battery voltage. A fully charged battery will measure 12.6 volts, or more, while a battery only 25% charged will measure 12.15 volts. Using an analog meter may result in inaccurate results.

For the battery to maintain its proper voltage level, the charging system must be functioning correctly. If in doubt about the condition of the charging system, have the system checked. Most automotive repair shops can test the system quickly and at a reasonable cost.

Wiring and Harness Connectors

The cars covered by this manual are equipped with electronic fuel injection and ignition systems that are controlled by central electronic control units. Many of the circuits operate on very low current and are very sensitive to increased resistance due to faulty or corroded wiring or connectors.

> **NOTE —**
> In most cases, faulty electronic control units are not the cause of driveability problems. These units are extremely durable and reliable. Actual failures are very rare. Driveability problems are more often caused by missing or incorrect signals to the control unit, or by other faulty components.

The electrical system is subject to corrosion, vibration, roadway elements and general wear. Because of this, the integrity and freedom from corrosion in the connections, wires, and switches, including all ground connections, is one of the most important conditions for trouble-free operation of the engine management systems.

Always make a thorough visual inspection of all wires and connectors, switches and fuses. Loose or damaged connectors can cause intermittent problems, especially at the small terminals in each control unit connector. In most cases, a visual inspection will detect any faults. If a connector shows no visible faults, but is still suspect, perform a voltage drop test at the connector. Even a small amount of corrosion in a connector can cause a large voltage drop to the circuit's load. See **ELECTRICAL SYSTEM** for more troubleshooting information.

Ground Connections

For any electrical circuit to work, it must make a complete path, beginning at the negative (–) battery terminal and ending at the positive (+) terminal. The negative (–) battery cable runs directly to the car's chassis. Therefore, connecting a wire to the chassis or any metal part bolted to the chassis provides a good ground path back to the negative (–) side of the battery.

Poor ground connections are one of the major sources of driveability problems. There are only a few main ground connections or points for the engine management systems. These ground points are a grouping of many wires crimped or welded into an eyelet that is then bolted to the car's chassis or metal parts. If any of these ground points are faulty, the voltage to the circuit will be reduced or even eliminated.

When checking ground wires, ground points, or ground straps, begin with a thorough visual inspection. Ground connections and wires can corrode, become loose, or break in areas that are not visible. To thoroughly check a circuit ground, check the voltage drop between the connector and a good ground source. Large voltage drops indicate too much resistance—that the connection is corroded, dirty or otherwise damaged. Clean or repair the connection and retest. Also check both battery terminals and all ground straps between the engine and the body for voltage drops.

NOTE —
For voltage drop tests and other general electrical troubleshooting information, see **ELECTRICAL SYSTEM**.

Fig. 3-5 through Fig. 3-10 show the main ground points for the engine management systems used on the cars covered by this manual.

Fig. 3-5. Ground point G103 for fuel injection system on 533i models.

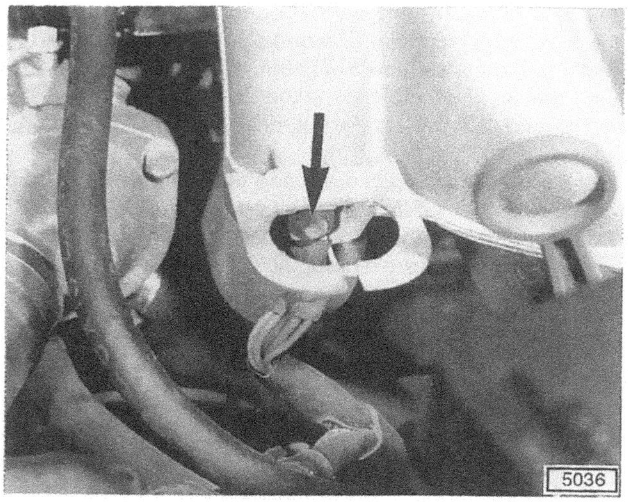

Fig. 3-6. Ground G103 (arrow) for fuel injection system on 1986 and later 528e models. Ground G103 for fuel injection system on 1982–1985 528e models is on left side of engine block, above starter motor.

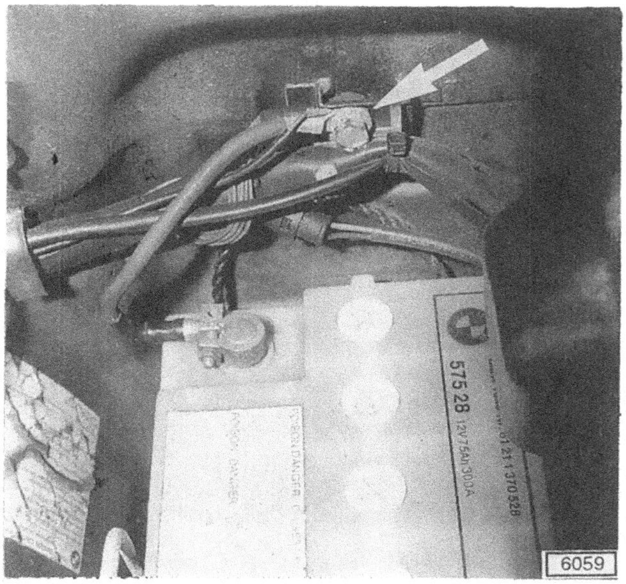

Fig. 3-7. Main battery ground G102 (arrow).

Fig. 3-8. Ground point G103 (arrow) on 535i models.

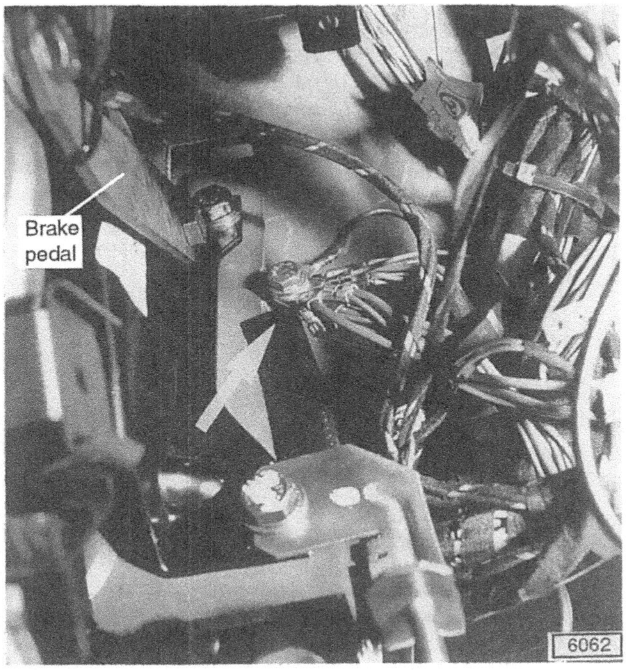

Fig. 3-9. Ground point G200 (arrow) for ignition switch, on-board computer, automatic transmission range switch, and start relay.

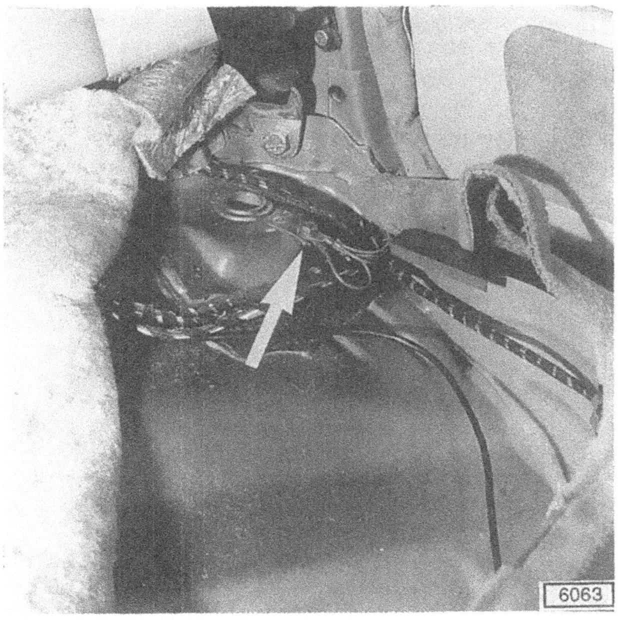

Fig. 3-10. Ground point G301 (arrow) for fuel pumps beneath rear seat bottom on driver's side.

3.5 Fuel System

3

For the engine to start quickly when cold or hot, run correctly throughout all operating conditions, and accelerate smoothly without hesitation, the fuel system must deliver a precise amount of fuel in relation to the amount of air that is drawn in by the engine.

Fuel Supply

To start and run, the engine needs an adequate supply of fuel. Fuel from the tank is supplied to the engine via two electric fuel pumps, a fuel filter, and the connecting fuel lines. If either the filter or a fuel line is restricted, the engine may not run properly. If the restriction is severe, or the main fuel pump is faulty, the engine may not start at all.

Power to run the fuel pump is controlled by a fuel pump relay mounted in the engine compartment. In order for the pump to run, the relay must be functioning correctly.

To check for a clogged fuel filter or a restricted fuel line, perform a fuel delivery rate test to check that the fuel pump is delivering enough fuel to the fuel injection system. For main fuel pump, fuel pump relay, and fuel pump delivery rate tests, see **FUEL INJECTION**. For transfer fuel pump testing, see **FUEL SUPPLY**.

Fuel Delivery Rate

The fuel injection system has the main function of delivering an optimum air-fuel mixture for all engine operating conditions. Basic fuel delivery is dependent on fuel pressure and the correctly functioning injectors. Fuel pressure is often overlooked when diagnosing driveability problems.

The fuel pressure from the fuel pump is controlled by a pressure regulator by allowing surplus fuel to return back to the fuel tank. See Fig. 3-11. The amount of fuel delivered to the engine is varied by changing the amount of time the electric pulsed-type fuel injectors remain open. A change in fuel pressure results in a change in the amount of fuel (or fuel mixture) that is delivered to the engine. Fuel pressure and fuel pressure regulator tests are described in **FUEL INJECTION**.

> **NOTE —**
> Fuel pressure tests require the use of a pressure gauge. If this equipment is not available, fuel pressure tests can be performed by an authorized BMW dealer or other qualified shop.

Fig. 3-11. Fuel pressure regulator (arrow) controls fuel pressure.

Correctly operating fuel injectors play a major role in fuel delivery. The injectors are switched on and off at the ground side of the connector. Positive (+) battery voltage is always present at the connector when the car is running. An injector can become clogged, it can completely fail or lose power and refuse to open, or it can short to ground and remain open whenever the engine is running. Checking if an injector is fundamentally working can be accomplished easily. Checking an injector's spray pattern is more difficult. See **FUEL INJECTION** for additional information.

> **NOTE —**
> High or low fuel pressure or a faulty injector will result in an incorrect fuel mixture and overall poor driveability. A lean mixture (too little fuel) can cause the engine to run poorly when cold or stumble upon acceleration. A rich mixture (too much fuel) can dilute the engine oil, foul the spark plugs, and cause a rough idle.

4. STARTING TROUBLESHOOTING

Starting difficulties can usually be grouped into one of two categories based on engine operating temperature. A cold engine that is difficult to start has different causes and remedies than an engine that is hard to start when warm. A third category, an engine starts but will not keep running, also has its own causes and remedies. Careful observation of the symptoms is the key to isolating and identifying starting problems.

4.1 No Start

Only a few requirements are necessary for an engine to start. An engine should start if it has fuel, a properly timed spark and sufficient compression. Make sure the battery is fully charged before troubleshooting an engine that will not start.

When an engine refuses to start, the first thing to check is the main relay. This relay powers the Motronic control unit, the fuel pumps, and the injectors. The main relay should click on when the key is turned on. If the main relay is not working correctly, the engine will not start. Fig. 4-1 shows the main relay location. A quick test for checking the main relay can be found in **FUEL INJECTION.**

> ***CAUTION —***
> *The main relay may be located in an adjacent position. To check if the main relay has been correctly identified, inspect the wires leading to the relay socket. There should be two large (4 mm) wires leading to the main relay socket.*

If the main relay is not faulty, test the reference (TDC) sensor next. This sensor gives the Motronic control unit its baseline timing mark for starting. If the sensor is not generating a signal to the control unit, the car will not start. The reference sensor can be easily checked as described in **IGNITION.**

> **NOTE —**
> On all models except 528e models with Motronic 1.1, the reference sensor is mounted in the transmission bellhousing and has a grey connector (next to the speed sensor with the black connector). On 528e models with Motronic 1.1, the reference sensor and speed sensor are combined into one unit and are mounted to the front of the engine near the vibration damper. See **IGNITION** for more information.

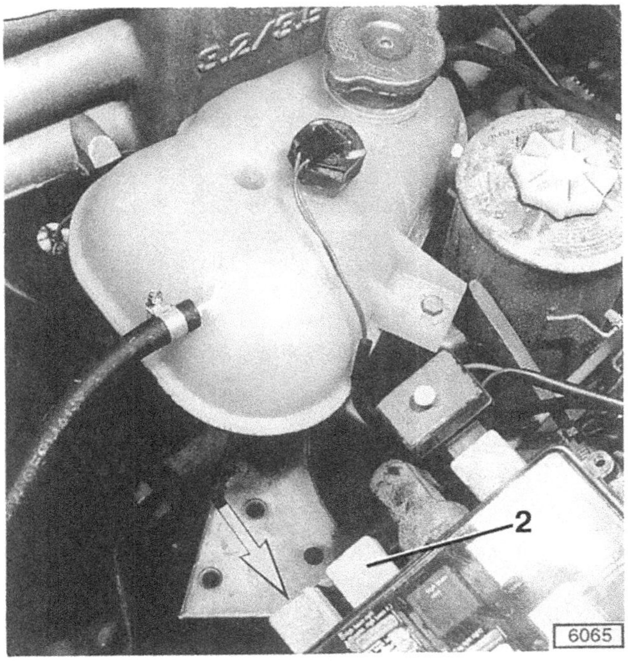

Fig. 4-1. Main relay (arrow) location on fuse/relay panel on most models covered by this manual. On 1985 and 1986 528e models, main relay is located in position (**2**).

If no faults are found up to this point, check to make sure the ignition system is producing a spark. See **IGNITION** for procedures on making an ignition system quick-test. If the ignition system is producing a spark, then the most likely causes are fuel related.

Next check that the fuel pump is operating. Turn the key on or actuate the starter while a helper listens for the fuel pump to come on. If the fuel pump is operating, check that the fuel injectors are getting fuel by making a fuel delivery test. Fuel pump checks and fuel rate delivery tests are described in detail in **FUEL INJECTION**. If no problems have been found up to this point, see **4.2 Hard to Start (cold)** below for further troubleshooting.

4.2 Hard to Start (cold)

Starting a cold engine has different fuel and air requirements than those of a warm engine. When the engine is cold, additional fuel and air are needed to maintain a steady idle and to overcome cold engine friction. On most models, ignition timing is also adapted for cold starting. If these systems used to adapt the fuel and ignition systems for cold starting are not operating, excessively long cranking times will result or the engine may not start at all.

Table b lists probable causes and corrective actions for cold starting problems. The boldface type indicates the section of the manual where the applicable test and repair procedures are found.

Table b. Cold Starting Troubleshooting

Symptom	Probable cause	Corrective action
1. Engine hard to start when cold	a. Cold start system not operating	a. Test cold start system. See **FUEL INJECTION**
	b. Fuel injectors clogged or not opening	b. Test fuel injectors and fuel injector wiring. See **FUEL INJECTION**
	c. Fuel pressure too low	c. Test fuel pressure. See **FUEL INJECTION**
	d. Air flow sensor flap binding or stuck in open position	d. Inspect air flow sensor flap for free movement. See **FUEL INJECTION**
	e. Coolant temperature sensor faulty or faulty sensor wiring	e. Test coolant temperature sensor. See **FUEL INJECTION**
	f. Reference signal missing	f. Faulty reference sensor or reference wheel or flywheel damaged. Reference pin missing. See **IGNITION**
	g. Throttle switch incorrectly adjusted or faulty	g. Check throttle switch. See **FUEL INJECTION**
	h. Large intake air leaks	h. Make thorough inspection of hoses, connections, duct work, and oil filler cap seal and oil dipstick seal
	i. Idle speed control system faulty	i. Test idle speed control system components. See **FUEL INJECTION**
	j. Poor fuel quality. Water in fuel	j. Replace fuel in tank, replace fuel filter. See **FUEL SUPPLY** and **LUBRICATION AND MAINTENANCE**
	k. Faulty Motronic control unit	k. Test control unit inputs. See **FUEL INJECTION**

4.3 Hard To Start (warm)

If the engine starts and idles well when cold, but is difficult or refuses to start when warm, the most probable cause is fuel related. Although the basic ignition system function can be eliminated as a source of trouble, the components that adapt ignition timing for varying operating conditions should not be overlooked.

Check the systems that adapt the engine to its cold running settings. If additional cold-start fuel is supplied to a warm engine, it will become flooded. When the engine reaches operating temperature, these systems should no longer be adapting the fuel system. **Table c** lists probable causes and corrective actions for warm starting problems. The boldface type indicates the section of the manual where the applicable test and repair procedures are found.

4.4 Starts But Will Not Keep Running

An engine that starts but then immediately stops is most likely due to one of three reasons.

The most common of these is a fuel pump that quits after the key is turned from the start position. In order for the fuel pump to continue to run after the key is turned from the start position, the Motronic control unit (which controls the fuel pump relay) needs a speed signal. If the speed signal is not present the fuel pump relay is switched off. That is why the fuel pump runs for only a few seconds when the key is first turned on. Check the speed sensor as described in **IGNITION**. Check the fuel pump relay electrical circuit as described in **FUEL SUPPLY.**

The second most common fault is caused by vacuum or air leaks. This unmeasured air can cause the air-fuel mixture to lean out to the point where the engine cannot maintain a steady idle. See **3.3 Air Flow Measurement and Vacuum Leaks** above for common sources of air leaks.

The last common fault is that the engine is relying on the additional fuel from the cold start system to keep running. Once this fuel is burned, the engine quits. Check for clogged fuel injectors that are not opening or low fuel pressure. Check the wiring to the injectors. Electrical signals to the injectors can be checked using a special LED tester. These testers are available from foreign automobile supply outlets.

NOTE —
- Complete engine management wiring schematics are shown in **FUEL INJECTION.**

- See **LUBRICATION AND MAINTENANCE** for information on fuel additives and clogged injectors.

- On 528e models with Motronic 1.1 engine management system, air leaks are not usually a cause of hard starting. The Motronic 1.1 engine management system is so adaptive that the air leak(s) would have to be very large (and probably easily visible) for the system not to adapt.

As a last resort, review the probable causes shown in **Table b** and **Table c**. If present in a lesser degree, many of these listed causes could also cause an engine to stop running immediately after starting.

Table c. Warm Starting Troubleshooting

Symptom	Probable cause	Corrective action
1. Engine hard to start when warm or hot	a. Fuel pump not operating when warm	a. Check for worn fuel pump by making fuel delivery test with pump at operating temperature. See **FUEL INJECTION**
	b. Fuel injector(s) leaking or stuck open	b. Test fuel injectors. See **FUEL INJECTION**
	c. Cold start system operating when warm	c. Test cold start system. See **FUEL INJECTION**
	d. Fuel pressure incorrect	d. Test fuel pressure. Inspect for leaks. See **FUEL INJECTION**
	e. Coolant temperature sensor faulty	e. Test coolant temperature sensor. See **FUEL INJECTION**
	f. Vapor lock (usually on hot days only)	f. Check fuel pressure. See **FUEL INJECTION**
	g. Evaporative emission system faulty	g. Test evaporative emission system. See **FUEL INJECTION**
	h. Idle speed control system faulty	h. Test idle speed control system components. See **FUEL INJECTION**
	i. Air flow sensor flap binding or stuck in open position	i. Inspect air flow sensor flap for free movement. See **FUEL INJECTION**
	j. Faulty Motronic control unit	j. Test electronic control unit inputs. See **FUEL INJECTION**

5. COLD RUNNING AND WARM-UP TROUBLESHOOTING

During engine warm-up, the engine requires a slightly richer mixture and a higher idle speed. This helps to overcome cold engine friction and also gives the engine extra torque to prevent stalling during take off or when selecting a drive position on models with automatic transmission.

A cold engine that accelerates poorly, hesitates or has poor off-idle characteristics can be fuel, ignition, or even emission controls related. Some of the most common causes are due to a lean air-fuel mixture as a result of intake air leaks or clogged fuel injectors. Although these are some of most probable causes, poor driveability during warm-up can have many other causes, including those that lead to an overly rich air-fuel mixture.

> **NOTE —**
> For an engine that idles poorly when cold, see **6. Idle Speed Troubleshooting**.

To ensure smooth running, good off-idle and acceleration characteristics, and overall driveability while the engine warms up, both fuel mixture and basic ignition timing need to be modified for cold engine operation.

Additional fuel is supplied to the engine by increasing the opening or pulse time of the injectors based on inputs from the coolant temperature sensor. See Fig. 5-1. To improve driveability, ignition timing is adapted for cold running based on inputs from the intake-air temperature sensor and the coolant temperature sensor.

Table d lists probable causes and corrective actions for cold-running and warm-up problems. The boldface type indicates the section of the manual where the applicable test and repair procedures are found.

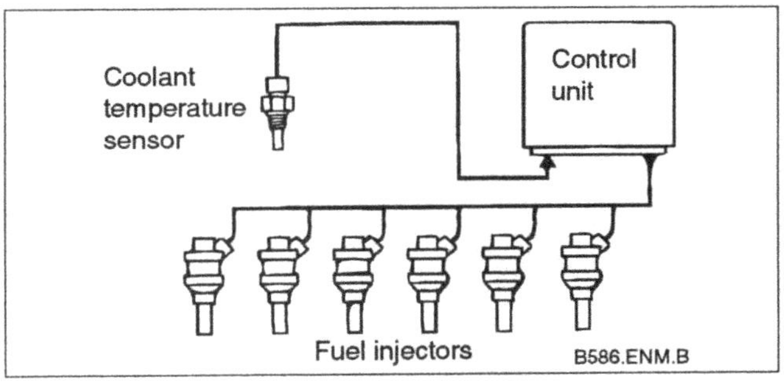

Fig. 5-1. Schematic of coolant temperature sensor used to adapt fuel system for warm-up enrichment. Injectors stay open longer when engine is cold based on cold engine signal from sensor. Courtesy Robert Bosch Corp.

Table d. Cold-Running and Warm-up Driveability Troubleshooting

Symptom	Probable cause	Corrective action
1. Engine runs poorly during warm-up, has poor acceleration, off-idle hesitation, and backfires	a. Intake air leaks	a. Make thorough inspection of hoses, connections, duct work, and oil filler cap seal and oil dipstick seal
	b. Fuel injector(s) clogged or not opening	b. Test fuel injectors and fuel injector wiring. See **FUEL INJECTION**
	c. Fuel pressure too low	c. Test fuel pressure. See **FUEL INJECTION**
	d. Air flow sensor flap binding	d. Inspect air flow sensor flap for free movement. See **FUEL INJECTION**
	e. Coolant temperature sensor faulty or faulty sensor wiring	e. Test coolant temperature sensor. See **FUEL INJECTION**
	f. Throttle switch incorrectly adjusted or faulty	f. Check throttle switch. See **FUEL INJECTION**
	g. Idle mixture incorrectly adjusted	g. Adjust idle mixture (except 528e models with Motronic 1.1). See **FUEL INJECTION**
	h. Idle mixture incorrect	h. Oxygen sensor faulty. Test oxygen sensor or replace if replacement interval exceeded. See **FUEL INJECTION**
	i. Idle speed control system faulty	i. Test idle speed control system components. See **FUEL INJECTION**
	j. Poor fuel quality. Water in fuel	j. Replace fuel in tank, replace fuel filter. See **FUEL SUPPLY** and **LUBRICATION AND MAINTENANCE**
	k. Cold start system operating all the time	k. Test cold start system. See **FUEL INJECTION**
	l. Faulty Motronic control unit	l. Test control unit inputs. See **FUEL INJECTION**

6. IDLE SPEED TROUBLESHOOTING

Erratic idle speed is one of the most common driveability problems encountered on the cars covered by this manual. An electronic idle speed system controls the idle speed for all engine operating conditions. The idle air stabilizer valve constantly adjusts the amount of air allowed to bypass the throttle plate to either increase or decrease engine speed. See Fig. 6-1.

Fig. 6-1. Idle air stabilizer valve (arrow) controls idle bypass air to regulate idle speed.

Idle speed is adapted based on various inputs from engine sensors and components. See Fig. 6-2.

NOTE —

- On all 533i models and all 528e models built up to March 1987, idle speed is controlled by a separate idle speed control unit. On all 535i models and 528e models built March 1987 and later (Motronic 1.1), idle speed is controlled by the Motronic control unit.

- Be sure the idle switch is correctly functioning as described in **FUEL INJECTION** before performing any idle speed troubleshooting.

6.1 Idle Speed Basics

Begin troubleshooting idle problems by visually inspecting for faulty wiring, especially checking the connectors at the idle air stabilizer valve, the injectors, the throttle switch, the oxygen sensor, the coolant temperature sensor, and the coolant temperature sensor (if applicable).

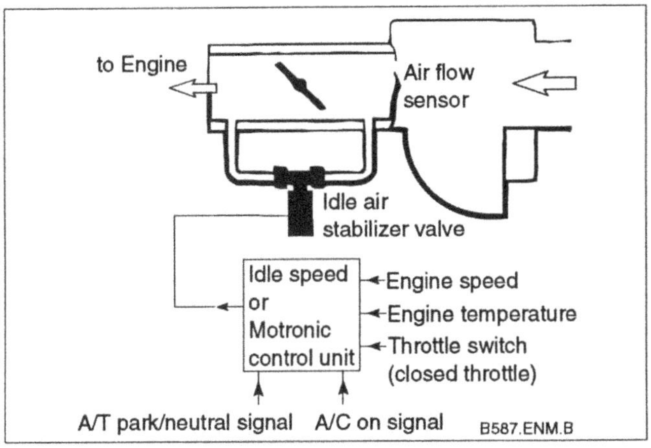

Fig. 6-2. Schematic of idle speed control system. Courtesy Robert Bosch Corp.

Carefully inspect for any air leaks, especially checking for cracks in the hoses at the idle air stabilizer valve, the large intake air duct, and the thin hose between the intake manifold and the pressure regulator. Make sure the oil filler cap and the dipstick are installed correctly. Check that the throttle switch and the throttle valve are correctly adjusted as described in **FUEL INJECTION**.

The engine management systems that control idle speed rely on two key inputs—a closed-throttle signal and an engine temperature signal. See Fig. 6-2 above. In addition to these signals, idle speed is further adapted based on engine loads, intake air temperature, and on some models with automatic transmission, outside air temperature.

NOTE —
If the test equipment is available, check that idle mixture (%CO) is within specifications. In addition, troubleshoot warm idle problems prior to working on cold idle problems. Cold idle is adjusted based on warm engine settings. Therefore, it is important to troubleshoot idle problems in this order.

Analysis of the observed symptoms is the key to isolating and identifying idle problems. Pay close attention to engine conditions whenever idle problems occur. For example, if the idle speed steadily increases as the engine warms up, check the sensors that give the idle speed control system its temperature information.

6.2 Cold Idle

If the engine idles poorly when cold but maintains a steady idle when warm, the most probable cause is that the air-fuel mixture is too lean for a cold engine. A lean mixture is caused by one of two conditions: either too much air or too little fuel. Excess air is mainly caused by intake air leaks. Insufficient fuel has many causes, such as clogged fuel injectors or incorrect fuel pressure.

NOTE —
The oxygen sensor can be ruled out as a source of cold idle problems. Until the engine reaches a specified temperature, the signal from the oxygen sensor is ignored.

As opposed to a lean mixture, an excessively rich mixture can also be the cause cold idle problems. If too much fuel is delivered to the engine, the spark plugs will not be able to burn the excess fuel. If the condition is severe enough, the engine will eventually stall as it warms up and refuse to start until it cools down. **Table e** lists some of the more common causes of cold idle problems.

To help determine if the idle mixture is lean or rich, observe the engine's idling characteristics. An engine that runs lean will normally be hard to start or may stall immediately after starting, is unresponsive with off-idle hesitation, and has poor acceleration characteristics. An engine that is running rich will have a steady stumbling or gallop when idling and may possibly stall after running for a minute or so due to fouled spark plugs. Another sure sign of a rich mixture is black exhaust smoke and strong fuel odors. As a quick check, blip the accelerator pedal to quickly raise the idle speed. If the idle speed falls below specifications when the pedal is released and then returns to normal, the engine is running too rich. If the engine hesitates or stalls as it tries to speed up, the mixture is too lean.

NOTE —
The condition of the spark plugs is a good indicator of combustion quality that can help to diagnose idle faults. See **LUBRICATION AND MAINTENANCE** for more specifics on spark plug appearance.

6.3 Warm-up Idle

When the engine is cold, the fuel system supplies the engine with additional fuel. As the engine warms up, this enrichment fuel is slowly cut back until the engine is at operating temperature. Engine temperature from the temperature sen-

Table e. Cold Idle Troubleshooting

Symptom	Probable cause	Corrective action
1. Engine idles poorly when cold (lean air-fuel mixture)	a. Faulty fuel injector seals (O-rings)	a. Replace injector seals. See **FUEL INJECTION**
	b. Faulty or cracked air hoses, vacuum hoses, air duct, and connections	b. Replace any faulty hoses or duct work. Tighten any loose connections
	c. Clogged or faulty fuel injectors	c. Test injectors. See **FUEL INJECTION**
	d. Fuel pressure too low	d. Check fuel pressure and fuel delivery rate. See **FUEL INJECTION**
	e. Coolant temperature sensor or switch faulty	e. Test coolant temperature sensor and or switch. See **FUEL INJECTION**
	f. Idle air stabilizer valve faulty or out of adjustment	f. Test idle air stabilizer valve. See **FUEL INJECTION**
	g. Valve cover gasket, front or rear crankshaft oil seal leaking or damaged	g. Check for leaking or damaged gaskets or oil seals. See **ENGINE—RECONDITIONING**
	h. Motronic control unit or idle speed control unit faulty	h. Test control unit inputs. See **FUEL INJECTION**
2. Engine idles poorly when cold (rich air-fuel mixture)	a. Injectors leaking or sticking open	a. Test injectors. See **FUEL INJECTION**
	b. Cold start system operating continuously	b. Test cold start system. See **FUEL INJECTION**
	c. Coolant temperature sensor or switch faulty	c. Test coolant temperature sensor and/or switch. See **FUEL INJECTION**
	d. Fuel pressure too high	d. Check fuel pressure. See **FUEL INJECTION**
	e. Intake air flow restricted	e. Check air flow path for obstructions. Replace dirty air filter. See **LUBRICATION AND MAINTENANCE**
	f. Weak ignition spark	f. Check ignition coil and spark intensity. See **IGNITION**
	g. Motronic control unit or idle speed control unit faulty	g. Test control unit inputs. See **FUEL INJECTION**

sor is the main signal that handles this warm-up enrichment. See Fig. 7-1. On some models, a coolant temperature switch also controls warm-up enrichment. Test these components as described in **FUEL INJECTION** first when experiencing warm-up driveability problems.

Fig. 6-3. Coolant temperature sensor (arrow) used to adapt engine management system for varying operating conditions.

If no faults are found with the coolant temperature sensor or switch, check to see when the oxygen sensor was last replaced. If the sensor is old, the sensor's signal may be incorrect during warm-up. See **3.2 Basic Adjustments** for more information on the oxygen sensor. Checking the accuracy of the oxygen sensor requires special test equipment. As a general rule, replace the sensor if replacement is due. See **LUBRICATION AND MAINTENANCE** for BMW's recommended maintenance schedules.

Other possible causes of poor idle during engine warm-up are clogged injectors and intake valve carbon deposits, although these problems will usually show up first during cold starting and cold idle. See **3.1 Engine** for more on clogged injectors and carbon deposits.

6.4 Idle at Operating Temperature

Warm idle speed problems can have numerous causes. The problems can range from a faulty coolant temperature sensor to a faulty fuel injection control unit. **Table f** lists probable causes and corrective actions for warm idle problems. Troubleshoot warm idle problems in the order listed in table. The boldface type indicates the section of the manual where the applicable test and repair procedures are found.

Table f. Warm Idle Troubleshooting

Symptom	Probable cause	Corrective action
1. Engine idles poorly when warm	a. Coolant temperature switch faulty (528e models built up to March 1987)	a. Test coolant temperature switch. See **FUEL INJECTION**
	b. Throttle switch incorrectly adjusted or faulty	b. Check throttle switch adjustment. See **FUEL INJECTION**
	c. Throttle plate incorrectly adjusted	c. Check throttle plate basic adjustment. See **FUEL INJECTION**
	d. Coolant temperature sensor faulty	d. Test coolant temperature sensor. See **FUEL INJECTION**
	e. Coolant level incorrect	e. Check coolant level. See **COOLING SYSTEM**
	f. Fuel injector(s) clogged, leaking or faulty	f. Test fuel injectors and fuel injector wiring. See **FUEL INJECTION**
	g. Idle air stabilizer valve out of adjustment or idle speed control system faulty	g. Test idle speed control system components. See **FUEL INJECTION**
	h. Oxygen sensor faulty	h. Test oxygen sensor output. See **FUEL INJECTION**
	i. Cold start system operating continuously	i. Test cold start system. See **FUEL INJECTION**
	j. Charcoal canister purge valve faulty	j. Test evaporative emission system. See **FUEL INJECTION**
	k. Intake air leaks	k. Make thorough inspection of hoses, connections, duct work, and oil filler cap seal and oil dipstick seal
	l. Fuel pressure regulator faulty or hose to regulator leaking or faulty	l. Test fuel pressure. See **FUEL INJECTION**
	m. Motronic control unit or idle speed control unit faulty	m. Test control unit inputs. See **FUEL INJECTION**

7. NORMAL WARM RUNNING TROUBLESHOOTING

The problems of normal warm running are very similar to those of warm idle. In most cases, warm engine driveability problems also affect warm engine idle. Additional causes of normal warm running that do not manifest themselves at idle are those of the more demanding operating conditions. Be sure the engine idles properly before troubleshooting the engine management systems for warm running conditions.

7.1 Rough Running/Misfiring

Begin troubleshooting an engine that runs poorly or misfires under all operating conditions and speeds by checking the main grounds and the electrical connections for the control units as described above under **3.4 Electrical System**. Make sure the throttle switch is correctly adjusted and properly functioning. Check that the air flow sensor flap is not binding and that there are no faults in the air flow sensor's potentiometer.

Remove the distributor cap and look for any oil-fouled or moisture-laden components. A faulty distributor cap dust shield O-ring can allow small amounts of engine oil to seep into the distributor. See **IGNITION** for more troubleshooting information.

If no faults can be found and all of the causes listed in **Table f** above have been eliminated, carbon deposits on the intake valves or injectors may be the cause of the problem. See **3.1 Engine** for more information on carbon deposits.

7.2 Poor Acceleration

During acceleration, the fuel mixture needs to be quickly enriched and the ignition timing needs to be adapted to prevent detonation and engine damage.

Three fuel system components handle fuel enrichment: the air flow sensor, the fuel pressure regulator, and the throttle switch (except 535i models with electronically-controlled transmission). When the throttle is quickly opened, the sudden rush of air past the air flow sensor causes the sensor flap to open quickly and over swing, sending a proportional electrical signal to the control unit. As the throttle plate is opened, the vacuum supplied to the top of the fuel pressure regulator is reduced and fuel pressure rises. If the pedal is fully depressed, a full-throttle signal is sent to the Motronic control unit. All of these components send additional signals to the control unit to increase fuel flow when the throttle is fully open. Check these fuel system components carefully when experiencing acceleration problems.

NOTE —
- On 1986 through 1988 535i models with automatic transmission, the full throttle signal to the Motronic control unit comes from the automatic transmission control unit.
- If no fuel system components are found to be faulty, the injectors may be clogged. Although clogged injectors are most pronounced during cold acceleration, slightly clogged injectors can cause flat spots and poor acceleration. See **LUBRICATION AND MAINTENANCE** for advice on clogged injectors.

If any ignition system components are faulty or worn, their ability to deliver a strong, precisely timed spark will be reduced. The engine may idle fine but miss and skip during acceleration due to a weak spark. Poor spark intensity at the plugs can be caused by worn or fouled plugs, old or faulty plug wires, or a faulty ignition coil, rotor, or distributor cap. Inspect the ignition components carefully and replace any that are worn. See **IGNITION.**

7.3 Poor Fuel Economy

High fuel consumption normally results when the air-fuel mixture is too rich. As a first step in diagnosing high fuel consumption, check the idle mixture. If the idle mixture cannot be adjusted to within specifications, check the components that adapt the air-fuel mixture for varying operating conditions.

NOTE —
High fuel consumption can also be caused by abnormal friction in the engine or drive train. For example, a faulty engine driven component, or a sticking parking brake.

The one component that constantly adjusts fuel mixture is the oxygen sensor. See Fig. 7-1. As an oxygen sensor ages, it loses its ability to accurately and quickly adjust fuel mixture. An oxygen sensor that fails will usually cause the engine to run rich. In addition, check the coolant temperature sensor and the coolant temperature switch as described in **FUEL INJECTION**. Check for a restricted air filter element.

NOTE —
Oxygen sensor testing procedures are covered in **FUEL INJECTION.**

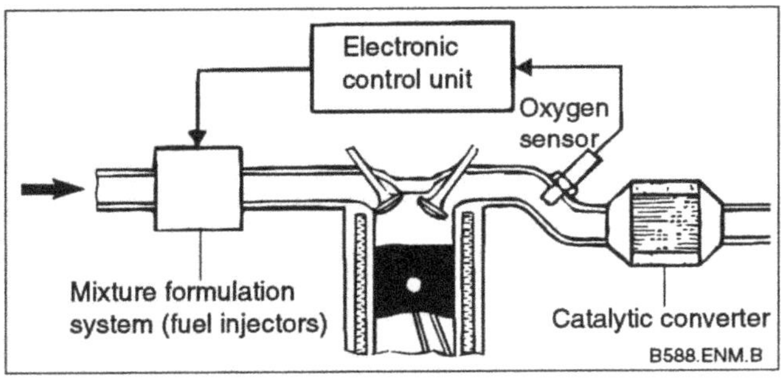

Fig. 7-1. Schematic of oxygen sensor circuit. Courtesy Robert Bosch Corp.

7.4 High Exhaust Emissions

Exhaust emission levels that are out of specification usually indicate an engine that is running poorly or is out of tune. For example, an engine that runs lean (low %CO) will usually exhibit cold starting and running problems. An engine that runs rich (high %CO) will usually show up as an increase in fuel consumption. A rich fuel mixture can also cause starting and running problems when the ambient air temperatures are high. Therefore, correcting engine driveability problems will usually return the exhaust emissions to within specifications.

Low exhaust emissions or a low %CO reading result when the basic air-fuel mixture is too lean. A lean fuel mixture will normally cause poor driveability and engine performance. Some of the side effects of a lean mixture are increased operating temperature and reduced spark plug life. The most common causes of low exhaust emissions are intake air leaks and clogged injectors. Review the information under the above headings to troubleshooting lean air-fuel mixtures.

High exhaust emissions or a high %CO reading results when the basic air-fuel mixture is too rich. Unless the ambient air temperatures are high, rich mixtures usually do not cause major engine driveability problems unless the mixture is excessively rich. In severe cases, the engine will run and then die once the spark plugs become fouled, and the engine will refuse to restart.

A rich mixture can cause serious engine damage if allowed to continue. A rich mixture will dilute the engine oil and wash the oil from the cylinder walls and friction surfaces, causing rapid engine wear. The catalytic converter could also be damaged due to the excess fuel. High exhaust emission should be corrected as soon as possible. The most common causes of high exhaust emissions are failed fuel delivery components, such as a stuck-open cold start valve or injector, a clogged air filter, or a faulty coolant temperature sensor. Review the troubleshooting information under the above headings when troubleshooting rich air-fuel mixtures.

Section 4

4

ENGINE — SERVICE AND REPAIR

Contents

TABLES

INTRODUCTION

This section of the manual covers the troubleshooting, service, and repair to the three 6-cylinder engines used in the 528e, the 533i, and the 535i models. All of the 5-series engines are front-mounted and water-cooled. The valves are operated by an overhead camshaft.

The information in this section of the manual is an engine service and repair guide both to car owners and to professional mechanics. For information on engine rebuilding or reconditioning, see **ENGINE—RECONDITIONING**.

1. GENERAL DESCRIPTION

The in-line overhead cam engine is water-cooled and transmits power through a piston-driven crankshaft. The cylinder block is made of cast iron with integral cylinders completely exposed on all sides to the coolant that circulates through the water jacket. A separate cast aluminum alloy cylinder head contains the camshaft and the entire valve train.

The engine is bolted to a bellhousing on the transmission and is inclined toward the right side of the engine compartment. This permits a lower hood line and lower center of gravity. The engine and the transmission are supported as a single unit by bonded rubber mounts that reduce the transfer of noise and vibration to the rest of the car.

The engine in the 528e has a displacement of 2693 cc (164 cu. in.). The 1982 through 1987 (built up to March 1987) 528e models are rated at 121 horsepower (SAE net). 528e models built from March 1987 and later are rated at 127 horsepower (SAE net). Owing to its small displacement compared to the displacement of the 533i and the 535i models, this engine is often referred to as the "baby six," the "small six," or the "eta engine."

The engine in the 533i has a displacement of 3210 cc (196 cu. in.). It is rated at 181 horsepower (SAE net). The engine in the 535i has a displacement of 3430 cc (209 cu. in.). It is rated at 182 horsepower (SAE net). These engines are often referred to as the "big six."

2. MAINTENANCE

BMW specifies the steps below to be carried out at particular times or mileage intervals for proper maintenance of the engine. A number in bold type indicates that the procedure is covered in this section, under that numbered heading. Information on other engine maintenance and on the prescribed maintenance intervals can be found in **LUBRICATION AND MAINTENANCE**.

1. Checking engine oil level
2. Changing engine oil and filter
3. Replacing spark plugs
4. Checking compression pressure **3.2**
5. Adjusting valve clearance **4.2**
6. Replacing cylinder head cover gasket (when adjusting valves) **4.1**

3. TROUBLESHOOTING

This troubleshooting section applies to problems affecting the basic engine assembly—the cylinder block, cylinder head, and their internal moving parts.

Only a few basic functions are required of the engine. The block, cylinder head, and their moving parts must fit together properly, operate smoothly, seal well enough to create and maintain compression, and keep pistons, valve train, and ignition properly timed. The problems discussed in this troubleshooting section are those that affect one or more of these functions.

Troubleshooting specifically for the lubrication system can be found in this section under **6. Lubrication System**. To troubleshoot overheating and other cooling system problems, see **COOLING SYSTEM**. Troubleshooting for other general starting and running problems can be found in **ENGINE MANAGEMENT—DRIVEABILITY**.

3.1 Basic Troubleshooting Principles

Begin troubleshooting with careful observation, keeping in mind the following questions:

How has the symptom developed? A symptom that develops quickly is probably caused by a problem that can be corrected by simple maintenance or repair. A symptom that has developed gradually over time, especially after fifty or sixty thousand miles, is more likely an indication of general wear and the need for more comprehensive overhaul work.

Is the symptom engine-speed dependent? A noise that is caused by an engine mechanical problem will be dependent mainly on engine speed, with similar symptoms regardless of changes in car speed. Noises that repeatedly occur only in a certain rpm range suggest a vibration problem. Noises that change with car speed are more likely due to drivetrain or running gear problems.

Is the symptom load dependent? Forces at work inside a running engine vary as the demand for power varies. Symptoms that are more severe during hard acceleration indicate certain kinds of problems. Symptoms that are more apparent at no load or high vacuum (example: coasting at high rpm) point to other problems. Note that higher engine loads also affect the fuel and ignition systems, which may be responsible for high-load performance problems.

Is the symptom temperature dependent? Does it only occur when the engine is cold? Does it change as the engine warms up? How? Metal parts expand and contract with changes in temperature. Clearances change. Oil viscosity and cooling system pressure change. In troubleshooting symptoms that change as the engine gets warm, look for an engine characteristic that changes with temperature.

Noise

In order to run smoothly under harsh conditions, the internal engine parts are made to precise dimensions, assembled with precision clearances, and lubricated by a pressurized oiling system.

Most unidentified engine noises result from clearances that have become too large due to worn parts, lack of adequate lubrication, or both. The importance of lubrication cannot be over-emphasized. For best results, troubleshooting engine noises should only be done when the oil and filter have been recently changed and the oil level is correct.

High-pitched metallic tapping noises are caused by relatively small, lightweight parts. These noises are most likely an indication of excessive clearances in the valve train. Valve train noise accompanied by burning oil (blue-gray smoke in the exhaust), particularly at start-up or when decelerating from high rpm, is an indication of worn valve guides that can only be remedied by recondition or replacement of the cylinder head.

In a high-mileage engine, a light metallic rattle or chatter under acceleration, accompanied by increased oil consumption and smoking, may indicate severely worn or broken piston rings. Since this diagnosis means overhaul or replacement of the engine, the problem should be further investigated with a compression or cylinder leakage test. A vacuum gauge is also helpful when diagnosing engine mechanical faults. See **3.2 Diagnostic Testing**.

Deep, metallic knocking sounds are caused by excessive clearances between heavier components. Closer analysis of the noise will often help identify the problem. Piston slap, caused by excessive piston skirt to cylinder wall clearance, is worse when the engine is cold and may be accompanied by increased oil consumption and reduced compression due to accelerated piston ring wear. A double knock, most pronounced at idle or low load, is due to excessive clearance at the piston pin and upper connecting rod bushing.

Crankshaft bearing problems produce a deep, hollow knock that is worst when the engine is warm. A noise that is very pronounced under load, perhaps louder during the transition from acceleration to coasting, is most likely caused by a damaged connecting rod bearing. Crankshaft main bearings produce a lower, dull knock. An intermittent knock, that may be most apparent when depressing or releasing the clutch, indicates excessive crankshaft end play. These problems seldom occur as isolated failures. They are almost always an indication of the overall engine condition that can only be properly corrected by complete engine overhaul or replacement. See **ENGINE— RECONDITIONING.**

Rumbling or groaning from the engine compartment may not indicate engine problems at all, but rather a worn bearing or bushing in an engine-driven accessory. They include the coolant pump, the alternator, the power steering pump, and the air conditioning compressor. To check these accessories, run the engine briefly with the drive belt disconnected and see if the noise has stopped. Once the drive belt is removed, turning the pulley and shaft by hand may also reveal a bad bearing or bushing. A properly functioning accessory should turn smoothly.

Fluid Leaks

Fluid leaking from and around the engine is most likely either oil, coolant, brake fluid, or power steering fluid from the power steering system and the brake booster system (533i and 535i models). Look for wet spots on the engine to help pinpoint the source. It may be helpful to start by cleaning the suspected area.

The most likely sources of engine oil leaks are the oil filter gasket, the crankcase oil seals, the cylinder head cover gaskets, or the oil pan gaskets. Because the crankcase is under slight vacuum when the engine is running, some oil leaks may not be apparent until the engine is turned off. The power steering system and the brake system are other sources of leaks near the engine.

Coolant is a mixture of water and anti-freeze, yellow-green in color or perhaps brown if the coolant is old. A pressure test of the cooling system is the best way to discover and pinpoint leaks. See **COOLING SYSTEM.**

Brake fluid is clear, perhaps slightly purple, and a little slippery. Look for wet spots around the master cylinder or brake lines. Especially check the flexible hoses near the wheels. See **BRAKES** for repair information.

Smoking

Smoke that is visible under the hood is usually either blue-gray smoke from burning oil, or white steam from the cooling system. Smoke in the exhaust indicates something getting into the combustion chamber and being burned. The color of the smoke identifies the contaminant.

Blue-gray smoke is from oil. Oil smoke, probably accompanied by increased oil consumption and oil residue on spark plugs, indicates that engine oil is getting past piston rings, valve guides, the cylinder head gasket, or some combination of the three. Use a compression test for diagnosis. See **3.2 Diagnostic Testing**.

In an older engine, compression pressures that are even but below specifications suggest worn piston rings and cylinder walls, and the need to overhaul or replace the engine. If smoking is most obvious under high engine vacuum, such as while coasting at high rpm, and compression pressures are within specifications, leaking valve guide oil seals or worn valve guides are a probable cause. See **ENGINE—RECONDITIONING** for repair information.

Oil smoke or steam appearing suddenly in the exhaust, along with low compression pressure in one cylinder or two adjoining cylinders, is very probably due to a failed cylinder head gasket. Look also for coolant loss, oil in the radiator, or water in the oil (that turns the oil an opaque, creamy brown). See **4.5 Removing and Installing Cylinder Head** for repair procedures.

Black smoke is caused by the engine getting too much fuel. See **ENGINE MANAGEMENT—DRIVEABILITY** for more troubleshooting information.

Excessive Oil Consumption

Some oil consumption is normal and indicates healthy flow and distribution of the vital lubricant in the engine. BMW states that the maximum allowable oil consumption for the engines covered by this manual is 1 quart per 390 miles. This is why the oil level must be checked, and occasionally corrected, between oil changes. Aside from leaks, increased oil consumption will usually be accompanied by some smoking, however slight, and the causes of excessive oil consumption are the same as those for oil smoke in the exhaust. As with smoking symptoms, gradual increases are caused by worn piston rings and/or valve guides. Sudden high oil consumption suggests broken rings or a failed cylinder head gasket. See **Smoking** above for more troubleshooting information.

Poor Fuel Consumption and Low Power

Poor fuel consumption and low power can suggest problems with the fuel or ignition systems, particularly on a low-mileage engine. On an engine with high mileage, low compression may be the cause.

Normal wear of the valves, piston rings, and cylinder walls decreases their ability to make a good seal. The engine becomes less efficient and has to work harder, using more fuel to produce the same amount of power. Engine condition can be evaluated with a compression test. See **3.2 Diagnostic Testing**.

Engine Not Running

An engine problem that affects ignition or valve timing may prevent the engine from starting or running. The camshaft drive belt or timing chain and sprockets are responsible for correctly timing the actions of the valves and the ignition system. A worn camshaft drive belt or timing chain may jump teeth, throwing off all the engine's timing functions, and still appear to be perfectly normal. Begin by doing a compression test. See **3.2 Diagnostic Testing**.

To check camshaft timing, see **4.3 Camshaft Drive Belt (528e models)** or **4.4 Camshaft Timing Chain (533i and 535i models)**. Other troubleshooting information for an engine that fails to start can be found in **ENGINE MANAGEMENT—DRIVEABILITY**.

In addition to engine timing, all of the Motronic—controlled engines covered by this manual rely on a Top Dead Center (TDC) signal from the reference sensor for starting. If this sensor is faulty, the engine will not start. See **ENGINE MANAGEMENT—DRIVEABILITY** for more information.

Table a lists symptoms of engine problems, their probable causes, and suggested corrective actions. The boldface numbers in the corrective action column indicate the heading in this section of the manual where the applicable test and repair procedures can be found.

Table a. Engine Troubleshooting

Symptom	Probable cause	Corrective action
1. Pinging or rattling noise under load, uphill or accelerating, especially from low speeds. Indicates detonation or pre-ignition	a. Fuel does not meet manufacturer's octane requirements b. Overheating c. Spark plugs damaged or wrong heat range d. Air-fuel mixture too lean	a. Switch to higher octane fuel. See **FUEL INJECTION** for fuel octane requirements b. See **COOLING SYSTEM** c. Replace spark plugs. See **IGNITION** d. See **ENGINE MANAGEMENT—DRIVEABILITY**

continued on next page

Table a. Engine Troubleshooting (continued)

Symptom	Probable cause	Corrective action
2. Screeching or squealing noise under load. Goes away when coasting. Indicates slipping V-belt	**a.** Loose, worn, or damaged V-belt(s) **b.** Excessive belt loads due to failed engine-driven component	**a.** Inspect, replace, or tighten belt(s). See **LUBRICATION AND MAINTENANCE** **b.** Locate and replace failed component. **3.2**
3. Growling or rumbling, varies with engine rpm. Bad bearing or bushing in an engine-driven accessory	**a.** Coolant pump worn **b.** Alternator worn **c.** Power steering pump worn **d.** Camshaft drive belt tensioner bearing worn (528e models only) **e.** Air conditioning compressor faulty	**a.** See **COOLING SYSTEM** **b.** See **ELECTRICAL SYSTEM** **c.** See **STEERING AND WHEEL ALIGNMENT** **d.** Replace belt tensioner and drive belt. **4.3** **e.** Replace compressor
4. Light metallic tapping noise, varies directly with engine speed. Oil warning light **not** illuminated	**a.** Low oil pressure and defective warning light circuit **b.** Valve clearance out of adjustment **c.** Defective rocker arm(s) or rocker arm shaft(s)	**a.** Check oil pressure. **6.** **b.** Adjust valve clearance. **4.2** **c.** Check rocker arms/shafts and replace as required. See **ENGINE—RECONDITIONING**
5. Light metallic knock, varies directly with engine speed. Oil warning light blinking or fully illuminated (may be most noticeable on hard stops or cornering) Indicates lack of sufficient oil supply	**a.** Low oil level **b.** Restricted (dirty) oil filter **c.** Insufficient oil pressure	**a.** Check and correct oil level. See **LUBRICATION AND MAINTENANCE** **b.** Change engine oil and filter. See **LUBRICATION AND MAINTENANCE** **c.** Check oil pressure. **6.**
6. Blue-gray exhaust smoke, oily spark plugs. Indicates oil burning in combustion chamber	**a.** Leaking valve stem oil seals **b.** Worn valve guides **c.** Worn, broken, or incorrectly installed pistons or piston rings	**a.** Replace valve stem oil seals. See **ENGINE—RECONDITIONING** **b.** Overhaul or replace cylinder head. See **ENGINE—RECONDITIONING** **c.** Overhaul or replace engine. See **ENGINE—RECONDITIONING**
7. Blue-gray smoke and/or white steam in exhaust	**a.** Failed cylinder head gasket (probably accompanied by low compression readings) See **3.2 Diagnostic Testing** **b.** Warped or cracked cylinder head (probably accompanied by low compression readings) See **3.2 Diagnostic Testing** **c.** Cracked cylinder block	**a.** Replace cylinder head gasket. **4.5** **b.** Resurface cylinder head or replace gasket. **4.5** or see **ENGINE—RECONDITIONING** **c.** Replace engine or short block. To remove engine see **5.**
8. Black exhaust smoke	**a.** Rich air/fuel mixture	**a.** See **ENGINE MANAGEMENT—DRIVEABILITY**
9. Engine runs badly, pops and backfires	**a.** Spark plug wires installed incorrectly **b.** Incorrect valve timing	**a.** Install wires correctly. See **IGNITION** **b.** Check camshaft drive belt or camshaft timing chain. **4.3** or **4.4**

continued on next page

Table a. Engine Troubleshooting (continued)

Symptom	Probable cause	Corrective action
10. Engine will not start or run. Starter operates, engine turns over at normal speed	a. Failed ignition system or fuel system b. Broken camshaft drive belt or timing chain c. Incorrect camshaft timing due to jumped or incorrectly installed belt or chain	a. See **ENGINE MANAGEMENT—DRIVEABILITY** b. Check camshaft or ignition rotor rotation as engine turns over. Install new camshaft drive belt or timing chain as necessary. **4.3** or **4.4** c. Check camshaft timing. Replace belt or chain and sprockets as necessary. **4.3** or **4.4**

3.2 Diagnostic Testing

The tests that follow can be used to help isolate engine problems, to better understand a problem before starting expensive repairs, or just to periodically check engine condition.

Compression Test

A test of compression pressures in the individual cylinders will tell a lot about the condition of the engine without the need for taking it apart. The test is relatively simple. It requires a compression tester, a spark plug wrench, and a few hand tools.

To obtain accurate results, the battery and starter must be capable of turning the engine at normal cranking speed. The area around the spark plugs or injectors should be clean, to avoid getting debris inside the engine when they are removed. Because engine temperature may affect compression, the most accurate results are obtained when the engine is at normal operating temperature.

To test compression:

1. Make sure the ignition is disabled. See **IGNITION** for information on how to disable the systems used on models covered by this manual.

2. On 533i and 535i models, remove the air filter housing together with the air flow sensor assembly. On 533i models, also remove the large rubber intake air boot. See Fig. 3-1.

3. Remove and label the spark plug wires from the plugs. Use care to pull only on the boot to avoid damage to the connectors. Remove and label the spark plugs.

   NOTE —
   Used spark plugs should be installed in the same cylinder from which they were removed.

4. Thread the compression tester into the first cylinder's spark plug hole, just tight enough to seal around the spark plug hole.

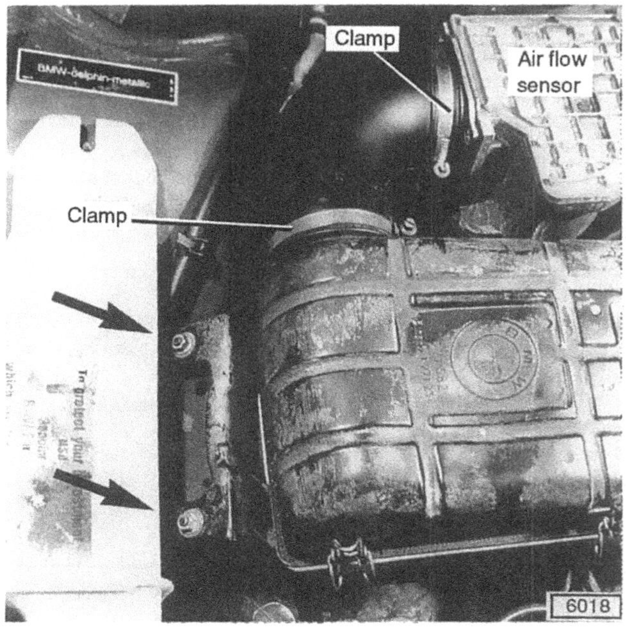

Fig. 3-1. Air filter housing mounting points (arrows) on 535i and 533i models.

5. With the transmission in neutral and the throttle held wide open, crank the engine with the starter. The gauge reading should increase with each engine revolution. Crank the engine about 4 to 5 revolutions. Record the value shown on the gauge.

   NOTE —
   The engine should be cranked an equal number of revolutions at each cylinder to obtain the most accurate readings.

6. Release the gauge pressure and thread the gauge out of the spark plug hole.

7. Repeat the test for each of the other cylinders. Record the data and compare with **Table b**.

Table b. Compression Pressure Specifications

Engine	Compression pressure
All	10–11 bar (142–156 psi)

8. Reinstall the spark plugs and the spark plug wires in their original locations. The remainder of installation is the reverse of removal.

Tightening torque
• spark plugs . . . . 20 to 30 Nm (14 to 22 ft-lb)

Low compression is evidence of poorly sealed combustion chambers. Generally, compression pressures that are relatively even but low indicate worn piston rings and/or cylinder walls. Erratic values tend to indicate valve leakage. Dramatic differences, such as good values in some cylinders and low values in one or two cylinders are the sign of a localized failure, probably of a head gasket. There are other tests that can further isolate the problem.

Wet Compression Test

To further help analyze the source of poor compression, repeat the compression test, this time with about a teaspoon of oil squirted into each cylinder. The oil will temporarily help seal between the piston rings and the cylinder wall, practically eliminating leakage past the rings for a short time. If this test yields higher compression readings than the "dry" compression test, there is probably leakage between the piston rings and cylinder walls, due either to wear or to broken piston rings. Little or no change in compression readings indicates other leakage, probably from the valves or a failed cylinder head gasket.

Leak-down Test

The most conclusive diagnosis of low compression symptoms requires a leak-down test. Using a special tester and a supply of compressed air, each cylinder is pressurized. The rate at which the air leaks out of the cylinder, as well as where the air leaks out, can accurately pinpoint the magnitude and location of the leakage. Any engine compression diagnosis that will require major disassembly should first be confirmed by the more accurate leak-down test. Because the test requires special equipment and experience, it may be desirable to have it done by a BMW dealer or other qualified repair shop.

Vacuum Gauge Test

A vacuum gauge can be a useful tool to diagnose a variety of engine problems. Care must be taken in interpreting the readings and the movements of the gauge needle. In many instances, the readings may indicate several problems and further testing may be required to isolate the exact problem. The vacuum gauge should be connected to a vacuum source on the intake manifold and the engine should be at operating temperature.

The engine vacuum gauge measures manifold vacuum. Manifold vacuum varies with different engine operating conditions and also with different engine problems. The manner in which the vacuum reading varies from the normal reading can indicate the type of engine problem. **Table c** lists vacuum gauge readings, their probable causes, and corrective actions.

Table c. Vacuum Gauge Readings

Vacuum gauge reading	Probable cause	Corrective action
1. High and steady reading (15-22 in./Hg.)	**a.** Normal performance. Engine in good condition	**a.** No corrective action required
2. Low and steady reading(10-15 in./Hg.)	**a.** Incorrect ignition timing (retarded) **b.** Incorrect valve timing **c.** Low compression **d.** Throttle valve sticking **e.** Leaking intake manifold gasket or fuel injector seals	**a.** Check ignition timing. See **IGNITION** **b.** Inspect camshaft timing. **4.3** or **4.4** **c.** Test compression as described above **d.** Remove intake air boot and check throttle movement. See **FUEL INJECTION** **e.** Inspect intake manifold gasket. Remove injectors and replace O-rings. See **ENGINE MANAGEMENT—DRIVEABILITY** or **FUEL INJECTION**

continued on next page

Table c. Vacuum Gauge Readings (continued)

Vacuum gauge reading	Probable cause	Corrective action
3. Very low but steady reading at idle (below 10 in./Hg.) Engine idles very rough and stalls	**a.** Large intake air leak	**a.** Visually inspect for faulty gaskets, hoses, or connections. Check the oil dipstick seal, the oil filler cap and the cylinder head cover gasket carefully
4. Gauge reading drifts or floats at idle	**a.** Minor intake air leak **b.** Air/fuel mixture incorrect (rich)	**a.** Visually inspect for intake air leaks **b.** Check air/fuel mixture. See **FUEL INJECTION**
5. Gauge reading fairly steady at idle but vibrates rapidly when engine speed is increased	**a.** Worn (weak) valve springs **b.** Ignition miss **c.** Faulty cylinder head gasket	**a.** Replace valve springs See **ENGINE—RECONDITIONING** **b.** Test ignition system. See IGNITION **c.** Test compression as described above
6. Gauge reading steady at idle but drops regularly	**a.** Sticking or faulty valve or incorrect valve clearance. (Needle drops when faulty valve operates) **b.** One or more spark plugs not firing **c.** Faulty head gasket	**a.** Adjust valve clearance. **4.2** If valve clearance is correct, inspect valves. See **ENGINE—RECONDITIONING** **b.** Replace faulty spark plugs or spark plug wires. See **IGNITION** **c.** Test compression as described above
7. Gauge reading vibrates rapidly at idle but steadies when engine speed is increased	**a.** Worn valve guides	**a.** Check valve guides and repair or replace as necessary. See **ENGINE—RECONDITIONING**
8. Gauge reading gradually drops to 0 in./Hg at idle	**a.** Plugged or restricted exhaust system	**a.** Check exhaust system for restrictions, especially check for a plugged catalytic converter

4. CYLINDER HEAD

The cylinder head can be removed from the cylinder block for repairs with the engine in the car. Since the camshaft bearing bores are integral with the head and are not split with removable caps, it is necessary to remove the cylinder head for most cylinder head repairs. Cylinder head service and repair is not overly complicated, but requires time and, in some instances, special tools.

> **NOTE —**
> If it is determined that the cylinder head will require significant reconditioning, a remanufactured cylinder head may be a good alternative. Remanufactured cylinder heads are available from an authorized BMW dealer.

4.1 Cylinder Head Cover and Gasket

Because the cylinder head cover gasket is deformed during installation, it is not reusable. It should be replaced anytime the cylinder head cover is removed and anytime there is evidence of leaks. A faulty gasket can also be the source of vacuum leaks and possibly an erratic idle caused by a lean mixture.

To replace cylinder head cover gasket (528e models):

1. Remove the intake manifold support bracket, and if applicable, the bracket for the engine sensors (1982 through 1985 models) or the bracket for the idle air stabilizer (1986 and 1987 models). See Fig. 4-1.

> ***CAUTION —***
> *On 1982 through 1985 models, mark the engine sensor wires before disconnecting them. If the connectors are installed incorrectly, the engine will not start.*

2. Remove the harness connectors from the cold-start valve and the idle air stabilizer valve.

> **NOTE —**
> The 528e models built after March 1987 are not equipped with a cold-start valve.

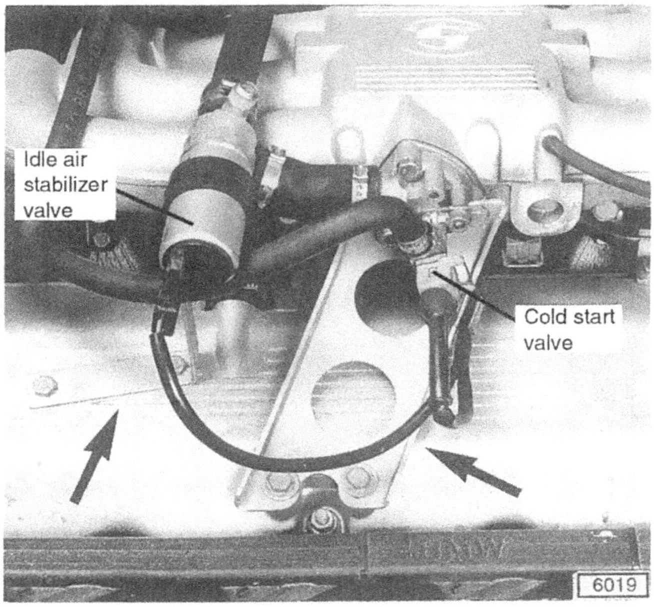

Fig. 4-1. Intake manifold and idle air stabilizer valve support brackets (arrows) on 1986 and 1987 528e models. Other models are similar.

3. Disconnect the breather hose from the cover. Remove the eight mounting nuts and the ground strap. See Fig. 4-2. Lift off the large ignition wire cover and place it out of the way.

Fig. 4-2. Breather hose and cylinder head cover mounting nuts (arrows).

4. Lift off the cylinder head cover and its gasket. If the gasket is stuck to the cylinder head, use a gasket removing tool or a dull knife to separate the gasket from the head.

> ***CAUTION —***
> *Use care when removing a stuck gasket. Damage to either surface can cause vacuum and oil leaks.*

5. Install a new gasket and the cylinder head cover.

6. Install the support bracket(s). Reconnect the harness connectors. Attach the ground strap under the rear mounting nut.

7. Tighten the mounting nuts in the sequence shown in Fig. 4-3. Reconnect the breather hose.

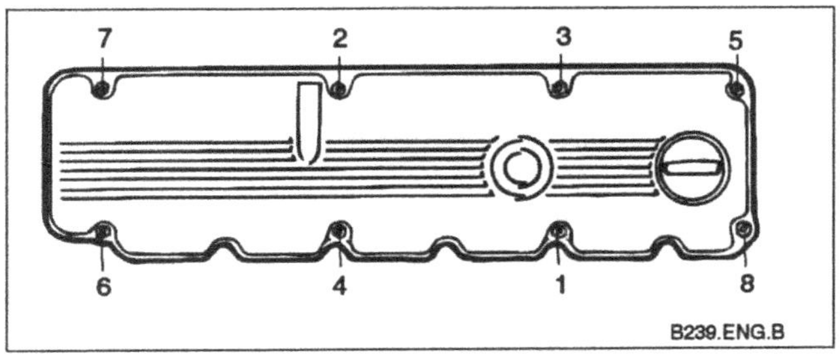

Fig. 4-3. Cylinder head cover mounting nut tightening sequence.

Tightening torque
• cylinder head cover mounting nuts. 9±1 Nm (80±9 in-lb)

To replace cylinder head cover gasket (533i and 535i models):

1. Disconnect the harness connector for the air flow sensor and unclip the harness, placing it out of the way.

2. Remove the hoses and fittings from the sides of the large of the intake air boot. Then separate the boot from the throttle valve by loosening the large clamp. See Fig. 4-4.

3. Remove the mounting nuts for the air cleaner housing, and remove the housing together with the air boot and the air flow sensor. See Fig. 4-5.

Fig. 4-4. Intake air boot being separated from throttle valve on 535i model.

Fig. 4-5. Air filter housing being removed.

4. Disconnect the breather hose from the cover. Remove the nuts and bolts shown in Fig. 4-6. Place the ignition wiring cover out of the way.

5. Lift off the cylinder head cover and its gasket. If the gasket is stuck to the cylinder head, use a gasket removing tool or a dull knife to separate the gasket from the head.

Fig. 4-6. Fasteners needed to be removed for cylinder head cover removal (arrows).

> ***CAUTION —***
> *Use care when removing a stuck gasket. Damage to either surface can cause vacuum and oil leaks.*

6. Install a new gasket and the cylinder head cover.

7. Installation is the reverse of removal. Be sure to reconnect all hoses and attach the ground strap under the rear mounting nut. Tighten the mounting nuts and bolts using the sequence shown above in Fig. 4-3.

Tightening torque
• cylinder head cover mounting nuts and bolts . . . . 9±1 Nm (80±9 in-lb)

4.2 Valve Adjustment

Valve adjustment should be done as part of scheduled maintenance. For a listing of recommended adjustment intervals, see **LUBRICATION AND MAINTENANCE**.

A special tool (BMW tool no. 11 3 070) is available to help turn the rocker arm eccentric, which is used to set valve clearance. As an alternative, a tool can be easily fabricated from stiff wire (approximate diameter of 3/32 in.). The valves can be adjusted with the engine hot or cold.

To adjust valve clearance:

1. Remove the cylinder head cover as described in **4.1 Cylinder Head Cover and Gasket**.

2. Using a socket wrench on the vibration damper nut, hand-turn the crankshaft clockwise until the No. 1 cylinder's camshaft lobes are pointing approximately downward (valves fully closed).

NOTE —
• On 533i and 535i models, it may be easier to reach the crankshaft vibration damper nut after removing the radiator cooling fan and plastic radiator cowling as described in **COOLING SYSTEM.**

• The no. 1 cylinder is the one closest to the radiator end of the engine.

• Crankshaft rotation can be done more easily if the spark plugs are first removed.

3. Measure the valve clearances of the no. 1 cylinder intake and exhaust valves using a feeler gauge. See Fig. 4-7. Compare the measured clearance values with the specifications listed in **Table d.**

Fig. 4-7. Valve clearance being measured. Feeler gauge is inserted between rocker arm eccentric and valve.

Table d. Valve Clearance Specifications

Model	Engine warm (coolant temp. above 176°F (80°C))	Engine cold (coolant temp. below 95°F (35°C))
528e	0.30 mm (0.012 in.)	0.25 mm (0.010 in.)
533i, 535i	0.35 mm (0.014 in.)	0.30 mm (0.012 in.)

NOTE —
Valve clearance specifications for both the intake valves and the exhaust valves are the same.

4. If the valve clearance is incorrect, loosen the setscrew nut (locknut) with a 10 mm box-end wrench. Rotate the eccentric with a stiff wire hook until the specified clearance is obtained and tighten the nut. See Fig. 4-8.

4

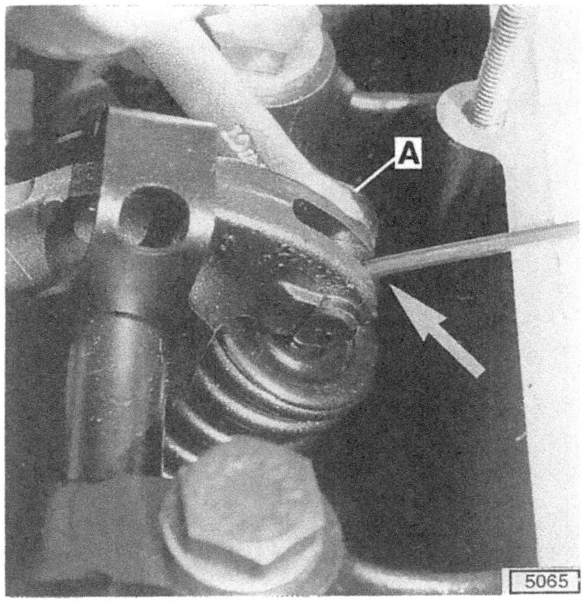

Fig. 4-8. Valve clearance being adjusted. Loosen setscrew nut (**A**), then insert stiff wire hook into hole in eccentric (arrow).

Tightening torque
• rocker arm eccentric locknut . . . . 10±1 Nm (89±9 in-lb)

5. Repeat the adjusting procedure for each pair of valves. Follow the sequence listed below. Rotate the crankshaft until the camshaft lobes for the next cylinder are pointing approximately downward (valves fully closed).

NOTE —
Rotate crankshaft approximately 1/3 of a turn (120°) between adjustments.

Valve adjustment sequence
• adjustment sequence (cylinder no.) . . . . 1—5—3—6—2—4

6. Recheck all clearances before installing the cylinder head cover as described in **4.1 Cylinder Head Cover and Gasket**. This will require two complete rotations of the crankshaft pulley.

4.3 Camshaft Drive Belt (528e models)

The 528e eta engine uses an overhead camshaft that is driven by a toothed rubber belt. See Fig. 4-9. Due to belt composition and high under-hood temperatures, the belt is subject to wear. BMW recommends that the belt be replaced at least every 60,000 miles, every four years, or anytime belt tension is relieved.

> ***CAUTION —***
> *Reusing a camshaft drive belt could result in overstretching of the belt, which can cause decreased belt life and possible engine damage.*

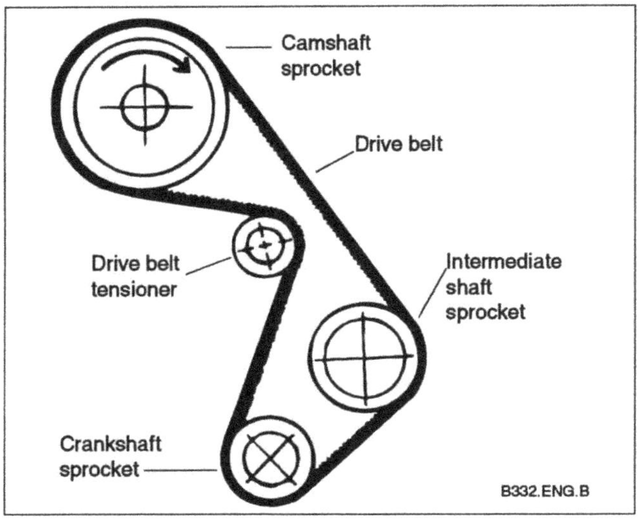

Fig. 4-9. Camshaft drive belt configuration with tensioner and sprockets. Arrow shows engine rotation direction.

Various versions of drive belt tensioner pulleys, belts, and sprockets were installed on the engines used in the 528e model. Some of the earlier versions of these parts are no longer available. During drive belt replacement, only parts containing the Z 127 markings should be installed and all parts should be carefully inspected. See Fig. 4-10. If an earlier version belt is being replaced, the tensioner pulley must also be replaced. Consult your authorized BMW dealer parts department for the latest parts information.

To remove camshaft drive belt (528e models):

1. With the engine cold, drain the coolant and remove the radiator cooling fan and the radiator. See **COOLING SYSTEM**.

> **NOTE —**
> Although it is possible to replace the camshaft drive belt without removing the radiator, the removal of the radiator greatly simplifies the task.

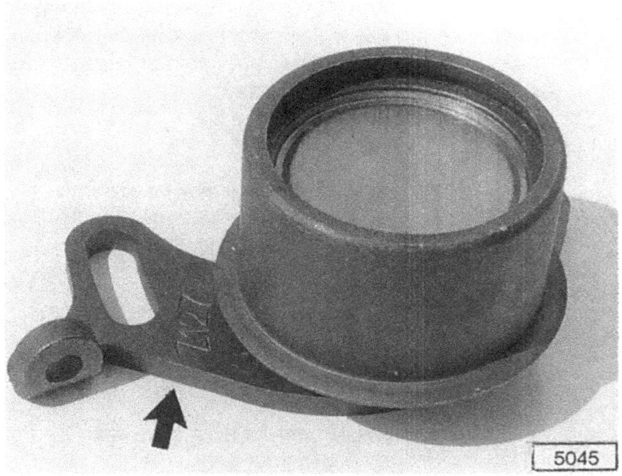

Fig. 4-10. Camshaft drive belt tensioner with stamp "Z 127" (arrow).

2. Remove the distributor cap, the ignition rotor, and the black protective cover from behind the rotor as described in **IGNITION**. Place the cap and wires out of the way.

> **NOTE —**
> • Label the spark plug wires so that they can be reinstalled in their original locations.
>
> • On models with screw-mounted ignition rotors, it is necessary to first remove the ignition rotor adaptor from the front of the camshaft before the black cover can be removed. See **IGNITION** for removal information. Reinstall the adaptor after removing the cover.

3. Remove the alternator V-belt, the air conditioning compressor V-belt, and the power steering pump V-belt as described in **LUBRICATION AND MAINTENANCE**.

4. Remove the two mounting bolts from the upper camshaft drive belt cover. See Fig. 4-11. Remove the cover and the engine hoisting bracket, if applicable.

5. Remove the upper alternator bracket by removing the mounting nut and loosening the alternator tensioning bolt. See Fig. 4-12.

6. Using a socket wrench (27 mm) on the center vibration damper (crankshaft pulley) bolt, rotate the crankshaft clockwise until the engine is at Top Dead Center (TDC or O/T) of the No. 1 cylinder.

Fig. 4-11. Upper camshaft drive belt cover mounting bolts (arrows).

Fig. 4-12. Upper alernator bracket and mounting nut (arrows).

NOTE —
At TDC on the compression stroke of the No. 1 cylinder, the "0/T" mark on the vibration damper will align with the notch on the lower camshaft drive belt cover. The arrow and notch on the camshaft drive belt sprocket will align with the mark stamped on the cylinder head. See Fig. 4-13.

Fig. 4-13. Left: Top Dead Center (TDC) mark on vibration damper aligned with mark on lower drive belt cover. **Right**: camshaft sprocket aligned with mark on cylinder head.

7. Remove the Top Dead Center (TDC) sensor from its clip as shown in Fig. 4-14.

Fig. 4-14. Top Dead Center (TDC) sensor.

8. On models built March 1987 and later, remove the reference sensor mounting bolt. Remove the sensor with the wiring from the mounting clips on the camshaft drive belt cover. See Fig. 4-15.

9. Remove the lower camshaft drive belt cover mounting bolt together with the cover.

Fig. 4-15. Reference sensor mounting bolt (arrow) on 528e models built March 1987 and later.

10. Remove the six bolts that hold the lower drive belt pulley and the vibration damper to the front of the crankshaft. Remove the pulley and the damper.

NOTE —
If the pulley and damper do not come loose easily, it may be necessary to pry gently with a screwdriver to free the parts from the crankshaft hub.

11. On early models with a two-piece crankshaft hub, hold the hub stationary and loosen the hub's center mounting bolt so it is threaded in three turns. Then using a suitable puller remove the hub from the crankshaft. See Fig. 4-16. Note the orientation of the woodruff key in the keyway. Replace the woodruff key if it is damaged.

NOTE —
- Models with a two-piece crankshaft hub are identified by the need to remove hub to remove the belt. On models with a one-piece hub, there is enough clearance to slide the belt over the hub.

- The two-piece crankshaft sprocket and hub mounting bolt is tightened to a torque of approximately 410 Nm (302 ft-lb). A heavy duty holding device, such as BMW tool no. 11 2 150, should be used to hold the sprocket. Use a heavy duty 3/4-inch drive socket and breaker bar to break the bolt free.

12. Loosen the upper bolt that secures the belt tensioner pulley to the front of the engine. See Fig. 4-17.

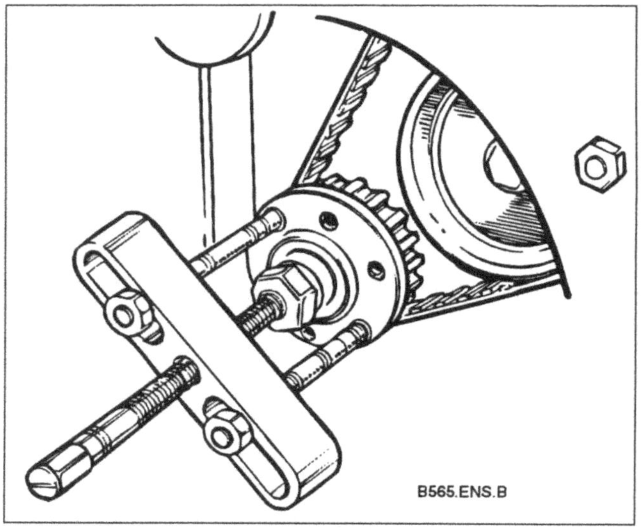

Fig. 4-16. Crankshaft hub being removed from front of crankshaft on models with two-piece hub. Center bolt is temporarily threaded in three turns.

Fig. 4-17. Camshaft drive belt tensioner upper mounting bolt (arrow).

13. Loosen the tensioner lower mounting nut at the base of the long stud.

14. Using a long screwdriver, push the tensioner in as far as it will go to relieve the belt tension and tighten the upper bolt. Remove the belt by peeling it off the sprockets.

CAUTION —
Tools should not be used to remove the belt. Once the belt is removed, do not disturb the camshaft or crankshaft position. Rotation of either shaft could result in engine damage.

15. Inspect the camshaft belt sprocket and the tensioner pulley. If the pulley catches, is noisy, or has radial (side-to-side) or axial (in-and-out) play, or does not spin freely, it must be replaced.

NOTE —
• On 528e models built between July 1986 and September 1986 the camshaft drive belt sprocket may break in service due to faulty design. Replacement is covered under warranty. Sprockets that have been replaced should be noted by a blue paint dot on the thermostat housing and a round label affixed to the driver's door pillar with the number "14" punched out. If in doubt, have the sprocket inspected by an authorized BMW dealer. All sprockets with stamped markings "W+P 2 86" must be replaced. See **ENGINE—RECONDITIONING** for camshaft sprocket removal procedures.

• The build date is on the driver's door, below the door latch.

• 1986 and later cars have belt tensioner pulleys that are marked "Z 127." 1985 and earlier cars may be equipped with an earlier version pulley. If there is no Z 127 stamp on the base of the belt tensioner pulley, it must be replaced with one so marked. See Fig. 4-10 shown above. The only replacement drive belt available is the Z 127 belt, which will only work with the newer Z 127 tensioner pulley.

To remove the tensioner, first remove the upper mounting bolt and swing the tensioner down until the spring tension is relieved. Then remove the lower mounting nut and the tensioner. When installing the new tensioner, make sure the spring assembly correctly engages the hole in the tensioner and the detent in the coolant pump. See Fig. 4-18.

Fig. 4-18. Camshaft drive belt tensioner and coolant pump assembly. Tensioner spring must engage hole in tensioner and detent in pump (arrows).

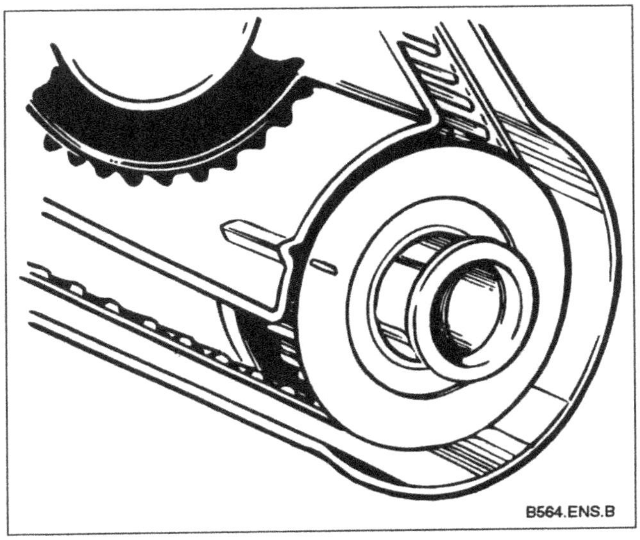

Fig. 4-19. Crankshaft hub mark aligned with notch on front engine cover.

To install camshaft drive belt (528e models):

1. Check that the crankshaft and camshaft timing marks are correctly aligned. See Fig. 4-13 above for camshaft alignment marks. To check the crankshaft position, check that the notch in the inner cover aligns with the mark on the sprocket. See Fig. 4-19.

2. Install the belt beginning at the crankshaft sprocket. Install the camshaft drive belt in a counterclockwise direction without slack, slipping the belt over the intermediate shaft and the camshaft sprockets, and the camshaft belt tensioner pulley.

CAUTION —
Do not use any tools to force the camshaft drive belt onto the pulleys.

3. With the belt correctly positioned on the sprockets, slowly loosen the camshaft belt tensioner pulley upper retaining bolt so that the belt is tensioned.

4. Using a socket wrench (27 mm) on the center crankshaft bolt, slowly rotate the engine clockwise through two complete revolutions (720°) until the timing marks are aligned again.

CAUTION —
If either the crankshaft or the camshaft are not perfectly aligned after turning the engine over, the drive belt tension will have to be released and the belt repositioned by repeating the above steps. It is permissible to rotate the camshaft drive gear by hand by one or two teeth in either direction to ensure proper alignment.

NOTE —
Rotating the engine automatically tensions the belt to the correct tightness.

5. Once the alignment of crankshaft and camshaft timing marks are correct, first torque the upper and then the lower tensioner mounting bolts. Tightening torques are given below.

The remainder of the installation is the reverse of removal. On models with a two-piece crankshaft hub, hold the crankshaft still using a holding device while the mounting bolt is torqued. Sprocket lettering should face toward the radiator.

CAUTION —
Be sure the woodruff key and crankshaft hub are correctly installed in the crankshaft keyway. If they are incorrectly installed the crankshaft will be destroyed.

Tightening torques

- camshaft drive belt tensioning bolt and nut . . . . 22±2 Nm (16±1 ft- lb)
- camshaft drive belt covers (upper and lower)
  M6 nuts or bolts . . . . 9 to 11 Nm (80 to 97 in-lb)
  M8 nuts or bolts . . . . 22±2 Nm (16±1 ft-lb)
- vibration damper to crankshaft hub
  M8 bolts . . . . 22±2 Nm (16±1 ft-lb)
- crankshaft hub to crankshaft center bolt (two-piece hub) . . . . 410±20 Nm (302±15 ft-lb)
- reference sensor (built March 1987 and later) . . . . 7±1 Nm (62±9 in-lb)
- ignition rotor adaptor to camshaft (with thread locking compound) . . . . . 60±5 Nm (44±4 ft-lb)
- camshaft sprocket to camshaft . . . . 65 to 70 Nm (48 to 52 ft-lb)

Install the V-belts removed earlier. Adjust V-belt tensions as described in **LUBRICATION AND MAINTENANCE**. Place a label in the engine compartment noting camshaft drive belt replacement date and odometer reading. Refill the cooling system as described in **COOLING SYSTEM**.

4.4 Camshaft Timing Chain (533i and 535i models)

The 533i and 535i engines use an overhead camshaft that is driven by a camshaft timing chain. The chain is driven by a sprocket on the front of the crankshaft. The timing chain is automatically tensioned by a chain tensioner and is lubricated by engine oil. The timing chain does not require maintenance. A worn timing chain and sprockets can be noisy and can cause erratic valve timing, leading to loss of performance. A noisy timing chain can also be caused by a faulty timing chain tensioner piston.

To remove camshaft timing chain (533i and 535i models):

1. Disconnect the negative (–) battery cable.

CAUTION —
BMW Anti-theft radios become inoperable whenever the battery is disconnected. See your owner's manual for more information.

2. Remove the cylinder head cover as described in **4.1 Cylinder Head Cover And Gasket**.

3. Partially drain the coolant and remove the thermostat housing. Remove the primary cooling fan from the front of the coolant pump. See **COOLING SYSTEM.**

4. Remove the distributor cap, rotor and black protective cover from behind the rotor as described in **IGNITION.**

NOTE —
On models with screw-mounted ignition rotors, it is necessary to first remove the ignition rotor adaptor from the front of the camshaft before the black cover can be removed. See **IGNITION** for removal information. Reinstall the adaptor after removing the cover.

5. Remove the timing chain tensioner. Be ready to catch the piston spring as the nut is removed. See Fig. 4-20. Pull the piston from its bore.

NOTE —
Check the removed piston for free movement of the check ball by shaking the piston. The ball should rattle freely. Blow air into the closed end of the piston. No air should pass. Blow air into the slotted guide end of the piston. Air should pass freely. If any faults are found, try disassembling and cleaning the piston. If the piston cannot be made to operate correctly, it should be replaced.

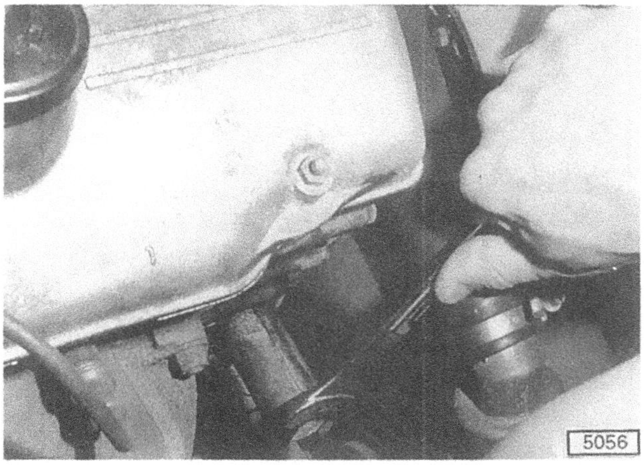

Fig. 4-20. Camshaft timing chain tensioner being removed. Remove tensioner slowly to relieve spring tension.

6. Remove the upper timing case cover mounting bolts. See Fig. 4-21. Carefully separate the cover from the cylinder head.

CAUTION —
Use care when removing the cover from the cylinder head gasket. If the cover is stuck, use a sharp knife to separate it from the gasket. If the cylinder head gasket becomes damaged, it will have to be replaced.

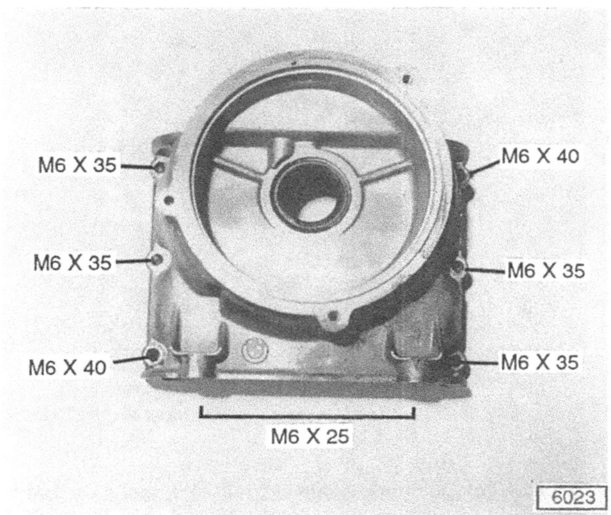

Fig. 4-21. Upper timing case mounting bolt holes.

7. Remove all V-belts as described in **LUBRICATION AND MAINTENANCE.**

8. Remove the coolant pump pulley from the pump. See **COOLING SYSTEM.**

9. Remove the alternator and its bracket as described in **BATTERY, STARTER, ALTERNATOR.**

10. Remove the power steering pump and its brackets as described in **STEERING AND WHEEL ALIGNMENT.** Using a stiff wire, support the power steering pump.

CAUTION —
Do not allow the power steering pump to hang from its hoses, as damage to the hoses, the fittings, or the pump may occur.

11. Remove the air conditioning compressor from its brackets by removing the mounting bolts. Using a stiff wire, support the compressor.

WARNING —
Do not remove the hoses from the air conditioner compressor. Doing so can lead to the expulsion of refrigerant under high pressure, leading to possible eye damage or frostbite.

12. Remove the bolts that hold the air conditioning compressor bracket to the cylinder block and remove the bracket.

13. Secure the vibration damper or the flywheel to prevent the crankshaft from turning. Using a 36 mm socket, loosen but do not remove the nut that holds the vibration damper to the crankshaft.

NOTE —
The crankshaft vibration damper mounting nut is tightened to a torque of approximately 440 Nm (325 ft-lb). A heavy duty flywheel or vibration damper holding device, such as BMW tool no. 11 2 200 (1983 to 1985 models) or BMW tool no. 11 2 220 (1986 to 1988 models), should be used to hold the vibration damper stationary while the nut is loosened. Use a heavy duty 3/4-inch drive socket and breaker bar to break the nut free.

14. Remove the holding device and rotate the crankshaft clockwise until the engine is at Top Dead Center (TDC or 0/T) on the compression stroke of the No.1 cylinder.

NOTE —
At TDC, the **0/T** mark on the vibration damper will be aligned with the cast boss on the lower camshaft timing chain cover. See Fig. 4-22.

15. Without disturbing the position of the crankshaft, remove the vibration damper mounting nut and the damper assembly from the front of the crankshaft. Remove the woodruff key from the crankshaft. Note the orientation of the woodruff key in the keyway. Replace the woodruff key if it is damaged.

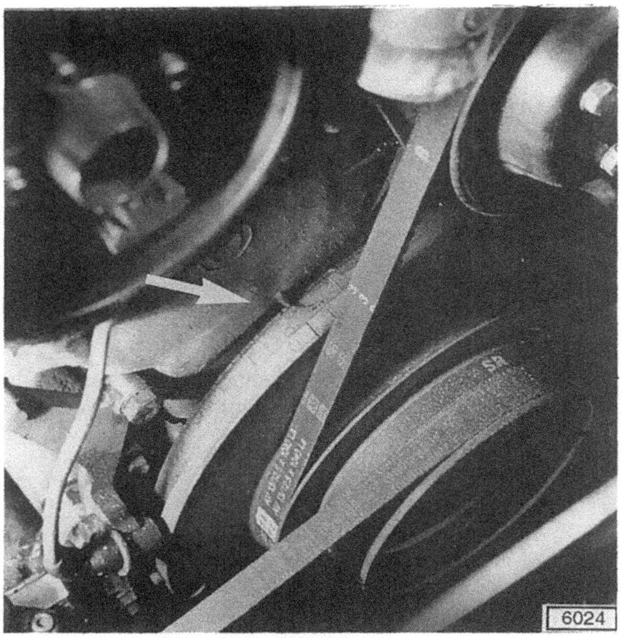

Fig. 4-22. 0/T (top dead center) mark on vibration damper aligned with mark on timing chain cover (arrow).

NOTE —
If the vibration damper is difficult to remove, use a puller to remove it from the crankshaft. Do not lever the damper off, as this may distort the harmonic vibration damper.

16. Remove the front mounting bolts from the lower timing chain cover.

17. Working from beneath the engine, remove the oil pan mounting bolts that are threaded into the bottom of the timing chain cover. See Fig. 4-23. Loosen the remainder of the oil pan mounting bolts. Using a sharp knife, separate the cover from the oil pan gasket.

18. Remove the bolts from the camshaft sprocket and pry the sprocket from the camshaft. See Fig. 4-24. Carefully remove the chain from the sprockets and tensioning rails.

CAUTION —
Do not rotate the crankshaft or the camshaft while the timing chain is removed, as bent valves may result.

NOTE —
Some models may be equipped with locking tabs on the camshaft sprocket mounting bolts. The locking tabs are no longer available from BMW and do not have to be reinstalled.

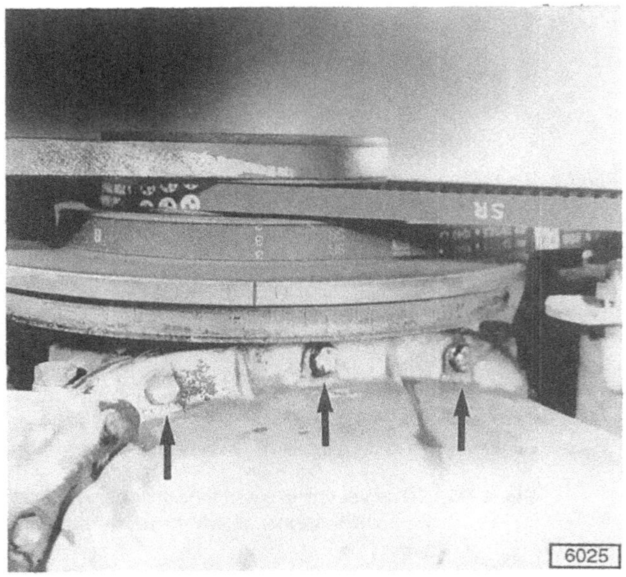

Fig. 4-23. Oil pan mounting bolts threaded into timing chain cover (arrows).

Fig. 4-24. Camshaft drive sprocket mounting bolts (arrows).

19. Inspect the timing chain sprockets. Sprockets that have worn or missing teeth should be replaced. The chain should also be replaced if the sprockets are excessively worn.

NOTE —
If the crankshaft sprocket requires replacement, the oil pan and the oil pump drive sprocket must be removed as described under **6. Lubrication System**. The crankshaft sprocket can be removed from the crankshaft using a puller after removing the woodruff key. Prior to installing the replacement sprocket, it must first be heated in an oil bath to a maximum temperature of 390°F (200°C). At this temperature, the sprocket can be driven onto the crankshaft while aligning the woodruff keyway. Due to the high temperatures required, we recommend that this be done by an authorized BMW dealer or other qualified shop.

20. Inspect the guide rail and tensioner rail for deep grooves caused by chain contact. The rails can be replaced after removing the circlips from the mounting pins. See Fig. 4-25. Always use new circlips.

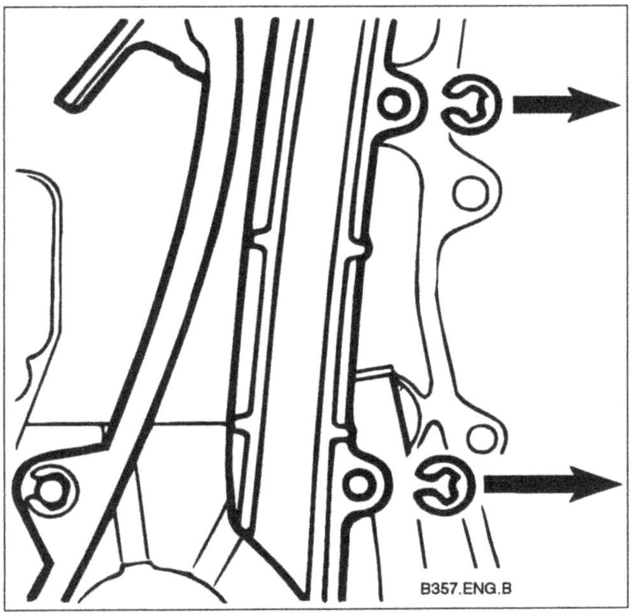

Fig. 4-25. Camshaft timing chain guide rail mounting circlips. Remove circlips in direction of arrows.

21. Inspect the crankshaft oil seal in the front of the lower timing chain cover. If the oil seal is hard, cracked, or otherwise damaged, it should be replaced as described in **ENGINE—RECONDITIONING**.

To install camshaft timing chain (533i and 535i models):

1. Temporarily install the lower timing chain cover and the vibration damper onto the crankshaft. Check that both the camshaft and the crankshaft are set at Top Dead Center (TDC). See Fig. 4-26.

Fig. 4-26. Camshaft and crankshaft TDC position. Camshaft sprocket mounting bolts should be straight up and down with the locating pin in the lower left-hand corner (**A**). 0/T mark on vibration damper should align with mark on front case (**B**).

2. Make sure the chain is properly engaged on the crankshaft sprocket. Install the camshaft sprocket to the chain so that it engages the camshaft flange correctly.

3. Install the sprocket onto the camshaft and tighten the mounting bolts. Recheck that the timing marks are still aligned.

Tightening torque
• camshaft sprocket mounting bolts . . . . 7 Nm (62 in-lb)

4. Remove the vibration damper and the timing chain cover.

5. Apply a light coat of gasket sealer to the lower camshaft timing chain cover gaskets. Install the gaskets on the cylinder block. Apply a light coating of sealant to the exposed portion of the oil pan gasket. Apply an extra amount of sealer to each corner of the oil pan/cylinder block surface area.

6. Apply a light coat of engine oil to the lip of the crankshaft oil seal in the lower timing chain cover. Install the cover and loosely install its mounting bolts. Be sure to install the reference sensor bracket beneath the upper mounting bolts.

NOTE —
Various sizes of bolts are used in the lower timing chain cover. Use Fig. 4-27 as a guide when installing the bolts.

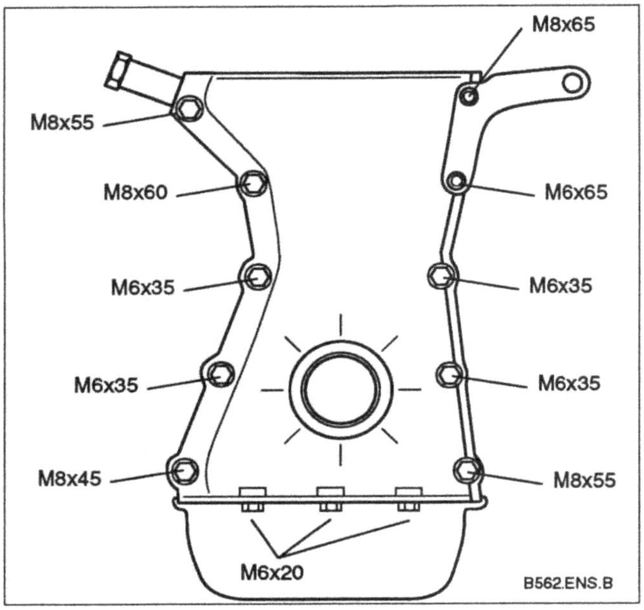

Fig. 4-27. Lower timing chain cover mounting bolt locations.

7. Loosely install the power steering bracket and the air conditioning bracket with their mounting bolts.

8. When all brackets and bolts are installed, tighten all the front cover and oil pan bolts in an alternating pattern.

Tightening torque	
• front timing chain cover	
M6 bolts	9 to 11 Nm (80 to 97 in-lb)
M8 bolts	22±2 Nm (16±1 ft-lb)

9. Install the camshaft chain tensioner piston, the spring, a new seal, and the chain tensioner securing nut. The tapered end of the spring should face the nut.

10. Bleed the camshaft chain tensioner piston by moving the tensioner rail back and forth slowly until oil runs out past the securing nut and resistance can be felt, then tighten the nut.

NOTE —
If the tensioner piston cannot be bled, the oil reservoir in the lower corner of the oil pan may be empty. Pour a small amount of oil down the side of the camshaft timing chain cover and repeat the process.

Tightening torque	
• camshaft timing chain tensioner nut	35±5 Nm (25±4 ft-lb)

11. Install the vibration damper to the crankshaft, while aligning the keyway with the key. While holding either the damper or the flywheel stationary, tighten the nut. Recheck that the TDC marks are still aligned. See Fig. 4-26 above.

Tightening torque	
• vibration damper to crankshaft	440±10 Nm (325±7 ft-lb)

12. Apply a light coat of sealer to the two upper camshaft timing chain cover gaskets. Set the gaskets into position on the cover. Apply a brush-on gasket sealer using enough around the small bores to fill them. See Fig. 4-28.

Fig. 4-28. Apply sealer (arrows) to fill bores.

13. Install the upper timing chain cover and hand-tighten the mounting bolts. First tighten the six bolts on the front of the cover then the two lower bolts

NOTE —
Upper timing chain cover mounting bolts vary in size. See Fig. 4-21 above for correct bolt orientation.

Tightening torques	
• front timing chain cover	
M6 bolts	9 to 11 Nm (80 to 97 in-lb)

14. Install the cylinder head cover as described in **4.1 Cylinder Head Cover and Gasket**.

15. Install the coolant pump pulley and fan as described in **COOLING SYSTEM**.

16. Install the air conditioning compressor and its bracket.

17. Install the power steering pump and its bracket as described in **STEERING AND WHEEL ALIGNMENT**.

18. Install the alternator and its bracket as described in **BATTERY, STARTER, ALTERNATOR**.

19. Install all belts as described in **LUBRICATION AND MAINTENANCE**.

20. Install the radiator, fan shroud, thermostat housing, and refill the cooling system as described in **COOLING SYSTEM**.

21. Install the ignition components as described in **IGNITION**. Be sure to use a thread-locking compound on the ignition rotor adaptor mounting bolts.

22. Reconnect the battery. Check all fluid levels.

4.5 Removing and Installing Cylinder Head

The cylinder head can be removed with the engine installed. Special tools are necessary to tighten the cylinder head bolts to a specified torque angle during installation.

> NOTE —
> On 528e models only, BMW recommends replacing all hex-head type cylinder head bolts with the new Torx® head ASA 14 bolts when ever the cylinder head is removed or any bolts are found to be damaged.

If a failed cylinder head gasket or warped head is suspected, a compression test, as described in **3. Troubleshooting** should be done before removing the cylinder head. A failed head gasket may be caused by a warped cylinder head. Overheating due to insufficient cooling is a major cause of cylinder head warpage. When replacing the cylinder head gasket, always check the cylinder head for straightness. The procedure for checking for warpage can be found in **ENGINE—RECONDITIONING**.

> ***CAUTION —***
> *To prevent the cylinder head from warping, the engine should be fully cool (not run for at least six hours, and preferably overnight) before removing the cylinder head.*

To remove cylinder head (528e models):

1. Disconnect the negative (–) battery cable.

> ***CAUTION —***
> *BMW anti-theft radios become inoperable whenever the battery is disconnected. See your owner's manual for more information.*

2. Drain the engine coolant and remove the coolant hoses from the front and rear of the cylinder head. Remove any hose retaining brackets and move the hoses out of the way. Remove the block drain from the engine block. See **COOLING SYSTEM**.

3. Unbolt and separate the front exhaust pipe from the manifold. Discard the mounting nuts.

> ***CAUTION —***
> *Do not reuse self-locking nuts. These nuts are designed to be used only once and should be replaced whenever they are removed.*

4. Disconnect the hoses, connectors, and fasteners shown in Fig. 4-29 below. Move all hoses and harnesses out of the way, cutting wire ties and removing brackets as necessary.

> NOTE —
> • Label all components, wires, and hoses before removing them as an aid to installation.
>
> • Some hoses may be secured with one-time hose clamps that are crimped with special pliers. The clamps may be pried loose with a small screwdriver. Replace the clamps with standard screw-type clamps.

5. Remove the two mounting nuts from the side of the air cleaner housing and remove the housing together with the air flow sensor.

6. Remove the harness connectors and components shown in Fig. 4-30. Move all harnesses and connectors out of the way.

7. Remove the distributor cap, spark plugs, and the distributor rotor as described in **IGNITION**.

> NOTE —
> Label the spark plug wires so that they can be reinstalled in their original locations.

8. Remove the cylinder head cover as described earlier under **4.1 Cylinder Head Cover and Gasket.**

9. Disconnect the wires from the alternator and from the starter motor as described in **BATTERY, STARTER, ALTERNATOR**.

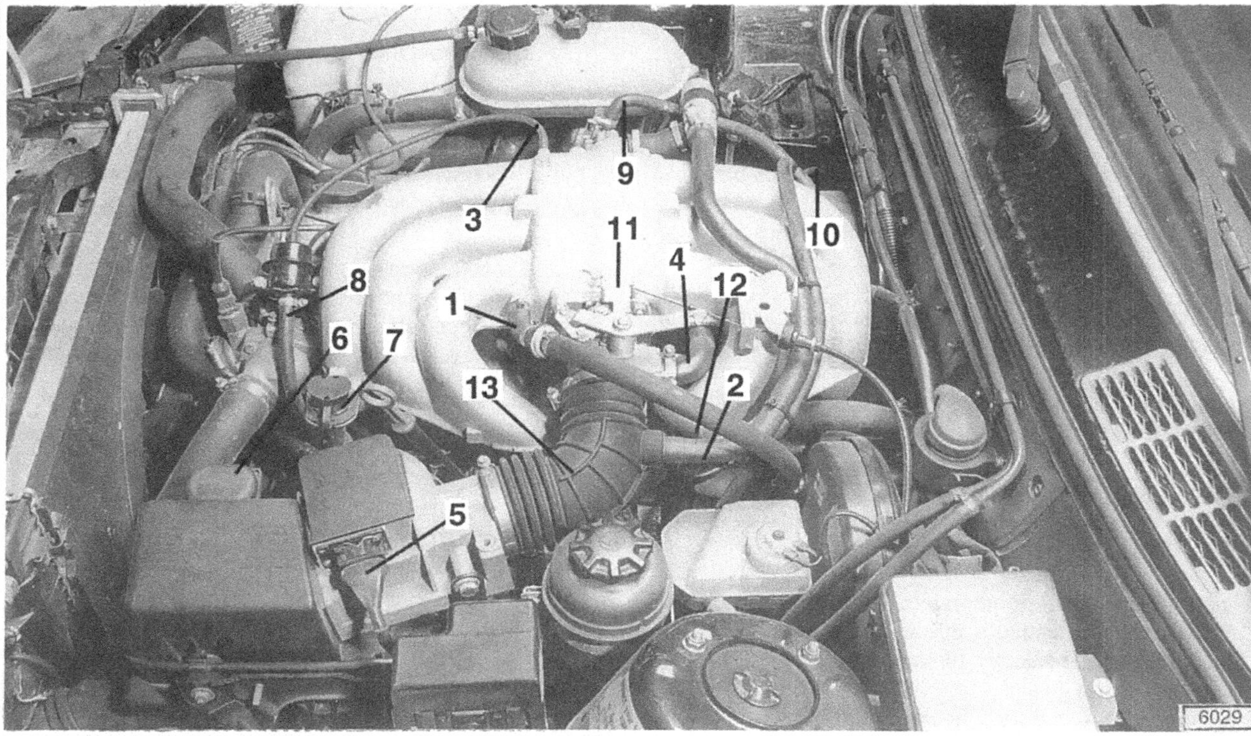

1. Vacuum hose
2. Vacuum hose
3. Vacuum hose
4. Vacuum hose
5. Air flow sensor connector
6. Altitude sensor connector
7. Diagnostic connector mounting bracket
8. Fuel return hose
9. Fuel supply hose
10. Fuel supply hose
11. Throttle, accelerator, and cruise control cables, if applicable
12. Coolant hoses
13. Intake air duct

Fig. 4-29. Components, hoses, and connectors needed to be removed before removing cylinder head on 528e model.

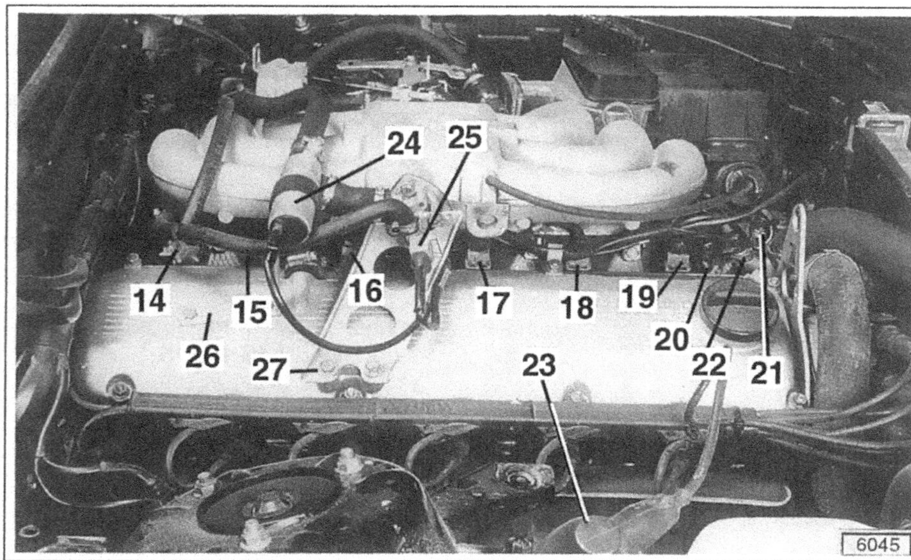

14. Fuel injector connector
15. Fuel injector connector
16. Fuel injector connector
17. Fuel injector connector
18. Fuel injector connector
19. Fuel injector connector
20. Coolant temperature sensor
21. Coolant temperature sender
22. Thermo-time switch connector
23. Ignition coil wiring
24. Idle stabilizer valve connector
25. Cold start valve connector, if applicable
26. Mounting bracket
27. Mounting bracket

Fig. 4-30. Harness connectors, wiring, and components needed to be removed before removing cylinder head on 528e model.

10. Remove the camshaft drive belt from the camshaft sprocket as described under **4.3 Camshaft Drive Belt (528e models)**.

11. Check carefully for any remaining connectors, wiring, hoses, brackets, clamps or components that will interfere with the removal of the cylinder head.

12. Gradually and evenly loosen and remove the cylinder head bolts using the sequence shown in Fig. 4-31 and remove the cylinder head. If the head is stuck, use a soft-faced mallet or pry gently with a wooden stick.

> **NOTE —**
> A crankcase ventilation tube is connected between the intake manifold and the crankcase breather assembly. The vent tube is spring loaded. A special tool is available from BMW (BMW tool no. 11 1 290) to compress the spring and hold the tube to the cylinder block while the cylinder head is removed.

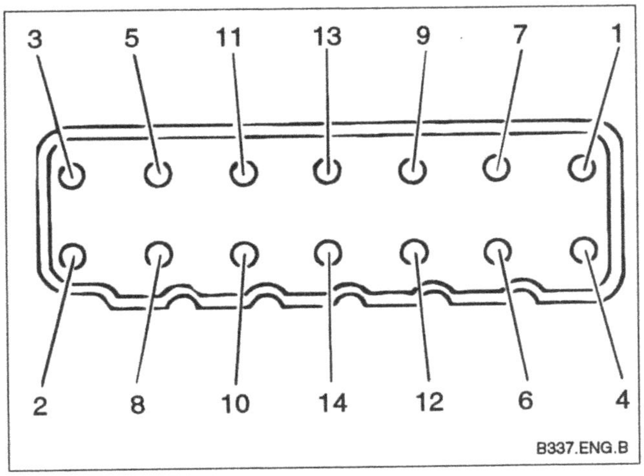

Fig. 4-31. Cylinder head bolt loosening sequence.

To install cylinder head (528e models):

1. Clean the cylinder head and the gasket surface of the cylinder block. Clean the threads of the head bolts and bolt holes with a thread chaser and remove all foreign matter and any liquid from the bolt holes. Avoid letting debris into the cylinders or oil passages in the cylinder block.

> ***CAUTION —***
> *Do not use a metal scraper or wire brush to clean the aluminum cylinder head or pistons. These tools may damage the cylinder head and pistons. Instead, use a solvent to soften carbon deposits and old sealing materials. If necessary, use a hard wooden or plastic scraper.*

2. Check the gasket surface of the cylinder head and the cylinder block for warpage as described **ENGINE—RECONDITIONING**.

3. Place a new cylinder head gasket on the surface of the cylinder block. The cylinder head gasket will fit correctly in only one orientation. The word "OBEN," found printed on the gasket, should face up.

> ***CAUTION —***
> *Cylinder head gaskets will make a reliable seal only once. Always use a new cylinder head gasket that has not been previously compressed by tightening the cylinder head bolts.*

> **NOTE —**
> The cylinder head gasket should have the 2.7 engine code stamped on the gasket. See Fig. 4-32. Failure to use the most up-to-date gasket could result in engine overheating and cylinder head damage. See an authorized BMW dealer for the latest parts information. In addition, if the cylinder head has been machined to correct for warpage, a special 0.3 mm thicker gasket should be installed. See **ENGINE—RECONDITIONING** for more information.

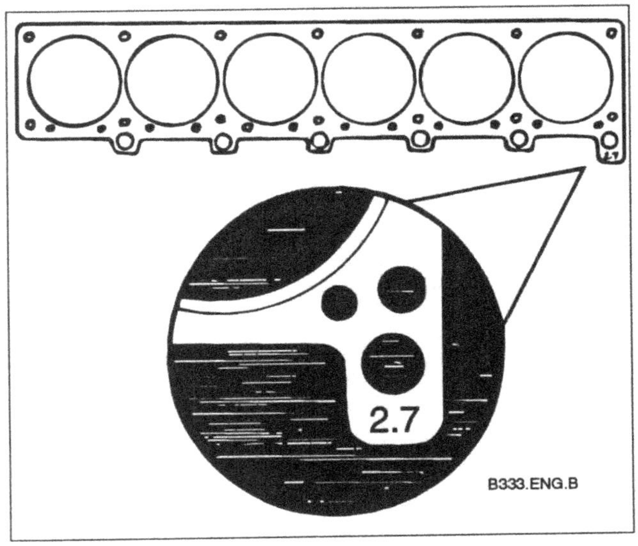

Fig. 4-32. Cylinder head replacement gasket code for 528e model (2.7 eta engine).

4. Place the cylinder head in position on the cylinder block. Check that the vent tube is correctly positioned. Lightly lubricate the cylinder head bolts. Loosely install the head bolts and their washers, then thread them in until they are finger tight. Guide the vent tube into its opening as the cylinder head bolts are tightened.

CAUTION —
BMW recommends replacing all hex-head cylinder head bolts with the new style Torx®head ASA 14 bolts whenever the cylinder head is removed or if any bolts are found to be faulty (such as a broken off bolt head). When replacing old style bolts with the cylinder head installed, remove and install one bolt at a time until all 14 are replaced.

NOTE —
To help guide the cylinder head onto the cylinder block, insert two 8 in. long by 3/8 in. round wooden dowels into two of the outermost head bolt holes. Thread in several bolts, then remove the dowels and install the remaining bolts.

5. Tighten the cylinder head bolts in the sequence shown in Fig. 4-33. The bolts should be tightened in three stages as listed below. The final stage(s) requires using a special tool (BMW tool no. 11 2 110) or a suitable protractor to tighten the bolts to a specified torque angle.

CAUTION —
The cylinder head bolt torque is critical to proper engine operation. The bolts must be tightened in the stages specified below.

NOTE —
When installing Torx®head bolts, the torque angles can be done with the engine cold. There is no specified waiting time or engine temperature.

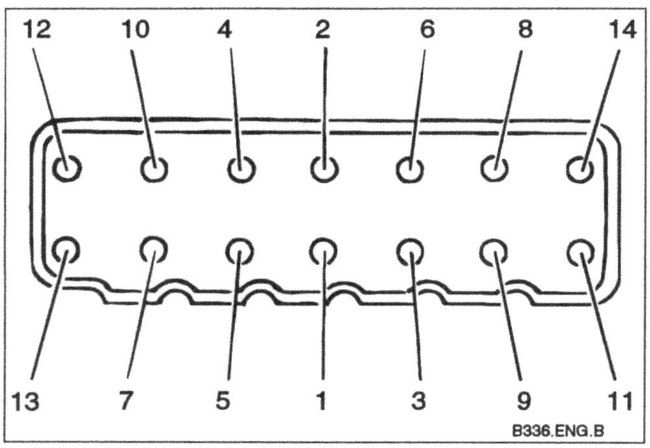

Fig. 4-33. Cylinder head bolt tightening sequence for 528e model.

Tightening torques	
• cylinder head bolts	
Torx®head bolts	
stage 1	30 Nm (22 ft-lb)
stage 2	90°
stage 3	90°
hex-head bolts	
stage 1	40+5 Nm (29+4 ft-lb)
then wait 15 minutes	
stage 2	60+5 Nm (44+4 ft-lb)
then run engine until fully warm (about 25 minutes)	
stage 3	25+5°

Installation of the remaining parts is the reverse of removal. Adjust the valve clearances as described under **4.2 Valve Adjustment**. Install the front pipe to the exhaust manifold with CRC® copper paste or equivalent on the mounting studs. Replace the gasket if necessary. Refill the cooling system as described in **COOLING SYSTEM.** Change the engine oil and filter as described in **LUBRICATION AND MAINTENANCE.** If applicable, adjust idle mixture as described in **FUEL INJECTION.** Adjust the accelerator cable as described in **FUEL INJECTION.** Adjust the transmission throttle cable as described in **AUTOMATIC TRANSMISSION.**

To remove cylinder head (533i and 535i models):

1. Disconnect the battery cables. Remove the smaller wires from the positive terminal.

CAUTION —
BMW anti-theft radios become inoperable whenever the battery is disconnected. See your owner's manual for more information.

2. Drain the engine coolant and remove the coolant hoses from the front and rear of the cylinder head. Disconnect the hose from the coolant reservoir. Remove any hose retaining brackets and move the hoses out of the way. Remove the block drain from the engine block. See **COOLING SYSTEM.**

3. Unbolt and separate the front exhaust pipe from the manifold. Discard the mounting nuts.

CAUTION —
Do not reuse self-locking nuts. These nuts are designed to be used only once and should be replaced whenever they are removed.

4. Remove the cover from the fuse/relay panel. Separate the connector and lift out the two relays shown in Fig. 4-34. Place the harness together with the relays on top of the engine.

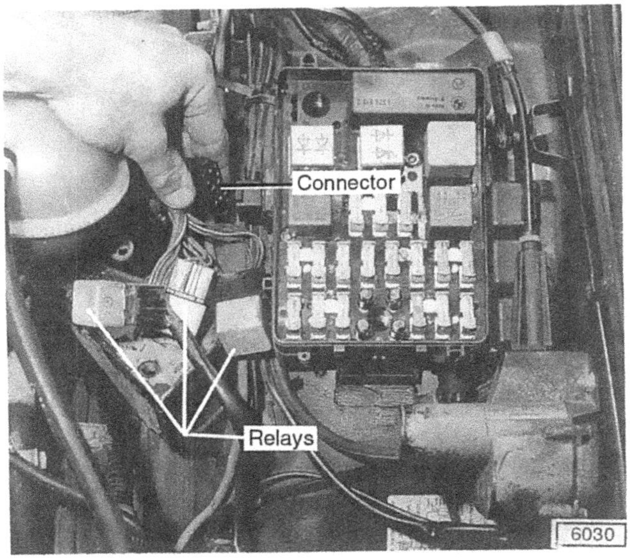

Fig. 4-34. Separate fuse/relay panel main harness connector and lift relays out of bracket. Set harness with relays on top of engine.

5. On 533i models, disconnect the hose, connectors, and fasteners shown in Fig. 4-35 below. Move all hoses and harnesses out of the way, cutting wire ties and removing brackets as necessary.

> **NOTE —**
> • Label all components, wires, and hoses before removing them as an aid to installation.
>
> • Some hoses may be secured with one-time hose clamps that are crimped with special pliers. The clamps may be pried open with a small screwdriver. Replace the clamps with standard screw-type clamps.

6. On 535i models, disconnect the hose, connectors, and fasteners shown in Fig. 4-36 below. Move all hoses and harnesses out of the way, cutting wire ties and removing brackets as necessary. Remove the air filter housing together with the air flow sensor.

1. Vacuum hose
2. Vacuum hose
3. Vacuum hose
4. Air flow sensor connector
5. Fuel return hose
6. Intake air duct clamp
7. Air filter housing mounting nuts
8. Idle air stabilizer valve connector

Fig. 4-35. Components, hoses, and connectors needed to be removed before removing cylinder head on 533i model.

1. Vacuum hose
2. Vacuum hose
3. Vacuum hose
4. Vacuum hose
5. Air flow sensor connector
6. Intake air duct clamp
7. Idle air stabilizer valve mounting nut
8. Idle air stabilizer valve connector
9. Air filter housing mounting nuts
10. Fuel return hose

Fig. 4-36. Components, hoses, and connectors needed to be removed before removing cylinder head on 535i model.

7. On all models, remove the harness connectors and components shown in Fig. 4-37. Move all harnesses and connectors out of the way

8. Remove the small vacuum hose (for the interior temperature sensor) from beneath the intake plenum.

9. Remove the intake plenum support bracket mounting bolt.

10. Remove the distributor cap, spark plugs, and the distributor rotor as described in **IGNITION**.

> **NOTE —**
> Label the spark plug wires so that they can be reinstalled in their original locations.

11. Remove the cylinder head cover as described earlier under **4.1 Cylinder Head Cover and Gasket.**

12. Disconnect the wires from the alternator and from the starter motor as described in **BATTERY, STARTER, ALTERNATOR**. Disconnect the oxygen sensor connector located beneath the car.

13. Remove the upper camshaft timing chain cover and remove the sprocket from the front of the camshaft as described in **4.4 Camshaft Timing Chain (533i and 535i models)**.

14. Check carefully for any remaining connectors, wiring, hoses, brackets, wire ties, clamps or components that will interfere with the removal of the cylinder head.

15. Gradually and evenly loosen and remove the cylinder head bolts using the sequence shown in Fig. 4-38 and remove the cylinder head. If the head is stuck, use a soft-faced mallet or pry gently with a wooden stick.

To install cylinder head (533i and 535i models):

1. Clean the cylinder head and the gasket surface of the cylinder block. Clean the threads of the head bolts and bolt holes with a thread chaser and remove all liquids and foreign matter from the bolt holes. Avoid letting debris into the cylinders or oil passages in the cylinder block.

> ***CAUTION —***
> *Do not use a metal scraper or wire brush to clean the aluminum cylinder head or pistons. These tools may damage the cylinder head or piston. Instead, use a solvent to soften carbon deposits and old sealing materials. If necessary, use a hard wooden or plastic scraper.*

2. Check the gasket surface of the cylinder head and the cylinder block for warpage as described in **ENGINE—RECONDITIONING.**

1. Fuel injector connector
2. Fuel injector connector
3. Fuel injector connector
4. Fuel injector connector
5. Fuel injector connector
6. Fuel injector connector
7. Coolant temperature sensor
8. Coolant temperature sender
9. Thermo-time switch connector
10. Ignition coil wiring
11. Cold start valve connector
12. Diagnostic connector bracket
13. Reference sensor connector
14. Speed sensor connector
15. Throttle, accelerator cable, and cruise control cable, where applicable
16. Mounting bracket
17. Mounting bracket
18. Fuel supply hose
19. Oil pressure switch connector
20. Throttle switch connector
21. Ground wire

Fig. 4-37. Harness connectors, wiring, and components needed to be removed before removing cylinder head on 533i and 535i models.

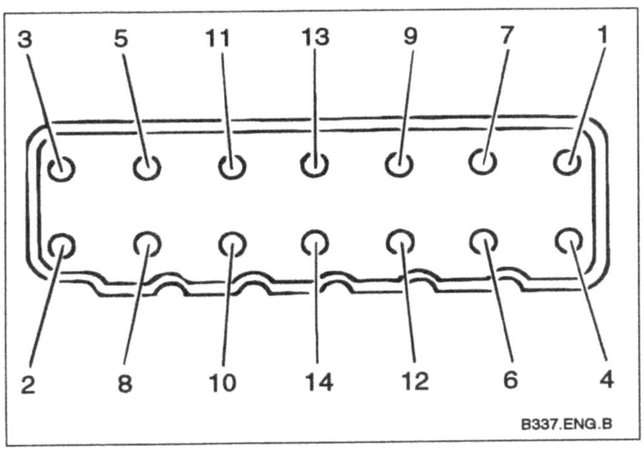

Fig. 4-38. Cylinder head bolt loosening sequence.

3. Place a new cylinder head gasket on the surface of the cylinder block. The cylinder head gasket will fit correctly in only one orientation. The word "OBEN," found printed on the gasket, should face up.

CAUTION —
Cylinder head gaskets will make a reliable seal only once. Always use a new cylinder head gasket that has not been previously compressed by tightening the cylinder head bolts.

NOTE —
Two different cylinder head gaskets are used for the 533i and 535i. An engine displacement code is stamped on the gasket flange as shown in Fig. 4-39. In addition, if the cylinder head has been machined to correct for warpage, a special 0.3 mm thicker gasket should be installed. See **ENGINE—RECONDITIONING** for more information.

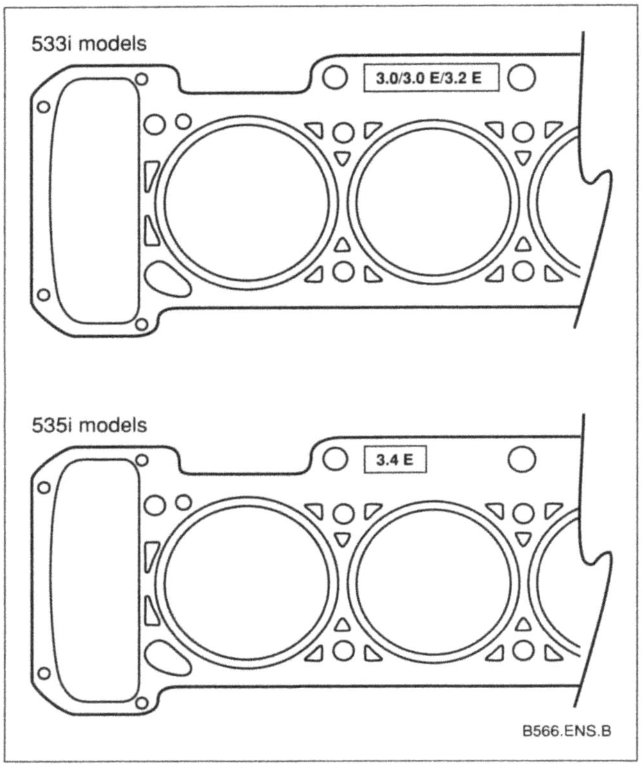

Fig. 4-39. Cylinder head replacement gasket code on flange of gasket.

4. Place the cylinder head in position on the cylinder block. Check that the rocker shafts are correctly positioned in their bores. Loosely install the head bolts and their washers, then thread them in until they are finger tight.

NOTE —
To help install the cylinder head, insert two 8 in. long by 3/8 in. round wooden dowels into two of the outermost head bolt holes in the block. Thread in several bolts, then remove the dowels and install the remaining bolts.

5. Tighten the cylinder head bolts in the sequence shown in Fig. 4-40. The bolts should be tightened in three stages. See below. The final stage requires using a special tool (BMW tool no. 11 2 110) or a suitable protractor to tighten the bolts to a specified torque angle.

CAUTION —
The cylinder head bolt torque is critical to proper engine operation. The bolts must be tightened in the stages specified below.

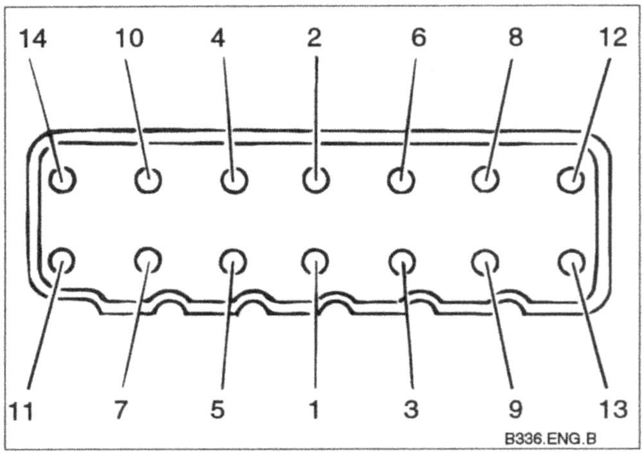

Fig. 4-40. Cylinder head bolt tightening sequence for 533i and 535i engine.

Tightening torques	
• cylinder head bolts	
stage 1	60±2 Nm (44±1 ft-lb)
then wait 20 minutes	
stage 2	80±2 Nm (59±1 ft-lb)
then run engine until fully warm (about 25 minutes)	
stage 3	35±5°

Installation of the remaining parts is the reverse of removal. Adjust the valve clearances as described under **4.2 Valve Adjustment**. Install the camshaft chain as described under **4.4 Camshaft Timing Chain (533i and 535i models)**. Install the front pipe to the exhaust manifold with CRC® copper paste or equivalent on the mounting studs. Replace the gasket if necessary. Refill and bleed the cooling system as described in **COOLING SYSTEM**. Change the engine oil and filter as described in **LUBRICATION AND MAINTENANCE**. Adjust idle speed and idle mixture as described in **FUEL INJECTION**. Adjust the accelerator cable as described in **FUEL INJECTION**. Adjust the transmission throttle cable as described in **AUTOMATIC TRANSMISSION**.

5. ENGINE

Before removing the engine it is first necessary to remove the transmission as described in **MANUAL TRANSMISSION AND CLUTCH** or **AUTOMATIC TRANSMISSION.** Remanufactured engines are available from an authorized BMW dealer.

Removing Engine

The air conditioning compressor and the power steering pump should be unbolted and set aside without disconnecting any of the refrigerant lines or the power steering fluid lines. The hood should either be removed from the car or supported in its fully open position. Drain the engine oil.

To remove engine (528e models):

1. Remove the transmission from the car. See **MANUAL TRANSMISSION AND CLUTCH** or **AUTOMATIC TRANSMISSION.**

2. Remove the front exhaust pipe from the exhaust manifold. See **EXHAUST SYSTEM.**

3. Remove the battery cables and then remove the battery from the car. Remove the small wire from the negative (–) battery terminal.

> ***CAUTION —***
> *BMW anti-theft radios become inoperable by disconnecting the battery cables. See your owner's manual for more information.*

4. Disconnect the hood support from the hood. Securely support the hood in its fully open position. As an alternative, mark the position of the hinges and remove the hood from the car. If applicable, disconnect the ground wire from the hood.

> **NOTE —**
> A special tool (BMW tool no. 51 2 110) to hold the hood in the fully open position is available from an authorized BMW dealer parts department.

5. Drain the engine coolant and remove the radiator. Disconnect the coolant hoses from the front of the engine. Disconnect the heater hoses from the heater core at the rear firewall. See **COOLING SYSTEM.**

6. Remove the V-belts from the air conditioning compressor and the power steering pump. Remove the compressor and the pump from their mounting brackets. Move the components out of the way as much as possible without distorting or damaging any lines. Use stiff wire to hang the components from the body.

7. Remove the cover from the fuse/relay panel. Separate the main connector and lift out the relays shown in Fig. 5-1. Place the harness together with the relays on top of the engine.

> **NOTE —**
> • Label all components, wires, and hoses before removing them as an aid to installation.
>
> • Some hoses may be secured with one-time hose clamps that are crimped with special pliers. The clamps may be pried loose with a small screwdriver. Replace the clamps with standard screw-type clamps.

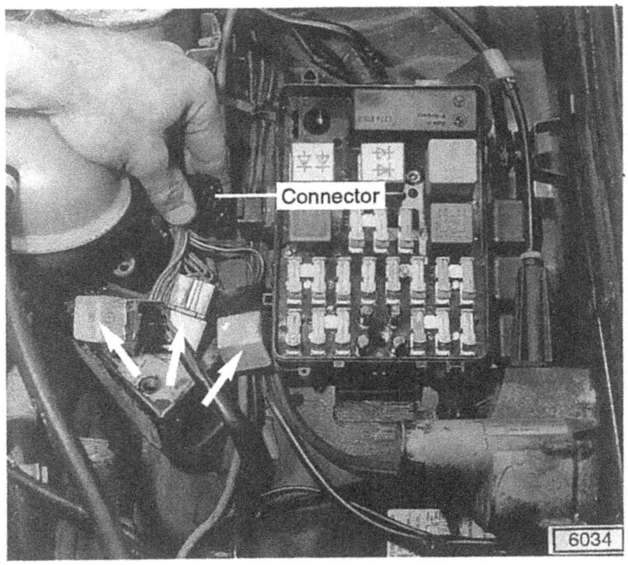

Fig. 5-1. Separate fuse/relay panel main harness connector and lift relays (arrows) out of brackets. Set harness with relays on top of engine.

8. Disconnect the hose, connectors, and fasteners shown in Fig. 5-2. Move all hoses and harnesses out of the way, cutting wire ties and removing brackets as necessary.

> ***WARNING —***
> *Fuel may be expelled under pressure. Do not smoke or work near heaters or other fire hazards. Keep a fire extinguisher handy. Wrap a cloth around fuel hoses to absorb leaking fuel before disconnecting them. Plug all open fuel lines.*

9. Working from above the glove compartment, disconnect the harness connectors from the idle speed control unit (1982 through 1987 models only) and the Motronic control unit. Disconnect the white (6-point) harness connector leading out of the Motronic wiring harness. Working from the engine compartment, pull the main wiring harness into the engine compartment. Cut any wire ties and remove any brackets holding the harness to the firewall. Place the harness on top of the engine.

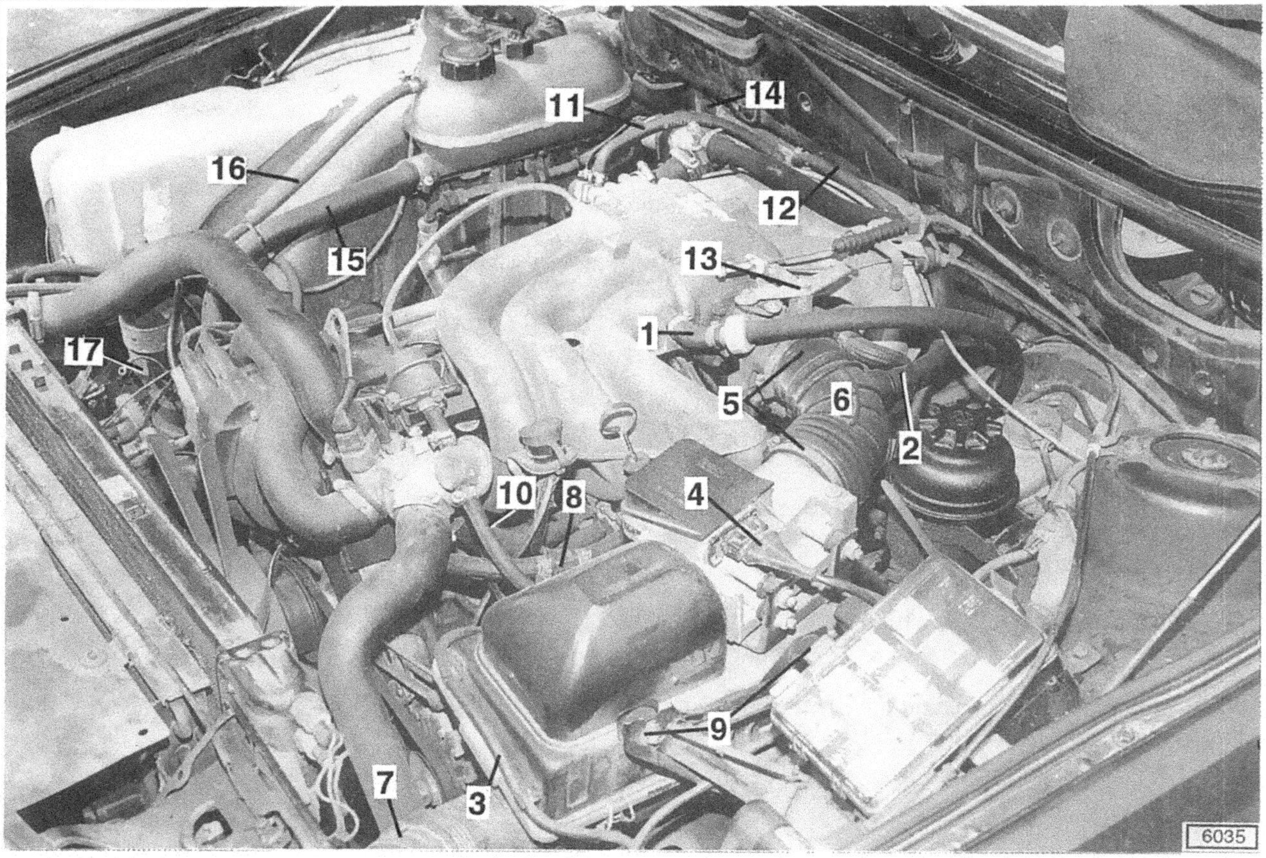

1. Vacuum hose
2. Vacuum hose
3. Vacuum hose
4. Air flow sensor connector
5. Intake air duct clamps
6. Intake air duct
7. Intake air hose
8. Altitude sensor connector
9. Air cleaner housing mounting nuts
10. Fuel return hose
11. Fuel supply hose
12. Fuel supply hose
13. Throttle, accelerator, and cruise control cables, if applicable
14. Ground strap, rear of valve cover
15. Coolant expansion tank hose
16. Coolant expansion tank hose
17. Ignition coil wiring

Fig. 5-2. Components, hoses, and connectors needed to be removed before removing engine on 528e model.

10. Working under the car, remove the nuts and ground strap from the engine mounts. See Fig. 5-3. Install an engine lifting device, using the front and rear lifting points. Carefully raise the engine out of the car, checking for any wiring, fuel lines, or mechanical parts that might become snagged as the engine is removed.

Installation is the reverse of removal. Engine mount tightening torques are listed below. Adjust the tension of all V-belts as described in **LUBRICATION AND MAINTENANCE.** Refill the cooling system as described in **COOLING SYSTEM.** Change the engine oil and filter as described in **LUBRICATION AND MAINTENANCE.** Install the front exhaust pipe as described in **EXHAUST SYSTEM.** Adjust the accelerator cable, check the idle speed and adjust idle mixture as described in **FUEL INJECTION.** Adjust the transmission throttle cable as described in **AUTOMATIC TRANSMISSION.** Adjust the cruise control cable so that there is 1.0 to 2.0 mm (0.04 to 0.08 in.) of play at the adjusting nut with the throttle closed. Check all fluid levels.

Tightening torques	
• engine mounts	
all M10 fasteners	43 to 48 Nm (32 to 35 ft-lb)
all M8 fasteners	22 to 24 Nm (16 to 18 ft-lb)

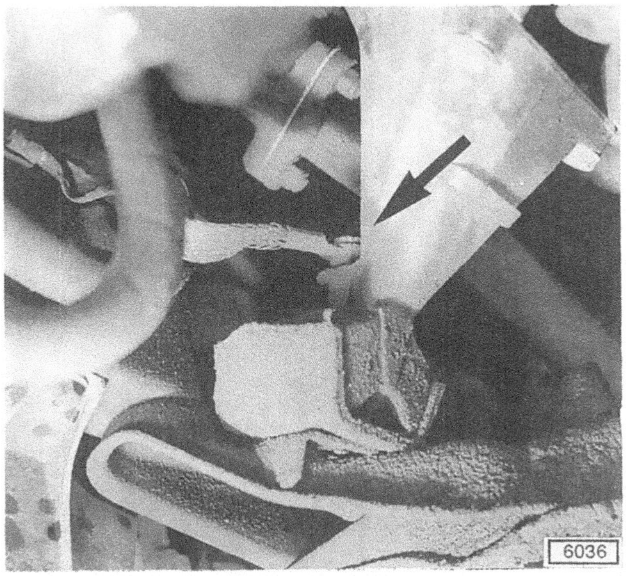

Fig. 5-3. Left engine mount nut and ground strap (arrow). Remove nuts from left and right mount.

To remove engine (533i and 535i models):

1. Remove the transmission from the car. See **MANUAL TRANSMISSION AND CLUTCH** or **AUTOMATIC TRANSMISSION.**

2. Remove the front exhaust pipe from the exhaust manifold. See **EXHAUST SYSTEM.**

3. Remove the battery cables and then remove the battery from the car. Remove the small wire(s) from the battery terminals. Drain the engine oil.

   CAUTION —
   BMW anti-theft radios become inoperable by disconnecting the battery cables. See your owner's manual for more information.

4. Disconnect the hood support from the hood. Securely support the hood in its fully open position. As an alternative, mark the position of the hinges and remove the hood from the car. If applicable, disconnect the ground wire from the hood.

   NOTE —
   A special tool (BMW tool no. 51 2 110) to hold the hood in the fully open position is available from an authorized BMW dealer parts department.

5. Drain the engine coolant and remove the radiator. Disconnect the coolant hoses from the front of the engine. Disconnect the heater hoses at the rear firewall. See **COOLING SYSTEM.**

6. Remove the V-belts from the air conditioning compressor and the power steering pump. Remove the compressor and the pump from their mounting brackets. Move the components out of the way as much as possible without distorting or damaging any lines. Use stiff wire to hang the components from the body. Disconnect the electrical connector from the A/C compressor.

4

7. Remove the cover from the fuse/relay panel. Separate the main connector and lift out the relays shown in Fig. 5-4. Place the harness together with the relays on top of the engine.

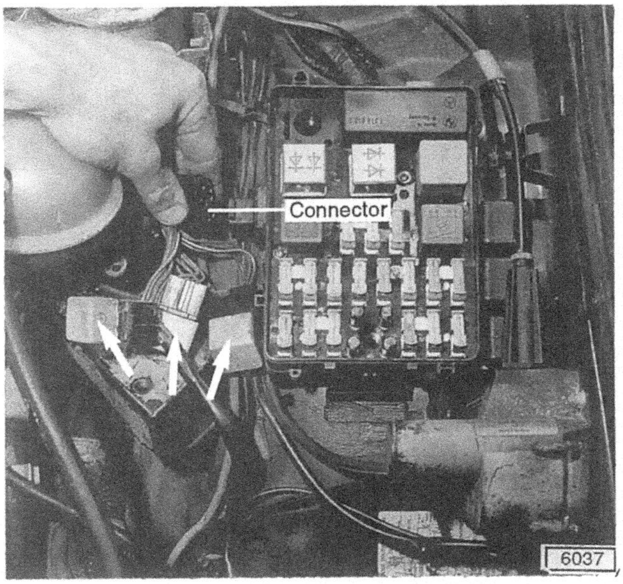

Fig. 5-4. Separate fuse/relay panel main harness connector and lift relays (arrows) out of brackets. Set harness with relays on top of engine.

NOTE —
• Label all components, wires, and hoses before removing them as an aid to installation.

• Some hoses may be secured with one-time hose clamps that are crimped with special pliers. The clamps may be pried loose with a small screwdriver. Replace the clamps with standard screw-type clamps.

8. On 533i models, disconnect the hoses, connectors, and fasteners shown in Fig. 5-5. Move all hoses and harnesses out of the way, cutting wire ties and removing brackets as necessary.

   WARNING —
   Fuel may be expelled under pressure. Do not smoke or work near heaters or other fire hazards. Keep a fire extinguisher handy. Wrap a cloth around fuel hoses to absorb leaking fuel before disconnecting them. Plug all open fuel lines.

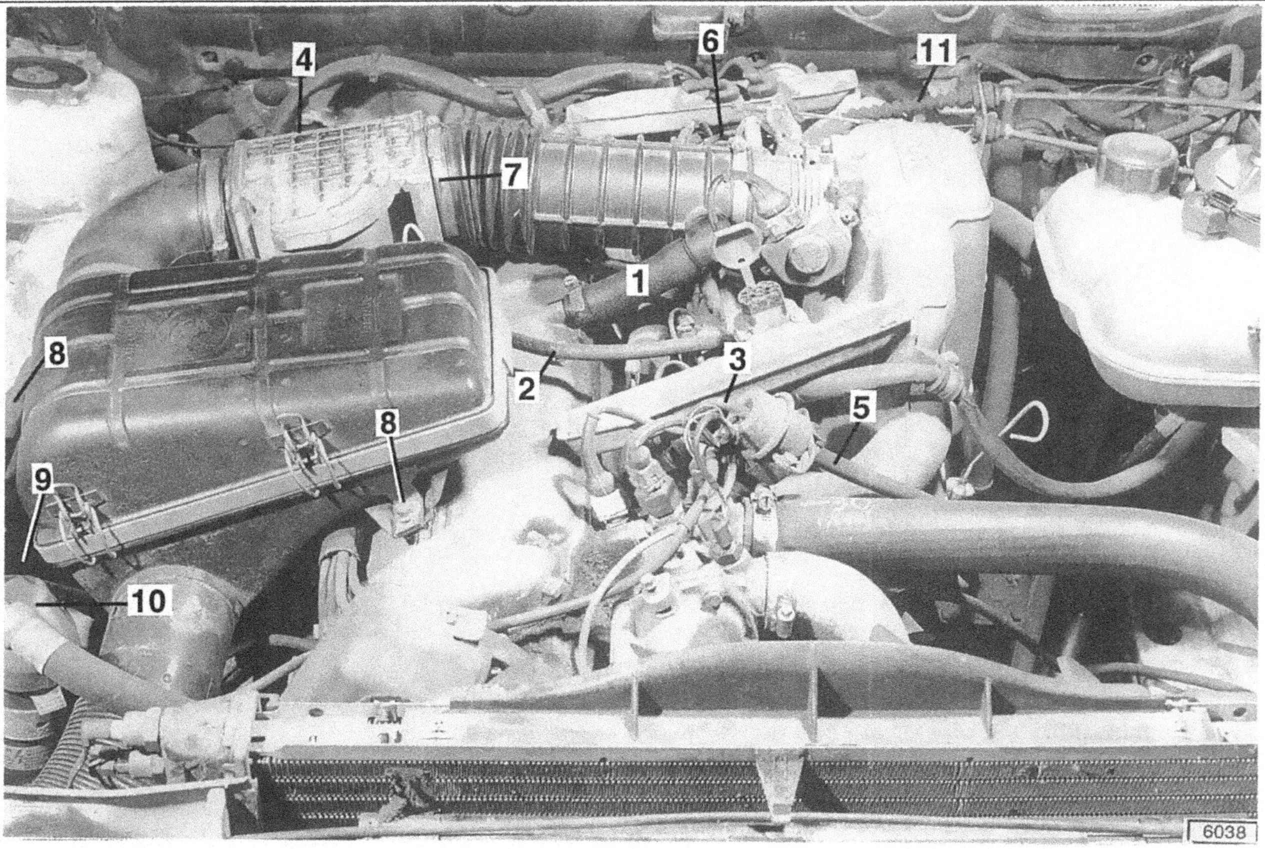

1. Vacuum hose
2. Vacuum hose
3. Vacuum hose
4. Air flow sensor connector
5. Fuel return hose
6. Fuel supply hose
7. Intake air duct clamp
8. Air filter housing mounting bolts
9. Charcoal canister purge valve connector
10. Ignition coil wiring
11. Accelerator, throttle and cruise control cable, where applicable

Fig. 5-5. Components, hoses, and connectors needed to be removed before removing engine on 533i model.

9. On 535i models, disconnect the hoses, connectors, and fasteners shown in Fig. 5-6 below. Move all hoses and harnesses out of the way, cutting wires ties and remove brackets as necessary.

10. Remove the air filter housing together with the air flow sensor.

11. On 535i models, disconnect the oxygen sensor harness connector from beneath the car, near the catalytic converter.

12. Disconnect the small vacuum hose (for interior temperature sensor) from the bottom of the intake air plenum.

13. Working from above the glove compartment, disconnect the harness connectors from the idle speed control unit (533i models only) and the Motronic control unit. Disconnect the white harness connector leading out of the Motronic wiring harness. See Fig. 5-7. Working from the engine compartment, pull the main wiring harness into the engine compartment. Cut any wire ties and remove any brackets holding the harness to the firewall. Place the harness on top of the engine.

14. Working under the car, remove the nuts and ground strap from the engine mounts. See Fig. 5-8.

15. Install an engine lifting device, using the front and rear lifting points. Carefully raise the engine out of the car, checking for any wiring, fuel lines, or mechanical parts that might become snagged as the engine is removed.

1. Charcoal canister hose
2. Air flow sensor connector
3. Intake air duct clamp
4. Air filter housing mounting nuts
5. Fuel return hose
6. Fuel supply hose
7. Ground strap
8. Ignition coil wiring
9. Accelerator, throttle and cruise control cable, where applicable
10. Air conditioning compressor connector

Fig. 5-6. Components, hoses, and connectors needed to be removed before removing engine on 535i model.

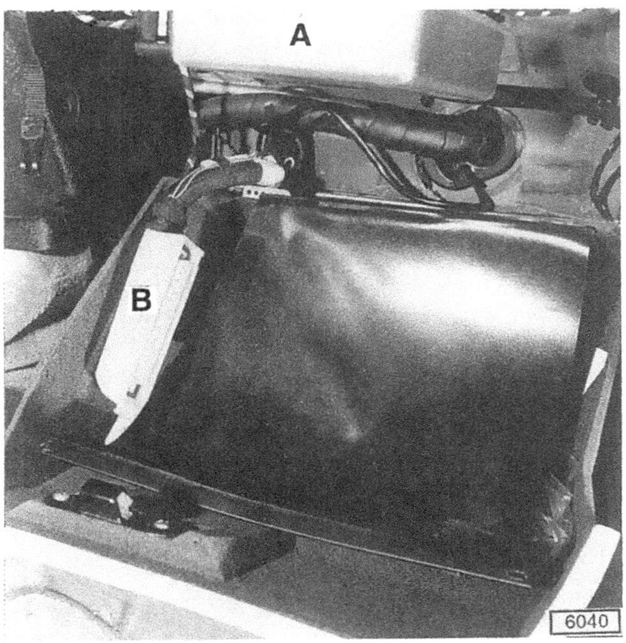

Fig. 5-7. Motronic control unit (**A**), and white harness connector (**B**) on 535i model. 533i model has additional idle speed control unit connector.

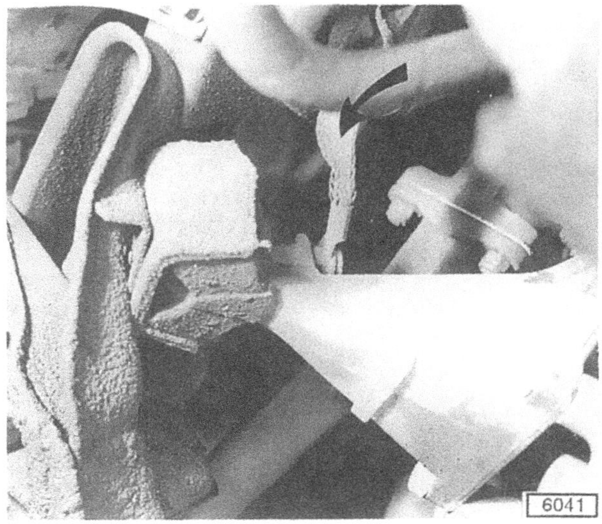

Fig. 5-8. Left engine mount nut and ground strap (arrow). Remove nuts from left and right mount.

Installation is the reverse of removal. Engine mount tightening torques are listed below. Adjust the tension of all V-belts as described in **LUBRICATION AND MAINTENANCE**. Refill the cooling system as described in **COOLING SYSTEM**. Change the engine oil and filter as described in **LUBRICATION AND MAINTENANCE**. Install the front exhaust pipe as described in **EXHAUST SYSTEM**. Adjust the accelerator cable, check the idle speed and adjust idle mixture as described in **FUEL INJECTION**. Adjust the transmission throttle cable as described in **AUTOMATIC TRANSMISSION**. Adjust the cruise control cable so that there is 1.0 to 2.0 mm (0.04 to 0.08 in.) of play at the adjusting nut with the throttle closed. Check all fluid levels.

Tightening torques	
• engine mounts	
all M10 fasteners.	43 to 48 Nm (32 to 35 ft-lb)
all M8 fasteners.	22 to 24 Nm (16 to 18 ft-lb)

6. LUBRICATION SYSTEM

The primary function of the lubrication system is to lubricate the internal moving parts of the engine. The circulation of oil also aids engine cooling. Proper lubrication requires a constant supply of oil, fed to the moving parts under pressure.

Engine lubrication is directly related to engine longevity. Therefore, its importance cannot be overemphasized. Change the engine oil and oil filter regularly, at least as often as specified by BMW's recommended maintenance intervals, and preferably more frequently.

This section covers inspection and repair of the lubrication system. Oil and oil filter change as well as oil specifications are covered in **LUBRICATION AND MAINTENANCE**.

6.1 Oil Pressure and Oil Level Warning Systems

All engines covered by this manual have an oil pressure warning system and an oil level warning system. To prevent serious engine damage, the oil pressure warning system warns the driver of insufficient oil pressure. The oil level warning system also warns the driver when the oil level drops below a safe level. Other safety features include a filter bypass, to guard against bursting the filter due to over pressure, and an oil pump pressure relief valve to prevent excessive system pressure.

> ***CAUTION —***
> *If the engine oil pressure or oil level warning indicator stays on after the engine is started, or flashes on while driving, always assume that there is insufficient oil pressure or the oil level is too low. Check the oil level and test the oil pressure before proceeding with tests of the warning system.*

Testing Oil Pressure

Test the oil pressure by removing the pressure switch and installing a pressure gauge. See Fig. 6-1 or 6-2. **Table e** lists oil pressure specifications. If testing shows low oil pressure, one or more of the following conditions is indicated: 1) worn or faulty oil pump 2) worn or faulty bearings or 3) severe engine wear. All of these conditions are serious and indicate the need for major repairs.

Fig. 6-1. Oil pressure switch (arrow) on 528e engine.

> **NOTE —**
> **Oil pressure specifications should only be checked with fresh oil of the correct type.**

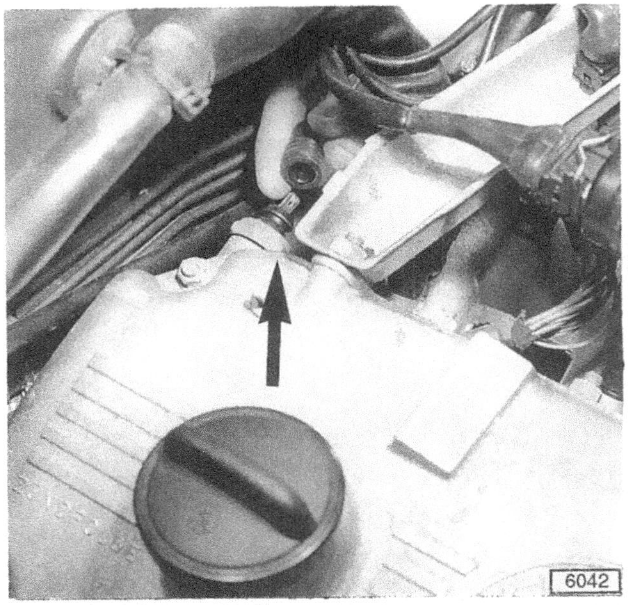

Fig. 6-2. Oil pressure switch on 533i and 535i engine.

Table e. Lubrication System Specifications

Oil pressure	
at idle	0.5 to 2.0 bar (7 to 29 psi)
at maximum engine speed	4.0 to 5.0 bar (58 to 72.5 psi)

Testing Oil Pressure Warning System

The oil pressure warning system consists of an oil pressure switch mounted in the oil circuit and an instrument panel warning light. When the ignition is turned on, a warning light comes on. When the engine is started and the oil pressure rises, the oil pressure switch opens and the warning light goes out. If this doesn't happen, check to make sure the oil level is correct before doing the tests listed below.

To test oil pressure warning system:

1. Turn the ignition switch on. The warning light on the instrument panel should light up. Remove the wire from the oil pressure switch (See Fig. 6-1 or Fig. 6-2 above). The light should go out.

> **NOTE —**
> If the light does not go out, the wiring to the switch is most likely grounded between the terminal and the warning light. See **ELECTRICAL SYSTEM** for electrical wiring troubleshooting information.

2. If the warning light does not light when the ignition is on, remove the terminal from the oil pressure switch and ground the wire to a clean metal surface.

NOTE —
If the warning light comes on, check the switch as described in the next step. If the warning light does not come on, the wiring circuit to the dash light or the light itself is faulty.

3. To test the switch, connect an ohmmeter between the switch terminal and ground. See Fig. 6-3. With the engine off, there should be continuity. With the engine running there should be no continuity. If any faults are found, the switch is faulty and should be replaced.

Fig. 6-3. Oil pressure switch being tested using an ohmmeter (shown schematically).

Tightening torque
• oil pressure switch. . . . . 30 to 40 Nm (22 to 30 ft-lb)

CAUTION —
If the oil pressure switch is not faulty and the light remains on while the engine is running, the oil pressure is too low. Do not operate the engine until the problem is corrected. The engine may be severely damaged.

Testing Oil Level Sensor

The oil level sensor is a device similar to the fuel gauge sender in the fuel tank. When the oil level falls below a safe level, the float assembly in the sensor unit completes the electrical circuit to the warning light on the active check control panel. Check that the oil level is correct before testing the sensor.

The oil level sensor is mounted in the bottom of the oil pan. The level incorporates two switches. The static switch senses the oil level when the engine is off and the ignition is on, and the dynamic switch senses the level when the engine is running. The oil level sensor can be tested using an ohmmeter.

4

To test the sensor, remove the negative terminal from the battery. Then disconnect the 17-pin connector from the side of the fuse/relay panel. See Fig. 6-4. With the engine off, there should be continuity through the static switch (terminal 2 and ground) and no continuity through the dynamic switch (terminal 13 and ground). Replace the sensor if it is faulty. Drain the engine oil before removing the sensor from the bottom of the oil pan. Inspect the sensor O-ring, replacing it if crushed or damaged.

NOTE —
To check the dynamic switch with the engine running, reconnect the fuse/relay panel connector and the battery terminal. Check the switch by probing terminal 13 from the rear of the connector. There should be no continuity between terminal 13 and ground with the engine running.

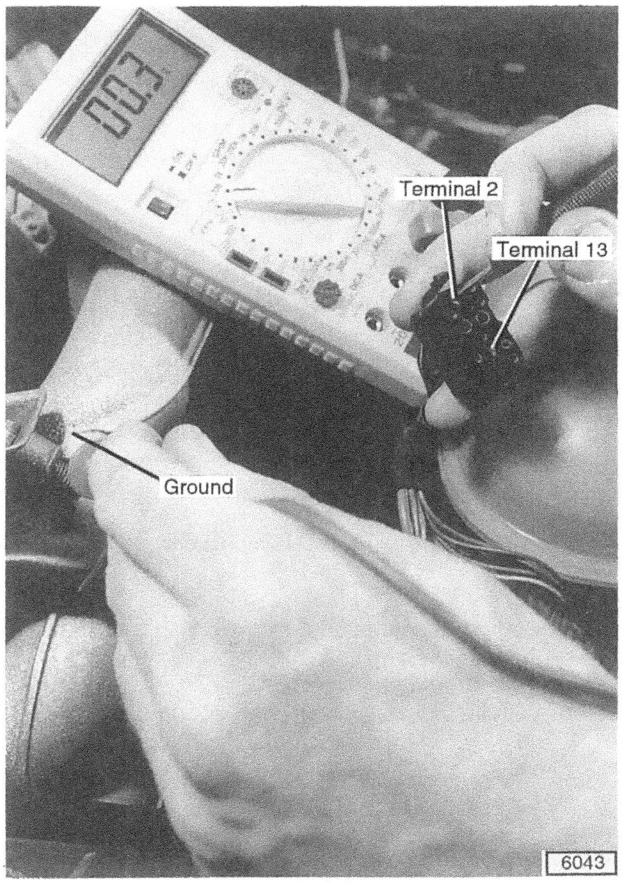

Fig. 6-4. Oil level sensor being tested at fuse/relay panel connector.

6.2 Oil Pan

The oil pan stores the oil used in lubrication and seals the bottom of the engine. The oil pump is located within the oil pan. On 528e models, the oil pan is easily removed with the engine installed in the car. On 533i and 535i models, oil pan removal is more difficult in that the engine needs to be lifted slightly to remove the oil pan.

To remove and install oil pan (528e model):

1. Drain the engine oil as described in **LUBRICATION AND MAINTENANCE**.

2. Disconnect the oil level sensor harness connector.

3. Remove the hex-head bolts and Torx®head bolts that secure the bellhousing reinforcement plate. See Fig. 6-5. Remove the plate.

Fig. 6-5. Bellhousing reinforcement plate mounting bolts (arrows).

4. Remove the oil pan retaining bolts and lower the oil pan.

5. Remove the three oil pump mounting bolts that hold the pump to the engine block.

6. Withdraw the oil pan with the oil pump.

> ***CAUTION —***
> *If the oil pan does not separate easily from the engine cylinder block, a few taps with a rubber mallet should break it free. Never pry between the oil pan and the engine cylinder block with a sharp instrument, as oil leaks can occur if either is scratched or scored.*

Installation is the reverse of removal. Thoroughly clean all the old gasket material from the mating surfaces. Install the oil pump as described in **6.3 Oil Pump**. Coat the mating surfaces on the timing case cover and the end cover with a sealing compound. Install a new gasket and the oil pan. Tighten all bolts evenly in stages. Reconnect the oil level sensor harness connector. Fill the engine with oil as described in **LUBRICATION AND MAINTENANCE**.

Tightening torques
• oil pan to engine block . . . . 9 to 11 Nm (80 to 97 in-lb)
• oil pump to engine block . . . . 22±2 Nm (16±1 ft-lb)

To remove and install oil pan (533i and 535i models):

1. Drain the engine oil as described in **LUBRICATION AND MAINTENANCE**.

2. Disconnect the oil level sensor harness connector.

3. Remove the hex-head bolts and Torx®head bolts that secure the bellhousing reinforcement plate. See Fig. 6-5 above. Remove the plate.

4. Remove the alternator and the alternator bracket as described in **BATTERY, STARTER, ALTERNATOR**.

5. Without disconnecting any fluid lines, remove the power steering pump and the power steering pump bracket. Support the pump from the body using stiff wire. See **STEERING AND WHEEL ALIGNMENT**.

6. Remove the lower part of the air conditioning compressor bracket.

7. Remove the engine mount retaining nuts. See Fig. 5-8 shown earlier. Using an engine lifting device at the front and rear engine lifting points, raise the engine as much as possible.

8. Remove the oil pan retaining bolts and lower the oil pan. Remove the two oil pump mounting bolts and lower the oil pump into the pan.

> ***CAUTION —***
> *If the oil pan does not separate easily from the engine cylinder block, a few taps with a rubber mallet should break it free. Never pry between the oil pan and the engine cylinder block with a sharp instrument, as oil leaks can occur if either is scratched or scored.*

9. Rotate the crankshaft so that no. 6 connecting rod journal is at its highest point (TDC) and withdraw the oil pan. If necessary, rotate the crankshaft to get extra clearance as the pan is being removed.

Installation is the reverse of removal. Thoroughly clean all the old gasket material from the mating surfaces. Coat the mating surfaces on the timing case cover and the end cover with a sealing compound. Install the oil pan using a new gasket. Tighten the oil pan mounting bolts evenly in stages. Reconnect the oil level sensor harness connector. Fill the engine with oil as described in **LUBRICATION AND MAINTENANCE**.

Tightening torque
• oil pan to engine block . . . . 9 to 11 Nm (80 to 97 in-lb)

6.3 Oil Pump

The oil pump is mounted to the bottom of the cylinder block inside the engine oil pan. When the engine is running, oil is drawn from the oil pan by the oil pump's pickup tube and circulated through the engine. On models with 533i and 535i engines, the oil pump is driven by a chain actuated by a gear on the front of the crankshaft. On 528e engines, the oil pump is driven by a gear on the intermediate shaft.

There is normally no need to remove the oil pump unless oil pressure is inadequate. Check the oil pressure as described above under **6.1 Oil Pressure and Oil Level Warning Systems**.

Removing and Installing Oil Pump
(528e models)

To remove the pump, remove the oil pan as described above. Once the pump is removed, remove the oil pump cover mounting bolt and remove and clean the oil filter screen. Insert the pump shaft and spin the pump gears. Remove the oil pump cover to check for any wear or scoring of the housing. If the gears spin with difficulty or any wear is present, the pump should be replaced. Install the oil pan as described in **6.2 Oil Pan**.

Removing and Installing Oil Pump
(533i and 535i models)

To remove the pump, first remove the oil pan as described above. Remove the large nut from the front of the oil pump sprocket mounting bolt and pull the sprocket off the pump. Remove the three bolts that secure the pump to the engine cylinder block and remove the pump, noting any shims between the pump and the block. See Fig. 6-6.

Fig. 6-6. Oil pump mounting bolts (arrows). Additional rear bolt not shown. Type of oil pump varies between models.

On models without chain tensioners (1983 through 1986 models), temporarily install the pump to check chain tension. A correctly tensioned chain will deflect slightly at the midpoint under light finger pressure. If the chain tension is incorrect, shims can be installed between the pump mounting points and the block to obtain the correct tension. Shims are available from an authorized BMW dealer parts department in two thicknesses, 0.1 mm and 0.3 mm. Install the shim so that the oil bore in the block is aligned with the hole in the shim. Install the oil pan as described in **6.2 Oil Pan**.

Tightening torques
• oil pump to engine block . . . . 22±2 Nm (16±1 ft-lb)
• chain sprocket to oil pump M10 nut . . . . 25 to 30 Nm (18 to 22 ft-lb)

7. TECHNICAL DATA

I. Tightening Torques

Camshaft sprocket to camshaft	
528e (bolts)	65–70 Nm (48–52 ft-lb)
Camshaft sprocket to camshaft flange	
533i, 535i (bolts)	7 Nm (62 in-lb)
Camshaft timing chain tensioner nut to front cover (533i, 535i)	35±5 Nm (26±4 ft-lb)
Crankshaft hub to crankshaft (bolt or nut)	
528e	410±20 Nm (302±15 ft-lb)
533i, 535i	440±10 Nm (325±7 ft-lb)
Cylinder head bolts	
528e models	
Torx®head bolts	
stage 1	30 Nm (22 ft-lb)
stage 2	90°
stage 3	90°
Hex-head bolts	
stage 1	40+5 Nm (29+4 ft-lb)
wait 15 minutes	
stage 2	60+5 Nm (44+4 ft-lb)
run engine until fully warm (about 25 minutes)	
stage 3	25+5°
533i and 535i models	
stage 1	60±2 Nm (43±1 ft-lb)
wait 20 minutes	
stage 2	80±2 Nm (58±1 ft-lb)
run engine until fully warm (about 25 minutes)	
stage 3	35±5°
Cylinder head cover to cylinder head (nuts and bolt)	9±1 Nm (89±9 in-lb)
Engine mounts to subframe and body	
M10 fasteners	43–48 Nm (32–35 ft-lb)
M8 fasteners	22–24 Nm (16–18 ft-lb)
Front timing chain cover to engine (533i, 535i)	
M6 bolts	9–11 Nm (80–97 in-lb)
M8 bolts	22±2 Nm (16±1 ft-lb).
Ignition rotor adaptor to camshaft (install with thread-locking compound)	
528e models	60±5 Nm (44±3 ft-lb)
533i, 535i models	23±1 Nm (17±0.5 ft-lb)
Oil pan to cylinder block	9–11 Nm (80–97 in-lb)
Oil pan to engine block	9–11 Nm (80–97 in-lb)
Oil pressure switch to cylinder head or cylinder block	30–40 Nm (22–30 ft-lb)
Oil pump to engine block	22±2 Nm (16±1 ft-lb)
Reference sensor to front cover (528e models built from March 1987 only)	7±1 Nm (62±9 in-lb)
Rocker arm eccentric locknut	10±1 Nm (7±0.5 ft-lb)
Spark plugs to cylinder head	20–30 Nm (15–22 ft-lb)
Timing chain sprocket to oil pump (533i, 535i models)	25 to 30 Nm (18 to 22 ft-lb)
Vibration damper to crankshaft hub	22±2 Nm (16±1 ft-lb)

Section 5

ENGINE—RECONDITIONING

5

Contents

TABLES

INTRODUCTION

This section contains engine reconditioning specifications for the three 6-cylinder engines used in the models covered by this manual. Much of the information contained here is intended to be used as a reconditioning guide for the professional or experienced automotive mechanic. Most of the operations described in this section require special equipment and experiences. If you lack the skills, tools, or a suitable workplace for reconditioning, we suggest you leave these repairs to an authorized BMW dealer or other qualified shop.

For information on engine and cylinder head removal, as well as other general engine troubleshooting, service and repair, see **ENGINE—SERVICE AND REPAIR.**

1. GENERAL DESCRIPTION

The 528e model has a 2.7 liter engine. This engine may sometimes be referred to as the "eta" engine or the "baby-six" engine. The 533i models has a 3.2 liter engine and the 535i models has a 3.4 liter engine. These two engines are virtually identical, with the only major difference being the bore size. These engines may sometimes be referred to as the "big-six."

1.1 Engine Components

Crankshaft and Bearings

The fully counterweighted crankshaft rotates in replaceable split-shell main bearings. A 2-piece center main bearing controls crankshaft end thrust. Flexible lip seals, pressed into light alloy seal carriers, are installed at both ends of the crankshaft to prevent oil leakage.

Connecting Rods and Pistons

The connecting rods are steel forgings. Replaceable split-shell bearings are installed at the crankshaft end and solid bushings at the piston pin end. The pistons are of the three-ring type with two upper compression rings and a lower one-piece oil scraper ring. Full-floating piston pins are retained at each end by circlips.

Cylinder Head

The cylinder head is an aluminum alloy casting. Replaceable valve guides are press-fit, while the bearing surfaces for the overhead camshaft and the rocker arm shafts are machined directly into the cylinder head casting.

Valve Train

An overhead camshaft operates the valves through rocker arms. The camshaft is chain-driven on the big-six engines (533i and 535i models) and belt driven on the baby-six engine (528e model). One end of each rocker serves as a cam follower and the other end contains the valve adjusting eccentric and its locknut.

1.2 Engine Specifications

Table a lists the major specifications for the engines covered by this manual. Some of this information may be useful when buying parts or making repairs.

Table a. Engine Specifications

Model	1982–1987 528e	1988 528e	1983–1984 533i	1985–1988 535i
No. of cylinders	6	6	6	6
Bore mm (in.)	84.0 (3.307)	84.0 (3.307)	89.0 (3.504)	92.0 (3.622)
Stroke mm (in.)	81.0 (3.189)	81.0 (3.189)	86.0 (3.386)	86.0 (3.386)
Displacement cc (cu. in.)	2693 (164)	2693 (164)	3210 (196)	3430 (209)
Compression ratio	9.0:1	8.5:1	8.8:1	8.0:1
Horsepower SAE net @ rpm	121@4250	127@4800	181@6000	182@5400
Torque lb-ft @ rpm SAE net	170@3250	170@3200	192@4000	214@4000

2. MAINTENANCE

For information on BMW's recommended engine maintenance, see **ENGINE—SERVICE AND REPAIR** or **LUBRICATION AND MAINTENANCE.**

3. TROUBLESHOOTING

Before disassembling the engine for reconditioning, make as many diagnostic tests as possible. Make a compression test, a leak-down test, and if the engine is running, a vacuum test. Each of these tests should help to assess the engine's condition while it is still assembled. For general engine troubleshooting information as well as diagnostics tests, see **ENGINE—SERVICE AND REPAIR.**

3.1 Basic Troubleshooting Principles

When inspecting the disassembled engine components for wear or damage, be sure all parts are clean and, if necessary, decarbonized. Damage such as a small crack in the cylinder head can only be detected when the head is shiny clean.

Where available, engine specifications are given by listing a maximum wear specification and/or a tolerance range. A tolerance range can be very helpful to the rebuilder when deciding if a component should be replaced. By calculating how much wear has taken place since new, the rebuilder can make a more informed decision, even if the part is within the allowable specifications.

The basic BMW six-cylinder engine has been in production for many years. This has allowed BMW engine repair facilities to identify common problems. Where possible, this information is included in the text.

4. CYLINDER HEAD

Most cylinder head work requires that it first be removed. Cylinder head removal is described in **ENGINE—SERVICE AND REPAIR.**

To prevent damaging the open valves while the cylinder head is on the workbench, place four 1/2" by 2" bolts into each corner of the head. Put the bolts in from the bottom of the cylinder head so that it rests on the bolt heads.

4.1 Disassembling Cylinder Head

The BMW cylinder head is somewhat difficult to disassemble owing to its design. The camshaft journals are integral with the head and do not have removable split caps. This means that valve spring tension is always bearing on the camshaft via the rocker arms and cannot be easily relieved. In addition, the rocker arm shafts are tightly fitted in their bores.

On 533i and 535i models, special BMW tools may be needed to compress the rocker arms and release the valve spring tension so that the camshaft can be pulled from the head. These tools are expensive, are usually only available through an authorized BMW dealer, and can damage the valves if not properly used.

On the 528e model, special spring-compressing tools are not needed, although other special tools may be required. If any of the special tools are not available, one alternative is to remove the cylinder head and have the head disassembled by an authorized BMW dealer.

To disassemble cylinder head (528e model):

1. Remove the socket head bolt from the camshaft sprocket using a 7 mm hex wrench. See Fig. 4-1. Remove the rotor adaptor and the sprocket from the camshaft.

NOTE —
It may be necessary to immobilize the camshaft drive belt sprocket while breaking the bolt loose. The old camshaft drive belt, which must be replaced, can be wrapped around the belt sprocket, then clamped with a pair of Vise-grips.

Fig. 4-1. Distributor rotor adaptor and camshaft sprocket mounting bolt being removed. Sprocket is being held stationary with a strip of leather and Vise-grips.

2. Remove the camshaft oil seal housing mounting bolts. Remove the housing by rotating it off the camshaft. See Fig. 4-2.

Fig. 4-2. Camshaft oil seal housing. Pull housing off while slowly turning back and forth (arrow).

3. Adjust the valves to maximum clearance by loosening the adjusting eccentric locknut. Rotate the eccentric until the adjusting hole is facing inward.

4. Remove the rocker arm retainers by lifting them straight off. Note that the straight leg of the retainer fits into the slot in the rocker arm shaft. Lift out the rocker arm shaft guide plate. See Fig. 4-3.

Fig. 4-3. Rocker arm retainer (**1**) and rocker arm shaft guide plate (**2**) on 528e cylinder head. Note orientation of retainer.

5. Remove the four rubber plugs from the front and rear of the cylinder head. See Fig. 4-4.

Fig. 4-4. Rubber retaining plug being removed from cylinder head.

6. Temporarily mount the distributor rotor adaptor to the front of the camshaft.

7. Using the rotor adaptor mounting bolt, rotate the camshaft clockwise until the no. 6 intake and exhaust valves are both open (overlapped). Slide the no. 1 exhaust-side rocker arm off the top of the valve. See Fig. 4-5.

8. Slowly rotate the camshaft while relieving the valve spring tension on the exhaust-side rocker arms and sliding the rocker arms off the top of their valves.

9. If applicable, remove the threaded plug from the front of the rocker arm shafts.

NOTE —
If the shaft turns while trying to remove the plug, place a cylinder head bolt in the no. 1 cylinder head bolt hole. This should hold the shaft stationary.

Fig. 4-5. Rocker arm positioned (arrow) for removal of exhaust-side rocker arm shaft.

10. Remove the exhaust-side rocker arm shaft from the front of the cylinder head by either driving it out using a drift or, on models with removable end plugs, pulling it out using a slide hammer that threads into the front of the shaft.

CAUTION —
If driving the shaft out, strike the drift using firm, hard blows. Tapping on the drift will cause the end of the shaft to mushroom, making it difficult to remove and possibly damaging the head. If the end of the shaft does become mushroomed, use a small grinding stone or equivalent to redress the end of the shaft. The shaft can be turned using Vise-grips to access the under side. Be sure to wrap a cloth or piece of leather on the shaft before applying the pliers.

NOTE —
- Two different versions of rocker arm shafts were installed in the engines covered by this manual—one with removable end plugs and one with welded end plugs. On shafts with removable plugs, a special BMW slide hammer (BMW tool no. 11 3 060) is designed to thread into the front of the shaft to facilitate shaft removal. On shafts with welded plugs, a special drift (BMW tool no. 11 3 050) is designed for this purpose. As an alternative, use a length of 9/16" bar stock.

- Label each rocker arm assembly as it is removed. Rocker arm assemblies should always be reinstalled in their original locations.

11. Remove the intake-side rocker arm shaft using the above procedure.

12. Carefully rotate the camshaft out of the head.

13. Remove the valves using a standard C-type valve spring compressor. Label each valve assembly as it is removed. Remove and discard the valve stem oil seals from the valve guides.

Installation is the reverse of removal. Prior to installation, check all components for wear as described below under **4.2 Cylinder Head Reconditioning.** Be sure all parts are installed in their original positions. When installing the rocker arm shafts, the notches for the guide plate and the small oil bores must face in and the large oil bores must face down towards the valve guides. See Fig. 4-6.

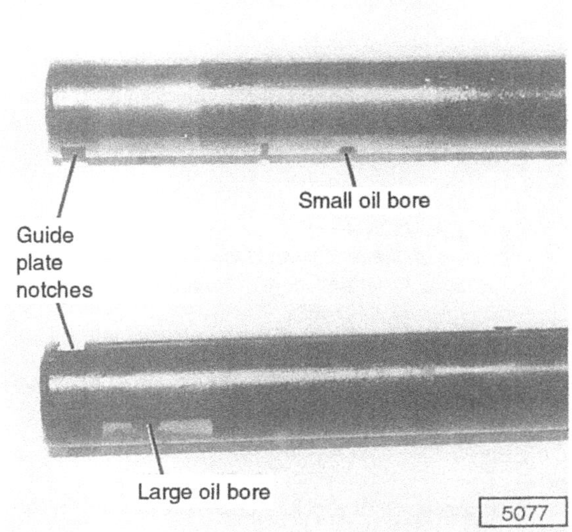

Fig. 4-6. Rocker arm shafts showing notches for front guide plate and oil bores.

Lubricate the camshaft prior to installation. Coat the camshaft oil seal lip and the housing O-ring with motor oil before installation. Install the camshaft sprocket and the distributor rotor adaptor so the guide pin correctly aligns with the guide hole in the sprocket and adaptor.

Tightening torques	
• oil sprayer bar fittings	6 to 8 Nm (53 to 71 in-lb)
• ignition rotor adaptor	65 to 70 Nm (48 to 52 ft-lb)

To disassemble cylinder head (533i and 535i models):

1. Remove the oil sprayer bar from the top of the head. Note the orientation of the spacer washers on the center retaining bolt.

2. Adjust the valves to maximum clearance by loosening the adjusting eccentric locknut. Rotate the eccentric until the adjusting hole points inward.

3. Remove the two camshaft guide plate mounting bolts from the front of the cylinder head. If necessary, rotate the camshaft until the cutouts in the camshaft flange are positioned in front of the bolts.

NOTE —
At this point, the valve spring pressure on the camshaft needs to be relieved. This can be done in one of two ways. The first method requires the BMW special tool (BMW tool no. 11 1 060), which is a large press that mounts over the top of the entire head and presses down on all of the rocker arms. The second method is to manually lever the necessary rocker arms down using screwdrivers or forked levers made of stiff steel. Once the pressure on the camshaft is relieved, it can be easily withdrawn from the head.

4. If the special tool (BMW tool no. 11 1 060) is available, install the tool and tighten the clamping nuts on the exhaust side first. Then tighten the intake side clamping nuts. Pull the camshaft out of the head.

CAUTION —
When tightening down the BMW special tool, be sure to tighten down the exhaust side first to prevent damaging the valves. In addition, the BMW special tool may not work on cylinder heads with a special non-factory camshaft. If a special camshaft is installed, watch the valves while slowly tightening down the tool. The valves must not contact each other.

5. If the special tool is not available, rotate the camshaft until the least number of rockers arms are exerting pressure on the camshaft. Using either screwdrivers or forked steel levers, relieve the pressure on the remaining rocker arms. See Fig. 4-7. Then pull the camshaft out of the head.

WARNING —
The valves are under considerable pressure from the valve springs. Use extreme care when depressing the rocker arms. Wear hand and eye protection.

6. Remove the rear cover plate from the back of the head. Note the location of the sealing washer and O-ring on one of the cover mounting bolts.

NOTE —
Always replace the rear cover sealing washer and O-ring to prevent oil leaks.

7. Remove the rocker arm retainers by sliding the rocker arms and thrust rings to the side and prying off the thin retaining rings. See Fig. 4-8.

Fig. 4-7. Rockers arms being depressed to relieve spring tension on camshaft (arrows).

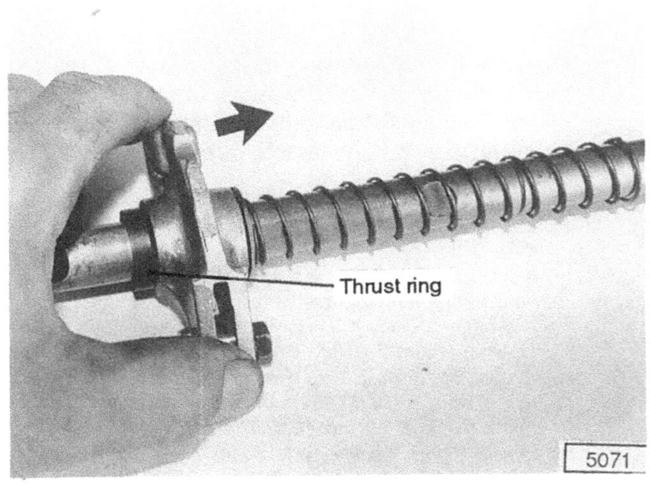

Fig. 4-8. Remove rocker arm retaining ring by sliding rocker arm and thrust ring to side. Pick off retaining ring using a small screwdriver.

8. If applicable, remove the threaded end plugs from the front and rear rocker arm shafts.

NOTE —
If the shaft turns while trying to remove the plug, place a cylinder head bolt in a cylinder head bolt hole. This will hold the shaft stationary.

9. Remove the rocker arm shafts by either driving them out using a long drift or, on models with removable end plugs, pulling them out using a slide hammer that threads into the threaded end of the rocker arm shaft.

CAUTION —
If driving the shaft out, strike the drift using firm, hard blows. Tapping on the drift will cause the end of the shaft to mushroom, making it difficult to remove and possibly damaging the head. If the end of the shaft does become mushroomed, use a small grinding stone or equivalent to re-dress the end of the shaft. The shaft can be turned using Vise-grips to access the under side. Be sure to wrap a cloth or piece of leather on the shaft before applying the pliers.

NOTE —
- Two different versions of rocker arm shafts were installed in the engines covered by this manual—one with removable end plugs and one with welded end plugs. On shafts with removable plugs, a special BMW slide hammer (BMW tool no. 11 3 060) is designed to thread into the front of the shaft to facilitate shaft removal. On shafts with welded end plugs, a special drift (BMW tool no. 11 3 050) is designed for this purpose. As an alternative, use a length of 9/16" bar stock.

- Label all parts as they are removed so they can be in reinstalled in their original positions. The 533i and 535i cylinder head uses a two-piece rocker arm shaft. The intake-side shafts are different from the exhaust side shafts.

10. Remove the valves using a standard C-type valve spring compressor. Label each valve assembly as it is removed. Remove and discard the valve stem oil seals from the valve guides.

Installation is the reverse of removal. Prior to installation, check all components for wear as described below under **4.2 Cylinder Head Reconditioning**. Be sure all parts are installed in their original positions. When installing the rocker arm shafts, the longer of the two shafts goes towards the front of the cylinder head and the plugged ends face out. Make sure the cutouts for the head bolts are correctly aligned and the oil holes for the rocker arms face the valve stems.

When installing the oil sprayer bar, make sure the arrow on the tube faces the front of the cylinder head and the center mounting bolt has a spacer on either side of the tube. See Fig. 4-9. If the arrow is not visible, turn the tube upside down and position it on the cylinder head. The oil holes should be directly above the cam lobes. It is also recommended that the center mounting bolt be installed with a small amount of blue Loctite and that the two crush washers be replaced.

CAUTION —
An incorrectly installed oil sprayer bar will cause severe and rapid camshaft wear.

NOTE —
If replacing damaged or worn rocker arm springs, BMW supplies only short springs that replace the older, longer springs. All springs are interchangeable.

Fig. 4-9. Oil sprayer bar correctly installed with arrow facing front (**A**).

Lubricate the camshaft prior to installation. If the camshaft is being replaced, transfer the camshaft flange parts from the old shaft to the new shaft.

Tightening torques
• oil sprayer bar mounting bolt . 11 to 13 Nm (97 to 115 in-lb)
• ignition rotor adaptor (install with Loctite 270 or equivalent) . . . . 23±1 Nm (17±0.5 ft-lb)
• flange to camshaft (large nut) . . . . 142±5 Nm (105±4 ft-lb)

4.2 Cylinder Head Reconditioning

Reconditioning procedures for the BMW cylinder heads are similar to those for most other modern water-cooled engines. For anyone with the proper tools and equipment and basic experience in cylinder head reconditioning, this section provides the specifications and special reconditioning information necessary to repair the cylinder heads covered by this manual.

If machine shop services are not readily available, one alternative is to install a remanufactured cylinder head. Remanufactured cylinder heads are available from an authorized BMW dealer parts department.

Cylinder Head Assembly

The disassembled cylinder head should be carefully inspected for warpage and cracks. Check the valve guides and valve seats as described later in this section.

Always decarbonize and clean the head before inspecting it. A high quality straightedge can be used to check for warpage. See Fig. 4-10. Visually inspect the cylinder head for cracks. If a cracked cylinder head is suspected and no cracks are detected through the visual inspection, have the head further tested for cracks by an authorized BMW dealer. A cracked cylinder head should be replaced.

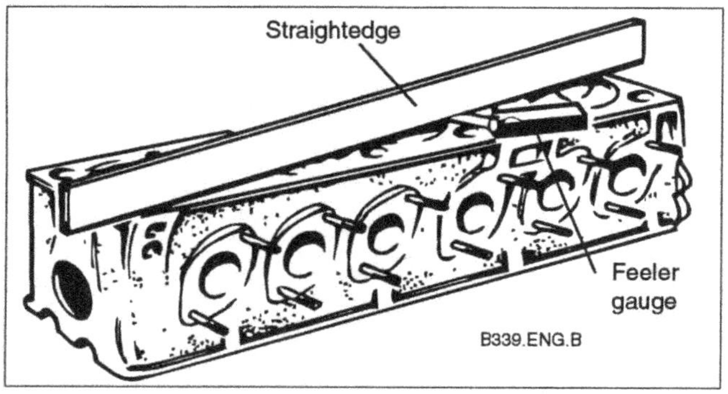

Fig. 4-10. Straightedge and feeler gauge being used to check straightness of cylinder head gasket surface.

A warped cylinder head can be machined provided no more than 0.3 mm (0.012 in.) of material is removed. If further machining is required, the head should be replaced. Removing more than this amount will reduce the size of the combustion chamber and adversely affect engine performance.

NOTE —
A 0.3 mm thicker gasket is available from an authorized BMW parts department for machined heads.

Before machining the head to correct for warpage, measure the total height of the cylinder head as shown in Fig. 4-11. **Table b** lists the minimum resurfacing height specifications. If the cylinder head height will not meet the minimum height dimension after machining the cylinder head should be replaced.

NOTE —
When machining cylinder heads on 533i and 535i models, the upper camshaft timing chain cover and the rear cover should be bolted to the cylinder head before machining.

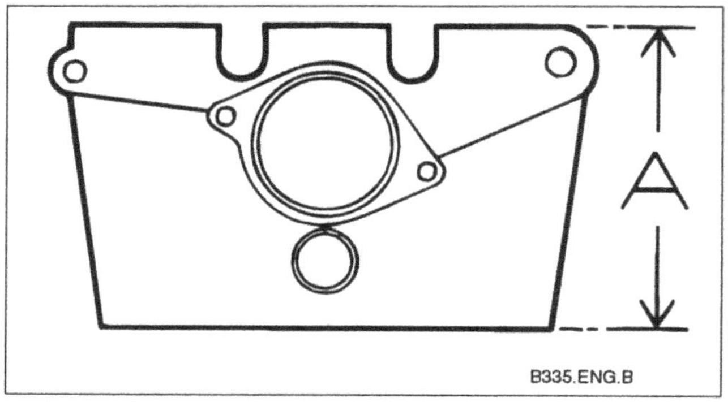

Fig. 4-11. Front view of cylinder head showing minimum resurfacing dimension (**A**).

Table b. Cylinder Head Resurfacing Specifications

Model	Minimum permissible height (dimension A)	
	new	after machining
528e	125.1±0.1 mm (4.925±.004 in.)	124.7 mm (4.909 in.)
533i and 535i	129.0±0.1 mm (5.079±.004 in.)	128.6 mm (5.063 in.)

Camshaft

Camshaft wear is usually caused by insufficient lubrication. On 533i and 535i models, the installation of the oil sprayer bar is critical—the bar is asymmetrical and must be installed only one way. See Fig. 4-9 above. On all models, a clogged or loose oil sprayer bar will reduce oil flow to the camshaft. Engine oil dilution, extended oil drain intervals, or high mileage are all additional causes of camshaft wear.

Visually inspect the removed camshaft lobes and journals for wear. Lightly lubricate the camshaft and place it in the cylinder head. Install the guide (thrust) plate. On 528e models, install the oil seal housing. Check that the camshaft turns smoothly. Using a feeler gauge, check the camshaft axial (thrust) play as shown in Fig. 4-12.

NOTE —
On cars experiencing severe camshaft wear, a special hardened camshaft may be available from BMW. Check with an authorized BMW dealer parts department for the latest parts information regarding BMW Service-Information Bulletin 11 07 88 (1767).

Fig. 4-12. Camshaft axial play (arrow) should be measured between guide plate and flange.

Camshaft axial play
• 528e model (maximum) . . . . 0.2 mm (0.008 in.)
• 533i and 535i model . . . . 0.03 to 0.18 mm (0.0012 to 0.0071 in.)

Rocker Arms and Rocker Arm Shafts

Rocker arms and rocker arm shafts are components that wear and are usually the cause of valve train noise. Although valve train noise may indicate worn rocker arms or rocker arm shafts, valve train noise may also be caused by incorrect valve clearances or other engine wear. Valve clearance adjustment is described in **ENGINE—SERVICE AND REPAIR**.

First check the rocker arm bushings for wear. If the lower part of the bushing is shiny and somewhat rough, the rocker arm is worn. A good rocker bushing surface will have a dull, smooth finish all around. Continue checking the rocker arms by installing them on a known good rocker arm shaft. There should be no noticeable play in any direction. The exhaust-side rocker arms are the most susceptible to wear owing to the design of the lubrication system. If the rocker arm or shafts show any visible signs of wear, it is highly recommended that these parts be replaced, especially on a high mileage engine.

Rocker arm wear
• radial play . . . . . . 0.016 to 0.052 mm (0.0006 to 0.0020 in.)

Valve Guides

Special tools and a press are required to replace valve guides. It is also necessary to heat the cylinder head and to chill the valve guides. Check valve guide wear with a new valve as shown in Fig. 4-13. Inspect the valve seats to ensure that the cylinder head can be reconditioned before installing new valve guides.

5

NOTE —
If valve guide wear is greater than 0.8 mm (0.031 in.), but less than 1.0 mm (0.039 in.), the valve guide may be reamed out to accept valves with oversized stems. If valves with oversized stems are installed, the valve seat must also be machined to accept the new valve.

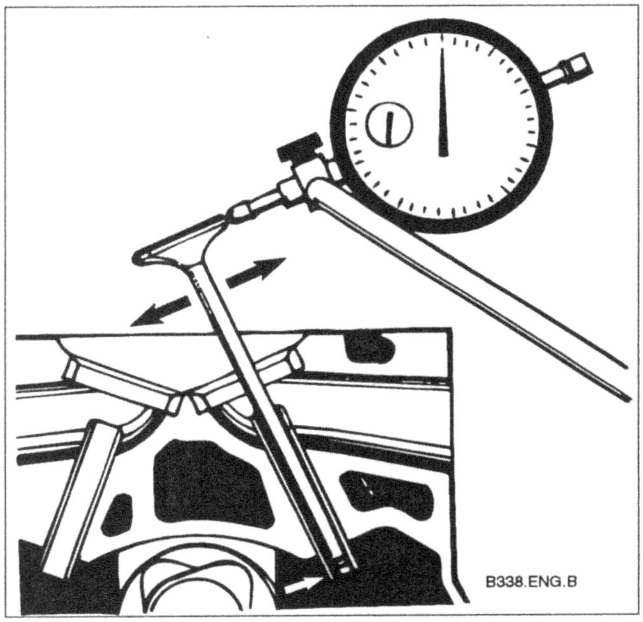

Fig. 4-13. Valve guide wear being checked with dial indicator. Insert new valve until stem end is flush with end of guide (white arrow).

Valve guide wear
• maximum play . . . . 0.8 mm (0.031 in.).

Worn valve guides are driven out from the camshaft side of the cylinder head. Remove valve guides at room temperature. Install new valve guides from the camshaft side of the cylinder head with the stepped end of the valve guide facing the camshaft. Valve guide specifications, including correct installation temperatures, are listed in **Table c**.

Table c. Valve Guide Specifications

Specifications	528e	533i and 535i
Valve guide wear, maximum (valve-tilt clearance measured with new valve)	0.8 mm (0.031 in.)	0.8 mm (0.031 in.)
Valve stem diameter		
standard	7.0 mm (0.275 in.)	8.0 mm (0.315 in.)
oversize 1	7.1 mm (0.279 in.)	8.1 mm (0.319 in.)
oversize 2	7.2 mm (0.283 in.)	8.2 mm (0.323 in.)
Valve guide inside diameter (tolerance per ISO allowance H7)		
standard	7.0 mm (0.275 in.)	8.0 mm (0.315 in.)
oversize 1	7.1 mm (0.279 in.)	8.1 mm (0.319 in.)
oversize 2	7.2 mm (0.283 in.)	8.2 mm (0.323 in.)
Valve guide outside diameter (tolerance per ISO allowance u6)		
standard		
old version	13.0 mm (0.5118 in.)	14.0 mm (0.5512 in.)
new version	13.2 mm (0.5197 in.)	NA
oversize 1		
old version	13.1 mm (0.5157 in.)	14.1 mm (0.5551 in.)
new version	13.3 mm (0.5236 in.)	NA
oversize 2		
old version	13.2 mm (0.5197 in.)	14.2 mm (0.5590 in.)
new version	13.4 mm (0.5276 in.)	NA
oversize 3	13.3 mm (0.5236 in.)	14.3 mm (0.5630 in.)
Valve guide bore diameter in cylinder head (tolerance per ISO allowance M7)		
standard		
old version	13.0 mm (0.5118 in.)	14.0 mm (0.5512 in.)
new version	13.2 mm (0.5197 in.)	NA

continued

Table c. Valve Guide Specifications (continued)

Specifications	528e	533i and 535i
Valve guide bore diameter in cylinder head (tolerance per ISO allowance M7) (cont'd)		
oversize 1		
old version	13.1 mm (0.5157 in.)	14.1 mm (0.5551 in.)
new version	13.3 mm (0.5236 in.)	NA
oversize 2		
old version	13.2 mm (0.5197 in.)	14.2 mm (0.5590 in.)
new version	13.4 mm (0.5276 in.)	NA
oversize 3	13.3 mm (0.5236 in.)	14.3 mm (0.5630 in.)
Valve guide installation temperature		
cylinder head	122°F (50°C)	122°F (50°C)
valve guide	–238°F (–150°C)	–238°F (–150°C)
Installed depth (height above cylinder head surface)	14.5 mm (0.5709 in.)	13.5 mm (0.5315 in.)
Special tools		
removal	BMW tool no. 11 1 330	BMW tool no. 11 1 100
installation	BMW tool no. 11 1 320	BMW tool no. 11 1 160

4.3 Valve Stem Oil Seals

The purpose of the valve stem oil seal is to prevent excess oil from entering the combustion chamber. The signs of faulty valve stem oil seals are 1) excessive oil consumption and 2) blue-gray smoke from the exhaust after starting and during sudden deceleration. For more information on excessive oil consumption and smoking, see **ENGINE—SERVICE AND REPAIR**.

Valve stem oil seal replacement requires that the cylinder head be fully disassembled as described above under **4.1 Disassembling Cylinder Head**. When installing the valve stem oil seals, the use of BMW tool no. 11 1 200 is highly recommended. Lubricate the new seals and install them using hand pressure only. Be sure to install the valve spring seat before installing the seal.

NOTE —
A BMW special tool (BMW tool no. 11 1 250) is available to remove the valve stem oil seals. As the seals will not be reused, it is also possible to carefully remove the valve stem seals with Vise-grips. Do not over tighten Vise-grips, as damage to the valve guide may occur.

Fig. 4-14. Valve stem oil seal being removed from valve guide using special removal tool.

Valve Seats

The valve seats should be resurfaced whenever new valves or valve guides are installed. Cutters or stones are required to resurface the seats. Always check the valve for leaks after reconditioning a valve seat as described below.

Table d lists valve seat dimensions. **Table e** lists valve seat replacement specifications, including correct installation temperatures. As with valve guides, replacing the valve seats requires heating the cylinder head and chilling the valve seat.

5

Table d. Valve Seat Dimensions

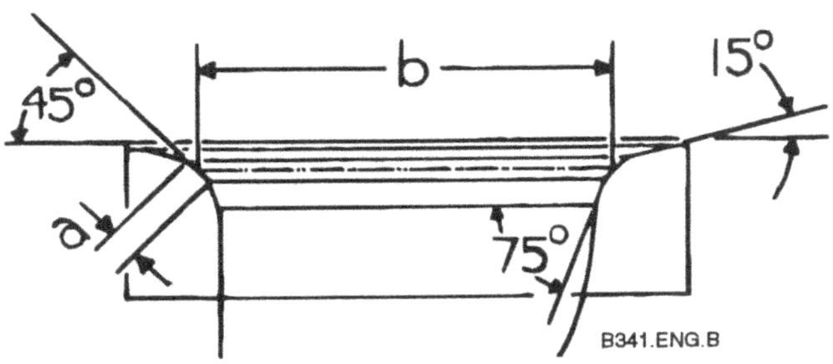

Specification	1982–1987 528e	1988 528e	1983–1988 533i, 535i
Valve seat width (**a**)			
intake	1.65±0.35 mm (0.065±0.014 in.)	1.65±0.35 mm (0.065±0.014 in.)	1.4±0.4 mm (0.055±0.016 in.)
exhaust	1.65±0.35 mm (0.065±0.014 in.)	1.65±0.35 mm (0.065±0.014 in.)	1.7±0.4 mm (0.067±0.016 in.)
Valve seat diameter (**b**)			
intake	38.6 mm (1.520 in.)	40.6 mm (1.598 in.)	44.6 mm (1.756 in.)
exhaust	32.6 mm (1.283 in.)	34.6 mm (1.362 in.)	36.6 mm (1.441 in.)

Table e. Valve Seat Replacement Specifications

Specification	1982–1987 528e	1988 528e	1983–1988 533i, 535i
Valve seat insert outside diameter (tolerance as per ISO allowance g6)			
intake			
standard			
old version	42.15 mm (1.6594 in.)	43.15 mm (1.6988 in.)	47.15 mm (1.8563 in.)
new version	NA	NA	48.15 mm (1.8957 in.)
oversize 0.2 mm			
old version	42.35 mm (1.6673 in.)	43.35 mm (1.7067 in.)	47.35 mm (1.8642 in.)
new version	NA	NA	48.35 mm (1.9035 in.)
oversize 0.4 mm			
old version	42.55 mm (1.6752 in.)	43.55 mm (1.7146 in.)	47.55 mm (1.8720 in.)
new version	NA	NA	48.55 mm (1.9114 in.)
exhaust			
standard	37.65 mm (1.4823 in.)	37.65 mm (1.4823 in.)	40.15 mm (1.5807 in.)
oversize 0.2 mm	37.85 mm (1.4902 in.)	37.85 mm (1.4902 in.)	40.35 mm (1.5886 in.)
oversize 0.4 mm	38.05 mm (1.4980 in.)	38.05 mm (1.4980 in.)	40.55 mm (1.5964 in.)
Valve seat bore diameter in cylinder head (tolerance as per ISO allowance H7)			
intake			
standard			
old version	42.00 mm (1.6535 in.)	43.00 mm (1.6929 in.)	47.00 mm (1.8504 in.)
new version	NA	NA	48.00 mm (1.8898 in.)
oversize 0.2 mm			
old version	42.20 mm (1.6614 in.)	43.20 mm (1.7008 in.)	47.20 mm (1.8583 in.)
new version	NA	NA	48.20 mm (1.8976 in.)
oversize 0.4 mm			
old version	42.40 mm (1.6693 in.)	43.40 mm (1.7086 in.)	47.40 mm (1.8661 in.)
new version	NA	NA	48.40 mm (1.9055 in.)
exhaust			
standard	37.50 mm (1.4764 in.)	37.50 mm (1.4764 in.)	40.00 mm (1.5748 in.)
oversize 0.2 mm	37.70 mm (1.4842 in.)	37.70 mm (1.4842 in.)	40.20 mm (1.5827 in.)
oversize 0.4 mm	37.90 mm (1.4921 in.)	37.90 mm (1.4921 in.)	40.40 mm (1.5905 in.)
Installation temperature			
cylinder head	122°F (50°C)	122°F (50°C)	122°F (50°C)
valve seat insert	–238°F (–150°C)	–238°F (–150°C)	–238°F (–150°C)

Valves

Valves should be machined using standard valve-grinding techniques. **Table f** lists valve specifications. Remove carbon from the valves using a wire brush or wire wheel.

Testing Valves for Leakage

The valves and their seats can be easily tested for leakage. With the camshaft and the rocker arm assemblies removed, install the valve assemblies and the spark plugs in each cylinder. Place the cylinder head on a workbench with the combustion chamber facing upward. Fill each combustion chamber with water. After fifteen minutes, check the level of the water. If the level of the water in any cylinder drops, that cylinder is not sealing properly.

5. CYLINDER BLOCK AND PISTONS

If the engine was removed as a complete assembly, remove the cylinder head as described in **ENGINE—SERVICE AND REPAIR**. Remove the clutch from the flywheel as described in **MANUAL TRANSMISSION AND CLUTCH**.

If machine shop services are not readily available, one alternative is to install a BMW remanufactured cylinder block (engine cylinder block complete with crankshaft, pistons and connecting rods installed, but without cylinder head or external components). A remanufactured cylinder block is available from an authorized BMW dealer parts department.

5.1 Disassembling Cylinder Block

Disassembly and assembly procedures for the cylinder blocks covered in this manual are similar to those for most other water-cooled, in-line engines. For anyone with the proper tools and equipment and basic experience in engine recondi-

Table f. Valve Specifications

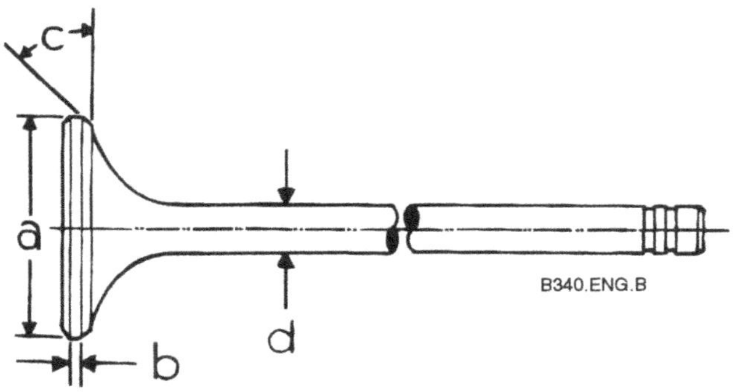

Specification	1982–1987 528e	1988 528e	1983–1988 533i,535i
Valve head diameter (**a**)			
intake	40 mm (1.575 in.)	42 mm (1.654 in.)	46 mm (1.811 in.)
exhaust	34 mm (1.339 in.)	36 mm (1.417 in.)	38 mm (1.496 in.)
Minimum valve head thickness (**b**)			
intake	1.3 mm (0.051 in.)	1.3 mm (0.051 in.)	1.3 mm (0.051 in.)
exhaust	2.0 mm (0.079 in.)	2.0 mm (0.079 in.)	2.0 mm (0.079 in.)
Valve face angle (**c**)	45°	45°	45°
Valve stem diameter (**d**)			
standard	7.0 mm (0.275 in.)	7.0 mm (0.275 in.)	8.0 mm (0.315 in.)
oversize 1	7.1 mm (0.279 in.)	7.1 mm (0.279 in.)	8.1 mm (0.319 in.)
oversize 2	7.2 mm (0.283 in.)	7.2 mm (0.283 in.)	8.2 mm (0.323 in.)

tioning, this section provides the specifications and special reconditioning information necessary to repair these BMW engines.

Thoroughly mark the position and orientation of all parts as they are removed. Connecting rods, rod caps, rod bearings, pistons, main bearing caps and main bearings are assembled in an exact location and orientation. This "assembly code" is crucial to proper engine operation. Use a punch or an electric engraving pencil to mark the parts as they are removed.

> ***CAUTION —***
> *On 528e models, the intermediate shaft bushing is not available from BMW. Do not remove the bushing as part of cylinder block disassembly.*

5.2 Cylinder Block and Piston Reconditioning

All of the disassembled parts should be thoroughly cleaned before inspection.

Connecting Rods and Pistons

If pistons, piston pins, piston rings, connecting rods, and bearings are to be reused, they should always be reinstalled in their exact positions. Parts should never be interchanged between cylinders.

Before removing the pistons, mark the cylinder numbers on connecting rods and connecting rod caps if they have not been previously marked at the factory. Mark each connecting rod's position on the crankshaft so that they are not reversed during installation. The matching numbers on the side of the connecting rod and rod cap should all be facing the same side of the engine (i.e., all marked numbers face exhaust side of engine). Mark the top of the pistons.

The piston pin should fit without any play. No specifications for the pin-to-rod fit are provided by BMW. The fit can be checked by pushing the pin through the rod by hand. There should be slight resistance as the pin moves through the rod. If there is any play at all, replace the connecting rod bushing or the connecting rod. If the piston pin is worn, the piston with pin will have to be replaced as an assembly. Piston pins are machined to match the piston. The piston pin circlips should be replaced with new circlips anytime they are removed.

When replacing the connecting rods, use only ones having the same weight class. The weight class is either stamped into the bottom of the connecting rod cap (alphanumeric code) or painted (colored dot) onto the side of the rod.

Each connecting rod should be matched to its cap with a stamped number code. The numbers are found on the side of the connecting rod and rod cap. When reinstalling the rods, be sure all matched numbers face the same side of the engine.

Table g lists connecting rod specifications. Connecting rods should be carefully checked for bend, twist and bearing bore out-of-roundness using connecting rod alignment tools.

Table g. Connecting Rod Specifications

Specification	528e	533i, 535i
Big end diameter (Bearing shells removed)		
standard no classification	NA	52.000–52.010 mm (2.0472–2.0476 in.)
double classification (red or blue paint dot on rod)		
red	48.000–48.008 mm (1.8898–1.8901 in.)	52.000–52.008 mm (2.0472–2.0476 in.)
blue	48.009–48.016 mm (1.8901–1.8904 in.)	52.009–52.016 mm (2.0476–2.0479 in.)
Connecting rod bushing		
outside diameter	24.060–24.100 mm (0.9472–0.9488 in.)	24.060–24.100 mm (0.9472–0.9488 in.)
inside diameter (nominal diameter 22.0 mm)	22.003–22.008 mm (0.8662–0.8664 in.)	22.003–22.008 mm (0.8662–0.8664 in.)
Maximum parallel deviation of connecting rod bores (bearing shells installed) at distance of 150 mm (5.905 in.)	0.04 mm (0.0016 in.)	0.04 mm (0.0016 in.)
Maximum deviation of weight between connecting rods (bearing shells removed)		
total	±4.0 grams (0.14 oz.)	±4.0 grams (0.14 oz.)
small end only	±2.0 grams (0.07 oz.)	±2.0 grams (0.07 oz.)
big end only	±2.0 grams (0.07 oz.)	±2.0 grams (0.07 oz.)
Connecting rod bolt torque	20 Nm (15 ft-lb) plus an additional 70°	52 to 57 Nm (38 to 42 ft-lb)

Piston specifications are listed in **Table h**. When replacing pistons, use only pistons having the same weight class. The piston weight class is stamped with a "+" or "–" on the piston top. The total weight difference between the pistons should not vary more than 10 grams (.35 oz.). The nominal piston diameter is also stamped on the crown. See Fig. 5-1. When installing the pistons, make sure the arrow on the piston top faces the front of the engine.

Table h. Piston Specifications

Specification	528e	533i	535i
Piston diameter (see Fig. 5-2)			
standard	83.98 mm (3.3063 in.)	88.97 mm (3.5027 in.)	91.97 mm (3.6209 in.)
special	84.06 mm (3.3094 in.)	89.05 mm (3.5059 in.)	92.05 mm (3.6240 in.)
oversize 1	84.23 mm (3.3161 in.)	89.22 mm (3.5126 in.)	92.22 mm (3.6307 in.)
oversize 2	84.48 mm (3.3260 in.)	89.47 mm (3.5224 in.)	92.47 mm (3.6405 in.)
Piston to cylinder clearance			
new	0.01–0.04 mm (.0004–.0016 in.)	0.02–0.05 mm (.0008–.0020 in.)	0.02–0.05 mm (.0008–.0020 in.)
wear limit	0.12 mm (0.0047 in.)	0.15 mm (0.0060 in.)	0.15 mm (0.0060 in.)
Weight difference between pistons (maximum permissible)	10 grams (.35 oz.)	10 grams (.35 oz.)	10 grams (.35 oz.)

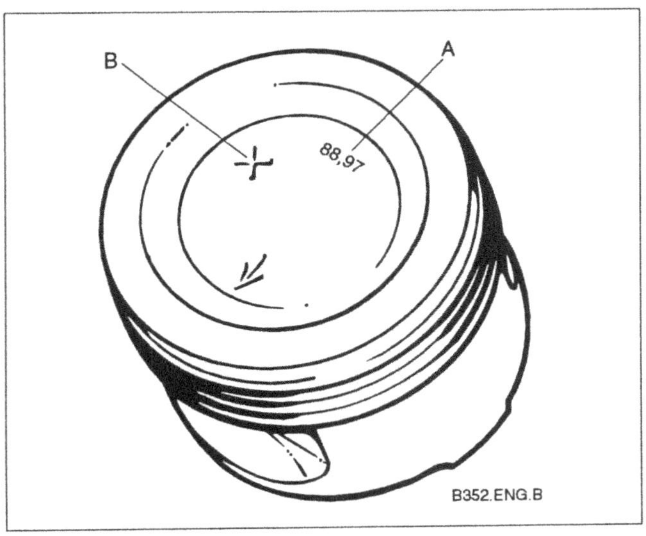

Fig. 5-1. Piston crown identification markings. Arrow should face front. Nominal piston diameter shown at **A**, and weight classification at **B**.

If the piston pin-to-piston clearance is excessive, the piston and the pin should be replaced as a set, as they are machined as a matched pair.

5

When measuring the piston diameter, make measurements at three places around the piston at the height indicated in Fig. 5-2, using the information listed in **Table i** or **Table j**.

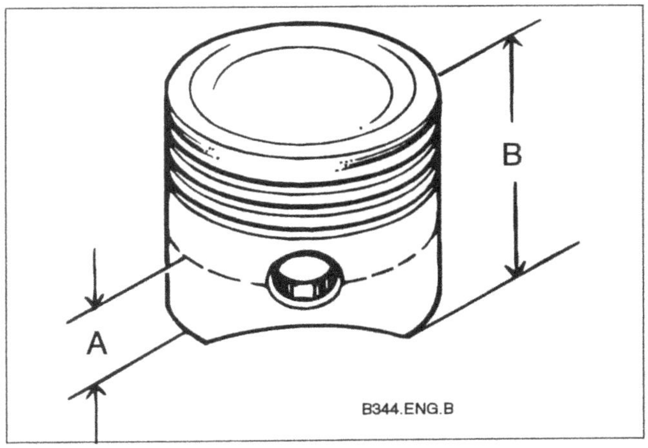

Fig. 5-2. Distance from bottom of piston skirt to measuring point (**A**) and total piston height (**B**) for determining piston diameter. Distance varies depending on model and piston manufacturer.

Table i. Piston Skirt Measuring Points (528e model)

Manufacturer	Total piston height (B)	Measurement A
Mahle	68.7 mm (2.705 in.)	8 mm (0.315 in.)
KS	68.7 mm (2.705 in.)	14 mm (0.551 in.)
Mahle or KS	77.7 mm (3.059 in.)	23 mm (0.905 in.)

Table j. Piston Skirt Measuring Points (533i, 535i models)

Manufacturer	Distance A
533i models	
Mahle	26.00 mm (1.024 in.)
KS	33.95 mm (1.337 in.)
535i models	
Mahle	14.00 mm (0.551 in.)

Cylinders

Measure cylinder bores at approximately the top, the middle, and the bottom of piston travel. Make all measurements at right angles (90°). Cylinder bore measuring points are shown in Fig. 5-3. Cylinder bore diameters are listed in **Table k**.

Table k. Cylinder Bore Specifications

Specification	528e	533i	535i
Cylinder bore diameter			
standard	$84.00^{+0.01}_{-0}$ mm ($3.3071^{+0.0004}_{-0}$ in.)	$89.00^{+0.01}_{-0}$ mm ($3.5039^{+0.0004}_{-0}$ in.)	$92.00^{+0.01}_{-0}$ mm ($3.6220^{+0.0004}_{-0}$ in.)
special	$84.08^{+0.01}_{-0}$ mm ($3.3102^{+0.0004}_{-0}$ in.)	$89.08^{+0.01}_{-0}$ mm ($3.5071^{+0.0004}_{-0}$ in.)	$92.08^{+0.01}_{-0}$ mm ($3.6252^{+0.0004}_{-0}$ in.)
oversize 1	$84.25^{+0.01}_{-0}$ mm ($3.3169^{+0.0004}_{-0}$ in.)	$89.25^{+0.01}_{-0}$ mm ($3.5138^{+0.0004}_{-0}$ in.)	$92.25^{+0.01}_{-0}$ mm ($3.6319^{+0.0004}_{-0}$ in.)
oversize 2	$84.50^{+0.01}_{-0}$ mm ($3.3268^{+0.0004}_{-0}$ in.)	$89.50^{+0.01}_{-0}$ mm ($3.5236^{+0.0004}_{-0}$ in.)	$92.50^{+0.01}_{-0}$ mm ($3.6417^{+0.0004}_{-0}$ in.)
Maximum out-of-round	0.03 mm (0.0012 in.)	0.01 mm (0.0004 in.)	0.01 mm (0.0004 in.)
Maximum conicity	0.02 mm (0.0008 in.)	0.01 mm (0.0004 in.)	0.01 mm (0.0004 in.)

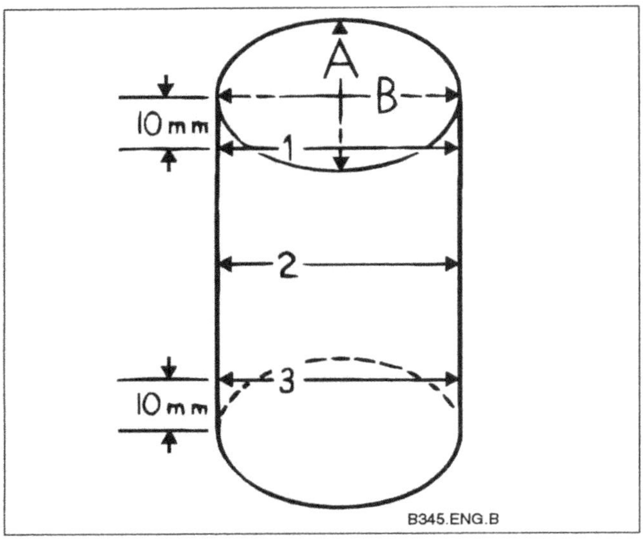

Fig. 5-3. Cylinder bore measuring points. Top (**1**) and bottom (**3**) measurements should be made at least 10 mm (3/8 in.) from ends of cylinder, first in direction **A** and then in direction **B**.

Fig. 5-4. Piston ring end gap being checked in cylinder block using feeler gauge.

Piston Rings

Piston ring end gaps are checked with the piston rings inserted approximately 15 mm (5/8 in.) from the top of the cylinder. See Fig. 5-4. The piston ring end gap specifications are listed in **Table l**. The gap should be checked after the final cylinder hone.

Table l. Piston Ring End Gap

Piston ring	528e	533i, 535i
Upper compression ring (top ring)	0.30–0.50 mm (0.012–0.020 in.)	0.30–0.50 mm (0.012–0.020 in.)
Lower compression ring (middle ring)	0.30–0.50 mm (0.012–0.020 in.)	0.20–0.40 mm (0.008–0.016 in.)
Oil ring (bottom ring)	0.25–0.50 mm (0.010–0.020 in.)	0.25–0.50 mm (0.010–0.020 in.)

Piston ring side clearance is checked using feeler gauges. Measure each ring in its original groove as shown in Fig. 5-5. Piston ring side clearance specifications are listed in **Table m**.

Fig. 5-5. Measuring piston ring side clearance with feeler gauge.

Table m. Piston Ring Side Clearance

Piston ring	528e	533i, 535i
Upper compression ring (top ring)	0.040–0.072 mm (0.0016–0.0028 in.)	0.050–0.082 mm (0.0020–0.0032 in.)
Lower compression ring (middle ring)	0.030–0.062 mm (0.0012–0.0024 in.)	0.040–0.072 mm (0.0016–0.0028 in.)
Oil ring (bottom ring)	0.020–0.042 mm (0.0008–0.0017 in.)	0.020–0.052 mm (0.0008–0.0020 in.)

Install the rings with the gaps offset from each other by 120°. Each ring should be installed so that the word "TOP" found on each ring faces up. Lightly coat the cylinders and the rings with engine oil before installation.

Crankshaft

To remove the crankshaft, both the rear crankshaft oil seal carrier and the front end cover must be removed from the ends of the engine cylinder block. Crankshaft main bearing caps must not be interchanged. Crankshaft main bearing shells, if they are to be reused, should only be installed in their original positions.

Crankshaft journal diameters are listed in **Table n**. Crankshaft clearance specifications are listed in **Table o**. Crankshaft bearings are available in standard sizes and undersizes according to a color coding system. A paint dot on the edge of the bearings, on the crankshaft, or on the base of the main bearing web of the cylinder block indicates the sizing classification of the crankshaft and bearings. The sizing colors available are red and blue for the "double classification," or yellow, green, and white for the "triple classification."

5

> ***CAUTION —***
> *On 528e models, do not interchange main bearing upper shells (in crankcase) with main bearing lower shells (in main bearing caps). The upper bearing shell must have two oil holes and a full oil groove (double or triple classification). The lower bearing shell can have either the type with the full oil groove and two oil holes (double classification) or the full-faced type with two crescent-shaped oil passages (triple classification). The full-faced bearing with crescent-shaped oil passages must never be used in the upper position.*

> **NOTE —**
> • On 528e models, double classification replacement bearings are no longer available from BMW. The triple classification bearings (yellow, green, and white) can be used in engines with double classification bearings. Replace red bearings with yellow bearings and blue bearings with green bearings.
>
> • On 533i and 535i models, the double classification replacement bearings (red and blue) are the only bearings available.

Crankshaft axial play (also called end play, float, or thrust) is controlled by a two-piece thrust bearing, which has raised edges to control the fore-and-aft movement of the crankshaft. The thrust bearing is the no. 6 main bearing. Axial play should be measured at the center main bearing journal with the crankshaft supported on rollers on the outer main bearing journals.

Table n. Crankshaft Journal Diameters

Specification	528e	533i, 535i
Main bearing journal diameter		
standard	(nominal dia. 60.00 mm)	(nominal dia. 60.00 mm)
red	59.980–59.990 mm (2.3614–2.3618 in.)	59.980–59.990 mm (2.3614–2.3618 in.)
blue	59.971–59.980 mm (2.3611–2.3614 in.)	59.971–59.980 mm (2.3611–2.3614 in.)
yellow	59.984–59.990 mm (2.3616–2.3618 in.)	59.984–59.990 mm (2.3616–2.3618 in.)
green	59.977–59.983 mm (2.3613–2.3615 in.)	59.977–59.983 mm (2.3613–2.3615 in.)
white	59.971–59.976 mm (2.3611–2.3613 in.)	59.971–59.976 mm (2.3611–2.3613 in.)
undersize 1	(nominal dia. 59.75 mm)	(nominal dia. 59.75 mm)
red	59.730–59.740 mm (2.3516–2.3520 in.)	59.730–59.740 mm (2.3516–2.3520 in.)
blue	59.721–59.730 mm (2.3513–2.3516 in.)	59.721–59.730 mm (2.3513–2.3516 in.)
yellow	59.734–59.740 mm (2.3517–2.3520 in.)	59.734–59.740 mm (2.3517–2.3520 in.)
green	59.727–59.733 mm (2.3515–2.3517 in.)	59.727–59.733 mm (2.3515–2.3517 in.)
white	59.721–59.726 mm (2.3512–2.3514 in.)	59.721–59.726 mm (2.3512–2.3514 in.)
undersize 2	(nominal dia. 59.50 mm)	(nominal dia. 54.50 mm)
red	59.480–59.490 mm (2.3417–2.3421 in.)	59.480–59.490 mm (2.3417–2.3421 in.)
blue	59.471–59.480 mm (2.3414–2.3417 in.)	59.471–59.480 mm (2.3414–2.3417 in.)
yellow	59.484–59.490 mm (2.3419–2.3421 in.)	59.484–59.490 mm (2.3419–2.3421 in.)
green	59.477–59.483 mm (2.3416–2.3418 in.)	59.477–59.483 mm (2.3416–2.3418 in.)
white	59.471–59.476 mm (2.3414–2.3416 in.)	59.471–59.476 mm (2.3414–2.3416 in.)
undersize 3	NA	(nominal dia. 59.25 mm)
red	NA	59.230–59.240 mm (2.3319–2.3323 in.)
blue	NA	59.221–59.230 mm (2.3316–2.3319 in.)
yellow	NA	59.234–59.240 mm (2.3320–2.3323 in.)
green	NA	59.227–59.233 mm (2.3318–2.3320 in.)
white	NA	59.221–59.226 mm (2.3315–2.3317 in.)

continued

Table n. Crankshaft Journal Diameters (continued)

Specification	528e	533i, 535i
Connecting rod journal diameter		
standard	(nominal dia. 45.00 mm) 44.975–44.991 mm (1.7707–1.7713 in.)	(nominal dia. 48.00 mm) 47.975–47.991 mm (1.8888–1.8894 in.)
undersize 1	(nominal dia. 44.75 mm) 44.725–44.741 mm (1.7608–1.7614 in.)	(nominal dia. 47.75 mm) 47.725–47.741 mm (1.8789–1.8795 in.)
undersize 2	(nominal dia. 44.50 mm) 44.475–44.491 mm (1.7510–1.7516 in.)	(nominal dia. 47.50 mm) 47.475–47.491 mm (1.8691–1.8697 in.)
Double classification		
standard	44.975–44.991 mm (1.7707–1.7713 in.)	47.975–47.991 mm (1.8888–1.8894 in.)
undersize 1	44.725–44.741 mm (1.7608–1.7615 in.)	47.725–47.741 mm (1.8789–1.8796 in.)
undersize 2	44.475–44.491 mm (1.7510–1.7516 in.)	47.475–47.491 mm (1.8691–1.8697 in.)
undersize 3	NA	47.225–47.241 mm (1.8592–1.8599 in.)
Crankshaft thrust bearing width		
standard	25.020–25.053 mm (0.9850–0.9863 in.)	30.025–30.064 mm (1.1821–1.1836 in.)
oversize 1	25.220–25.253 mm (0.9929–0.9942 in.)	30.225–30.264 mm (1.1900–1.1915 in.)
oversize 2	25.420–25.453 mm (1.0008–1.0021 in.)	30.425–30.464 mm (1.1978–1.1994 in.)
oversize 3	NA	30.625–30.664 mm (1.2057–1.2072 in.)

Table o. Crankshaft Clearances

Crankshaft main bearing radial clearance (Plastigage®)	
red or blue classification	0.030–0.070 mm (0.0012–0.0028 in.)
yellow, green, or white classification	0.020–0.046 mm (0.0008–0.0018 in.)
Connecting rod bearing radial clearance (Plastigage®)	
no classification	0.030–0.070 mm (0.0012–0.0028 in.)
double classification	0.020–0.055 mm (0.0008–0.0022 in.)

continued on next page

Table o. Crankshaft Clearances (continued)

Crankshaft axial clearance	
528e model	0.080–0.163 mm (0.0031–0.0064 in.)
533i, 535i models	0.085–0.174 mm (0.0033–0.0069 in.)
Maximum permissible crankshaft runout	
528e model	0.15 mm (0.006 in.)
533i, 535i model	0.10 mm (0.004 in.)

The crankshaft should only be replaced with a new or reconditioned BMW crankshaft. BMW reconditioned crankshafts come supplied with bearings. Although many machine shops are capable of regrinding crankshafts, BMW crankshafts are specially heat treated at the factory. BMW recommends that crankshafts be reconditioned only at the factory. See your authorized dealer parts department for the most up-to-date information regarding crankshafts. Crankshaft and cylinder block tightening torques are listed below.

> **NOTE —**
> When replacing the crankshaft on models with manual transmission, be sure to install a new transmission input shaft pilot bearing in the end of the crankshaft. Also check the bearing before installing the crankshaft. If the bearing does not roll smoothly, feels gritty, or has any play, it should be replaced. Pack the area behind the bearing with about one gram of grease before installing the new bearing.

Tightening torques

- rear crankshaft oil seal carrier bolts
  - M6 bolts . . . . 9±1 Nm (80±9 in-lb)
  - M8 bolts . . . . 22±2 Nm (16±1 ft-lb)
- front end cover mounting bolts
  - M6 bolts . . . . 9±1 Nm (80±9 in-lb)
  - M8 bolts . . . . 22±2 Nm (16±1 ft-lb)
- main bearing cap bolt . . . . 58 to 63 Nm (43 to 46 ft-lb)
- oil pan bolts . . . . 9 to 11 Nm (80 to 97 in-lb)

Flywheel and Drive Plate

Replacement of the flywheel or the drive plate are similar operations. When removing the mounting bolts, hold the ring gear stationary using a holding fixture. Flywheels and drive-plates should be inspected for excessive runout using a dial indicator. The maximum allowable runout is listed below in **Table p**. If the runout exceeds the maximum, the flywheel or drive plate can be machined, provided it remains within the specifications listed in the table. Inspect the flange height after machining the flywheel. If the flange height has been reduced to zero clearance during machining, the flange surface of the flywheel must be machined until some clearance exists. See Fig. 5-6. BMW does not list specifications for this flange height.

5

> **NOTE —**
> • On 528e models, two different types of flywheels are installed on the models covered by this manual. Models built up to May 1986 have a conventional plate-type flywheel. All models built after this date have a dual mass flywheel. The dual mass flywheel incorporates a spring damper system. This spring damper system replaces the spring system in the clutch disc. The dual mass flywheel is designed to reduce body and transmission noise. Dual mass flywheels can be serviced as described above. Separate replacement parts for the dual mass flywheel are not available.
>
> • Replace cracked flywheels.

Table p. Flywheel Specifications

Maximum axial runout (measured at outer diameter)	0.10 mm (0.004 in.)
Minimum flywheel thickness	
528e models	
1984	25.0 mm (.984 in.)
1985–1988	32.0 mm (1.260 in.)
533i,535i models 1984–1985	26.5 mm (1.043 in.)
Starter ring gear replacement temperature (manual transmission only)	395° to 445°F (200° to 230°C)

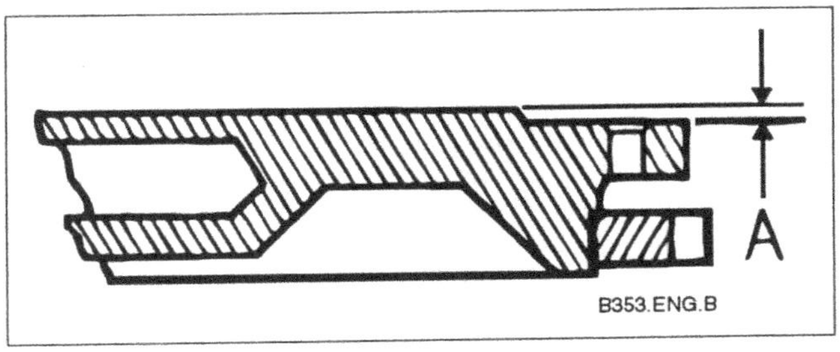

Fig. 5-6. Schematic front view of removed flywheel. Inspect flange height **A** after machining flywheel.

Installation of the flywheel and the drive plate is the reverse of removal. Be sure all bolt holes are clean and use new bolts.

CAUTION —
Flywheel and drive plate stretch-type mounting bolts must not be reused. If not already applied to the new bolts, install a thread locking compound (Loctite® 270 or equivalent) to the bolt threads. Tighten the bolts using the sequence shown in Fig. 5-7 below.

Tightening torque
• flywheel to crankshaft . . . . 105±7 Nm (77±5 ft-lb)

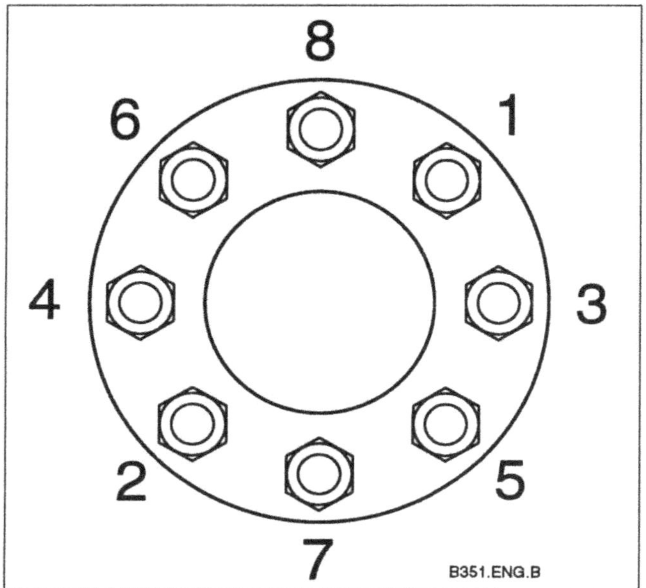

Fig. 5-7. Flywheel or drive plate mounting bolt tightening sequence.

If the starter ring gear (models with manual transmission only) is damaged, it may be replaced. To remove the old gear, drill a 6 mm (0.23 in.) hole about 8 mm deep (0.315 in.) at a tooth gap in the gear. Using a chisel, break the ring gear at the drilled point. To install the new gear, heat the ring gear to 395° to 445°F (200° to 230°C). Using a brass mandrel, drive the ring gear into place on the flywheel, with the beveled edge of the gear facing the engine.

6. CYLINDER BLOCK OIL SEALS

On all engines covered by this manual, the front cylinder block oil seal(s) can be replaced with the engine installed. Replacement of the rear oil seal requires that the engine and transmission be separated.

6.1 Replacing Cylinder Block Oil Seals

Front Crankshaft Oil Seals

On 533i and 535i models, the front crankshaft oil seal can be replaced without removing the timing chain cover or the timing chain. On the 528e model, it is necessary to remove the camshaft drive belt and the lower drive belt sprocket before replacing the crankshaft oil seal or the intermediate shaft oil seal. Camshaft drive belt removal is described in **ENGINE—SERVICE AND REPAIR.**

To replace front crankshaft and intermediate shaft oil seals (528e model):

Requirements:

- Camshaft drive belt removed. See **ENGINE—SERVICE AND REPAIR**

1. Remove the crankshaft sprocket, if applicable. See Fig. 6-1.

NOTE —
• On models with a two-piece crankshaft hub, the crankshaft sprocket is removed as part of the camshaft drive belt removal procedure.

• It is always good practice to replace the camshaft drive belt if it has been removed.

• The crankshaft sprocket and hub mounting bolt is tightened to a torque of approximately 410 Nm (302 ft-lb). A heavy duty holding device, such as BMW tool no. 11 2 150, should be used to hold the sprocket. Use a heavy duty 3/4-inch drive socket and breaker bar to break the bolt free.

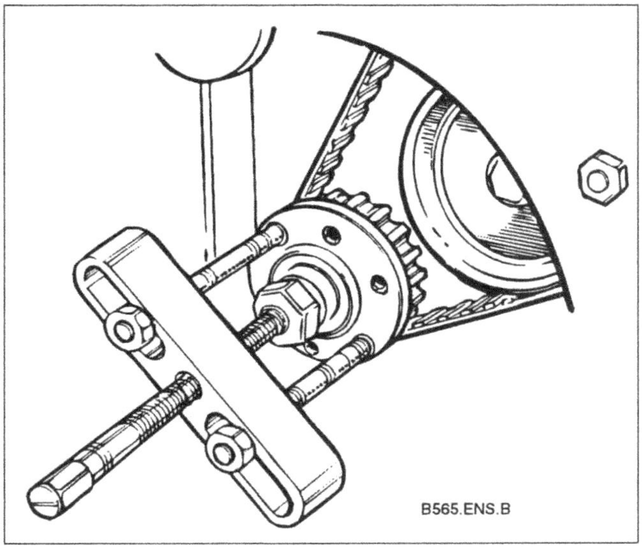

Fig. 6-1. Crankshaft sprocket being removed using puller.

2. While holding the intermediate shaft sprocket stationary with a holding device, loosen the bolt that secures the sprocket to the shaft. Remove the sprocket from the shaft using a suitable puller.

CAUTION —
Before removing the intermediate shaft sprocket mounting bolt, inspect for oil leakage at the bolt and on the sprocket. If signs of oil leakage are present, the center internal sealing plug in the intermediate shaft may be faulty or missing. A leaking intermediate shaft should be replaced. Always replace a drive belt that has been contaminated with oil.

NOTE —
• A special device for holding the intermediate shaft sprocket stationary (BMW tool no. 11 2 190) is available from an authorized BMW dealer parts department.

• The intermediate shaft can be replaced once the block front-end cover and the shaft guide plate are removed. Intermediate shaft bearings are not replaceable.

3. Working beneath the car, remove the three oil pan bolts that thread into the front end cover. Loosen but do not remove the remaining oil pan bolts. Remove the six bolts that secure the front end cover to the cylinder block. Pry the cover from the engine, being careful not to tear the oil pan gasket.

NOTE —
If the front end cover is stuck to the oil pan gasket, use a sharp knife to cut it free.

4. Press the oil seals out of the front end cover. Press the new seals into the cover until they are approximately 1 to 2 mm (0.04 to 0.08 in.) below the surface of the cover.

NOTE —
The old seals were originally installed flush to the cover surface. The new seals should be installed slightly deeper so that they do not ride on the same area of the shafts. This will allow for a good seal despite any wear.

5. Apply a light coating of silicone sealer to the portion of the oil pan gasket that mates to the front end cover. Apply a light coat of engine oil to the seals. Using a new front end cover gasket, install the cover.

CAUTION —
Do not use too much sealer as it can leak into the oil pan and plug small oil passages when circulated with the engine oil. This can lead to oil starvation and engine damage.

NOTE —
• To protect the oil seals during installation of the front end cover, BMW recommends using special tools (BMW tool no. 24 1 040 and 24 1 050) to cover the shafts. As an alternative, the shafts may be wrapped with plastic tape and coated with a small amount of engine oil.

• Permatex® 77B or Wurth® DP-300 sealers can be used to seal the cover to the gasket.

The remainder of the installation is reverse of removal. Install the camshaft drive belt as described in **ENGINE—SERVICE AND REPAIR.** Be sure the centering pin is installed in the intermediate shaft and the woodruff key is correctly installed in the crankshaft.

NOTE —
Most torque wrenches cannot handle the torque setting of the large crankshaft sprocket mounting bolt. Special arrangements should be made to tighten the bolt to the proper torque using a heavy-duty torque wrench.

Tightening torques	
• front oil seal cover	
M6 bolts	9 to 11 Nm (80 to 97 in-lb)
M8 bolts	22±2 Nm (16±1 ft-lb)
• oil pan mounting bolts	9 to 11 Nm (80 to 97 in-lb)
• intermediate shaft sprocket	60±5 Nm (44±4 ft-lb)
• vibration damper to crankshaft hub	
M8 bolts	23±1 Nm (17±1 ft-lb)
• crankshaft hub to crankshaft	410±20 Nm (302±15 ft-lb)

5

To replace front crankshaft oil seal (533i, 535i models):

Requirements:

- All engine V-belts removed. See **LUBRICATION AND MAINTENANCE**
- Vibration damper removed from crankshaft. See **ENGINE—SERVICE AND REPAIR**

1. Use an appropriate seal remover to pry the seal from the front cover. Note the installed depth of the seal before removing. See Fig. 6-2. As an alternative, use a screwdriver to carefully pry the seal from its housing.

CAUTION —
Be careful not to mar the surface of the housing when removing the seal.

NOTE —
The crankshaft sprocket and hub mounting bolt is tightened to a torque of approximately 440 Nm (325 ft-lb). A heavy duty holding device, such as BMW tool no. 11 2 100 (cars with manual transmission) or 11 2 120 (cars with automatic transmission), should be used to hold the flywheel. Use a heavy duty 3/4-inch drive socket and breaker bar to break the bolt free.

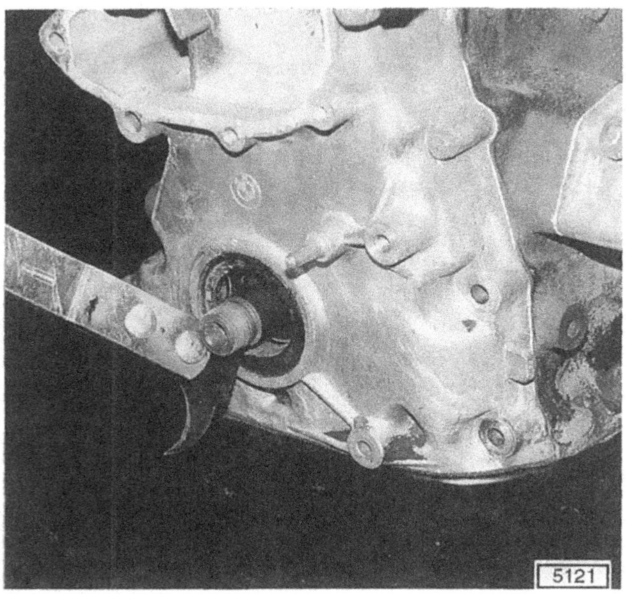

Fig. 6-2. Crankshaft oil seal being removed.

2. Lubricate the nose of the crankshaft and the inside of the oil seal with engine oil. Install the seal, driving it in until the seal is flush with the timing chain cover.

CAUTION —
- *To prevent the oil seal from being torn as it is placed over the crankshaft use tape around the threads on the front of the crankshaft.*

- *Do not attempt to hammer directly on the surface of the seal, as its rubber cover might be torn or the metal housing may be bent. A damaged oil seal should be replaced. Do not drive the seal in beyond its correct position. There is no shoulder in the housing.*

Installation of the remainder of the components is the reverse of removal. Make sure the woodruff key is correctly positioned in the crankshaft. Install all the V-belts as described in **LUBRICATION AND MAINTENANCE**.

Tightening torque	
• crankshaft pulley vibration damper	440±10 Nm (325±7 ft-lb)

NOTE —
Most torque wrenches cannot handle the torque setting of the large crankshaft sprocket mounting nut. Special arrangements should be made to tighten the bolt to the proper torque using a heavy-duty torque wrench.

Replacing Rear Crankshaft Oil Seal

Replacing the rear crankshaft oil seal requires that the engine be separated from the transmission and that the flywheel or drive plate be removed. These removal and installation procedures are described in **MANUAL TRANSMISSION AND CLUTCH** or **AUTOMATIC TRANSMISSION**.

Once the seal is exposed, remove the mounting bolts for the oil seal carrier as shown in Fig. 6-3. Loosen, but do not remove, the oil pan mounting bolts and remove the oil seal carrier. With the carrier removed, the old seal can be pressed out. Press the new seal in until it is 1 to 2 mm (0.04 to 0.08 in.) below the surface of the carrier surface.

NOTE —
- The old seals were originally installed flush to the cover surface. The new seals should be installed slightly deeper so that they do not ride on the same area of the shafts. This will allow for a good seal despite any wear.

- If the oil seal carrier is stuck to the oil pan gasket, free the cover from the gasket using a sharp knife.

When installing the oil seal carrier, apply a light coating of silicone sealer to the portion of the oil pan gasket that mates to the carrier. Apply a light coat of engine oil to the lip of the seal.

Fig. 6-3. Rear crankshaft oil seal carrier mounting bolts (arrows). Transmission, clutch assembly, and flywheel removed from engine.

Install a new gasket and the carriers. See Fig. 5-7 above for flywheel/driveplate bolt torque sequence.

> ***CAUTION —***
> *Do not use too much sealer as it can leak into the oil pan and plug small oil passages when circulated with the engine oil. This can lead to oil starvation and engine damage.*

Tightening torques

- crankshaft rear oil seal carrier
  - M6 bolts . . . . 9±1 Nm (80±9 in-lb)
  - M8 bolts . . . . 22±2 Nm (16±1 ft-lb)
- oil pan mounting bolts . . . . 9 to 11 Nm (80 to 97 in-lb)
- flywheel to crankshaft (new bolts with Loctite® 270) . . . . 105±7 Nm (77±5 ft-lb)

7. TECHNICAL DATA

I. Tightening Torques

Camshaft flange nut
- to camshaft (533i, 535i) . . . . 142±5 Nm (105±4 ft-lb)

Camshaft drive belt covers to engine (528e)
- M6 nuts or bolts . . . . 9–11 Nm (80–97 in-lb)
- M8 nuts or bolts . . . . 22±2 Nm (16±1 ft-lb)

Camshaft sprocket to camshaft (bolt)
- 528e . . . . 65–70 Nm (48–52 ft-lb)
- 533i, 535i . . . . 9 Nm (80 in-lb)

Cylinder head cover to
- cylinder head (nuts and bolt) . . . . 9±1 Nm (80±9 in-lb)

Connecting rod cap to connecting rod (nut or bolt)
- 528e
  - stage 1 . . . . 20 Nm (15 ft-lb)
  - stage 2 . . . . additional 70°
- 533i, 535i . . . . 52–57 Nm (38–42 ft-lb)

Crankshaft hub to crankshaft
- 528e (bolt) . . . . 410±20 Nm (302±15 ft-lb)
- 533i, 535i (nut) . . . . 440±10 Nm (325±7 ft-lb)

Cylinder head to cylinder block (bolts)
- 528e models
  - Torx head bolts
    - stage 1 . . . . 30 Nm (22 ft-lb)
    - stage 2 . . . . 90°
    - stage 3 . . . . 90°
  - Hex-head bolts
    - stage 1 . . . . 40+5 Nm (30+4 ft-lb)
    - wait 15 minutes
    - stage 2 . . . . 60+5 Nm (44+4 ft-lb)
    - run engine until fully warm (about 25 minutes)
    - stage 3 . . . . 25+5°
- 533i and 535i models
  - stage 1 . . . . 60±2 Nm (44±1 ft-lb)
  - wait 20 minutes
  - stage 2 . . . . 80±2 Nm (59±1 ft-lb)
  - run engine until fully warm (about 25 minutes)
  - stage 3 . . . . 35±5°

Engine mounts to subframe and body
- M10 fasteners . . . . 43–48 Nm (32–35 ft-lb)
- M8 fasteners . . . . 22–24 Nm (16–18 ft-lb)
- Flywheel to crankshaft (install new bolts with Loctite 270 or equivalent) . . . . 105±7 Nm (77±5 ft-lb)

Front engine cover with oil seal(s) to cylinder block (bolts)
- M6 . . . . 9±1 Nm (80±9 in-lb)
- M8 . . . . 22±2 Nm (16±1 ft-lb)

Ignition rotor adaptor to camshaft (bolt)
- 528e . . . . 65–70 Nm (48–52 ft-lb)
- 533i, 535i (install with Loctite 270 or equivalent) . . . . 23±1 Nm (17±0.5 ft-lb)

Intermediate shaft sprocket
- to intermediate shaft (bolt) . . . . 60±5 Nm (44±4 ft-lb)

Main bearing cap
- to cylinder block (bolt) . . . . 58–63 Nm (43–46 ft-lb)

Oil pan to cylinder block (bolts) . . . . 9–11 Nm (80–97 in-lb)

Oil pump to engine block (bolts . . . . 22±2 Nm (16±1 ft-lb)

Oil pump sprocket
- to oil pump (M10 nut) . . . . 25–30 Nm (18–22 ft-lb)

Oil sprayer bar to cylinder head
- 528e (flare fittings) . . . . 6–8 Nm (53–71 in-lb)
- 533i,535i (hollow banjo bolt) . . . . 11–13 Nm (97–115 in-lb)

Rear crankshaft oil seal carrier to cylinder block (bolts)
- M6 . . . . 9±1 Nm (80±9 in-lb)
- M8 . . . . 22±2 Nm (16±1 ft-lb)

Spark plugs to cylinder head . . . . 20–30 Nm (15–22 ft-lb)

Vibration damper to crankshaft hub . . . . 22±2 Nm (16±1 ft-lb)

Section 6

FUEL SUPPLY

6

Contents

TABLES

INTRODUCTION

The fuel supply system has two functions: 1) it stores the fuel, and 2) it delivers the fuel to the fuel injection system. This section covers repair information specifically for the fuel pumps, tank, filter, and fuel lines. Information on the fuel injection system and emission controls are covered in **FUEL INJECTION.** General fuel system troubleshooting to be used as an aid in diagnosing problems can be found in **ENGINE MANAGEMENT—DRIVEABILITY.**

1. GENERAL DESCRIPTION

Each of the cars covered by this manual is equipped with a fuel tank, two fuel pumps, a fuel level transmitter, a fuel filter, and connecting lines.

2. MAINTENANCE

BMW specifies the procedures below to be carried out at particular time or mileage intervals for proper maintenance of the fuel system. Information on these steps and the prescribed maintenance intervals can be found in **LUBRICATION AND MAINTENANCE.**

1. Inspecting fuel lines, tank, filler cap and filter for tight fit and leaks
2. Replacing the fuel filter

3. TROUBLESHOOTING

Poor driveability or faulty running may have a variety of causes. The fault may lie with the fuel system, the ignition system, parts of the emission control system, or a combination of the three. Because of the interrelated functions of these systems it is often difficult to know where to begin looking for problems. For this reason, effective troubleshooting must always consider these systems as one major system.

This section of the manual covers those tests and repairs that apply specifically to the fuel supply system.

Please observe the following cautions and warnings when performing any service or repair on the fuel system.

WARNING —
- *The fuel system is designed to retain pressure even when the ignition is off. When working with the fuel system, loosen the fuel lines very slowly to allow residual fuel pressure to dissipate gradually. Avoid spraying fuel.*
- *Before beginning any work on the fuel system, place a fire extinguisher in the vicinity of the work area.*
- *Fuel is highly flammable. When working around fuel, do not disconnect any wires that could cause electrical sparks. Do not smoke or work near heaters or other fire hazards.*
- *Always unscrew the fuel tank cap to release pressure in the tank before working on the tank or lines.*
- *Do not use a work light near any fuel. Fuel may spray onto the hot bulb causing a fire.*
- *Make sure the work area is properly ventilated.*

CAUTION —
- *Before making any electrical tests with the ignition turned on, disable the ignition system as described in **IGNITION**. Be sure the battery is disconnected when replacing components.*
- *To prevent damage to the ignition system or the electronic fuel system components, including the control unit, always connect and disconnect wires and test equipment with the ignition off.*
- *Absolute cleanliness is essential when working with the fuel system. Even a minute particle of dirt can cause trouble if it reaches an injector. Thoroughly clean the fuel line unions before disconnecting any of the lines.*
- *Use only clean tools. Keep removed components clean and sealed or covered with a clean, lint-free cloth, especially if completion of the repair is delayed.*
- *When replacing parts, install only new, clean components.*
- *Avoid nearby use of compressed air, and do not move the car while the fuel system is open.*
- *Avoid using high pressure compressed air to blow out lines and components. High pressure can rupture internal seals and gaskets.*
- *Always replace seals and O-rings.*

4. FUEL TANK AND LINES

The fuel tank is located at the rear of the car beneath the luggage compartment. The tank capacity is 63 liters (16.6 gallons). Mounted with the fuel tank are the fuel gauge sending unit, the fuel transfer pump, connecting lines for the evaporative emission control system and expansion tank, and a fuel supply and return line. Some later models have a fuel cooler in the return fuel line.

Removing and Installing Fuel Tank

The fuel tank must be completely emptied into a storage unit or another car's fuel tank before removing it from the car. Use an approved transfer pump and container. Be sure the container is large enough to hold the volume of fuel that is to be pumped out of the tank. The fuel can be pumped out via a hose inserted in the filler neck. As an alternative, the tank can be drained through the drain plug in the bottom of the tank.

> ***WARNING —***
> - *Always have a fire extinguisher handy when working on the fuel system.*
>
> - *Before removing the tank, be sure that all hot components such as the exhaust system are completely cooled down.*

To remove fuel tank:

1. Disconnect the negative (–) battery cable. Remove the fuel tank filler cap.

   > ***CAUTION —***
   > *BMW anti-theft radios can be rendered useless by disconnecting the battery. See your owner's manual for more information.*

2. Empty the fuel tank using an approved method as described above.

3. Working in the luggage compartment, remove the floor carpeting. Then remove the three screws for the round access cover above the tank and pry the cover off.

4. Remove the two hoses and two harness connectors from the top of the tank transfer pump/sending unit assembly. See Fig. 4-1.

5. Working from within the filler door, remove the fuel tank cap and pry out the rubber sealing gasket, which is over the filler neck. Then remove the three expansion tank vent hoses.

6. Disconnect the two rubber exhaust hangers.

Fig. 4-1. Fuel tank transfer pump/sending unit assembly.

> **NOTE —**
> Inspect the two rubber exhaust hangers and replace them if they are hard or cracked.

7. Remove the exhaust system rear support bolt. Slowly lower the rear of the exhaust system and secure with wire.

   > ***CAUTION —***
   > *Do not let the exhaust system hang without support. The pipes or catalytic converter may be damaged.*

8. While supporting the tank, remove the fuel tank mounting bolts and strap around the perimeter of the tank. Carefully lower the tank from the car.

Installation is the reverse of removal. Always use new seals, gaskets, O-rings, and hose clamps. Inspect all rubber parts including hoses and replace any that are worn, chafed, or cracked. Inspect heat shields and replace if fatigued, cracked, brittle, or corroded.

Tightening torques	
• fuel tank to body (bolts)	20 to 22 Nm (15 to 16 ft-lb)
• heat shield to fuel tank (bolts)	9 to 10 Nm (80 to 89 in-lb)
• fuel tank drain plug (with new seal ring)	21 to 25 Nm (15 to 18 ft-lb)
• hose clamps 10mm to 16mm diameter	2.0±0.3 Nm (18±3 in-lb)

Fuel Level Sending Unit

The fuel level sending unit is mounted in the fuel tank as part of the transfer pump assembly. The unit has a float connected to a variable resistance for fuel level, and a switch that closes to turn on the low fuel light. The unit cannot be repaired. When replacing the sending unit always replace hose clamps, gaskets and O-rings.

> ***WARNING —***
> *When removing the fuel pump, the transfer pump or the fuel gauge sending unit, the fuel tank must be drained. If the tank is not drained, fuel may be spilled.*

To remove and install level sending unit:

1. Disconnect the negative (–) battery cable. Drain the fuel from the tank as described above in **Removing and Installing Fuel Tank**.

> ***CAUTION —***
> *BMW anti-theft radios can be rendered useless by disconnecting the battery. See your owner's manual for more information.*

2. While working in the trunk, remove the access cover from above the fuel tank transfer pump/sending unit assembly.

3. Remove the harness connector from the sending unit and remove the four sending unit mounting bolts. See Fig. 4-2.

> ***WARNING —***
> *Always unscrew the fuel tank cap to release pressure in the tank before working on the tank or lines.*

4. Slowly withdraw the sending unit from the tank, allowing fuel to drain off. Discard the O-ring.

> ***WARNING —***
> *Fuel may be spilled. Do not disconnect any wires that could cause electrical sparks. Do not smoke or work near heaters or other fire hazards.*

5. Install the fuel gauge sending unit with a new O-ring and tighten the mounting nuts.

> **NOTE —**
> New fuel level sending units have a packing guard for protection during shipping. This guard must be removed prior to installing the unit.

Fig. 4-2. Fuel level sending unit mounting nuts (arrows).

Tightening torques	
• fuel gauge sending unit nuts	1.8 to 2.0 Nm (16 to 18 in-lb)
• transfer pump mounting bolts	5.9 Nm (52 in-lb)

6. Reconnect the harness connector. Fill the tank and check for leaks. Install the access cover and its mounting screws. Reconnect the negative (–) battery cable.

Test the removed sending unit using an ohmmeter connected across the terminals shown in Fig. 4-3. See **Table a** for test results.

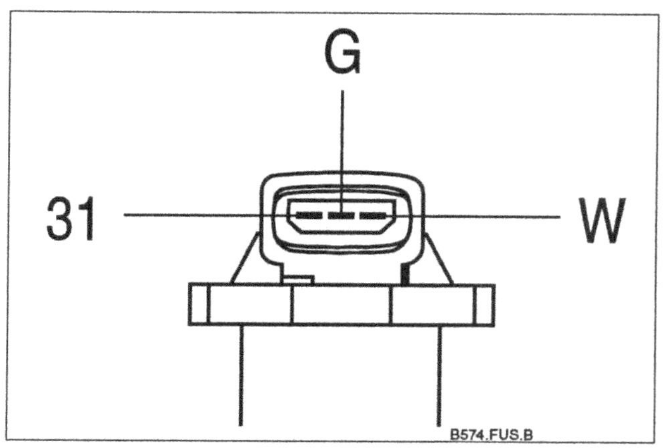

Fig. 4-3. Fuel gauge sending unit terminal identification.

CAUTION —
Always install a new O-ring any time the sending unit is removed.

Table a. Fuel Gauge Sending Unit Specifications

Connector terminals	Float position	Test result (ohms)
G and 31	slowly moving from EMPTY to FULL	resistance continuously changing (no breaks)
	EMPTY	71.7±2.3 ohms
	FULL	3.2±0.7 ohms
W and 31	EMPTY (low fuel warning)	continuity

5. FUEL PUMPS

All models covered by this manual have two electric fuel pumps. A transfer pump, mounted in the fuel tank, supplies fuel at low pressure to the main fuel pump. This is to help reduce fuel starvation and vapor lock. The main fuel pump is mounted beneath the car near the passenger's side rear wheel and supplies fuel at high pressure to the fuel injection system.

During starting, the pumps run as long as the ignition switch is in the start position and continue to run once the engine starts. A failure of the transfer pump will not necessarily prevent the engine from running, but may cause starting problems or fuel starvation at the main pump, particularly in summer weather. If an electrical system fault interrupts power to the main fuel pump, however, the engine will not run at all.

Fuel pressure and rate of fuel delivery are the important fuel-pump performance measurements. Transfer pump tests are covered in this section. See **FUEL INJECTION** for information on testing the main fuel pump and system fuel pressures.

5.1 Fuses and Relays for Fuel Pumps

Both the transfer pump and the main fuel pump are electric, and operated by a fuel pump relay located on the main fuse panel. The fuel pumps are protected by fuse no. 1 (16 amp) located on the main fuse panel. An additional main relay supplies power to the Motronic control unit and the fuel pump relay.

If both pumps are not running, the cause may be a faulty fuse or a faulty relay. In cold weather, water in the fuel may freeze in the pump, causing the circuit to overload and the fuse to fail. Troubleshooting of any fuel pump fault should begin with checking the fuel pump fuse and the fuel pump relay. The main relay should also be checked.

To quickly check if the fuel pump is coming on, actuate the starter while an assistant listens for the running pumps. To accurately hear the transfer pump it may be necessary to remove the access cover as described below in **5.2 Transfer Pump**.

Operating Fuel Pumps for Tests

The fuel pump relay is actuated (closed) via a ground signal from the Motronic control unit. The fuel pumps operate for a few seconds when the ignition is first turned on, and continue to operate only if the engine is running. To operate the pumps longer for fuel system tests without the engine running, the relay can be bypassed to power the pumps directly.

6

One method is to remove the fuel pump relay and connect two sockets on the fuse/relay panel with a fused jumper wire. The sockets used are those corresponding to relay terminals 30 and 87. Fig. 5-1 shows the location of the fuel pump relay.

CAUTION —
- *Fuse and relay locations are subject to change and may vary due to production line changes. Use care when troubleshooting the electrical system at the fuse/relay panel. To resolve problems in identifying a relay, see an authorized BMW dealer.*
- *Do not jumper or ground any other sockets or wires. Damage may result to the control units.*

NOTE —
The jumper wire should be 1.5 mm (14 ga.) and include an in-line fuse holder with a 16 amp fuse. To avoid fuse/relay panel damage from repeated connecting and disconnecting, also include a toggle switch. A heavy-duty jumper, BMW tool no. 61 3 050, is also available from an authorized BMW dealer.

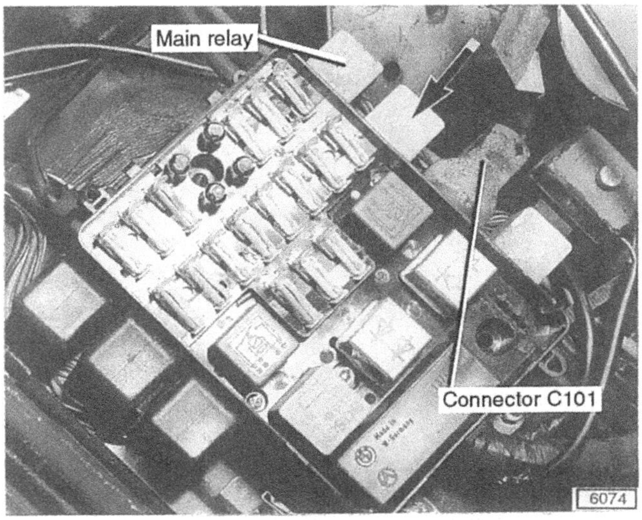

Fig. 5-1. Typical fuel pump relay location (arrow) on fuse/relay panel.

With the ignition off and the fuel pump relay removed, use the jumper (be sure toggle switch is off) to bridge the sockets (terminals 30 and 87) in the relay panel. Turn on the toggle switch to operate the fuel pumps.

NOTE —
On 1982 528e models only, the ignition switch must also be on.

If the pumps do not run with the jumper in place of the relay, the fault could be in the fuel pump or the electrical wiring to the pump. Make pump electrical tests as described below in **5.2 Transfer Pump** and **5.3 Main Fuel Pump**. For fuel pump circuit electrical diagrams, see **FUEL INJECTION**. After completing the tests, remove the jumper wire and turn off the ignition (where applicable).

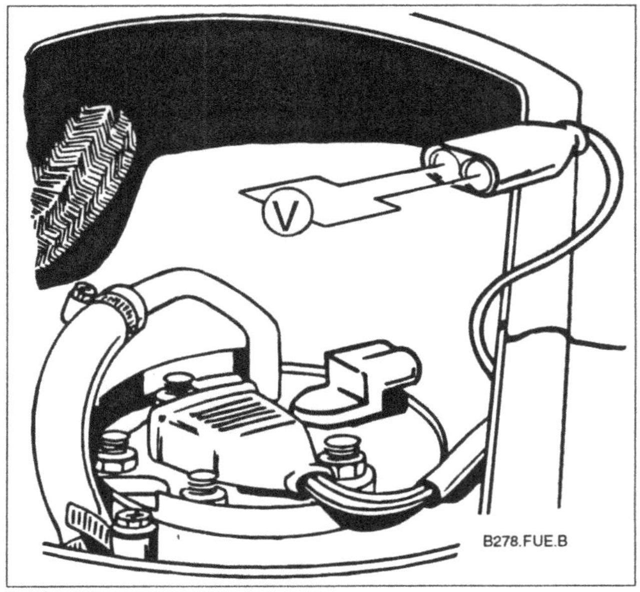

Fig. 5-2. Voltage supply to transfer pump being checked at harness connector (shown schematically).

5.2 Transfer Pump

The pump is cooled and lubricated by fuel and may therefore be damaged if allowed to run dry or if the fuel pickup strainer/filter becomes blocked. Access to the transfer pump is from the top of the fuel tank through the luggage compartment.

WARNING —
When removing the transfer pump or the fuel gauge sending unit, the fuel tank must be drained. If the tank is not drained, fuel may be spilled.

To test transfer pump:

1. Working from the luggage compartment, remove the floor carpeting. Then remove the three access cover mounting screws.

2. Jump power to the transfer pump to run it as described in **5.1 Fuses and Relays for Fuel Pumps**. Both the transfer pump and the main pump should operate. Disconnect the jumper wire when finished.

   NOTE —
   If the transfer pump does not run, disconnect the transfer pump harness connector. With the jumper wire connected as described in step 2 above, check for voltage across the connector terminals. See Fig. 5-2. If there is voltage, the transfer pump is probably faulty and should be replaced. If there is no voltage, trace the wiring and check for breaks or shorts to ground. For fuel pump circuit wiring diagrams, see **FUEL INJECTION.**

3. Check the transfer pump fuel delivery pressure. Remove the fuel tank filler cap. Remove the transfer pump output hose and connect a fuel pressure gauge to the hose fitting. See Fig. 5-3.

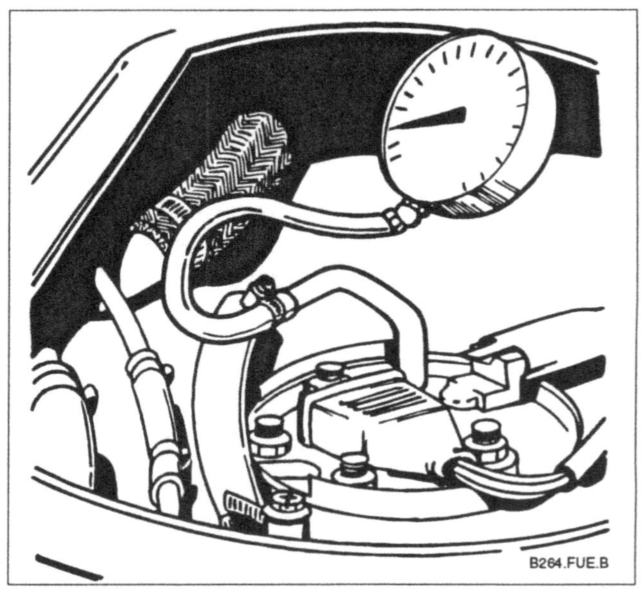

Fig. 5-3. Transfer pump fuel pressure being measured with fuel pressure gauge.

NOTE —
Make sure all electrical connectors are connected.

4. Operate the fuel pump as described under **5.1 Fuses and Relays for Fuel Pumps.**

Transfer pump pressure
• at 68°F (20°C) . . . . 0.3 bar (4.3 psi)

5. If the pressure is low, check the pump's power consumption with the pump running. Connect an ammeter as shown in Fig. 5-4.

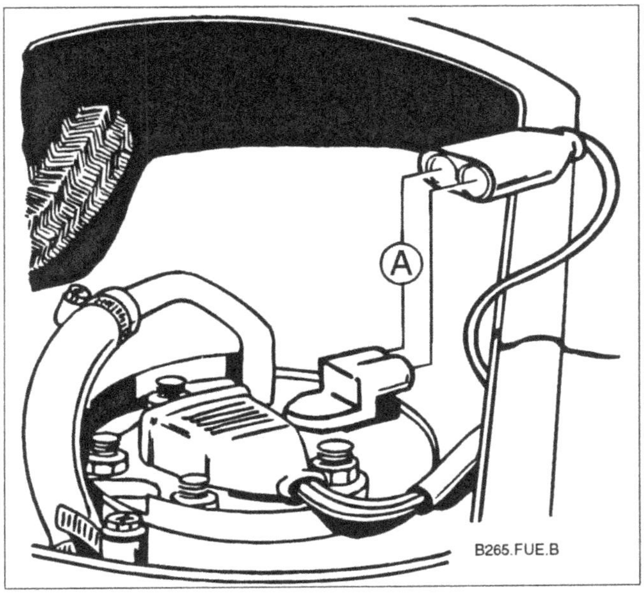

Fig. 5-4. Transfer pump current consumption being tested. Connect ammeter between one connector terminal and its corresponding terminal on pump. Jumper other terminal using a length of insulated wire.

Transfer pump current
• current draw at 13 volts. . . . . 1.4 amps or less.

NOTE —
To achieve accurate test results, the battery voltage at the connector must be 13 volts. Charge the battery if necessary. If the battery voltage is too high, turn on the headlights until the correct voltage is obtained.

6. If the current is too high, remove the transfer pump as described below and inspect the pickup strainer/filter for blockage. Recheck fuel pressure and current draw. If test results are still incorrect, the transfer pump is faulty and should be replaced.

To remove and install transfer pump:

1. Disconnect the negative (–) battery cable. Drain the fuel from the tank as described in **4. Fuel Tank and Lines**.

CAUTION —
BMW anti-theft radios can be rendered useless by disconnecting the battery. See your owner's manual for more information.

2. Remove the transfer pump/sending unit assembly access cover and the harness connectors as described earlier.

3. Remove and plug the transfer pump/sending unit assembly fuel hoses and discard the hose clamps.

WARNING —
- *Fuel will be spilled. Do not disconnect any wires that could cause electrical sparks. Do not smoke or work near heaters or other fire hazards. Have an approved fire extinguisher handy.*
- *Always remove the fuel tank cap to release any pressure in the tank before working on the tank or lines.*

4. Remove the four mounting nuts from the top of the fuel gauge sending unit. Slowly remove the sending unit from the tank, allowing fuel to drain off. Discard the O-ring.

5. Remove the bolts securing the transfer pump assembly. See Fig. 5-5. Remove the assembly from the fuel tank. Discard the O-ring.

Fig. 5-5. Fuel tank transfer pump mounting bolts (arrows). On some models there may be six bolts.

6. Inspect and, if necessary, clean the strainer/filter. See Fig. 5-6.

7. Lubricate the new O-ring with a small amount of fuel, then place the assembly with the O-ring into the tank and tighten the bolts evenly. Install the fuel gauge sending unit with a new O-ring and tighten the mounting nuts.

6

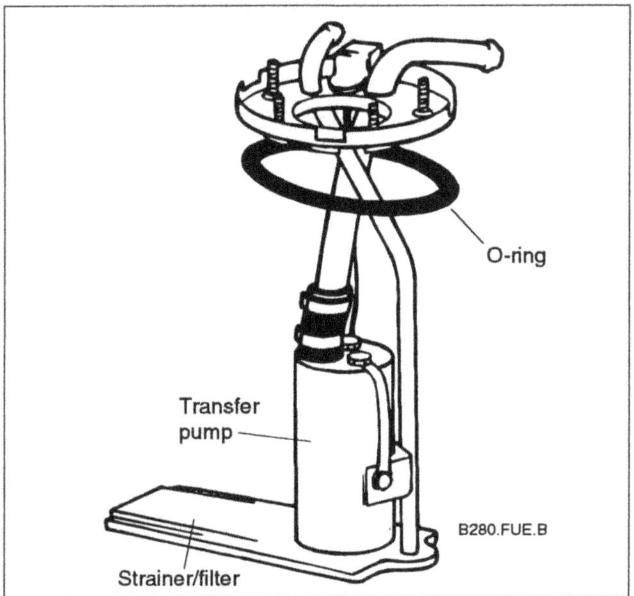

Fig. 5-6. Transfer pump, shown with fuel gauge sending unit removed from assembly.

Tightening torques
• fuel gauge sending unit to transfer pump assembly (nuts). . . . . . .1.8 to 2.0 Nm (16 to 18 in-lb)
• transfer pump assembly to fuel tank (bolts) . . . . 5.9 Nm (52 in-lb)

8. Using new hose clamps, connect the fuel hoses. Reconnect the harness connectors. Reinstall the filler cap. Reconnect the negative (–) battery cable. Run the fuel pump as described under **5.1 Fuses and Relays for Fuel Pumps** and check for leaks. Install the access cover and its mounting screws.

5.3 Main Fuel Pump

The main fuel pump delivers fuel at high pressure to the fuel injection system. A pressure regulator maintains system pressure. The quantity of fuel supplied exceeds demand, so excess fuel returns to the fuel tank via a return line. This constant recirculation of fuel in the system keeps the fuel cooler and helps prevent fuel starvation due to vapor lock (air bubbles created by fuel vaporizing in the lines). See **FUEL INJECTION** for more information on main fuel pressure and the pressure regulator.

Main Fuel Pump Electrical Tests

These tests are for use in finding out why the main fuel pump does not run or why the fuel delivery rate is below specifications. The tests check the fuel pump relay, power to the fuel pump, and fuel pump current draw.

The fuel pump relay is actually a part of two separate circuits. One circuit powers the fuel pump, as the relay acts as a switch to turn the pump on and off. The other circuit is the one that energizes the relay with power.

The first step in troubleshooting the fuel pump circuit is to check the fuel pump fuse no. 1 (16 amp). Replace a failed fuse and test the fuel pump again. If the fuel pump fuse is good, test the fuel pump circuit as described below.

> ***CAUTION —***
> *Fuse and relay locations are subject to change and may vary due to production line changes. Use care when troubleshooting the electrical system at the fuse/relay panel. To resolve problems in identifying a relay, see an authorized BMW dealer.*

To test main fuel pump circuit:

1. Remove the fuel pump relay. See Fig. 5-1 above. Identify the relay terminal numbers on the bottom of the relay.

> ***WARNING —***
> *The battery negative (–) cable must be disconnected when disconnecting or connecting electrical connections. Sparks may ignite spilled fuel. Keep an approved fire extinguisher handy.*

> ***CAUTION —***
> *The fuel pump relay may be located in an adjacent position. To check if the fuel pump relay has been correctly identified, check the wire colors leading to the fuel pump relay socket. See **FUEL INJECTION** for fuel pump circuit wiring diagrams and wire colors.*

2. Use a test light or voltmeter to check for voltage between the socket corresponding to relay terminal 30(+) and ground. There should be approximately battery voltage.

> **NOTE —**
> On 1982 528e models the key must be in the run or start position for voltage to be present between terminal 30 and ground. On these models only, the wire color at terminal 30 is green/yellow.

If voltage is not present between socket 30 and ground, power is not reaching the fuel pump relay from the battery. Check the red wire between the battery and the relay socket.

If no faults are found and the fuel pump will run only with the jumper connected, as described in **5.1 Fuses and Relays for Fuel Pumps**, then the relay itself is probably faulty.

If there is power to the fuse/relay panel but the pump still will not run with the jumper connected, check for voltage at the fuel pump. Pull back the two rubber boots that protect the fuel pump electrical connectors.

With the relay removed and the jumper wire connected, check for voltage between the fuel pump connectors as shown in Fig. 5-7. If there is voltage and the connectors are not loose or corroded, then the fuel pump is faulty. If there is no voltage, look for a faulty wire or connection between the fuse/relay panel and the fuel pump connector. Also check the fuel pump ground.

> ***WARNING —***
> *The battery negative (–) cable should be disconnected when disconnecting or connecting electrical connections. Sparks may ignite spilled fuel. Keep an approved fire extinguisher handy.*

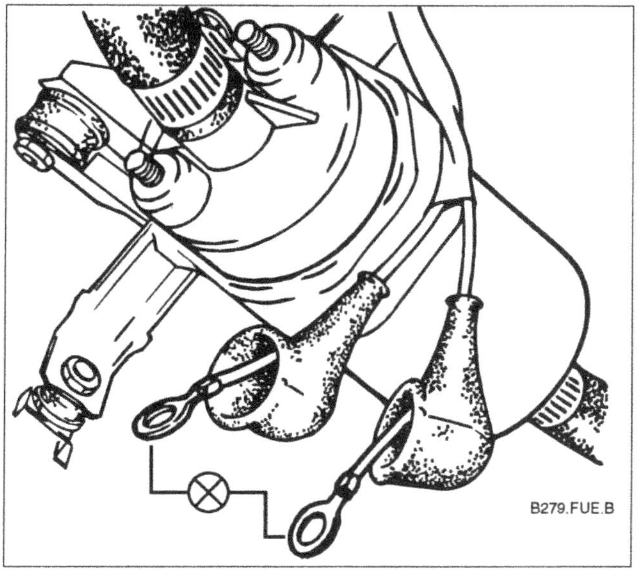

Fig. 5-7. Voltage supply to fuel pump being checked with test light (shown schematically).

To test main fuel pump current draw:

1. Pull back the two protective rubber boots from the pump's connectors. Remove the nut and disconnect the green/violet wire from the fuel pump.

2. Connect an ammeter between the disconnected wire and the fuel pump terminal. The other wire remains connected to the pump. See Fig. 5-8.

> ***CAUTION —***
> *Do not allow the test leads to short to ground.*

3. Use a jumper wire to bypass the fuel pump relay as described in **5.1 Fuses and Relays for Fuel Pumps**.

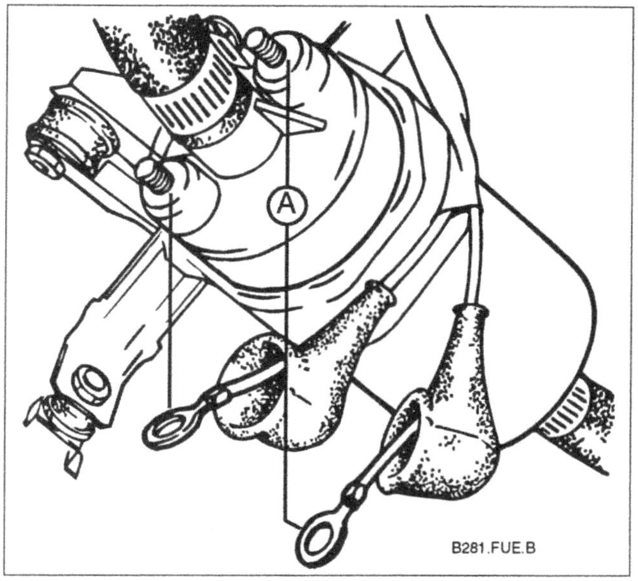

Fig. 5-8. Schematic view of ammeter connected between fuel pump connector and fuel pump.

4. Compare the ammeter reading with the specification listed in **Table b**.

Table b. Main Fuel Pump Current Specifications

Fuel pump code number	Maximum current consumption
0 580 464 013 (up to 1986 models)	6.5 amps
0 580 464 032 (1986 and later models)	5.0 amps

A higher than normal power consumption usually indicates a worn fuel pump, which may cause intermittent fuel starvation due to pump overheating and seizure. The only remedy is pump replacement.

Replacing Main Fuel Pump

To replace the fuel pump, a large container for catching fuel and a clamping device to pinch off the hoses on either side of the pump are the only special tools needed. For more information on replacing the fuel filter see **LUBRICATION AND MAINTENANCE**.

> **NOTE —**
> • Make sure that the Bosch code number on the replacement pump matches the code number on the old pump. See **Table b** above for more information.
>
> • It is good insurance to replace the fuel filter when a new fuel pump is installed.

6

To replace main fuel pump:

1. Disconnect the negative (–) battery cable. Remove the filler cap to release tank pressure.

> ***CAUTION —***
> *BMW anti-theft radios can be rendered useless by disconnecting the battery. See your owner's manual for more information.*

2. Working beneath the car, pull back the two rubber boots that protect the pump's electrical connectors. Remove the two nuts and disconnect the wires from the pump. Thoroughly clean the fuel line unions.

3. Temporarily pinch shut the supply line hose that comes from the transfer pump and the outlet hose to the fuel injection. See Fig. 5-9. Place a large container beneath the fuel pump for catching fuel.

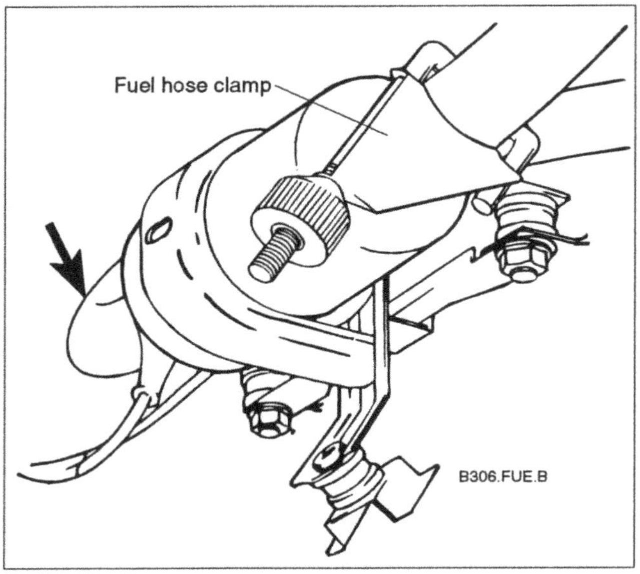

Fig. 5-9. Fuel pump inlet hose being pinched shut with special clamp. Also pinch shut the outlet hose to the fuel injection (arrow) with another clamp.

> **NOTE —**
> BMW special tool no. 13 3 010 should be used to clamp the fuel lines. These tools are available from an authorized BMW dealer.

4. Loosen the supply line and output line hose clamps. Remove the fuel hoses. Discard the hose clamps.

> ***WARNING —***
> *Fuel will be discharged. Do not disconnect any wires that could cause electrical sparks. Do not smoke or work near heaters, or other fire hazards such as a hot exhaust. Keep an approved fire extinguisher handy.*

5. Remove the bracket clamping nuts and bolts and remove the pump and bracket from the car. See Fig. 5-10. Unbolt the yoke bracket securing the pump and filter.

> **NOTE —**
> Inspect all rubber components including the bracket mounts. Replace any that are worn, cracked, or chafed.

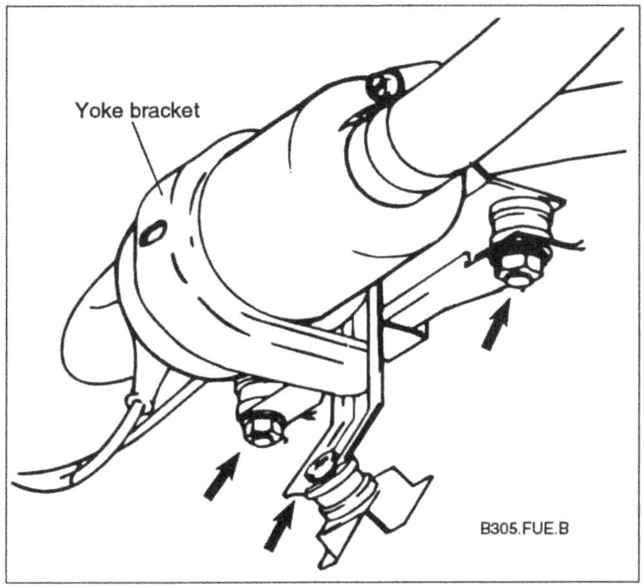

Fig. 5-10. Typical fuel pump bracket mounting points (arrows). Fuel filter (not shown) is also mounted to the bracket.

6. Place the new pump with its new rubber ring, in the mounting bracket and install. Tighten the mounting bolt.

Tightening torque	
• fuel pump clamp bolt	6 to 7 Nm (53 to 62 in-lb)

7. Reconnect the fuel lines using new hose clamps. Reconnect the electrical wires and tighten the nuts. Reposition the rubber boots over the connectors.

Tightening torques	
• electrical connections to fuel pump (nut).............	1.5 to 2.0 (13 to 18 in-lb)
• hose clamps 10mm to 16mm diameter......	2.0±0.3 Nm (18±2.5 in-lb)

8. Reconnect the negative (–) battery cable. Remove the special hose clamps. Be sure to run the engine and check for leaks.

Section 7

FUEL INJECTION

7

Contents

TABLES

2 FUEL INJECTION

INTRODUCTION

The fuel injection system has three main functions: 1) it delivers fuel to the engine 2) it admits filtered air to the engine to be mixed with fuel and 3) it mixes fuel and air in precise proportions and delivers the mixture to the cylinders as a combustible vapor.

The fuel injection system injects atomized fuel into the intake air stream under pressure. This method of active fuel metering means that the fuel mixture entering the engine can be controlled more precisely and used more efficiently, yielding improved driveability, fuel economy, and performance.

This section covers repair information specifically for the fuel injection system and emission controls. Information on the fuel pump, fuel tank, and fuel lines is covered in **FUEL SUPPLY.** General fuel system troubleshooting to be used as an aid in diagnosing problems can be found in **ENGINE MANAGEMENT—DRIVEABILITY**.

1. GENERAL DESCRIPTION

Each of the cars covered by this manual is equipped with an electronic fuel injection system. Fuel is delivered to the engine under pressure, then metered by precise electronic control in proportion to measured air flow. Though there are some minor differences between the systems installed on the various models, all are Bosch Motronic Engine Management Systems.

Motronic Engine Management System

The Bosch Motronic Engine Management System is electronically controlled. Sensors supply information about air flow, engine temperature, and other operating conditions to the central electronic control unit. The control unit then operates the solenoid-type fuel injectors and adjusts ignition spark and timing. Because the system controls both fuel injection and ignition it is called an Engine Management System. For more information on timing and repairs to the ignition system, see **IGNITION.**

Two versions of the Motronic engine management system are used on the cars covered by this manual, a "basic" Motronic system and Motronic 1.1. Motronic 1.1 is a more sophisticated version of the basic system, using a control unit that features a more advanced adaptive control circuitry. Where the basic Motronic system compares data from its sensors to baseline values in memory and makes adjustments, Motronic 1.1 actually changes its baseline values in the control unit memory based on how that particular car is operating.

Many of the testing and adjusting procedures are the same for both Motronic systems. Unless the text clearly states otherwise, the procedures in this section apply to all systems. Read each procedure carefully to see the differences that may apply only to certain models.

Applications–Identifying Features

Most of the cars covered by this manual are equipped with the basic Motronic engine management system. All 528e models built March 1987 and later are equipped with Motronic 1.1.

A distinguishing feature of Motronic 1.1 is the 20-pin diagnostic connector located near the engine oil dipstick. See Fig. 1-1. All other models use a 15-pin connector. Motronic 1.1 uses a single, front-engine mounted speed/reference sensor, while all other models use two separate transmission bellhousing-mounted sensors. In addition, the early Motronic system uses a 35-pin connector at the Motronic control unit, while the later Motronic 1.1 system uses a 55-pin connector.

NOTE —
On 528e models the diagnostic connector is mounted to the intake manifold, in front of oil level dipstick. On 533i and 535i models, the diagnostic connector is mounted to the front intake manifold support bracket, near the fuel pressure regulator.

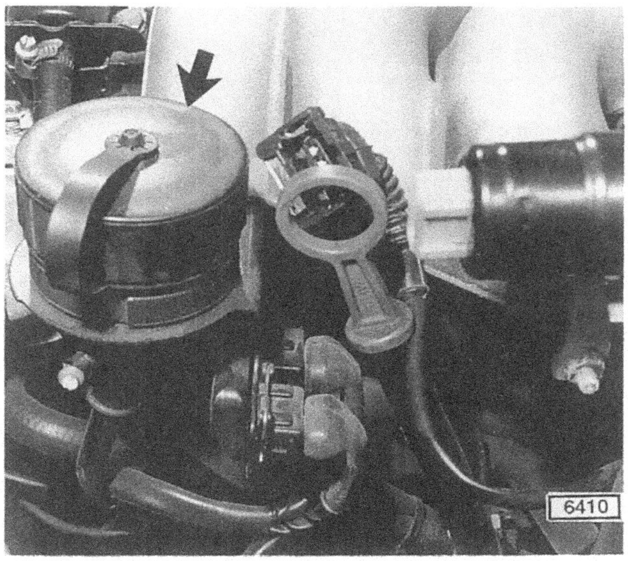

Fig. 1-1. 20-pin Diagnostic connector (arrow) used on 528e models with Motronic 1.1. All other models have smaller 15-pin connector.

Fig. 1-2 and 1-3 show the fuel system component locations on the cars covered by this manual. **Table a** lists the models covered by this manual and their corresponding fuel injection systems.

NOTE —
A car's build date is listed on the driver's door, below the door latch.

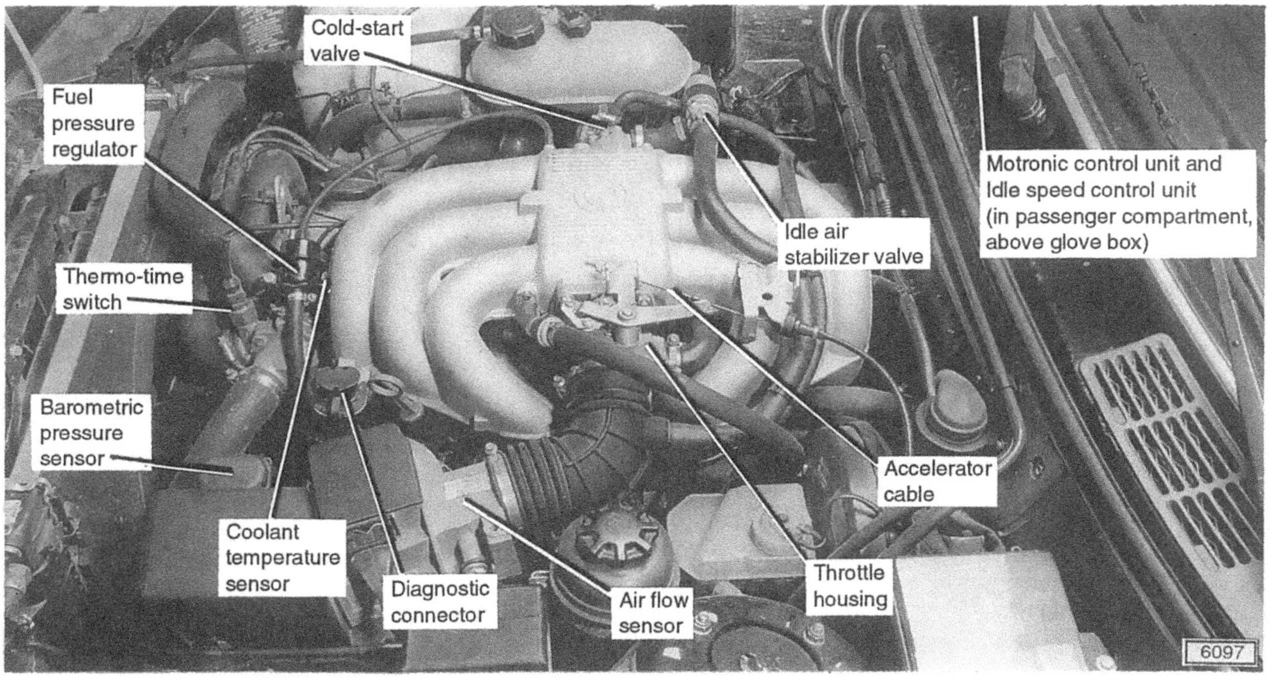

Fig. 1-2. Motronic engine management system components used on 528e models. Early Motronic system shown. Motronic 1.1 system is similar.

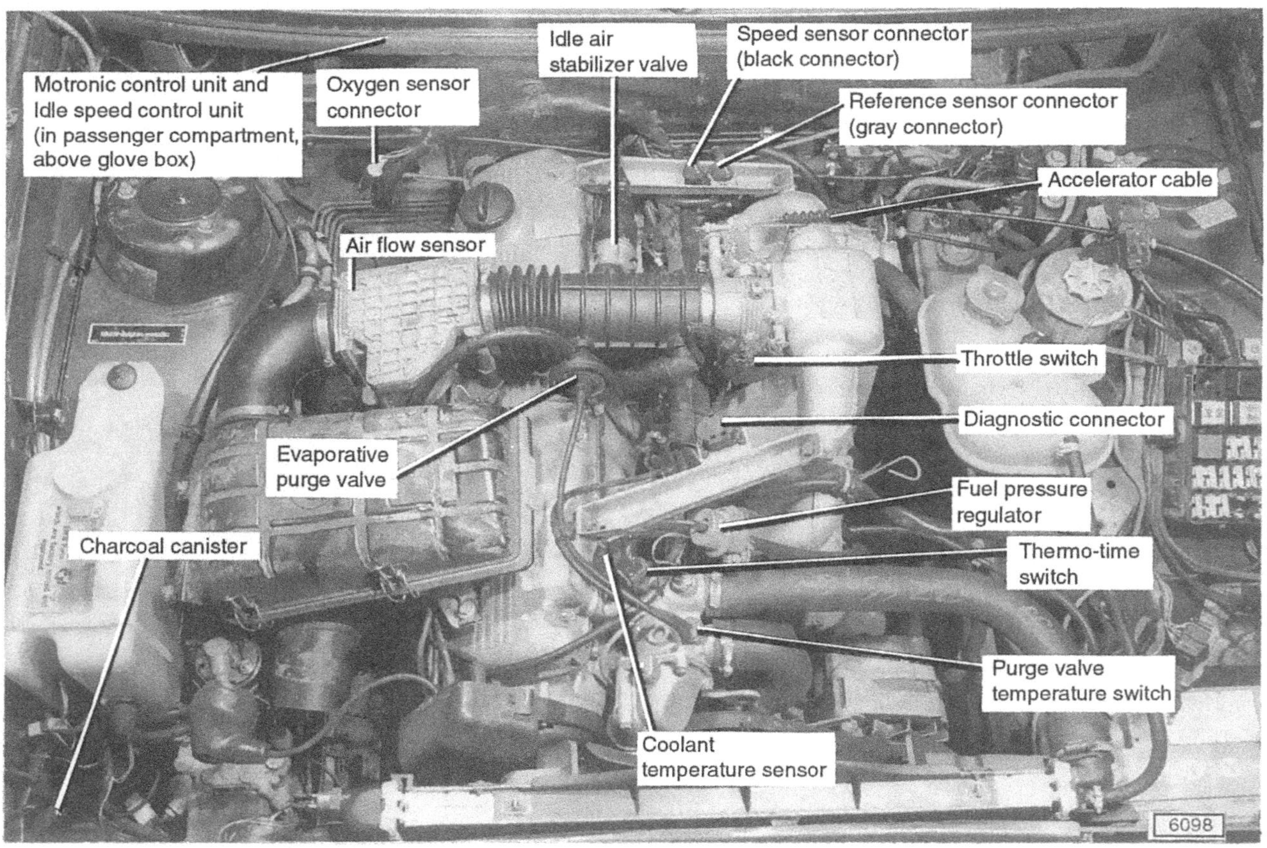

Fig. 1-2. Motronic engine management system components used on 535i models. 533i models are similar.

Table a. Fuel Injection System Applications

Model and year	Fuel injection system
528e 1982 through 1987 (built up to March 1987)	Motronic
528e 1987 and 1988 (built from March 1987)	Motronic 1.1
533i (all)	Motronic
535i (all)	Motronic

2. MAINTENANCE

BMW specifies the procedures below to be carried out at particular time or mileage intervals for proper maintenance of the fuel system. Information on these steps and the prescribed maintenance intervals can be found in **LUBRICATION AND MAINTENANCE**.

1. Replacing the oxygen sensor
2. Servicing the air filter
3. Replacing the fuel filter
4. Lubricating throttle linkage and throttle bearings
5. Inspecting fuel hoses

3. TROUBLESHOOTING

Poor driveability or faulty running may have a variety of causes. The fault may lie with the fuel system, the ignition system, parts of the emission control system, or a combination of all three. Because of the interrelated functions of these systems it is often difficult to know where to begin looking for problems. For this reason, effective troubleshooting must always consider these systems as one major system called Engine Management.

This section of the manual covers those tests, adjustments, and repairs that apply specifically to the fuel injection system. For troubleshooting information addressing engine management and car-running problems and their likely causes, see **ENGINE MANAGEMENT—DRIVEABILITY**.

Please observe the following cautions and warnings when performing any service or repair on the fuel system.

WARNING —
- *The fuel system is designed to retain pressure even when the ignition is off. When working with the fuel system, loosen the fuel lines very slowly to allow residual fuel pressure to dissipate gradually. Avoid spraying fuel.*

WARNING —
- *Fuel is highly flammable. When working around fuel, do not disconnect any wires that could cause electrical sparks. Do not smoke or work near heaters or other fire hazards. Keep an approved fire extinguisher handy.*
- *There are lethal voltages present at the ignition system when the engine is running or the key is on.*

CAUTION —
- *Before making any electrical tests with the ignition turned on, disable the ignition system as described in* ***IGNITION****.*
- *To prevent damage to the ignition system or the electronic fuel system components, including the control unit, always connect and disconnect wires and test equipment with the ignition off.*
- *Absolute cleanliness is essential when working with the fuel system. Even a minute particle of dirt can cause trouble if it reaches an injector. Thoroughly clean the fuel line unions before disconnecting any of the lines.*
- *Use only clean tools. Keep removed components clean and sealed or covered with a clean, lint-free cloth, especially if completion of the repair is delayed.*
- *When replacing parts, install only new, clean components.*
- *Avoid nearby use of compressed air, and do not move the car while the fuel system is open.*
- *Do not use compressed air above 40 psi to blow out lines or components. Internal damage may result to components.*
- *Always replace seals and O-rings.*

7

4. EMISSION CONTROLS

Fig. 4-1 shows a schematic of the typical emission control systems used on the cars covered by this manual. The emission controls include the evaporative emission system and the oxygen sensor system. The evaporative emission system provides venting for the fuel tank while at the same time trapping the fuel vapors that would otherwise be vented into the atmosphere. The oxygen sensor system provides the fuel injection system with feedback information about combustion efficiency.

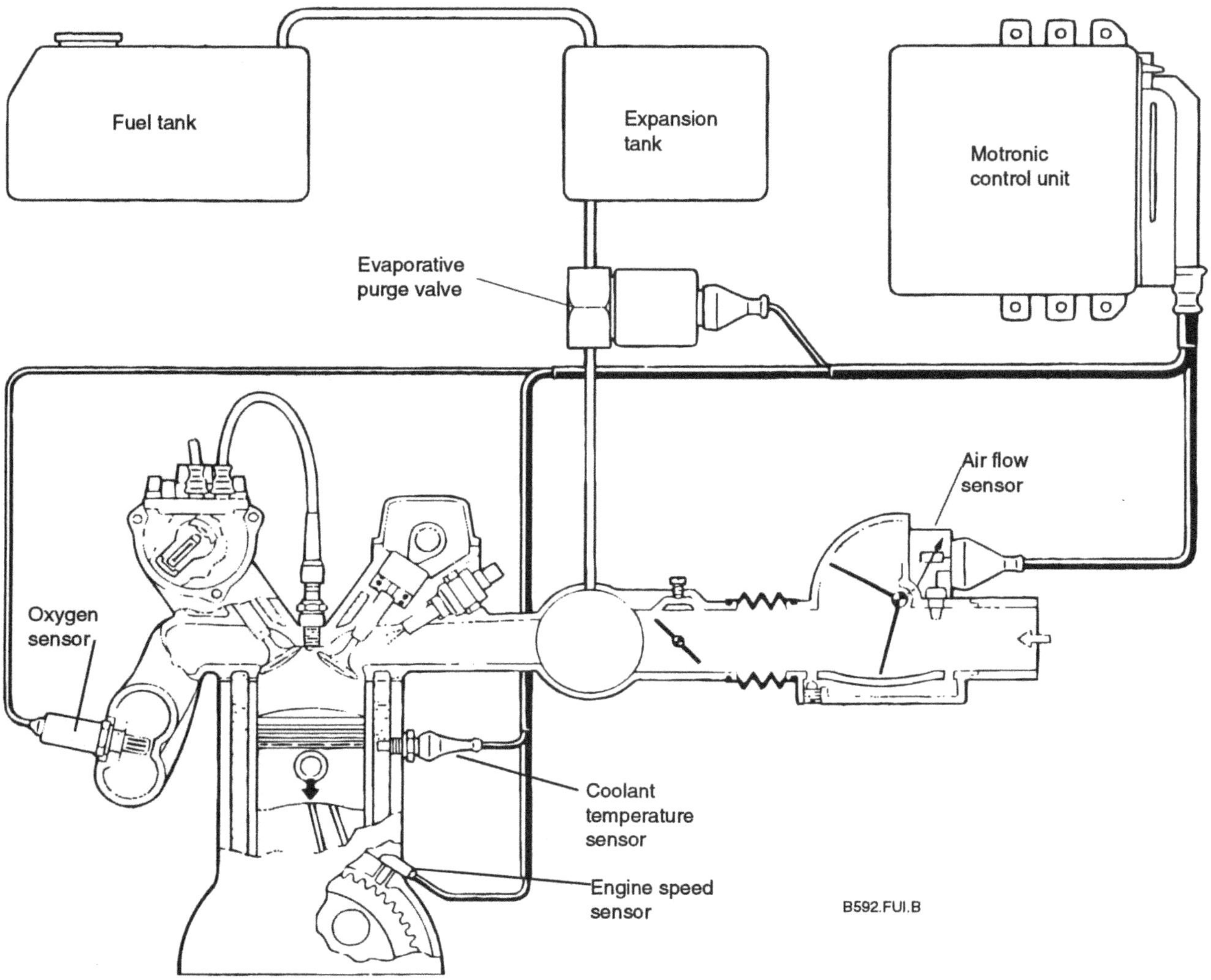

Fig. 4-1. Typical emission controls used on BMW 5-series cars. Courtesy Robert Bosch Corp.

4.1 Evaporative Emission Controls

The evaporative emission system collects fuel vapors from the fuel tank in a charcoal canister. During certain engine operating conditions, the canister is purged, drawing the fuel vapors into the engine to be burned.

The main components of the system are the charcoal canister, the bypass or purge valve (if applicable), and an expansion tank (near the fuel tank filler neck in trunk). The charcoal canister collects and stores fuel vapors from the tank when the engine is idling or stopped. During normal driving, the fuel vapors are drawn into the engine.

A faulty evaporative emission control system can cause hard warm engine starting, erratic idle, or poor acceleration if the fuel vapors are drawn into the engine at the wrong time.

NOTE —
A faulty evaporative emission control system usually affects driveability only when the engine is warm and/or outside temperatures are high, such as during prolonged idling on a hot day. As a general rule, the evaporative emissions system is not normally the cause of cold engine running problems.

Testing Charcoal Canister Purge Valve

The purge valve is tested by trying to pass air through it to determine when it is open or closed. Two types of purge valves are used on the models covered by this manual, an electrically-operated valve and a pneumatically-operated valve. See **Table b** for valve applications.

The purge valve can be easily found by following the hose from the charcoal canister (behind the right-front headlight assembly) up to the throttle housing. If installed, the valve will be located in the middle or end of this hose.

Table b. Evaporative Purge Valve Applications

Model and year	Type of Valve
528e 1982 through 1984	Electrically-operated*
528e 1985 through 1987 (built up to March 1987)	Not equipped**
528e 1987 through 1988 (built March 1987 and later)	Electrically-operated
533i (all)	Electrically-operated
535i (all)	Pneumatically-operated

*1982–1984 528e models may have had the purge valve removed and the purge line re-routed as part of an authorized BMW dealer field fix to help correct a rough idle condition. In addition, these models may have had a small purge control valve (BMW Part No. 11 63 1 707 770) installed to further correct this problem. Do not confuse the electric purge valve with the pneumatic control valve. Check the vacuum hose routing diagram on the inside (front edge) of the hood. See Fig. 4-2.

**528e models built before March 1987 may have a small restrictor installed in the hose between the charcoal canister and the throttle housing. This restrictor is not a purge control valve and is designed to eliminate poor throttle response during warm engine acceleration.

Fig. 4-2. This 528e vacuum routing hose diagram shows that the evaporative emissions system has been modified according to a BMW-authorized field fix no. 84-2.7V5-2. The original system had a purge valve, which was later removed.

To test the vacuum-actuated purge valve (535i models), remove the vent lines from the valve and blow into the larger port of the valve. It should be closed and not pass any air. Apply vacuum to the small vacuum hose connection on the purge valve. When vacuum is applied, the valve should be open and air should be able to pass through. Replace a purge valve that fails either of these tests.

To test the electrically-operated valve (528e and 533i models), first remove the valve from the car. See Fig. 4-3. Do not disconnect the harness connector. Turn the ignition on and check that the valve makes an audible click. If the valve does not click, remove the harness connector. Using a voltmeter, check for voltage between the connector terminals. With the ignition on, there should be approximately battery voltage. If there is voltage, the valve is faulty and should be replaced. If there is no voltage, then there is a either a fault in the wiring to the Motronic control unit or the control unit or purge valve relay (if equipped) is faulty.

NOTE —
On 528e models with Motronic 1.1, the purge valve is controlled by the Motronic control unit and a purge valve relay is not used.

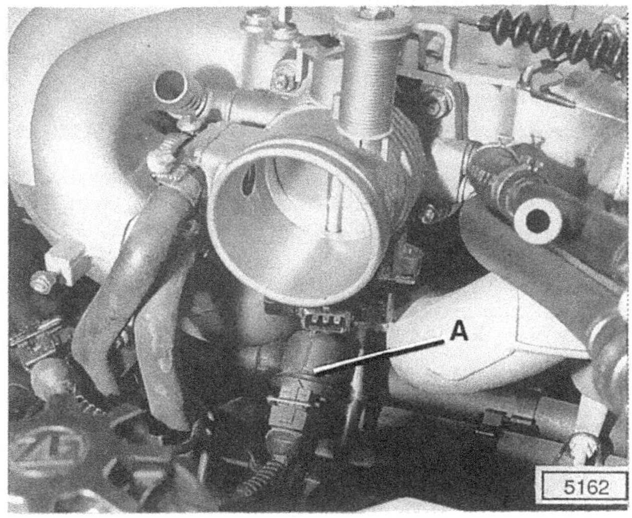

Fig. 4-3. Charcoal canister purge valve (**A**) on 528e models with Motronic 1.1. On other 528e models and 533i models, the valve is located in front of the windshield washer reservoir on right front fender.

On 535i models, a purge valve vacuum switch prevents purging of the charcoal canister until the engine has warmed up. See Fig. 4-4. The switch is immersed in engine coolant and opens when the coolant reaches a specified temperature. Check the switch by trying to pass air through it with the engine cold and then with the engine warm. Air should pass through the valve when the engine is warm and no air should pass through the valve (valve closed) when the engine is cold. Replace a faulty switch.

NOTE —
Replacing the evaporative purge valve vacuum switch may require partial draining of the cooling system, See **COOLING SYSTEM** for draining and filling procedures.

Fig. 4-4. Evaporative purge valve vacuum switch (arrow) used on 535i models. The switch prevents fuel vapors from being drawn into the engine until it is warmed up.

Purge valve vacuum switch tests	
• switch opens	above 43±3° C (109±5° F)
• switch closes	below 33±3° C (91±5° F)

Tightening torque	
• purge valve vacuum switch	15 Nm (11 ft-lb)

4.2 Oxygen Sensor System

The oxygen sensor system provides the fuel injection system with feedback information about combustion efficiency. It does this by reacting to the oxygen content in the exhaust.

The exhaust-mounted oxygen sensor is constructed of ceramic material coated with platinum. One surface is exposed to the exhaust gas, while the other is exposed to the atmosphere. The difference in oxygen content between the two surfaces causes a chemical reaction that generates a low-voltage electrical signal (100–1000mv).

The signal from the oxygen sensor is monitored by the Motronic control unit. This signal is just one of many inputs the Motronic control unit receives and processes in order to properly control the air-fuel mixture.

Since the oxygen sensor system relies on low-voltage signals, it is very sensitive to contamination or poor connections. Before doing any testing, make sure that all electrical contacts are clean and dry. This may cure system problems easily and preclude the need for time-consuming testing.

Oxygen sensor replacement is a maintenance procedure scheduled at a specified time or mileage interval that varies depending on model. See **LUBRICATION AND MAINTENANCE** for more information on recommended service intervals and for oxygen sensor replacement.

> ***CAUTION —***
> *• Fuel injection additives or lubricating sprays containing silicone can contaminate the oxygen sensor. Check with an authorized BMW dealer for approved fuel additives and sprays.*
>
> *• Handle the oxygen sensor with care. Hitting or dropping the oxygen sensor will damage it.*

> **NOTE —**
> Emission controls, including the oxygen sensor system, are covered by an extended warranty. See **LUBRICATION AND MAINTENANCE** for maintenance requirements. Consult an authorized BMW dealer about warranty coverage before beginning any repairs.

Testing Oxygen Sensor System

Because of the closed-loop interaction between the fuel injection system and the oxygen sensor, accurate testing of the sensor requires special equipment. There are, however, a few simple tests that can help determine whether the sensor is functioning correctly. These tests require an accurate voltmeter. Two types of sensors—heated and non-heated— are installed on the 5-series cars:

Oxygen sensor applications	
• 1982 to 1987 528e (ex. Motronic 1.1)	single-wire (non-heated)
• 1987 to 1988 528e (Motronic 1.1)	4-wire (heated)
• 1983 to 1984 533i	single-wire (non-heated)
• 1985 to 1988 535i	3-wire (heated)

This is a sensitive measurement. In the interest of accuracy, the engine must be fully warmed up, the exhaust system must be free of leaks, and all electrical consumers (cooling fan, air conditioning, lights, etc.) must be off.

To test oxygen sensor:

1. With the ignition off, disconnect the oxygen sensor connector. See Fig. 4-5.

Fig. 4-5. Oxygen sensor connector (arrow) location on all 533i models, 535i models, and all 528e models built before March 1987.

NOTE —
On 528e models with Motronic 1.1 (built March 1897 and later), the oxygen sensor connector is located beneath the coolant reservoir, mounted to the wheel well in the rear of the engine compartment.

2. Using a voltmeter (0 to 1 VDC scale), connect the positive (+) test lead to the oxygen sensor signal wire and the negative (-) test lead to a good ground. **Table c** lists the correct test terminals for the various models and sensor types. If necessary, peel back the rubber boot to identify the wire colors specified in the table.

CAUTION —
Connect and disconnect test leads only with the ignition off.

NOTE —
• On most 1985 and 1986 535i models, the connector for the oxygen sensor heater circuit is beneath the car above a small protective cover, directly below the driver's seat.

• On 1985 and 1986 535i models that have had the sensor replaced, a sensor with a longer cable should have been installed. The connector for the new sensor should be located in the engine compartment, in the same location as all of the other cars listed and shown in Fig. 4-5.

3. Start the engine and let it idle. After a maximum of two minutes the oxygen sensor's output voltage should be fluctuating, indicating correct oxygen sensor function.

WARNING —
Exhaust manifolds and pipes can be hot enough to cause serious burns. When working near hot pipes or mufflers, use heavy gloves and other appropriate protection.

4. With the engine running, loosen the oil filler cap or pull up on the oil level dipstick to admit unmeasured intake air and simulate a lean running condition. The oxygen sensor's output voltage should drop.

Lack of a fluctuating voltage signal from the sensor indicates a problem, perhaps only that the sensor is not hot enough. This problem occurs mainly at idle, and is more likely to occur on cars with unheated oxygen sensors. Accelerate the engine several times or run it at fast idle for a few minutes to increase sensor temperature, then check the output signal again.

If the voltage signal does not fluctuate or if there is no change when the dipstick is withdrawn, check for a faulty sensor wire. Also check for a bad ground connection. See **7. Fuel System Electrical Circuits** for additional electrical circuit information. If no wiring faults can be found, the sensor is probably faulty and should be replaced.

Table c. Oxygen Sensor System Test Connections

Model (model year)	Sensor type	Sensor output signal—VDC (test at sensor connector)	Heater power supply (approx. 12 VDC) (test at harness connector)
533i (all), 528e (built up to March 1987)	Unheated (single wire)	black wire (+) and ground (–)	N/A
528e (built from March 1987)	Heated (four-wire)	black wire (+) and ground (–)	grey/blue wire (+) and brown wire (–)*
535i (all)	Heated (three-wire)	black wire (+) and ground (–)	green/blue wire (+) and brown wire (–)

*528e models with 4-wire sensors have two brown (ground) wires—one for the heater circuit and one for the Motronic control unit. Be sure to make the test between the grey/blue wire and both brown wires when checking the voltage supply to the heater.

On cars with heated oxygen sensors, low sensor temperature may be caused by a faulty heater circuit. Check the heater circuit as described below before replacing the oxygen sensor.

Checking Oxygen Sensor Heater Circuit

If the heated oxygen sensor function is in doubt, particularly at idle speeds, check the oxygen sensor heater circuit at the harness side of the oxygen sensor connector. See **Table c** above. With the engine running, there should be battery voltage (approximately 12 VDC) to the sensor's heater. If not, check the oxygen sensor heater relay or the voltage source to the relay. See **7. Fuel System Electrical Circuits** for additional electrical wiring information. The oxygen sensor heater relay is shown in Fig. 4-6.

Fig. 4-6. Oxygen sensor heater relay (arrow).

CAUTION —
Connect and disconnect test leads only with the ignition off.

NOTE —
The oxygen sensor heater relay may be located in the adjacent relay position. To determine the exact position of the relay, check the wire colors leading into the bottom of the relay. The connector that has a green/blue wire leading to it is the oxygen sensor heater relay.

5. FUEL DELIVERY

While the engine management system controls the amount of fuel metered to the engine, the fuel delivery system supplies the fuel to the engine management system. The fuel delivery system includes the fuel pumps, the pressure regulator, and the fuel injectors.

The main fuel pump supplies fuel to the engine management system and engine and creates fuel pressure. The pressure regulator controls the fuel pressure. It is a diaphragm-type regulator, which means that when fuel pressure reaches a certain point a diaphragm deflects to open a valve. See Fig. 5-1. This recirculates excess fuel back to the fuel tank and maintains the desired pressure.

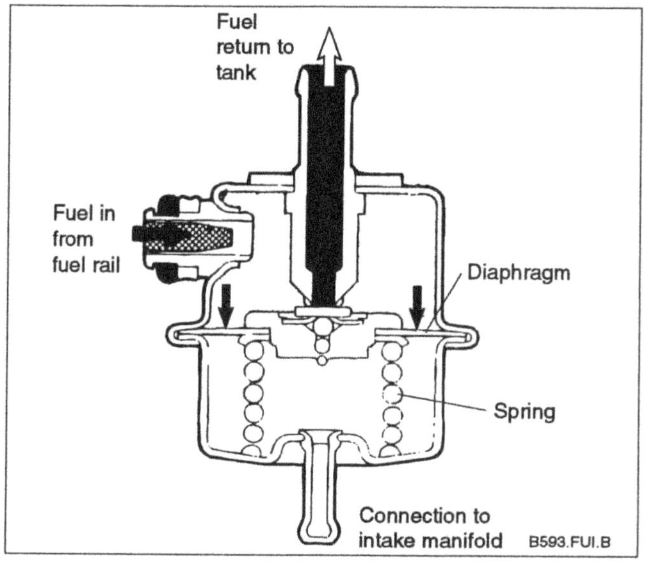

Fig. 5-1. Schematic view of fuel pressure regulator. Fuel pressure deflects diaphragm to return fuel to tank when pressure reaches desired limit. Courtesy Robert Bosch Corp.

A vacuum connection to the intake manifold ensures that the difference between fuel pressure and manifold pressure is constant. Manifold pressure changes act on the regulator diaphragm to increase or decrease fuel pressure. As a result, fuel pressure does not affect fuel metering as engine load changes. The amount of injector opening time is the only factor that regulates fuel metering.

NOTE —
A cracked or leaking pressure-regulator vacuum hose may cause an erratic idle.

The pressure regulator is mounted on the end of the fuel rail. See Fig. 5-2. Fuel flows into the fuel rail first, and then into the pressure regulator. In addition to controlling system pressure, it acts as a one-way check valve to maintain slight residual pressure in the system after the engine is turned off.

Fig. 5-2. Fuel pressure regulator (arrow). Vacuum hose connection to engine intake manifold adjusts fuel pressure based on engine load to keep difference between fuel pressure and manifold pressure constant.

5.1 Fuel Pressure Tests and Specifications

Although the quantity of fuel delivered to the engine is controlled primarily by how long the injectors are open, fuel pressure also influences fuel delivery. To eliminate the influence of fuel pressure variations, the fuel pressure is regulated by a diaphragm-type pressure regulator mounted to the end of the fuel rail. Pulsation dampeners are installed to reduce noise and pressure pulsations caused by the opening and closing injectors and by the fuel pump.

CAUTION —
Cleanliness is essential when working with fuel system components. Before disconnecting any fuel lines, thoroughly clean the unions. Use clean tools.

System Pressure and Residual Pressure

System pressure is the pressure value that is maintained in the system by the pressure regulator. System pressure is not adjustable. The regulator should be replaced if system pressure is not within specifications when tested

To avoid fuel vaporization and hard starting (vapor lock) when the engine is hot, the system is designed to retain slight fuel pressure after the engine has been turned off. This residual pressure is maintained by a check valve in the pressure regulator and at the main fuel pump outlet. The fuel pump check valve is not serviceable as an individual part.

WARNING —
Fuel will be discharged. Do not disconnect any wires that could cause electrical sparks. Do not smoke or work near heaters or other fire hazards. Keep an approved fire extinguisher handy.

NOTE —
The fuel pressure gauge must have a range of at least 0 to 5 bar (0 to 75 psi) and must be securely connected to prevent it from coming loose under pressure.

7

To check system pressure:

1. On 528e models with basic Motronic (built up to March 1987), remove the supply hose from the cold-start valve. Using a small length of fuel hose, two hose clamps, and a T-fitting, connect a fuel gauge between the cold-start valve and the supply line. See Fig. 5-3.

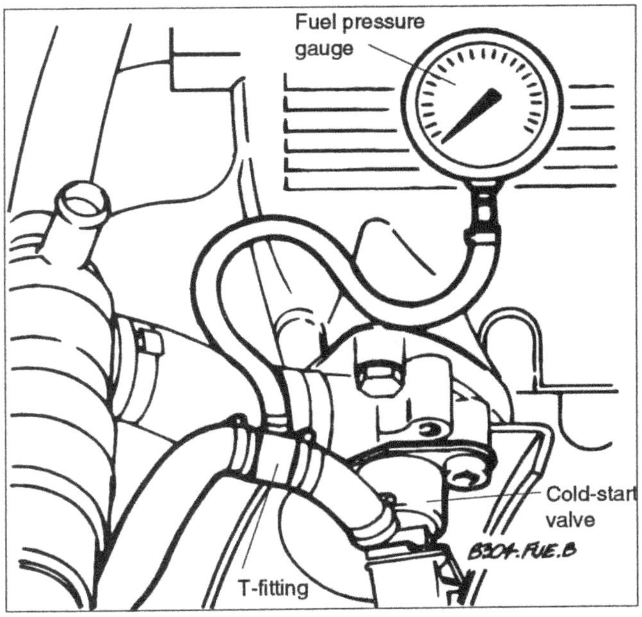

Fig. 5-3. Fuel pressure gauge connected between the cold-start valve and cold-start valve supply line on 528e models with cold start valve.

2. On all 533i, 535i, and 528e models with Motronic 1.1, disconnect the fuel supply line at the firewall end of the fuel rail. Using a small length of fuel hose, two hose clamps, and a T-fitting, connect a fuel gauge between the fuel rail and the supply line.

3. Operate the fuel pump by bypassing the fuel pump relay as described in **FUEL SUPPLY**.

4. Check that the fuel pressure corresponds to the specifications listed in **Table d.**

Table d. Fuel Pressure Specifications

Model	Fuel pressure
528e, 533i	2.5±0.05 bar (36.3±0.7 psi)
535i	3.0±0.06 bar (43.5±0.9 psi)

If the system pressure is too high, check the return line from the pressure regulator to the tank. Check for kinks in the hose. Blow compressed air through the line to check for blockages. If no faults can be found, the pressure regulator is faulty and should be replaced.

CAUTION —
Do not use compressed air above 40 psi to blow out lines or components. Internal damage to components may result.

If the system pressure is too low, repeat the pressure test while gradually clamping off the return hose at the fuel pressure regulator as shown in Fig. 5-4. The pressure should rise to at least 3.2 bar (46.4 psi) or possibly higher. If it does, then the pressure regulator is faulty and should be replaced.

Fig. 5-4. Fuel pressure regulator return hose being clamped shut to check fuel pump delivery pressure.

If the pressure is still low, check visually for leaks in fuel lines or at unions. Leaks can also be due to a leaking cold-start valve or injector. If no leaks can be found, test the fuel pump delivery rate as described below under **5.2 Fuel Delivery Rate and Specifications.**

To check fuel pressure regulator response to engine load:

1. Reinstall the fuel pump relay. Start the engine and allow it to idle. Check that the regulating pressure is as specified in **Table e.**

Table e. Fuel Regulating Pressure

Model	Regulating pressure
528e, 533i	2.3 to 2.7 bar (33.4 to 39 psi)
535i	2.8 to 3.2 bar (40.6 to 46.4 psi)

2. With the engine idling, remove the vacuum hose from the regulator. The pressure should increase.

3. Reconnect the hose and check that the pressure returns to the lower regulating pressure.

If any faults are found, either the fuel pressure regulator or the hose (and vacuum supply) is faulty and should be replaced. When finished testing, disconnect the gauge and reconnect the fuel line to the cold-start valve or the fuel rail. Tighten the hose clamp.

Replacing Pressure Regulator

The fuel pressure regulator attaches to the fuel rail with two bolts. Disconnect the vacuum hose and the fuel return hose as shown in Fig. 5-5. Remove the bolts and pull the regulator from the fuel rail. Inspect the O-ring for damage and replace it if necessary. Installation is the reverse of removal. Be sure the code number on the side of the old pressure regulator matches that of the new regulator.

WARNING —
Fuel will be discharged. Do not disconnect any wires that could cause electrical sparks. Do not smoke or work near heaters or other fire hazards. Keep an approved fire extinguisher handy.

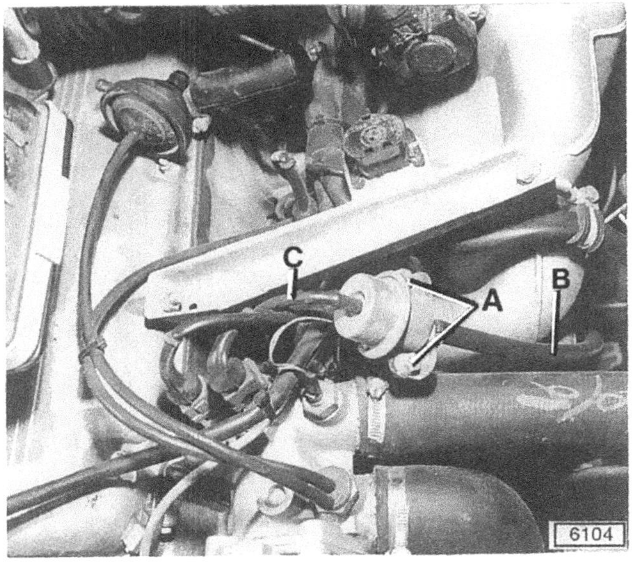

Fig. 5-5. Fuel pressure regulator mounting bolts (**A**), return fuel line (**B**), and vacuum line (**C**).

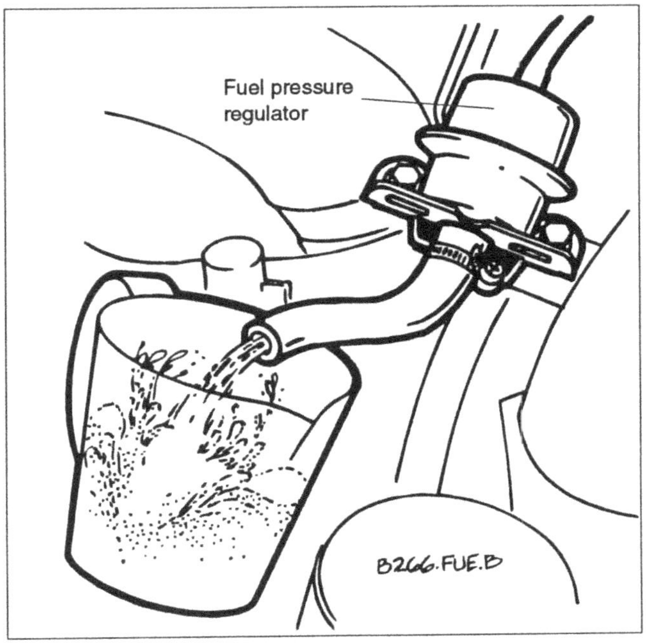

Fig. 5-6. Fuel delivery rate being measured at fuel pressure regulator return line.

5.2 Fuel Delivery Rate and Specifications

The main fuel pump, which is mounted underneath the rear of the car, supplies the fuel system with the correct quantity of fuel under pressure. If the system pressure, as tested above, is not as specified, the fuel pump delivery rate should be checked. A worn fuel pump can cause low fuel pressure and quantity, reusulting in driveability problems.

To test fuel delivery rate:

1. Disconnect the fuel return hose from the pressure regulator. Connect a length of hose to the pressure regulator and place the open end into a container. See Fig. 5-6.

2. Run the fuel pump by bypassing the fuel pump relay as described in **FUEL SUPPLY**.

3. Check that the quantity of fuel delivered corresponds to the specifications listed in **Table f**.

NOTE —
• When making a fuel delivery test, make sure the battery is fully charged (12±0.5 VDC). Low battery voltage can result in incorrect test results.

• When making a fuel delivery test, it is good practice to do the test with a new fuel filter installed. A plugged filter can result in incorrect test results. To change the filter, see **LUBRICATION AND MAINTENANCE**.

Table f. Fuel Delivery Rate Specifications

Model	Fuel quantity (fuel pump running for exactly one minute)
1982 through 1985 (all)	at least 2.0 liters (2.2 quarts)
1986 through 1988 (all)	at least 1.8 liters (1.9 quarts)

If the amount of fuel delivered is too low, check that the fuel pressure regulator is operating correctly and that the fuel pressure is as specified. See **5.1 Fuel Pressure Test and Specifications.** If no faults are found with the pressure regulator or fuel pressure, test the fuel pump as described in **FUEL SUPPLY.**

5.3 Fuel Injectors

The electrically-operated fuel injectors are turned on and off by signals from the Motronic control unit. The injectors are connected to a common fuel supply, called the fuel rail. See Fig. 5-7.

In each injector, a solenoid opens a needle valve to spray fuel into the intake port. See Fig. 5-8. The frequency at which the injectors open is synchronized to the engine's speed. The quantity of fuel delivered to the engine is determined by the amount of time the injectors remain open. The injectors are not repairable and must be replaced if found to be faulty.

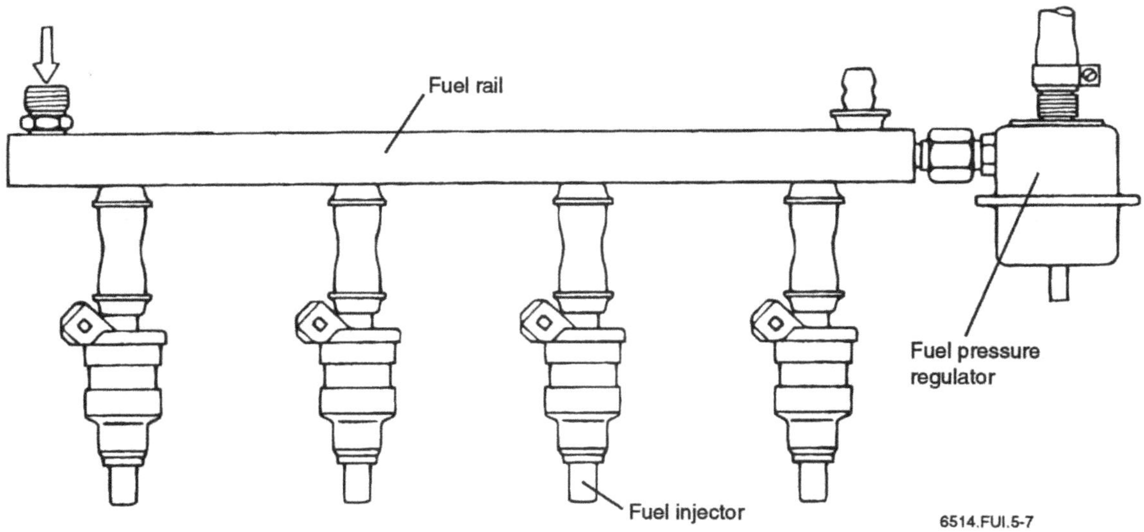

Fig. 5-7. Fuel injector and fuel rail assembly. Fuel rail is supplied with fuel from pump. Fuel pressure regulator controls fuel pressure in the fuel rail. Courtesy Robert Bosch Corp.

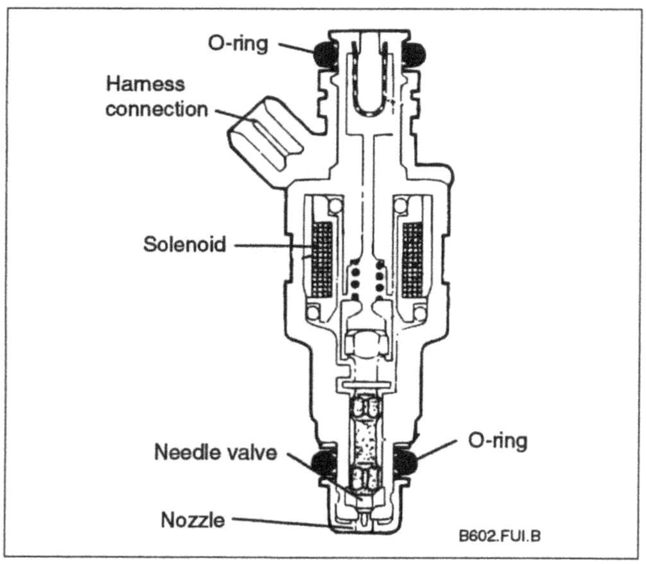

Fig. 5-8. Cutaway view of pulsed-type fuel injector. Courtesy Robert Bosch Corp.

Faulty or clogged injectors may cause rough running. Carbon deposits that form at the tips of the injectors reduce fuel flow, which can cause erratic idle and poor performance, especially when accelerating a cold engine.

If the O-rings that seal the injectors are cracked or seated poorly, rough running may also result. Air leaking past the O-rings is not measured by the fuel injection system, and it will lean the mixture. The injectors must be removed as described below to replace the O-rings.

Testing Injectors

Injector tests include electrical checks for resistance, power, ground and signal from the control unit, and flow tests to make sure that the injectors are not clogged.

Most electrical tests can be performed using a volt/ohmmeter. In addition, inexpensive LED fuel injector testers that plug into the injector harness connector to check the electrical signal to the injector are available from many parts sources.

Flow tests require specialized equipment that opens each injector for a precise amount of time to measure the amount of fuel delivery. Most authorized BMW dealers can perform an injector balance test to locate clogged or faulty injectors. For more information on cleaning and preventing clogged injectors, see **LUBRICATION AND MAINTENANCE**.

To quickly check if an injector is electrically functioning, start the engine and place a screwdriver or a finger on the injector body. A slight vibration or buzzing sound indicates that the injector is functioning.

If the injector does not vibrate, turn the engine off and disconnect the electrical connector from the suspected injector. Using an ohmmeter, check the injector's resistance. The resistance should be within the specifications listed in **Table g**. If the resistance is not as specified, the injector is faulty and should be replaced.

If the injector's resistance is within specifications, but the injector does not operate as described above, check for positive (+) battery voltage at the connector. There should be battery voltage between the red/white wire and ground when the ignition is on. If no faults are found, check for continuity between the connector's ground wire (brown wire with colored stripe)

Table g. Fuel Injector Specifications

Model and date of manufacture	Injector code no. (BMW Part No.)	Color of plug connection	Color of injector shield	Coil resistance in ohms at 68° F (20° C)
528e models built up to March 1987	0 280 150 126 (13 64 1 273 271)	light grey	orange	2 to 3
528e models built March 1987 and later	0 280 150 716 (13 64 1 726 989)	white	yellow	14.5 to 17.5
533i models	0 280 150 201 (13 64 1 273 272)	dark grey	dark grey	2 to 3
535i models built up to Jan. 1986	0 280 150 203 (13 64 1 276 149)	yellow	yellow	14.5 to 17.5
535i models built Jan. 1986 and later	0 280 150 714 (13 64 1 706 176)	yellow	yellow	14.5 to 17.5

and the control unit. If no wiring faults can be found, the control unit may be faulty. See **6.4 Electrical Tests** for control unit terminal identification and additional electrical tests. See **7. Fuel System Electrical Circuits** for engine management wiring diagrams.

> ***WARNING —***
> *To prevent the engine from starting, disable the ignition system before operating the starter. See **IGNITION** for more information.*

Fig. 5-9. Fuel rail mounting bolts (two of four shown) on 535i model. Other models are similar.

Removing and Installing Injectors

The fuel injectors are removed by first partially removing the complete injector/fuel rail assembly and then unclipping the injectors from the fuel rail. On most models, a single injector can be removed without removing the entire fuel rail.

To remove:

1. Disconnect the negative (–) battery cable.
2. Disconnect the electrical connectors from the injectors.
3. Remove the fuel rail mounting bolts. See Fig. 5-9.
4. Carefully pull the fuel rail assembly up and away from the cylinder head until the injectors are completely removed from the intake ports.
5. Use a small screwdriver to pry off the injector's retaining clip. See Fig. 5-10. Then pull the injector out of the fuel rail.

> ***WARNING —***
> *Fuel will be discharged. Do not disconnect any wires that could cause electrical sparks. Do not smoke or work near heaters or other fire hazards. Keep an approved fire extinguisher handy.*

Fig. 5-10. Fuel injector mounting clip being removed (arrow).

Installation is the reverse of removal. Replace injector O-rings that are hard, cracked, or otherwise damaged. Apply a light coat of gear oil (SAE 90 weight) to the injector O-rings to aid in installation. Check that the injector's electrical connections are correctly fitted and that the injectors are fully seated prior to installing the fuel rail mounting bolts. Be sure to reconnect the negative (–) battery cable.

Tightening torque
• fuel rail mounting bolts ....9 to 11 Nm (80 to 97 in-lb)

CAUTION —
Use only gear oil to lubricate the O-rings when installing the injectors. Other lubricants can clog and contaminate the injectors.

5.4 Accelerator Cable

The accelerator cable links the accelerator pedal to the throttle lever on the throttle housing. Cars with non-electronically-controlled automatic transmission also have an additional accelerator cable, called the transmission accelerator cable. It connects the throttle lever to the transmission to make the transmission responsive to throttle position.

CAUTION —
*The throttle plate adjusting screw is factory-set and should not be used to correct idle speed. For more information on idle speed adjustments, see the appropriate heading under **6. Motronic Engine Management System**.*

NOTE —
• Because the adjustment of the transmission accelerator cable affects the operation of the automatic transmission, adjustment of this cable is covered in **AUTOMATIC TRANSMISSION**.

• Cars with an electronically-controlled transmission use a throttle position sensor to communicate throttle position to the automatic transmission control unit. The automatic transmission control unit signals the Motronic control unit when the throttle is in the full-open position.

To adjust accelerator cable:

1. Check the throttle cable adjustment by checking the amount of freeplay at the cable end. With the accelerator pedal in the rest position, there should be a small amount of play. If necessary, turn the adjusting nut to obtain the correct clearance. See Fig. 5-11.

NOTE —
Make sure that the cable is slack by pulling lightly on the cable end.

Fig. 5-11. Correct amount of clearance in accelerator cable (dimension **A**) at rest. Adjust cable by turning plastic adjusting nut. On some models, it is necessary to first loosen a small locknut.

Accelerator cable clearance
• cable freeplay at rest position (**Dimension A**)........ about 1 to 2 mm (0.04 to 0.08 in.)
• full throttle stop clearance (**Dimension B**)............... about 0.5 mm (0.020 in.)

2. Have a helper push the accelerator pedal down until it reaches its full-throttle position. There should be a small amount of clearance between the full-throttle stop and the throttle lever. See Fig. 5-12.

3. If the clearance is incorrect, adjust the height of the pedal stop or kickdown stop/switch by loosening the locknut and turning the stop/switch until the specified clearance is obtained. See Fig. 5-13.

Tightening Torque
• pedal stop or kickdown switch (locknut)... 9 Nm (80 in-lb.)

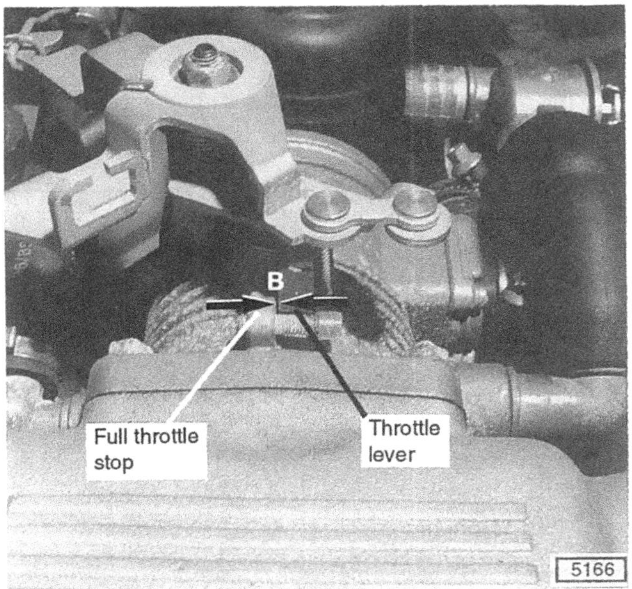

Fig. 5-12. Correct amount of clearance at full-throttle stop (dimension **B**). 528e engine shown, other models are similar.

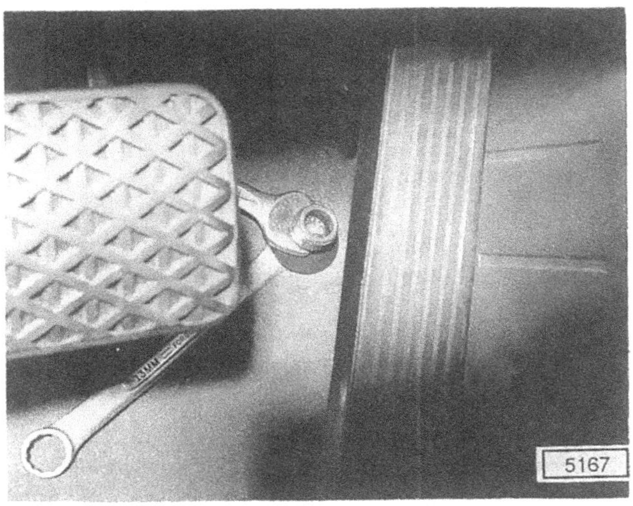

Fig. 5-13. Accelerator pedal stop/switch locknut being loosened using two open end wrenches. Rotate stop/switch until there is about 0.5 mm (0.02 in.) of clearance.

NOTE —
- Rotating the pedal stop or kickdown switch 1 1/2 turns results in a 0.5 mm (0.02 in.) change at the full-throttle stop.

- On models with electronically controlled transmissions (code letters 4 HP 22 EH), disconnect the electrical connector from the kickdown switch before rotating it. See **AUTOMATIC TRANSMISSION** for more information.

4. Recheck all adjustments after adjusting the accelerator pedal stop or kickdown switch.

5. On models with automatic transmission, check the adjustment of the transmission accelerator cable as described in **AUTOMATIC TRANSMISSION**.

To remove and install accelerator cable:

1. Pry off the plastic retaining clip from the throttle lever and remove the cable end from the lever. Remove the plastic clip from the cable. See Fig. 5-14.

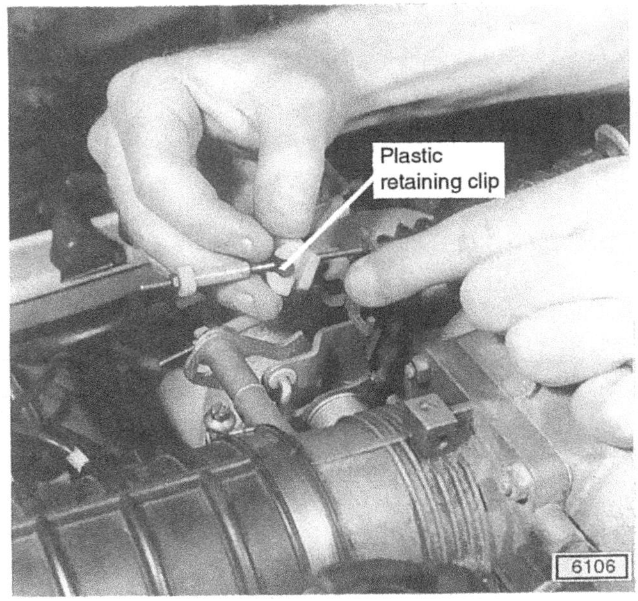

Fig. 5-14. Accelerator cable being removed from throttle lever.

2. Remove the cable together with the rubber grommet from the bracket on the intake manifold.

3. Working from inside the car, remove the lower instrument panel trim.

4. Pull the cable end from its ball socket in the accelerator pedal assembly.

5. Working from inside the car at the firewall, compress the locking tabs on the firewall grommet and press the grommet into the engine compartment.

6. Working from the engine compartment, pull the cable through the firewall.

7. Installation is the reverse of removal. Adjust the new cable as described above. Depress the accelerator pedal several times to check for binding before driving.

7

6. MOTRONIC ENGINE MANAGEMENT SYSTEM

Fig. 6-1 shows the main components of the Motronic engine management system (also known as Digital Motor Electronics (DME)). The Motronic system is similar to the widely installed L-Jetronic fuel injection system. The major difference is that the single Motronic control unit also controls the ignition speed functions. The Motronic control unit continuously collects various engine sensor signals to determine the optimum ignition timing point and injection quantity (air-fuel mixture).

In the Motronic system, basic fuel metering is determined by engine rpm and by the volume and temperature of the air entering the engine. Fuel is metered to the engine via the fuel injectors, which are electrically pulsed on and off by the control unit. The control unit receives air flow volume and temperature data from the air flow sensor and engine rpm from the pulse or speed sensor.

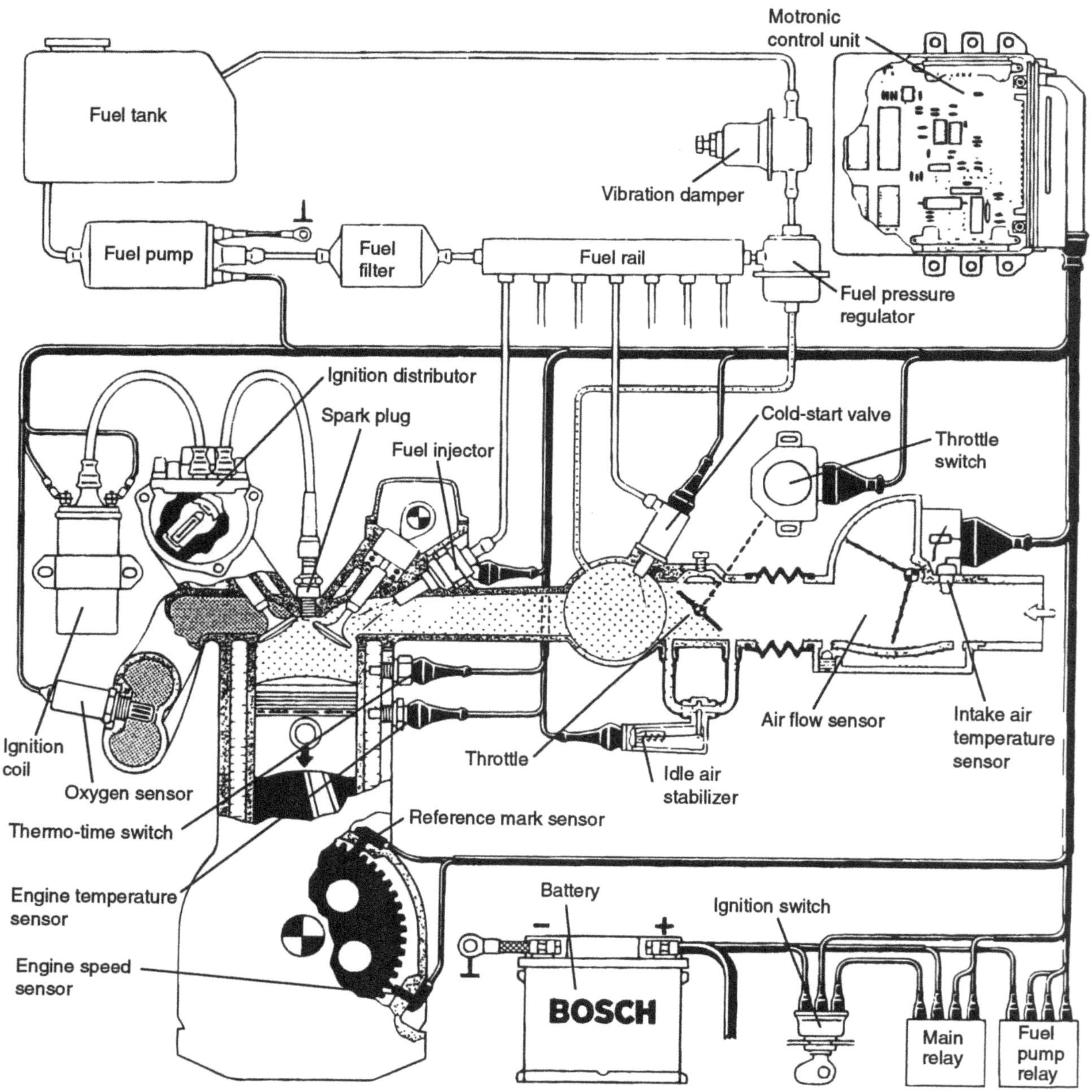

Fig. 6-1. Schematic view of Motronic engine management system. Motronic 1.1 system is similar. Courtesy Robert Bosch Corp.

Six injectors, one for each cylinder, are triggered on and off at a rate governed by engine rpm. Fuel mixture is controlled by the length of the electrical impulse from the control unit to the injector—the length of time that the injectors are open. The injectors are mounted to a common fuel supply, called the fuel rail. Fuel pressure is regulated by a fuel pressure regulator mounted to the end of the fuel rail.

Information on fuel injectors can be found in **5.3 Fuel Injectors**. Information on the fuel pressure regulator as well as fuel pressure tests are covered separately in **5.1 Fuel Pressure Tests and Specifications**.

The control unit makes additional adjustments to the amount of fuel metered by the injectors based on inputs from various engine sensors. The oxygen sensor system provides the control unit with information on engine combustion efficiency. A throttle switch and/or throttle position sensor informs the control unit when the throttle is in the fully open or fully closed position. A coolant temperature sensor provides the control unit with engine temperature. On some models, a barometric pressure switch or sensor adjusts the fuel mixture based on changes in altitude. For testing and repair information on the oxygen sensor system, see **4.2 Oxygen Sensor System**.

NOTE —
On 535i models with electronically-controlled automatic transmission (4 HP 22 EH), the full throttle signal to the Motronic control unit comes from the throttle position sensor via the automatic transmission control unit.

Two versions of the Motronic engine management system are installed on the 5-series models covered by this manual. The first is the "basic" Motronic system just described. The second is Motronic 1.1, a newer version of the basic Motronic. The differences between the two systems are mainly incorporated in the circuitry. Both systems have adaptive abilities that adjust the system to compensate for things such as engine wear and vacuum leaks.

On Motronic 1.1, the system has additional and wider adaptive features that change the system's base settings in the control unit's memory. In addition, Motronic 1.1 injectors are controlled in two sets of three cylinders each (on basic Motronic, all injectors are controlled as a single set). This allows for a more precise fuel mixture. An induction sensor is connected to the no. 6 cylinder spark plug wire as a timing reference signal for the control unit. Visible Motronic 1.1 differences include a single, front-engine mounted reference/ speed sensor (earlier system uses two sensors at the transmission bellhousing), a 20-pin diagnostic connector (earlier models use a 15-pin connector) and a 55-pin Motronic control unit harness connector (earlier models use a 35-pin connector).

For specific model applications, see **1. General Description**. The information in this repair section applies to both Motronic systems unless the text specifically mentions an exception.

In order for the system to operate properly, basic idle speed, ignition timing, and idle mixture (% CO) must be correct. The throttle switch must be functioning correctly. Idle speed and ignition timing are electronically controlled and are therefore not adjustable. In addition, idle mixture is not adjustable on models with Motronic 1.1. If any of these basic settings are incorrect, all other test results will be misleading. See **6.3 Idle Speed** and **6.6 Idle Specifications (rpm and % CO)** for more information.

The Motronic system has been designed so that most electrical components can be tested using a low-current LED test light and a digital multimeter. Most of these tests can be performed right at the control unit harness connector using a special test harness adapter.

7

On-board Diagnostics
(Motronic 1.1 only)

An On-Board Diagnostic system is incorporated into the Motronic 1.1 control unit. This system is able to detect certain emissions-related engine management malfunctions. When a malfunction is detected, the Check Engine warning light in the instrument cluster will come on and flash in coded intervals representing the faults(s).

These fault codes can be read when the key is in the On position. The code(s) begin after three seconds and each flash is approximately one second apart. If more than one fault has been detected, each code will be separated by a 3 second pause. When all fault codes have been displayed, there will be another 3 second pause and then the check engine light will come on and stay on. To read the codes again, simply turn the key off and then on again. After 3 seconds, the codes will be repeated.

Table h lists the fault codes, their probable causes and corrective actions. The boldface numbers in the corrective action column indicate the headings in this section or the sections in the manual where the applicable test and repair procedures can be found. After the faults have been corrected, the fault memory can be erased by starting the engine at least five times.

NOTE —
- Fault codes that are not emissions-related are stored in the internal memory of the control unit and can only be read using the special BMW Service Tester. The check engine light does not come on for these malfunctions.
- The flashing of the check engine light is shown graphically in the table. Each peak in the graph indicates one flash.

Table h. 528e Motronic 1.1 On-Board Diagnostics Fault Codes

Code number, meaning, and number of light flashes	Possible reason for malfunction	Corrective action
Engine check light on / Engine check light off Code 1: Air flow sensor malfunction	Air flow sensor flap binding or air flow sensor potentiometer faulty	Test air flow sensor. **6.1 Air Flow Measurement**
Engine check light on / Engine check light off Code 2: Oxygen sensor malfunction or oxygen sensor parameters exceeded	Oxygen sensor faulty or fuel mixture too lean or too rich, due to faulty fuel or ignition system components causing oxygen sensor to exceed its operating parameters	Test oxygen sensor. **4.2 Oxygen Sensor System**. Check for intake air leaks or reasons for rich mixture. See **ENGINE MANAGEMENT—DRIVEABILITY**
Engine check light on / Engine check light off Code 3: Coolant temperature sensor malfunction	Coolant temperature sensor faulty, or engine running too hot or too cold due to faulty cooling system component	Test coolant temperature sensor. **6.2 Cold Start and Cold Running Enrichment.** Test cooling system components. See **COOLING SYSTEM**
Engine check light on / Engine check light off Code 4: Throttle switch malfunction	Throttle switch incorrectly adjusted or faulty	Test throttle switch. **6.3 Idle Speed**

6.1 Air Flow Measurement

The amount of air entering the engine is measured by the air flow sensor. See Fig. 6-2. As intake air is drawn past the air flow sensor flap, the spring-tensioned flap opens and in turn actuates a potentiometer inside the sensor. The varying resistance of the potentiometer provides a varying voltage signal (approximately 0–5 volts) to the control unit that is proportional to air flow. The air flow sensor also contains an intake air temperature sensor that measures the temperature of the incoming air.

The throttle is operated by the accelerator pedal via the accelerator cable, and controls the amount of air drawn in by the engine. The throttle is adjusted during manufacture and does not require any routine adjustment. The throttle can be readjusted if the factory settings have been tampered with. A throttle switch and/or sensor (535i models with electronic automatic transmission) is mounted to the throttle housing. The throttle switch and/or sensor sends a voltage signal to the control unit whenever the throttle is fully closed or fully open.

NOTE —
On 535i models with an electronically-controlled automatic transmission (4 HP 22 EH), the closed throttle position signal comes directly from a small throttle switch (idle position switch). The full throttle position signal comes from the automatic transmission control unit via the throttle position sensor. See **6.3 Idle Speed** for more information.

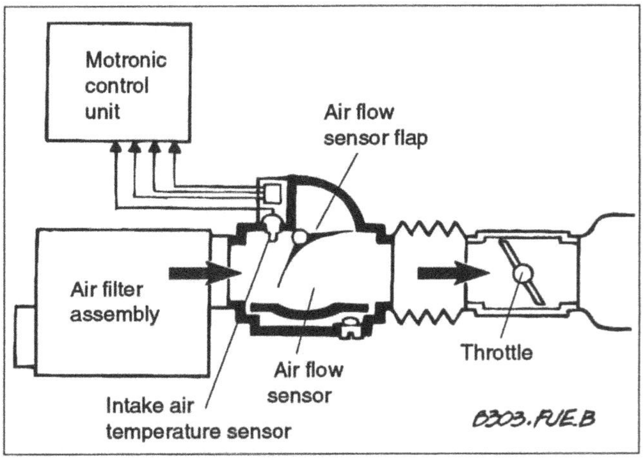

Fig. 6-2. Schematic of air flow sensor and related components. Arrows indicate direction of air flow.

Air Flow Sensor

The sensor's mechanical operation depends upon the free movement of the flap inside the main air passage. To check the flap movement, loosen the clamps, remove the rubber intake air duct, and move the flap through its range of travel. See Fig. 6-3.

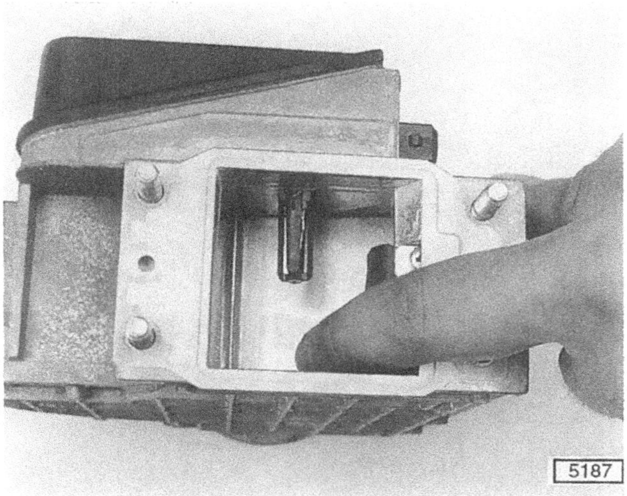

Fig. 6-3. Air flow sensor flap being checked. Move flap through entire range of travel to check for binding and smooth operation.

If the flap binds at any point, remove the intake air flow sensor, as described later under this heading, and check for any foreign material that may be interfering with the flap's movement. Check that the flap returns to the closed position smoothly. If the flap cannot be made to move freely, or if the flap's spring tension is uneven, the air flow sensor is faulty and should be replaced.

> **NOTE —**
> No electrical test specifications are available for the air flow sensor installed on models covered by this manual. Check the air flow sensor's input signals from the control unit as described under **6.4 Electrical Tests**.

Throttle Basic Adjustment

The factory-set throttle adjusting screw is not used to adjust idle speed. Its only function is to provide a mechanical stop for the linkage to prevent the throttle plate from contacting the inside of the throttle housing and causing wear. It should be adjusted only if the factory setting has been tampered with. Check for a slight clearance between the throttle housing and the throttle plate as shown in Fig. 6-4. If the clearance is incorrect, the throttle should be adjusted.

On 528e models, correct a faulty throttle adjustment by removing the tamper proof cap from the throttle lever stop screw. See Fig. 6-5. Turn the screw counterclockwise until the plate contacts the throttle housing, then turn the screw clockwise just until the until the throttle plate is no longer contacting the housing. For a more accurate setting, use a .0015 in. feeler gauge between the throttle plate and the throttle housing. Apply paint or lacquer to the screw's threads to lock it in place then install the tamper-proof cap.

Fig. 6-4. Throttle plate clearance between housing and plate (arrow). Throttle plate should not contact housing.

7

> **NOTE —**
> Before checking or adjusting the throttle plate clearance, clean the throttle plate and housing with a commercial carburetor cleaner and a clean cloth.

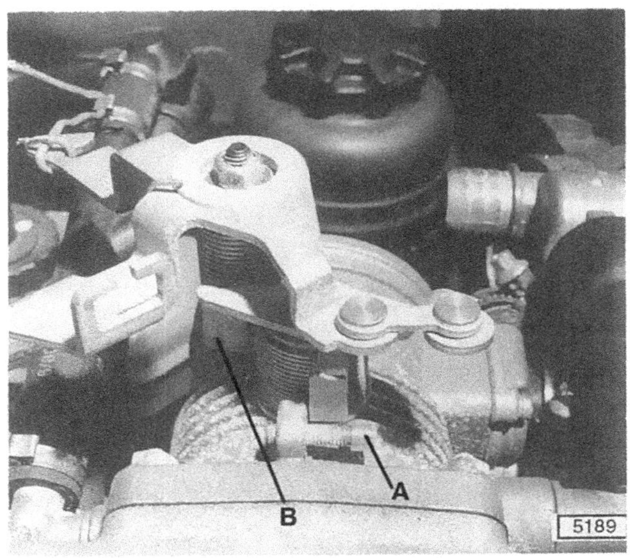

Fig. 6-5. Throttle stop adjusting screw with tamper-proof cap (**A**) and throttle lever (**B**) on 528e models.

On 533i and 535i models, correct a faulty throttle adjustment by loosening the throttle stop clamping nut. See Fig. 6-6. Then rotate the stop just until throttle plate no longer contacts the throttle housing. Tighten the clamping nut.

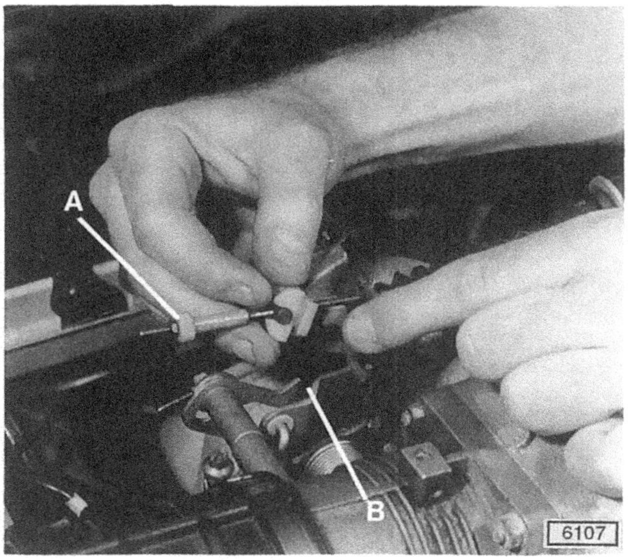

Fig. 6-6. Throttle stop clamping nut (**A**) and throttle stop (**B**) on 533i and 535i models.

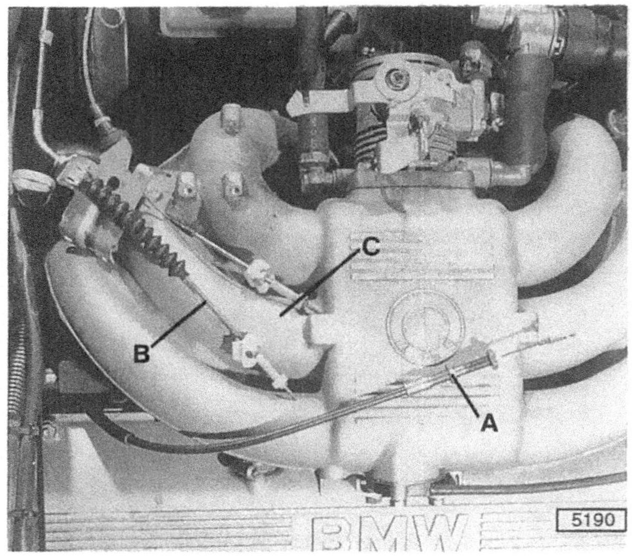

Fig. 6-7. Automatic transmission accelerator cable (**A**), cruise control cable (**B**), and accelerator cable (**C**) on 528e models. Other models are similar.

NOTE —
Throttle switch and/or throttle position sensor adjustment should be checked anytime the throttle is adjusted. See **6.3 Idle Speed**.

Removing and Installing Throttle Housing

If the throttle housing is being removed for the purpose of adjusting the throttle switch (528e models only), first check that the throttle basic adjustment is correct before removing the housing. See **Throttle Basic Adjustment** above.

NOTE —
On 528e models, removing the throttle housing requires that some of the engine coolant be drained as described in **COOLING SYSTEM**.

To remove and install throttle housing:

1. Disconnect the cables from the throttle levers. See Fig. 6-7.

2. On 528e models, drain approximately 2 quarts of coolant from the cooling system.

3. Remove the large hose clamps and remove the rubber intake-air duct from the throttle housing.

4. On 528e models, disconnect the crankcase ventilation hose from the housing. On 528e models with Motronic 1.1, separate the hose from the idle air stabilizer valve. See Fig. 6-8.

5. Disconnect the harness connector from the throttle switch.

Fig. 6-8. Throttle housing, showing intake duct removed on 528e models with Motronic 1.1. Hose for idle air stabilizer valve shown at (**A**) and crankcase ventilation at (**B**).

6. On 528e models with Motronic 1.1, disconnect the connector from the evaporative purge valve beneath the throttle housing.

7. On 528e models, loosen the hose clamps and remove the coolant hoses from the housing. See Fig. 6-9.

8. Remove the four mounting bolts from the housing and withdraw the housing from the intake manifold. See Fig. 6-10.

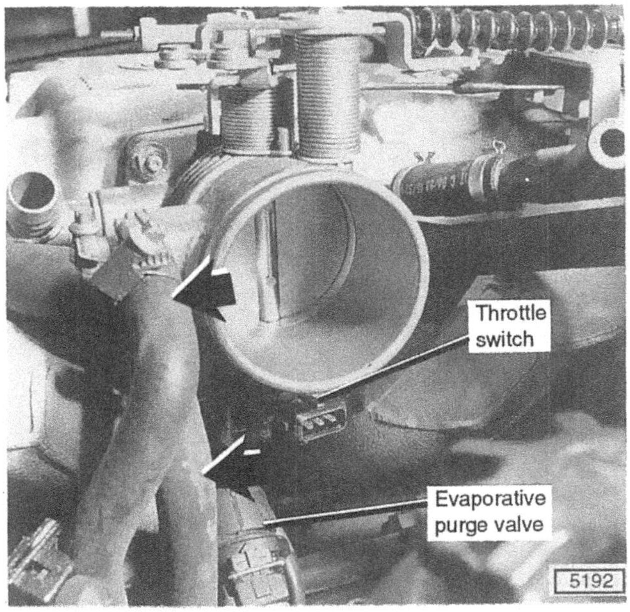

Fig. 6-9. Coolant hoses to be removed from throttle housing (arrows).

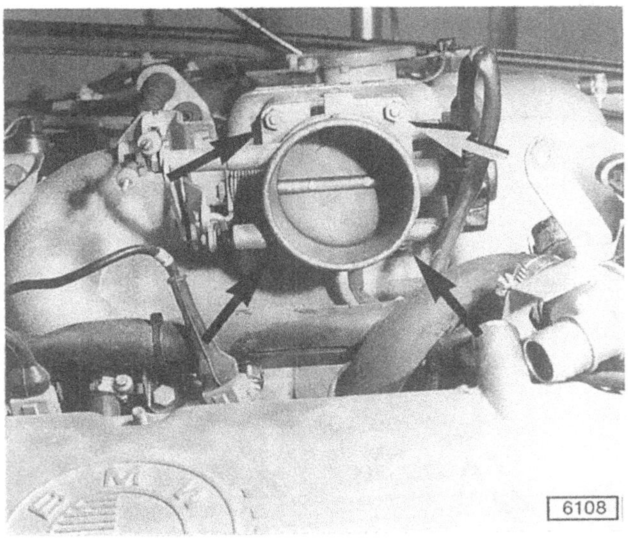

Fig. 6-10. Throttle housing mounting bolts (arrows) on 533i and 535i models. 528e models are similar.

Installation is the reverse of removal. Replace the gasket between the housing and the intake manifold. Adjust the accelerator cable as described earlier under **4.6 Accelerator Cable**. Adjust the automatic transmission accelerator cable as described in **AUTOMATIC TRANSMISSION**. Reconnect the cruise control cable. On 528e models, refill and bleed the cooling system as described under **COOLING SYSTEM**. Lubricate the cables and throttle shafts as described in **LUBRICATION AND MAINTENANCE**. Before driving the car, press the accelerator pedal a few times to check that the cables and throttle are not binding.

6.2 Cold Start and Cold Running Enrichment

When an engine is cold, additional fuel is needed for starting and during warmup. Two methods are used on models with Motronic fuel injection.

On all models except those with Motronic 1.1, an electrically operated cold-start valve sprays extra fuel into the intake manifold for several seconds during starting.

On models with Motronic 1.1, an auxiliary cold-start system is not used. Instead, cold-start enrichment is handled in the same away as cold running enrichment. Based on the signal from the coolant temperature sensor, additional fuel is supplied to the engine during starting by increasing injector opening time.

7

On all Motronic models, to ensure smooth running during warmup, the coolant temperature sensor, and on some models, the coolant temperature switch, give the control unit information on engine temperature. The control unit, in turn, makes an adjustment to injector opening time. For additional cold running driveability and fuel economy, the throttle housing is heated by engine coolant. This system warms the air entering the engine in cold weather to reduce fuel condensation and improve mixture formation.

Cold-Start Valve

The cold-start valve is controlled by the thermo-time switch. Below a certain engine coolant temperature, the thermo-time switch is closed, allowing power to reach the cold-start valve and open it when the starter is actuated. When the switch warms up from engine heat it opens, and the valve does not operate during starting. To further limit valve operation and prevent flooding, an electric current also warms the switch and opens the circuit after a few seconds. The thermo-time switch is located in the coolant outlet housing. See Fig. 6-11. The cold-start valve is shown in Fig. 6-12 and Fig. 6-13.

If the cold-start valve fails to inject fuel during starting, it will be difficult or impossible to start the engine. If the cold-start valve leaks or stays open too long, the engine may receive extra fuel at the wrong time and become flooded, especially if the engine is hot.

To test cold-start valve:

1. Make sure engine coolant temperature is below 86° F (30° C). Preferably the engine should sit for several hours.

2. Remove the two internal-hex-head screws holding the cold-start valve to the intake manifold. Without disconnecting the fuel line or the harness connector, remove the cold-start valve from the intake manifold.

3. Disable the ignition system as described **IGNITION**.

Fig. 6-11. Thermo-time switch (arrow) with brown harness connector on 533i and 535i models. Early 528e models are similar.

Fig. 6-12. Cold-start valve (arrow) controlled by thermo-time switch is used to provide extra fuel for cold starting. 528e models shown.

4. Wipe the cold-start valve nozzle dry. Point the valve into a container, then have a helper actuate the starter. See Fig. 6-14. The valve should spray in an even, cone-shaped pattern until the thermo-time switch interrupts the circuit, up to a maximum of eight seconds. An irregular spray pattern indicates a dirty or faulty cold-start valve. If the valve does not spray, test the thermo-time switch as described below.

WARNING —
Fuel will be discharged. Do not disconnect any wires that could cause electrical sparks. Do not smoke or work near heaters or other fire hazards. Keep an approved fire extinguisher handy.

Fig. 6-13. Cold start valve (arrow) on 533i and 535i models.

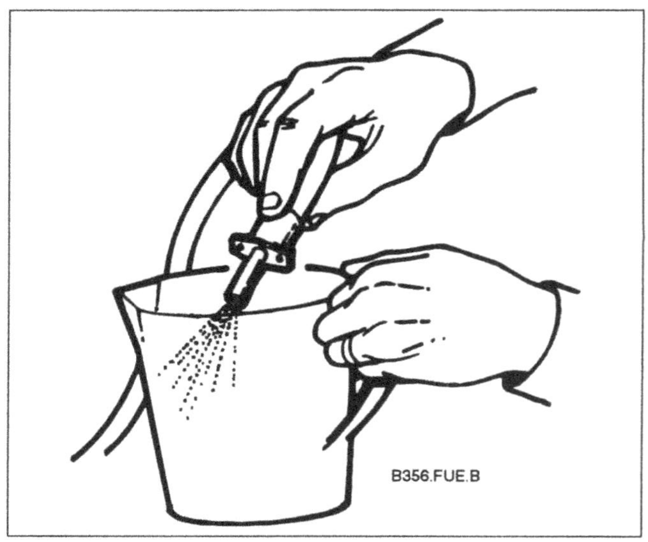

Fig. 6-14. Cold-start valve operation being checked.

5. Wipe the nozzle dry. The valve should not drip for at least a minute. If it does, it is faulty and should be replaced.

To test thermo-time switch:

1. Make sure engine coolant temperature is below the "switch closed" temperature. See specifications listed below. Preferably the engine should sit for several hours.

   NOTE —
   There are "freeze sprays" available from electronic supply houses that can be used to chill the coolant temperature switch to a closed position.

2. Disable the ignition system as described in **IGNITION**.

3. Disconnect the cold-start valve harness connector and connect a test light or voltmeter across the connector terminals, as shown in Fig. 6-15.

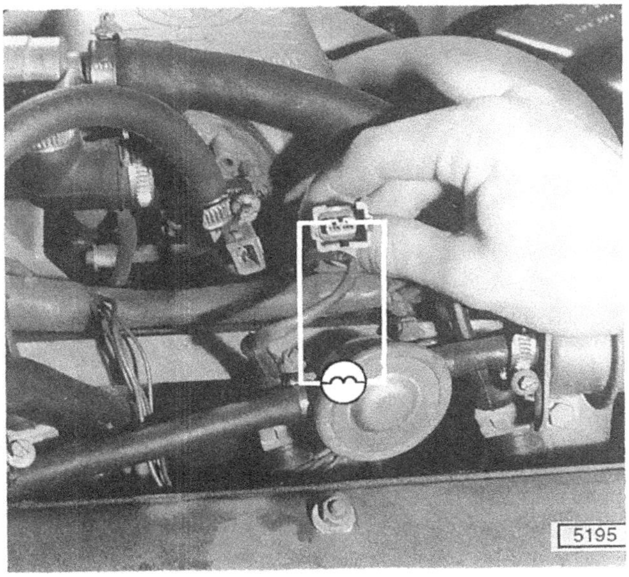

Fig. 6-15. Schematic view of test light connected across cold-start valve harness connector.

4. Actuate the starter while observing the test light. The test light should light for a few seconds, then go out.

   NOTE —
   The amount of time that the test light stays on depends on coolant temperature and cranking time.

If the test light does not light, it is either because voltage is not reaching the connector due to faulty wiring, or because the path to ground (through the thermo-time switch) is interrupted by a faulty switch.

Check for voltage at the removed thermo-time switch harness connector. There should be battery voltage between terminal G (black/yellow wire) and ground when the starter is actuated.

Check for a faulty switch by checking for continuity between terminal W (terminal corresponding to black/red or white wire) and ground. At coolant temperatures at or below the specified switch closed temperature listed below, there should be continuity to ground (switch closed). At coolant temperatures at or above the specified open temperature, there should be a specified resistance as listed below (switch open). If any faults are found, the switch should be replaced.

NOTE —
• Three versions of thermo-time switches are used on the cars covered by this manual. Each thermo-time switch is stamped with the nominal opening temperature and the maximum length of time that the switch remains closed. For example, a switch stamped **35° C 8s** would open at temperatures greater than 95° F (35° C) and would stay closed for a maximum of eight seconds and then open.

• On 1982 528e models and 1983 533i models, excessively long cranking time may develop at ambient temperatures above 59° F (15° C). If this problem occurs, replace the 15° C thermo-time switch with a 35° C switch.

Thermo-time switch tests	
59° F (15° C) 8s switch	
• test condition	below 50° F (10° C)
resistance between terminal G and switch housing (ground)	50 to 70 ohms
resistance between terminal W and switch housing (ground)	0 ohms
• test condition	above 68° F (20° C)
resistance between terminal G and switch housing (ground)	50 to 70 ohms
resistance between terminal W and switch housing (ground)	infinite ohms
95° F (35° C) 8s switch	
• test condition	below 86° F (30° C)
resistance between terminal G and switch housing (ground)	25 to 40 ohms
resistance between terminal W and switch housing (ground)	0 ohms
• test condition	above 105° F (40° C)
resistance between terminal G and switch housing (ground)	50 to 80 ohms
resistance between terminal W and switch housing (ground)	infinite ohms
95° F (35° C) 12s switch	
• test condition	below 86° F (30° C)
resistance between terminal G and switch housing (ground)	25 to 80 ohms
resistance between terminal W and switch housing (ground)	0 ohms
• test condition	above 105° F (40° C)
resistance between terminal G and switch housing (ground)	50 to 100 ohms
resistance between terminal W and switch housing (ground)	infinite ohms

Replacing the thermo-time switch will require draining and replacing some of the engine coolant. See **COOLING SYSTEM**. Use a thread sealant when installing the new switch.

Tightening torque
• thermo-time switch to coolant outlet . . . . 20 to 25 Nm (15 to 18 ft-lb)

Coolant Temperature Sensor

The coolant temperature sensor is located in the cylinder head coolant outlet, next to the thermo-time switch. See Fig. 6-16. The signal from this sensor is the main temperature input to the Motronic control unit and the idle speed control unit (if applicable). To test the sensor, disconnect the harness connector and measure the resistance across the sensor's terminals. The proper resistance value depends on engine coolant temperature as listed in **Table i.**

Fig. 6-16. Coolant temperature sensor on 533i and 535i models (**A**) and 528e models (**B**).

Table i. Coolant Temperature Sensor Test Values

Model	Test temperature	Resistance (ohms)
528e (except with Motronic 1.1), 533i	14±2° F (–10±1° C)	7000–11600
	68±2° F (20±1° C)	2100–2900
	176±2° F (80±1° C)	270–400
528e (with Motronic 1.1), 535i	14±2° F (–10±1° C)	8200–10500
	68±2° F (20±1° C)	2200–2700
	176±2° F (80±1° C)	300–360

If the resistance of the sensor is incorrect, it is faulty and should be replaced. Replacing the sensor will require draining and replacing some of the engine coolant. See **COOLING SYSTEM.**

Coolant Temperature Switch
(528e models built up to March 1987)

The coolant temperature switch is located in the cylinder head coolant outlet, next to the thermo-time switch. The switch turns a ground circuit for idle control on and off, depending on coolant temperature. The coolant temperature switch can be identified by its two single push-on connectors. See Fig. 6-17.

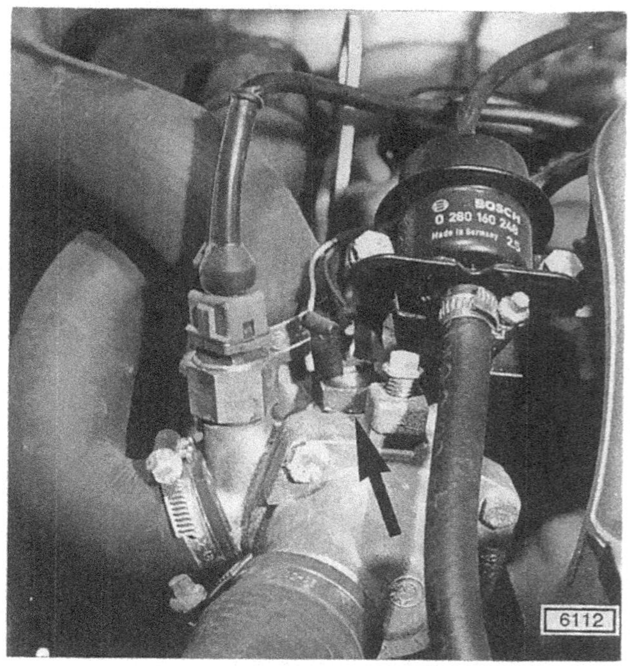

Fig. 6-17. Coolant temperature switch (arrow) used on all 528e models built up to March 1987.

To test the switch, disconnect both wires from the switch. Using an ohmmeter, check for continuity across the switch terminals at the temperatures listed below.

Coolant temperature switch tests	
• 86° F (30° C)	switch open (no continuity)
• 118° F (48° C)	switch closed (continuity)

If any faults are found, the switch should be replaced. Replacing the switch will require draining and replacing some of the engine coolant. See **COOLING SYSTEM**.

6.3 Idle Speed

Engine idle speed is controlled by an electronic idle stabilization system. In this system, an idle air stabilizer valve regulates a small amount of air that bypasses the throttle plate. By this method, idle speed is more accurate and reliable as it compensates for varying engine loads and operating conditions.

On 1982 through 1987 (built up to March 1987) 528e and all 533i models, the idle air stabilizer valve is controlled by a separate idle speed control unit. See Fig. 6-18. On 535i, and 528e models with Motronic 1.1 (built after March 1987), the idle air stabilizer valve is controlled by the Motronic control unit.

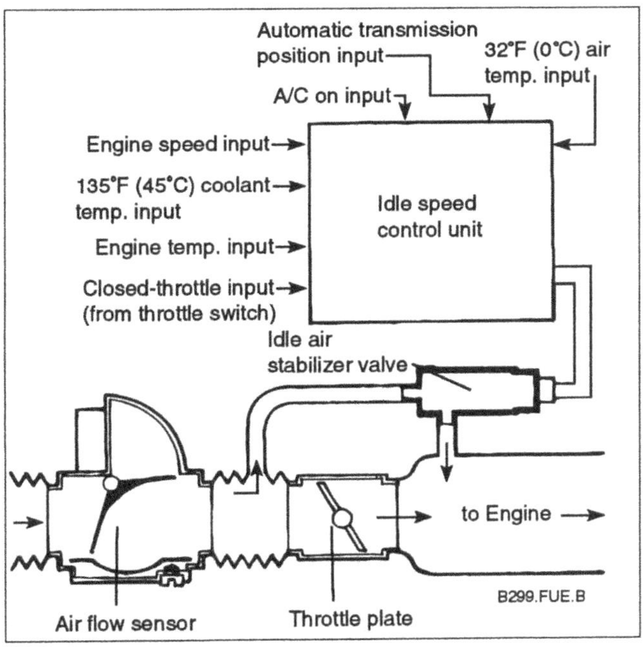

Fig. 6-18. Schematic view of idle air stabilization system installed on models with separate idle speed control unit.

NOTE —
Both the Motronic control unit and the idle speed control unit are located above the glove box.

Idle speed cannot be adjusted, as it is electronically regulated by the idle stabilization system. On some early models, an adjustment screw in the idle air stabilizer valve can be used to adjust the valve's base setting (middle of its adaptive range), but changing the position of this screw should have no effect on idle speed. The valve normally does not require adjustment unless the factory setting has been tampered with. An incorrect or erratic idle speed is usually caused by a faulty idle speed stabilization system. See **ENGINE MANAGEMENT—DRIVEABILITY** for more information on troubleshooting idle speed problems.

7

Idle mixture should be checked before troubleshooting the idle stabilization system. See **6.6 Idle Specifications (rpm and % CO)**. If the equipment needed to do this work is not available, we suggest turning the job over to an authorized BMW dealer or other qualified repair shop. In a properly equipped shop, these checks and any applicable adjustments can be made quickly, accurately, and at reasonable cost.

When checking the idle stabilization system, the following requirements apply:

1. The engine must be at normal operating temperature (oil temperature at approximately 140° F (60° C)).
2. All electrical accessories should be off, including the auxiliary radiator cooling fan. Make adjustments only when the fan is not on.
3. The throttle switch must be operating correctly.
4. The accelerator cable and the basic throttle must be correctly adjusted.
5. The exhaust system must be free of leaks.
6. There must be no engine vacuum leaks.
7. The oxygen sensor must be operating correctly.

Connect a tachometer according to the instrument manufacturer's instructions in order to accurately measure rpm. The ignition signal lead from the tachometer should be connected to terminal 1 of the coil. See **8. Technical Data** for idle speed checking specifications.

CAUTION —
The ignition must be off before disconnecting or connecting any electrical connections.

Idle Air Stabilizer Valve

The electronic idle air stabilizer valve controls the amount of air allowed to bypass the throttle to either increase or decrease idle speed based on varying operating conditions. Whenever the throttle switch is closed, the valve receives voltage from the electronic control unit.

To quickly check that the valve is functioning, turn the ignition on or start the engine. The valve should vibrate and hum slightly. If the valve is not operating, check that the idle switch is operating correctly as described below. If the valve is functioning but idle speed is erratic, test the valve as described below.

NOTE —
Two different idle stabilizer valves are installed on the cars covered by this manual. On 1982 through 1987 (up to March 1987) 528e and all 533i models, the stabilizer valve has two wires. On all 535i models and 1987 (from March 1987) and 1988 528e models, the stabilizer valve has three wires. Testing procedures vary depending on the type of valve.

On two- or three-wire valves, first check the valve's resistance. If any faults are found, the valve should be replaced. If no faults are found continue troubleshooting as described below.

Idle air stabilizer valve resistance	
• two-wire valve	9 to 10 ohms (approximate)
• three-wire valve	
between outer terminals	40 ohms (approximate)
between center terminal and each outer terminal	20 ohms (approximate)

On two-wire valves, remove the hoses and the electrical connector from the valve. Carefully, using fused jumper wires, apply battery voltage to the valve's terminals as shown in Fig. 6-19. The valve's piston should close when voltage is applied and open when voltage is removed. If any faults are found, the valve is faulty and should be replaced. If no faults are found, check the control signal to the valve as described below in **Idle Speed Control**.

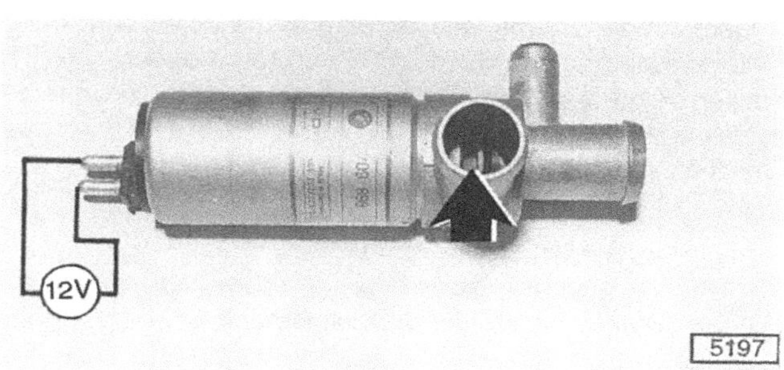

Fig. 6-19. Schematic view of battery voltage being applied to two-wire idle air stabilizer valve. Valve piston (arrow) should close when voltage is applied.

On three-wire valves, remove the two hoses and the electrical connector from the valve. Withdraw the valve from the rubber mount. Check that the valve's piston moves freely by quickly rotating the valve back and forth. If no faults are found, reinstall the valve and reconnect the wiring. Check the control signal to the valve as described below in **Idle Speed Control**.

NOTE —
The idle air stabilizer valve piston can be cleaned with a carburetor spray cleaner if there is carbon build-up.

Idle Speed Control

If no faults are detected with the idle air stabilizer valve, but the idle speed is not within specifications, check the idle stabilization function of the control unit.

On models with two-wire idle stabilizer valves, check the signal from the idle speed control unit by measuring the current to the valve. Fig. 6-20 shows an ammeter (0–1000 mA range) correctly connected to the idle air stabilizer valve. With the engine idling at operating temperature, the current reading should fluctuate within the range listed below.

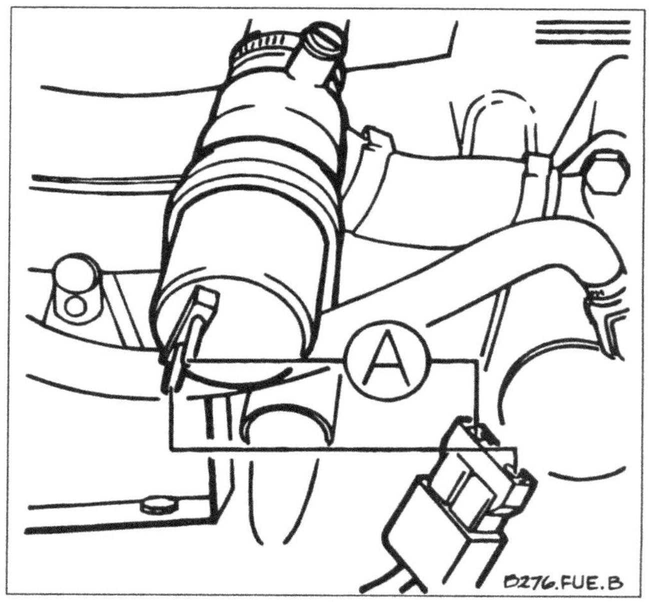

Fig. 6-20. Schematic view of ammeter connected to idle air stabilizer valve to measure current from control unit, on models with 2-wire valves.

Two-wire idle speed stabilizer valve control current	
• checking value	400 to 500 mA

NOTE —
Current will fluctuate significantly if the engine is cold, if the coolant temperature switch or coolant temperature sensor is faulty, if there is an intake air leak, or if any electrical accessories are on.

If there is no current reading, check the wires between the idle speed control unit and the valve. If the current is not as specified, adjust the base setting of the idle air stabilizer valve as described below.

To adjust idle air stabilizer valve (two wire valve with external adjusting screw):

1. Allow the engine to reach operating temperature.
2. Connect a tachometer according to the instrument manufacturer's instructions in order to accurately measure rpm.
3. Connect an ammeter to the valve as shown above in Fig. 6-20 and slowly turn the valve's bypass screw until the current reading is correct.

Two-wire idle air stabilizer valve current
• adjusting . . . . 460±10 mA at 700±50 rpm.

If the valve cannot be correctly adjusted, disconnect the battery then remove the idle speed control unit from above the glove compartment and disconnect its harness connector. Working at the harness connector, test the inputs to the control unit. Fig. 6-21 shows the connector terminals. **Table j** lists the electrical tests. If no electrical faults can be found, the control unit is probably faulty and should be replaced.

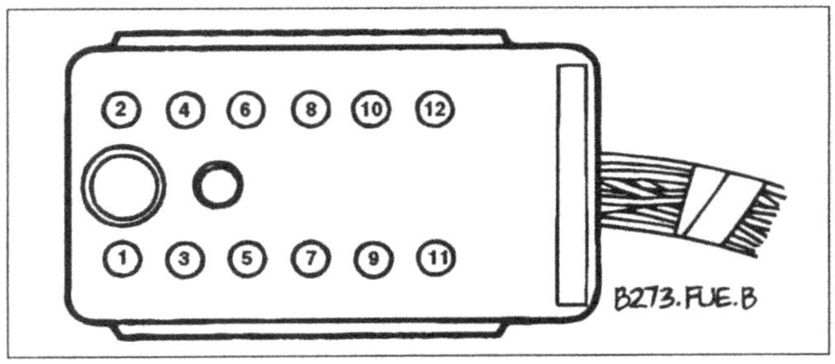

Fig. 6-21. Idle speed control unit harness connector terminal identification. Terminal numbers are also printed on connector. Remove trim panel above glove compartment to reach control unit.

Table j. Idle Speed Control Unit Electrical Tests (1982 through 1987 (up to March 1987) 528e and 533i models)

Component or circuit being tested	Test terminals on harness connector	Test conditions	Correct test value
Voltage supply to control unit	2 and ground	ignition ON	battery voltage (approximately 12 VDC)
Main ground	4 and ground		continuity
Coolant temp. switch	2 (+) and 6 (–)	ignition ON, coolant temp. above 118° F (48° C)	battery voltage (approximately 12 VDC)
		ignition ON, coolant temp. below 86° F (30° C)	no voltage
Air temp. switch	10 (+) and 4 (–)	ignition ON, air temp. below 18° F (–8° C)	battery voltage (approximately 12 VDC)
		ignition ON, air temp. above 39° F (4° C)	no voltage
Aut. trans. range switch	7 (+) and 4 (–)	ignition ON, selector lever in Neutral position	battery voltage (approximately 12 VDC)
	8 (+) and 4 (–)	ignition ON, selector lever in Park position	battery voltage (approximately 12 VDC)
Throttle switch	12 (+) and 4(–)	ignition OFF, accelerator pedal in rest position	continuity
		ignition OFF, accelerator pedal slightly depressed	no continuity
Air cond. ON signal	9 (+) and 4 (–)	ignition ON, air cond. ON	battery voltage (approximately 12 VDC)
Engine temp.	11 (+) and 4 (–)		resistance varies with temperature. Test coolant temperature sensor. See **6.2 Cold Start and Cold Running Enrichment**
Idle air stabilizer valve	1 and 5	ignition OFF	9–10 ohms at 73±9° F (23±5° C)
Engine speed (rpm)	3 and ground	LED test light connected between terminal 3 and ground. starter actuated	LED must flicker

NOTE —
528e models built before January 1985 may experience poor warm engine driveability after restarting the engine when the air conditioner is on and when outside temperatures are above 68° F. If these symptoms are present, disconnect the harness connector at the idle speed control unit and peel back the rubber boot from the connector. Insulate the wire end and tape it back to the harness. Reposition the boot and reconnect the connector.

On models with three-wire valves, check idle speed control by starting the engine and allowing it to reach operating temperature. Then turn off the engine and all electrical accessories. Remove the hose so that the piston can be seen and turn the ignition on. The valve's piston should move to approximately halfway across the valve's opening. See Fig. 6-22.

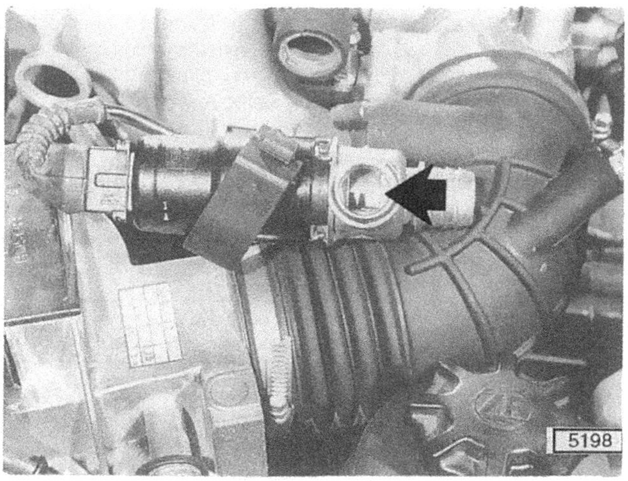

Fig. 6-22. Three-wire idle stabilizer valve piston (arrow) in center position with harness connector connected and ignition on.

NOTE —
Idle stabilizer valve piston position can vary significantly if the engine is cold, if the coolant temperature sensor is faulty, if the throttle switch is open, or if any electrical accessories are on.

If the piston position is incorrect, check for voltage at the harness connector with the ignition on. See Fig. 6-23. There should be battery voltage (approximately 12 VDC).

If there is no voltage, check power to the connector. See **7.1 Wiring Schematics** for more information. If there is power and the valve still does not operate correctly, then either the wiring to the control unit or the control unit itself is faulty.

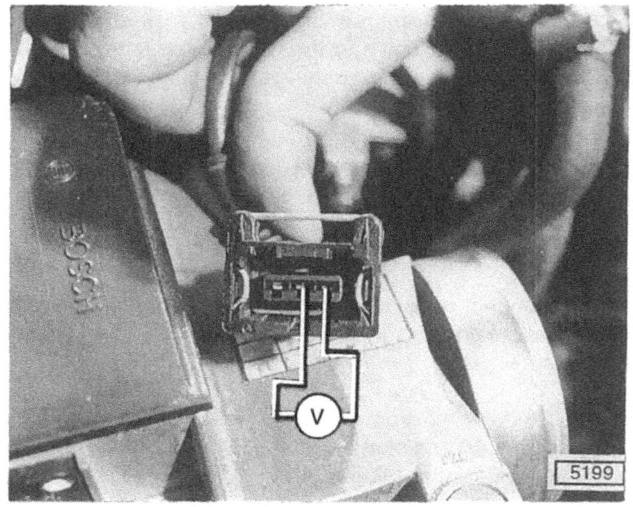

Fig. 6-23. Voltmeter (shown schematically) being used to test idle stabilizer valve voltage signal from the control unit on models with 3-wire valves.

Idle and Full Throttle Signal

The throttle position signal to the Motronic and idle speed control units is handled in one of two ways depending on the model and type of transmission installed.

On all models except 1986 through 1988 535i models, a combined idle and full throttle switch is used. On 1986 through 1988 535i models with electronically-controlled automatic transmission, a separate idle position switch is used for the closed throttle signal, while a throttle position sensor (potentiometer) is used for the full throttle signal. Both switch types and the position sensor are mounted to the throttle housing.

NOTE —
On 1986 through 1988 535i models with automatic transmission, the throttle position sensor is wired directly to the automatic transmission control unit. When the throttle is in the full open position, the automatic transmission control unit sends a "full throttle" signal to the Motronic control unit.

An incorrectly adjusted switch can cause an erratic idle speed. The switch can be easily checked with the throttle housing installed

NOTE —
A digital multimeter should be used to check and adjust the throttle switch(es) and throttle position sensor. Some analog meters may not be sensitive enough.

Before testing the throttle switch or idle position switch, check that it is receiving voltage. Disconnect the harness connector. With the ignition on, there should be 5 volts (VDC) between the center terminal and either of the outer terminals of the harness connector. If voltage is not present, test the electronic control unit inputs as described under **6.4 Electrical Tests**.

When testing the throttle switch, remove the harness connector from the switch and check for continuity at the switch terminals shown in Fig. 6-24. When testing the idle position switch, remove the two harness connectors from the switch when checking for continuity. See Fig. 6-25.

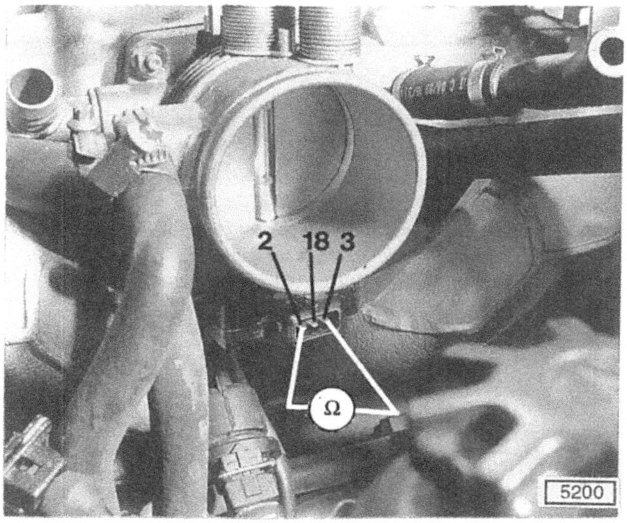

Fig. 6-24. Idle and full throttle switch showing terminal identification. Throttle switch from 528e shown.

Test the throttle switch and idle position switch using the information shown in **Table k.** If either switch does not operate as specified, it should be adjusted. With the switch mounting screws loosened, use an ohmmeter connected across the switch terminals. With the throttle in the rest position, rotate the switch just until there is continuity. Tighten the mounting screws. Check the adjustment of the switch as described above. Seal the mounting screws using paint or lacquer.

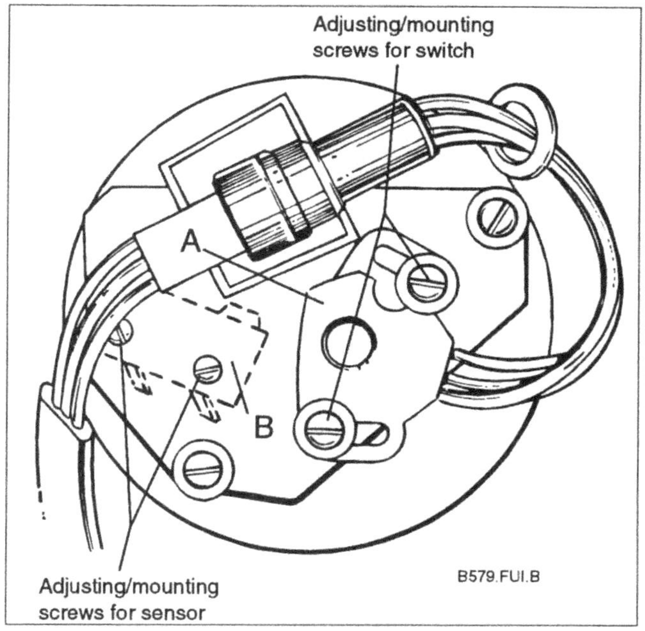

Fig. 6-25. Idle position switch (**A**) and throttle position sensor (**B**) used on 1986 through 1988 535i models with automatic transmission.

NOTE —
On 528e models, the throttle housing will have to be removed to reach the throttle switch mounting screws. Throttle housing removal is described under **6.1 Air Flow Measurement.**

Check the throttle position sensor with the ignition on. Make the tests listed in **Table l** working from the rear of the throttle position sensor harness connector shown in Fig. 6-25 above.

NOTE —
When testing the throttle position sensor, the harness connector must remain connected.

Table k. Throttle Switch and Idle Position Switch Tests

Component	Test conditions	Test terminals	Correct test results
Throttle switch (Harness connector disconnected at switch, ignition OFF. Measure resistance at switch terminals)	throttle closed throttle partially open throttle fully open	2 and 18 2 and 18, 3 and 18 3 and 18	continuity no continuity continuity
Idle position switch (535i models with electronic A/T only) (Wire connectors disconnected at switch, ignition OFF. Measure resistance at switch side of connector)	throttle closed throttle partially open	1 and 2 1 and 2	continuity no continuity

Table I. Throttle Position Sensor Tests (1986-1988 535i with automatic transmission)

Component	Test terminals (wire colors)	Test conditions	Correct test results
Throttle position sensor (Measure voltage at sensor connector (connector remains connected). Probe rear of connector with voltmeter test leads.)	3 (black wire) and ground	Ignition ON, throttle closed	5 VDC (approximate)
	3 (black wire) and 1 (brown wire)		5 VDC (approximate)
	2 (yellow wire) and 1 (brown wire)	Ignition ON, throttle slowly depressed to full open position	0.7 VDC (throttle closed) to 4.78* VDC (throttle fully open) (Check that voltage increases evenly as the throttle is depressed to the full open position)

*The wide open throttle voltage should be .22 volts less than the stabilized voltage. Stabilized voltage is measured between terminals 3 and 1.

If the sensor does not operate as specified, it should be adjusted. With the sensor mounting screws loosened, connect a voltmeter between terminals 2 (yellow wire) and 3 (black wire). With the accelerator pedal fully depressed (including kick-down), rotate the switch just until there is 0.22±.04 volts less than the stabilized voltage. (Stabilized voltage is measured between terminals 3 and 1.) Tighten the mounting screws. Check the adjustment of the sensor as described above. Seal the mounting screws using paint or lacquer. If the sensor cannot be correctly adjusted, it should be replaced.

6.4 Electrical Tests

These voltage and continuity tests can be used to help determine whether there are faults in the wiring or components that provide information to the Motronic control unit. If all inputs are found to be correct and the system still does not perform as specified, the control unit itself may be faulty.

As a general rule, a complete absence of the voltage or continuity specified in the tests suggests an open circuit in the wiring harness. Test results that differ from the specified values do not necessarily mean that components are faulty. Check carefully for connections that are loose, corroded or otherwise damaged. Before testing, disable the ignition system as described in **IGNITION.**

With the battery disconnected, remove the Motronic control unit connector retaining screw, if applicable, and disconnect the connector. The connector pivots at one end. Use a multimeter to check for voltage or continuity at the connector terminals. Use care to avoid spreading the small connector terminals with the meter probes. For best and safest results, make a set of test leads as shown in Fig. 6-26. The Motronic control unit is located above the glove compartment. See Fig. 6-27. Test terminals are identified in Fig. 6-28. Tests are listed in **Table m** and **Table n**.

CAUTION —
- *Use only a digital voltmeter-ohmmeter, or multimeter with high input impedance. The electrical characteristics of other types of test equipment may cause inaccurate results or damage to the electronic components. For more information, see **FUNDAMENTALS**.*

- *The ignition must be off before disconnecting or connecting any electrical connections.*

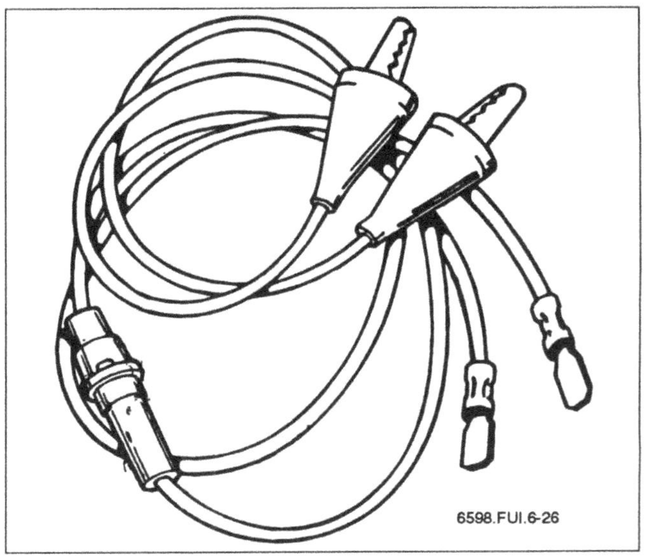

Fig. 6-26. Test leads made with flat connectors, alligator clips, and in-line fuse. Test leads are used to make contact without damaging control unit connector terminals.

Fig. 6-27. Motronic control unit and connector. Retaining screw and lock (arrows) used on some models. Control unit is above glove box.

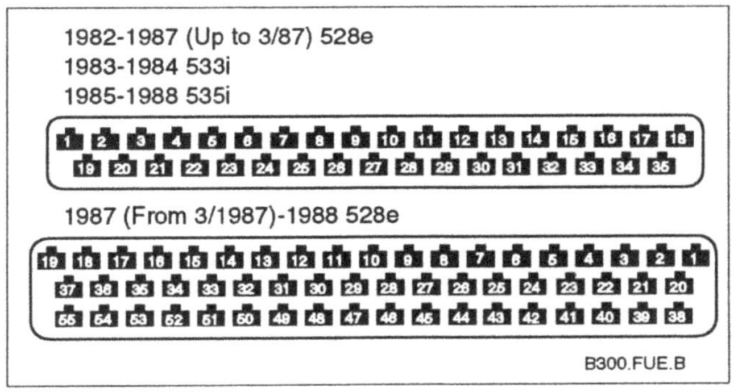

Fig. 6-28. Terminal identification for 35-point Motronic control unit connector (top) and 55-point Motronic 1.1 control unit connector (bottom).

NOTE —
For accurate test results, battery voltage at the connector must be 12 to 13 volts. Charge the battery if necessary. If voltage is too high, turn on the headlights to lower voltage.

Table m. Motronic Control Unit Harness Electrical Tests (1982 through 1987 (up to March 1987) 528e models, all 533i and 535i models)

Component or circuit	Test terminals (at 35-point connector unless otherwise specified)	Test conditions	Correct test value
Voltage supply to control unit	18 and ground 35 and ground	ignition ON	battery voltage (approximately 12 VDC)
Main grounds	5 and ground 16 and ground 17 and ground 19 and ground	ignition OFF	continuity
Fuel pump relay control	20 and ground	ignition ON	fuel pumps run (audibly) when terminal 20 is touched to ground
Starter input (terminal 50)	4 and ground	actuate starter	8 VDC (minimum)
Throttle switch (except 1986-1988 535i models with A/T)	2 and ground 2 and ground 3 and ground	throttle closed open throttle slowly to half-open position throttle fully open	continuity no continuity continuity
Idle position switch (1986-1988 535i models with A/T only)	2 and ground	throttle closed open throttle slowly to half-open position	continuity no continuity
Idle air stabilizer valve (535i models only)	34 and 33		40 ohms resistance
Engine speed sensor	8 and 27	ignition OFF	960±96 ohms resistance
Reference sensor	25 and 26	ignition OFF	960±96 ohms resistance
Coolant temperature sensor	13 and ground	ignition OFF	resistance varies with temperature. See **6.2**
Fuel injector control (injectors 1, 2, and 3)	14 and ground	ignition ON	fuel injectors (1, 2, and 3) click when terminal 14 is touched to ground
Fuel injector control (injectors 4, 5, and 6)	15 and ground	ignition ON	fuel injectors (4, 5, and 6) click when terminal 15 is touched to ground

continued on next page

Table m. Motronic Control Unit Harness Electrical Tests (1982 through 1987 (up to March 1987) 528e models, all 533i and 535i models) (continued)

Component or circuit	Test terminals (at 35-point connector unless otherwise specified)	Test conditions	Correct test value
Air flow sensor	7 and 9	move sensor flap by hand or actuate starter	resistance must fluctuate
Oxygen sensor	24 and ground	separate oxygen sensor connector and connect green wire to ground	continuity
Evaporative purge valve control (1982-1984 528e only)	31 and ground	ignition ON	purge valve clicks when terminal 31 is touched to ground
Air conditioning ON signal (535i only)	29 and ground	ignition on, A/C switch on	battery voltage (approximately 12 VDC)

Table n. Motronic 1.1 Control Unit Harness Electrical Tests (1987 (from March 1987) and 1988 528e models)

Component or circuit	Test terminals (at 55-point connector unless otherwise specified)	Test conditions	Correct test value
Main grounds	2 and ground 14 and ground 19 and ground 24 and ground		continuity
Voltage supply to control unit	18 and ground	Ignition OFF	battery voltage (approximately 12 VDC)
Voltage supply to control unit	27 and ground	Ignition ON	battery voltage (approximately 12 VDC)
Voltage supply to control unit	37 and ground	connect terminal 36 to ground with fused jumper wire	battery voltage (approximately 12 VDC)
Main relay control	36 and ground		Main relay must click on when terminal 36 is touched to ground
Fuel pump relay control	3 and ground	connect terminal 36 to ground with fused jumper wire	Fuel pumps run (audibly) when terminal 3 is touched to ground
Throttle switch	52 and ground	throttle closed open throttle slowly to half-open position throttle fully open	continuity no continuity continuity
Fuel injector control (injectors 1, 3, and 5)	16 and ground	connect terminal 36 to ground with fused jumper wire	fuel injectors (1, 3, and 5) click when terminal 16 is touched to ground
Fuel injector control (injectors 2, 4, and 6)	17 and ground	connect terminal 36 to ground with fused jumper wire	fuel injectors (2, 4, and 6) click when terminal 17 is touched to ground
Pulse sensor	48 and 47		540±54 ohms resistance
Coolant temperature sensor	45 and ground		resistance varies with temperature. See **6.2**
Idle air stabilizer valve	4 and 22		40 ohms resistance
Air flow sensor	7 and 12	move sensor flap by hand or actuate starter	Resistance must fluctuate

continued on next page

Table n. Motronic 1.1 Control Unit Harness Electrical Tests (1987 (from March 1987) and 1988 528e models) (continued)

Component or circuit	Test terminals (at 55-point connector unless otherwise specified)	Test conditions	Correct test value
Evaporative emissions purge valve	5 and ground	connect terminal 36 to ground with fused jumper wire	purge valve clicks on when terminal 5 is touched to ground
Cylinder identification sensor	8 and 31		less that 1.0 ohm resistance
Air conditioning on signal	40 and ground 41 and ground	air conditioning switch ON; ignition ON; ambient temperature above 60° F (16° C)	battery voltage (approximately 12 VDC)
Automatic transmission park/neutral signal	42 and ground	selector lever in park or neutral position, ignition ON	battery voltage (approximately 12 VDC)

Testing Main Relay

The Motronic control unit, the fuel injectors, and the fuel pump relay all receive battery voltage via the main relay. On some models the idle air stabilizer valve and the evaporative emission purge valve are also powered by the main relay. If the relay fails, the car will not start. Below is a quick test to help determine if the relay is faulty.

On all models except those with Motronic 1.1, disconnect a harness connector from one of the fuel injectors. On models with Motronic 1.1, disconnect the harness connector from the idle air stabilizer valve. Check for voltage between the red/-white wire of the connector and ground with the ignition on. If battery voltage is present, the relay is probably working.

If voltage is not present, check the voltage supply and wiring to the relay connector first before eliminating the main relay as a source of trouble. Remove the main relay from the relay panel. See Fig. 6-29. Check for voltage between terminal 30 and ground. If battery voltage is not present, check the large red wire between the relay socket and the battery. Turn the ignition key on and check for voltage between terminal 86 and ground. If battery voltage is present and no faults are found in the wires or the connectors, the relay is faulty and should be replaced. For more information see **7. Fuel System Electrical Circuits**.

> ***CAUTION —***
> *The main relay may be located in an adjacent position. To check if the main relay has been correctly identified, inspect the wires leading to the relay socket. There should be two large (4 mm) wires leading to the main relay socket.*

> **NOTE —**
> The on-board computer powers the main relay terminal 86.

Fig. 6-29. Main relay (arrow) on side of fuse/relay panel. On 1985 and 1986 528e models, main relay is located in position (2).

6.5 Idle Specifications (rpm and % CO)

Idle speed and ignition timing are electronically controlled and are therefore not adjustable. In addition, idle mixture is not adjustable on models with Motronic 1.1. However, checking idle speed and idle mixture (% CO) together is a good way to determine the general state of condition of the Motronic system.

The Motronic fuel management systems installed on the cars covered by this manual are adaptive. In other words, if any of the above operating parameters are not within specified limits, the system will electronically self-correct, within limits, until the correct settings are obtained.

Making these checks requires special test equipment. If the equipment necessary to accurately perform this work is not available, we suggest turning the job over to an authorized BMW dealer or other qualified repair shop. In a properly equipped shop, these checks can be made quickly, accurately, and at reasonable cost.

When checking idle mixture, the following requirements apply:

1. The engine must be at normal operating temperature (oil temperature at approximately 140° F (60° C)).

2. All electrical accessories should be off (including the auxiliary radiator cooling fan—make adjustments only when the fan is not on).

3. The throttle switch (or idle position switch) must be operating correctly.

4. The exhaust system must be free of leaks.

5. There must be no engine vacuum leaks.

6. The valves must be correctly adjusted.

7. The oxygen sensor must be operating correctly.

NOTE —
CO readings may be affected by the presence of unburned gasoline and combustion by-products in the engine oil. For the most accurate CO settings, always change the engine oil and filter before making any adjustments.

To check and adjust idle mixture (%CO) (except 528e models with Motronic 1.1):

1. On early 528e and 533i models with electric purge valve, disconnect from the charcoal canister the small hose that runs from the canister to the purge valve. Do not plug the hose. Disconnect the electrical connector from the charcoal canister purge valve.

NOTE —
1982 through 1984 528e and 533i models were originally equipped with an electric purge valve. The purge valve is mounted on the right (passenger) side fender, near the windshield washer reservoir. As part of an authorized BMW dealer field fix, this valve may have been removed.

2. Remove the CO-tap bolt from the exhaust manifold. See Fig. 6-30. Install a threaded adapter with a pipe fitting and connect the hose of the exhaust gas analyzer to the adapter.

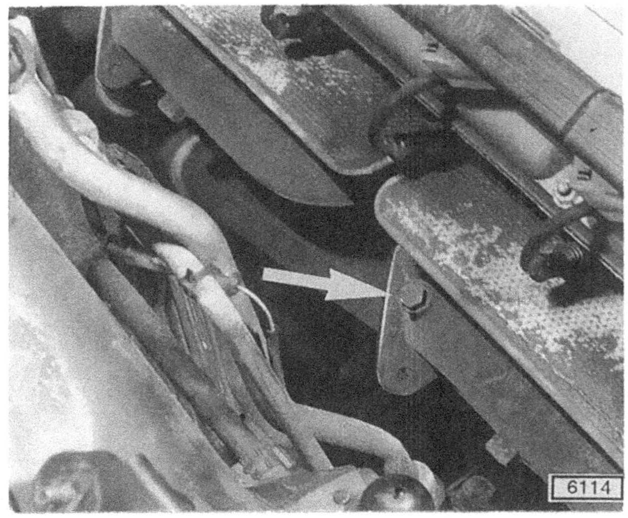

Fig. 6-30. Exhaust manifold CO-tap bolt (arrow). Remove bolt and connect exhaust gas analyzer with adapter.

NOTE —
• Exhaust gas content should only be measured at the exhaust manifold. Checking the exhaust gas content at the tailpipe is checking the gas after it has been cleaned by the catalytic converter.

• An exhaust manifold adapter (BMW tool no. 13 0 100) is available from an authorized BMW dealer parts department.

3. Disconnect the oxygen sensor harness connector. See Fig. 6-31.

Fig. 6-31. Typical location of oxygen sensor harness connector (arrow).

4. On 535i models, briefly disconnect, then reconnect the negative (–) battery cable.

NOTE —
Disconnecting the negative battery terminal on 535i models will cancel any electronic adaptive corrections made by the Motronic control unit.

5. Start the engine and run it to normal operating temperature. Remove the anti-tampering plug. With the engine idling, turn the idle mixture screw using a 5 mm hex wrench. **Table o** lists idle specifications.

NOTE —
• Turning the CO screw counterclockwise makes the mixture leaner (lower % CO). Turning it clockwise makes the mixture richer (higher % CO).

• The idle mixture screw can only be adjusted after removing the anti-tampering plug in the air flow sensor housing. See Fig. 6-32. The air flow sensor anti-tampering plug is removed by drilling a 2.5 mm (3/32 in.) hole into the plug and threading a screw into the plug so that it can be extracted with pliers.

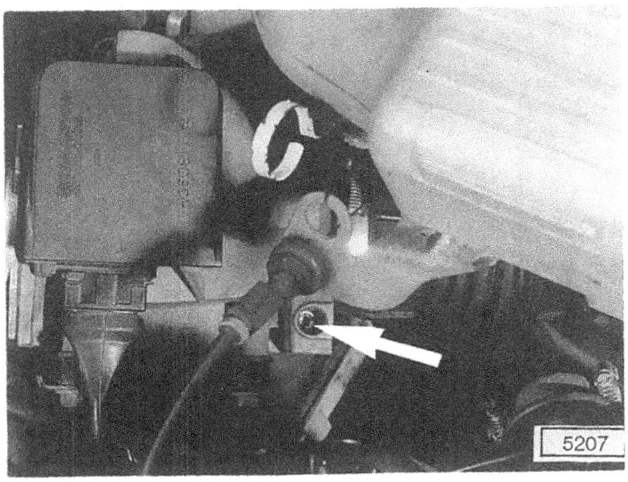

Fig. 6-32. Air flow sensor idle mixture screw (arrow), beneath anti-tampering plug.

Table o. Idle Specifications (except 528e models with Motronic 1.1)

Specification	Value
Idle speed 528e, 533i 535i	 700 rpm±50 800 rpm±50
% CO (all)	0.2–1.2%

NOTE —
Check the idle mixture (or oxygen sensor voltage) with and without the oxygen sensor connected before making any adjustments. The idle mixture (oxygen sensor voltage) should be almost the same in both cases. If there is a big difference, the CO screw may be misadjusted or there may be other faults such as a vacuum leak or a faulty injector. See **ENGINE MANAGEMENT—DRIVEABILITY** for more troubleshooting information.

6. Turn the engine off and disconnect all test equipment.

7. Install the exhaust manifold tap bolt. Reconnect the oxygen sensor harness connector. Replace the anti-tampering plug.

7

7. Fuel System Electrical Circuits

The electrical information under this numbered heading can be used as a guide in diagnosing fuel system electrical problems. When troubleshooting Motronic electrical problems, tight and corrosion-free ground connections, including the negative terminal of the battery and the ground straps, are essential to trouble-free operation. For additional electrical troubleshooting information and tips, see **ELECTRICAL SYSTEM** For more Motronic-specific troubleshooting, including ground locations, see **ENGINE MANAGEMENT—DRIVEABILITY.**

7.1 Wiring Schematics

Figure 7-1, 7-2, and 7-3 are schematic representations of the fuel system electrical circuits used on the cars covered by this manual. Each diagram's figure caption specifies all model years and models that are covered by that diagram. A complete index of the component, ground, and splice locations that appear on the diagrams below can be found directly after the wiring schematics under **7.2 Harness Connector, Ground, and Splice Locations.**

CAUTION —
The information under this heading is intended to be used as a guide in troubleshooting fuel system electrical problems. Some wire colors and sizes as well as components may vary slightly from year to year and from model to model.

NOTE —
A full listing of the wiring and component symbols used in the fuel system diagrams as well as general instructions on using the wiring diagrams can be found in **ELECTRICAL SYSTEM.**

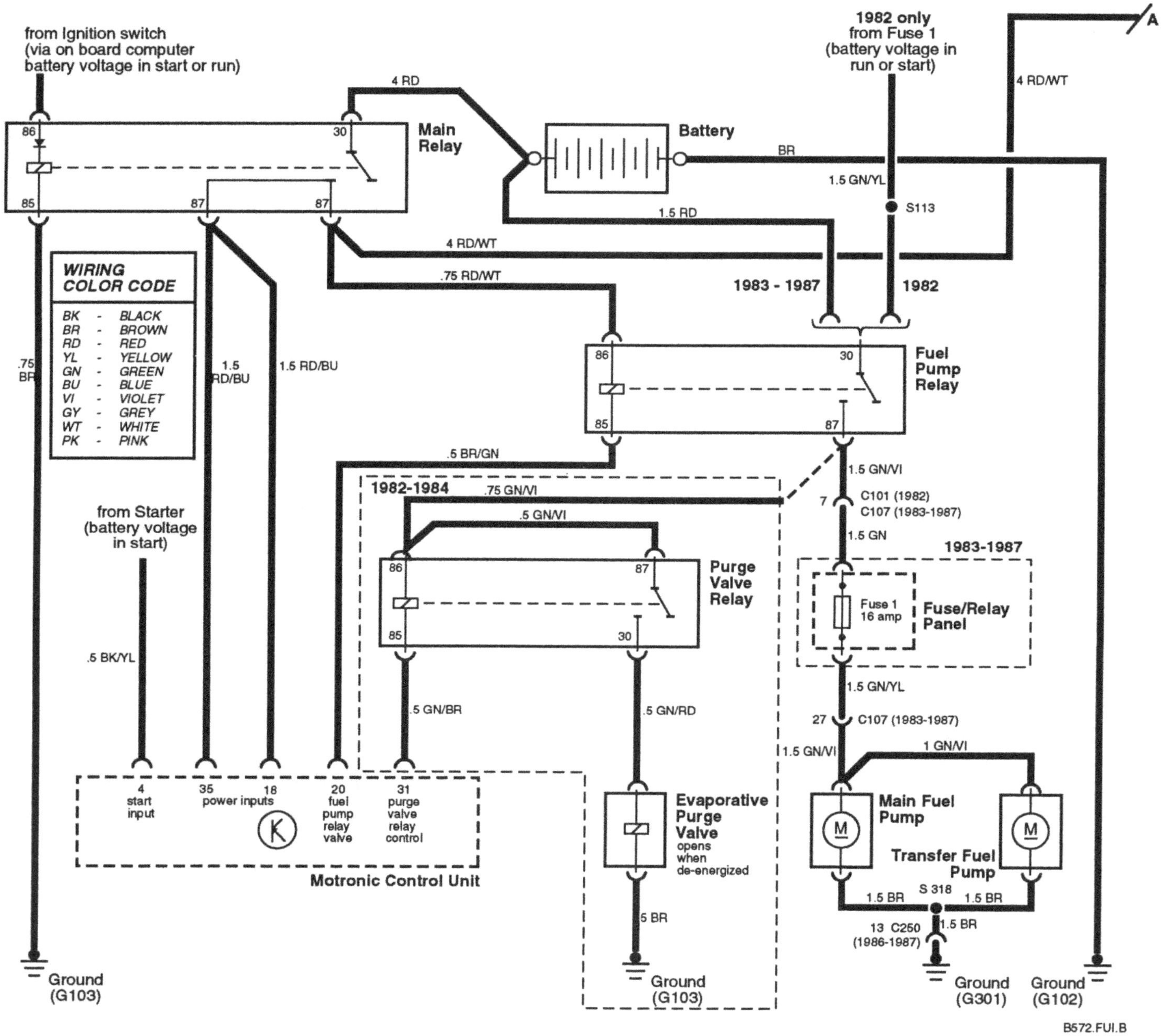

Fig. 7-1. Schematic wiring diagram for 1982 through 1987 (built up to March 1987) 528e models and 1983 and 1984 533i models.

continued on next page

WIRING COLOR CODE		
BK	-	*BLACK*
BR	-	*BROWN*
RD	-	*RED*
YL	-	*YELLOW*
GN	-	*GREEN*
BU	-	*BLUE*
VI	-	*VIOLET*
GY	-	*GREY*
WT	-	*WHITE*
PK	-	*PINK*

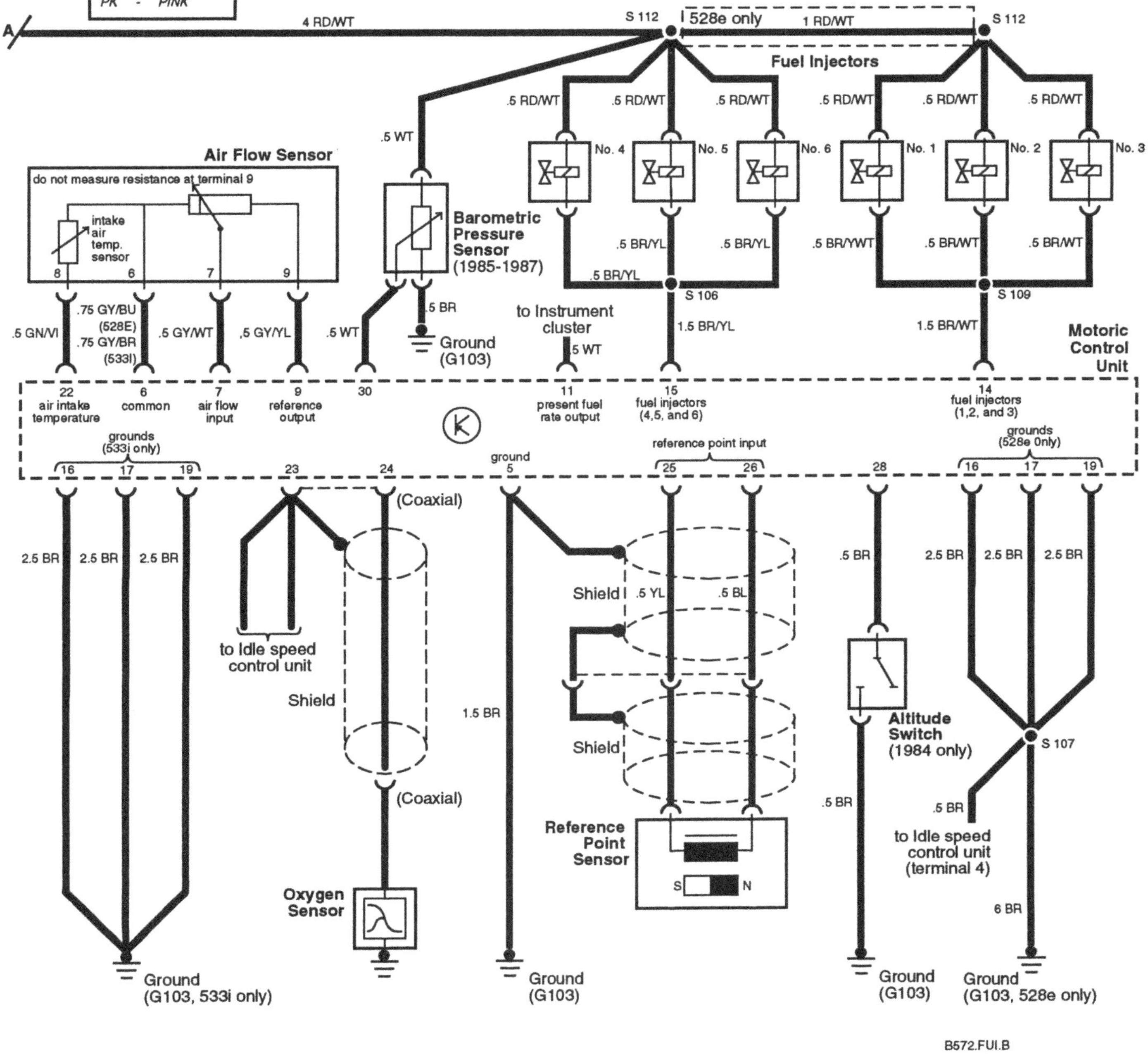

continued on next page

7

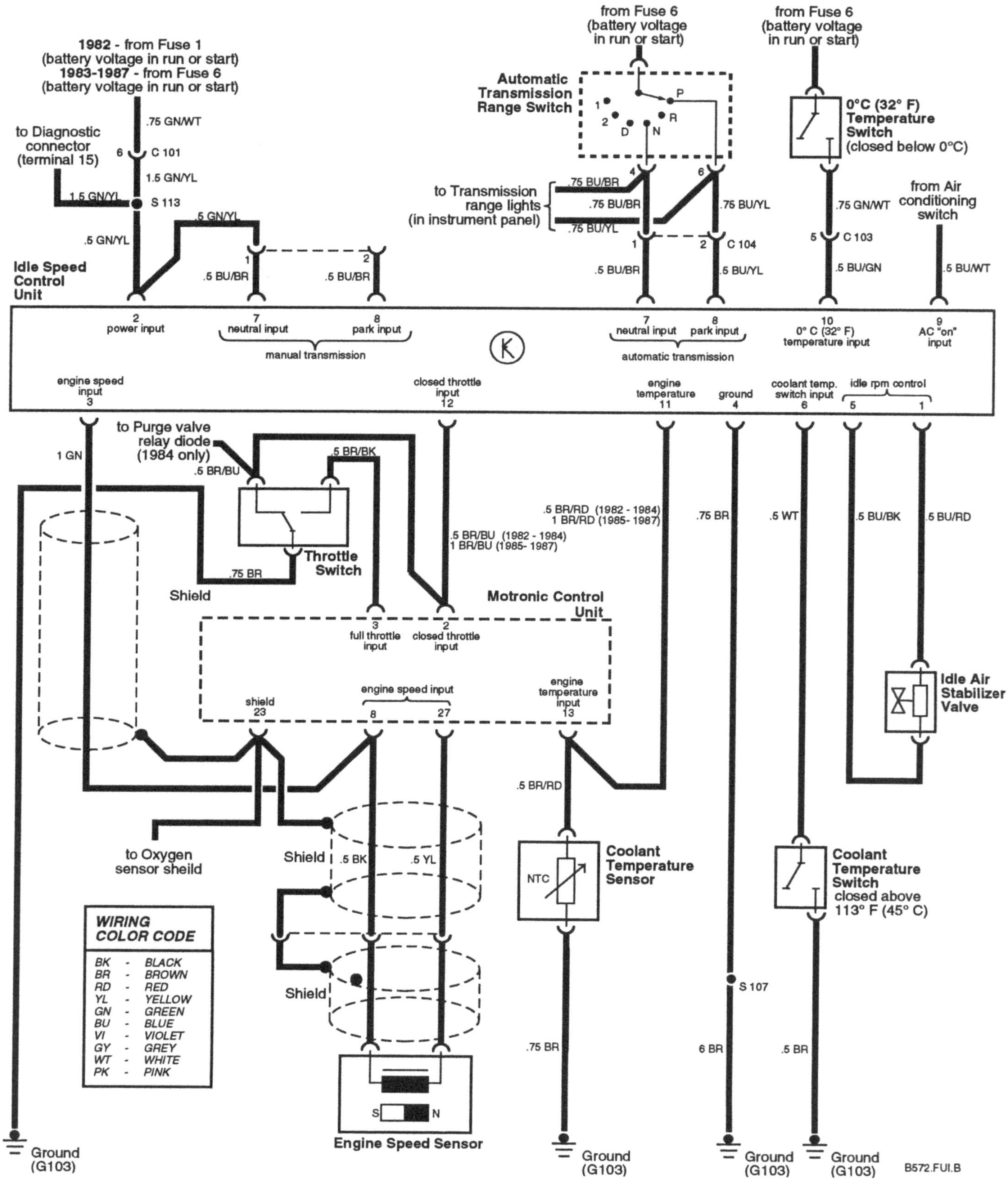

Fig. 7-1. (continued)

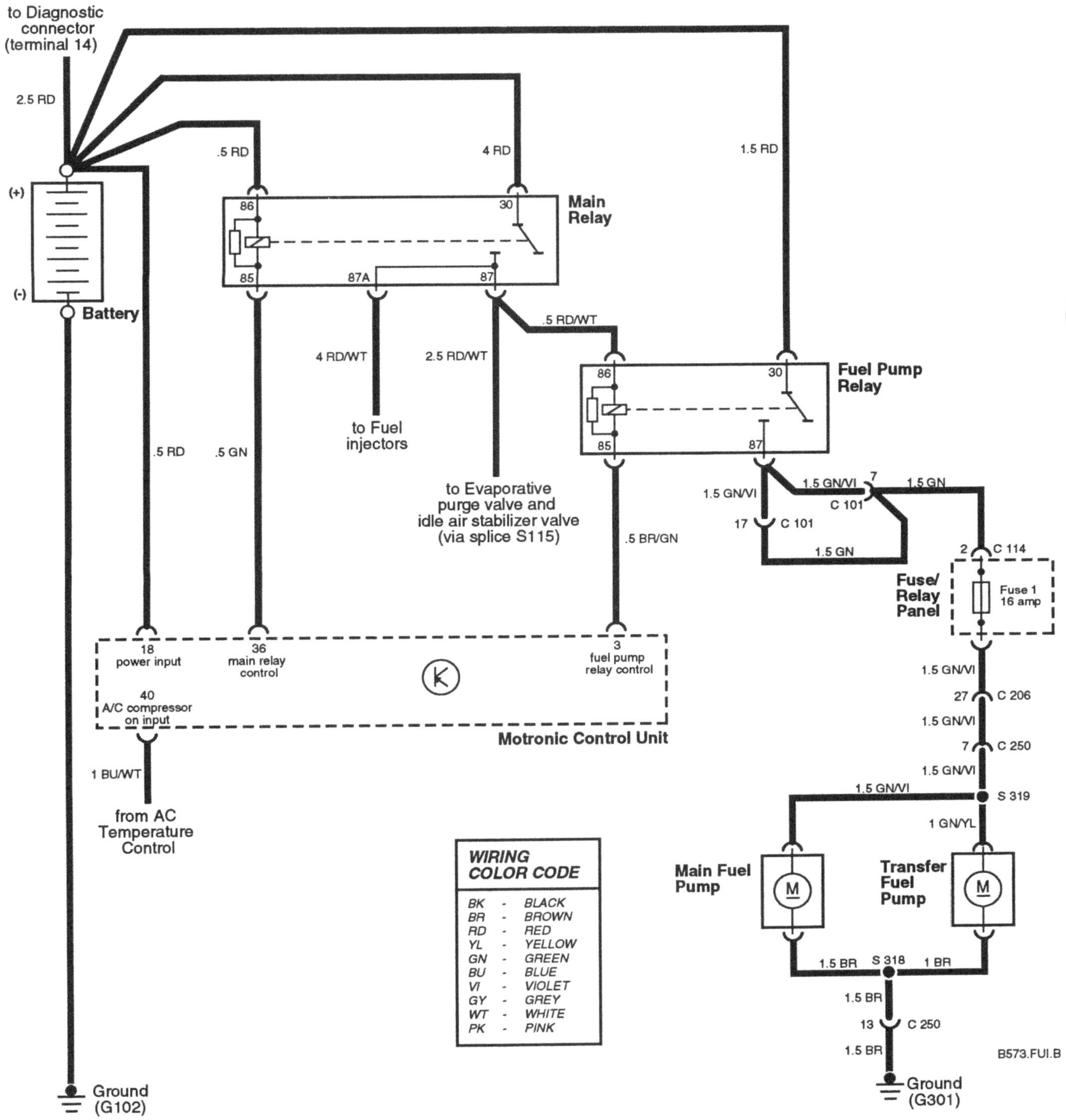

continued on next page

Fig. 7-2. Schematic wiring diagram for 1987 (built from March 1987) and 1988 528e models.

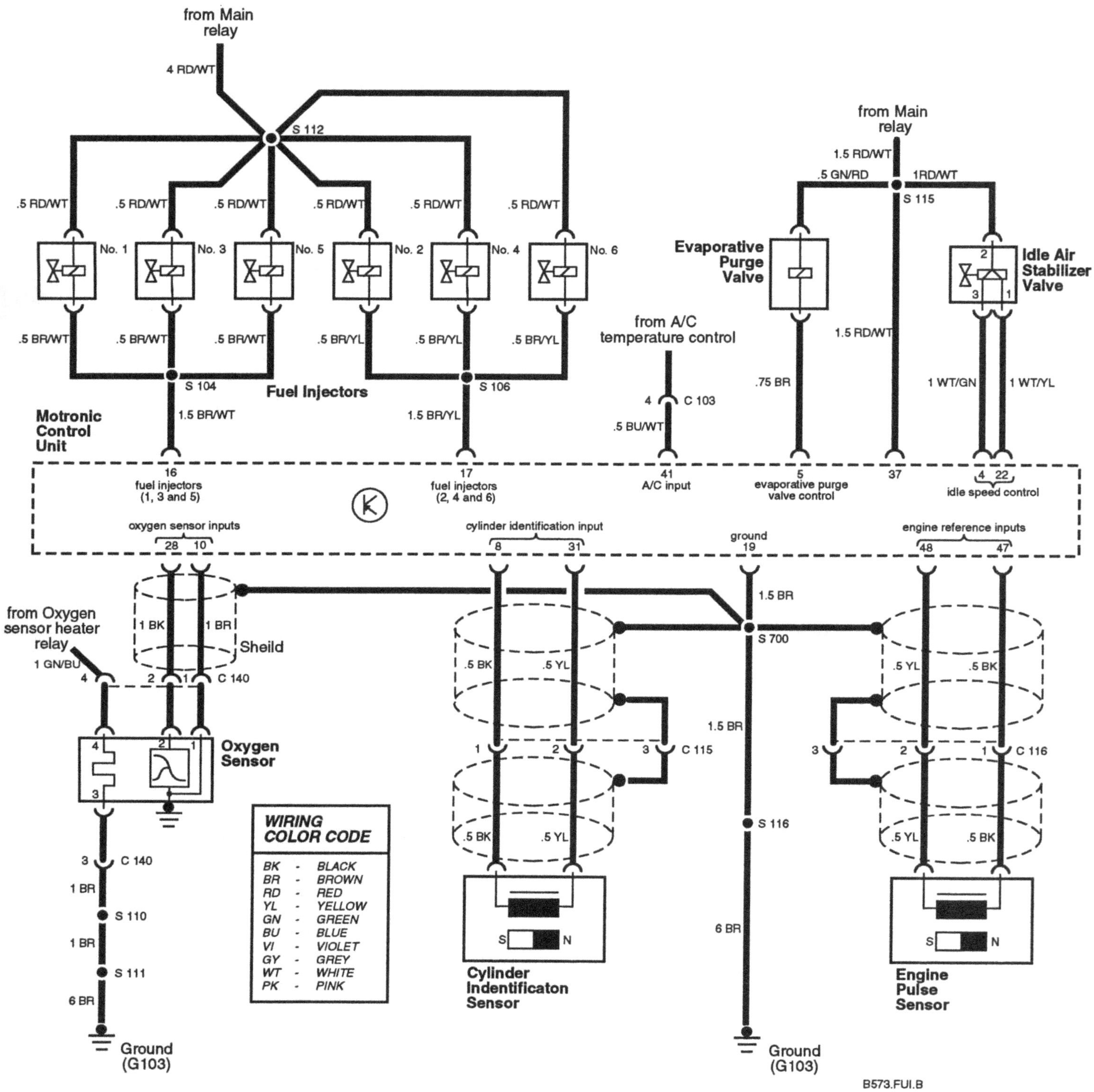

Fig. 7-2. (continued)

continued on next page

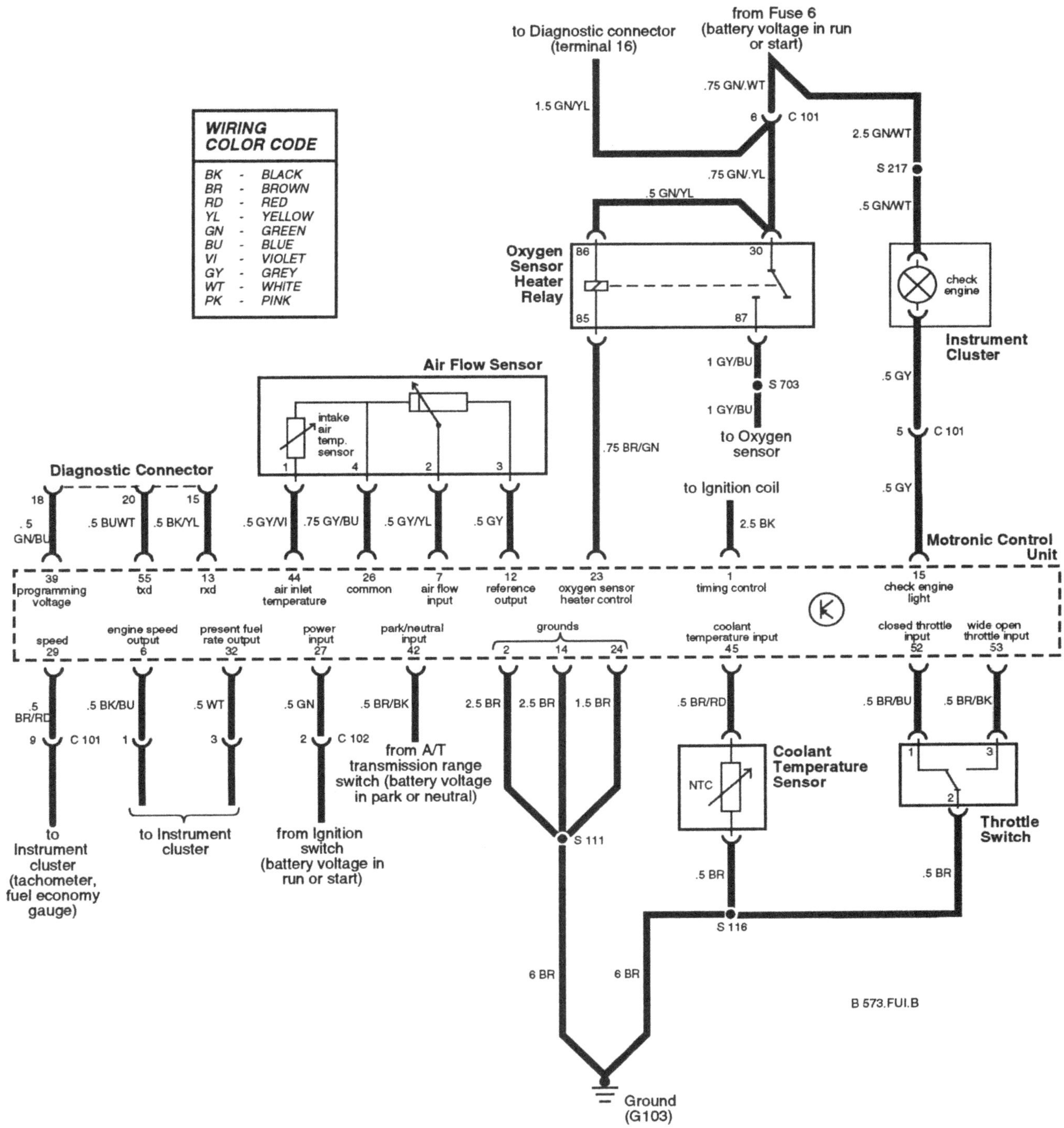

Fig. 7-2. (continued)

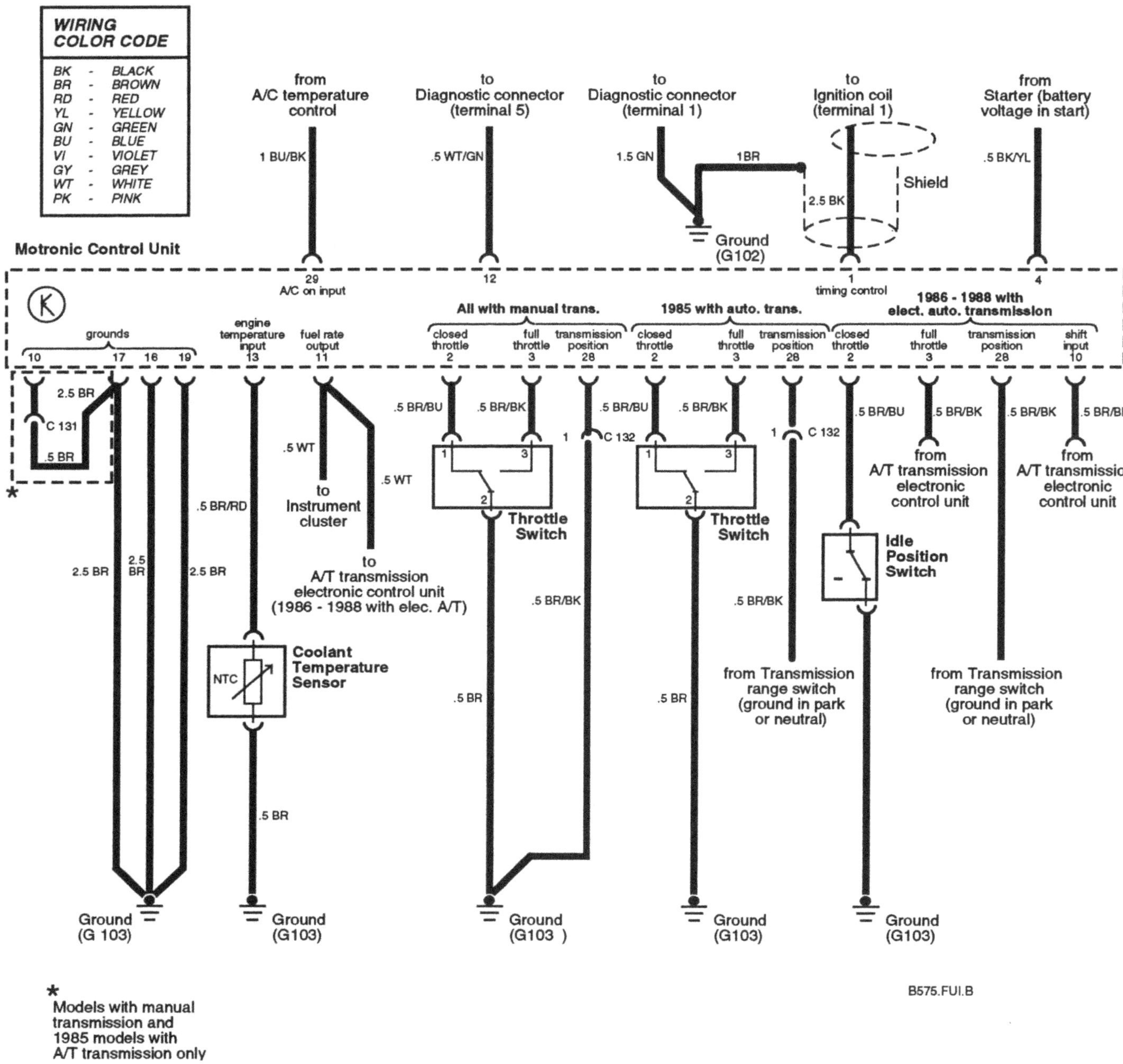

Fig. 7-3. Schematic wiring diagram for 1985 through 1988 535i models.

continued on next page

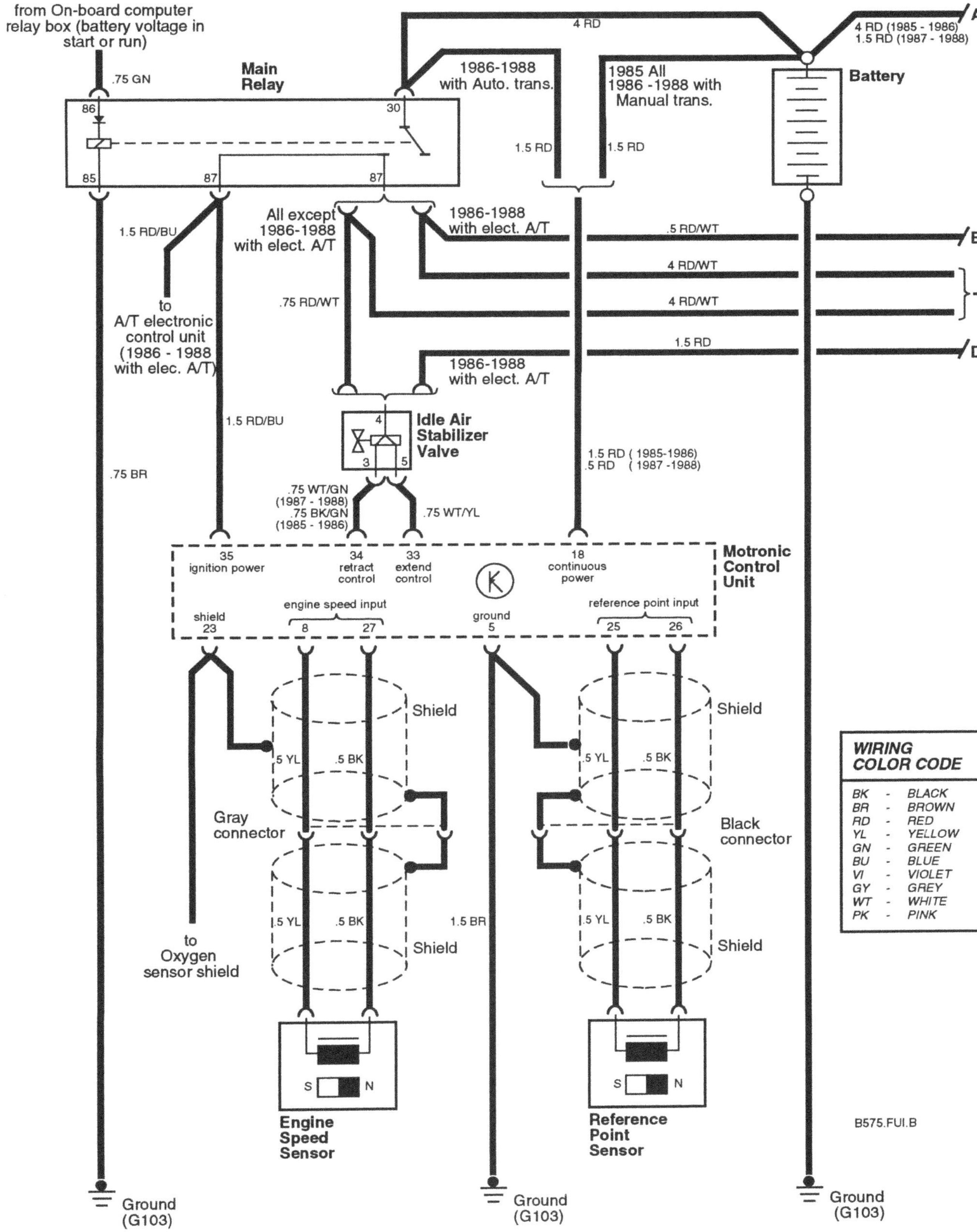

WIRING COLOR CODE		
BK	-	BLACK
BR	-	BROWN
RD	-	RED
YL	-	YELLOW
GN	-	GREEN
BU	-	BLUE
VI	-	VIOLET
GY	-	GREY
WT	-	WHITE
PK	-	PINK

continued on next page

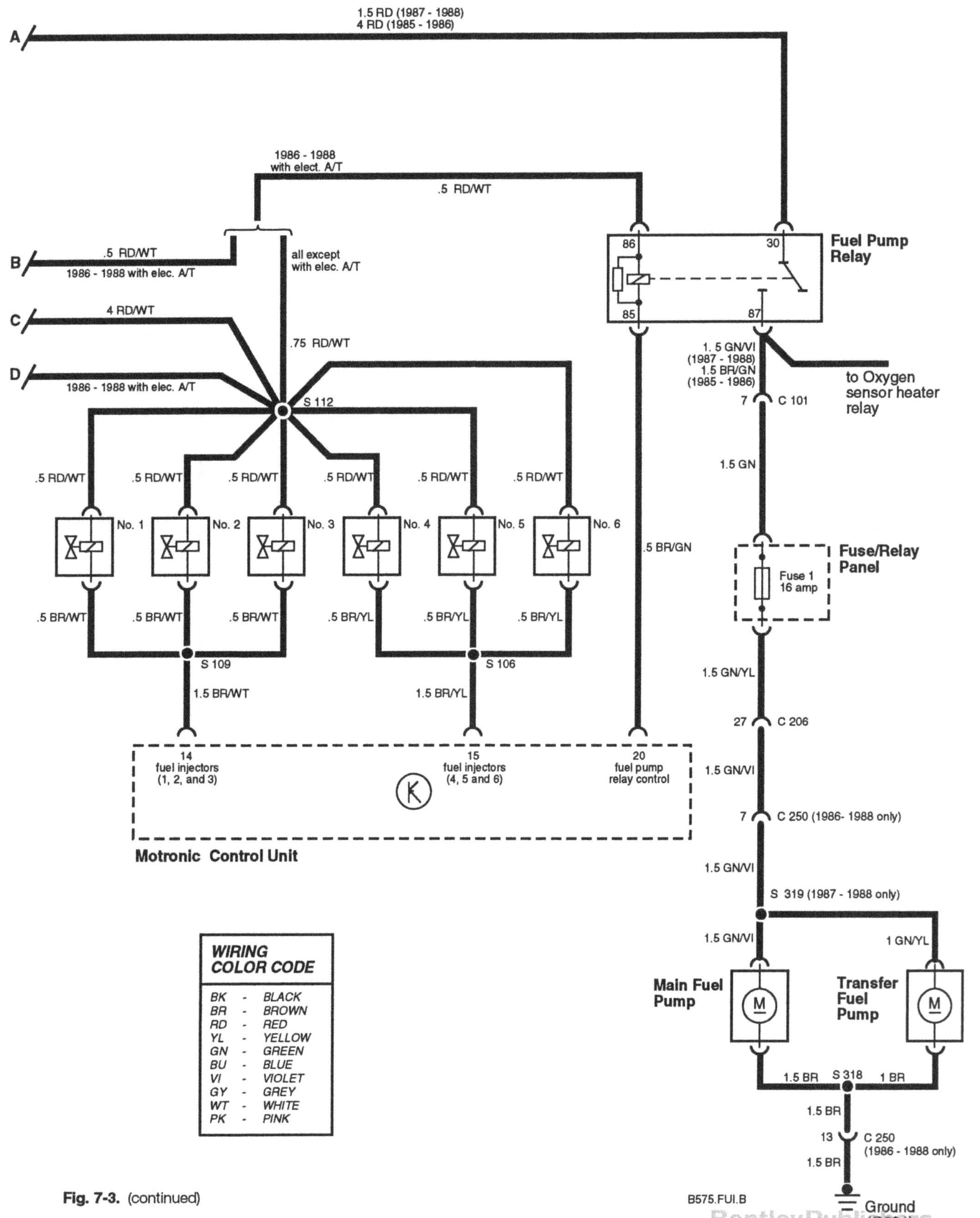

WIRING COLOR CODE		
BK	-	BLACK
BR	-	BROWN
RD	-	RED
YL	-	YELLOW
GN	-	GREEN
BU	-	BLUE
VI	-	VIOLET
GY	-	GREY
WT	-	WHITE
PK	-	PINK

Fig. 7-3. (continued)

continued on next page

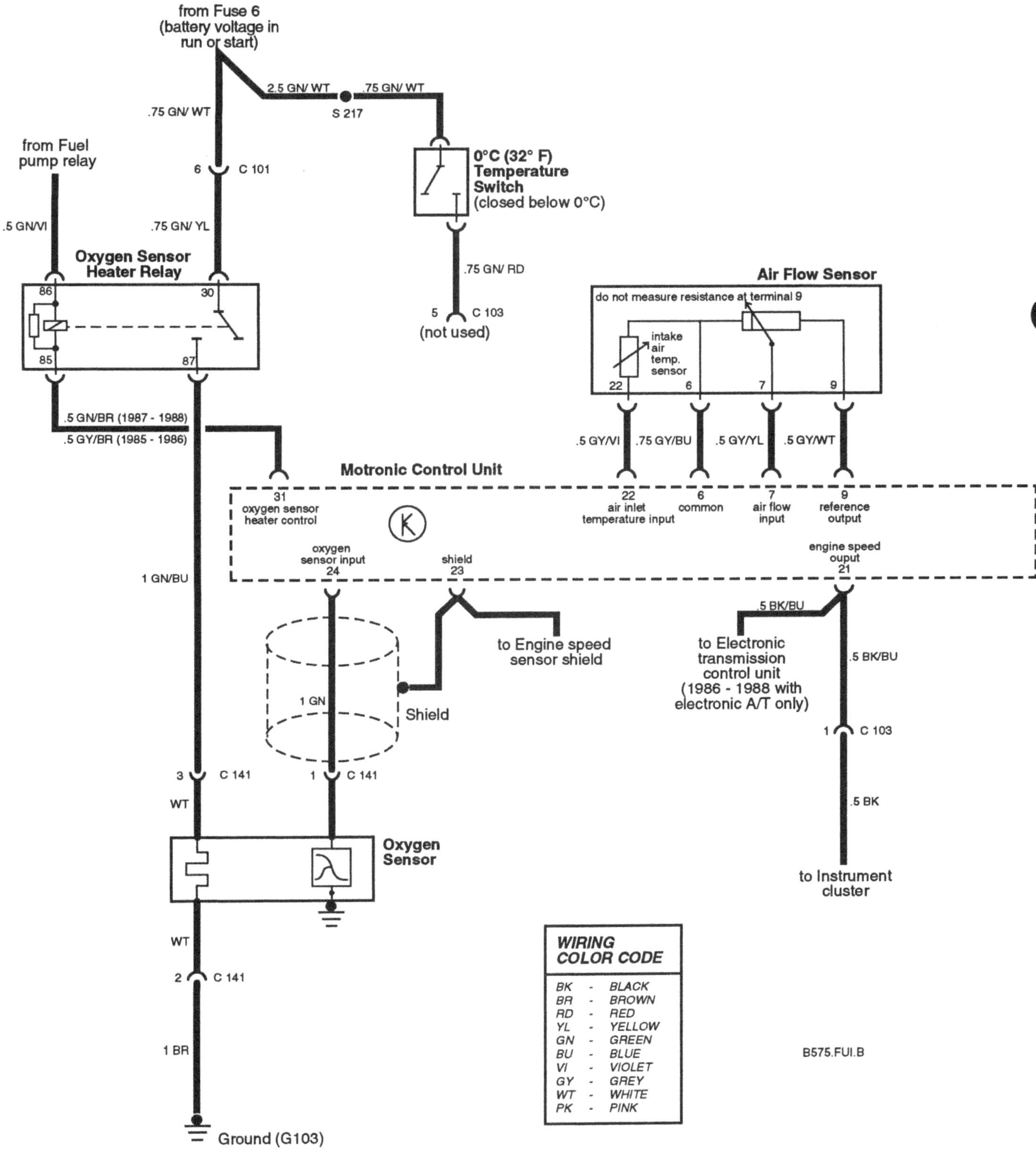

Fig. 7-3. (continued)

7.2 Harness Connector, Ground, and Splice Locations

Table p, **Table q**, and **Table r** list the locations of the various harness connectors, grounds, and splices (welded connections in the wiring harness). For the exact positions of the fuses and relays, see **ELECTRICAL SYSTEM**

Table p. Wiring Harness Connector Locations

Connector	Location
C101	On side of fuse/relay panel (17-pin)
C102	Under fuse/relay panel (2-pin)
C103	Right side under dash, near Motronic control unit (6-pin)
C104	Right side under dash, near Motronic control unit (2-pin)
C107	In engine compartment, left front (2-pin)
C114	Under fuse/relay panel (8-pin)
C115	Left side of engine compartment near diagnostic connector (3-pin)
C116	Left side of engine compartment near diagnostic connector (3-pin)
C131	Right side under dash, near Motronic control unit (1-pin)
C132	Right side under dash, near Motronic control unit (1-pin)
C140	Right side of engine compartment at rear (4-pin)
C141	Rear of engine, near exhaust manifold (3-pin)
C206	Left side under dash, on main connector bracket (29-pin)
C250	Below rear seat— center (13-pin)

Table q. Wiring Harness Ground Locations

Ground point	Location
G102	Main ground above battery post
G103 (1982-1985 528e)	Main engine ground, on rear side of engine above starter
G103 (1986-1988 528e)	Main engine ground, on front of engine under diagnostic connector
G103 (533i, 535i)	Main engine ground, on rear of engine above starter
G200	Above brake pedal cluster
G301	Under rear seat— left side

Table r. Wiring Harness Splice Locations

Splice (welded connection in wiring harness)	Harness or approximate location
S104	Front of engine compartment
S106	Engine compartment, near firewall
S107	Engine compartment, near right side firewall
S109	Engine harness,
S110	Engine harness
S111	Engine compartment, near firewall
S112	Engine compartment, near firewall
S113	Engine compartment, left side rear
S115	Engine compartment, near firewall
S116	Engine compartment, near firewall
S117	Engine compartment, left side
S217	Under left side of dash
S318	Rear harness, under rear seat
S319	Rear harness, under rear seat
S700 (1988 528e)	Under right side of dash
S703 (1988 528e)	Under right side of dash

8. TECHNICAL DATA

I. Motronic Fuel Injection Specifications

Fuel pump delivery rate with fuel pump operated for exactly one minute
- 1982–1985 models . . . . at least 2.0 liters (2.2 quarts)
- 1986–1988 models . . . . at least 1.8 liters (1.9 quarts)

System fuel pressure
- 528e, 533i models . . . . 2.5±0.05 bar (36.3±0.7 psi)
- 535i models . . . . 3.0±0.06 bar (43.5±0.9 psi)

System regulating pressure
- 528e, 533i models . . . . 2.3–2.7 bar (33.4–39 psi)
- 535i models . . . . 2.8–3.2 bar (40.6–46.4 psi)

Idle speed (non-adjustable)
- 528e, 533i models . . . . 700±50 rpm
- 535i models . . . . 800±50 rpm

Idle mixture
- All . . . . 0.2–1.2 %CO

Section 8

IGNITION

8

Contents

INTRODUCTION

The ignition systems installed on the cars covered by this manual incorporate many components that maximize engine performance and reliability in the face of increasing restrictions on engine exhaust emissions. The ignition components are really part of the Motronic engine management system. For more information on the Motronic system, see **FUEL INJECTION**.

1. GENERAL DESCRIPTION

The ignition system provides each spark plug with a precisely timed high-voltage charge to ignite the air-fuel mixture in the combustion chamber. The system also makes adjustments to the ignition timing in response to changes in engine speed and load.

The high-voltage charge is created by the ignition coil. In the primary circuit, battery voltage is applied to the ignition coil to charge it. When the primary circuit is open, the coil discharges its high voltage. The secondary ignition circuit—the rotor, the distributor cap, the spark plug wires, and the spark plugs—distributes the high voltage to the cylinders to ignite the air-fuel mixture. Ignition timing refers to the position of the piston in the cylinder when the coil discharges.

In Motronic systems, also known as Digital Motor Electronics (DME), all ignition functions and fuel injection control functions are controlled by a single electronic control unit. Ignition timing and fuel control are based on inputs the control unit receives for engine load, engine speed, ignition quality, coolant temperature, intake-air temperature, and on some models altitude (barometric pressure). The only function that the distributor serves is to distribute the high voltage to the individual spark plugs. The distributor is an integral part of the cylinder head and there is no mechanical spark advance system. See Fig. 1-1.

Motronic Identifying Features

There are two versions of Motronic systems used on cars covered by this manual. A quick check to determine whether the car has Motronic or Motronic 1.1 is to look at the diagnostic connector on the engine. Motronic versions use a 15 pin connector, Motronic 1.1 versions use a 20 pin connector. For more information see **FUEL INJECTION**.

Engine management system applications

- 533i, 535i models . . . . .Motronic
- 528e models built up to March 1987. . . . . .Motronic
- 528e models built from March 1987 . . . . .Motronic 1.1

> **NOTE —**
> A car's build date is listed on the driver's door, below the door latch.

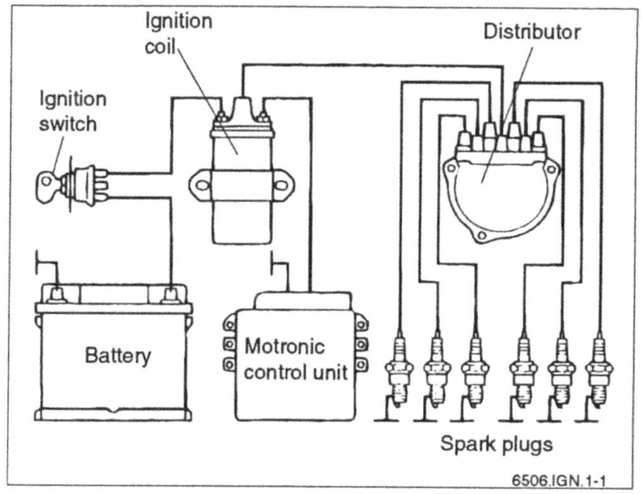

Fig. 1-1. Schematic view of ignition system on Motronic engine-management system. Courtesy Robert Bosch Corp.

2. MAINTENANCE

BMW specifies the maintenance steps below to be carried out at particular time or mileage intervals for proper maintenance of the ignition system. A number in bold type indicates that the procedure is covered in this section, under that numbered heading. Information on other ignition system maintenance and on the prescribed maintenance intervals can be found in **LUBRICATION AND MAINTENANCE**.

1. Replacing spark plugs
2. Inspecting distributor cap, rotor, and spark plug wires

3. TROUBLESHOOTING

Poor driveability may have a variety of causes. The fault may lie with the ignition system, the fuel system, parts of the emission control system, or a combination of the three. Because of the interrelated functions of these systems and their effects on each other, it is often difficult to know where to begin looking for problems. For this reason, effective troubleshooting should always consider these systems in unison, as one major system.

This troubleshooting section applies to starting and running problems caused specifically by faults in the ignition system, including the coil, the distributor cap and rotor, and the spark plug wires. A complete failure of the ignition system to produce spark at the spark plugs is self-evident. For other problems such as rough idle, misfiring, or poor starting, however, the cause is not so clear. For troubleshooting engine management and the way the car runs, see **ENGINE MANAGEMENT—DRIVEABILITY**. There, you will be referred back to the appropriate part of this section for further tests and repairs.

3.1 Basic Troubleshooting Principles

The ignition system's function is to provide a properly timed high-voltage spark.

On Motronic ignition systems, the reference sensor determines the crankshaft (piston) position and gives the electronic system its baseline ignition timing. Ignition timing is then controlled by the Motronic control unit based on engine load, engine speed, engine temperature, and throttle position.

An engine that starts and runs indicates that the ignition system is fundamentally working—delivering voltage to the spark plugs. A hard-starting or poor-running engine, however, may indicate a problem with how well the spark is delivered. A faulty coil, cracked or deteriorated spark plug wires, a worn or cracked distributor cap or rotor, and worn or fouled spark plugs are all causes of reduced spark intensity and inefficient combustion.

WARNING —
For cars with catalytic converters, inefficient combustion can cause serious problems. The poorly burned mixture can overload the catalytic converter with raw fuel, leading to converter overheating or plugging. An overheated catalytic converter can be a fire hazard.

An engine that has good cranking speed but will not even begin to start may indicate a complete failure of the system to produce spark. Inspect the ignition system visually. Make sure the spark plug wires have not been interchanged. Ignition firing order is described under **4.1 Distributor Cap and Rotor**.

If no faults are located, make a basic check to see if spark is being produced as described below under **Quick-Check of Ignition System**. This will be the most important first troubleshooting step. If a strong spark is observed, then the failure to start is due to another cause, perhaps no fuel being delivered to the engine.

The Motronic ignition system contains very sensitive electronic components. To protect the system, and for general safety, the following cautions should be observed during any ignition system troubleshooting, maintenance, or repair work.

WARNING —
Do not touch or disconnect any of the high tension cables from the coil, distributor, or spark plugs while the engine is running or being cranked by the starter. Fatal voltages are present.

CAUTION —
- *Connect or disconnect ignition system wires, multiple connectors, and ignition test equipment leads only while the ignition is off.*

CAUTION —
- *Switch multimeter functions or measurement ranges only with the test probes disconnected.*
- *Before operating the starter without starting the engine (as when making a compression test), always disable the ignition. See **3.5 Disabling Ignition System** for detailed information.*
- *Do not disconnect terminal 4 (center terminal) from the coil or remove the distributor cap to disable the ignition.*
- *Do not connect test instruments with a 12-volt supply to terminal 15 (+) of the ignition coil. The voltage backflow can damage the Motronic control unit. In general, make test connections only as specified by BMW, as described in this manual, or as described by the instrument's manufacturer.*
- *Do not disconnect the battery while the engine is running.*
- *The ignition systems covered by this manual use a special shielded rotor with 1000 ohms resistance. Take care to install the correct part. Do not substitute any other BMW or Bosch part.*
- *Do not quick-charge the battery (for boost starting) for longer than one minute, and do not exceed 16.5 volts at the battery with the boosting cables attached. Wait at least one minute before boosting the battery a second time. On models equipped with on-board computers, remove the computer fuses (no. 5, no. 6, no. 12) prior to quick-charging to prevent damaging the computer.*
- *Do not wash the engine while it is running, or any time the ignition is switched on.*
- *Disconnect the battery when doing any electric welding on the vehicle or charging the battery.*
- *Do not try to start the engine of a car which has been heated above 176°F (80°C), for example, in a paint drying booth, until allowing it to cool to normal temperature.*
- *Do not conduct ignition system tests with a test lamp that uses a normal incandescent bulb. The high electrical consumption of these test lamps may damage the electronic components.*
- *Do not connect terminal 1 of the coil to ground as a means of preventing the engine from starting (for example, when installing or servicing anti-theft devices).*

Test Equipment

Many of the tests of ignition system components require the use of high-impedance test equipment to prevent damage to the electrical components. A high impedance digital multimeter should be used for all voltage and resistance tests. An LED test light should be used in place of an incandescent-type test lamp.

Many tests require checking for voltage, continuity, or resistance at the terminals of the components' harness connectors. The blunt tips of a multimeter's probes can spread open the small connector terminals and cause poor connections. To prevent damage, use flat male connectors to probe the harness connector terminals.

Quick-Check of Ignition System

If the engine does not start, the most fundamental step in troubleshooting the ignition system is to determine whether or not the system is making a spark at the spark plug. If no spark is present, then more detailed testing of the ignition system is necessary.

To check for spark, turn the ignition off and remove a connector from one of the spark plugs. Connect it to a known good spark plug, preferably a new plug. Do not hold the spark plug or its connector, even if using insulated pliers. Position the plug so that the outer electrode is grounded on the engine.

> ***WARNING —***
> *The ignition systems installed on the cars covered by this manual are high-energy systems operating in a dangerous voltage range which could prove to be fatal if exposed terminals or live parts are contacted. Use extreme caution when working on a vehicle with the ignition on or the engine running.*

> ***CAUTION —***
> *Any test set-up other than the one described above may cause damage or inconclusive tests.*

While a helper actuates the starter, look and listen for spark in the spark plug gap. A bright blue spark indicates a healthy ignition system. A yellow-orange spark is weaker and indicates that, while spark is present and the system is functioning, it is not operating at peak efficiency. Check the condition of the ignition system components as described in **3.2 Ignition System Visual Inspection** and replace any faulty components.

> ***WARNING —***
> *If ignition system failure is not the problem, the engine may start during this test. Be prepared to turn off the ignition immediately. Also, running the engine with a spark plug wire disconnected may damage the catalytic converter.*

> **NOTE —**
> Before checking the ignition system when there is no spark or a weak spark, make sure that the battery is fully charged. See **BATTERY, STARTER, ALTERNATOR**.

If there is no spark, test for primary voltage at the ignition coil. Connect a voltmeter between terminal 15 (+) of the ignition coil and ground (a clean, bare metal part of the engine or chassis). See Fig. 3-1. When the ignition is turned on, there should be battery voltage at the terminal. If battery voltage is not present, there is either a fault in the wire between terminal 15 and the ignition switch, in the ignition switch itself, in the board computer (where applicable), or in the wiring from the battery to the ignition switch. See **4.4 Ignition System Electrical Circuits** for ignition system wiring schematics.

> **NOTE —**
> The wiring from the battery to terminal 15 (+) of the coil (via the ignition switch) is not fused. Short circuits can destroy components. Be careful when testing this circuit.

If no faults have been detected up to this point but there is still no spark or a weak spark, refer to **Table a** for more troubleshooting information. Bold numbers in the corrective action column refer to numbered headings in this section where repair information is located. If the coil is receiving voltage, or if a strong spark is observed but the engine still will not start, refer to **ENGINE MANAGEMENT—DRIVEABILITY** for more troubleshooting information.

Table a. Ignition System Troubleshooting

Symptom	Probable cause	Corrective action
1. No spark or weak spark observed during spark test	a. Wet or damp distributor cap and/or spark plug wires	a. Remove cap and wires. Dry and reinstall. **3.3**, **3.4**
	b. Faulty wires or connectors (primary circuit)	b. Inspect and repair as needed
	c. Weak or faulty coil	c. Test and replace as needed. **3.3**
	d. Defective spark plug wires	d. Test and replace as needed. **3.3**
	e. Worn or fouled spark plugs	e. Replace spark plugs. See **LUBRICATION AND MAINTENANCE**
	f. Faulty reference or pulse sensor	f. Test and replace as needed. **4.4**
	g. Faulty Motronic control unit	g. Test and replace as needed. See **FUEL INJECTION**

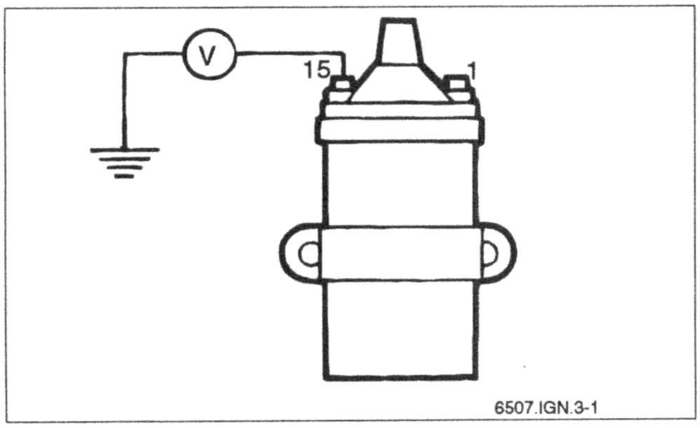

Fig. 3-1. Primary voltage to coil being checked with a voltmeter (shown schematically) between terminal 15 and ground. Coil terminal numbers are found on coil.

3.2 Ignition System Visual Inspection

The spark plug wires, the distributor cap, and the distributor rotor are subject to wear and electrical breakdown which will impair their ability to deliver a crisply timed and powerful spark. Many of these conditions are most easily detected by a thorough visual inspection. Dirt and moisture on any of these components are also potential causes of poor spark at the spark plugs.

To inspect the distributor cap and rotor, first remove the cap as described in **4.1 Distributor Cap and Rotor**. Inspect the contacts inside the distributor cap and at the tip of the rotor for corrosion, wear, or pitting. See Fig. 3-2. Parts with corroded contacts can be cleaned and reused, but if there is wear, pitting, or heavy corrosion, replacement is highly recommended. The center black carbon brush inside the cap should spring back when compressed.

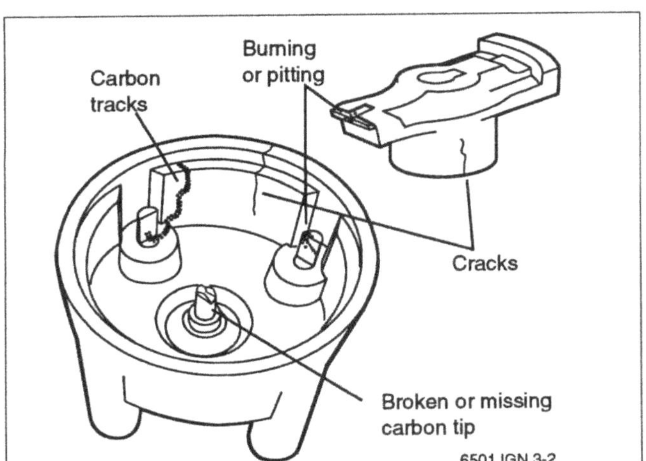

Fig. 3-2. Distributor cap and rotor. Inspect cap and rotor at contact points.

Cracks or carbon tracks in the distributor cap may cause shorts to ground. The cracks may be fine and difficult to see. Check carefully, especially around the contacts. Carbon tracks are the faint black lines, usually running between two contacts or to ground, left over from high-voltage arcing. A distributor cap that shows any sign of cracks or carbon tracking should be replaced.

Inspect the black dust shield mounted behind the rotor. See Fig. 3-3. Excessive oil residue on the shield, rotor, or cap can lead to engine misfire and poor running under load. If any signs of oil residue are found on the shield, carefully inspect the camshaft oil seal. See **ENGINE—RECONDITIONING** for oil seal replacement.

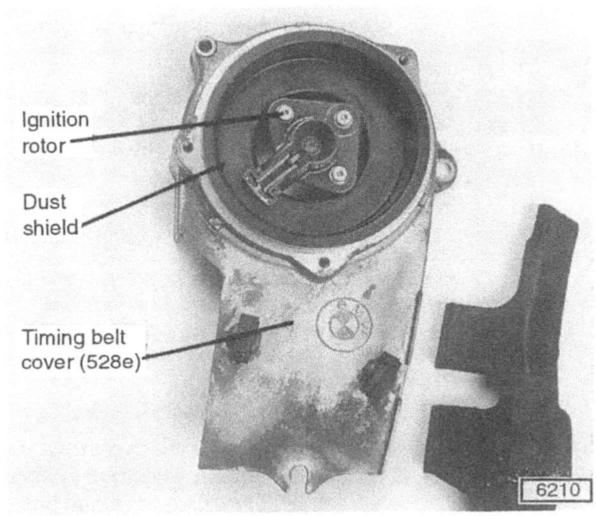

Fig. 3-3. Ignition rotor and dust shield on 528e models. On 528e models, front camshaft oil seal is behind timing belt sprocket. On 533i and 535i models, the oils seal is mounted in the front timing chain cover.

To visually check the spark plug wires, gently bend them in several places. This will expose cracks in the insulation which may cause spark "leaks." Peel back the rubber boots and check them for pliancy and the ability to seal out dirt and moisture. Wires that are cracked, oil-soaked or dry and brittle should be replaced.

For a quick-check of distributor cap and spark plug wire condition, listen for the sound of voltage arcing or watch while the engine runs at night. In darkness, the arc of high voltage to ground because of a crack in the cap or a poorly insulated wire may be visible as a blue spark.

The coil should be closely examined for cracks, burns, carbon tracks, or any leaking fluid. The coil tower, terminal 4, should be clean and dry. If necessary, remove the coil for cleaning and closer examination. Check that the wiring at the coil top is routed as shown in Fig. 3-4. Loosen the nuts and reposition the wires if necessary.

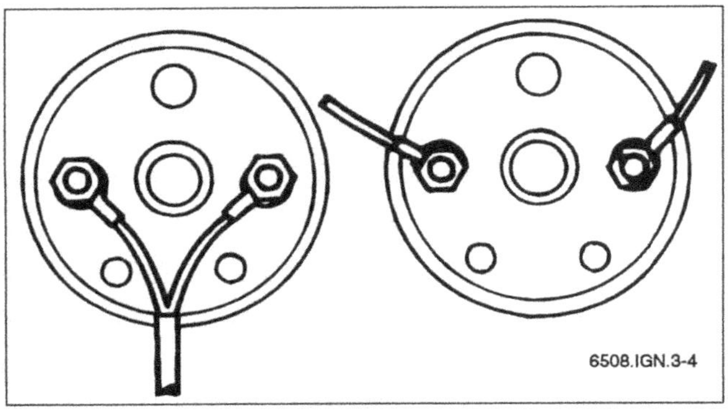

Fig. 3-4. Coil wiring correctly positioned (left) and incorrectly positioned (right). Wires terminals should not touch or come close to metal coil housing.

Inspect all primary wires and connections for any corrosion or damage. Clean or repair any faults found. In these sensitive electronic ignition systems, corroded or loose connections may interfere with the ignition function.

NOTE —
On models with Motronic 1.1, a failed Motronic control unit can sometimes damage the ignition coil. Be sure to check the ignition coil anytime a faulty Motronic 1.1 control unit is replaced and a no-start condition still exists.

Inspect the reference and speed sensor(s), their connectors, and wiring. Check for dents or cracks in the sensors, or chafed spots in the wiring. On older cars, damage to the sensors from foreign material in the bellhousing can destroy a sensor. For more information on the sensors and their locations, including testing procedures, see **4.3 Reference Sensor and Speed Sensor**.

3.3 Testing Coil and Spark Plug Wires

Use an ohmmeter to test the ignition coil primary and secondary resistance with the wires removed from the coil. See Fig. 3-5. Resistance values are listed in **Table b**. Replace any coil which has higher primary or secondary resistance.

Table b. Ignition Resistance Specifications

Test	Terminals	Resistance
Coil primary resistance	1 (–) and 15 (+)	0.50 ohms
Coil secondary resistance	15 (+) and 4 (center tower)	6000 ohms
Spark plug end connectors		5000±1000 ohms
Spark plug wires		0 ohms (approx.)
Distributor cap suppression connectors		1000±200 ohms
Rotor		1000 ohms

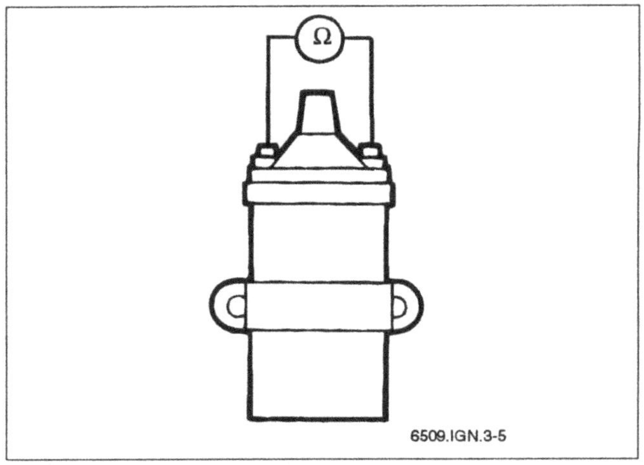

Fig. 3-5. Primary coil resistance being measured with an ohmmeter (shown schematically) between terminals 1 and 15. Measurement of secondary resistance is similar.

To check each spark plug wire, check the resistance of each spark plug end connector as well as the wire itself. To check the ends, carefully probe either side of the spark plug end. Check the shielded connectors at the distributor cap end of the wire and the coil wire ends using the same method. Check the resistance of the rotor as shown in Fig. 3-6. Correct values are listed in **Table b**.

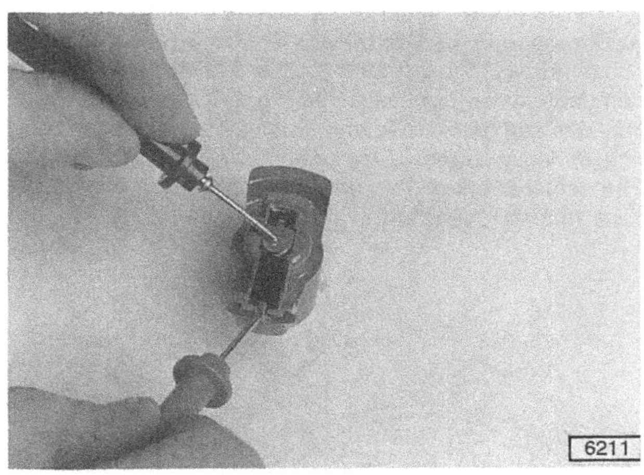

Fig. 3-6. Resistance of rotor being checked with ohmmeter. Push-on rotor from early model shown.

CAUTION —
To avoid damaging the distributor cap, do not wiggle the connectors when removing them. If necessary, twist to loosen. Then, pull straight out from the cap.

If the measured resistance is too high, the wire assembly should be replaced. Also check for corrosion at the connections. Wires or connectors with too much resistance should be replaced.

NOTE —
Individual connectors are available as replacement parts from an authorized BMW dealer. Special tools are required to correctly install the replacement ends to the wires, therefore it may be more economical to replace the complete wire assembly.

3.4 Testing Distributor Cap

To check the distributor cap, first remove them as described in **4.1 Distributor Cap and Rotor**. Be sure to label the spark plug wires and their location when disconnecting them from the cap. Check the resistance of the cap between the tower and its matching contact inside of the cap. The resistance should be nearly zero ohms. If the measured resistance is too high the distributor cap should be replaced.

3.5 Disabling Ignition System

The ignition system should be disabled when making certain tests such as compression checks or starter current draw tests.

On all models except 528e models built from March 1987 and later, remove the main relay to disable the ignition system. See Fig. 3-7. Check for a 4mm red wire to terminal 30 of the main relay to verify the exact location of the main relay. See **ELECTRICAL SYSTEM** for more information.

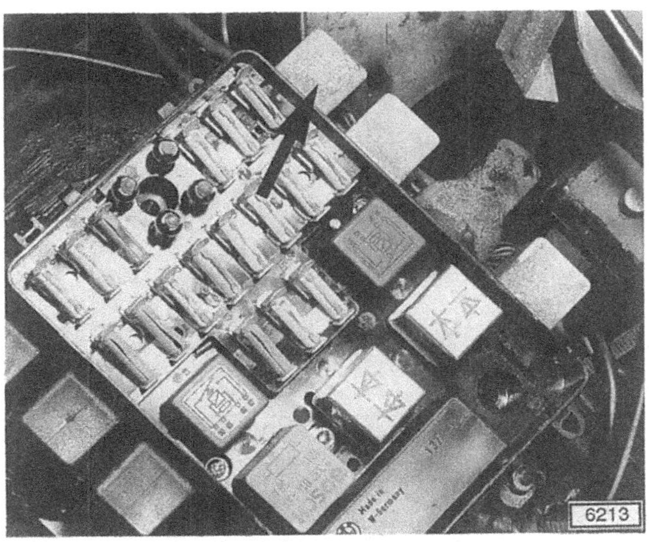

Fig. 3-7. Typical location of main relay (arrow) to be removed to disable ignition system (all models except 528e's built from March 1987 and later). Main relay location can vary.

On 528e models built from March 1987 (with Motronic 1.1), remove the wire from terminal 15 of the coil with the ignition off. Completely insulate the wire end with electrical tape.

CAUTION —
The wire at terminal 15 of the coil is not fuse protected. Always disconnect the wires at the coil only with the ignition switch in the off position.

NOTE —
A car's build date is listed on the driver's door, below the door latch.

4. MOTRONIC (DME) IGNITION SYSTEM

This heading covers the removal and installation of the distributor cap and rotor, and the spark plug wires. Inspection and testing procedures for these components are covered in **3. Troubleshooting**.

4.1 Distributor Cap and Rotor

The distributor cap and rotor are generally replaced as a part of normal maintenance. See **LUBRICATION AND MAINTENANCE** for recommended replacement intervals.

Removing and Installing Distributor Cap, Rotor, and Spark Plug Wires

These components all carry high voltage to the spark plugs and proper engine performance depends on getting the best possible spark at the spark plug. Worn and corroded contacts or poor insulation which allows the spark to short to ground are the primary reasons for replacement of these components.

Each spark plug wire leads from a specific terminal on the distributor cap to a specific spark plug. This order is known as the ignition firing order. When removing the wires, label their positions so that they can be reinstalled in the proper places. If the wires get mixed up, see **Firing Order** below.

Before the distributor cap can be removed, the radiator cooling fan and the fan shroud should be removed as described in **COOLING SYSTEM**. If the fan and shroud are not removed, it may be difficult to remove the distributor cap and rotor mounting screws.

The distributor cap is held in place with three screws. See Fig. 4-1. To replace the cap remove the two plastic shields then remove the three screws. Temporarily leave the spark plug wires in the old cap. Mount the new cap and tighten the screws. Change over one wire at a time from the old cap to the new one.

NOTE —
The cap can only be mounted in one position. Make sure the cap is properly seated before tightening the mounting screws.

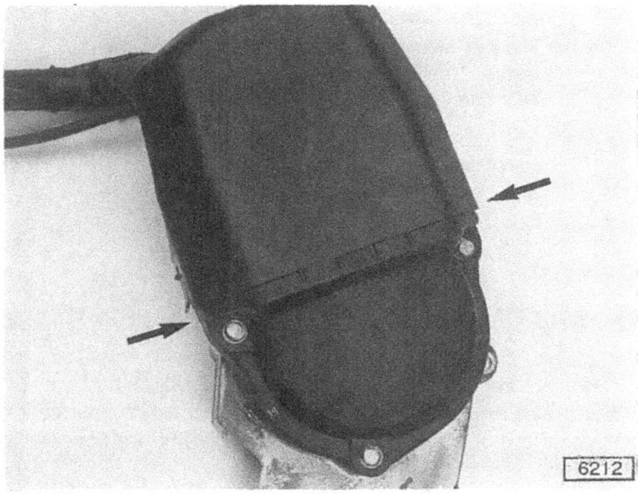

Fig. 4-1. Distributor cap protective cover. Unclip cap at points indicated by arrows. Also shown are the distributor cap mounting screws.

Two types of rotors are used on these ignition systems, screw-mounted and push-on. On screw-mounted rotors, use a 3 mm hex wrench to remove the three rotor mounting screws to remove the rotor. Remove the dust shield and inspect the large O-ring on the rear of the shield. See Fig. 4-2. If it is crushed or damaged, it should be replaced. Installation of the rotor is the reverse of removal. Be careful not to over tighten the rotor mounting screws.

NOTE —
The rotor can only be mounted in one position. On push-on style rotors, align cutout in the rotor adaptor with the notch in the rotor and push the rotor onto the shaft.

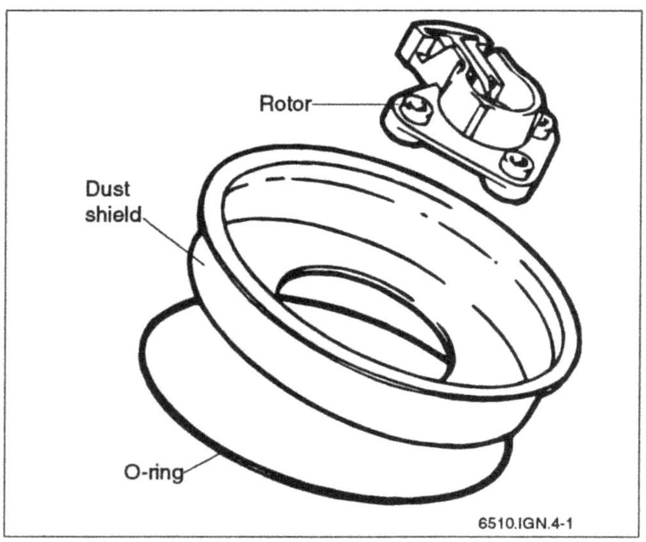

Fig. 4-2. Ignition rotor, distributor cap dust shield and O-ring.

Tightening torque
• rotor mounting screws . . . . 2.8±0.2 Nm (25±1.8 in-lb)

NOTE —
It is recommended that the cap and rotor be replaced as a set.

The spark plug wires are removed from the distributor cap and spark plugs by pulling straight up. Pull only on the connector, not on the wire itself. On stuck wires, twisting the connector slightly or peeling back the rubber boot from the distributor cap tower may ease removal.

CAUTION —
Wiggling the connectors from side to side when removing the spark plug wires from the distributor cap may damage the cap.

NOTE —
• On 528e models built from March 1987 and later, there is an inductive pickup on the no. 6 spark plug wire. The signal from this sensor is used to sequence the fuel injectors. See **FUEL INJECTION** for more information.

• A car's build date should be listed on the driver's side door pillar.

Firing Order

Spark plug wires must be installed so that the spark plugs fire in the following order: 1-5-3-6-2-4. The rotor turns in a clockwise direction when viewed from the front with the cap removed. Fig. 4-3 shows the correct routing of the spark plugs wires in the distributor cap.

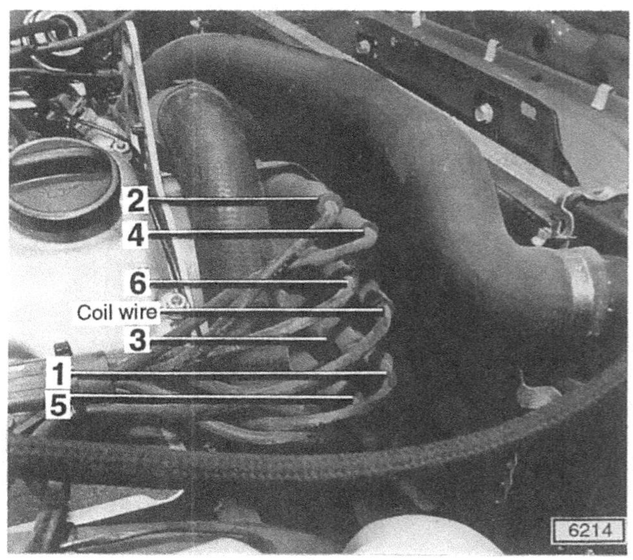

Fig. 4-3. Spark plugs wires correctly installed in distributor cap.

NOTE —
Cylinder no. 1 is at the front of the engine (closest to the radiator).

4.2 Ignition Timing

The ignition point is controlled according various inputs to the control unit. Inputs such as engine load, engine speed, temperature, altitude, and throttle position are used to determine the optimum ignition point.

Ignition timing is electronically controlled and is not adjustable on Motronic systems. The initial baseline ignition point is determined based on the crankshaft position during starting. This is signalled by a reference sensor. Once the engine is running, the ignition point is continually changed based on the various inputs to the control unit. Engine speed is signalled by a speed sensor. A Motronic ignition characteristic map illustrating all the possible ignition points is shown in Fig. 4-4. A map similar to the one shown is digitally stored in the Motronic control unit.

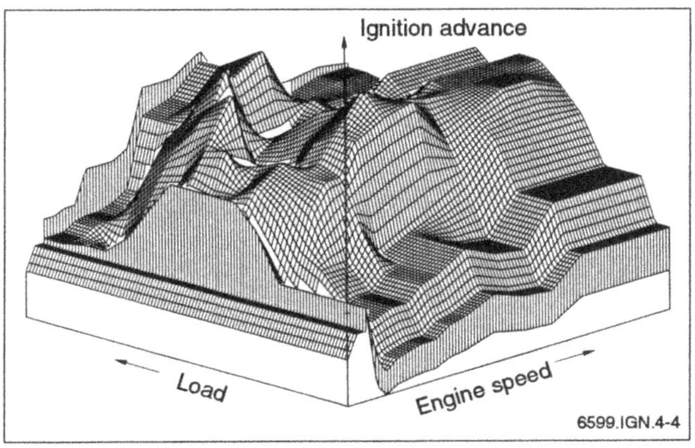

Fig. 4-4. Motronic ignition characteristic map. Courtesy Robert Bosch Corp.

4.3 Reference Sensor and Speed Sensor

On all 533i, 535i, and on 528e models built up to March 1987, two separate sensors supply the control unit with engine speed and crankshaft position signals. The sensors are mounted on the driver's side of the transmission bellhousing. The speed sensor uses the flywheel teeth to determine engine speed. The reference sensor determines the crankshaft's position, or angle, when a raised pin on the flywheel passes the sensor. See Fig. 4-5.

NOTE —
A Top Dead Center (TDC) position sensor mounted on the front of the engine is for use with the BMW Service-test unit. This sensor is not linked to the Motronic system or the engine in any way and does not affect ignition timing or the way the engine runs.

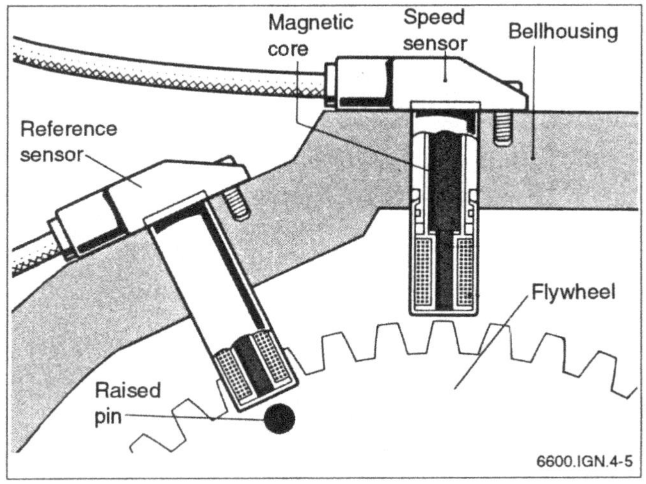

Fig. 4-5. Schematic of reference sensor and speed sensor on all 533i, all 535i, and 528e models built up to March 1987. Courtesy Robert Bosch Corp.

528e models built from March 1987 and later are equipped with Motronic 1.1 which uses a single front-mounted pulse sensor to sense engine speed and crankshaft position. The pulse sensor is mounted on the front of the engine and reads a pulse wheel mounted on the front of the crankshaft. Engine speed is determined by the rate at which the wheel's teeth pass the sensor. Crankshaft position is determined by the missing-teeth gap on the pulse wheel. See Fig. 4-6.

Fig. 4-6. Pulse sensor (arrow) on 528e models with Motronic 1.1 (built from March 1987 and later).

NOTE —
A car's build date is listed on the driver's door, below the door latch.

If the control unit does not receive a crankshaft position signal from the sensor, the engine will not start. If the engine consistently misses at constant speeds, test the speed sensor. Motronic electrical tests other than those described below can be found in **FUEL INJECTION**. When testing the sensors, the temperature should be approximately 77°F (25°C) to obtain the most accurate results.

Testing Reference and Speed Sensors
(533i, 535i and 528e built up to March 1987)

Disconnect the two electrical connectors. See Fig. 4-7, Fig. 4-8 or Fig. 4-9. The gray connector is for the reference mark sensor and the black connector is for the speed sensor.

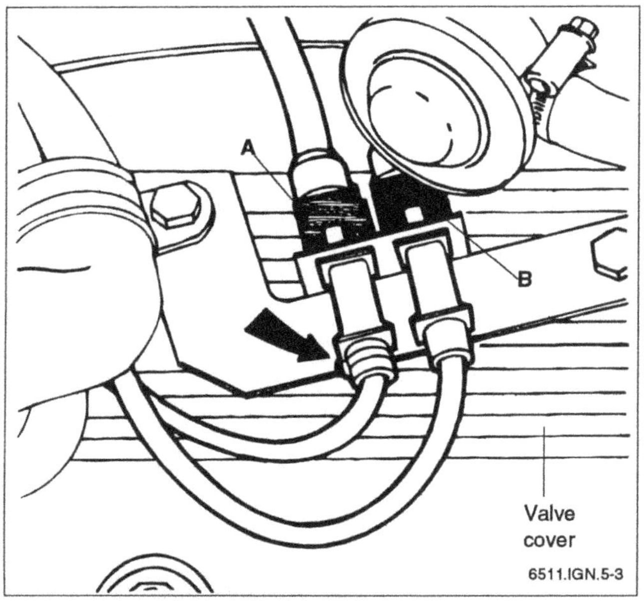

Fig. 4-7. Gray reference sensor connector (**A**) and black speed sensor connector (**B**) for 1982 through 1985 528e models. Reference sensor also marked with a ring (arrow).

Using an ohmmeter, check the resistance between terminal 1 (yellow wire) and terminal 2 (black wire) on the sensor side end of each connector. Also check the resistance between terminal 3 (shield) and either terminal 1 or terminal 2. See Fig. 4-10.

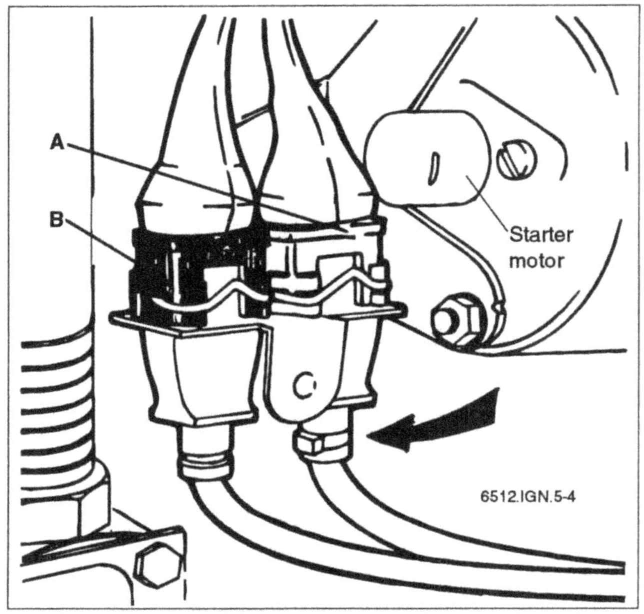

Fig. 4-8. Gray reference sensor connector (**A**) and black speed sensor connector (**B**) for 1986 through March 1987 528e models. Reference sensor also marked with a ring (arrow).

Fig. 4-9. Gray reference sensor connector (**A**) and black speed sensor connector (**B**) for 533i and 535i models.

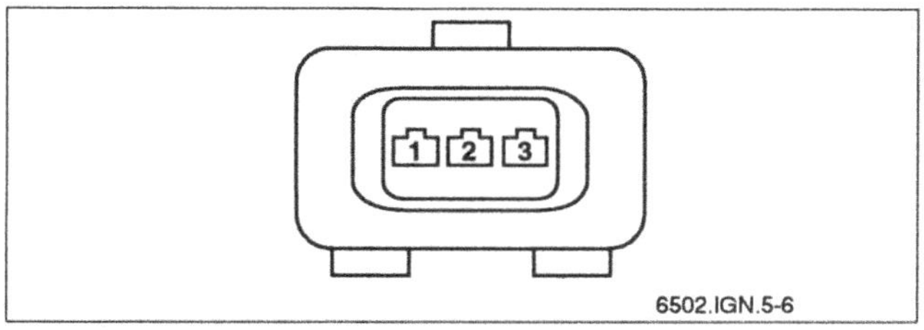

Fig. 4-10. Terminal identification for speed and reference sensor connectors.

Speed or reference sensor resistance	
• terminals 1 and 2	960±96 ohms
• terminal 1 and 3	100,000 ohms minimum
• terminal 2 and 3	100,000 ohms minimum

When replacing a sensor, be sure the connectors are not interchanged, and that the sensor is placed in the correct position in the bellhousing. The bellhousing is marked with a **B** for the reference sensor, and a **D** for the speed sensor. See Fig. 4-11. Tighten the sensor mounting screw. Be careful not to overtighten the mounting screw.

NOTE —
If the sensors are reversed in the bellhousing the engine will not start.

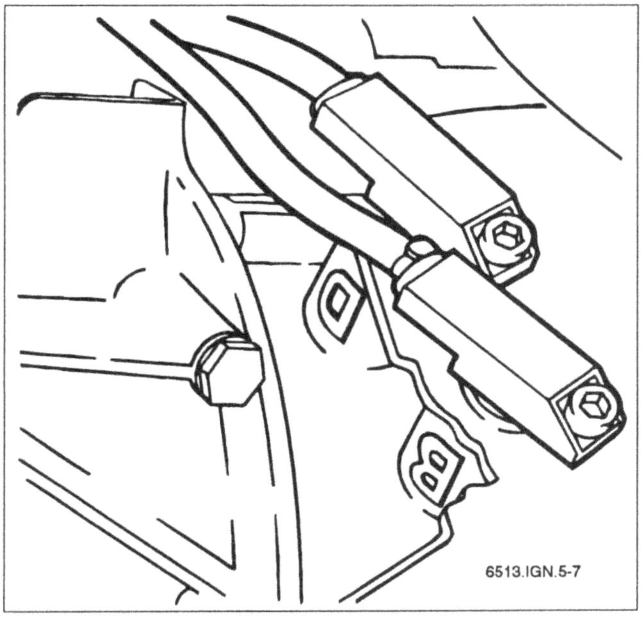

Fig. 4-11. Bellhousing reference sensor position (**B**) and speed sensor position (**D**).

Tightening torque	
• speed and reference sensor screws	7±1 Nm (62±9 in-lb)

If no faults are found with the reference sensor, inspect the raised pin on the flywheel. Using a socket wrench on the front of the crankshaft vibration damper (pulley), hand-turn the engine clockwise until the pin is visible through the bellhousing timing check hole. If the pin is missing or damaged, it should be replaced with an original pin available from BMW. Remove the socket wrench when finished.

Testing Pulse Sensor
(528e models built from March 1987)

Disconnect the electrical connector shown in Fig. 4-12. Using an ohmmeter, check the resistance between terminal 1 (yellow wire) and terminal 2 (black wire) on the sensor side end of the connector.

8

Fig. 4-12. Pulse sensor connector (arrow) for 528e models with Motronic 1.1 (built from March 1987).

Pulse sensor specifications	
• pulse sensor resistance	540±54 ohms (approximately)
• pulse sensor distance, tip to toothed wheel	1.0±0.3 mm (0.04±0.01 in.)

If the resistance is not correct, the sensor is faulty and should be replaced. Remove the sensor mounting bolt using a 5 mm hex wrench. When installing the sensor position the tip to the correct distance from the toothed wheel. Tighten the sensor mounting screw. Be careful not to overtighten the mounting screw. Be sure the wiring is correctly routed through the protective covering. See Fig. 4-13.

Fig. 4-13. Pulse sensor wiring routed through protective covering (arrow).

Tightening torque
• pulse sensor screw . . . . .7±1 Nm (62±9 in-lb)

4.4 Ignition System Electrical Circuits

This section contains wiring diagrams of the ignition systems for the cars covered by this manual. Fig. 4-14 is schematic of the system used on all 533i and 535i models, and 528e models built up to March 1987. Fig 4-15 is a schematic of the system used on 528e models built from March 1987 and later (with Motronic 1.1).

Table c and **Table d** list the locations of the various harness connectors and grounds used throughout the circuit diagrams contained in this section. For the exact positions of the fuses and relays, as well as splices (welded connections in the wiring harness) see **ELECTRICAL SYSTEM**. The Motronic control unit is located in the passenger compartment behind the glove box. See **FUEL INJECTION** for additional Motronic engine management circuit diagrams.

WARNING —
Do not touch or disconnect any of the high tension cables from the coil, distributor, or spark plugs while the engine is running or being cranked by the starter. Fatal voltages are present.

Table c. Wiring Harness Connector Locations

Connector	Location
C101	On side of fuse/relay panel (17-pin)
C116	Left side of engine compartment, below diagnostic connector (3-pin)
C200	To left of steering column (10-pin)

Table d. Wiring Harness Ground Locations

Ground point	Location
G102	Main ground above battery post
G103 (1982-1985 528e)	Main engine ground, on front of engine under diagnostic connector
G103 (1986-1988 528e)	Main engine ground, on rear side of engine above starter
G103 (533i, 535i)	Main engine ground, on rear top of engine

5. TECHNICAL DATA

I. Motronic (DME) Ignition System Specifications

Ignition coil code number	
up to March 1983	0 221 122 032 (blue label)
from March 1983	0 221 118 335 (yellow label)
Firing order	1 - 5 - 3 - 6 - 2 - 4
Spark plugs	
Bosch	WR9LS
Spark plug gap	0.7+0.1 mm (0.028+0.004 in.)
Spark plug tightening torque	20–30 Nm (15–22 ft-lb)
Reference or speed sensor coil resistance (ex. Motronic 1.1)	960±96 ohms
Pulse sensor coil resistance (Motronic 1.1)	540±54 ohms
Ignition rotor tightening torque	2.8±0.2 Nm (25±1.8 in-lb)
Reference, speed, or pulse sensor tightening torque	7±1 Nm (62±9 in-lb)

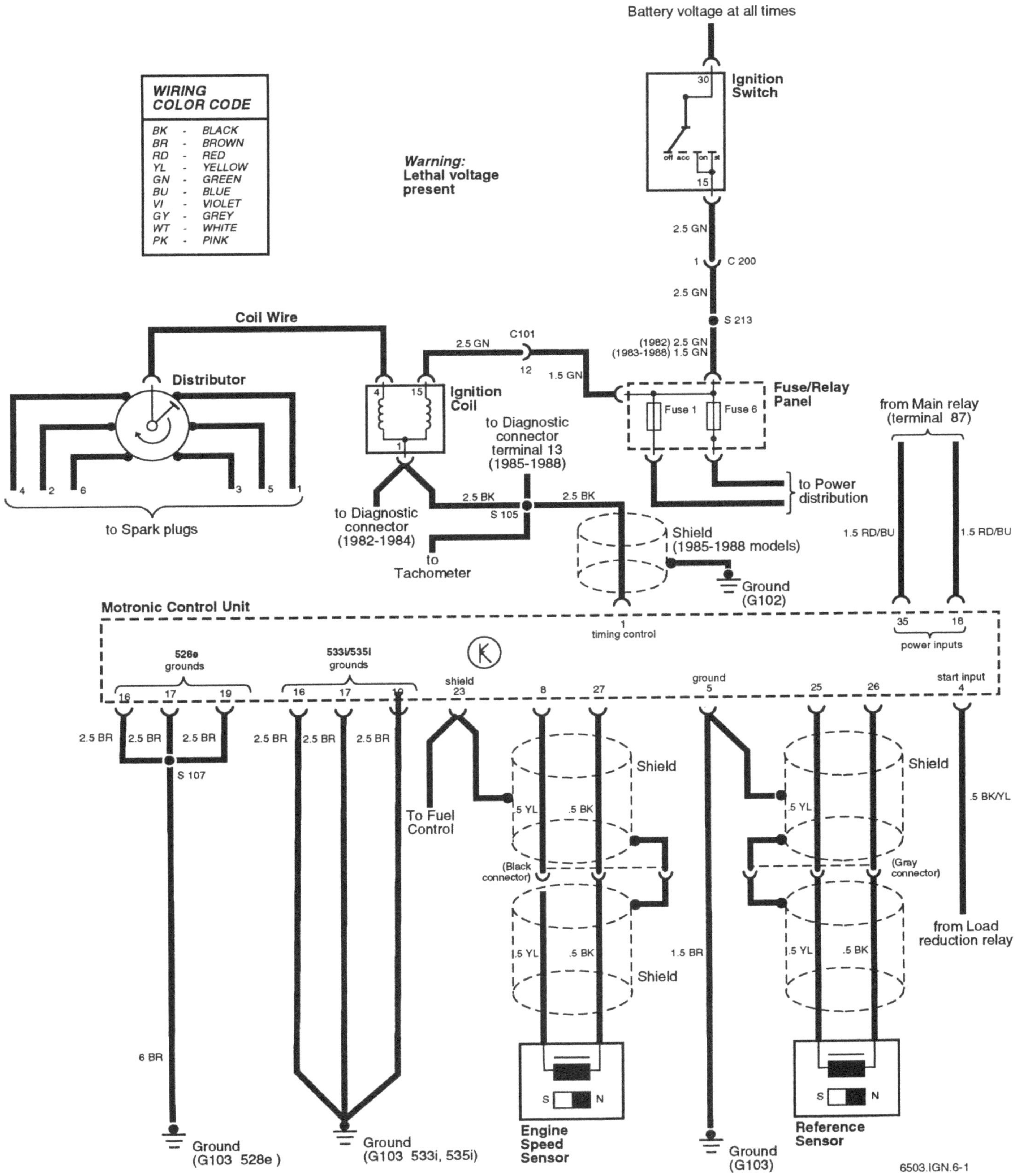

Fig. 4-14. Motronic ignition system for 533i and 535i models, and 528e models built up to March 1987.

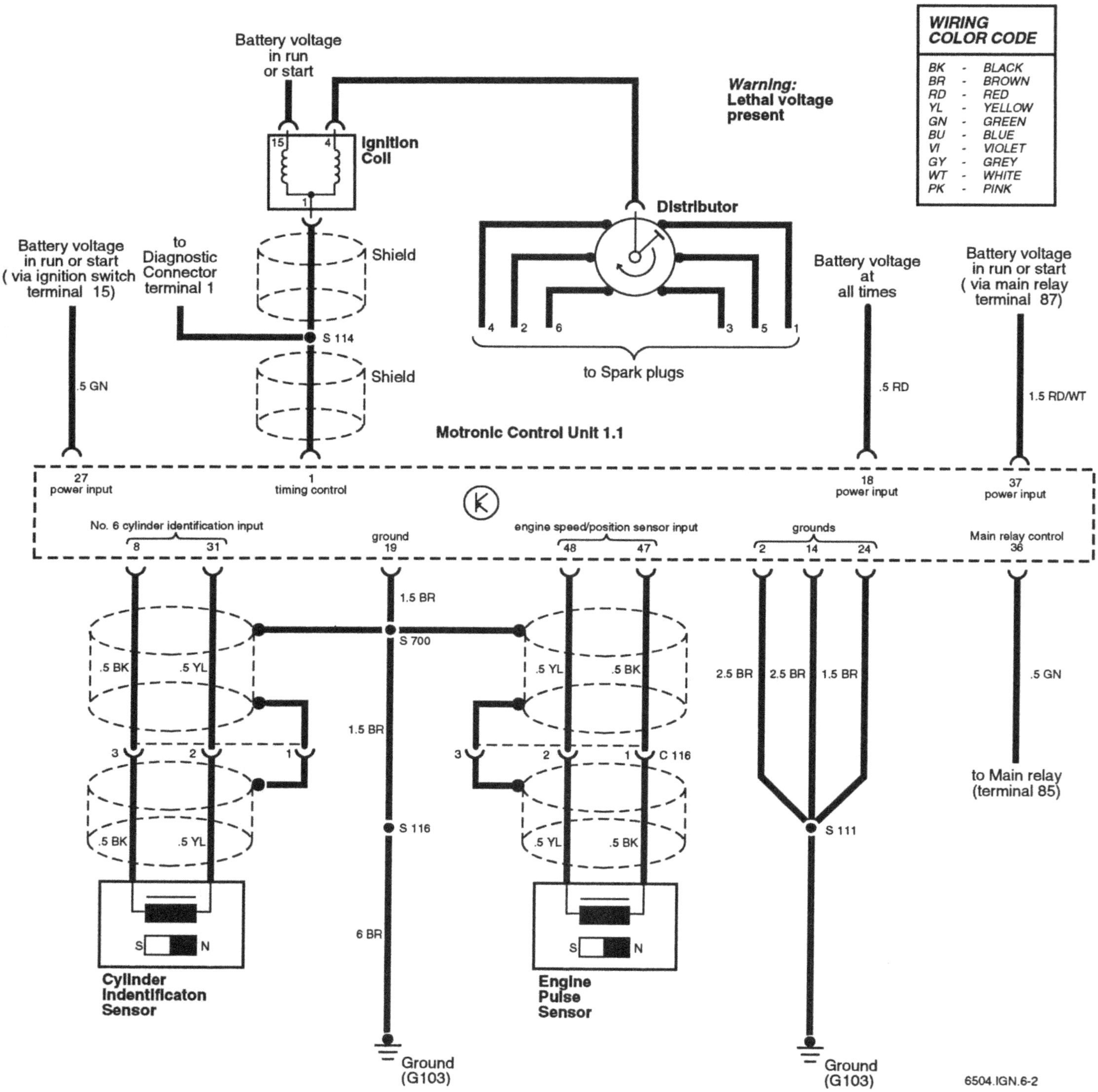

Fig. 4-15. Motronic 1.1 ignition system for 528e models built from March 1987.

Section 9

BATTERY, STARTER, ALTERNATOR

9

Contents

INTRODUCTION

This section covers the battery, starter, and alternator components of the electrical system. For additional electrical troubleshooting information that may apply to these components, see **ELECTRICAL SYSTEM.**

1. GENERAL DESCRIPTION

The battery, starter, and alternator are the three major components of the electrical system. The battery supplies electrical power when the engine is not running, and also supplies power to start the engine. The starter is an electric motor that rotates the engine via the flywheel, when the key is in the start position. The alternator converts some of the engine's mechanical energy into electrical energy whenever the engine is running.

The starter and the alternator are not protected by fuses. To prevent accidental shorts that might blow a fuse or damage wires and electrical components, the negative (–) battery cable should always be disconnected before working on the electrical system. On models equipped with anti-theft radios, the anti-theft circuitry is activated when power to the battery is removed, such as when the battery cables are removed. Be sure to have the anti-theft code on hand before removing the battery terminals or the radio fuse.

A brief description of the battery, starter, and charging system is presented here for familiarization with the system. The components are discussed in greater detail later in this section.

Voltage and Polarity

The cars covered by this manual have a 12-volt, direct current (DC), negative-ground electrical system. The voltage regulator maintains the voltage in the system at approximately the 12-volt rating of the battery, and all circuits are grounded by direct or indirect connection to the negative (–) terminal of the battery.

Battery

On all models covered by this manual, the battery is located in the engine compartment, behind the left headlight assembly.

The six-cell, 12-volt lead-acid battery capacity is rated by ampere/hours (Ah) and cold cranking amps (CCA). The Ah rating is determined by the average amount of current the battery can deliver over time without dropping below a specified voltage. The CCA is determined by the battery's ability to deliver starting current at 0°F (–18°C).

Starter

The starter and its attached solenoid are mounted to the left rear (driver's) side of the engine, below the intake manifold. To maximize the amount of current available to the starter, load-reduction relays interrupt the voltage to many of the electrical accessories when the ignition key is in the START position.

Charging System

The charging system consists of a belt-driven alternator and a voltage regulator. The voltage regulator, which is mounted on the alternator, also serves as the alternator brush holder. The alternator may be one of three output ratings depending on the production date of the car.

2. MAINTENANCE

No routine lubrication of the starter or alternator is required. The following electrical system maintenance is described in **LUBRICATION AND MAINTENANCE**.

1. Checking the battery

2. Checking alternator V-belt

3. TROUBLESHOOTING

This heading describes specific troubleshooting information for the battery, the starter, and the charging system.

3.1 Battery, Starter and Charging System Troubleshooting

The causes of a dead battery or an inoperative starter are often related. Before troubleshooting any battery-related problems, check that all ground connections are tight and free of corrosion. The battery ground is shown in Fig. 3-1. The engine ground is shown in Fig. 3-2.

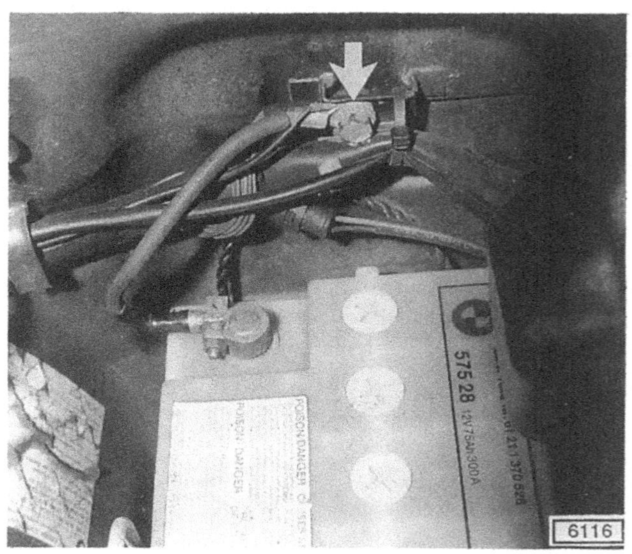

Fig. 3-1. Battery to chassis ground connection (G102) shown at arrow.

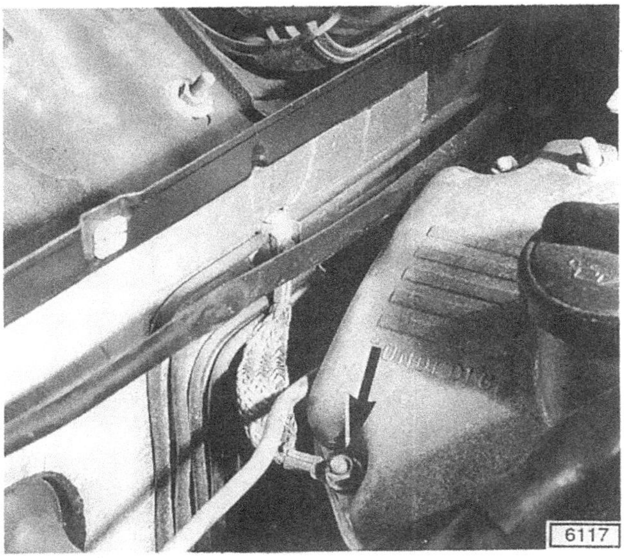

Fig. 3-2. Engine ground strap runs from cylinder head to chassis.

Table a lists symptoms of trouble with the battery, starter, and charging system, their probable causes, and suggested corrective actions. The bold numbers in the corrective action column refer to the numbered headings in this section where the suggested repairs can be found.

4. BATTERY

The battery produces electricity chemically. Lead plates inside the battery interact with a mixture of sulfuric acid and water, called electrolyte, to create a flow of current.

Low-maintenance batteries have been supplied as original equipment in the cars covered by this manual. These batteries have vented filler caps and may require distilled water to be added periodically to maintain the electrolyte level. Under normal operation the battery is maintenance-free. The electrolyte does not need to be periodically checked, and normal charging can be done without removing the caps.

Temperature affects the efficiency of the battery. The current-producing capacity of a battery at 5°F (–15°C) is only half its capacity at 68°F (20°C), and partly-discharged batteries can freeze due to the higher proportion of water in the electrolyte. A frozen battery produces no current, but can usually be restored when thawed.

For more information on battery maintenance, see **LUBRICATION AND MAINTENANCE.** For information on starting with jumper cables, see **FUNDAMENTALS**.

9

Table a. Battery, Starter and Charging System Troubleshooting

Symptom	Probable cause	Corrective action
1. Engine cranks slowly or not at all, solenoid clicks when starter is operated	**a.** Battery cables loose, dirty, or corroded **b.** Battery discharged **c.** Body ground strap loose, dirty, or corroded **d.** Poor connection or no voltage at starter motor terminal 30 **e.** Starter motor or solenoid faulty	**a.** Clean or replace cables. See **LUBRICATION AND MAINTENANCE** **b.** Charge battery, test and replace if necessary. **4.1** **c.** Inspect ground strap, clean, tighten, and replace as necessary **d.** Check connection, test for voltage at starter. **5.1**. Test automatic transmission range switch and start relay (auto. trans. only). See **ELECTRICAL SYSTEM** **e.** Test starter. **5.1**
2. Battery will not stay charged (more than a few days)	**a.** Shorted circuit draining battery **b.** Alternator V-belt loose or damaged **c.** Battery faulty **d.** Battery cables loose, dirty, or corroded **e.** Alternator or voltage regulator faulty	**a.** Test for current drain. **6.2** **b.** Inspect alternator V-belt. See **LUBRICATION AND MAINTENANCE** **c.** Test battery and replace if necessary. **4.1** **d.** Clean or replace cables. See **LUBRICATION AND MAINTENANCE** **e.** Test alternator and regulator. **6.2**
3. Battery losing water	**a.** Battery being overcharged	**a.** Test voltage regulator. **6.2**
4. Lights dim, light intensity varies with engine speed	**a.** Alternator V-belt loose or damaged **b.** Alternator or voltage regulator faulty **c.** Body ground straps loose, dirty, or corroded	**a.** Inspect alternator V-belt. See **LUBRICATION AND MAINTENANCE** **b.** Test alternator and voltage regulator. **6.2** **c.** Inspect ground straps, clean, tighten, and replace as required

WARNING —
- *Wear goggles, rubber gloves, and a rubber apron when working with battery electrolyte. Electrolyte contains sulfuric acid and can cause severe burns. If electrolyte is spilled on your skin or clothing, flush the area at once with large quantities of water. If electrolyte gets into your eyes, bathe them with large quantities of clean water for several minutes and call a physician.*

- *Batteries generate explosive gasses. Keep sparks and open flame away. Do not smoke.*

CAUTION —
Replace batteries if the case is cracked or leaking. Electrolyte can damage the car. If electrolyte is spilled, clean the area with a solution of baking soda and water.

4.1 Testing Battery

Battery testing determines the state of battery charge. The most common methods are open-circuit and load voltage testing. Batteries that have filler caps can also be tested by checking the specific gravity of the electrolyte. The specific gravity test checks the amount of acid in the electrolyte as an indication of battery charge. Inexpensive specific gravity testers are available at most auto supply stores.

Open-Circuit Voltage Test

An open-circuit voltage test checks battery voltage by connecting an accurate digital voltmeter to the battery posts after disconnecting the battery ground cable. Before making an open-circuit voltage test on a battery, first load the battery with 15 amps for one minute, for example by turning on the headlights without the engine running. Open-circuit voltage levels and their corresponding percentages of charge are in **Table b**.

CAUTION —
Never operate the engine with the battery disconnected.

Table b. Open-Circuit Voltage and Battery Charge

Open-circuit voltage	State of charge
12.6 V or more	Fully charged
12.4 V	75% charged
12.2 V	50% charged
12.0 V	25% charged
11.7 V or less	Fully discharged

The battery is in satisfactory condition if the open-circuit voltage is at least 12.4 volts. If the open-circuit voltage is at this level or above, but the battery still lacks power for starting, make a load voltage test to determine the battery's service condition. If the open-circuit voltage is below 12.4 volts, recharge the battery. If the battery cannot be recharged to at least 75%, it should be replaced.

Load Voltage Testing

A load voltage battery test is made by connecting a specific resistive load to the battery terminals and then measuring voltage. The test requires a special tester and can generally be performed quickly and inexpensively by an authorized BMW dealer or qualified repair facility

4.2 Battery Charging

Discharged batteries can be recharged using a battery charger, but a battery can never be charged to a voltage in excess of that which it is capable of producing electro-chemically. Prolonged charging causes gassing that will evaporate the electrolyte to a level that can damage the battery.

Always read and follow the instructions provided by the battery charger's manufacturer. Do not use a charger if the instructions are not available. **Table c** lists charging rates and times that should be followed when charging batteries.

WARNING —
- *The gasses given off by the battery during charging are explosive. Do not smoke. Keep open flames away from the top of the battery, and prevent electrical sparks by turning off the battery charger before connecting or disconnecting it.*

CAUTION —
- *Always allow a frozen battery to thaw before attempting to recharge it.*

- *Always disconnect the battery cables when using a battery charger. This will prevent damage to solid-state components.*

- *On models equipped with on-board computers, remove the on-board computer fuses no. 5, 6, 9, and 12 prior to fast-charging to prevent damaging the on-board computer module.*

Table c. Battery Charging Specifications

Charging rate (low-maintenance batteries)	Specific gravity*	Approximate charging time
Fast charge (at 80% to 90% of battery's capacity, example: 44 to 50 amperes for a 55-ampere hour battery)	1.150 or less 1.150 to 1.175 1.175 to 1.200 1.200 to 1.225	1 hour 3/4 hour 1/2 hour 1/4 hour
Slow charge (at 10% of battery's capacity, example: 5.5 amperes for a 55-ampere hour battery)	Above 1.225	Slow charge only, to a specific gravity of 1.250 to 1.265

*Check only on batteries with filler caps.

5. STARTING SYSTEM

When the ignition key is in the START position, battery power energizes a solenoid switch on the starter. The solenoid then engages the starter's drive pinion with the ring gear on the engine flywheel or drive plate. To minimize wear and stress on the drive pinion and ring gear, the solenoid does not switch current to the starter until the drive pinion is fully engaged. Also, the drive pinion has an overrunning clutch to prevent the starter from being driven by the engine.

On cars with automatic transmission, an additional relay controls power to the starter. The relay is only energized when the shift lever is in the "P" or "N" position. The neutral/park/back-up light switch controls power to the relay. The switch is located below the shift lever. The relay is located under the left side of the instrument cluster.

Three versions of starters are installed on the cars covered by this manual. **Table d** lists the starter rating and model production date.

Table d. Starter Applications

Model	Production date	Starter rating
528e	models built up to December 1986	1.1 kW
	models built from December 1986	1.4 kW
533i, 535i	—	1.5 kW

5.1 Starting System Troubleshooting

The battery and its cables should be in good condition when troubleshooting the starter. If in doubt about the battery, see **4. Battery**. Starter efficiency is affected by engine oil viscosity. This is especially true in cold weather. See **LUBRICATION AND MAINTENANCE.**

Troubleshooting information for the starting system appears in **Table e**. The bold numbers in the corrective action column refer to headings in this section where the repair procedures are described. Fig. 5-1 is a wiring schematic of the starting system that can be used as a guide when troubleshooting.

9

Table e. Starting System Troubleshooting

Symptom	Probable cause	Corrective action
1. Starter does not operate when ignition switch is turned to START	**a.** Ignition switch or wire leading from ignition switch to solenoid faulty (less than 8 volts to solenoid switch)	**a.** Test for voltage at terminal 50 of solenoid switch with ignition switch at START. If not at least 8 volts, test for voltage at terminal 50 of ignition switch with switch at START. Replace ignition switch (See **STEERING AND WHEEL ALIGNMENT**) or eliminate open circuit between ignition switch and solenoid switch
	b. Solenoid switch faulty (less than 8 volts to starter motor)	**b.** Test for voltage at field-winding connecting strap with ignition at START. (See Fig. 5-3 below). If not at least 8 volts, replace solenoid
	c. Starter motor faulty	**c.** Test for voltage at field-winding connecting strap with ignition at START. If at least 8 volts, repair or replace starter motor. **5.2**
	d. Automatic transmission range switch or start relay faulty (models with automatic transmission only)	**d.** Test switch. See **ELECTRICAL SYSTEM**
2. Starter turns slowly or fails to turn engine	**a.** Dirty, loose, or corroded starter connections	**a.** Remove, clean, and tighten connections. **5.2**
	b. Dirty, loose, or corroded ground strap between engine and body	**b.** Remove and clean or replace strap
	c. Starter worn or faulty	**c.** Repair or replace starter. **5.2**
3. Starter makes unusual noise, turns erratically, or fails to turn	**a.** Drive pinion defective	**a.** Repair or replace starter. **5.2**
	b. Flywheel or driveplate ring gear damaged	**b.** Replace flywheel or driveplate. See **MANUAL TRANSMISSION AND CLUTCH** or **AUTOMATIC TRANSMISSION**
4. Starter operates, but does not turn engine	**a.** Starter drive pinion or armature shaft faulty	**a.** Repair or replace starter. **5.2**
	b. Solenoid switch mechanism faulty	**b.** Replace starter solenoid switch. **5.2**

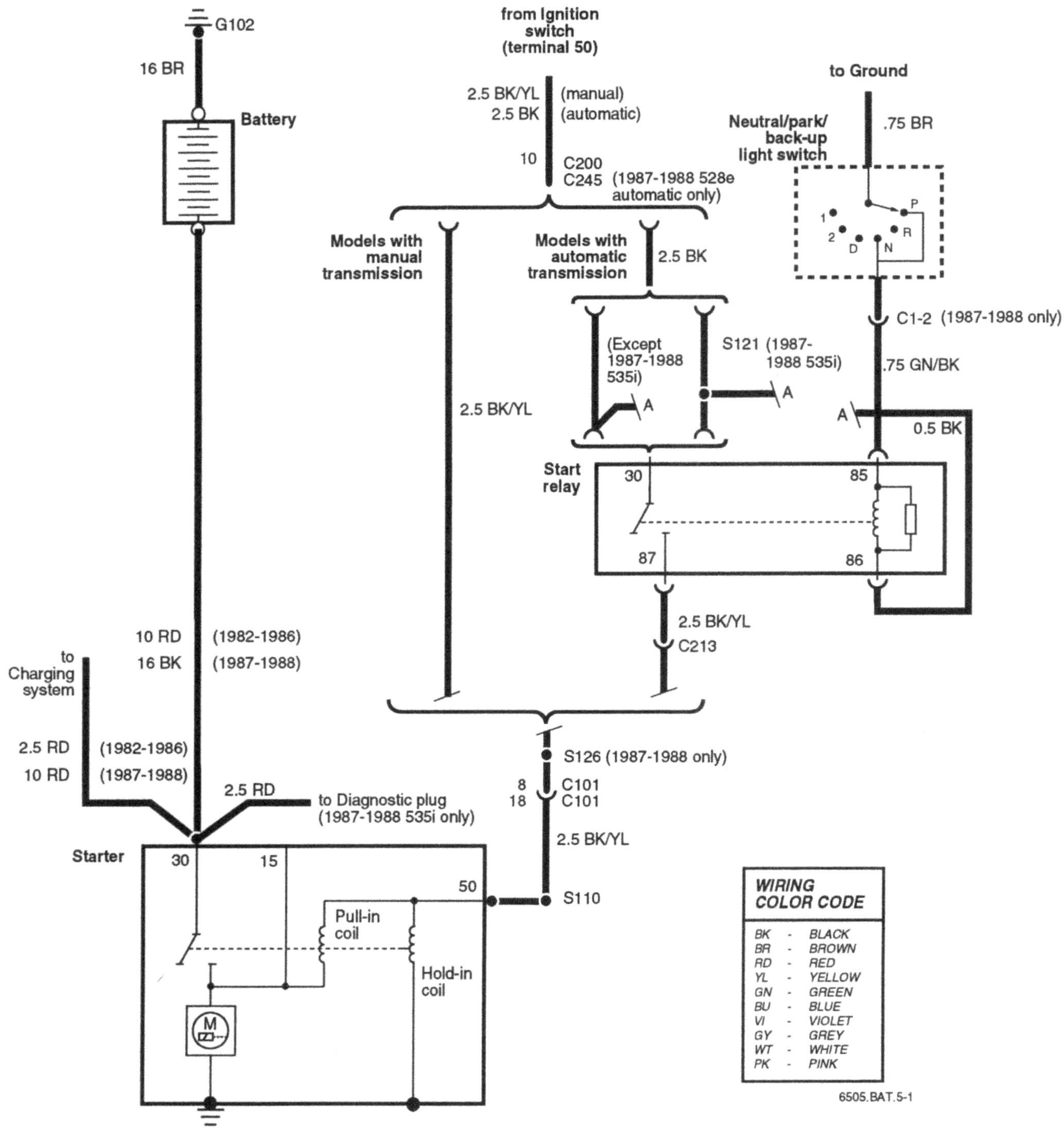

Fig. 5-1. Starting system circuit diagram. Circuit may vary slightly on some models.

5.2 Removing and Installing Starter

The starter is wired directly to the positive (+) battery terminal without fuse protection. To prevent shorts and electrical damage, always disconnect the negative (–) battery cable before removing the starter. The starter solenoid can be replaced separately once the starter is removed.

To remove starter (528e models):

1. Disconnect the negative (–) battery cable.

2. Remove the air cleaner housing and air flow sensor mounting nuts. Loosen the hose clamp on the rubber bellows. Disconnect the electrical connector from the air flow sensor and the altitude compensator, if applicable. Remove the air cleaner assembly and air flow sensor. See Fig. 5-2.

Fig. 5-2. Air cleaner housing and air flow sensor mounting nuts (arrows).

3. Disconnect the wires from the starter solenoid. See Fig. 5-3.

4. Remove the bolt and two nuts from the starter support bracket and remove the bracket. See Fig. 5-4.

5. Remove the bolts that hold the starter to the transmission bellhousing and remove the starter.

NOTE —
The use of a half-moon or starter box-end wrench or a 10 inch extension bar and deep-well socket will facilitate the removal of the mounting bolts.

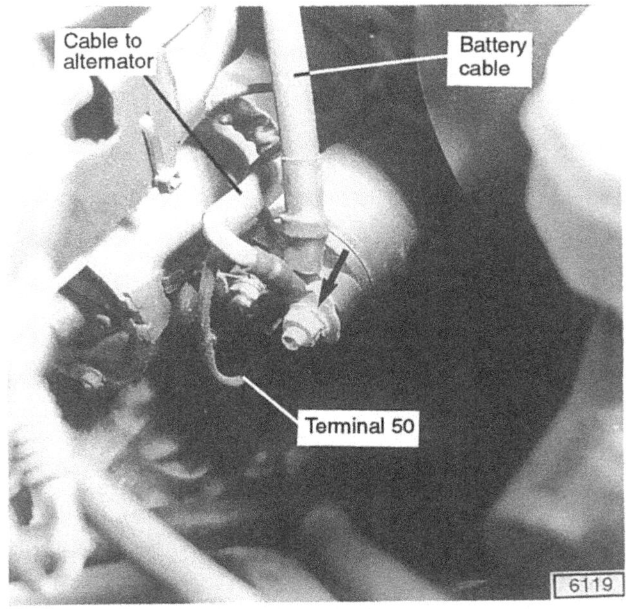

Fig. 5-3. Starter wiring at solenoid. Remove wires shown at arrow. Some models have an additional wire (not shown).

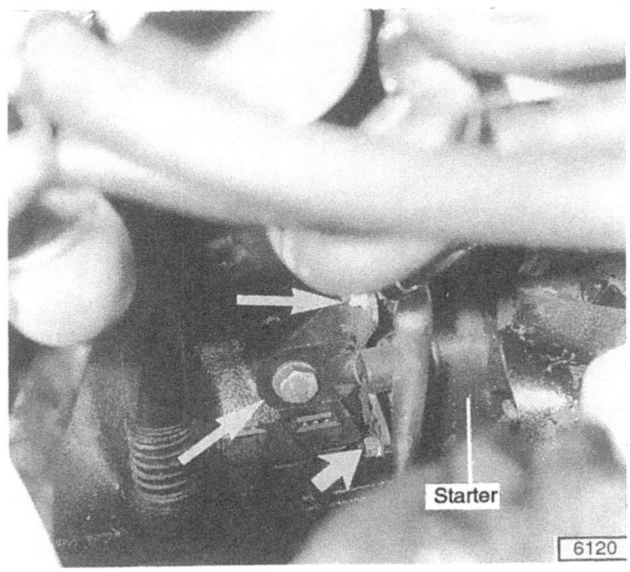

Fig. 5-4. Starter bracket mounting bolt and nuts (arrows).

Inspect the starter and solenoid. Replace or repair the starter if the drive pinion teeth are worn or broken. Also check the teeth on the flywheel of drive plate. Installation is the reverse of removal. Install the starter bolts to the proper torque. Install the starter wires as shown above in Fig. 5-3.

Tightening torques
• starter mounting bolts. . . . . 47 to 50 Nm (35 to 37 ft-lb)
• nut: battery cable to starter . . . 8 to 12 Nm (71 to 106 in-lb)

CAUTION —
Connect wires to the proper terminals. Incorrect installation may damage the electrical system.

To remove starter (533i, 535i models):

1. Disconnect the negative (–) battery cable.

2. Drain the engine coolant as described in **COOLING SYSTEM.**

3. Remove the expansion tank mounting bolts and position the tank out of the way. See Fig. 5-5.

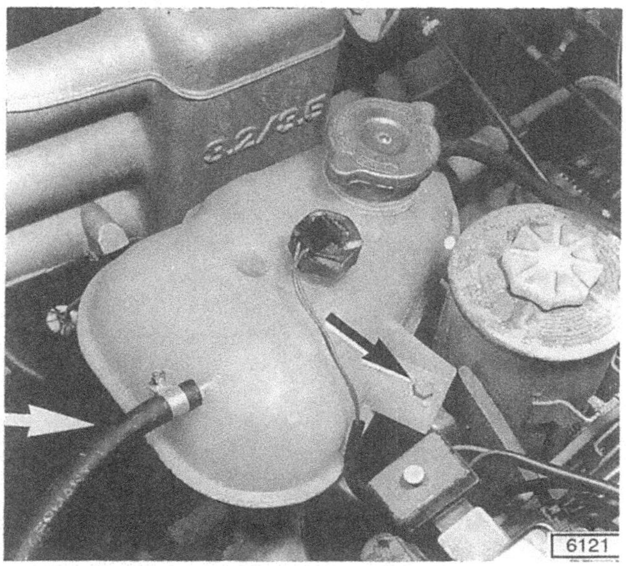

Fig. 5-5. Coolant expansion tank mounting bolts (arrows).

4. Remove one of the engine-to-heater core hoses for access to the top starter mounting bolt. See Fig. 5-6.

5. Disconnect the starter wires. See Fig. 5-7.

6. Remove the bolts that hold the starter to the transmission bellhousing and remove the starter.

NOTE —
The use of a half-moon or starter box-end wrench or a 10 inch extension bar and deep-well socket will facilitate the removal of the mounting bolts.

Inspect the starter and solenoid. Replace or repair the starter if the drive pinion teeth are worn or broken. Also check the teeth on the flywheel of drive plate. Installation is the reverse of removal. Install the starter bolts to the proper torque. Install the starter wires as shown in Fig. 5-7. Refill the cooling system as described under **COOLING SYSTEM.**

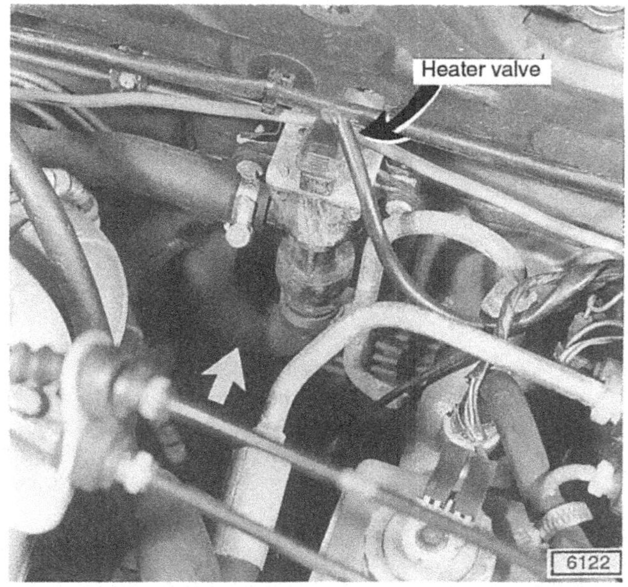

Fig. 5-6. Hose coming from heater core (arrow) to be removed to reach starter mounting bolt.

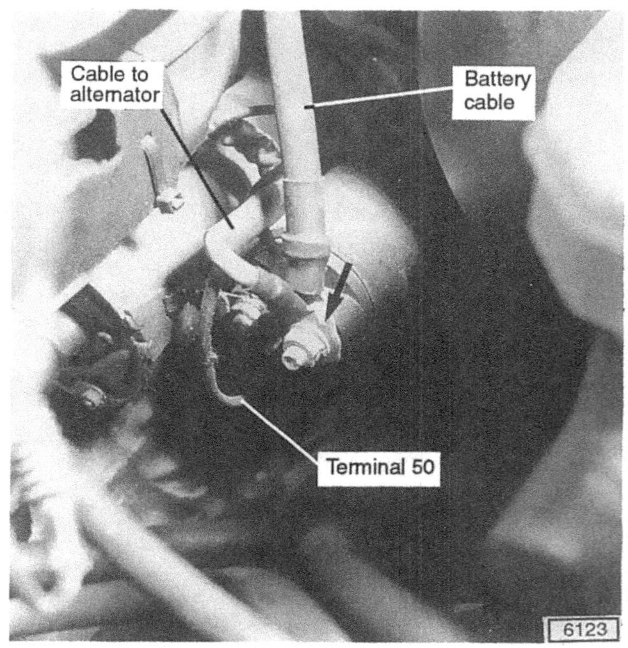

Fig. 5-7. Starter wiring at solenoid (arrow). Remove all wires. Some models have an additional wire (not shown).

Tightening torques
• starter mounting bolts. . . . . 47 to 50 Nm (35 to 37 ft-lb)
• battery cable to starter (nut) . . . 8 to 12 Nm (71 to 106 in-lb)

CAUTION —
Connect wires to the proper terminals. Incorrect installation may damage the electrical system.

Replacing Solenoid Switch

A faulty solenoid switch can be replaced separately. First remove the starter as described above. Remove the nut from the field-winding connecting strap and remove the strap from the solenoid. Remove the three screws holding the solenoid to the starter. Remove the solenoid and its spring.

NOTE —
The solenoid mounting screws are tightly fitted and may be difficult to remove. To help loosen the screws, use an impact driver and a slotted bit.

Installation is the reverse of removal. Apply a sealer to the screw heads after installing the solenoid.

6. CHARGING SYSTEM

The charging system provides the current necessary to keep the battery charged and to operate the car electrical accessories. The system includes an alternator driven from the engine crankshaft by a V-belt to generate the charging current, and a voltage regulator to control the charging current.

To prevent damage to the alternator or regulator when making tests or repairs, make all connections with negative (–) to negative, and positive (+) to positive unless directed otherwise. Even momentary contact with a conductor of the wrong polarity can damage the alternator's diodes. Make certain that the battery negative (–) cable is securely fastened, and that the cable to terminal 30 on the starter is connected to the battery's positive (+) terminal.

CAUTION —
*Never operate the engine with the battery disconnected. Never operate the alternator with its output terminal (B+ or 30) disconnected and the other terminals connected. Never short, bridge, or ground any terminals of the charging system except as specifically described in **6.1 Charging System Troubleshooting**.*

The charging capacity of the alternator installed on each car depends on the type and number of electrical accessories. The alternator rating can be found on its housing. **Table f** lists the alternator ratings.

NOTE —
Rated output of the alternator when measured at the battery with the engine running can be 10 to 15 amps less than specified in **table f**. The difference is accounted for by electrical consumers at the power distribution box and is a normal condition.

Table f. Alternator Output Ratings and Model Production Dates

Model	Production date	Alternator rating
528e	models built up to March 1983	65 A
	models built from March 1983 to Jan. 1986	80 A
	models built from Jan. 1986	90 A
533i, 535i	models built up to Jan. 1986	80 A
	models built from Jan. 1986	90 A

The voltage regulator is mounted to the back of the alternator housing. Voltage regulators and alternators are available as replacement parts from an authorized BMW dealer. BMW remanufactured alternators are also available.

9

6.1 Charging System Troubleshooting

Charging system trouble is indicated by an illuminated alternator warning light on the instrument panel, or by an under or overcharged battery.

The alternator generates electrical current by electrical induction. That is, a magnetic field is placed in motion to induce a current in a stationary coil. When the engine is running and the alternator is spinning, part of the current it produces energizes its electromagnetic field. When starting, some other current must be provided to initially energize the field and begin the current generating process. This current is provided by the battery through the alternator warning light in the instrument cluster.

NOTE —
On 1982 through 1986 models, a burned out instrument cluster alternator warning light will prevent the alternator from charging. On 1987 and later models, a resistor is wired in parallel with the alternator warning light. This resistor will allow current to reach the alternator D+ terminal even if the warning light is burned out. See Fig. 6-1.

As soon as the alternator's output equals the battery's voltage, the light goes out. Normally, the warning light should be off when the ignition is off and the engine is stopped. The light should only come on when the ignition is turned on (current to the rotor) and go out again when the engine is started and the battery is being charged.

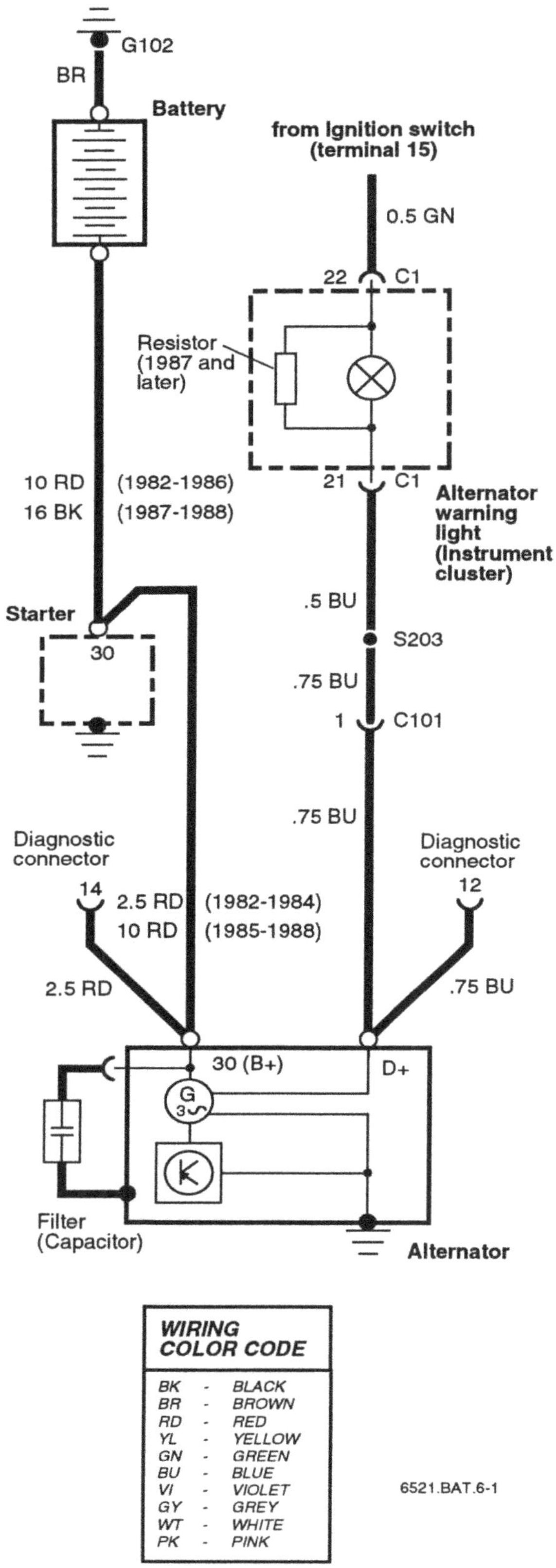

Fig. 6-1. Simplified charging system circuit.

Table g describes symptoms of trouble indicated by the warning light, lists tests and probable causes for the problem, and suggests corrective actions. The bold numbers in the corrective action column refer to the numbered headings in this section where the suggested repairs can be found.

> ***WARNING —***
> *Disconnect the battery negative (–) cable before disconnecting any wires from the rear of the alternator. Battery voltage is wired directly to the alternator without fuse protection. Reconnect the battery cable after all wires have been safely disconnected and insulated.*

An undercharged battery is usually associated with starting trouble. Again, make sure that the battery is in good condition and capable of accepting a full charge before blaming the charging system. Causes of an undercharged battery are: the simultaneous use of many electrical accessories for long periods of time, leaving accessories or lights in operation with the engine stopped, frequent long periods of starter usage, frequent short-trip driving and improper alternator V-belt condition or tension.

Broken or frayed charging system wiring or corroded connections at the D+ and B+ (30) terminals of the alternator, as well as worn, corroded, or loose battery cable connections will also prevent adequate charging or increase charging time. Chassis ground connections must be clean and tight.

6.2 In-Car Testing of Charging System

The tests described here will help determine the cause of charging system trouble. The battery should be fully charged and the alternator V-belt correctly tensioned. If in doubt about either, test the battery as described in **4.1 Testing Battery**, and adjust the V-belt as described in **6.3 Removing and Installing Alternator and Voltage Regulator**. The charging system warning light should be operating correctly. All electrical connections should be clean and tight. Replace wires that are hard or cracked.

> ***WARNING —***
> *Disconnect the battery negative (–) cable before disconnecting any wires from the rear of the alternator. Battery voltage is wired directly to the alternator without fuse protection. Reconnect the battery cable after all wires have been safely disconnected and insulated.*

A general test of charging system output can be made with an accurate digital voltmeter. The most accurate testing is done using an alternator and regulator tester that applies a high-current load to the alternator. Conclusive tests using this equipment can be made inexpensively by an authorized BMW dealer or other qualified shop.

Table g. Warning Light Troubleshooting

Symptom	Test and probable cause	Corrective action
1. Ignition off, engine not running, warning light glowing or on	**(TEST)** Disconnect blue wire (D+) from alternator **a.** Light goes out: Faulty alternator diodes **b.** Light does not go out: Short to ground in wiring harness or wiring connector	**a.** Repair or replace alternator. **6.3** **b.** Repair or replace faulty wiring
2. Ignition on, engine not running, warning light off	**a.** Battery fully discharged **b.** Bulb burned out **(TEST)** Disconnect blue wire (D+) from alternator. With battery connected and ignition on, touch blue wire to ground **c.** Light does not come on: Faulty bulb socket, open circuit between socket and terminal 15 of ignition switch, or open circuit between blue wire (D+) on alternator and instrument cluster **d.** Light comes on: Loose connection between regulator and alternator or loose connection between brushes and regulator **e.** Light comes on, no faults with regulator: Internal alternator faults or faulty regulator	**a.** Charge battery. **4.2** **b.** Remove and test bulb. Replace faulty bulb. See **ELECTRICAL SYSTEM** **c.** Replace instrument cluster printed circuit board. Repair wires or connections. See **ELECTRICAL SYSTEM** **d.** Inspect brushes. Correct loose connections. **6.3** **e.** Repair or replace alternator or voltage regulator. **6.3**
3. Engine running at any speed, warning light stays on	**a.** Loose or broken alternator V-belt **(TEST)** Disconnect red wire (B+) from alternator. Do not short wire. Using a voltmeter, test between red wire and ground with battery terminal connected **b.** No voltage to alternator: open circuit between red wire on alternator and starter or between starter and battery positive (+) pole **c.** Exciter diodes burned out **d.** Faulty regulator or faulty alternator windings **e.** High voltage drop between red alternator positive (+) wire and starter due to broken, loose, or corroded wires	**a.** Replace or adjust V-belt. See **LUBRICATION AND MAINTENANCE** **b.** Repair wire or connections between alternator and starter **c.** Repair or replace alternator. **6.3** **d.** Test charging system and replace faulty components as needed. **6.2** **e.** Repair wires or connections

CAUTION —
An alternator must never be run with the battery disconnected. This will severely damage the alternator, the voltage regulator, or electronic components. The alternator wiring should be secure.

Testing Alternator and Regulator

Start the engine and run it at about 1500 rpm with all electrical accessories turned off. Set the voltmeter to the DCV scale and measure the voltage between the positive and negative terminals of the battery. Make sure that the tester is connected to clean areas of the terminals.

Alternator charging voltage	
• at 1500 rpm	13.5 to 14.2 volts (DC)

A reading much higher than 14.2 volts most likely indicates a faulty voltage regulator. A reading below 13.5 volts means that the battery is not being adequately charged. This may be due to either a faulty V-belt, regulator, or alternator. To determine which, keep the car running at about 1500 rpm and turn on all electrical accessories, then check voltage across the battery. A reading that is the same as the first most likely indicates a faulty regulator. A reading that is lower most likely indicates a faulty alternator.

Testing Current Drain

If tests show that the alternator and regulator are operating correctly, but the battery still continually runs down, there may be a short in the electrical system causing a continuous current drain. Remember that the clock and some other components require constant current.

To test for current drain, turn off the ignition switch, the radio, and all lights. Turn on the switches for the blower motor, the air conditioning, the rear defogger and the heated seats. This will determine if the unloader relay is faulty. Disconnect the negative (–) cable from the battery and connect a fused ammeter (0 to 1 A DC scale) between the cable and the negative post.

If the ammeter indicates current flow, some electrical accessory or a short in the electrical system is draining current from the battery. Isolate the faulty circuit by removing and replacing the fuses one at a time until the ammeter reading goes down to nearly zero. This should be the faulty circuit. If no fault is found in this way, a fault may exist in the components without fuses, such as the alternator, the starter motor, and the instrument cluster. Disconnect and insulate the items one at a time. If a low current drain is still detected, repeat the test using a fused ammeter set to the 0 to 200 mA DC range.

> **NOTE —**
> Large current drains can be detected using a test light connected in series with the negative cable and negative battery post.

Noisy Alternator

Alternator noises are usually mechanical in origin, but a high, soft whistling sound may be produced by an alternator that is overcharging because of a faulty regulator diode. The same sound may be heard if there is a shorted diode placing an abnormal electrical strain on the alternator. Some alternators make this sound when operating normally at maximum output.

Alternator mechanical noises are usually the result of misalignment between the V-belt and the pulley, a loose or broken pulley, worn bearings, or a bent rotor shaft. Check for bad bearings by removing the V-belt as described below and rotating the alternator pulley by hand. If the shaft grinds or grates, the alternator should be replaced.

6.3 Removing and Installing Alternator and Voltage Regulator

The alternator is connected to battery current even when the ignition is turned off. To prevent shorts and electrical damage, always disconnect the battery negative (–) cable before removing the alternator.

To remove:

1. Disconnect the negative (–) battery cable.

2. On 528e models, remove the air cleaner assembly and air flow sensor mounting nuts. Disconnect the hose clamp securing the rubber bellows. Disconnect the electrical connector from the air flow sensor and the altitude compensator, if applicable. See Fig. 5-2, above. Remove the air cleaner and air flow sensor.

3. Remove the two mounting nuts and wiring from the back of the alternator. See Fig. 6-2.

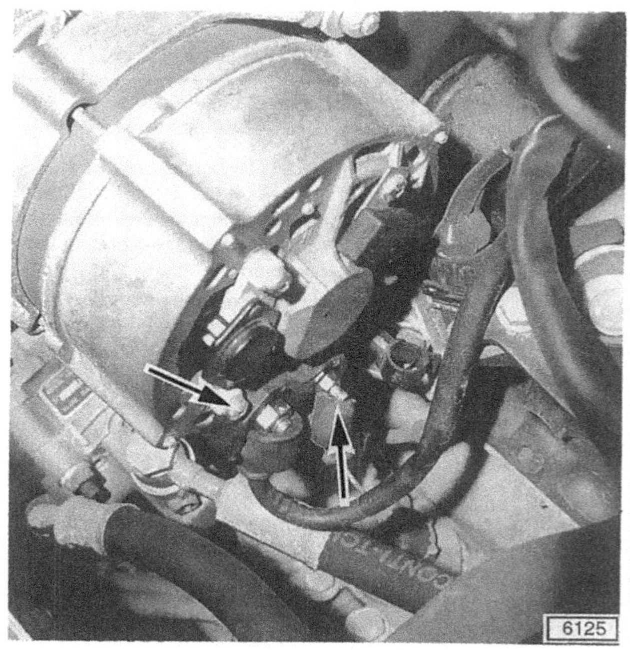

Fig. 6-2. Alternator wiring mounting nuts and wiring (arrows).

4. Loosen the upper alternator mounting bolt. See Fig. 6-3. Loosen the lower mounting bolt. See Fig. 6-4. Loosen the nut that holds the alternator belt adjusting bracket to the front of the engine.

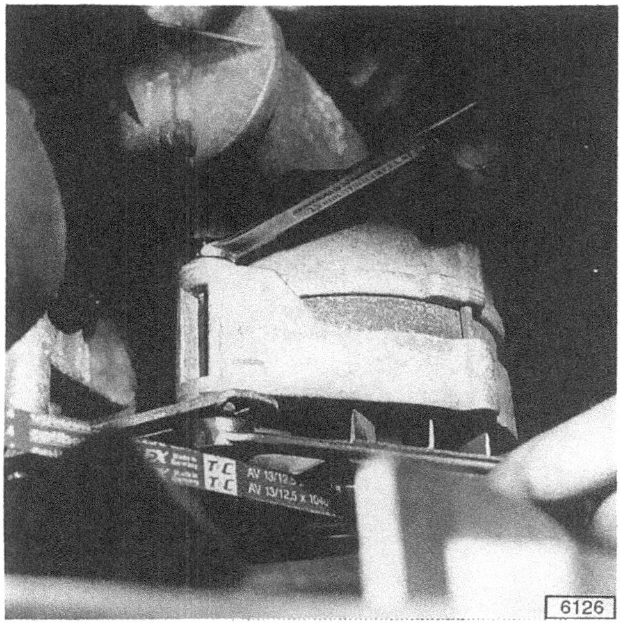

Fig. 6-3. Upper alternator mounting bolt and tensioning gear nut being loosened.

Fig. 6-4. Lower alternator mounting bolt being loosened.

5. Push the alternator as far as possible toward the engine, then take the V-belt off the alternator pulley.

6. Remove the lower alternator mounting bolt, then remove the upper alternator mounting bolt and remove the alternator from the engine.

Installation is the reverse of removal. Use a torque wrench and a crowfoot wrench to tension the V-belt. Turn the tensioning gear nut counterclockwise with the torque wrench. Hold the wrench steady and tighten the nut on the rear of the alternator. Tighten all other mounting nuts and reconnect the ground wire if applicable.

Tightening torques
• alternator wiring large nut (B+ terminal, red wire) . . . . . 7.5 to 8.0 Nm (66 to 71 in-lb) smaller nut (D+ terminal, blue wire). . . . . 1.6 to 2.3 Nm (14 to 20 in-lb)
• tensioning gear nut (for belt tensioning). . . . 7 Nm (62 in-lb)
• tensioning gear locking nut (rear of alternator) . . . . 24 Nm (18 ft-lb)

Brushes and Voltage Regulator

9

The brushes contact the alternator's slip rings to supply the current that magnetizes the alternator rotor. The voltage regulator maintains a nominal voltage in the electrical system by feeding excess output from the alternator back to ground. In addition, it regulates the amount of current supplied to the alternator rotor, turning it on and off as needed. If either component is faulty, the result will be an over or undercharged battery. The brushes wear under normal use, and will eventually need to be replaced.

The brushes and regulator are mounted on the rear of the alternator. The brushes and regulator can be replaced with the alternator installed provided the negative (–) battery cable is disconnected. The regulator and brush holder is mounted with two screws. Check the brush length as shown in Fig. 6-5.

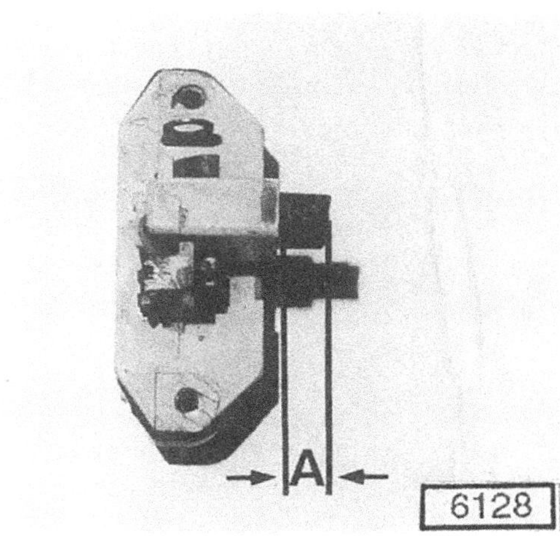

Fig. 6-5. Alternator brush length being checked (**A**). Replace brushes worn below minimum.

Alternator brush length (A)
• minimum. . . . . 5 mm (7/32 in.)

Section 10

COOLING SYSTEM

Contents

INTRODUCTION

The engines covered by this manual are liquid-cooled, and rely on a closed system of circulating coolant to maintain an even engine temperature and help transfer heat away from the engine. To provide adequate cooling over a range of temperature conditions, to lubricate the system's moving parts, and to prevent the buildup of mineral deposits and other contaminants, the coolant recommended for use year-round is a mixture of phosphate-free anti-freeze and clean water.

Proper care of the cooling system is easy. Simple preventive maintenance can keep the system operating at its best and help prevent temperature-related problems from shortening engine life.

1. GENERAL DESCRIPTION

Fig. 1-1 and Fig. 1-2 are schematic views of the cooling systems used on the cars covered by this manual.

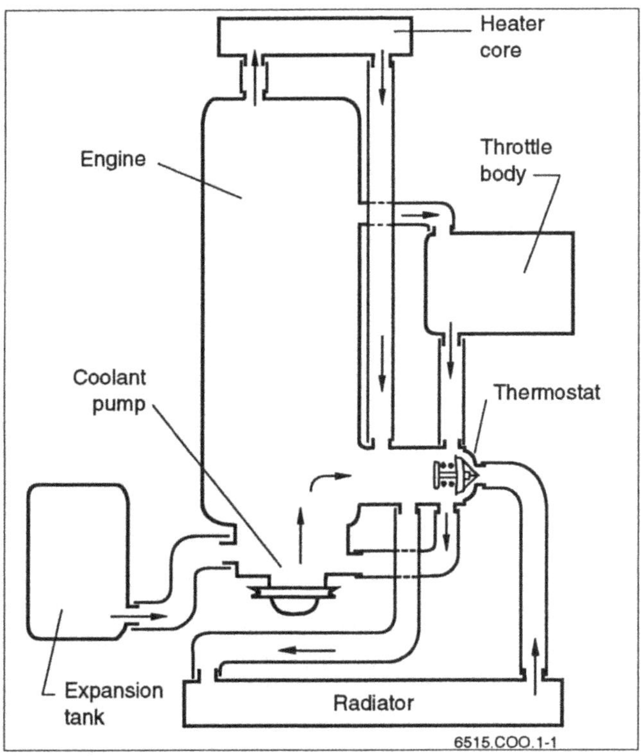

Fig. 1-1. Schematic view of cooling system components and hose routing for 528e model. Arrows show coolant flow.

Coolant Pump and Thermostat

A centrifugal-type coolant pump is mounted to the front of the engine block. The pump is crankshaft-driven by a V-belt and circulates coolant through the system whenever the engine is running.

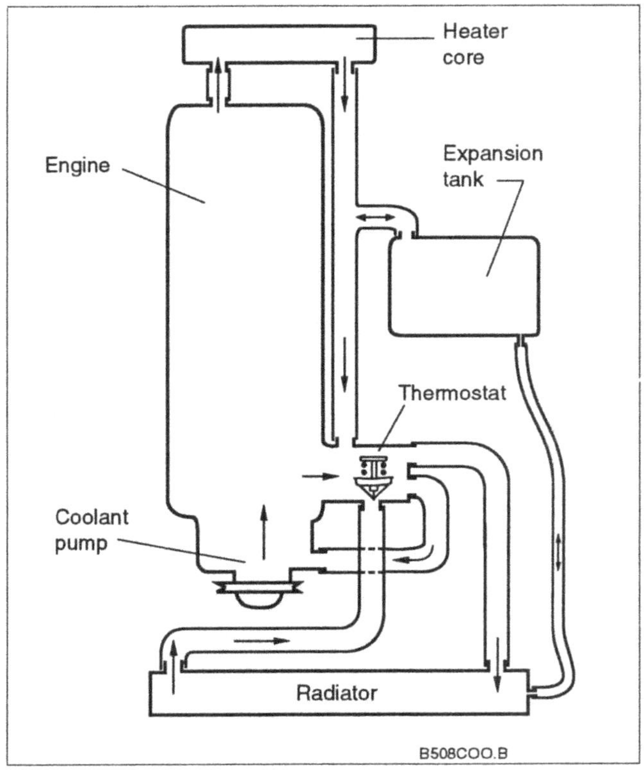

Fig. 1-2. Schematic view of cooling system components and hose routing for 533i and 535i models. Arrows show coolant flow.

A thermostat controls the flow of coolant into the radiator. When the engine is cold the thermostat is closed, so coolant bypasses the radiator, recirculating from the engine directly back to the pump. When the engine reaches operating temperature, the thermostat opens and coolant circulates through the whole system, including the radiator.

Radiator and Expansion Tank

The radiator is constructed of an aluminum core and plastic side tanks. On cars with automatic transmission, the radiator also has an integral transmission fluid cooler in the right side tank. A translucent expansion tank, or overflow reservoir provides for coolant expansion at higher temperatures and easy monitoring of coolant level. Depending on model, either a bayonet or screw-on cap is used on the expansion tank.

Radiator Cooling Fan

The primary cooling fan is mounted to the front of the coolant pump through a fan clutch. The fan clutch is a viscous coupling that controls the speed of the fan based on engine compartment temperature. An additional electric auxiliary cooling fan is mounted in front of the radiator and operates independent of the engine. Power to the auxiliary fan is controlled by coolant temperature switches mounted on the radiator, one switch for low speed and one for high speed. As coolant temperature increases, the switches close to start the cooling fan. When coolant temperature is in the correct range, the switches open. An additional circuit powers the auxiliary cooling fan whenever the air conditioning is on.

2. MAINTENANCE

BMW specifies the following steps for proper maintenance of the cooling system, to be carried out at particular time or mileage intervals. For information on the prescribed maintenance intervals and maintenance steps, see **LUBRICATION AND MAINTENANCE**.

1. Checking coolant level and anti-freeze concentration
2. Inspecting coolant pump V-belt tension and condition
3. Inspecting coolant hoses for leaks, tightness of hose clamps
4. Replacing engine coolant

3. TROUBLESHOOTING

This troubleshooting section applies to problems affecting the cooling system, which includes the components that store, pump, and regulate circulation of the coolant.

Problems associated with the operation and repair of the heating system are covered in **HEATING AND AIR CONDITIONING**. Removal and replacement of the coolant temperature gauge is covered in **ELECTRICAL SYSTEM**.

Overheating problems may also be caused by an engine fault that allows hot combustion gases to leak into the cooling system. See **ENGINE—SERVICE AND REPAIR** for more information.

3.1 Basic Troubleshooting Principles

When investigating the cause of overheating or coolant loss, begin with a visual inspection of the system. Check coolant level and check for evidence of coolant leaks. Leaks can occur at any place in the cooling system where there is a bolted housing or hose connection.

The closed cooling system becomes pressurized at normal operating temperature, and maintaining this pressure is an important system function. Leaks may prevent the system from becoming pressurized and allow the coolant to boil at a lower temperature. If visual evidence is inconclusive, a cooling system pressure test will determine whether the system leaks, and may help to indicate the source. See **3.2 Diagnostic Tests**.

If the cooling system is full of coolant and holds pressure, the next most probable cause of overheating is poor coolant circulation caused by a broken V-belt, a failed thermostat, a pinched or restricted hose, or a clogged system. In warm weather, virtually all clogs are caused by neglect of the coolant, or by the addition of substances to the coolant that are not recommended. In cold weather, the cooling system may also be clogged by frozen coolant resulting from an inadequate amount of anti-freeze.

The engine-driven coolant pump is subject to wear, just as any other moving engine part. Complete failure of the pump to circulate coolant is unusual, but excessive wear may cause noisy operation or coolant leaks at the pump shaft.

An otherwise sound cooling system may still overheat, particularly with prolonged idling, due to a failure of either the primary cooling fan or the auxiliary cooling fan.

The primary cooling fan is controlled by a temperature-dependent fan clutch. See Fig. 3-1. A failed fan clutch may affect air flow through the radiator, resulting in overheating or possibly overcooling. See **4.3 Coolant Pump** for more information on the fan clutch.

10

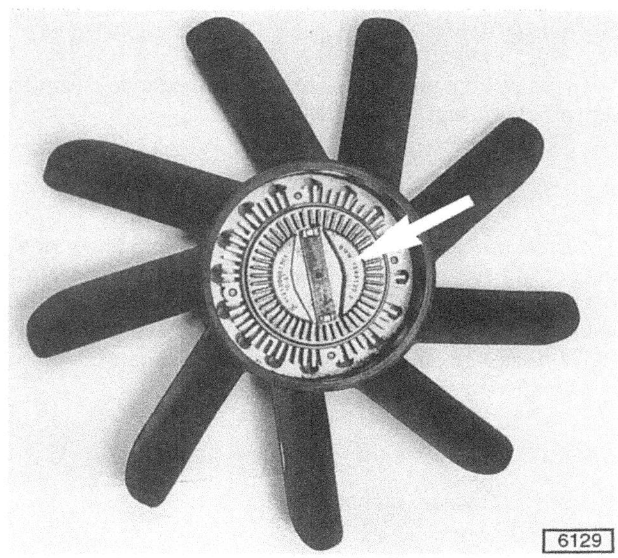

Fig. 3-1. Primary cooling fan and clutch (arrow) assembly.

The electrically operated auxiliary cooling fan should switch on and off according to engine coolant temperature. If the auxiliary fan does not operate properly, see **4.4 Auxiliary Cooling Fan** for testing information.

Table a lists overheating and underheating symptoms, their probable causes, and suggested corrective actions. The bold numbers in the corrective action column refer to headings in this section where the suggested repairs are described.

Table a. Cooling System Troubleshooting

Symptom	Probable cause	Corrective action
1. Engine overheats	a. Low coolant level	a. Fill cooling system. Pressure-test system to check for leaks. **3.2**
	b. Burst hose	b. Replace hose. **4.1**
	c. Radiator hose restricted (faulty lower hose may collapse at high engine speeds)	c. Replace hose. **4.1**
	d. V-belt loose or broken	d. Inspect V-belt and adjust or replace as necessary. See **LUBRICATION AND MAINTENANCE**
	e. Faulty thermostat	e. Remove and test thermostat and replace if faulty. **4.2**
	f. Auxiliary fan not switching on	f. Test coolant switches and fan. Replace faulty part. **4.4**
	g. Faulty cap on expansion tank	g. Pressure-test cap and replace if faulty. **3.2**
	h. Clogged radiator	h. Clean or replace radiator. **4.5**
	i. Coolant pump faulty	i. Inspect coolant pump and replace if faulty. **4.3**
2. Temperature gauge reads low (heater output inadequate)	a. Faulty thermostat	a. Remove and test thermostat, replace if faulty. **4.2**
	b. Auxiliary cooling fan not switching off	b. Replace coolant temperature switches for fan. **4.4**
	c. Cooling fan clutch hub seized	c. Inspect clutch and replace if faulty. **4.3**
3. Temperature gauge reads low (heater output normal)	a. Faulty temperature gauge or sending unit	a. Test temperature gauge and sending unit. Replace faulty part. **3.2**

3.2 Diagnostic Tests

These system tests are used to help diagnose cooling system problems and isolate their causes.

Pressure Testing

A pressure test will help find any leaks and show whether the cooling system can maintain pressure. If the system cannot maintain pressure, the engine will overheat more easily. If you do not have access to the necessary tools, an authorized BMW dealer or other qualified repair shop can perform this test inexpensively.

Requirements:

- cooling system pressure tester
- engine warm

With the engine at normal operating temperature, pressurize the system. See Fig. 3-2. Loss of pressure indicates leaks that may also be detected by the seepage of coolant. If the pressure drops rapidly and there is no sign of coolant leakage, the cylinder head gasket may be faulty. See **ENGINE—SERVICE AND REPAIR** for more information.

The expansion tank cap, either bayonet type or screw-on type, can also be tested using a pressure tester with the correct adapter.

Fig. 3-2. Pressure tester shown installed on expansion tank to pressurize cooling system.

Cooling system pressure tests
• maximum test pressure .... 1.5 bar (21.7 psi)
• maximum pressure drop. . . 0.1 bar (1.5 psi) after 2 minutes

CAUTION —
Exceeding the specified test pressure could damage the radiator or other cooling system components.

NOTE —
On screw-on type caps, the expansion tank cap opening pressure is molded onto the top of the cap. See Fig. 3-3.

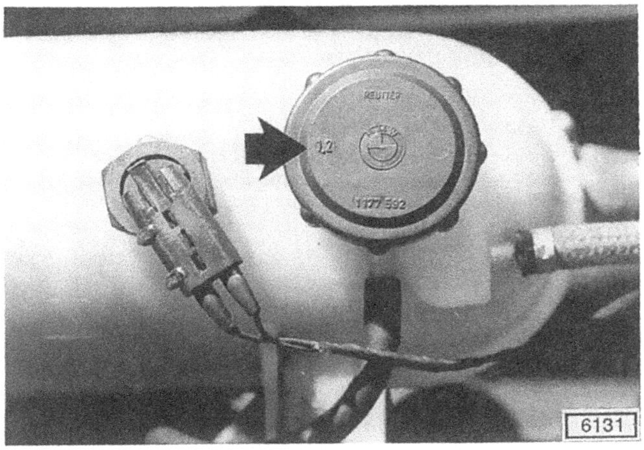

Fig. 3-3. Expansion tank cap showing opening pressure (arrow) on screw-on type cap.

On screw-on type caps, check the vacuum valve on the back of the cap by lifting up on the valve. See Fig. 3-4. The vacuum valve should spring back into position on its seat. Inspect the cap's gasket. A faulty cap or a damaged cap gasket should be replaced.

Fig. 3-4. Inside of screw-on expansion tank cap showing location of vacuum valve (arrow).

Temperature Gauge and Sending Unit Check

A quick, easy test of the coolant temperature sender can determine whether the coolant temperature gauge is functioning correctly. The checks described below should be made at the coolant temperature sender. The sender is mounted in the thermostat housing. See Fig. 3-5.

If the gauge needle remains at the rest position with the engine warm, disconnect the wire from the sending unit and ground the wire to the switch housing. See Fig. 3-6. Connecting the sender wire to ground simulates a high engine temperature to the gauge. Turn the ignition on. If the gauge needle moves upward, the sending unit is faulty. If the gauge does not respond, the wiring to the gauge is broken (open circuit) or the gauge itself is faulty.

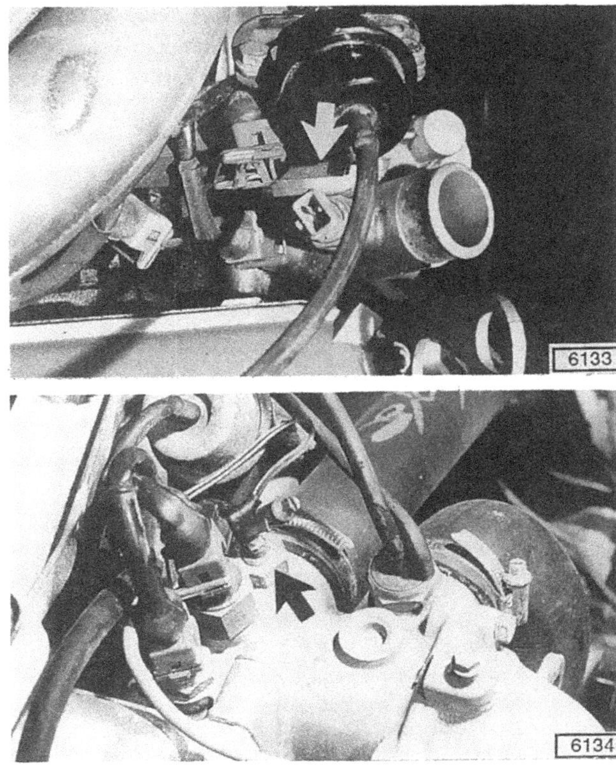

Fig. 3-5. Coolant temperature sender (brown/white wire) on 528e (top), and 533i and 535i models (bottom).

Fig. 3-6. Coolant gauge temperature sender wire being touched to ground (arrow).

If the gauge needle reads too high when the engine is cold, disconnect the wire from the sender. Do not let the wire touch ground. Turn the ignition on. If the gauge needle position does not change, the wiring or the gauge is shorted to ground. If the gauge needle drops, the sender is faulty and should be replaced.

When replacing a faulty coolant temperature sender, the gasket ring on the sender should also be replaced.

Tightening torque	
• temperature sending unit	18±1 Nm (13±0.5 ft-lb)

4. COOLING SYSTEM SERVICE

Most cooling system repairs are easy and require relatively little time. Always plan to replace gaskets and seals and have them on hand before beginning.

> ***WARNING —***
> - *At normal operating temperature the cooling system is pressurized. Allow the system to cool as long as possible before opening—a minimum of an hour—then release the cap very slowly to allow safe release of pressure.*
> - *Releasing cooling system pressure lowers the coolant's boiling point, and the coolant may boil suddenly. Use heavy gloves and wear eye and face protection to guard against scalding.*
> - *Use extreme care when draining and disposing of engine coolant. Coolant is poisonous and lethal to pets. Pets are attracted to coolant because of its sweet smell and taste.*

> ***CAUTION —***
> *Avoid adding cold water to the coolant while the engine is hot or overheated. If it is necessary to add coolant to a hot system, do so only with the engine running and coolant pump turning.*

4.1 Coolant and Hoses

The coolant level and the hoses should be inspected periodically as described in **LUBRICATION AND MAINTENANCE**. Hoses deteriorate with time, and regular inspection will help prevent unexpected failure.

> ***CAUTION —***
> *BMW recommends the use of phosphate-free anti-freeze to avoid the formation of harmful, clogging deposits in the cooling system. Use of anti-freeze containing phosphates is considered by BMW to be harmful to the cooling system.*

Draining and Filling Cooling System

Draining the coolant is a first step in almost all cooling system repairs. Coolant can be reused provided it is drained into a clean pan. New coolant is recommended every 2 years. Always mix anti-freeze with clean water. Distilled water is best because of its reduced mineral content. You will also need a new coolant drain plug gasket.

> ***CAUTION —***
> *Do not reuse the coolant when replacing damaged engine parts. Contaminated coolant may damage the engine or cooling system.*

To drain and fill cooling system:

1. Remove the expansion tank cap.
2. Set the temperature control lever in the passenger compartment to Warm.
3. Remove the radiator drain plug or lower radiator hose. See Fig. 4-1.

Fig. 4-1. Location of radiator drain plug (arrow) in bottom of radiator.

4. Remove the engine block drain plug from the engine block. See Fig. 4-2. The drain plug is located on the exhaust side of the block, beneath number six cylinder.
5. Reinstall and tighten the drain plugs. Use a new gasket on the engine block drain plug.

Tightening torques	
• radiator drain plug	2.0 to 3.0 Nm (18 to 27 in-lb)
• engine block drain plug (including new gasket)	50 +6/-0 Nm (36 +4/-0 ft-lb)

Fig. 4-2. Typical location of coolant drain plug on engine block.

6. Using a coolant mixture of 50% anti-freeze and 50% water, fill the system slowly so that air is allowed to escape. Cooling system capacities are given in **Table b**.

Table b. Cooling System Capacities

Model	Approximate capacity in liters (quarts)
528e	
1982–1983 models	12.0 (12.7)
1984–1988 models	11.0 (11.6)
533i, 535i	12.0 (12.7)
Coolant type: phosphate-free ethylene glycol	

Air may become trapped in the system during filling. Trapped air can prevent good coolant circulation. The system should be bled to remove trapped air.

To bleed cooling system:

1. Add coolant until the level reaches the mark on the coolant expansion tank shown in Fig. 4-3.
2. Loosen the 8 mm bleeder screw on the thermostat housing and add coolant to the expansion tank until it spills from the screw. See Fig. 4-4.
3. Tighten the screw and start the engine.
4. Set the temperature control lever in the passenger compartment to the Warm position.
5. With the engine idling at normal operating temperature, loosen the bleeder screw until the coolant spilling out is free from air bubbles, then tighten the screw.
6. After the engine has cooled, recheck the coolant level and add coolant as necessary.

Fig. 4-3. Fill mark on coolant expansion tank (arrow). Coolant level should be at mark with engine cold.

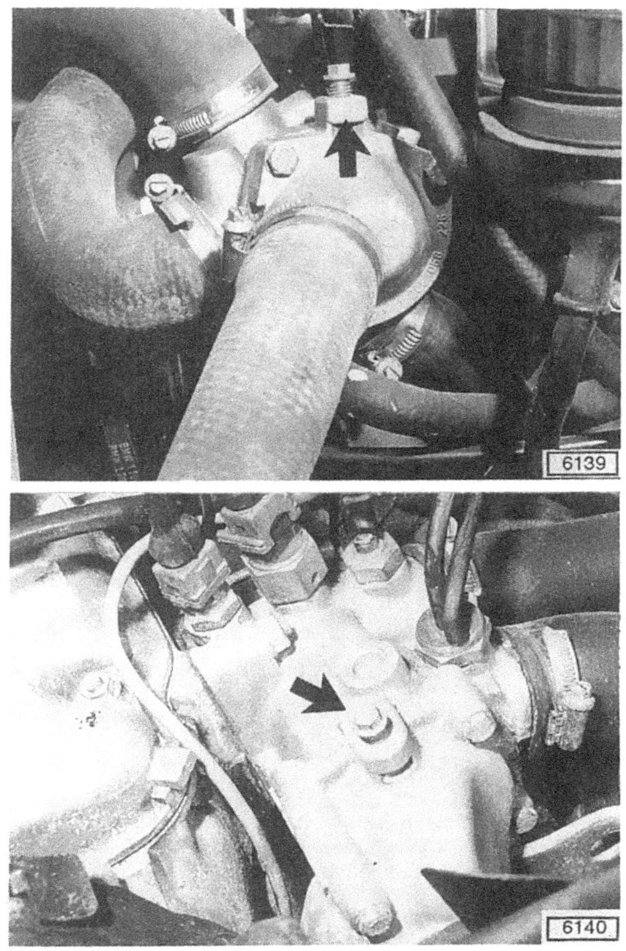

Fig. 4-4. Location of bleeder screw (arrow) on thermostat housing of 528e model (top) and 533i and 535i models (bottom).

Tightening torque
• cooling system bleeder screw. . .6 to 10 Nm (53 to 89 in-lb)

To replace a hose:

1. Drain the coolant as described above.
2. Remove the hose. Using a screwdriver, loosen each hose clamp and slide the clamps away from the hose ends.

> **NOTE —**
> If a radiator hose is stuck to the radiator connection, cut the old hose off the connection as shown in Fig. 4-5. Prying the hose loose may damage the connection or the radiator.

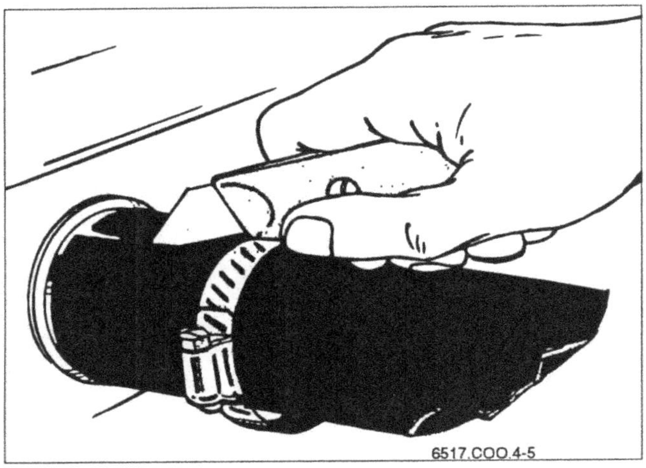

Fig. 4-5. Stuck hose being removed by cutting.

3. Clean the hose connections and remove any bits of old hose and sealer.
4. Install the new hose. Slide the loose clamps over the hose and slide the hose ends over the connections.
5. Position and tighten the clamps. Place the clamp as near the bead as possible and at least 4 mm (5/32 in.) from the hose end, as shown in Fig. 4-6. Tighten the clamps enough to compress the hose firmly around the connections.

> ***CAUTION —***
> *Do not overtighten clamps. Tighten just enough to seal. Overtightening may damage the hose and cause premature failure.*

6. Refill the radiator as described in **Draining and Filling Cooling System** above. Run the engine until warm. Check for leaks. If necessary, retighten the hose clamps. Check once more after the engine has cooled.

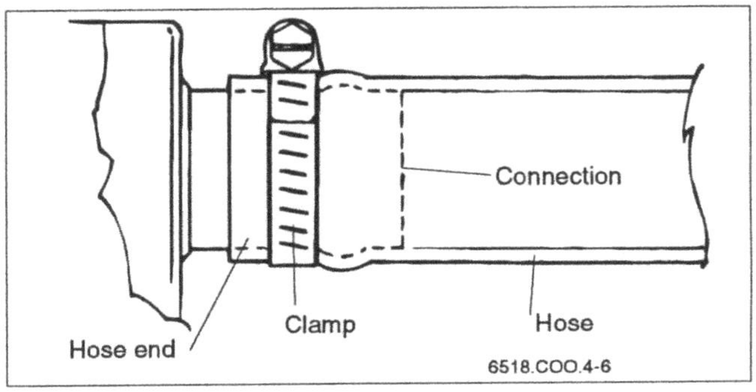

Fig. 4-6. Hose clamp correctly installed on hose end.

4.2 Thermostat

The thermostat controls coolant temperature by regulating coolant flow to the radiator. A thermostat that is stuck open will cause the engine to warm up slowly and run below normal temperature at highway speed. A thermostat that is stuck closed will restrict coolant flow to the radiator and cause overheating.

> **NOTE —**
> To quickly check if the thermostat is opening and if coolant is circulating through the radiator, allow a cold engine to reach operating temperature (temperature gauge needle approximately centered). Feel the upper radiator hose. If the hose is hot to the touch, the coolant is circulating. If the hose is not hot, either the thermostat is not opening or the radiator is plugged.

Removing and Installing Thermostat

The thermostat needs to be removed for testing and replacement. The engine should not be operated without a thermostat.

Requirements:

- new O-ring (528e)
- new gasket and O-ring (533i, 535i models)

To remove and install thermostat:

1. Drain the coolant as described in **Draining and Filling Cooling System**.
2. Remove the thermostat housing bolts. See Fig. 4-7.
3. Separate the housing from the coolant outlet and remove thermostat, O-ring and, on 533i and 535i models, the gasket. See Fig. 4-8.
4. Position the new thermostat in the outlet. See Fig. 4-9.

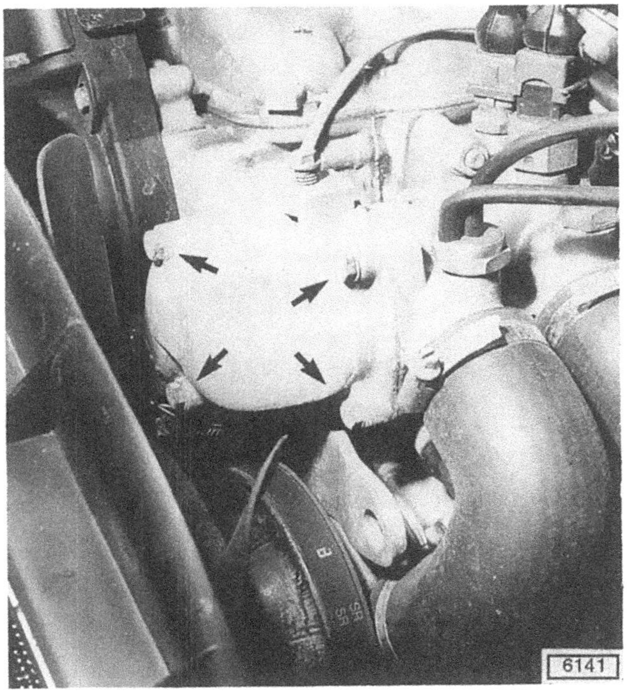

Fig. 4-7. Thermostat housing mounting bolts (arrows) on 533i and 535i models. Housing on 528e model is similar.

Fig. 4-8. Coolant outlet (on cylinder head) with thermostat housing, thermostat, and O-ring removed. 528e model shown.

5. Install the thermostat housing and new O-ring and gasket (where applicable).

Fig. 4-9. Thermostat positioned in coolant outlet (528e model shown). On thermostats with arrow on metal flange (**A**), position thermostat with arrow up.

NOTE —
Two types of thermostats are used on the models covered in this manual. On 1986 and later models, a larger diameter thermostat is used as compared to the thermostat used on earlier models. The two thermostats are not interchangeable.

Tightening torque	
• thermostat housing mounting bolts	9±1 Nm (80±9 in-lb)

6. Refill and bleed the cooling system as described in **4.1 Coolant and Hoses**.

Testing Thermostat

Test the thermostat when it is removed from the engine. It is also wise to perform this simple test to a new thermostat before installation. If any faults are found, the thermostat should be replaced.

Heat the thermostat in a pan of water while monitoring the temperature with a thermometer.

Thermostat opening temperature	
• all	176°F (80°C)

4.3 Coolant Pump

The coolant pump is replaced as a unit. Although individual parts are available, special tools and press equipment are necessary for rebuilding.

> **NOTE —**
> Always plan to replace gaskets and seals and to have them on hand before beginning repairs.

Inspecting and Replacing Coolant Pump

To inspect the coolant pump, first loosen the V-belt so that the pulley and pump shaft can be turned freely. Firmly grasp opposite sides of the fan and check for play in all directions. Rotate the pulley and check that the shaft runs smoothly. Inspect the pump for signs of coolant leakage.

The coolant provides some lubrication for the pump shaft, so an occasional drop of coolant leaking from the pump is acceptable. A larger leak, a loose pump shaft, or a shaft that is noisy and turns roughly indicates a worn shaft bushing. The pump is faulty and should be replaced. The pump replacement procedure differs depending on model. Also, plan to replace the V-belt if it is worn or damaged. For information on checking, replacing and adjusting the V-belt, see **LUBRICATION AND MAINTENANCE**.

To remove coolant pump (528e models):

1. Drain the cooling system as described in **4.1 Coolant and Hoses.**

2. Remove the fan clutch assembly as described later in this section under **Inspecting and Replacing Primary Fan Clutch**.

3. Place light pressure on the V-belt to hold the pulley, then loosen the four pulley mounting bolts and remove the V-belt and the pulley.

4. Loosen the hose clamps and remove the two hoses from the pump.

5. Remove the distributor cap with wire assembly, the ignition rotor, and the dust shield as described in **IGNITION**.

6. If the engine has a timing reference sensor (528e models built March 1987 and later), remove the sensor from its mounting bracket. See Fig. 4-10. Remove the sensor's wiring from the front of the camshaft drive belt cover.

7. Loosen the upper alternator mounting bracket nut. See Fig. 4-11.

Fig. 4-10. Timing reference sensor and internal-hex-head screw (arrow) on some 528e models. Use 5 mm hex wrench to remove screw.

Fig. 4-11. Upper alternator mounting bracket nut (arrow) on camshaft drive belt cover.

8. Remove the two camshaft drive belt cover mounting bolts. Remove the cover together with the side rubber piece. See Fig. 4-12.

9. Loosen but do not remove all three coolant pump mounting bolts. Remove the two bolts indicated in Fig. 4-13. Pivot the pump down as far as it will go and remove the camshaft drive belt tensioner spring and pin. Remove the remaining lower bolt together with the pump.

> ***CAUTION —***
> *Do not loosen the camshaft drive belt tensioner bolts. The valves can be damaged if the pistons are accidentally moved while belt tension is relieved.*

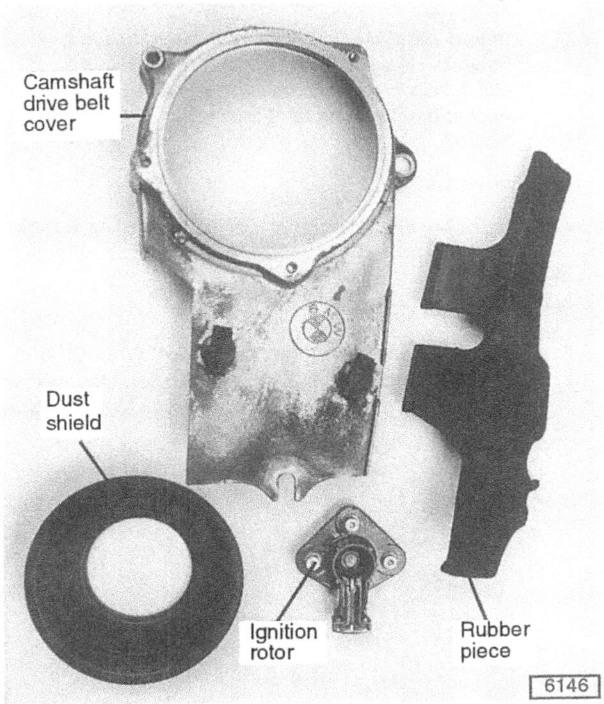

Fig. 4-12. Camshaft drive belt cover and rubber piece shown removed. Dust shield and ignition rotor also shown.

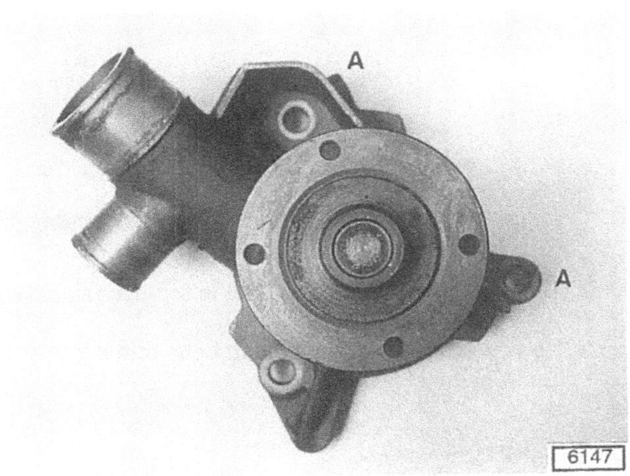

Fig. 4-13. When removing coolant pump, remove mounting bolts at **A** first.

NOTE —
It may be a good time to inspect the camshaft drive belt condition and tension while it is exposed. See **ENGINE—SERVICE AND REPAIR** for more information.

Installation is the reverse of removal. Be sure the camshaft drive belt tensioner pin correctly engages the notch in the coolant pump. See Fig. 4-14. Replace all gaskets. Install and tension the V-belt as described in **LUBRICATION AND**

Fig. 4-14. Coolant pump notch (arrow) for camshaft drive belt tensioner spring.

MAINTENANCE. Fill and bleed the cooling system as described in **4.1 Coolant and Hoses.**

Tightening torques (528e models)
• coolant pump mounting bolts . . . . 22±2 Nm (16±1 ft-lb)
• pulley mounting bolts . . . . 9±1 Nm (80±9 in-lb)

To remove coolant pump (533i, 535i models):

1. Drain the cooling system as described in **4.1 Coolant and Hoses.**

2. Remove the fan clutch assembly as described below in **Inspecting and Replacing Primary Fan Clutch.**

3. Placing light pressure on the V-belt to hold the pulley, loosen the four pulley mounting bolts. Remove the V-belt and remove the pulley.

4. Remove the coolant hose from the pump.

5. Remove the six coolant pump mounting bolts, remove the engine lifting bracket, and remove the pump. See Fig. 4-15.

Installation is the reverse of removal. Use a new gasket between the coolant pump and engine block. Install and tension the V-belt as described in **LUBRICATION AND MAINTENANCE**. Install the fan assembly as described below. Fill and bleed the system as described in **4.1 Coolant and Hoses.**

Tightening torques (533i, 535i models)
• coolant pump mounting bolts . . . . 9±1 Nm (80±9 in-lb)
• pulley mounting bolts . . . . 9±1 Nm (80±9 in-lb)

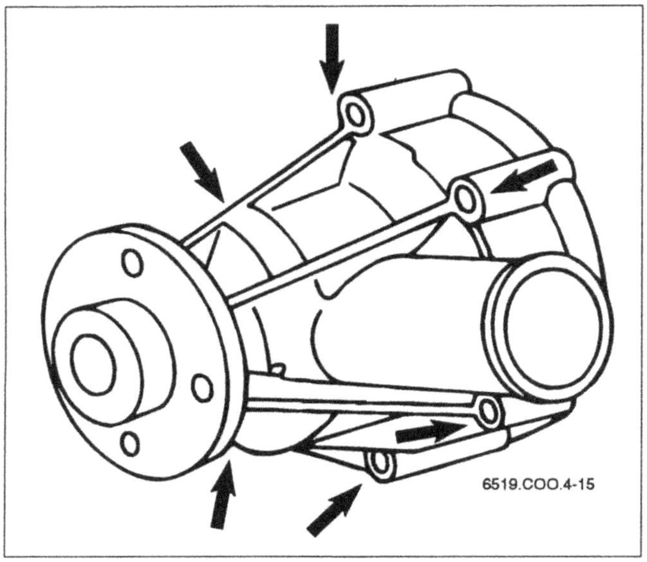

Fig. 4-15. Coolant pump mounting bolt locations (arrows) on 533i, 535i models.

Inspecting and Replacing Primary Fan Clutch

Inspect the fan clutch by spinning the fan with the engine off and cold. See Fig. 4-16. The fan should spin on the clutch with slight resistance. Check for signs of oil leaking from the clutch. If the fan cannot be turned by hand or if there are signs of oil leakage, the clutch is faulty and should be replaced.

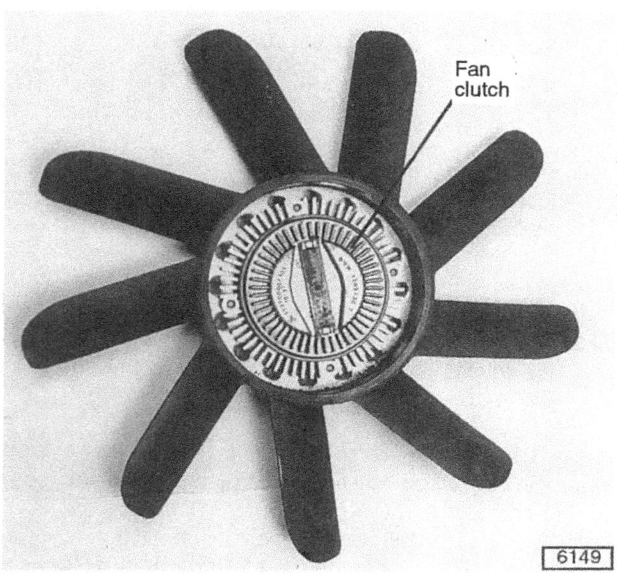

Fig. 4-16. Radiator cooling fan and clutch assembly.

NOTE —
In very cold weather, perform this test inside a heated work area. Cold temperatures cause the clutch fluid to thicken, making a good fan clutch appear bad by being hard to turn.

Requirements:

- 32 mm open-end wrench or BMW tool no. 11 5 040

To remove fan clutch:

1. Place the 32 mm wrench on the fan clutch nut. See Fig. 4-17. Strike the wrench quickly in a clockwise direction (as viewed from the front of the car) to loosen the nut.

Fig. 4-17. Nut (left-hand thread) on fan clutch (arrow).

2. Turn the left-hand threaded nut to remove fan clutch.

3. Remove the four fan mounting bolts and separate the fan from the clutch.

NOTE —
Always store the removed fan clutch assembly in an upright position. Failure to store the clutch assembly upright may result in a loss of clutch fluid.

Installation is the reverse of removal. Install and tension the V-belt as described in **LUBRICATION AND MAINTENANCE**.

Tightening torques
• fan-to-clutch mounting bolts . . . . 9±1 Nm (80±9 in-lb)
• fan clutch nut to coolant pump (left-hand thread) without BMW tool no. 11 5 040 . . . . 40 Nm (29 ft-lb) with BMW tool no. 11 5 040 . . . . 30 Nm (22 ft-lb)

NOTE —
BMW tool no. 11 5 040 is used in conjunction with a torque wrench to accurately torque the fan clutch to the coolant pump. See Fig. 4-18. Because the tool provides increased leverage, it multiplies the torque.

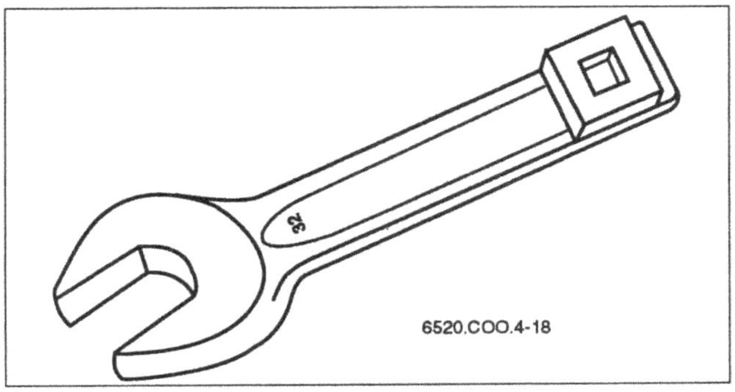

Fig. 4-18. 32 mm open end wrench (BMW tool no. 11 5 040).

4.4 Auxiliary Cooling Fan

The auxiliary cooling fan provides additional air flow through the radiator. It is controlled by temperature switches and a relay, and operates only when coolant temperature exceeds a predetermined level or whenever the air conditioning is on. If the fan does not operate properly, the engine may overheat. The fan assembly is mounted in front of the radiator. The switches are mounted on the side of the radiator. See Fig. 4-19.

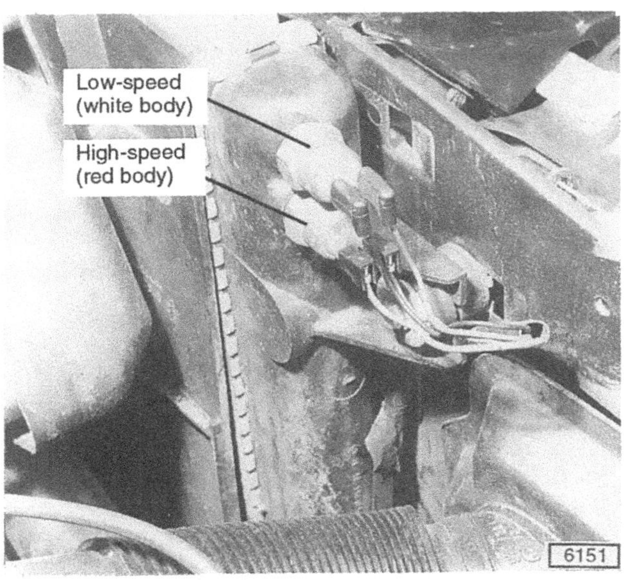

Fig. 4-19. Coolant temperature switches for auxiliary radiator cooling fan on 533i and 535i models. Switches on 528e models are located on upper left corner.

If a faulty thermostat or a restriction in the system is not allowing the coolant to circulate through the radiator to warm up the coolant temperature switches, the switches will not close and the fan will not run. Before making the tests described below, make sure the thermostat is operating correctly as described under **4.2 Thermostat**.

NOTE —
A temperature gauge that is not indicating normal operating temperature may indicate a faulty (stuck open) thermostat.

Testing Cooling Fan and Coolant Temperature Switches

For greatest safety, cooling fan and coolant temperature switch tests should be performed with the air conditioner off and with the engine cold. See **Table c** for switching temperatures.

WARNING —
Use extreme caution when testing the cooling fan and coolant temperature switches. Keep hands and wires clear of the fan blades. The cooling fan can run any time the ignition is ON.

10

Table c. Auxiliary Cooling Fan Switching Temperatures

Fan speed	Switch body color	Temperature switch closes at
Low	White	196°F (91°C)
High	Red	210°F (98°C)

If the coolant is circulating at normal operating temperature, and the auxiliary cooling fan does not run, test the fan as follows. Disconnect the wires from the suspect radiator temperature switch and jumper the wires. This will simulate a closed switch. **Table d** gives the correct test results. If the fan runs only when powered directly by the jumpered connector, the radiator temperature switch should be replaced. Use a new gasket ring for the switch.

Table d. Auxiliary Cooling Fan Temperature Switch Tests

Wires jumpered	Test conditions	Test results
Green/brown and black (white-body switch)	Ignition ON	Fan runs on low speed
Green/brown and green/yellow (red-body switch)	Ignition ON	Fan runs on high speed

Tightening torque
• coolant temperature switches .... 8±1 Nm (71±9 in-lb)

If the fan does not run when powered directly, check for battery voltage at the connectors with the ignition on. If battery voltage is not present at either of the connectors, check fuse no. 17 (low-speed) and fuse no. 14 (high-speed). If no faults are found, check power and ground for the the low-speed and high-speed relays with the ignition On. See Fig. 4-20.

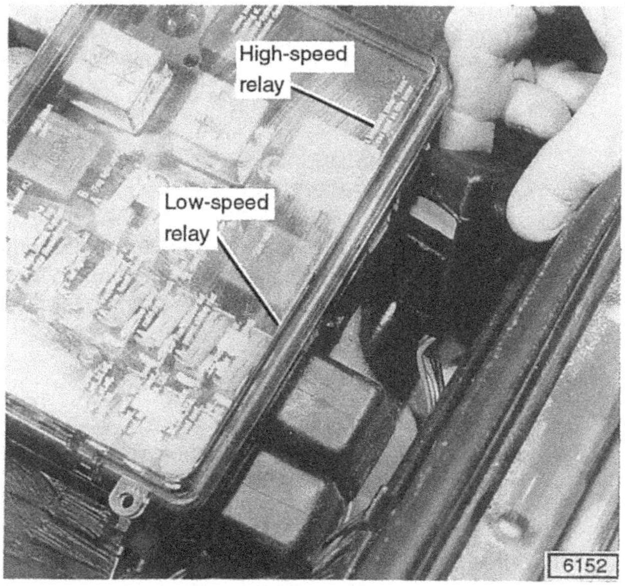

Fig. 4-20. Auxiliary cooling fan low-speed and high-speed relays. High-speed relay removed from bracket.

NOTE —
Relay positions may vary between models. The high-speed relay has two red/green wires leading to terminal 30 of the relay. The low-speed relay has a single red/green wire leading to terminal 30 of the relay.

Fix any wiring faults. If there is power and ground and the fan still does not run with the relays installed, then the relays are probably faulty.

If the fan operates only on high speed and no electrical faults have been found up to this point, check the resistance of the low-speed resistor. The resistor is mounted on the auxiliary fan housing. See Fig. 4-21.

Test specification
• low-speed cooling fan resistor. . . . . 0.6 ohms

To remove auxiliary cooling fan:

1. Remove the screws securing the trim panel in front of the radiator. See Fig. 4-22.

Fig. 4-21. Auxiliary cooling fan low-speed resistor and electrical connectors (arrows).

Fig. 4-22. Trim panel retaining screw being removed.

2. Remove the three nuts from the fan assembly. See Fig. 4-23.

3. Disconnect the fan's electrical connectors and remove the fan housing from the car.

Installation is the reverse of removal. If the fan is being replaced, transfer the old fan housing and resistor onto the new fan.

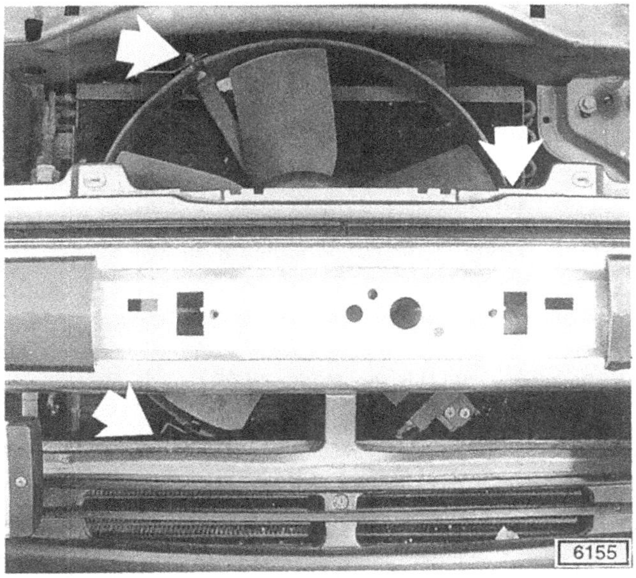

Fig. 4-23. Auxiliary cooling fan mounting nuts (arrows).

4.5 Radiator

The radiator fins should not be blocked with dirt or debris, and should be firmly mounted. Excess vibration due to loose, broken, or missing fasteners may damage the radiator. Clean the radiator fins using low-pressure water or compressed air. Blow from the engine side out.

If the engine overheats and no other cooling system tests indicate trouble, the radiator may have some plugged passages that are restricting coolant flow. This does not necessarily mean that the radiator must be replaced. In many cases the radiator can be chemically cleaned by a qualified radiator repair shop to relieve the plugging.

NOTE —
Repairing radiators of aluminum and plastic construction used on some BMW models requires some specialized knowledge. Choose a shop with the necessary experience.

Removing and Installing Radiator

To remove radiator:

1. Remove the expansion tank cap and drain the cooling system as described under **4.1 Coolant and Hoses.**

2. Disconnect the upper and lower radiator hoses and the small hose to the expansion tank.

3. Remove the auxiliary cooling fan and fan clutch assembly as described above.

4. Disconnect the electrical connectors from the radiator coolant temperature switches and from the auxiliary cooling fan.

5. If an engine splash guard is installed, remove it from beneath the car.

6. Remove the screws securing the plastic fan shroud. See Fig. 4-24. Pull the fan shroud up and off the radiator and position the shroud over the cooling fan, towards the engine.

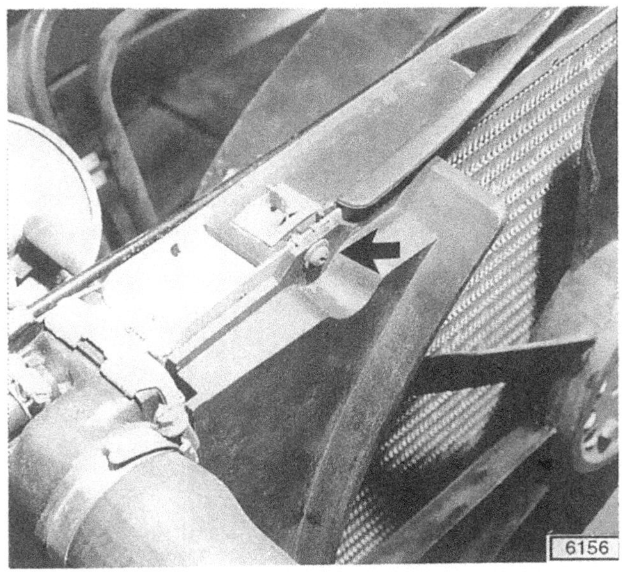

Fig. 4-24. Radiator shroud mounting screw (arrow). Remove screws from both sides and slide shroud back towards engine.

7. On models with automatic transmission, thoroughly clean the two ATF cooler line unions at the radiator and then disconnect the lines. See Fig. 4-25.

NOTE —
Plug the cooler lines and the radiator openings to keep them clean.

8. Remove the radiator mounting bolts from either side of the radiator. See Fig. 4-26. Lift the radiator out from the top.

Installation is the reverse of removal. Inspect the rubber lower radiator mountings and replace any that are damaged. Fill the cooling system as described in **4.1 Coolant and Hoses**. On models with automatic transmission, check the level of the automatic transmission fluid (ATF). See **AUTOMATIC TRANSMISSION**.

10

Fig. 4-25. Typical BMW automatic transmission cooler line at radiator (arrow).

Fig. 4-26. Removing left radiator mounting bolt. The right mounting bolt (not shown) must also be removed.

Tightening torques
• radiator mounting screws . . . . . 9 +1/-0 Nm (80 +9/-0) in-lb)
• transmission cooler lines. . . . . . 18 +3/-0 Nm (13 +2/-0 ft-lb)

Flushing Radiator

If the radiator has been contaminated with engine oil, BMW recommends that the radiator and expansion tank be flushed with Solvethane®, available from an authorized BMW dealer.

WARNING —
Solvethane® is poisonous. Wear safety glasses and gloves. Always follow the manufacturer's safety precautions listed on the container. Dispose of Solvethane® properly when flushing is complete.

CAUTION —
Solvethane® is damaging to rubber seals, gaskets and hoses. Do not allow it to enter the cooling system or contact rubber parts.

NOTE —
Although flushing the radiator and cooling system is part of scheduled maintenance, the procedure below applies only to oil-contaminated radiators. For information on routine flushing of the cooling system, see **LUBRICATION AND MAINTENANCE**.

To flush the system, remove the radiator as described above. Pour 4 to 6 pints (2 to 3 liters) of Solvethane® into the radiator. Cover any openings and vigorously shake the radiator. After approximately 2 to 3 minutes, drain the Solvethane® and reinstall the radiator. Flush using hot water until there are no signs of oil in the water. Repeat the process for the expansion tank. Finally, fill the system with coolant as described in **4.1 Coolant and Hoses**.

5. TECHNICAL DATA

I. Cooling System Specifications

Cooling system capacity	
528e (1982–1983)	12.0 liters (12.7 qts.)
528e (1984–1988)	11.0 liters (11.6 qts.)
533i	12.0 liters (12.7 qts.)
535i	12.0 liters (12.7 qts.)
Coolant type: phosphate-free ethylene glycol	

II. Tightening Torques

Automatic transmission cooler lines to radiator	18 +3/-0 Nm (13 +2/-0 ft-lb)
Coolant fan clutch to coolant pump (left-hand thread)	
with special tool (BMW Part No. 11 5 040)	30 Nm (22 ft-lb)
without special tool	40 Nm (29 ft-lb)
Coolant primary fan to coolant fan clutch	9±1 Nm (80±9 in-lb)
Coolant pump pulley to coolant pump (bolt)	9±1 Nm (80±9 in-lb)
Coolant pump to engine block	
533i, 535i models	9±1 Nm (80±9 in-lb)
528e model	22±2 Nm (16±1 ft-lb)
Coolant temperature sending unit to cylinder head water outlet	18±1 Nm (13±1 ft-lb)
Coolant temperature switches to radiator	8 Nm (71 in-lb)
Cooling system bleeder screw	6–10 Nm (54–84 in-lb)
Engine block (coolant) drain plug	50 +6/-0 Nm (36 +4/-0 ft-lb)
Radiator drain plug	2.0–3.0 Nm (18–27 in-lb)
Thermostat housing bolts	9±1 Nm (80±9 in-lb)
Upper radiator mounting	9±1 Nm (80±9 in-lb)

Section 11

EXHAUST SYSTEM

Contents

INTRODUCTION

The exhaust system has two main functions: first, to guide the rapidly expanding gasses of combustion out of the engine and away from the passenger compartment, and second, to quiet and cool the exhaust. The catalytic converter removes most of the pollutants that are the normal by-products of combustion.

Proper exhaust system function depends on each component being free from holes, with airtight seals at all joints. Emission control system function depends on proper adjustment of the fuel injection system and the ignition system, as well as chemical treatment of the exhaust gasses by the catalytic converter.

Exhaust system components are subjected to vibration and extreme temperature and pressure, as well as road hazards. The exhaust system is designed to be maintenance free, but regular inspection is warranted due to these harsh operating conditions. The only scheduled emission control maintenance is replacement of the oxygen sensor at specified time and mileage intervals.

This section covers maintenance, troubleshooting, and repair of the exhaust system. For information on testing the oxygen sensor and other emission controls, see **FUEL INJECTION**.

1. GENERAL DESCRIPTION

1.1 Exhaust System

The basic exhaust system components are the exhaust manifold, the front pipe with integral catalytic converter, and the rear pipe with muffler(s). See Fig. 1-1.

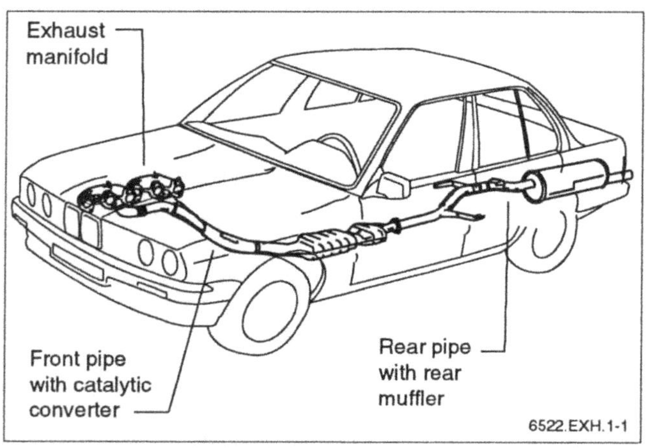

Fig. 1-1. Typical two-piece BMW exhaust system. Exact configuration varies from model to model.

Two individual exhaust manifolds are mounted to the cylinder head and channel exhaust gasses from the individual exhaust ports into the front pipe. The manifolds are made of cast iron.

The front end of the system is supported by attachment to the exhaust manifolds, and by a mounting bracket on the transmission. The rear part of the system is suspended by rubber retaining rings or hangers from the underbody of the car. The retaining rings provide positive but non-rigid mounting to allow some expansion and contraction of the system and to help isolate noise and vibration from the body.

Catalytic Converter

The catalytic converter is similar in appearance to a small muffler and is integral with the front pipe assembly. Its honeycombed ceramic core contains hundreds of tiny passages whose surfaces are coated with precious metal catalysts. The catalysts promote chemical reactions in the exhaust gasses that reduce the quantity of harmful pollutants in the exhaust. Under normal conditions the catalytic converter does not require replacement.

Oxygen Sensor

The catalytic converter reduces emissions most efficiently when the percentage of oxygen in the exhaust falls within a narrow range. The oxygen sensor monitors the exhaust gas and provides feedback about combustion efficiency to the fuel injection system.

2. MAINTENANCE

BMW specifies the maintenance steps below to be carried out at particular time or mileage intervals for proper maintenance of the exhaust system and emission controls. A number in bold type indicates that the procedure is covered in this section, under that numbered heading. Information on other exhaust system and emission control maintenance and on the prescribed maintenance intervals can be found in **LUBRICATION AND MAINTENANCE**.

1. Inspecting exhaust system
2. Replacing oxygen sensor **5.**

BMW recommends inspection of the exhaust system during each scheduled maintenance, but it is also a good idea to inspect it whenever other repair work allows access to the underside of the car. The system should always be inspected immediately if it becomes unusually noisy, if road hazard damage has occurred, or if exhaust odor is detected inside the car.

3. TROUBLESHOOTING

This troubleshooting section covers the exhaust system, including the exhaust manifolds and pipes, the muffler, and the catalytic converter.

Problems such as exhaust leakage or excessive noise almost certainly indicate faults in the exhaust system. Other problems such as poor performance, rough running, or increased emissions may have more complex causes. For help in troubleshooting running and performance problems, see **ENGINE MANAGEMENT—DRIVEABILITY**. For help in troubleshooting oxygen sensor problems, see **FUEL INJECTION**.

3.1 Basic Troubleshooting Principles

As with any troubleshooting, analysis of the observed symptoms is the key to isolating and identifying exhaust system problems.

In the case of the exhaust system, damage may be due to striking a road hazard. Noise is the main indicator of exhaust system problems. A gradual increase in noise level is most likely an indication of the general deterioration of the whole system, due to corrosion for example. More extensive repair or complete replacement, may be necessary.

The correct amount of back pressure helps the engine produce power smoothly over a wide range of engine speeds. Excessive back pressure due to a failed or damaged component may cause poor driveability, rough idling, or stalling. The catalytic converter with its small passages is especially susceptible to plugging if it gets overheated, if the fuel injection or ignition systems are faulty, or if the car is run on leaded fuel.

Table a lists common exhaust system problems, their probable causes, and suggested corrective actions. The numbers in bold type in the corrective action column refer to the numbered headings in this section of the manual where the suggested repairs are described.

WARNING —
- *Toxic exhaust gases are colorless and odorless. Do not run the engine in a non-ventilated area. Repair any structural damage to the body or exhaust system immediately.*

- *The catalytic converter operates at a very high temperature and can cause severe burns.*

11

Table a. Exhaust Troubleshooting

Symptom	Probable cause	Corrective action
1. Hissing, rumbling, loud noise during acceleration	**a.** Exhaust system leaks	**a.** Examine all system joints. See **LUBRICATION AND MAINTENANCE**. Replace faulty components. **4.1**
	b. Internally damaged muffler or catalytic converter	**b.** Replace faulty components. **4.1**
	c. Exhaust system out of alignment	**c.** Re-align exhaust system. **4.1**
2. Exhaust system rattles	**a.** Exhaust system out of alignment	**a.** Re-align exhaust system. **4.1**
	b. Missing or broken exhaust system rubber retaining ring	**b.** Replace retaining ring(s). **4.1**
3. Reduced power, poor mileage, hesitation on initial acceleration, rough idle	**a.** Excessive back pressure (damaged muffler or pipes or plugged catalytic converter)	**a.** Inspect components and replace as required. **4.1, 4.2**
	b. Oxygen sensor faulty	**b.** Test oxygen sensor function, replace faulty components. See **FUEL INJECTION**

4. EXHAUST SYSTEM

The main exhaust system components are bolted together at welded flanges. Gaskets are used at each flange and between the exhaust manifolds and the cylinder head. The entire system is mounted to the underbody using brackets and rubber retaining rings. Fig. 4-1, Fig. 4-2, Fig. 4-3, and Fig. 4-4 show the exhaust systems used on the various models covered by this manual. When mounting the exhaust pipe the exhaust manifold, tighten the mounting nuts in two stages. Tightening torques for all exhaust system fasteners are listed in **6. Technical Data.** If the exhaust manifold is removed, be sure to replace the mounting nuts. These nuts are only designed to be used once.

NOTE —
A car's build date is listed on the driver's door, below the door latch.

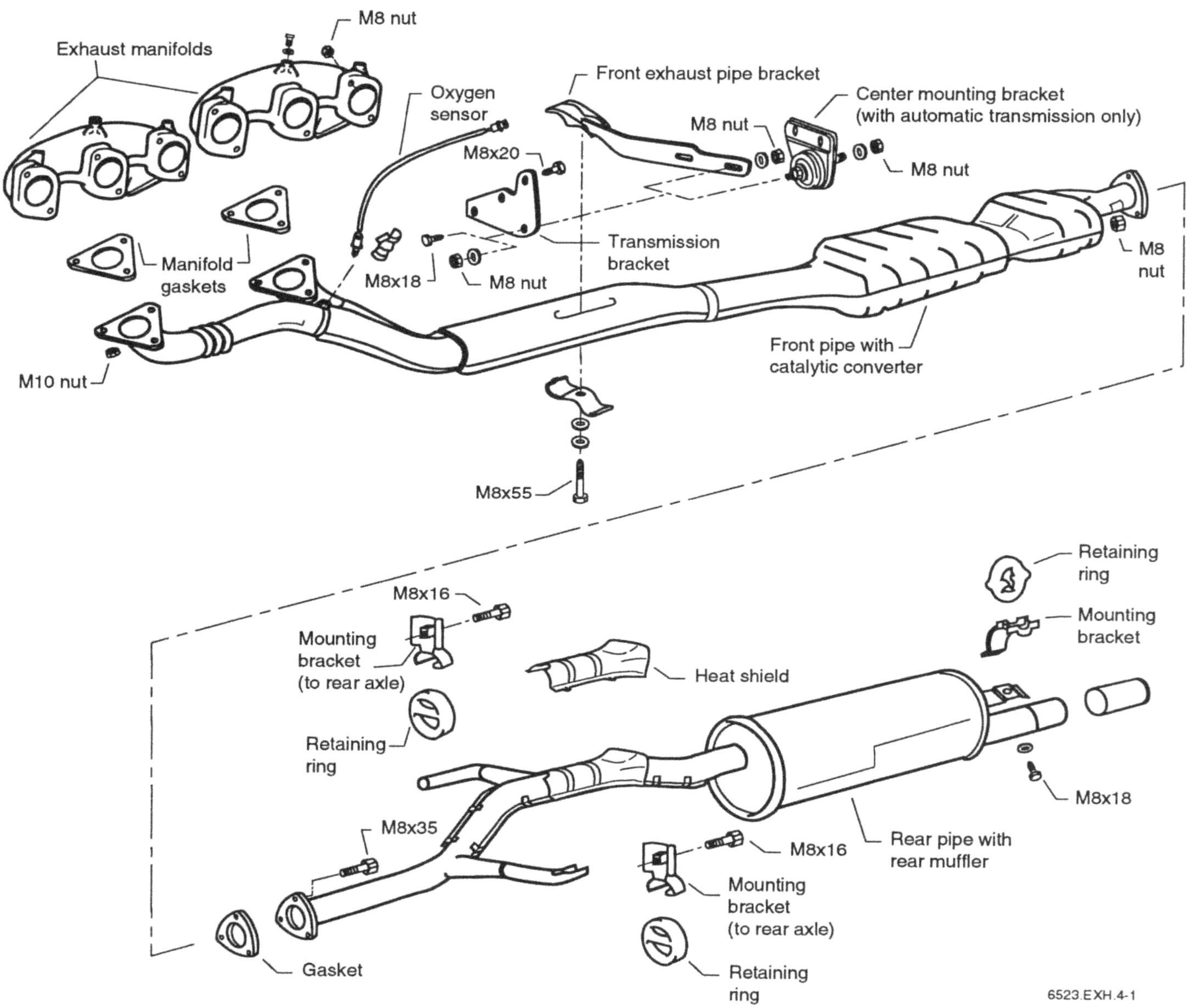

Fig. 4-1. Exploded view of exhaust system on 528e models built up to March 1987. See **6. Technical Data** for tightening torques.

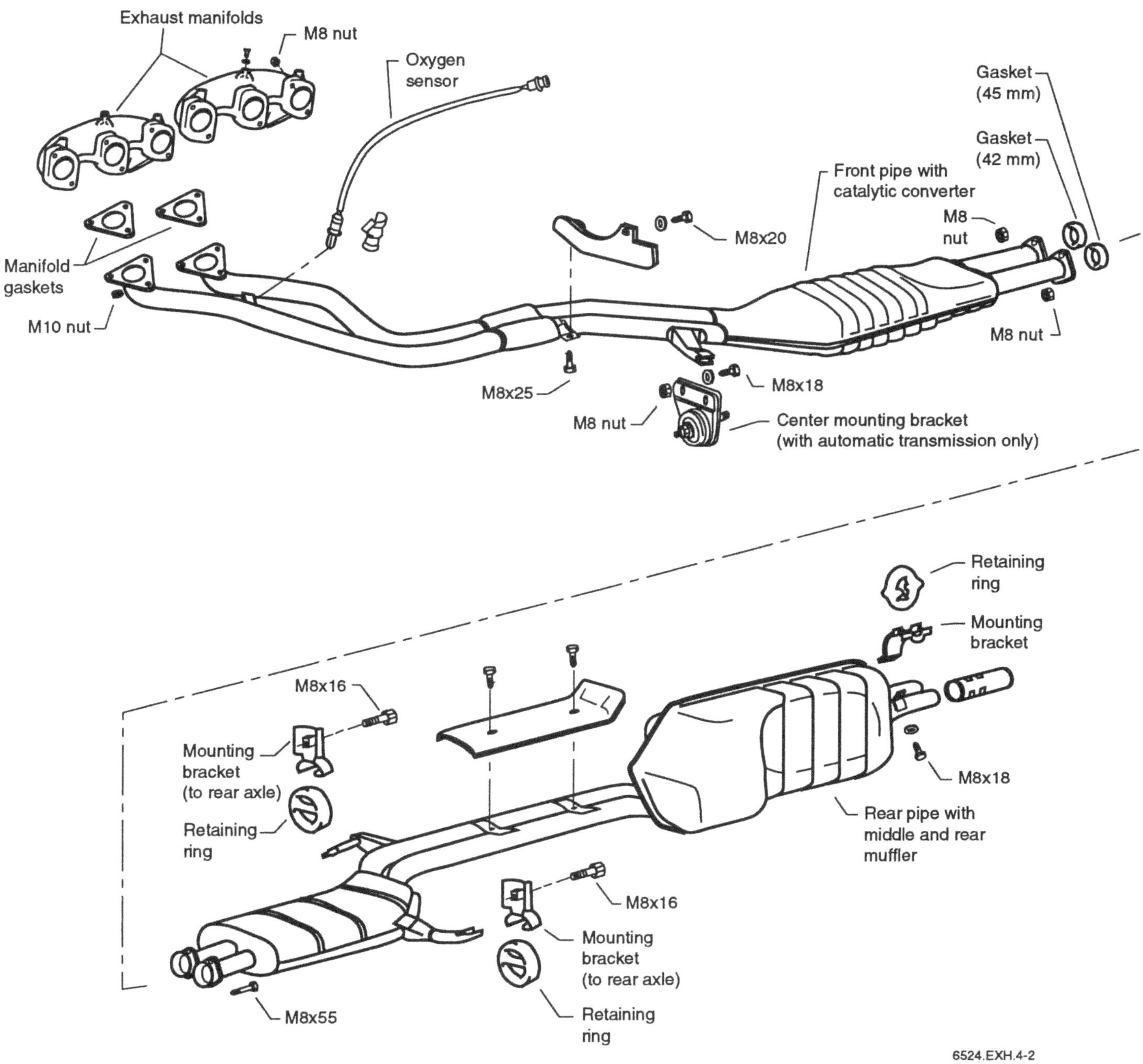

Fig. 4-2. Exploded view of exhaust system on 528e models built from March 1987. See **6. Technical Data** for tightening torques.

6 EXHAUST SYSTEM

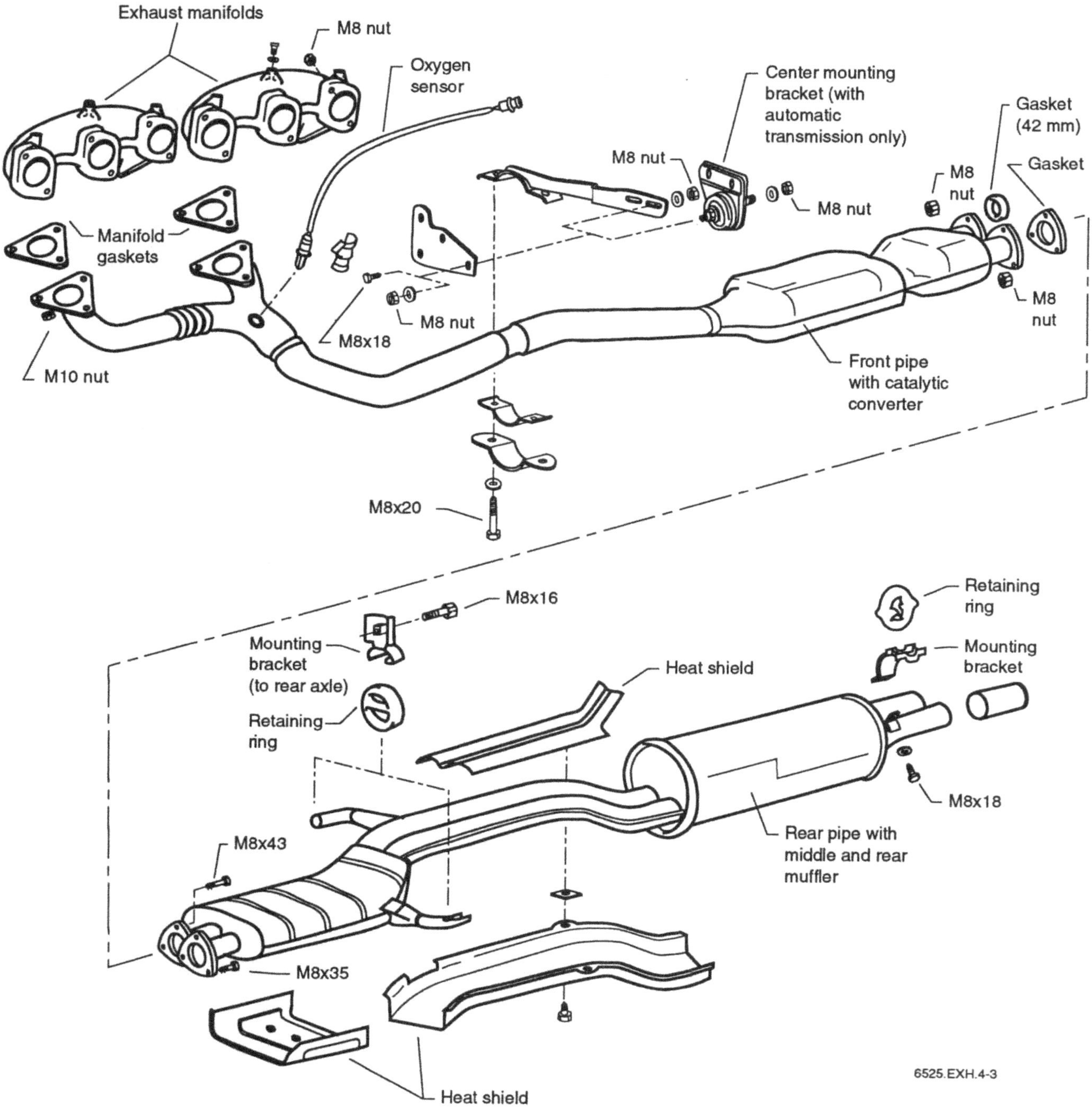

Fig. 4-3. Exploded view of exhaust system on 533i models. See **6. Technical Data** for tightening torques.

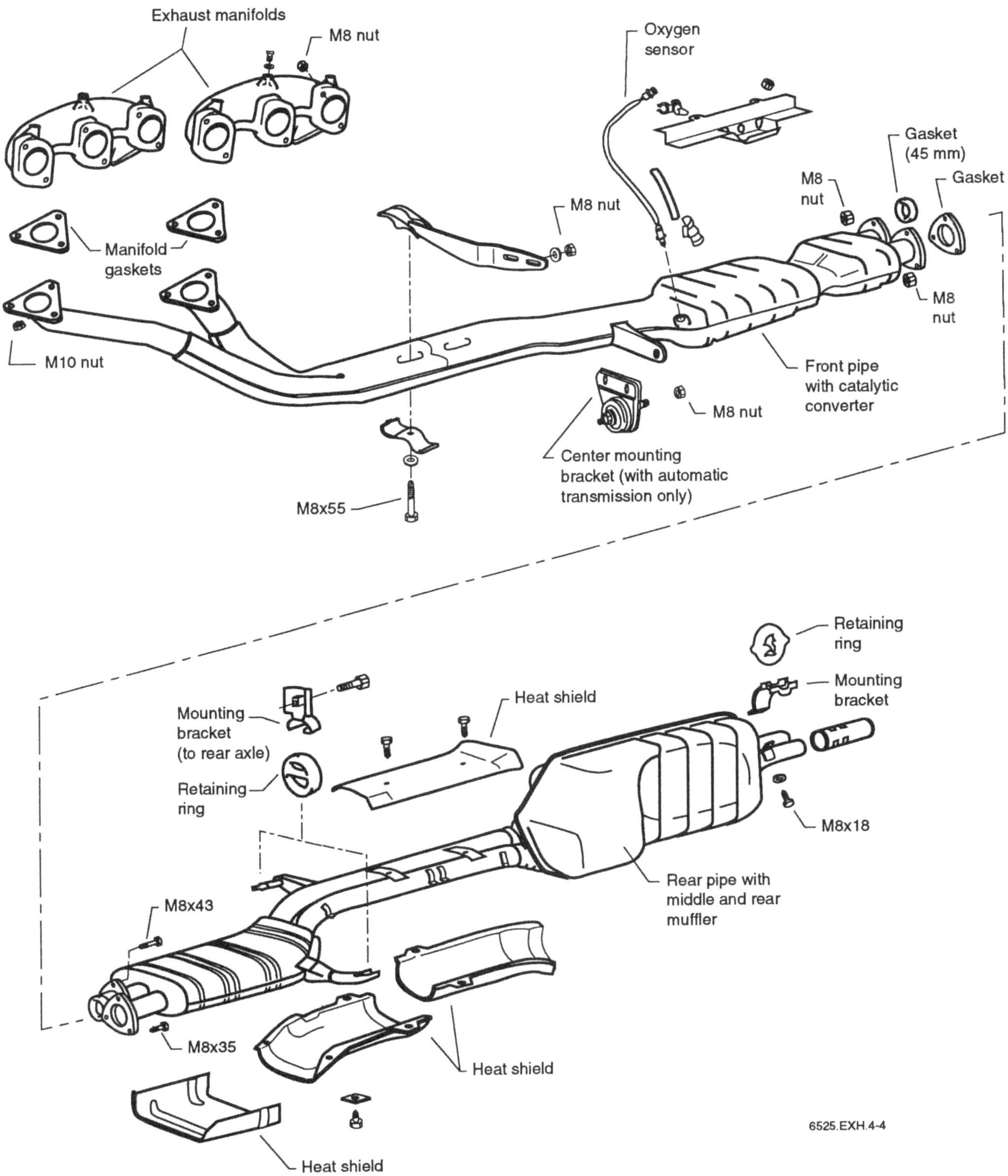

Fig. 4-4. Exploded view of exhaust system on 535i models. See **6. Technical Data** for tightening torques.

4.1 Removing and Installing Exhaust System

The pipes, mufflers, heat-shields, and catalytic converter are corrosion resistant, but they are still prone to deterioration. If any one part of the system is perforated by rust, it is very likely that other parts are similarly affected, or soon will be.

When replacing only part of the system, make sure that the new parts will fit properly with the old before removal. Parts from different sources may not mate properly. Genuine BMW replacement parts will probably mate best with the original parts.

> **NOTE —**
> Use of non-original exhaust system components can change system back pressure and may affect fuel mixture. It may be necessary to check and adjust the fuel injection system after installing any non-original exhaust pipes or mufflers. See **FUEL INJECTION**.

New fasteners, rubber hangers and clamps are always recommended. The chance of getting the old ones off undamaged and in reusable condition is slim. Gaskets should always be replaced whenever the flanged joints are disconnected.

> ***WARNING —***
> *To avoid injury, wear eye protection and heavy gloves when working on the rusty parts of the old exhaust system. Do not use flammable workshop chemicals near a hot catalytic converter. They may ignite.*

Removing Exhaust System

Individual components can be removed without completely dismantling the system. The rubber retaining rings allow some movement, enough to remove front components without removing the rear also. A penetrating oil applied to all bolts several hours in advance of beginning the work will make removal easier. Be sure to transfer all heat shields.

> **NOTE —**
> • On cars with automatic transmission, the bolts securing the front stabilizer to the body must be removed. This will allow enough clearance to remove the front exhaust pipe assembly.
>
> • When removing and installing the exhaust manifolds on 528e models built from March 1987 and later, be sure to mark front and rear manifolds. The manifolds are different and are not interchangeable.

When removing the front exhaust pipe from the car, be sure the oxygen sensor's electrical connector is disconnected. If the front pipe is to be replaced, the oxygen sensor can be transferred from one pipe to the other. For more information on removing and installing the oxygen sensor, see **5. Oxygen Sensor**.

Installing Exhaust System

Loosely install and assemble the complete system including heat shields, then evenly tighten the clamps and mounting bolts to their final torque values. Make sure that no part of the exhaust system contacts any part of the car body. Anti-seize compound used on all threaded fasteners will extend service life and make any future replacement easier.

> ***CAUTION —***
> *Do not let anti-seize compound or penetrating oil come in contact with the slit portion of the oxygen sensor body. The oxygen sensor will be destroyed.*

> **NOTE —**
> • For exhaust system fastener torque values see **6. Technical Data**.
>
> • On cars with automatic transmission, install the front stabilizer to the body and tighten the bolts.

Tightening torque
• front stabilizer to body . . . . 22 Nm (16 ft-lb)

Adjust the clearance between the tailpipe and the body using the mounting bracket that attaches to the rear axle. See Fig. 4-5 and Fig. 4-6. When the clearance is correct tighten the bolt.

Fig. 4-5. Clearance (**a**) between tailpipe and body.

Tightening torque
• mounting bracket bolt. . . . . 22 to 24 Nm (16 to 17 ft-lb)

After the system is aligned, tighten the remaining bolts and clamps. After completing the installation, start the engine and

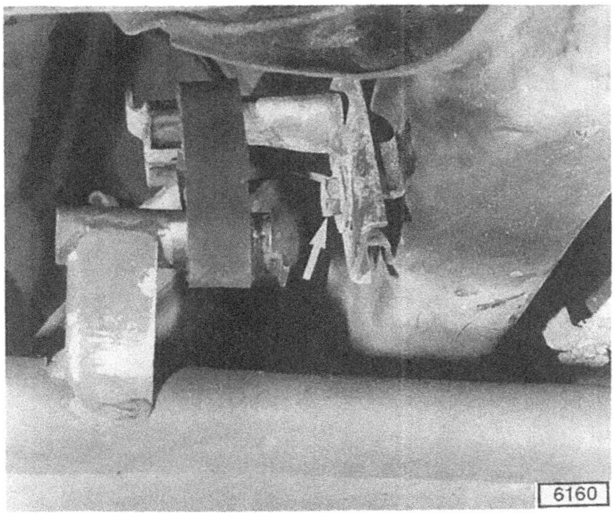

Fig. 4-6. Mounting bracket attached to rear axle. Loosen bolt (arrow) to adjust clearance between tailpipe and body.

check for any exhaust leakage. Some slight smoking and odor is normal as the new parts become hot for the first time. System alignment can be checked by driving the car over a rough road and listening for sounds of the exhaust system striking the underbody.

> **NOTE —**
> As a quick-check for detecting system leaks, start the engine and hold a non-flammable rag to cover the rear pipe outlet(s). If any hissing noises are heard under the car, the system has leaks. Loosen and readjust the system.

4.2 Catalytic Converter

All models covered by this manual are equipped with a three-way catalytic converter in the exhaust system. Fig. 4-7 shows the internal construction of the catalytic converter, designed to maximize surface area of the ceramic coated catalysts and promote the chemical reactions that convert harmful compounds into nitrogen, carbon dioxide, water, and heat.

The catalytic converter is designed to be maintenance free, and under normal operating conditions it should last at least 60,000 miles (96,000 km). Improper operating conditions can cause thermal breakdown of the catalytic material and leave the converter partially plugged or inoperative. Some of these conditions are: incorrect ignition timing, incorrect fuel metering, engine misfiring, prolonged idling, prolonged high load such as towing, and, the use of leaded gasoline.

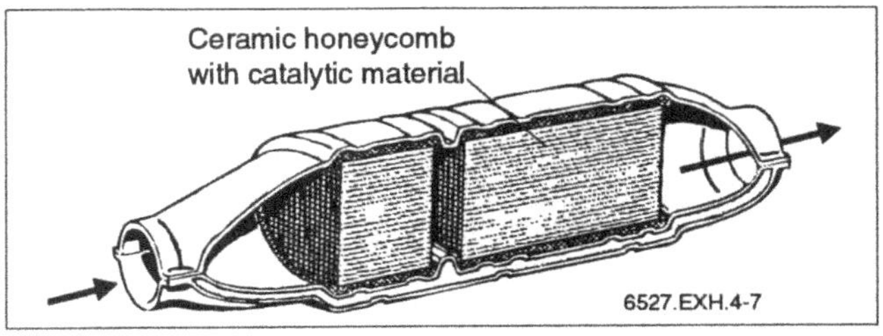

Fig. 4-7. Cutaway view of catalytic converter. Arrows indicate exhaust flow.

> ***WARNING —***
> *Do not operate the starter for more than 1 to 3 minutes if the engine fails to start. Excessive cranking may allow raw fuel to enter the converter, creating a fire hazard and potentially damaging the catalytic converter.*

> ***CAUTION —***
> *Carburetor and fuel injection additives or sprays can destroy the oxygen sensor and catalytic converter. Check with an authorized BMW dealer for approved fuel additives.*

11

Reduced power, stalling at idle, rattles in the exhaust system, and high exhaust emissions measured at the tailpipe are all possible indications of a faulty catalytic converter. The converter can be removed and visually inspected if converter failure is suspected. See **Checking Catalytic Converter** below.

Removing and Installing Catalytic Converter

The catalytic converter is one of the main components of the exhaust system and can be removed and installed as described in **4.1 Removing and Installing Exhaust System.**

Checking Catalytic Converter

Since the catalytic converter is an integral part of the front exhaust pipe, opportunity for visual inspection is limited. The only realistic test is to remove the front exhaust pipe and check for evidence of physical damage. Remove the pipe and hold it vertically with the outlet end down. Firmly tap the end on a block of wood, then turn the pipe over and tap the other end in the same manner. A knocking sound from inside the converter housing probably indicates that the ceramic core has become dislodged, and that the converter should be replaced.

5. OXYGEN SENSOR

The oxygen sensor provides the fuel injection system with feedback information about combustion efficiency by reacting to the oxygen content in the exhaust. Oxygen sensor replacement is a maintenance procedure scheduled at a specified time or mileage interval that varies depending on model year. See **LUBRICATION AND MAINTENANCE** for more information on recommended service intervals and servicing the oxygen sensor.

NOTE —
Emission controls, including the oxygen sensor, are covered by an extended warranty. See **LUBRICATION AND MAINTENANCE** for maintenance requirements. Consult an authorized BMW dealer about warranty coverage before beginning any repairs.

The signal from the oxygen sensor is monitored by the fuel system's electronic control unit. For more information on the operation of the oxygen sensor and the rest of the fuel injection system, see **FUEL INJECTION**.

Replacing Oxygen Sensor

To replace an oxygen sensor, disconnect the wiring to the sensor and use a wrench on the hex portion of the sensor housing. If not already present, apply an anti-seize compound to the threads of the new sensor. Install the new sensor, reconnect the wiring. See Fig. 5-1.

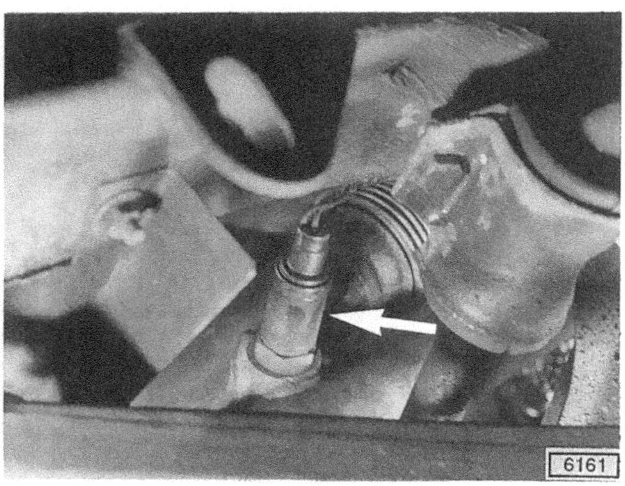

Fig. 5-1. Typical location of oxygen sensor (arrow).

Tightening torque
• oxygen sensor . . . . 55±5 Nm (41±4 ft-lb)

CAUTION —
• Do not let anti-seize compound or penetrating oil come in contact with the slit portion of the oxygen sensor or else the sensor will be damaged.

• Carburetor and fuel injection additives or sprays can destroy the oxygen sensor and catalytic converter. Check with an authorized BMW dealer for approved fuel additives.

• Handle the oxygen sensor with care. Hitting or dropping the oxygen sensor will damage it.

NOTE —
• Special tools are available from most auto supply stores to remove and install the oxygen sensor. The tool allows the sensor's wire to be positioned out of the way.

• On 528e and 533i models, see **ELECTRICAL SYSTEM** for turning out the oxygen sensor warning light.

6. TECHNICAL DATA

I. Tightening Torques

Exhaust manifold to cylinder head	
(nut)	22–24 Nm (16–17 ft-lb)
Front exhaust pipe to rear exhaust pipe	
(nut or bolt)	22–24 Nm (16–17 ft–lb)
Front exhaust pipe to exhaust manifold (nut)	
1st stage	30–35 Nm (22–26 ft-lb)
2nd stage	50–55 Nm (37–41 ft-lb)
Front exhaust pipe clamp to exhaust pipe	
(nut or bolt)	22–24 Nm (16–17 ft-lb)
Front exhaust pipe bracket to transmission bracket	
(nut or bolt)	22–24 Nm (16–17 ft-lb)
Front pipe mounting bracket to transmission bracket (rubber mounts)	
M8 bolts	22–24 Nm (16–17 ft-lb)
Front stabilizer to body	22 Nm (16 ft-lb)
Heat-shield to exhaust system	
(self-tapping screw)	7–8 Nm (62–71 in-lb)
Oxygen sensor to exhaust manifold	
or exhaust pipe	55±5 Nm (40±3 ft-lb)
Rear muffler clamp to rear muffler maximum permissible	
(clamping bolt)	14 Nm (10 ft-lb)
Rear muffler mounting bracket to rear axle	
(bolt)	22–24 Nm (16–17 ft-lb)

Section 12

MANUAL TRANSMISSION AND CLUTCH

Contents

INTRODUCTION

This section covers removal and overhaul procedures for the clutch, and removal procedures for the manual transmission. From the transmission, power is transmitted to the drive wheels by the driveshaft and final drive. Service and repair of the driveshaft and final drive are covered in **DRIVESHAFT AND FINAL DRIVE.** Repairs to the internal parts of the transmission are covered in **MANUAL TRANSMISSION OVERHAUL.**

1. GENERAL DESCRIPTION

The cars covered by this manual are equipped with one of three Getrag five-speed manual transmissions. The 260/5, 260/6, and the 265/6. Both 260 transmissions have an integral bellhousing-transmission case. The 265 has a detachable bellhousing. All three transmissions are fully synchronized in all gears and are housed in lightweight alloy cases. Fig. 1-1 shows a typical BMW five-speed manual transmission.

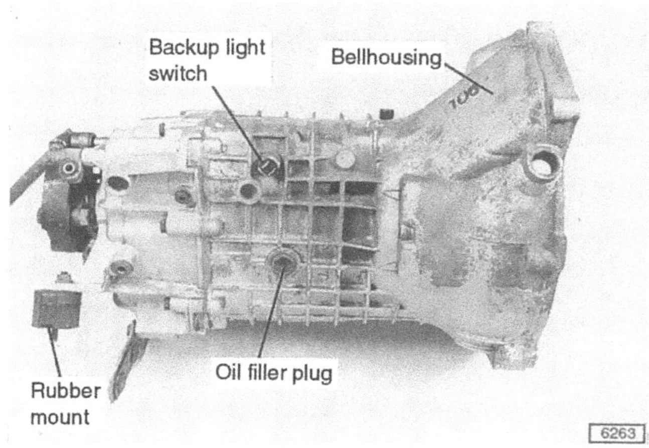

Fig. 1-1. Transmission case of five-speed Getrag 260 manual transmission. Other transmissions are similar.

1.1 Transmission

The transmission case, which contains the transmission gear train, is a lightweight alloy die casting with an integral bellhousing. The transmission gears are of the constant-mesh type with balk ring synchronizers. Each gear (1st through 5th) is actually a mating pair of gears, and constant-mesh simply means the mating gears are always meshing, as shown in Fig. 1-2.

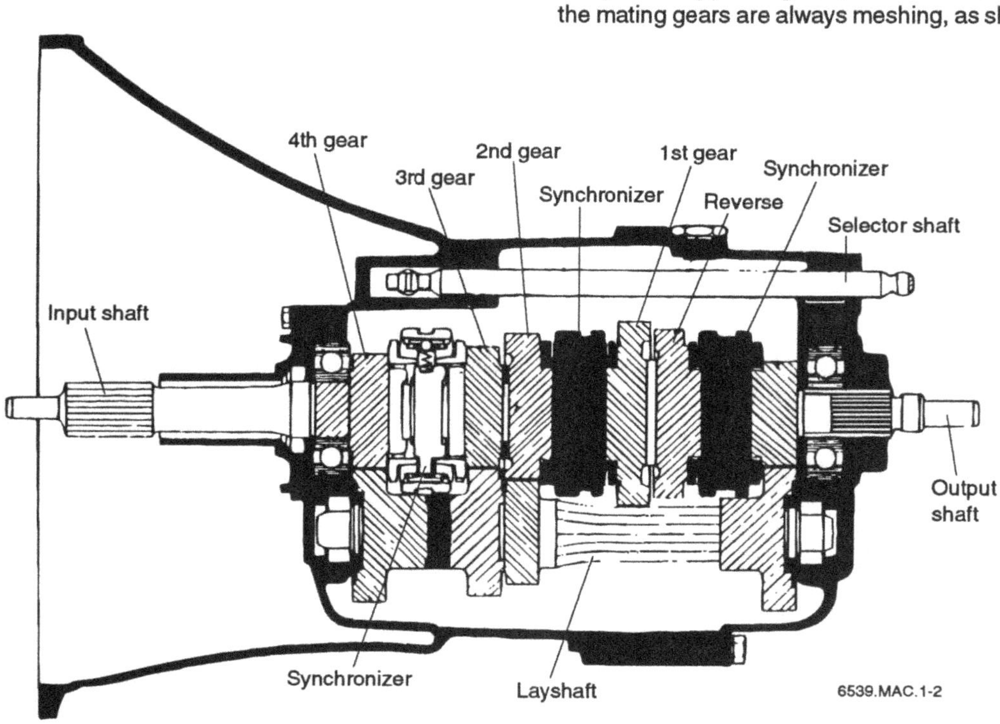

Fig. 1-2. Cross section of manual transmission showing position of gear train.

The 4th gear is mounted on the input shaft and rotates the layshaft at engine speed. The other forward gears and reverse gear are mounted in bearings with their mating gears on the layshaft, and the gears freewheel until engaged. The synchronizers are splined and mounted on the output shaft and always rotate at rear wheel speed.

Gear selection is made using the gearshift lever in the passenger compartment, which is linked to the transmission by the gearshift linkage. When shifting, the synchronizers match the speed of the selected gear to the speed of the output shaft, then lock the selected gear into rotation with the output shaft. This eases shifting, minimizes wear, and helps prevent damage to the gears.

1.2 Clutch

The transmission bellhousing houses the clutch components. The clutch is hydraulically operated. When the clutch pedal is depressed, the clutch master cylinder generates hydraulic pressure to activate the slave cylinder on the transmission. The slave cylinder moves the clutch release lever and release bearing to engage or disengage the clutch.

WARNING —
The clutch disc contains asbestos fibers. Asbestos materials can cause asbestosis. Always wear an approved respirator and protective clothing when handling components containing asbestos. Do not use compressed air, do not grind, heat, weld, or sand on or near any asbestos materials. Only approved cleaning equipment should be used to service the clutch disc or areas containing asbestos dust or asbestos fibers.

1.3 Identification Codes and Specifications.

Transmissions are identified by a manufacturer's stamp and by code numbers and letters. The stamp and the code numbers are typically in front of the mounting for the clutch slave cylinder as shown in Fig. 1-3 and Fig. 1-4. **Table a** lists manual transmission gear ratios.

Table a. Manual Transmission Gear Ratios

Transmission type	Getrag 260/5	Getrag 260/6	Getrag 265
Gear ratios			
1st gear	3.83	3.83	3.82
2nd gear	2.20	2.20	2.20
3rd gear	1.40	1.40	1.40
4th gear	1.00	1.00	1.00
5th gear	0.81	0.81	0.81
Reverse gear	3.46	3.46	3.71

Fig. 1-3. Manufacturer's stamp.

Fig. 1-4. Typical location of manufacturer's code numbers (arrow).

2. MAINTENANCE

BMW specifies the maintenance steps below to be carried out at particular time or mileage intervals for proper maintenance of the clutch and transmission. Information on clutch and transmission maintenance and on the prescribed maintenance intervals can be found in **LUBRICATION AND MAINTENANCE**.

1. Checking clutch master cylinder fluid level
2. Replacing clutch fluid. See **BRAKES**

3. Checking transmission oil level
4. Checking clutch disc for wear
5. Changing transmission oil

NOTE —
Checking clutch freeplay is not necessary. The clutch system is hydraulic and is self-adjusting.

3. TROUBLESHOOTING

This troubleshooting section applies to problems affecting the transmission—including the gearshift mechanism and the clutch.

The source of most problems is apparent from the symptoms. For example, difficulty in engaging a gear or imprecise shifting are transmission problems. Other symptoms, such as loss of performance, vibration or shuddering when releasing the clutch, or noises from the area of the transmission, are less specific.

Poor performance may be evidence of a slipping clutch, but it may also be caused by an engine problem. Vibration suggests a faulty clutch, but might also indicate a driveline problem.

Noises may be the result of a failure in the gear train, or of the driveshaft and final drive. What appears to be a transmission oil leak may be engine oil leaking from a faulty rear crankshaft oil seal especially if the leak is near the bottom of the transmission bellhousing. See **ENGINE—RECONDITIONING** for information on crankshaft oil seals. See **DRIVESHAFT AND FINAL DRIVE** for information on the driveshaft.

3.1 Basic Troubleshooting Principles

Transmission problems fall into two categories: those that can be fixed by external adjustments, and those that require disassembly of the transmission. Problems that at first appear to be caused by internal faults, such as gear shifting difficulty or noisy operation, can often be corrected externally.

Begin any transmission troubleshooting with a thorough visual inspection, both in the engine compartment and from beneath. Check all parts of the gearshift mechanism for wear that might cause misalignment and shifting difficulty. Look for wet spots that may indicate oil leaks. Low oil level, or the wrong type of oil may be the cause of hard shifting or noise. Accurate pinpointing of leaks may require that the suspected area be cleaned and reinspected.

Correct leaks and oil level before acting on suspected internal problems. The gearshift mechanism is covered in **4. Shift Mechanism**. Replacement of transmission oil seals is covered in **MANUAL TRANSMISSION OVERHAUL**. Checking and correcting oil level is covered in **LUBRICATION AND MAINTENANCE**.

As with most other troubleshooting, careful observation of the symptoms is the key to isolating and identifying transmission problems. A road test is an important step. Determining whether the problem is present in all gears, only during acceleration, when the clutch is engaged, or in some other special conditions may help isolate the source of the problem.

Table b lists manual transmission and clutch symptoms, their probable causes, and recommended corrective actions. The numbers in bold type in the corrective action column refer to the numbered headings in this section where the suggested repairs are described.

Table b. Manual Transmission Troubleshooting

Symptom	Probable cause	Corrective action
1. Difficult or noisy shifting	**a.** Clutch not fully releasing	**a.** Air or leaks in clutch hydraulic system. Bleed system or replace master and slave cylinders. **5.1**, **5.2.** Incorrect pedal adjustment. Check and correct pedal adjustment. **5.2**
	b. Clutch disc binding on transmission input shaft	**b.** Remove transmission and inspect splines of clutch disc and input shaft. If necessary, replace clutch disc. **5.3**
	c. Worn gearshift linkage, or loose shift console	**c.** Check shift mechanism and tighten shift console. If necessary, replace worn parts. **4.1**
	d. Low gear oil level	**d.** Check for transmission oil leaks. Check and correct oil level if needed. See **LUBRICATION AND MAINTENANCE**
	e. Incorrect gear oil for ambient temperature	**e.** Drain oil and replace with oil of correct type and viscosity. See **LUBRICATION AND MAINTENANCE**
	f. Input shaft bearing in flywheel seized or sticky	**f.** Replace ball bearing assembly. See **ENGINE—RECONDITIONING**
	g. Worn or damaged internal gear train components	**g.** Inspect internal transmission components and replace faulty parts. See **MANUAL TRANSMISSION OVERHAUL**
	h. Not waiting long enough before shifting (Reverse)	**h.** Wait at least 3-4 seconds after depressing clutch pedal before shifting into reverse

continued on next page

Table b. Manual Transmission Troubleshooting (continued)

Symptom	Probable cause	Corrective action
2. Transmission noisy	a. Insufficient gear oil b. Worn or damaged internal gear train components c. Clutch dampening springs broken	a. Check for transmission oil leaks. Check and correct oil level if needed. See **LUBRICATION AND MAINTENANCE** b. Inspect internal transmission components and replace faulty parts. See **MANUAL TRANSMISSION OVERHAUL** c. Inspect clutch components and replace faulty parts. **5.3**
3. Grinding noise when shifting (1st or Reverse gears)	a. Idle speed too high b. Not waiting long enough before shifting (Reverse) c. Worn or damaged internal gear train components	a. Adjust idle speed. See **FUEL SUPPLY** b. Wait at least 3-4 seconds after depressing clutch pedal before shifting into reverse c. Inspect internal transmission components and replace faulty parts. See **MANUAL TRANSMISSION OVERHAUL**
4. Transmission fails to engage a gear or jumps out of gear	a. Worn gearshift linkage or loose shift console b. Transmission output flange loose c. Worn or damaged internal gear train components	a. Check shift mechanism and tighten shift console. If necessary, replace worn parts. **4.1** b. Check output flange. See **DRIVESHAFT AND FINAL DRIVE** c. Inspect internal transmission components and replace faulty parts. See **MANUAL TRANSMISSION OVERHAUL**
5. Poor acceleration, clutch slipping on hills when accelerating	a. Clutch friction surfaces worn or burnt b. Clutch not fully engaging c. Clutch disc, pressure plate, or flywheel oil soaked	a. Inspect clutch components and replace faulty parts. **5.3** b. Inspect clutch disc for binding on input shaft. Check condition of flywheel bearing for input shaft. Check condition of release bearing and release lever. Replace faulty parts. **5.3**. Check for binding in clutch hydraulic system at slave cylinder and master cylinder pushrods. **5.2** c. Inspect clutch components and engine and transmission oil seals. If necessary, clean pressure plate and flywheel. Replace clutch disc and faulty oil seals. **5.3**
6. Clutch grabs or chatters when the pedal is released	a. Faulty clutch release system b. Clutch disc binding on transmission input shaft c. Input shaft binding to flywheel d. Contaminated (fluid soaked) or glazed (overheated) clutch disc e. Faulty engine/transmission mounting	a. Check condition of release bearing and release lever. Replace faulty parts. **5.3**. Check for binding in clutch hydraulic system. **5.2** b. Remove transmission and inspect splines of clutch disc and input shaft. If necessary, replace clutch disc. **5.3** c. Inspect and replace input shaft ball bearing assembly. See **ENGINE—RECONDITIONING** d. Inspect clutch disc and replace faulty parts. **5.3** e. Check engine and transmission mounts for oil contamination. Replace if necessary. See **ENGINE—SERVICE AND REPAIR**

12

3.2 Diagnostic Tests

Most internal transmission problems cannot be accurately diagnosed unless the unit is removed and disassembled. However, a quick test can be used to determine whether the clutch is performing satisfactorily or slipping.

To quick-check clutch performance, start the engine and set the parking brake. Depress the clutch pedal and place the gearshift lever in 3rd or 4th gear, then slightly accelerate the engine and slowly release the clutch pedal. The engine should immediately stall, indicating that the clutch is engaging properly and will not slip enough to allow the engine to continue to run. If the engine stalls slowly, or does not stall at all, the clutch is most likely slipping. See **5. Clutch** for more information.

4. SHIFT MECHANISM

The basic shift mechanism on the cars covered by this manual is a shift lever and shift rod connected to the transmission selector shaft. See Fig. 4-1. The shift lever swivels in a ball-and-socket mounted in the shift console. The shift rod is carried in bushings at either end to dampen transmission vibration.

There are two versions of the shift mechanism. One has a sheet-metal shift console and one has an aluminum shift console. The version with sheet-metal shift console is installed on most models through 1986.

On sheet-metal consoles, the shift lever is mounted in a two-piece plastic bearing secured by a circlip. At the front, the console is bolted to the transmission case.

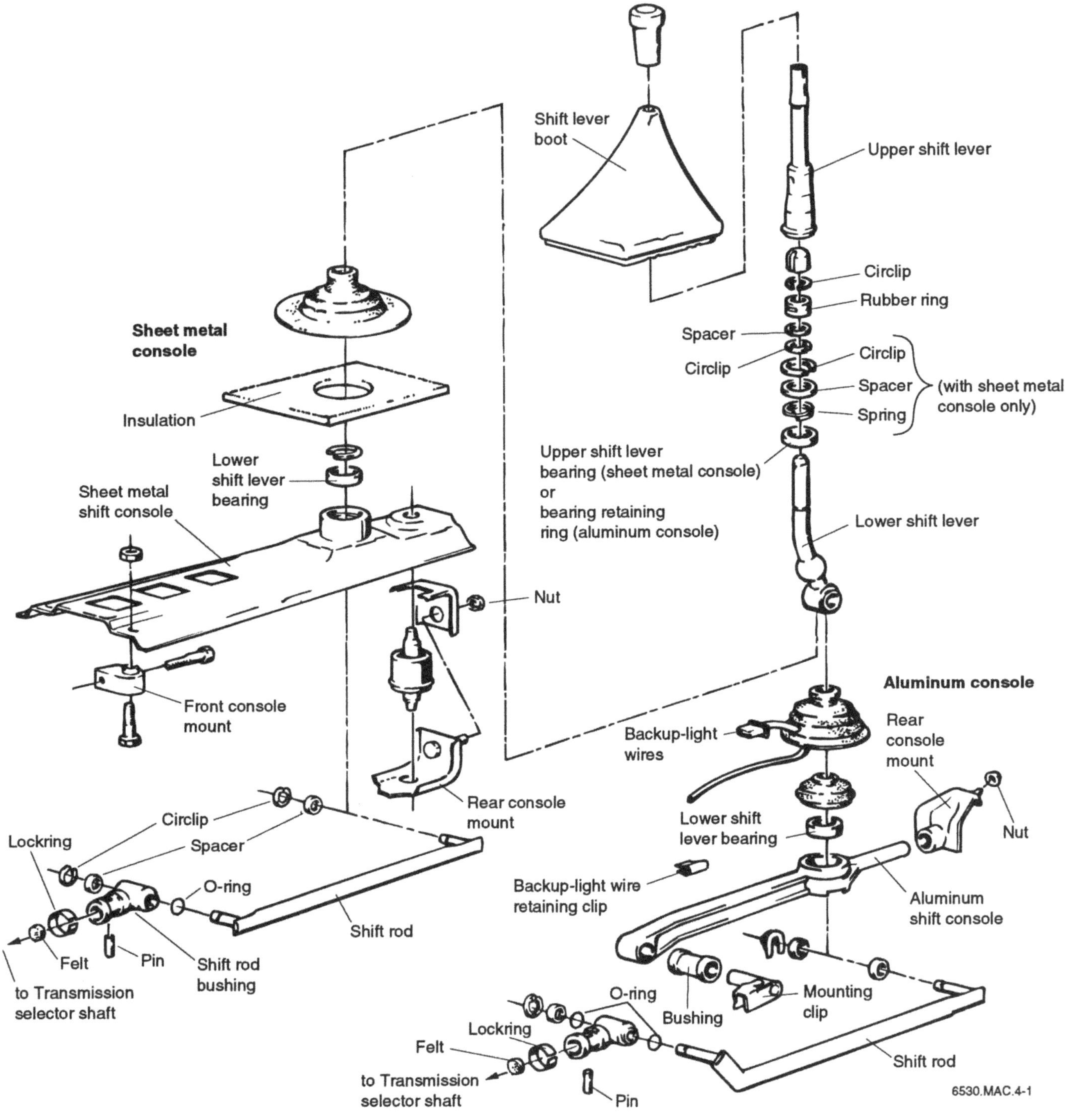

Fig. 4-1. Exploded view of shift mechanism.

On aluminum consoles, the shift lever and its bearing are held in the frame by a plastic retaining ring that snaps into place. The front of the console is mounted to the transmission case by a clip.

The shift mechanism is not adjustable. For any shifting problems, the gearshift bearing and shift-rod bushings should be inspected for wear or for a lack of lubrication that might affect smooth or accurate shifting. The shift console should also be checked for secure mounting and bushing wear. All bearings and bushings are available as replacement parts.

4.1 Disassembling and Assembling Shift Mechanism

The shift mechanism itself is easily removed and installed, but in most models complete access to the mechanism requires removing the exhaust system, the exhaust heat shield, and the driveshaft. See **EXHAUST SYSTEM** and **DRIVESHAFT AND FINAL DRIVE** for more information.

On models with the aluminum shift console, removal of the shift lever may require the use of a special tabbed tool to unlock the plastic retaining ring.

On models with the sheet-metal console, special bolts coated with a locking compound are used to hold the console to the transmission. This prevents the bolts from loosening and being stripped by the shifting action.

When assembling the shift rod or shift lever bearing, lubricate all joints and pivot points with molybdenum disulfide grease (Molykote Longterm 2 or equivalent).

To remove and install shift lever and shift rod:

1. Put the transmission in reverse.

2. Pull the shift knob up and off of the shift lever. Remove the shift boot and the sound insulating felt underneath it.

3. Remove the rubber dust cover. On models with aluminum shift console, it is necessary to first disconnect the plug for the backup light switch.

4. Working underneath the car, disconnect the shift lever from the shift rod by removing the circlip as shown in Fig. 4-2.

5. On models with sheet-metal shift console, remove the circlip from the top of shift lever bearing shown in Fig. 4-3 and lift out the shift lever.

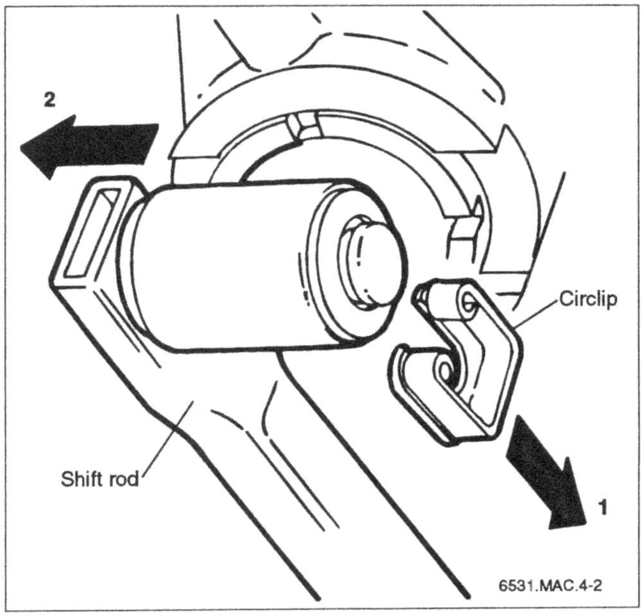

Fig. 4-2. Pull off shift rod circlip (arrow **1**) and then disengage shift rod from shift lever bushing (arrow **2**).

Fig. 4-3. Shift lever circlip being removed on models with sheet-metal shift console.

6. On models with aluminum shift console, work underneath the car to turn the bearing retaining ring 90° (1/4 turn) counterclockwise. See Fig. 4-4. Disengage the ring and lift out the shift lever.

NOTE —
A special tool (BMW tool no. 25 1 100) is available to unlock the bearing retaining ring. As an alternative, a needle-nose pliers or a screwdriver can be used to engage the locking tabs of the bearing retaining ring.

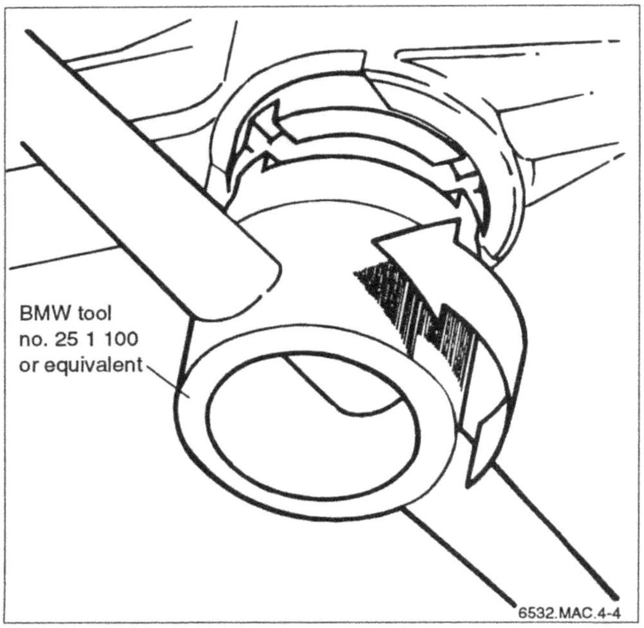

Fig. 4-4. Shift lever retaining ring being removed on models with aluminum shift console.

7. Remove the shift rod from the transmission selector shaft. This is done by pulling off the bushing lockring, then using a suitable drift to drive out the pin. See Fig. 4-5.

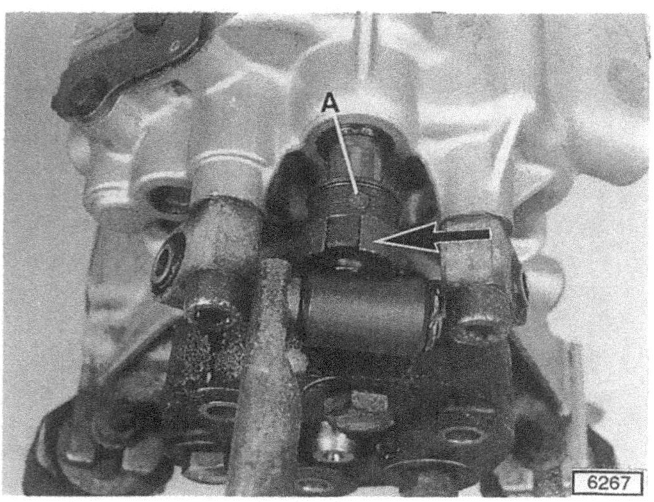

Fig. 4-5. Shift rod bushing lockring (arrow), and pin (**A**). With transmission installed, drive pin out from below. (Transmission shown removed.)

NOTE —
• Note the installed orientation of the shift rod so that it can be reinstalled the same way. Depending on the transmission, the shift rod may be on the right or left side of the bushing.

• On models with sheet-metal shift console, it may be easier to drive the pin out with the selector shaft in 3rd or 5th gear.

Installation is the reverse of removal. Check the condition of the felt in the shift rod bushing and replace it if it is torn or damaged. Drive the bushing pin in from the bottom. Remove any old grease from the shift lever bearing and lubricate it with molybdenum disulfide grease.

On models with sheet-metal shift console, the shift lever bearing should be lubricated with the same grease. On models with aluminum shift console, align the tabs on the plastic retaining ring with the openings in the shift console, as shown in Fig. 4-6, and press in the retaining ring until it clicks twice.

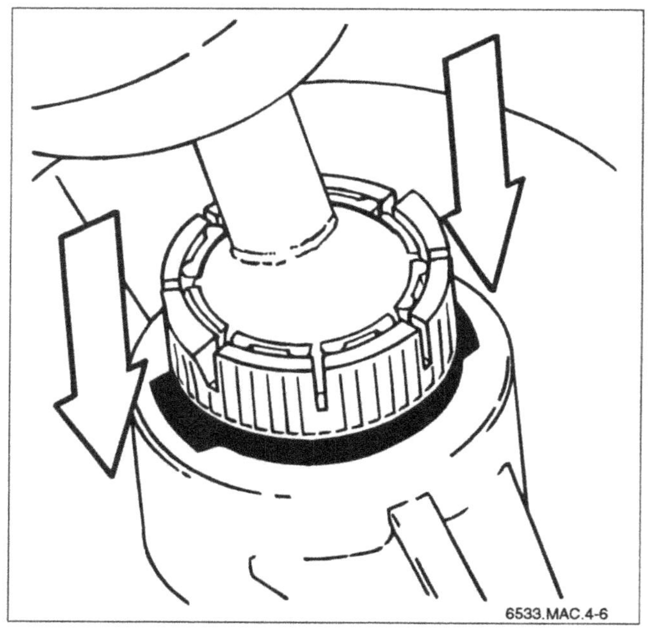

Fig. 4-6. Shift lever being installed on models with aluminum shift console.

To remove and install shift console:

1. Remove the shift lever and shift rod as described above.

2. Disconnect the wires for the backup lights from the transmission. On models with sheet-metal console also disconnect the plug on the top of the console. On models with aluminum console, remove the wiring clips on the console and remove the wires.

3. Working under the car, remove the shift console rear mounting nut, shown in Fig. 4-7. Removal of the driveshaft and loosening of the transmission mount bolts may be necessary to gain access.

> ***WARNING —***
> *Lowering the transmission requires a transmission jack or a floor jack with transmission adaptor, and jack stands to support the car. Use extreme caution when working beneath the car and lowering the transmission.*

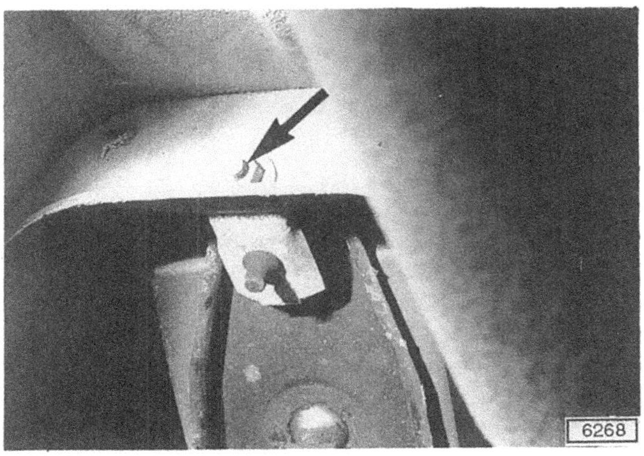

Fig. 4-7. Shift console rear mounting nut (arrow). Model with sheet-metal console shown. Aluminum console is similar.

4. On models with sheet-metal console, remove the front console-to-transmission mounting bolts shown in Fig. 4-8 and remove the console.

Fig. 4-8. Shift console-to-transmission mounting bolts (arrows) for sheet-metal console. (Transmission shown removed.)

> **NOTE —**
> The bolts may be difficult to remove since they are installed with a locking adhesive.

5. On models with aluminum shift console, support the transmission with a jack and unbolt the rear transmission support from the body. Then lower the transmission so that it rests on the front cross member. Remove the front mounting clip for the console as shown in Fig. 4-9 and remove the console.

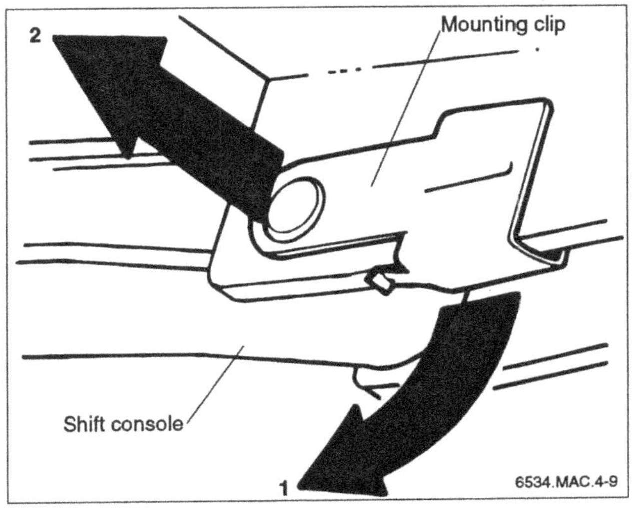

Fig. 4-9. Front mounting clip for aluminum shift console. Remove by first unclipping (**1**), and then sliding pin and clip out (**2**).

Installation is the reverse of removal. Torque the rear mounting nut. On models with sheet-metal console, use new front console mounting bolts with locking compound. Torque all fasteners.

Tightening torques	
• console rear mounting nut	11 Nm (97 in-lb)
• sheet-metal console front mounting bolts	23 Nm (17 ft-lb)
• transmission support bolts (aluminum console)	21.5 Nm (16 ft-lb)

5. CLUTCH

Fig. 5-1 is a schematic view of the hydraulically-operated clutch. The friction clutch disc is clamped by spring pressure between the diaphragm-type pressure plate assembly and the engine flywheel. The splined hub of the clutch disc rides on the transmission input shaft, while the pressure plate assembly is bolted to the engine flywheel.

Depressing the clutch pedal operates the master cylinder and generates hydraulic pressure. This pressure forces the slave cylinder pushrod out to move the release lever and release bearing. As the release bearing presses on the pressure-plate release levers, the clamping pressure is eased on the clutch disc, disengaging engine power from the transmission.

Two basic types of clutch discs are installed, depending on the type of flywheel. On models with a conventional plate-type flywheel, the clutch disc has integral cushion springs and dampening springs. The cushion springs are between the friction surfaces of the disc and reduce the shock when the clutch is engaged. The dampening springs are in the center hub and absorb the rotating power pulses of the engine. On models with dual-mass flywheels, the dampening springs are integrated into the flywheel. For more information on flywheels, see **ENGINE—RECONDITIONING**.

Except for repairs and maintenance to the clutch hydraulic system, inspection or repair of the clutch assembly requires removing the transmission. When the transmission is removed, the flywheel and the clutch assembly remain bolted to the crankshaft, while the release bearing and release lever remain mounted on the transmission. Fig. 5-2 shows the clutch components that are usually replaced when the clutch is overhauled.

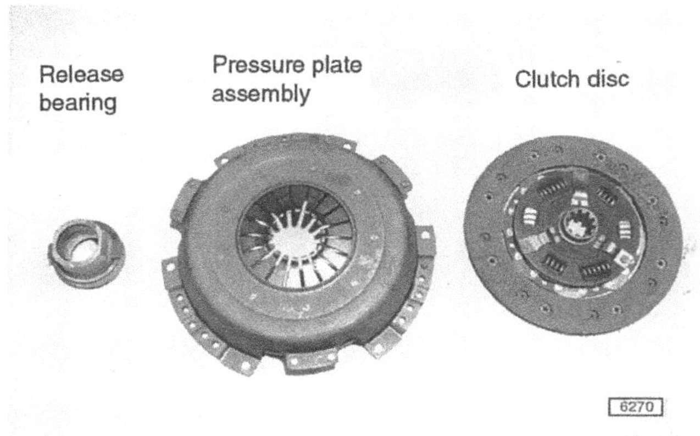

Fig. 5-2. Clutch components that are usually replaced when the clutch is overhauled.

The clutch release system is not adjustable. The clutch slave cylinder automatically compensates for normal wear of the clutch disc. The clutch disc should be replaced when worn beyond acceptable limits. Checking clutch disc wear is a regular maintenance procedure and is covered in **LUBRICATION AND MAINTENANCE**.

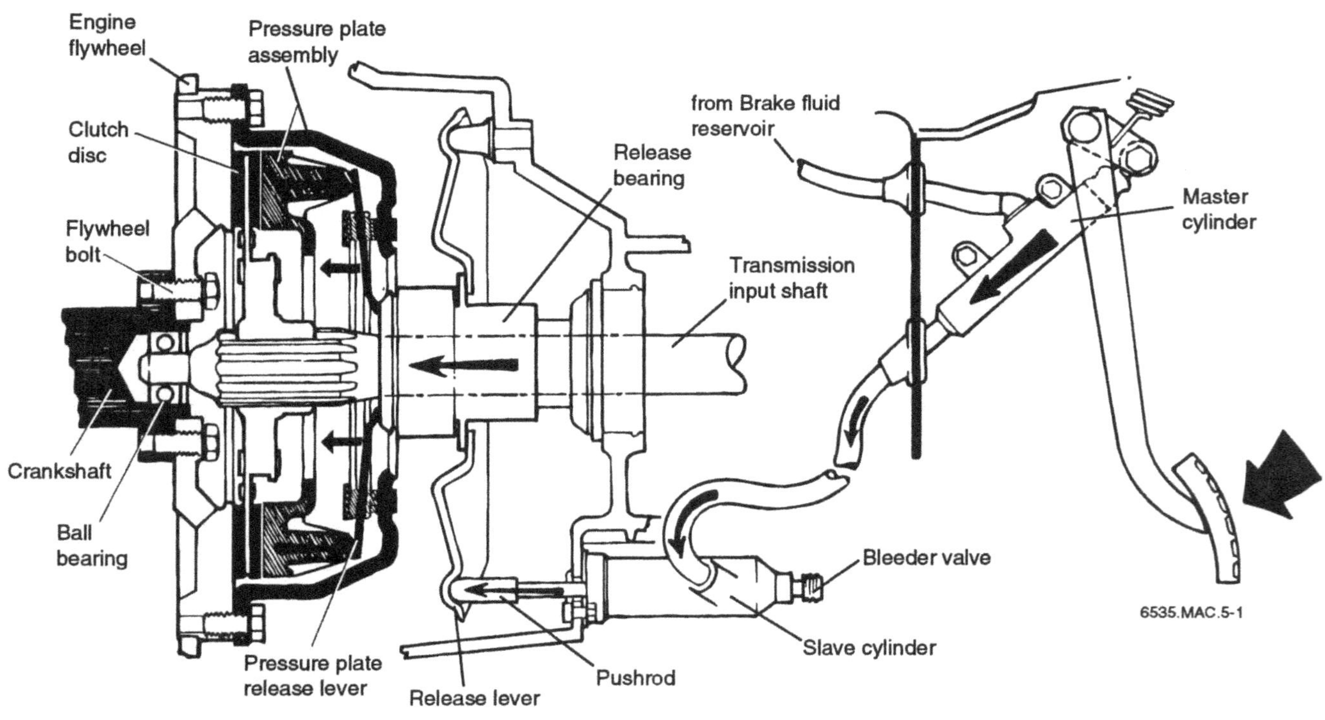

Fig. 5-1. Schematic view of hydraulic clutch components.

Failure of the clutch to fully disengage may be caused by inadequate travel or misalignment of the slave-cylinder pushrod, by a bent master-cylinder pushrod, by the clutch disc binding on the transmission input shaft, or by the input shaft seized to the flywheel ball bearing. Bleed the clutch as described in **5.1 Bleeding Clutch Hydraulic System**, and then check the slave and master cylinders as described in **5.2 Clutch Master Cylinder and Slave Cylinder**. Inspection of the input shaft and clutch disc splines requires removing the transmission, as described in **6.1 Removing and Installing Transmission**.

The clutch shares a common fluid reservoir with the brake system. The same problems that typically affect brake operation—air in the lines and moisture in the fluid—can also affect clutch operation. BMW does not specify any particular interval for replacement of the clutch hydraulic fluid, but the publisher recommends changing the fluid whenever the brake fluid is changed. This should provide adequate protection against corrosion in the clutch master and slave cylinders.

5.1 Bleeding Clutch Hydraulic System

Bleeding the clutch hydraulic system is necessary to remove any air from the lines that may affect the function of the clutch slave cylinder, and to drain the system when changing the fluid.

Bleeding the clutch requires a pressure bleeder and adaptor. Aerated or contaminated fluid is expelled from the system and replaced with new, clean fluid. The pressure bleeder is used to displace fluid in the line after the bleeder screw on the slave cylinder is open. For more information on bleeding methods and on handling brake fluid, see **BRAKES**.

To bleed the clutch, remove the cap and float assembly from the brake master cylinder reservoir. Attach the bleeding equipment to the reservoir, then open the bleeder screw on the slave cylinder. Pump the clutch pedal a few times, then shut the bleeder screw when the fluid is clear and has no air bubbles. Remove the bleeding equipment, check and top up the fluid level in the reservoir and reinstall the float and the cap.

Brake fluid requirement	
• all models	DOT 4

If you suspect that there is still some air in the lines, remove the slave cylinder from the transmission as described below. Push the pushrod all the way in, then release it slowly. This will force any remaining air in the fluid back into the fluid reservoir. When finished, check the fluid level and top it up if necessary.

CAUTION —
Do not operate the clutch with the slave cylinder removed from the transmission. The slave cylinder may be damaged if the pushrod is forced out of the cylinder.

5.2 Clutch Master Cylinder and Slave Cylinder

Although seal rebuild kits are available for both the master and slave cylinders, replacing leaking master or slave cylinders as complete units is usually preferred. In addition, rebuilding these units successfully depends on their internal condition, which can only be determined after the unit is removed and fully disassembled.

Checking Slave Cylinder Pushrod Travel

Inadequate travel of the slave cylinder pushrod may prevent the clutch from fully disengaging when the clutch pedal is depressed, leading to difficult shifting.

To check pushrod travel, use a strip of stiff metal that is thin enough to fit between the slave cylinder and the transmission bellhousing. Push the strip through the slot and hold it against the slave cylinder pushrod. See Fig. 5-3. Have an assistant depress the clutch pedal and release it. The metal strip should scribe a mark on the pushrod, indicating its travel. Remove the slave cylinder as described below and check the mark. Pushrod travel specifications are listed in below.

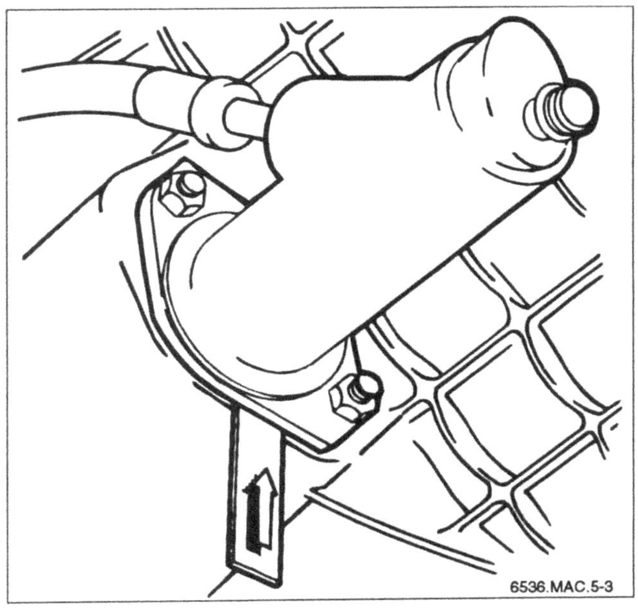

Fig. 5-3. Slave cylinder pushrod travel being checked with metal strip.

Specification	
• slave cylinder pushrod travel	23 mm (0.905 in.)

Inadequate pushrod travel may be due to air in the system, a leaky seal, or the clutch disc binding on the transmission input shaft. Reinstall the slave cylinder, bleed the system as described in **5.1 Bleeding Clutch Hydraulic System**, and recheck travel.

If pushrod travel is still inadequate after bleeding, hold clutch pedal down and check that the pushrod stays fully extended for 15 to 30 seconds. If not, there may be an internal leak. Check the master and slave cylinder seals and hose connections for leaks. Check the adjustment of the clutch pedal as described below. Finally, check the clutch disc and input shaft as described in **5.3 Removing and Installing Clutch**.

Removing and Installing Clutch Master Cylinder and Slave Cylinder

Incorrect installation of either the master cylinder pushrod or the slave cylinder pushrod may cause the clutch release mechanism to bind. This will prevent the clutch pedal from returning or cause the release action to feel jerky. It may even prevent the clutch pedal from being depressed.

To remove master cylinder:

1. Remove the brake fluid reservoir cap and, using a clean syringe, remove the brake fluid until the level is below the hose connection for the clutch master cylinder.

> ***WARNING —***
> *Brake fluid is poisonous. Do not siphon brake fluid with your mouth. Wear hand and eye protection. Do not allow brake fluid to enter the bloodstream through cuts or scratches.*

2. Disconnect the hose from the master cylinder, where it projects through the firewall. Be prepared to catch any leaking fluid.

3. Remove the lower left instrument panel trim.

4. Remove the bolt that connects the master cylinder pushrod to the clutch pedal. See Fig. 5-4.

5. Remove the two bolts that hold the master cylinder to the clutch pedal bracket, disconnect the fluid pipe from the top of the master cylinder, and remove the master cylinder.

> ***CAUTION —***
> *Brake fluid is damaging to paint. Be prepared to catch any fluid that leaks from the master cylinder and fill pipe. Any brake fluid that spills on the car should be cleaned off immediately.*

To install master cylinder:

1. Position the master cylinder against the pedal bracket and install the mounting bolts. Reattach the filler pipe.

Tightening torque
• clutch master cylinder mounting bolts . . . . . . 9 Nm (80 in-lb)

Fig. 5-4. Bolt (arrow) that connects master cylinder pushrod to clutch pedal.

2. Install the bolt that holds the pushrod to the clutch pedal and tighten it.

Tightening torque
• pushrod to clutch pedal bolt . . . . 21 Nm (15 ft-lb).

> **NOTE —**
> On models with an over-center pedal helper spring, shown in Fig. 5-5, make sure the spring is engaged in its bracket before installing the pushrod bolt.

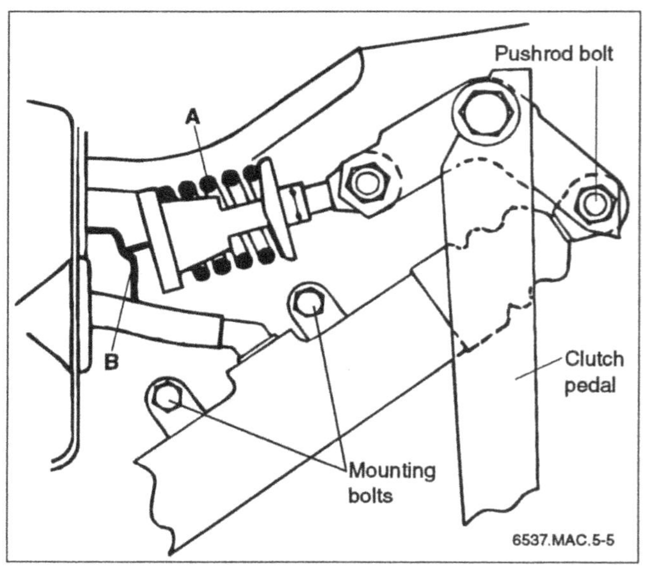

Fig. 5-5. *Clutch pedal over-center helper spring* (**A**) *and spring guide* (**B**).

3. Reconnect the fluid pipe and torque the fitting.

Tightening torque
• fluid pipe . . . . 13 to 16 Nm (10 to 12 ft-lb)

4. Check the adjustment of the clutch pedal as shown in Fig. 5-6. The specification is found below.

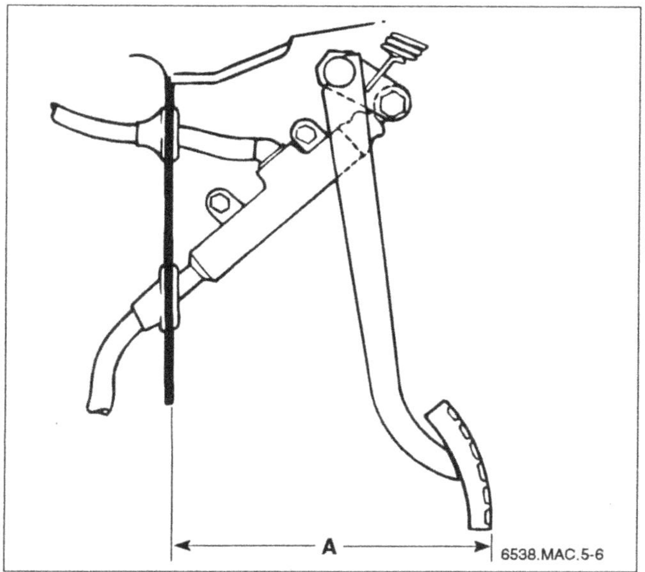

Fig. 5-6. Clutch pedal adjustment is checked at dimension **A**, from firewall to clutch pedal face.

Clutch pedal adjustment
• pedal face to firewall distance. . . . . 253 +8/–3 mm (10 +$^1/_3$/–$^1/_8$ in.)

5. If the adjustment is incorrect, loosen the pushrod nut and bolt. The bolt is eccentric and rotating it should change the adjustment.

6. Some models also have an adjustable pushrod. If the correct adjustment cannot be achieved by turning the eccentric bolt, loosen the pushrod locknut shown in Fig. 5-7, and turn the pushrod until the correct adjustment is achieved. Torque the locknut.

CAUTION —
• Do not screw the pushrod all the way in. This may cause the locknut to jam against the clutch pedal during operation and break the master cylinder pushrod. If pushrod length is changed, check the locknut clearance before fully depressing the clutch pedal.

• Do not over-torque the pushrod locknut. If the threads are stripped the master cylinder may jam, causing clutch failure.

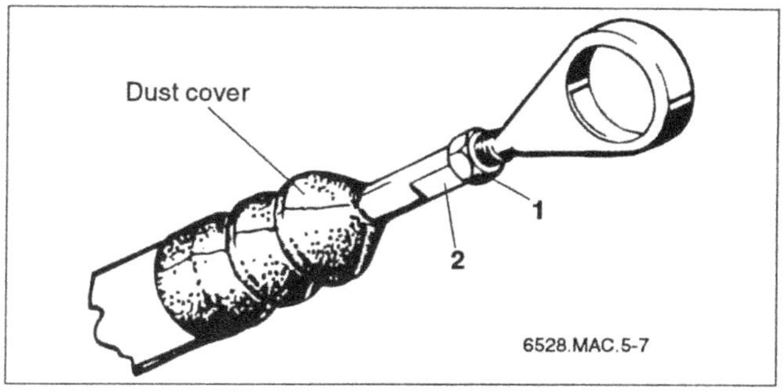

Fig. 5-7. Master cylinder pushrod locknut (**1**) and pushrod (**2**). Slide dust cover back and loosen locknut. Use wrench on slotted part of pushrod to adjust pushrod length.

Tightening torque
• clutch master cylinder pushrod locknut . . . . 4.5 to 6 Nm (40 to 53 in-lb)

7. Reinstall the instrument panel trim.

8. Refill the brake fluid reservoir and then bleed the clutch hydraulic system as described in **5.1 Bleeding Clutch Hydraulic System**.

To remove, inspect, and install slave cylinder:

1. Remove the brake fluid reservoir cap. Using a clean syringe, remove brake fluid until the level is below the hose connection for the master cylinder.

WARNING —
Brake fluid is poisonous. Do not siphon brake fluid with your mouth. Wear safety glasses when working with brake fluid, and wear rubber gloves to prevent brake fluid from entering the bloodstream through cuts or scratches. Always dispose of old fluid carefully.

2. Break the fluid pipe loose. Unbolt the slave cylinder from the transmission, then disconnect the fluid pipe and catch any excess fluid in a container.

3. Inspect the tip of the pushrod and the release lever for wear. The pushrod should be worn only on the tip.

NOTE —
The release lever should show wear only in the pushrod recess. Wear on the side of the pushrod tip, or wear on the release lever other than in the pushrod recess indicates that the pushrod was misaligned.

4. Check the slave cylinder for leaks. If any leaks are present around the pushrod seal, the slave cylinder should be replaced.

5. Lightly coat the plastic pushrod tip with molybdenum disulfide grease (Molykote Longterm 2 or equivalent) and make sure the tip engages the recess in the clutch release lever.

NOTE —
When installing the slave cylinder, the bleeder valve faces down.

6. Torque the mounting nuts.

Tightening torques
• slave cylinder mounting nuts. . . . . 24 Nm (18 ft-lb)
• fluid pipe connection . . . . 13 to 16 Nm (10 to 12 ft-lb)

7. Refill the brake fluid reservoir and then bleed the clutch hydraulic system as described in **5.1 Bleeding Clutch Hydraulic System**.

5.3 Removing and Installing Clutch

The transmission must be removed from the engine to reach the clutch. See **6.1 Removing and Installing Transmission.** It is normally recommended that the pressure plate and release bearing be replaced when a new clutch disc is installed. If the clutch pressure plate is removed from the flywheel, new pressure plate bolts should be used to reinstall it.

The flywheel and flywheel input shaft ball bearing should be carefully inspected when replacing the clutch. Replacing the clutch disc without replacing other worn components may accelerate clutch disc wear.

An inexpensive alignment tool—commonly called a clutch pilot tool or clutch arbor—is used to center the clutch disc during installation. Use of this tool or its equivalent will greatly aid the installation of the transmission. Clutch pilot tools are available from many aftermarket suppliers or from BMW (Part No. 21 2 100).

BMW recommends lubrication of some parts when reassembling the clutch. The BMW-recommended lubricants are noted in the procedures below. If there are signs of oil at the bottom of the transmission, carefully inspect the rear crankshaft oil seal while the clutch is removed. A faulty oil seal should be replaced. See **ENGINE—RECONDITIONING** for information on the crankshaft oil seal.

To remove clutch:

1. Remove the transmission from the engine as described in **6.1 Removing and Installing Transmission**.

2. Remove the pressure plate and clutch disc. The bolts are shown in Fig. 5-8. Loosen the bolts evenly, one turn at a time, until the clutch pressure is relieved, then remove the pressure plate and clutch disc.

WARNING —
The clutch disc contains asbestos fibers. Asbestos materials can cause asbestosis. Always wear an approved respirator and protective clothing when handling components containing asbestos. Do not use compressed air, do not grind, heat, weld, or sand on or near any asbestos materials. Only approved cleaning equipment should be used to service the clutch disc or areas containing asbestos dust or asbestos fibers.

NOTE —
A holding fixture—a BMW special tool or an equivalent—is needed to hold the flywheel while the six pressure-plate bolts are removed. BMW has two tools available: 11 2 160 for 533i/535i models and 11 2 170 for the 528e.

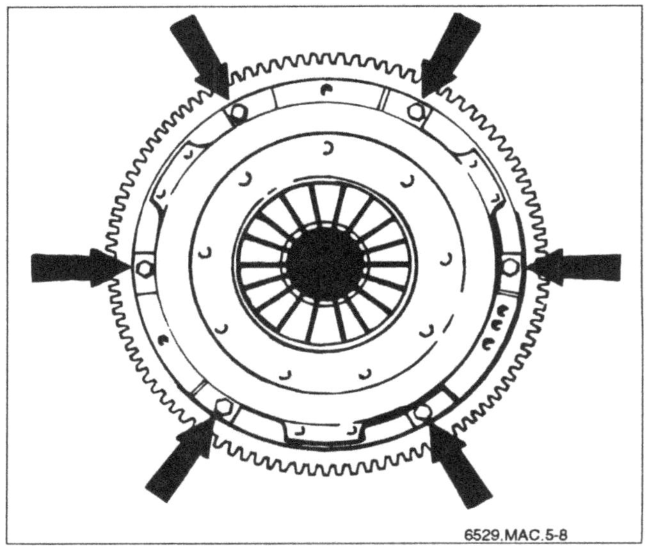

Fig. 5-8. Pressure plate bolts (arrows) to be removed for clutch removal. Always use new bolts when installing pressure plate.

3. Pull the release bearing off of the transmission input shaft. Remove the release lever by unclipping the retaining spring as shown in Fig. 5-9.

To inspect and install:

1. Inspect the clutch pressure plate. Check for loose rivets and for bent or misaligned release levers. See Fig. 5-10. The levers should be almost parallel with each other. **Table c** lists the correct specification. Lay a straightedge across the friction surface of the pressure plate to check that the surface is flat. See Fig. 5-11. Inspect the surface for cracks, scoring, discoloration due to heat or oil contamination, or other damage.

Fig. 5-9. Release lever retaining spring tabs (arrows). Squeeze tabs to remove spring.

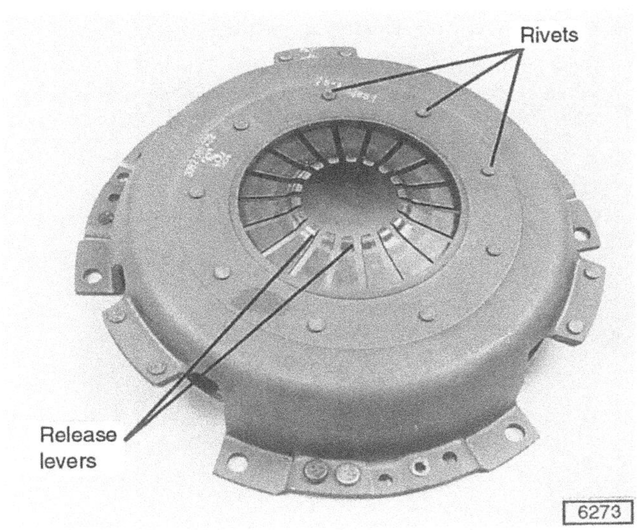

Fig. 5-10. Clutch pressure plate rivets and release lever tips to be checked.

Table c. Clutch Specifications

Release lever tips, maximum deviation from parallel	0.60 mm (0.024 in.)
Clutch disc runout, maximum allowable	0.50 mm (0.020 in.)
Clutch disc thickness, minimum allowable	7.5 mm (0.295 in.)

Fig. 5-11. Clutch pressure plate friction surface being checked for flatness using straightedge.

NOTE —
- A clutch pressure plate that shows any of the signs of damage described above or is not flat should be replaced.
- New pressure plates are coated with a corrosion inhibitor for storage. Use a grease-free solvent to clean the friction surface of new pressure plates before installation.

2. Clean the flywheel friction surface and inspect it for wear, cracks, and grooves. Check for loose or worn guide pins. Replace a flywheel with any of these faults. See **ENGINE—RECONDITIONING**.

3. Check the thickness, runout, and general condition of the clutch disc. If runout or thickness are not as specified above in **Table c**, or if there is any evidence of contamination by oil, the disc should be replaced.

NOTE —
The clutch disc is a relatively low-cost part that commonly wears out and requires replacement. Many experienced technicians routinely install a new clutch disc anytime the transmission is removed.

4. Inspect the splines of the clutch disc and the transmission input shaft. Check that the clutch disc is free to slide on the shaft. If the clutch disc is in any way unserviceable, it should be replaced. Clean all traces of grease from the shaft and clutch splines.

NOTE —
Always apply a light coat of Microlube®261 lubricant to the input shaft and clutch disc splines before reassembly.

5. Inspect the release bearing for smooth operation. Check for wear on the bearing face. See Fig. 5-12 and **Table d**. Replace the bearing if it feels rough or is worn beyond specification.

Fig. 5-12. Clutch release bearing should be checked for wear at dimension **A** and dimension **B**.

Table d. Release Bearing Specifications

Dimension A	
Fichtel & Sachs	49±0.40 mm (1.929±0.016 in.)
SKF	51.5±1.0 mm (2.028±0.039 in.)
Dimension B (both)	25±0.25 mm (0.984±0.010 in.)

NOTE —
Beginning in March 1984, the Getrag 260 0105 **090** transmission was replaced with a 260 0105 **091** transmission on 533i models. If the 090 transmission is being replaced by the 091 a new release bearing (BMW Part No. 21 51 1 204 525) will be required. This release bearing is longer—the original shorter release bearing will not work with the newer 091 transmission.

6. Check the grooved input shaft ball bearing in the center of the crankshaft. Replace it if it does not rotate smoothly.

7. Position the clutch disc and pressure plate assembly against the engine flywheel and start the mounting bolts in their holes. Be sure the correct side of the disc faces the flywheel. Leave the bolts loose enough so that the clutch disc can still move. Make sure the pressure plate is aligned on the guide pins. See Fig. 5-13.

CAUTION —
Always use new pressure plate mounting bolts. Once torqued, they are deformed and should not be reused.

8. Using the clutch pilot tool, center the clutch disc. Tighten the pressure plate mounting bolts evenly until the clutch disc is firmly held in place.

Fig. 5-13. Pressure plate guide pins (arrows). Note use of pilot tool to align clutch disc.

9. Remove the centering tool and finish torquing the bolts in a uniform pattern.

Tightening torques	
• pressure plate bolts	
(8.8 grade)	22 to 24 Nm (16 to 18 ft-lb)
(10.9 grade)	30 to 35 Nm (22 to 26 ft-lb)

NOTE —
• The bolt grade should be marked on the bolt head.

• BMW recommends replacing 8.8 grade bolts with 10.9 grade bolts (Part No. 07 11 9 913 612) when installing the pressure plate.

10. Using Microlube GL 261, lightly lubricate the clutch disc splines and the transmission input shaft splines.

CAUTION —
BMW recommends using Microlube GL 261 on the clutch disc and transmission input shaft splines. Otherwise the clutch disc may bind on the input shaft and cause hard shifting. This lubricant is available under the BMW Part No. 81 22 9 407 436.

11. Use molybdenum disulfide grease (Molykote Longterm 2 or equivalent) to pack the release bearing lubricating groove and to coat the guides. See Fig. 5-14. Grease the release lever pivot points and install the release lever as shown in Fig. 5-15, and then the release bearing, making sure the guides are flush against the release lever.

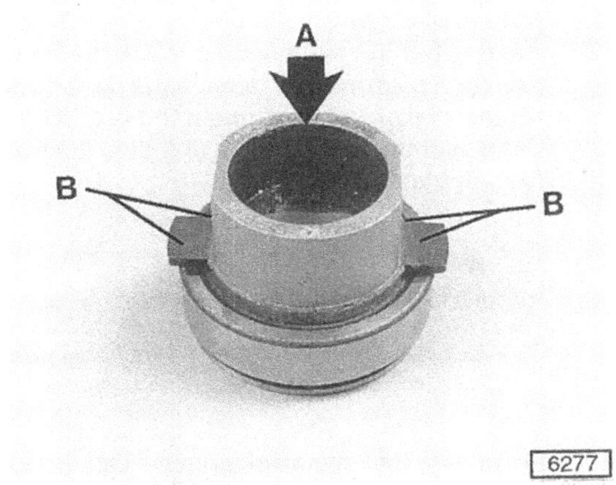

Fig. 5-14. Release bearing lubricating groove (**A**) and guides (**B**) to be greased.

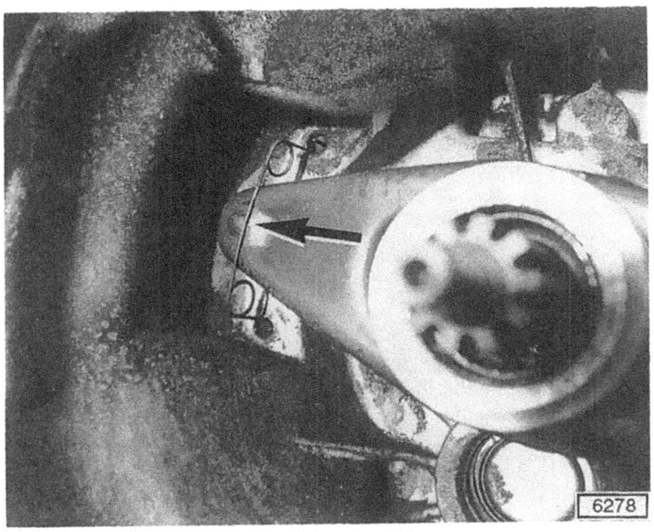

Fig. 5-15. Release lever installed position. Note location of retaining clip (arrow).

12. Remove the flywheel holding fixture, if used, and reinstall the transmission as described in **6.1 Removing and Installing Transmission**.

> **NOTE —**
> To help align the input shaft in the clutch disc, rotate the output shaft slightly as the transmission is being installed.

6. TRANSMISSION REMOVAL AND INSTALLATION

This section covers the removal and installation of the transmission only. Internal transmission repairs and replacement of oil seals are covered in **MANUAL TRANSMISSION OVERHAUL**.

Testing of the reverse light switch, located on the side of the transmission case, is covered in **ELECTRICAL SYSTEM**. It is not adjustable. Replacement is the only remedy for a faulty switch.

6.1 Removing and Installing Transmission

The transmission must be removed from the car for most internal repairs or for replacement. In the procedure below, the transmission is separated from the engine, supported on a jack, and removed from below.

Removal of the transmission requires a transmission jack or a floor jack with transmission adaptor, and jack stands to support the car. Use extreme caution when working beneath the car and lowering the transmission.

Some models use Torx® head bolts around the bellhousing. Be sure to have the appropriate tools on hand before starting the job.

12

To remove:

1. Disconnect the negative (–) battery cable. Disconnect the electrical connectors from the reverse light switch on the side of the transmission.

2. Remove the exhaust system as described in **EXHAUST SYSTEM**.

> **NOTE —**
> It is possible to disconnect only the exhaust downpipe from the exhaust manifold, and the exhaust hanger from the transmission support, and to then push the exhaust system to the side and out of the way. Be sure to disconnect the oxygen sensor wire(s) first. The transmission removal procedure is somewhat easier, though, if the exhaust system is removed completely.

3. Remove the driveshaft as described in **DRIVESHAFT AND FINAL DRIVE**.

4. Disconnect the shift rod and shift console from the transmission as described in **4.1 Disassembling and Assembling Shift Mechanism**.

5. Mark and remove the TDC and reference sensors from the bellhousing as described in **IGNITION**.

NOTE —
Label the location of the sensors as they are removed. If the sensors are not installed in their original locations, the engine will not start.

6. Remove the clutch slave cylinder from the transmission as described in **5.2 Clutch Master Cylinder and Slave Cylinder**. Do not disconnect the fluid hose. Instead, suspend the slave cylinder from the body with a short piece of wire.

CAUTION —
Do not operate the clutch with the slave cylinder removed from the transmission. The slave cylinder may be damaged if the pushrod is forced out of the cylinder.

7. Support the transmission from below with the jack and remove the transmission support mounting nuts shown in Fig. 6-1. Then lower the transmission/engine assembly so that it rests on the front crossmember.

Fig. 6-1. Transmission support nuts (arrows) to be removed.

8. Remove the bolts that hold the transmission to the engine. Some bolts may have a Torx® head and require the use of a special Torx® socket. Note the length and location of each bolt.

CAUTION —
At no time should the weight of the transmission be supported by the transmission input shaft. Such a load will damage the clutch and transmission components.

9. Separate the transmission from the engine, taking care not to place any strain on the transmission input shaft, and either lower the jack or raise the car until there is enough clearance to remove the transmission. Pull the transmission out from under the car. Mark the exact location of the dowels used to locate the transmission to the engine.

WARNING —
Make sure the car is stable and well supported at all times during the removal procedure. Use jack stands that are designed for the purpose. A floor jack alone is not adequate support.

Installation is the reverse of removal. Lightly lubricate the input shaft splines with Microlube GL 261, then carefully position the transmission on the engine. The splines of the input shaft and the splines of the clutch disc may not align exactly. If not, use a wrench on the crankshaft pulley to rotate the crankshaft. Be sure to replace the dowels that locate the transmission to the engine with new ones. Install them in there original positions. Refer to **Table e** for torque values.

CAUTION —
BMW recommends using Microlube GL 261 on the clutch disc and transmission input shaft splines. Otherwise the clutch disc may bind on the input shaft and cause hard shifting. This lubricant is available under the BMW Part No. 81 22 9 407 436.

NOTE —
Washers should always be used with Torx®-head bolts. Installing bolts without washers may cause the bolts to corrode and seal tightly against the transmission case, making them difficult to loosen.

Table e. Transmission Installation Tightening Torques

Bolt	Torque
Transmission to engine (hex-head)	M8: 22 to 27 Nm (16 to 20 ft-lb) M10: 47 to 51 Nm (35 to 38 ft-lb) M12: 66 to 82 Nm (49 to 60 ft-lb)
(Torx®-head)	M8: 20 to 24 Nm (15 to 18 ft-lb) M10: 38 to 47 Nm (28 to 35 ft-lb) M12: 64 to 80 Nm (47 to 59 ft-lb)
Rear transmission support to body (bolt)	22 to 24 Nm (16 to 18 ft-lb)
Rubber mount to transmission or bracket (nut)	43 to 48 Nm (32 to 35 ft-lb)
Transmission drain plug/fill plug	40 to 60 Nm (30 to 44 ft-lb)

If the transmission was drained, refill it with the appropriate lubricant before starting or towing the car. See **LUBRICATION AND MAINTENANCE** for more information.

7. TECHNICAL DATA

I. Tightening Torques

Aluminum shift console to transmission (bolts)	21.5 Nm (16 ft-lb)
Clutch master cylinder to pedal bracket (bolts)	9 Nm (80 in-lb)
Clutch master cylinder pushrod clevis to clutch pedal (shoulder bolt)	21 Nm (15 ft-lb).
Clutch master cylinder pushrod to pushrod clevis (locknut)	4.5–6 Nm (40–53 in-lb
Clutch pressure plate to flywheel (stretch bolts)	
8.8 grade	22–24 Nm (16–17 ft-lb)
10.9 grade	30–35 Nm (22–25 ft-lb)
Clutch slave cylinder to transmission (nuts)	24 Nm (18 ft-lb)
Hydraulic fluid pipe connection to clutch master cylinder or slave cylinder	13–16 Nm (10–12 ft-lb)
Shift console rear mount to body (nut)	11 Nm (97 in-lb)
Sheet-metal shift console to transmission (bolts)	23 Nm (17 ft-lb)
Rear transmission support to body (bolt)	22–24 Nm (16–18 ft-lb)
Rear transmission support rubber mount to transmission or support (nut)	43–48 Nm (32–35 ft-lb)
Transmission to engine	
hex-head bolt	
M8	22–27 Nm (16–20 ft-lb)
M10	47–51 Nm (35–38 ft-lb)
M12	66–82 Nm (49 to 60 ft-lb)
Torx®-head	
M8	20–24 Nm (15–18 ft-lb)
M10	38–47 Nm (28–35 ft-lb)
M12	64–80 Nm (47–59 ft-lb)
Transmission drain plug/fill plug	40–60 Nm (30–44 ft-lb)

12

Section 13

MANUAL TRANSMISSION OVERHAUL

Contents

INTRODUCTION

This section covers the internal repair to both Getrag 260 and Getrag 265 model transmissions. For information on removing and installing the transmission, see **MANUAL TRANSMISSION AND CLUTCH**.

Repairs to the internal parts of the transmission require special tools and knowledge. If you lack the skills and tools, or a suitable workplace for servicing the transmission, we suggest you leave these repairs to an authorized BMW dealer or other qualified shop. Completely reconditioned transmissions are available from an authorized BMW dealer as an exchange unit. See an authorized BMW dealer for correct applications and parts information.

Though you may not have the skills and knowledge for doing actual internal repairs, it may be possible to save some of the expense of professional repair by removing and installing the transmission yourself, using the procedures described in **MANUAL TRANSMISSION AND CLUTCH**. It is important to realize, however, that a partially disassembled transmission may be a problem for a mechanic. We strongly advise against taking the transmission apart to begin any repair that cannot be properly finished.

1. GENERAL DESCRIPTION

The cars covered by this manual are equipped with one of three Getrag five-speed manual transmissions, the 260/5, 260/6, and the 265/6. Both 260 transmissions have an integral bellhousing-transmission case. The 265 transmission has a detachable bellhousing. The 260/6 transmission can be identified by the long ribbed cooling fins cast into the bottom portion of the case. The 260/5 transmission does not have these ribs.

1.1 Transmission

The transmission case, which contains the transmission gear train, is a lightweight alloy die casting. The transmission gears are of the constant-mesh type with balk ring synchronizers. Each forward gear (1st through 5th) is actually a mating pair of gears, and constant-mesh simply means the mating gears are always meshing, as shown in Fig. 1-1.

The 4th gear is mounted on the input shaft and rotates the layshaft at engine speed. The 1st through 3rd and 5th gears are mounted in bearings with their mating gears on the lay-

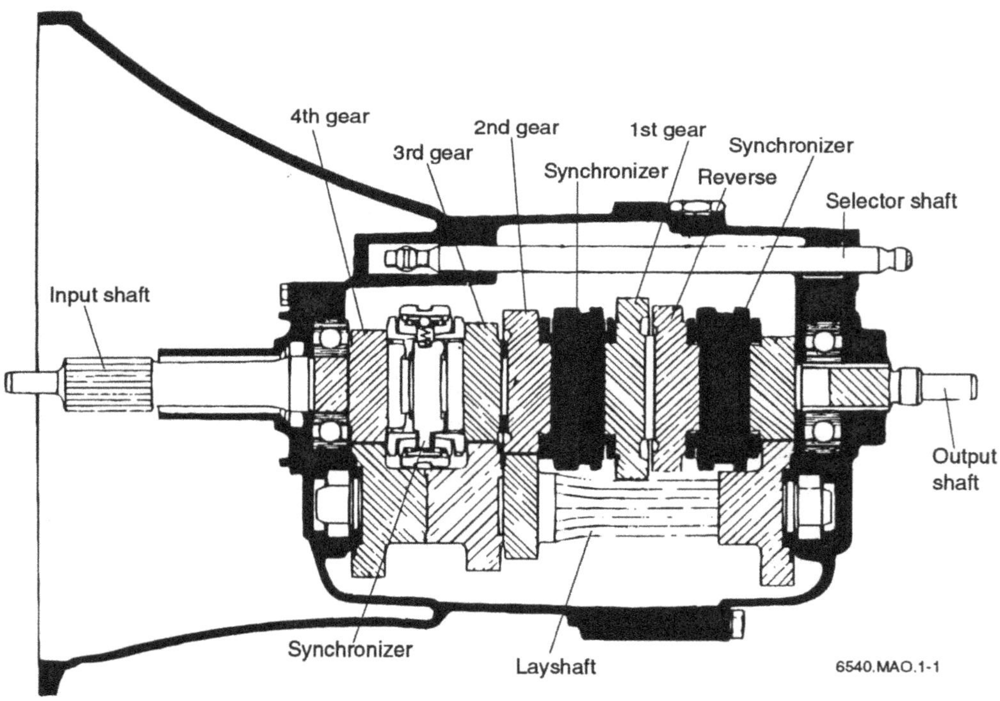

Fig. 1-1. Cross section of manual transmission showing position of gear train.

shaft, and the gears freewheel until engaged. The synchronizer assemblies are splined and mounted on the output shaft and always rotate at rear wheel speed.

Gear selection is made using the gearshift lever in the passenger compartment, which is linked to the transmission by the gearshift linkage. When shifting, the synchronizer rings match the speed of the selected gear to the speed of the output shaft, then the synchronizer hub locks the selected gear into rotation with the output shaft. This eases shifting, minimizes wear, and helps prevent gear damage.

1.2 Clutch

The transmission bellhousing houses the clutch components. For more information on the clutch components see **MANUAL TRANSMISSION AND CLUTCH**

WARNING —
The clutch disc contains asbestos fibers. Asbestos materials can cause asbestosis. Always wear a respirator and protective clothing when handling components containing asbestos. Do not use compressed air, do not grind, heat, weld, or sand on or near any asbestos materials. Only approved respirators and cleaning equipment should be used to service the clutch disc or areas containing asbestos or asbestos fibers.

1.3 Identification Codes and Specifications

Transmissions are identified by manufacturer's stamp and by code numbers and letters. The manufacturer's stamp is on the case, near the mounting for the clutch slave cylinder. For more information see **MANUAL TRANSMISSION AND CLUTCH. Table a** lists the transmissions installed in the models covered by this manual.

NOTE —
Although each of the three transmissions covered by this manual will bolt up to any of the engines used in the 5-series models, the driveshaft lengths vary between transmission types.

Table a. Manual Transmission Applications

Year	528e	533i	535i
1982	Getrag 265/6	N/A	N/A
1983	Getrag 260/5	Getrag 260/5	N/A
1984	Getrag 260/5	Getrag 260/5	N/A
1985	Getrag 260/5	N/A	Getrag 265/6
1986	Getrag 260/5	N/A	Getrag 260/6
1987	Getrag 260/5	N/A	Getrag 260/6
1988	Getrag 260/5	N/A	Getrag 260/6

2. MAINTENANCE

Information on clutch and transmission maintenance and on the prescribed maintenance intervals can be found in **MANUAL TRANSMISSION AND CLUTCH** and in **LUBRICATION AND MAINTENANCE**.

3. TROUBLESHOOTING

Troubleshooting is covered in detail in **MANUAL TRANSMISSION AND CLUTCH**.

To avoid removing the transmission unnecessarily, check the gearshift mechanism for wear and correct leaks and oil level before acting on suspected internal problems. The gearshift mechanism is covered in **MANUAL TRANSMISSION AND CLUTCH**.

As with any type of troubleshooting, careful observation of the symptoms is the key to isolating and identifying transmission problems. A road test is an important step. Determining whether the problem is present in all gears, only during acceleration, when the clutch is engaged, or in some other special conditions may help isolate the source of the problem.

3.1 Diagnostic Tests

Most internal transmission problems cannot be accurately diagnosed unless the unit is removed and disassembled. However, disconnecting the driveshaft at the transmission and running the car through the gears can usually eliminate driveshaft and final drive components as a source of noise or vibration.

4. TRANSMISSION SERVICE— GETRAG 260

This section covers internal transmission repairs, including replacement of the transmission case oil seals, bushings, and bearings. Read the procedures completely before starting any repair job under this heading. Special tools and measuring equipment are necessary for certain tasks. Thoroughly clean the outside of the transmission before beginning any disassembly work. This will help keep abrasive dirt from getting into the working components as they are disassembled.

4.1 Removing and Installing Gear Train (Getrag 260)

Removal of the gear train requires special tools to press the gear assembly out of the transmission case. Also covered under this heading is the removal of the shift forks, shafts and shift rods. Installation of the gear train requires new circlips, rolled pins, and spacers. The circlips and spacers set axial play of the input shaft. The rolled pins retain the shift forks to the shift rods.

To remove gear train (Getrag 260):

1. Remove the transmission as described in **MANUAL TRANSMISSION AND CLUTCH**.

> ***WARNING —***
> • *The clutch disc contains asbestos fibers. Asbestos materials can cause asbestosis. Always wear a respirator and protective clothing when handling components containing asbestos. Do not use compressed air, do not grind, heat, weld, or sand on or near any asbestos materials. Only approved respirators and cleaning equipment should be used to service the clutch disc or areas containing asbestos or asbestos fibers.*
>
> • *When pressing or pulling is required, only use a press or puller with the correct capacity. Do not use power tools, such as air impact tools, on the press or pulling equipment.*

2. Mount the transmission on a stand or place it on a work bench. Drain the transmission oil into a 4-quart pan by removing the fill plug and the drain plug shown in Fig. 4-1. Reinstall the plugs loosely.

> **NOTE —**
> Carefully observe the color of the transmission oil and metal filings. This may help pinpoint component wear.

3. Remove the clutch release lever, spring and release bearing. Remove the clutch release bearing guide sleeve mounting bolts and pull the sleeve from the transmission. Remove and label any spacers from behind the sleeve. Remove and label the input shaft circlip and the spacer from the front of the input shaft. See Fig. 4-2.

> **NOTE —**
> On Getrag 260 transmissions with two-piece guide sleeves, remove only the outer flange mounting bolts.

4. Unscrew the back-up light switch and pry out the plug for the selector shaft lockpin. Remove the spring and the lockpin from the transmission. Note the position of the lockpin when removing. See Fig. 4-3.

Fig. 4-1. Transmission fill plug (**A**) and transmission drain plug (**B**).

Fig. 4-2. Input shaft circlip (**A**) and spacer (**B**) on Getrag 260 transmission.

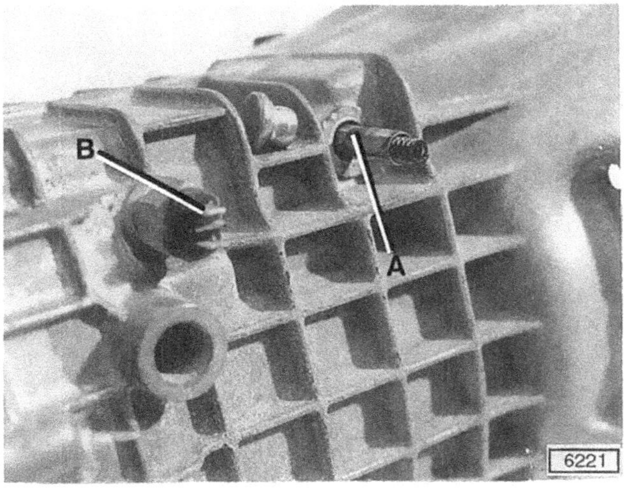

Fig. 4-3. Selector shaft lockpin and spring (**A**) and back-up light switch (**B**).

5. Working from the rear of the transmission, drive out the top guide pin and remove the case mounting bolts. See Fig. 4-4.

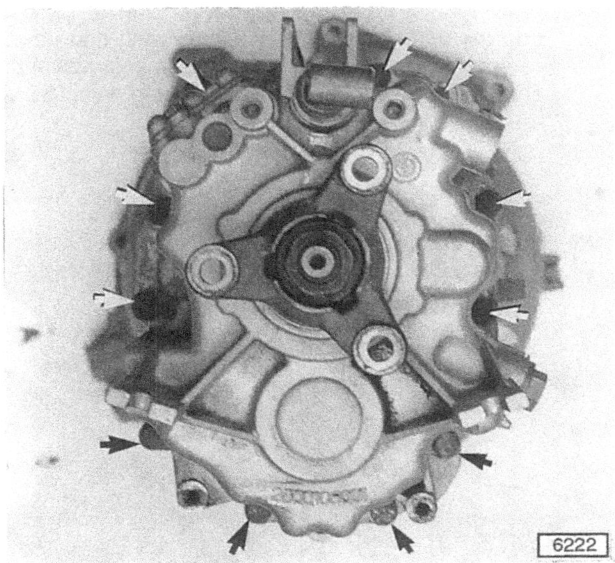

Fig. 4-4. Transmission case mounting bolts and guide pins (arrows).

6. Remove the reverse gear shaft retaining bolt from the front transmission case. See Fig. 4-5.

7. Separate the front transmission case from the rear case by pressing the input shaft out of the front case using BMW special tools. Remove the roller bearing from the top of the layshaft.

Fig. 4-5. Reverse gear shaft retaining bolt (arrow) in front transmission case.

CAUTION —
Do not attempt to separate the case by driving on the front of the input shaft or the front case casting. The front case casting is not designed to take this type of a load. The housing may crack and break.

NOTE —
BMW special tools (BMW tool no. 23 1 460 and BMW tool no. 33 1 301) or equivalents must be used. The special tool is bolted to the guide sleeve surface of the front case. The sleeve part of the tool extends above the tip of the input shaft. A bolt is then threaded through the top of the tool to bear on the input shaft, and the input shaft bearing is pulled off together with the front case. See Fig. 5-7 below.

8. At the rear of the transmission, pry out the drive flange lockplate from the mounting nut. Using a deep thin-walled 30 mm socket (BMW tool no. 23 1 210 or equivalent), remove the drive flange nut while holding the flange stationary. Remove the flange from the transmission output shaft.

NOTE —
• A thin-wall socket can be made by grinding the outer surface of an ordinary socket.

• If the drive flange cannot be easily removed, it may be necessary to use a puller.

9. Remove the reverse-gear shaft retaining bolt from the side of the rear transmission case. Remove the reverse-gear retaining bracket and its mounting bolt. See Fig. 4-6.

Fig. 4-6. Reverse gear shaft retaining bolt (**A**), retaining bracket (**B**), and bracket mounting bolt (**C**).

10. Pull the shaft out of the reverse gear and then remove the gear from the case. See Fig. 4-7.

Fig. 4-7. Reverse gear shaft with roller bearing being withdrawn from transmission case.

11. Pull out the lockpin for the shift operating lever and slide the lever off the shift operating shaft. Pull the shift operating shaft out of the transmission. See Fig. 4-8.

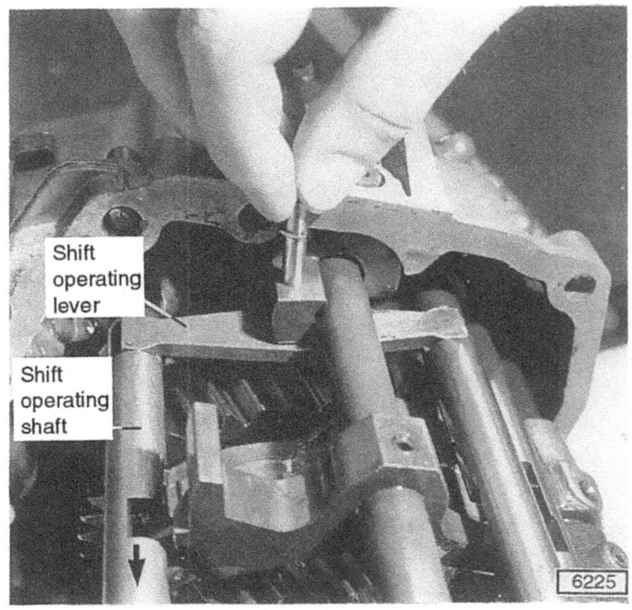

Fig. 4-8. Shift operating lever lockpin being removed. Slide out shaft in direction of arrow.

12. Hand-turn the input shaft while engaging 4th gear. Support the underside of the selector shaft and drive the selector shaft operating lever rolled pin in just until the shaft can be pulled from the operating lever. See Fig. 4-9.

CAUTION —
Do not drive the rolled pin all the way out. Pull on the selector shaft so that the rolled pin is between 2nd and 3rd gears. Otherwise the selector shaft cannot be removed.

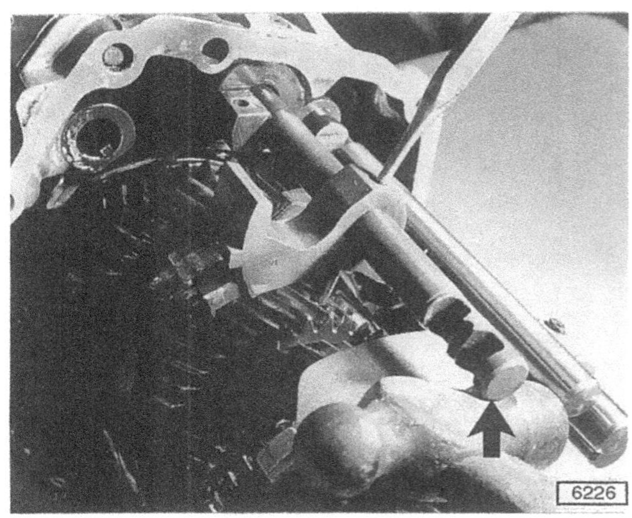

Fig. 4-9. Selector shaft operating lever rolled pin being driven in. Support end of shaft (arrow) while driving pin.

13. Pull the selector shaft from the rear of the transmission. Be ready to catch the four roller bearings on the end of the shaft. Remove the selector shaft oil seal from the shaft. Remove the selector shaft operating lever.

> **NOTE —**
> It may be necessary to tap on the end of the shaft to release the oil seal from the transmission case. Use only a soft-faced hammer when driving the shaft.

14. Remove the shift rod detent ball end plate and remove the three springs and balls. Note the length of each spring. Carefully pry out the shift rod detent ball end plug. See Fig. 4-10.

> **NOTE —**
> The end plug can be easily removed by first pivoting the plug in its bore. Use a small drift to tap on one side of the plug until it is on its side, then use small needle-nose pliers to remove the plug.

Fig. 4-10. Shift rod detent ball end plate (**A**), springs (**B**), and end plug (**C**). Detent balls are under springs.

15. Carefully drive out the rolled pin from the 3rd/4th gear shift fork. Pull the 3rd/4th gear selector rod out of the shift fork. Be sure the small pin in the end of the rod does not fall out as the rod is being removed. See Fig. 4-11. Remove the two detent balls from either side of the end plug bore.

16. Pull on the 1st/2nd gear shift rod to engage 2nd gear. Pull on the 5th/reverse gear shift rod to engage reverse gear.

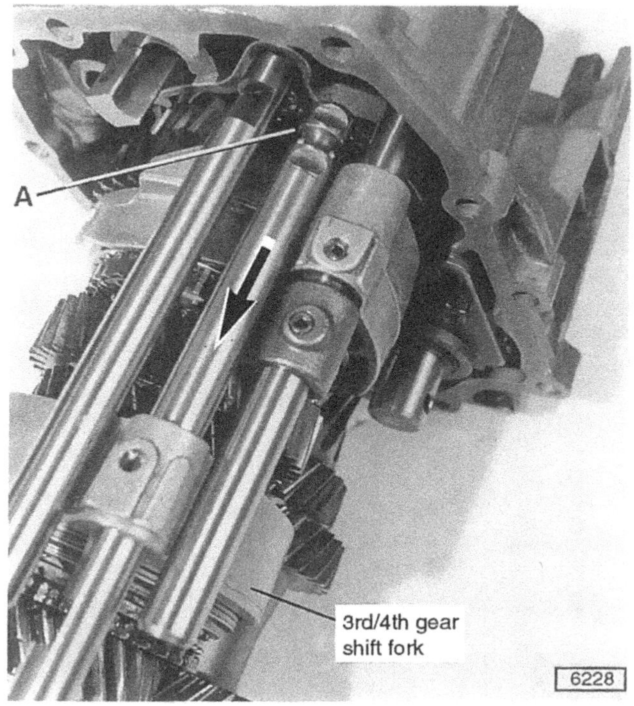

Fig. 4-11. 3rd/4th gear shift rod being removed. Arrow shows direction of removal. Note small pin (**A**) in end of rod.

17. Using a large puller, press the input shaft, the output shaft, and the layshaft assembly out of the transmission case. See Fig. 4-12.

13

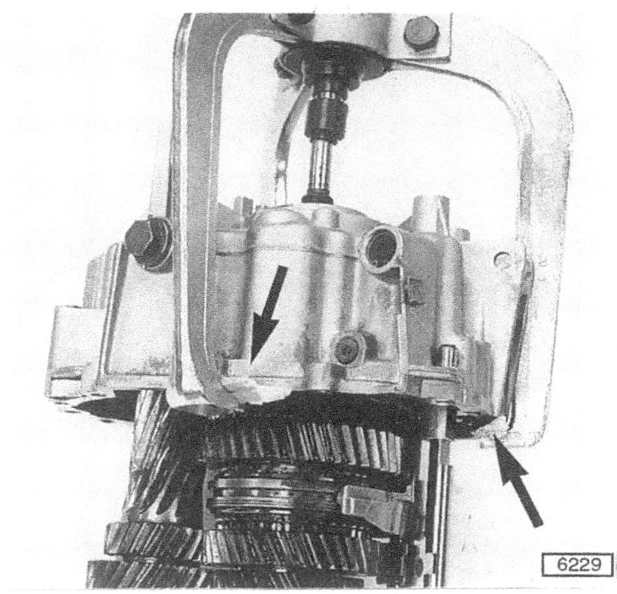

Fig. 4-12. Transmission gear assembly being removed from transmission rear case using puller. Place small blocks of wood beneath the jaws of the puller (arrows) to protect case.

CAUTION —
Place small blocks of wood under the jaws of the puller to prevent damaging the case sealing surface. Check to make sure that the selector rods, forks, and the layshaft clear the jaws of the puller as the gear assembly is being removed.

18. Remove the layshaft roller bearing from the rear case.

To install gear train:

1. Inspect all bearings and bushings in the transmission case. Replace any that are damaged as described below under **6. Inspecting Transmission Components.** Inspect the thrust washer on the end of the output shaft and replace it if scored or damaged.

2. Drive out the input shaft roller bearing from the front case.

3. Position the 3rd/4th gear shift fork, the 1st/2nd gear shift fork and rod, and the 5th/reverse gear shift fork and rod onto the gear train assembly. See Fig. 4-13. Engage 2nd gear and reverse gear.

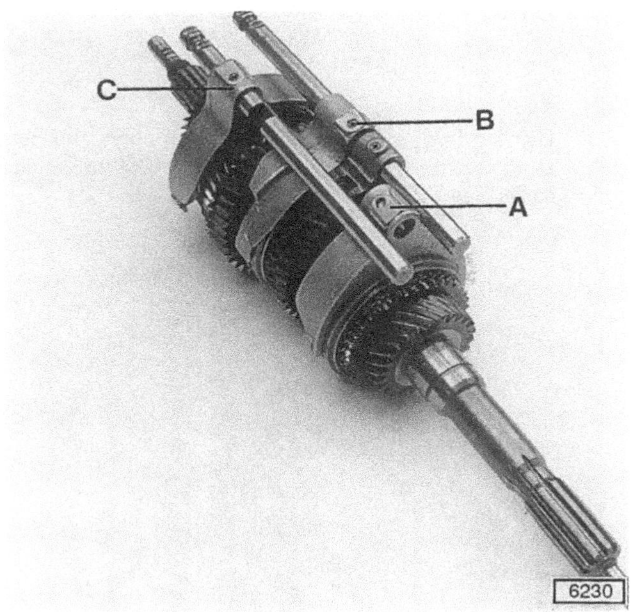

Fig. 4-13. 3rd/4th gear shift fork (**A**), 1st/2nd gear shift fork and rod (**B**), and 5th/reverse gear shift fork and rod (**C**) correctly positioned on gear train.

4. Apply grease to the layshaft roller bearing and then install it into the rear case so that the side with the larger diameter cage is facing up. See Fig. 4-14.

Fig. 4-14. Layshaft roller bearing position in rear transmission housing with large side of cage facing up (arrow).

5. Using a hot air blower, heat the output shaft bearing's inner race to approximately 176°F (80°C). Place the layshaft on the gear train, then install the gear train assembly into the rear transmission case while aligning the shift rods and layshaft. Using an appropriate puller, draw the output shaft into the rear case until it is firmly seated.

NOTE —
- The BMW special tools used to pull the output shaft onto the output shaft bearing are BMW tool no. 23 2 150, 23 2 160 and BMW tool no. 23 1 300.

- It may be necessary to have a helper support the shift fork assembly and the layshaft against the gear train until the assembly is seated in the rear case.

6. Apply a small amount of grease to the two end plug detent balls. Install the two balls through the plug opening and drive in a new end plug. See Fig. 4-15.

7. Position the shift rods in their neutral positions. Apply a small amount of grease to the 3rd/4th gear shift rod pin and install it into the hole in the end of the rod. Install the rod through the shift fork so that the cutouts in the rod are facing up. See Fig. 4-16. Push the shift rod in until it is fully seated.

8. While supporting the end of the 3rd/4th gear shift rod, drive in a new 3rd/4th gear shift-fork rolled pin.

Fig. 4-15. Shift rod detent balls being installed into transmission rear case. Arrows indicate position of balls.

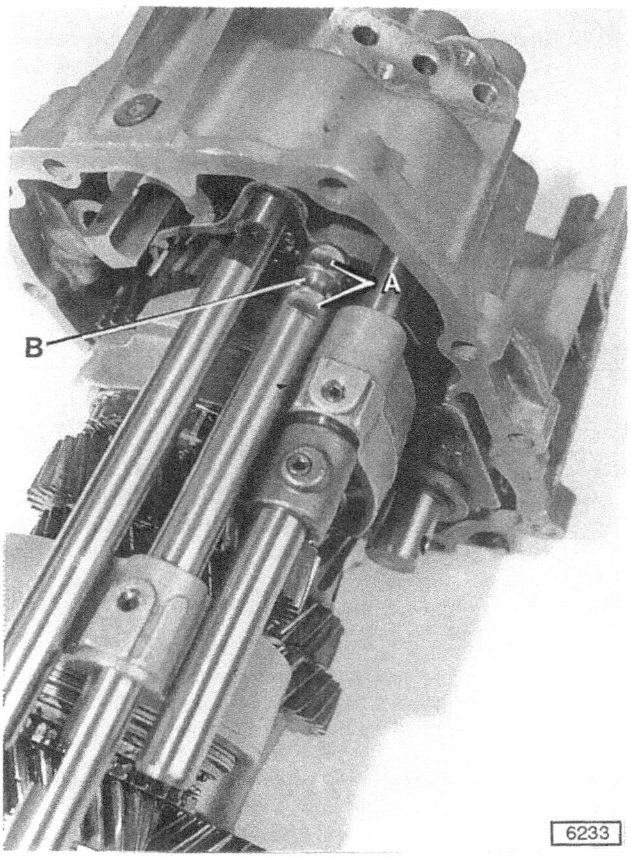

Fig. 4-16. 3rd/4th gear shift rod cutouts (**A**) and pin (**B**) correctly positioned.

9. Install the three detent balls and the springs into the shift rod bores shown in Fig. 4-17. Apply gasket sealer (Loctite® 573 or equivalent) to the sealing surface of the detent ball end plate and install the plate.

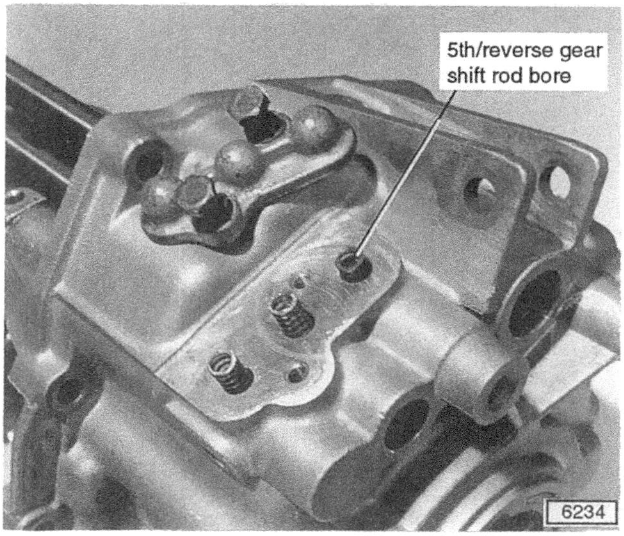

Fig. 4-17. Shift rod detent balls and springs installed into transmission case. Detent balls are under springs.

Tightening torque
• detent ball end plate . . . . 10 Nm (89 in-lb)

13

CAUTION —
Some transmissions may be equipped with one short spring and two longer, equal length springs. On these transmissions, install the short spring into the bore for the 5th/reverse gear shift rod (bore closest to the selector shaft). The other two springs should be installed in the remaining bores.

Spring length
• short spring, 5th/reverse detent bore . . 15.9 mm (0.626 in.)

10. Insert the rolled pin into the selector shaft operating lever. Install the small roller bearings on the selector shaft using a small amount of grease to hold them in place. Slide the selector shaft through the bore in the rear case while aligning the roller bearings in the case and operating lever. See Fig. 4-18.

11. While supporting the end of the selector shaft, drive in the operating lever rolled pin. Lubricate the new selector shaft oil seal sealing lip with oil and install it onto the shaft, driving it into position until it is firmly seated.

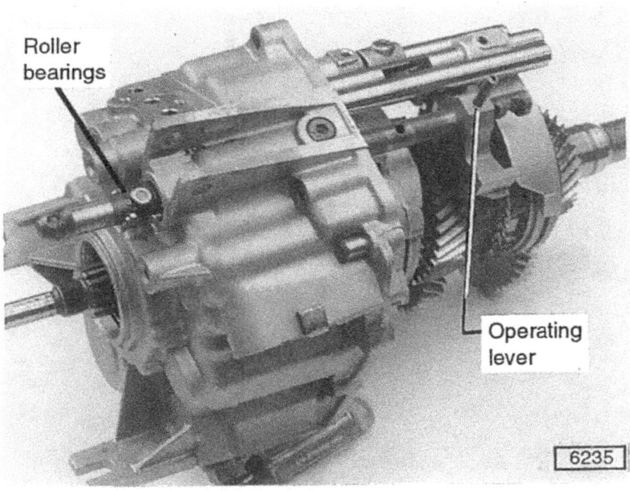

Fig. 4-18. Selector shaft with roller bearings being installed.

12. Install the shift operating shaft and shift operating lever. Position the shift operating lever so that groove in the top of the lever is facing up and towards the shift operating shaft, then install the lockpin. See Fig. 4-19.

Fig. 4-19. Shift operating shaft (**A**) and shift operating lever (**B**) correctly installed in transmission. Groove in lever should be up and facing left (arrow).

13. Thoroughly clean the reverse gear shaft sealing surface in the rear transmission case, then apply a gasket sealer (Loctite® 573 or equivalent) to the area. See Fig. 4-20.

Fig. 4-20. Area of rear transmission case (arrow) requiring gasket sealer.

14. Install the reverse gear in the case and then install the shaft with the needle bearing. Using a bolt locking compound, install and torque the shaft retaining bolt.

Tightening torque
• reverse gear shaft retaining bolt. . . . . 25 Nm (18 ft-lb)

15. Install the reverse gear shaft retaining bracket and bolt. While applying outward pressure on the bracket, torque the bolt. See Fig. 4-21.

Tightening torque
• reverse gear retaining bracket bolt. . . . . . . 25 Nm (18 ft-lb)

16. Install the output flange onto the output shaft. Apply a sealer (Loctite® 270 or Hylogrip) to the bearing surface of the nut and torque the nut.

> **NOTE —**
> Tighten the nut in two stages. First, tighten the nut to fully seat the flange. Then loosen the nut and retighten it. Install a new lockplate over the nut and bend the tab into the groove.

Tightening torque
• output flange nut, 1st stage tightening. . 170 Nm (125 ft-lb)
• output flange nut, 2nd stage tightening (after loosening) . . . . 120 Nm (89 ft-lb)

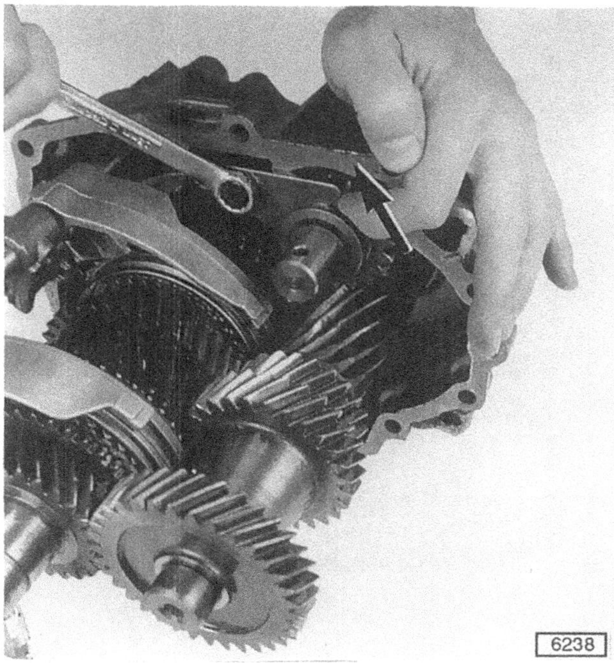

Fig. 4-21. Reverse gear shaft retaining bracket correctly positioned. Apply outward pressure (arrow) and then tighten bolt.

17. Install the layshaft roller bearing onto the layshaft so that the larger diameter part of the bearing's cage is facing the layshaft. Apply grease to the bearing's rollers and press them tight to the layshaft.

18. Thoroughly clean the front case, rear case, and reverse gear sealing surfaces. Apply gasket sealer (Loctite® 573 or equivalent) to the rear case and reverse gear sealing surface areas. See Fig. 4-22.

Fig. 4-22. Area of front transmission case (arrows) requiring gasket sealer.

NOTE —
- On 260/5 transmissions (528e, 533i models), the input shaft roller bearing should be removed from the front case when installing the case as described in step 19 below.

- On 260/6 transmissions (535i models), the input shaft roller bearing should be installed in the case and the bearing inner race should be heated prior to installing the case as described in step 19.

Installation temperature
• input shaft bearing inner race. . . . . 176°F (80°C)

19. Install the front transmission case into position. Using a screwdriver through the oil filler plug hole, align the layshaft as the case is lowered into position. See Fig. 4-23. Install the guide pin and install and torque the case mounting bolts.

NOTE —
On 260/6 transmissions, pull up on the input shaft as the case is lowered into position so the bearing is firmly seated.

Tightening torque
• transmission case mounting bolts . . . . 25 Nm (18 ft-lb)

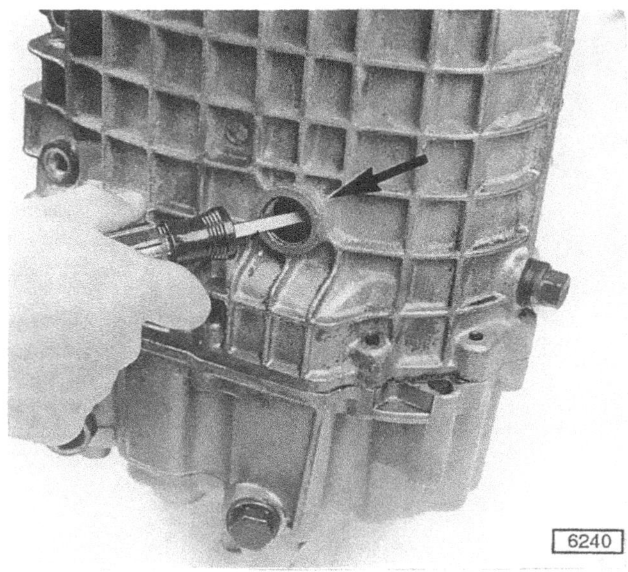

Fig. 4-23. Screwdriver being used to guide layshaft into bearing shell (arrow) as front case is installed.

20. Using a bolt locking compound, install the reverse gear shaft mounting bolt into the side of the transmission and torque the bolt.

Tightening torque
• reverse gear shaft mounting bolt. . . . . 25 Nm (18 ft-lb)

21. Insert the selector shaft lockpin and spring into the transmission side and install the plug. Install and torque the backup light switch. See Fig. 4-3 above.

Tightening torque
• back-up light switch. . . . . 20 Nm max. (15 ft-lb max.)

22. On 260/5 transmissions (528e, 533i models), heat the input shaft bearing area of the front case and the bearing's inner race. Position the bearing into the case so that the protruding side of the bearing's inner race faces the gear train. See Fig. 4-24. Carefully drive the bearing into the case and until it is firmly seated.

Installation temperature
• bearing inner race and case bearing area . . . 176°F (80°C)

Fig. 4-24. Input shaft bearing showing protruding part of inner race (arrow). Protruding part faces gear train.

23. Install the spacer and the new circlip onto the input shaft. See Fig. 4-25. While pulling on the input shaft, check the shaft's axial play between the spacer and the circlip using a feeler gauge. Selective size circlips are listed in **Table b**.

Axial play
• input shaft to circlip . . . . .0 to 0.09 mm (0 to 0.0035 in.)

NOTE —
Circlips can be ordered from an authorized BMW dealer parts department.

Fig. 4-25. Input shaft circlip (**A**) and spacer (**B**).

Table b. Selective Circlips for Input Shaft Bearing and Spacers for Clutch Guide Sleeve (Getrag 260)

Location	Dimension and thickness in mm
Input shaft bearing	35x2.0 (circlip) 35x2.1 (circlip) 35x2.2 (circlip) 35x2.3 (circlip)
Clutch guide sleeve	80x66x0.3 (spacer) 80x66x0.4 (spacer) 80x66x0.5 (spacer)

24. Using a vernier caliper, measure and record the input shaft bearing depth in the case. See Fig. 4-26. Then measure and record the height of the protrusion on the back of the clutch guide sleeve. See Fig. 4-27. Subtract the protrusion height from the depth, then select the correct spacer(s) from **Table b**.

NOTE —
Spacers can be ordered from an authorized BMW dealer parts department.

Clearance
• clutch guide sleeve to input shaft bearing . . . . 0 to 0.09 mm (0 to 0.0035 in.)

Fig. 4-26. Input shaft bearing depth in case (dimension **A**).

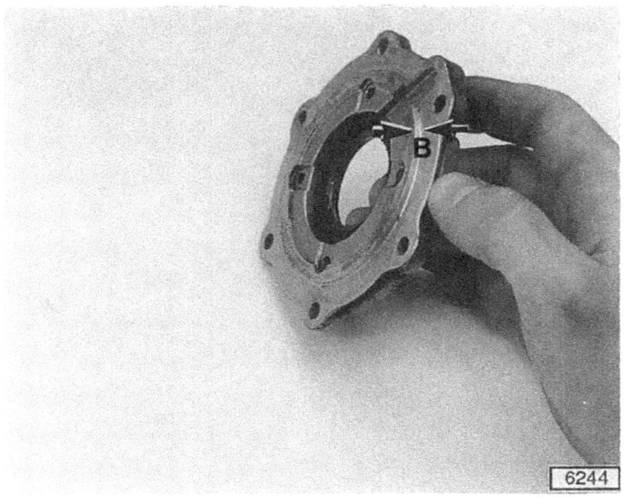

Fig. 4-27. Clutch guide sleeve protrusion height (dimension **B**).

25. Lubricate the input shaft oil seal lip with oil. Apply a gasket sealer (Loctite® 573 or equivalent) to the sealing surface of the guide sleeve and the bolts. Install the sleeve with the correct spacer(s). See Fig. 4-28. Tighten the mounting bolts. Install the clutch release lever, spring and bearing. Tighten the drain plug, fill the transmission with oil and tighten the fill plug. **Table c** gives final assembly and installation tightening torques.

NOTE —
When installing the guide sleeve, align the oil bore in the case with the oil groove in the sleeve. Some models may have an arrow cast into the sleeve. If applicable, align the arrow so that it is pointing down towards middle of layshaft bearing cover.

Fig. 4-28. Clutch guide seal oil groove and transmission case oil bore (arrows).

Table c. Gear Train Installation Tightening Torques (Getrag 260)

Back-up light switch to transmission case	20 Nm max. (15 ft-lb max.)
Clutch guide sleeve to front transmission case	10 Nm (89 in-lb)
Detent ball and spring locking plate to transmission case (bolt)	10 Nm (89 in-lb)
Drive Flange to output shaft (nut)	
initial torque	170 Nm (125 ft-lb)
final torque (after loosening nut)	120 Nm (89 ft-lb)
Front transmission case to rear case (bolt)	25 Nm (18 ft-lb)
Reverse gear shaft to transmission case (bolt)	25 Nm (18 ft-lb)
Reverse gear shaft retaining bracket to transmission case (bolt)	25 Nm (18 ft-lb)

4.2 Disassembling and Assembling Gear Train (Getrag 260)

Disassembly of the gear train requires a hydraulic press or other suitable pullers. Special press bed adaptors are available from an authorized BMW dealer to adapt the press bed to the Getrag gear train. Each of the three synchronizer guide sleeves are tightly fitted to the output shaft and need to be pressed off.

Assembly of the gear train requires new circlips and possibly new selective spacers. The circlips and the spacers correctly set axial play of the three synchronizer guide sleeves. Inspect the synchronizer rings before removing the gear clusters as described under **6. Inspecting Transmission Components.** Label all parts as they are removed from the shafts. Many parts are similar in appearance but can only be reinstalled in one position. Fig. 4-29 is an assembly view of the gear train components.

To disassemble gear train:

1. Remove the gear train from the transmission case as described in **4.1 Removing and Installing Gear Train**.

2. Pull the input shaft from the front of the output shaft. Remove the 4th gear brass synchronizer ring and the roller bearing. See Fig. 4-30.

NOTE —
If the synchronizer rings are going to be reinstalled, be sure to mark their gear positions as they are removed. Used synchronizer rings must be reinstalled in their exact position.

3. On 260/6 transmissions (535i models) remove the output shaft inner race from the output shaft using a bearing splitter. Note the small locating ball below the inner race.

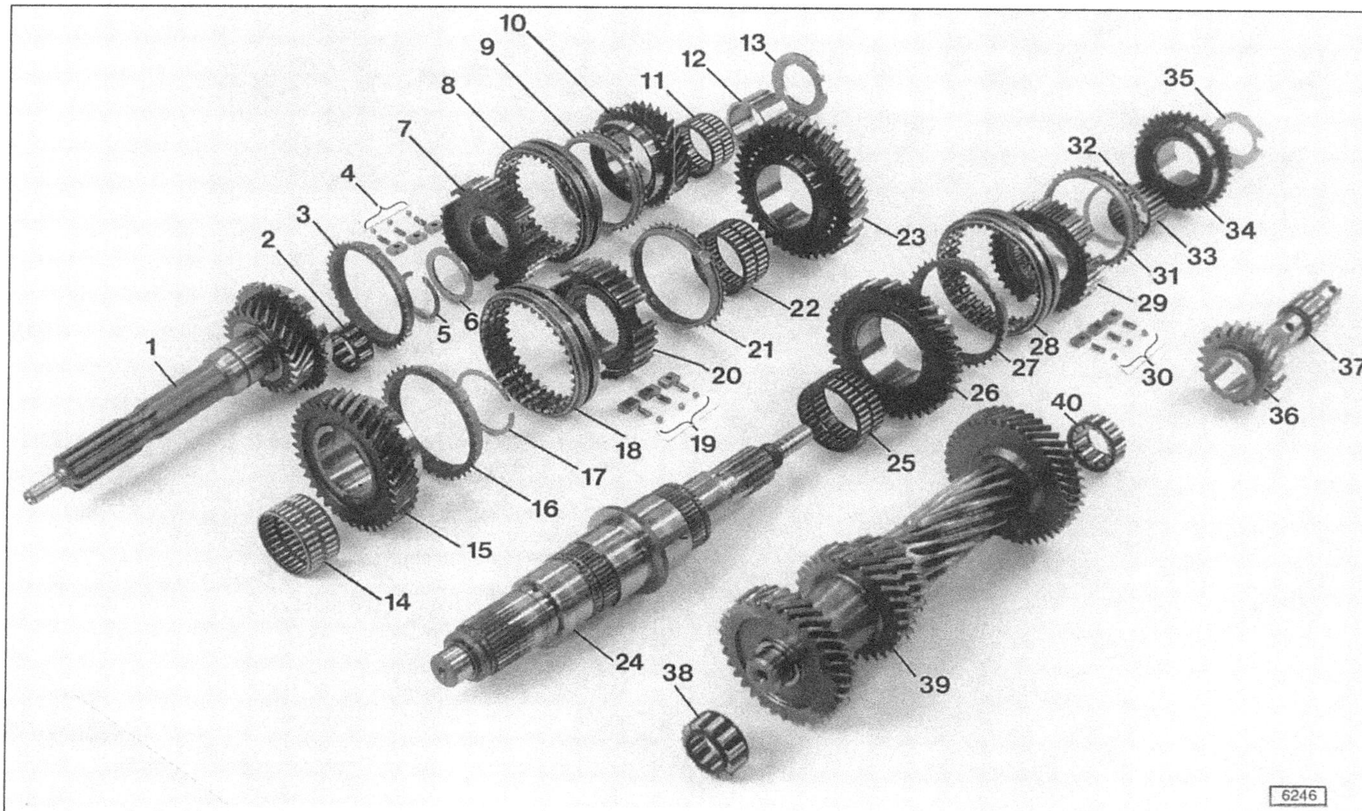

1. Input shaft with integral 4th gear
2. Roller bearing
3. 4th gear synchronizer ring
4. Pressure pieces (dogs), springs, and balls
5. Circlip
6. Spacer
7. 3rd/4th gear guide sleeve
8. 3rd/4th gear operating sleeve
9. 3rd gear synchronizer ring
10. 3rd gear
11. 3rd gear needle bearing
12. Spacer sleeve*
13. Thrust washer and ball*
14. 2nd gear needle bearing
15. 2nd gear
16. 2nd gear synchronizer ring
17. Circlip
18. 1st/2nd gear operating sleeve
19. Pressure pieces (dogs), springs, and balls
20. 1st/2nd gear guide sleeve
21. 1st gear synchronizer ring
22. 1st gear needle bearing
23. 1st gear
24. Output shaft
25. Reverse gear needle bearing
26. Reverse gear
27. Reverse gear synchronizer ring
28. 5th/reverse gear operating sleeve
29. 5th/reverse gear guide sleeve
30. Pressure pieces (dogs), springs, and balls
31. 5th gear synchronizer ring
32. Circlip
33. 5th gear needle bearing
34. 5th gear
35. Thrust washer**
36. Reverse idler gear
37. Reverse idler gear shaft
38. Layshaft roller bearing
39. Layshaft
40. Layshaft roller bearing

* Some transmissions may be equipped with a one-piece spacer sleeve and thrust washer.
**535i models with 260/6 transmissions have thrust washer and ball.

Fig. 4-29. Getrag 260 gear train components.

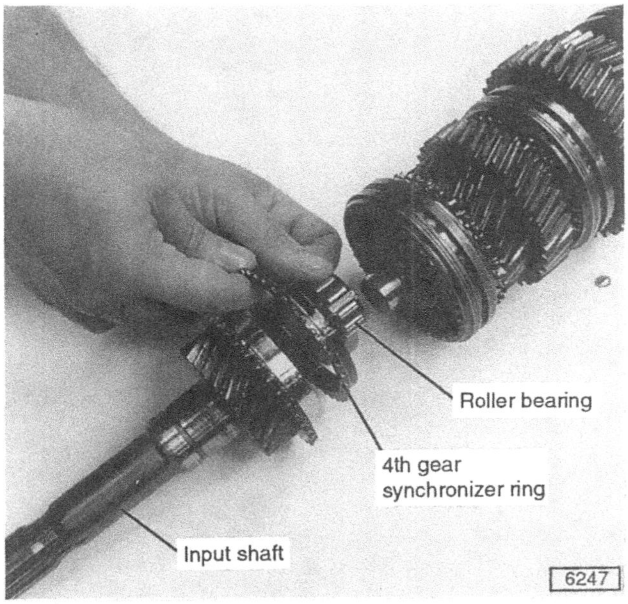

Fig. 4-30. Input shaft with 4th gear, 4th gear synchronizer ring, and roller bearing being removed.

4. Working from the rear of the output shaft, pull off 5th gear, the 5th gear synchronizer ring and its thrust washer with ball (where applicable). Remove the 5th gear needle bearing. See Fig. 4-31.

Fig. 4-31. 5th gear, 5th gear synchronizer ring, thrust washer with ball (where applicable) (arrow) and needle bearing being removed from rear of output shaft.

5. Working from the front of the shaft, remove the circlip and the spacer from in front of the 3rd/4th gear synchronizer guide sleeve.

CAUTION —
Do not reuse circlips.

6. Press off in a single operation the 3rd/4th gear guide sleeve with operating sleeve, the 3rd gear synchronizer ring, 3rd gear, the bearing sleeve and needle bearing, and 2nd gear. See Fig. 4-32. If applicable, remove the thrust washer and small ball from in front of 2nd gear. Pull the needle bearing from the shaft.

NOTE —

- Use an appropriate press bed adaptor (BMW tool no. 23 2 080) or gear puller on the back side of 2nd gear when pressing the gear cluster off the shaft. Pressing-off force should not exceed 3.7 tons.

- Two types of bearing sleeves are installed between 2nd and 3rd gear. On early transmissions, a one piece spacer/thrust washer is installed. On later transmissions, the thrust washer and sleeve are separate parts. On two piece models, the thrust washer is located on the shaft with a small ball. See Fig. 4-33.

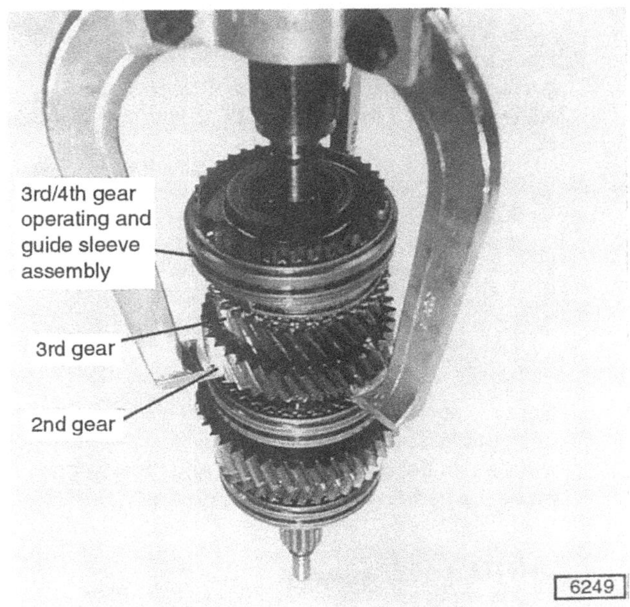

Fig. 4-32. 3rd gear cluster and 2nd gear being removed from output shaft.

13

Fig. 4-33. Thrust washer and ball between 2nd and 3rd gear. Ball in shaft (arrow) prevents washer from rotating.

7. Remove the circlip from the front of the 1st/2nd gear guide sleeve. Press off the 1st/2nd gear guide sleeve with operating sleeve and 1st gear. Remove the needle bearing.

NOTE —
Use an appropriate press bed adaptor (BMW tool no. 23 1 490) or gear puller on the back side of 1st gear when pressing the gear cluster off the shaft. Pressing off force should not exceed 3.7 tons.

8. Remove the circlip from the front of the 5th/reverse gear guide sleeve. Press off 5th/reverse gear guide sleeve with the operating sleeve and reverse gear. Remove the needle bearing. See Fig. 4-34.

NOTE —
Use an appropriate press bed adaptor (BMW tool no. 23 2 080) or gear puller on the back side of reverse gear when pressing the guide sleeve off the shaft. Pressing off force should not exceed 3.7 tons on the Getrag 260 transmission.

To assemble gear train:

1. Assemble the synchronizer guide sleeves to the operating sleeves. Position the dogs so that the beveled edge or stepped side of dog is facing the operating sleeve. See Fig. 4-35.

Fig. 4-34. 5th/reverse gear guide sleeve assembly being removed from output shaft.

Fig. 4-35. Guide sleeve operating dog correctly installed with beveled edge (arrow) facing operating sleeve. On early transmissions, dog with stepped side faces operating sleeve.

NOTE —
Two types of guide sleeve dogs are installed in the Getrag transmissions covered by this manual. On early transmissions, a stepped dog is installed. On later transmissions, a dog with beveled edges is installed. On early transmissions with the stepped dog, make sure that the balls align with the hole in the operating sleeve.

NOTE —
On 260/5 transmissions built from May 1985 and later, assemble the 3rd/4th gear operating sleeve to the guide sleeve so that the two grooves in the operating sleeve are on the opposite side from the longer collar on the guide sleeve hub. See Fig. 4-36.

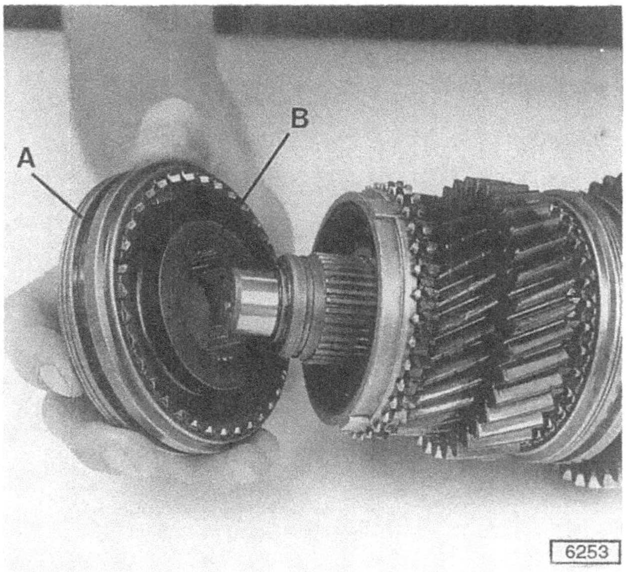

Fig. 4-36. Operating sleeve grooves (**A**) and longer collar on guide sleeve hub (**B**).

2. Install the reverse gear needle bearing, reverse gear, the reverse gear synchronizer ring, and the operating and guide sleeve assembly onto the output shaft. Align the tabs (blocks) on the synchronizer ring with the openings in the guide sleeve before pressing the sleeve onto the shaft. See Fig. 4-37.

CAUTION —
Pressing on force should not exceed 2.5 tons.

NOTE —
• On 260/5 transmissions built from May 1985 and later, install the 5th/reverse gear operating sleeve so that the side of the sleeve with the groove is facing 5th gear (stepped side of the sleeve faces reverse gear).

• On 260/6 transmissions, install the 5th/reverse gear operating sleeve so that the stepped side of the sleeve is facing reverse gear.

• There are two types of synchronizer rings in the Getrag 260 transmissions. One type of ring is for 1st and reverse gears and another type is for 2nd through 5th gears. The two types of rings should not be interchanged. See Fig. 4-38.

Fig. 4-37. Reverse gear assembly being installed on output shaft.

Fig. 4-38. Two types of synchronizer rings used on 260/5 transmissions built since May 1985 and all 260/6 transmissions.

3. Check the 5th/reverse gear guide sleeve axial play by inserting the old circlip and a feeler gauge into the groove in the shaft. See Fig. 4-39. Select a new circlip based on the information listed in **Table d** below so that there is no play.

CAUTION —
Do not reuse circlips.

NOTE —
- To determine the thickness of the circlip to install, add the thickness of the old circlip and the thickness of the feeler gauge that just fits in the groove. Select the circlip that is closest to this value but not over.

- Circlips can be ordered from an authorized BMW dealer parts department.

Fig. 4-39. 5th/reverse gear guide sleeve axial play being measured.

Table d. 5th/Reverse Gear and 1st/2nd Gear Guide Sleeve Selective Circlips (Getrag 260)

Dimension and thickness in mm
52x44x1.80
52x44x1.85
52x44x1.90
52x44x1.95
52x44x2.00

4. Install the 1st gear needle bearing, 1st gear, the 1st gear synchronizer ring, and the operating and guide sleeve assembly onto the output shaft. Align the tabs on the synchronizer ring with the openings in the guide sleeve before pressing the sleeve onto the shaft.

NOTE —
- On 260/5 transmissions built from May 1985 and later, install the 1st/2nd gear operating sleeve so that the side of the sleeve with the grooves is facing 2nd gear (stepped side of the sleeve faces 1st gear). See Fig. 4-40.

- On 260/6 transmissions, install the 1st/2nd gear operating sleeve so that the stepped side of the sleeve is facing 1st gear.

Fig. 4-40. 1st/2nd gear guide sleeve assembly correctly installed on late model (260/5 from May 1985 and later) transmissions. Grooves in operating sleeve should face 2nd gear. Stepped side of operating sleeve (if applicable) should face 1st gear.

5. Check the 1st/2nd gear guide sleeve axial play. See Fig. 4-39 above. Select a new circlip based on the information listed in **Table d** above.

CAUTION —
Do not reuse circlips.

6. Install the 2nd gear needle bearing, the 2nd gear synchronizer ring, and 2nd gear.

CAUTION —
Check that the bearing surface on the output shaft is slightly above the 2nd gear shoulder on the guide sleeve. See Fig. 4-41. If there is no protrusion, check that the circlip for the 1st/2nd gear guide sleeve below 2nd gear is correctly installed.

Fig. 4-41. Output shaft bearing surface protruding above second gear shoulder (arrow).

7. On models with a two-piece bearing sleeve, install the thrust washer and ball to the shaft while aligning the notch in the washer with the ball in the shaft. See Fig. 4-33 above.

8. Heat the bearing sleeve for 3rd gear to approximately 176°F (80°C) using a hot air blower and drive it onto the shaft.

9. Install the 3rd gear needle bearing, 3rd gear, and the 3rd gear synchronizer ring.

10. Install the 3rd/4th gear operating and guide sleeve assembly so that the longer collar on the guide sleeve hub is facing 3rd gear. See Fig. 4-42. Align the tabs on the synchronizer ring with the openings in the guide sleeve before pressing the sleeve onto the shaft.

NOTE —
- Pressing on force should not exceed 2.1 tons on the Getrag 260 transmission.

- On 260/5 transmissions built from May 1985 and later, install the 3rd/4th gear operating sleeve so that the side of the sleeve with the grooves is facing 4th gear.

11. Check the 3rd/4th gear guide sleeve axial play. See Fig. 4-39 above. Install a new circlip on the shaft using the information given in **Table e** below.

CAUTION —
Do not reuse circlips.

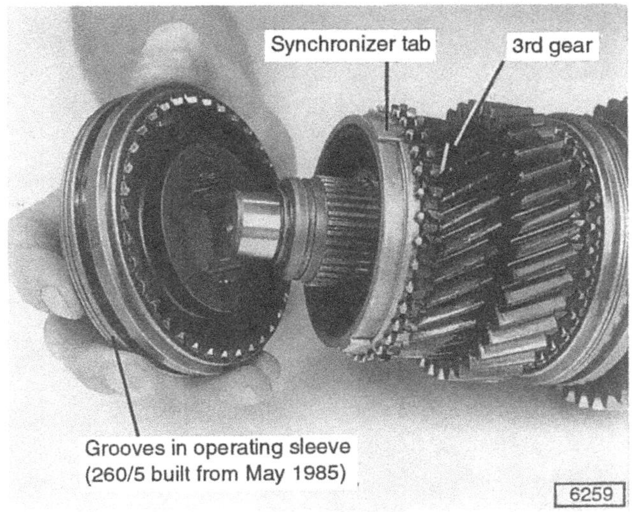

Fig. 4-42. 3rd/4th gear operating sleeve assembly being installed on output shaft with long collar (arrow) facing 3rd gear.

Table e. 3rd/4th Gear Guide Sleeve Selective Circlips (Getrag 260)

Dimension and thickness in mm
52x44x1.80
52x44x1.85
52x44x1.90
52x44x1.95
52x44x2.00

13

12. Install the input shaft needle bearing onto the output shaft and install the output shaft.

4.3 Transmission Oil Seals (Getrag 260)

Replacing the transmission oil seals can be done without any major disassembly of the transmission, and two of the three seals can be replaced with the unit installed in the car.

Low gear oil level due to a faulty oil seal may cause problems such as hard shifting, jumping out of gear, and transmission noise. Oil at the bottom of the bellhousing may also be due to a leaking rear crankshaft seal.

Before assuming that the seals are at fault, check the transmission vent on the top of the case. A clogged or damaged vent can cause the pressure inside the case to increase to a point where the lubricant is forced past the seals. The vent is shown in Fig. 4-43.

Fig. 4-43. Manual transmission vent (arrow). Check vent for free movement up and down.

To replace output shaft oil seal (transmission installed):

1. Remove the exhaust system as described in **EXHAUST SYSTEM.**

2. Disconnect the driveshaft and, where applicable, the vibration damper from the transmission. See **DRIVESHAFT AND FINAL DRIVE.**

3. Bend back the lockplate tabs shown in Fig. 4-44, and remove the lockplate. Hold the flange steady and remove the collar nut using a deep 30 mm thin-walled socket.

   NOTE —
   A thin-wall socket can be made by grinding the outer surface of an ordinary socket.

4. Remove the output flange from the output shaft. If necessary use a puller to remove the flange.

5. Using a seal remover or a small screwdriver, pry out the old seal. Lubricate the new seal with oil and carefully drive it into place.

   CAUTION —
   Be careful not to damage the transmission case housing when prying out the old seal. A damaged case can result in oil leaks.

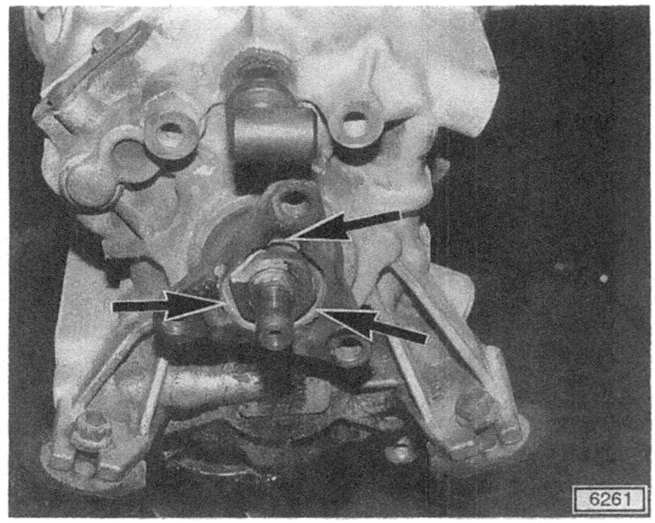

Fig. 4-44. Lockplate tabs (arrows). Bend tabs out of groove to remove lockplate and collar nut.

6. Install the output flange. Coat the bearing surface of the collar nut with sealer and install the nut, torquing it to the specifications listed below.

   NOTE —
   • BMW recommends the use of a sealer such as Loctite® 270, Hylogrip, or Curil K2 when installing the collar nut to prevent transmission oil leaks. Such a sealant is available from your BMW dealer.

   • The drive flange nut must be tightened in two stages. The first stage firmly seats the flange and any bearings. The final is done after first loosening the nut.

Tightening torque	
• output flange nut, 1st stage tightening	170 Nm (125 ft-lb)
• output flange nut, 2nd stage tightening (after loosening)	120 Nm (89 ft-lb)

7. Reinstall the lockplate and bend the tabs into the grooves as shown above in Fig. 4-44.

8. Reinstall the driveshaft as described in **DRIVESHAFT AND FINAL DRIVE.**

9. Reinstall the exhaust system as described in **EXHAUST SYSTEM.**

To replace selector shaft oil seal (transmission installed):

1. Remove the exhaust system as described in **EXHAUST SYSTEM.**

2. Remove the driveshaft or disconnect the flexible coupling as described in **DRIVESHAFT AND FINAL DRIVE.**

3. Bend back the lockplate tabs shown above in Fig. 4-44, and remove the lockplate. Hold the flange steady and remove the collar nut.

4. Remove the output flange from the output shaft. If necessary use a puller to remove the flange.

5. Disconnect the selector rod from the selector shaft by prying back the spring band and driving in the lockpin.

NOTE —
Engaging 3rd gear will help to gain access to lockpin.

6. Using a pick or small screwdriver, pry out the seal as shown in Fig. 4-45. Lubricate the new seal with oil and drive it into place.

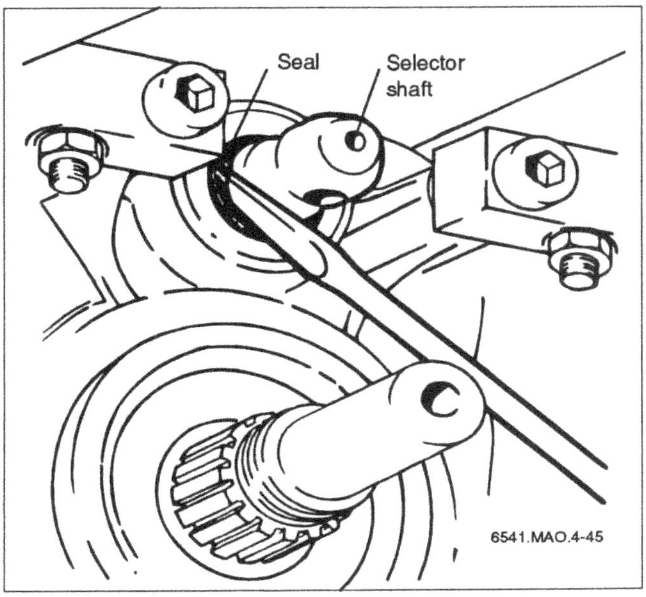

Fig. 4-45. Selector shaft oil seal being pried out with screwdriver. Take care not damage selector shaft or transmission housing.

7. Reinstall the selector rod to the selector shaft, drive in the lockpin, and install the spring band.

8. Reinstall the driveshaft flange.

CAUTION —
The tightening torque on the drive flange collar nut is critical. Tighten the collar nut as described above under the procedure for replacing the output shaft oil seal.

9. Reinstall the driveshaft as described in **DRIVESHAFT AND FINAL DRIVE.**

10. Reinstall the exhaust system as described in **EXHAUST SYSTEM.**

To replace input shaft oil seal (transmission removed):

1. Remove the transmission as described in **MANUAL TRANSMISSION AND CLUTCH.**

2. Remove the clutch release bearing and release lever from within the bellhousing.

3. Remove the bolts for the clutch release guide sleeve, shown in Fig. 4-46, and remove the sleeve.

NOTE —
The guide sleeve is sealed in place. On transmissions with two-piece guide sleeves, remove only the outer bolts when removing the sleeve.

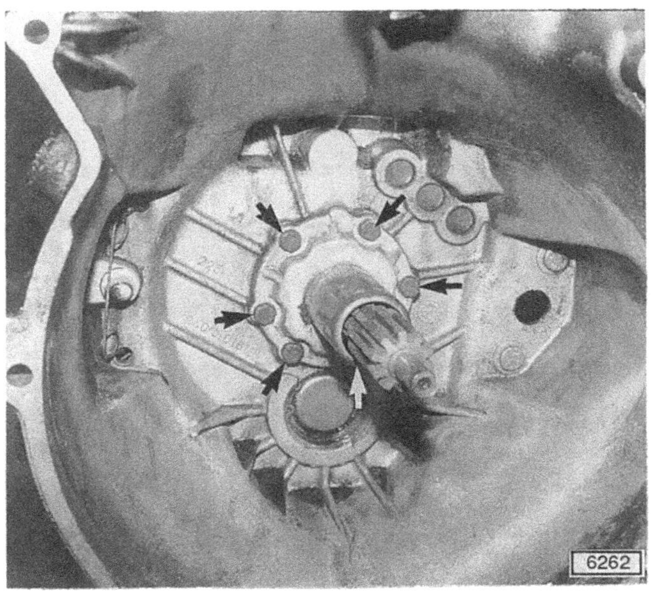

Fig. 4-46. Guide sleeve bolts (arrows) to be removed. Getrag 260 transmission is similar.

4. Pry the old seal out of the guide sleeve, lubricate the new seal with oil, and drive it into place.

CAUTION —
Be careful not to mar the housing when removing the seal.

5. Clean the mounting bolts and sealing surface of the guide sleeve. Apply sealer (Loctite® 573 or equivalent) to the guide sleeve sealing surface. Reinstall the guide sleeve and spacer(s).

> ***CAUTION —***
> *On Getrag 260 transmissions, make sure that the oil groove in the sleeve is aligned with the oil bore in the transmission case. See Fig. 4-28 shown earlier.*

> **NOTE —**
> A different spacer may be required if either the bearing or the guide sleeve are replaced. See **4.1 Removing and Installing Gear Train** for more information.

6. Coat the guide sleeve bolt shoulders with sealing compound and install the bolts.

Tightening torque
• Clutch release bearing guide sleeve bolts (M6). 10 Nm (89 in-lb)

7. Reinstall the clutch release lever, spring, and release bearing. Reinstall the transmission as described in **MANUAL TRANSMISSION AND CLUTCH.**

5. TRANSMISSION SERVICE — GETRAG 265

Read the procedures completely before starting any repair job under this heading. Special tools and measuring equipment are necessary for certain tasks. Thoroughly clean the outside of the transmission before beginning any disassembly work. This will help keep abrasive dirt from getting into the working components as they are disassembled. Label all parts as they are removed. Many parts are similar in appearance but must be reinstalled only in one position.

If the transmission is to be disassembled, drain the gear oil before removing it from the car as described in **LUBRICATION AND MAINTENANCE.** Fig. 5-1 (next page) is an exploded view of the Getrag 265 gear train, which can be used during the disassembly and assembly procedures described below.

> **NOTE —**
> Carefully observe the color of the transmission oil and metal filings. This may help pinpoint component wear.

5.1 Removing and Installing Front and Rear Cases (Getrag 265)

Special press tools are required to remove the cases. In addition, new parts (snap ring, circlip, and possibly new shims) will be required when installing the cases. The snap ring and circlip become distorted and stretched as they are removed. The shims are used to correctly set clearances between bearings and transmission shaft. The shims may be reused if the measured clearances are within the specifications described below. If the clearances are excessive, new shims will be required. New seals— such as the selector shaft seal will also need to be replaced.

To remove front and rear cases:

1. Remove the transmission as described in **MANUAL TRANSMISSION AND CLUTCH.**

> ***WARNING —***
> • *The clutch disc contains asbestos fibers. Asbestos materials can cause asbestosis. Always wear a respirator and protective clothing when handling components containing asbestos. Do not use compressed air, do not grind, heat, weld, or sand on or near any asbestos materials. Only approved respirators and cleaning equipment should be used to service the clutch disc or areas containing asbestos or asbestos fibers.*
>
> • *When pressing or pulling is required, only use a press or puller with the correct capacity. Do not use power tools, such as air impact tools, on the press or pulling equipment.*

2. Mount the transmission on a stand or place it on a work bench.

3. Remove the clutch release lever, spring, and release bearing from the input shaft.

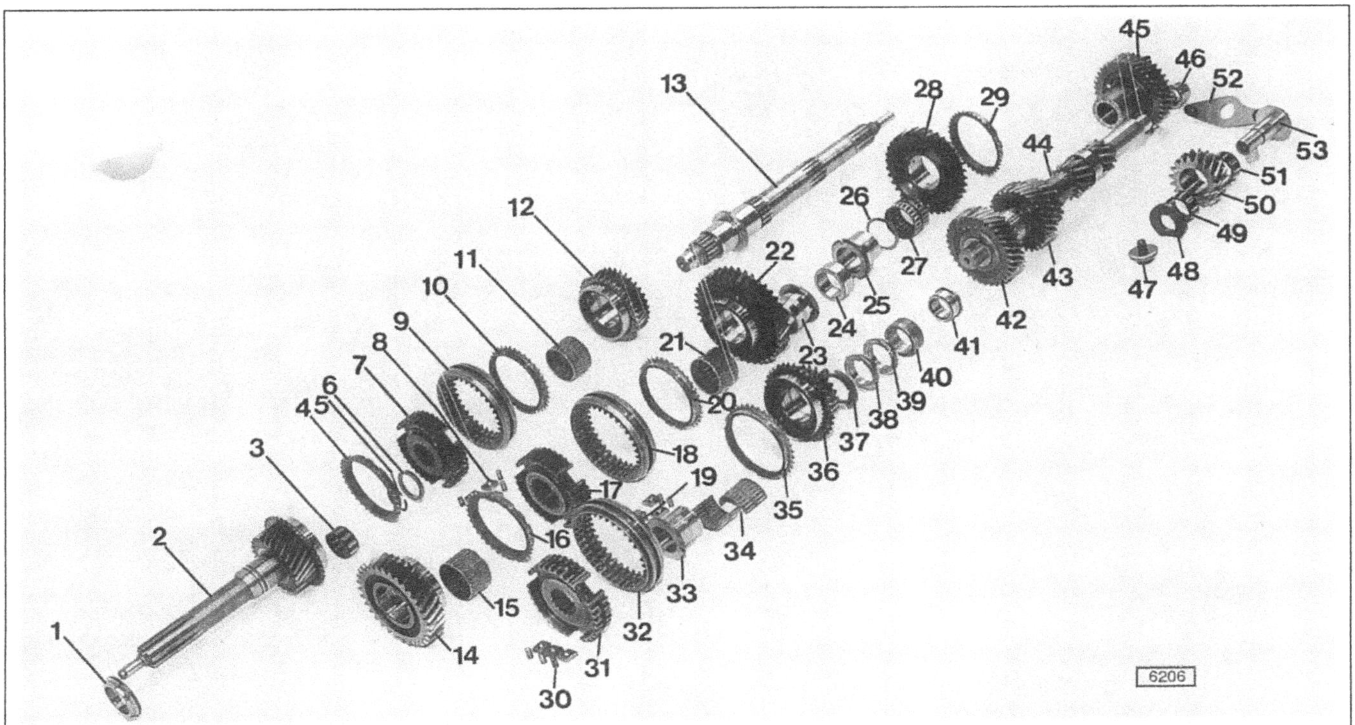

1. Input shaft bearing inner race
2. Input shaft with integral 4th gear
3. Input shaft roller bearing
4. 4th gear synchronizer ring
5. Circlip
6. Spacer
7. 3rd/4th gear guide sleeve
8. Pressure pieces (dogs), springs, and balls (6.5 mm)
9. 3rd/4th gear operating sleeve
10. 3rd gear synchronizer ring
11. 3rd gear needle bearing
12. 3rd gear
13. Output shaft
14. 2nd gear
15. 2nd gear needle bearing
16. 2nd gear synchronizer ring
17. 1st/2nd gear guide sleeve
18. 1st/2nd gear operating sleeve
19. Pressure pieces (dogs), springs, and balls (6.5 mm)
20. 1st gear synchronizer ring
21. 1st gear needle bearing
22. 1st gear
23. 1st gear inner bearing sleeve
24. Intermediate case bearing inner race
25. Reverse gear inner bearing sleeve
26. Shim
27. Reverse gear needle bearing
28. Reverse gear
29. Reverse gear synchronizer ring
30. Pressure pieces (dogs), springs, and balls (6.5 mm)
31. 5th/reverse gear guide sleeve
32. 5th/reverse gear operating sleeve
33. 5th gear inner bearing sleeve
34. 5th gear needle bearing halves
35. 5th gear synchronizer ring
36. 5th gear
37. Thrust washer with location ball (4 mm)
38. Output shaft bearing inner race (inner)
39. Output shaft bearing inner race (outer)
40. Output shaft sleeve or speedometer drive pinion
41. Layshaft inner bearing race
42. Layshaft 4th gear (can be pressed off layshaft)
43. Layshaft 3rd gear (can be pressed off, retained by snapring)
44. Layshaft
45. Layshaft 5th gear
46. Layshaft inner bearing race
47. Reverse gear idler shaft mounting bolt
48. Reverse gear idler shaft cutout washer
49. Reverse gear idler shaft roller bearing
50. Reverse idler gear
51. Reverse gear idler shaft roller bearing
52. Reverse gear idler shaft retainer plate
53. Reverse gear idler shaft

Fig. 5-1. Getrag 265 gear train components.

13

4. Remove the bellhousing after removing the nuts shown in Fig. 5-2.

5. Remove the clutch release bearing guide sleeve mounting bolts and pull the sleeve from the transmission. See Fig. 5-3.

6. Remove and label the spacer(s) from behind the guide sleeve. Remove the input shaft snap ring and the upper spacer. See Fig. 5-4. Remove and label the input shaft circlip and the lower spacer.

7. Working at the rear of the transmission, engage either 1st, 3rd, or 5th gear. Then pull back the spring loaded sleeve and drive out the retaining pin while supporting the selector shaft from below. Remove selector rod joint.

8. Unscrew the back-up light switch. Pry off the selector shaft lockpin plug and remove the spring and the lockpin from the transmission. See Fig. 5-5.

Fig. 5-2. Bellhousing being unbolted from front case. Remove nuts (arrows) securing bellhousing to transmission from both side of case.

Fig. 5-3. Clutch release bearing guide sleeve bolts to be removed.

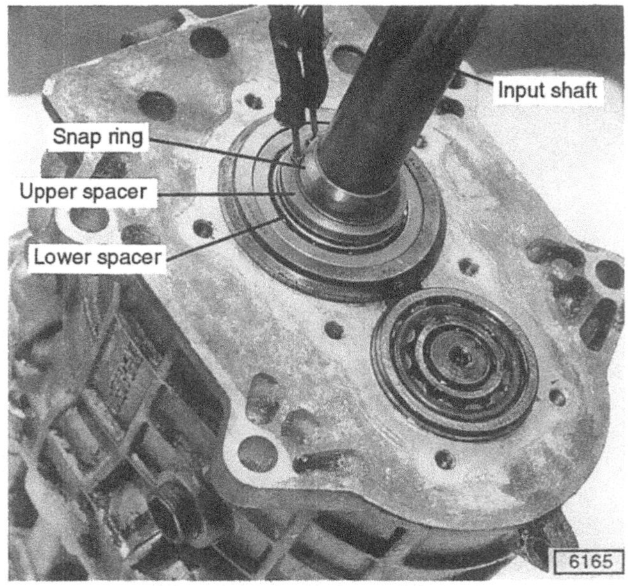

Fig. 5-4. Input shaft snap ring being removed. Circlip (not shown) below top spacer. Lower spacer should also be removed.

Fig. 5-5. Selector shaft lockpin and spring (arrow) being removed.

9. Working at the middle of the transmission, drive out the top guide pin for the front-to-intermediate case sections. See Fig. 5-6. Remove the front-to-intermediate case mounting bolts.

Fig. 5-6. Transmission case guide pin being driven out.

Fig. 5-7. Front case being pressed off of input shaft using pressing tool.

10. Separate the front transmission case from the intermediate case by pressing the input shaft out of the front case.

> ***CAUTION —***
> *Do not attempt to separate the case by driving on the front of the input shaft or the front case. The front case is not designed to take this type of a load. The case may crack and break.*

> **NOTE —**
> BMW special tools (BMW tool no. 23 1 170 and BMW tool no. 33 1 301) or an equivalent should be used to separate the case. The special tool is bolted to the guide sleeve surface of the front case. The sleeve part of the tool extends above the tip of the input shaft. A bolt is then threaded through the top of the tool to bear on the input shaft, and the input shaft is pressed out of the bearing and front case. As an alternative, a tool made of black iron pipe fixtures can be fabricated. See Fig. 5-7. The threaded pipe pieces are available from most hardware stores or plumbing supply houses.

11. Working at the output shaft, remove the output flange nut lockplate. Secure the output flange and remove the flange collar nut. Remove the output flange from the transmission. It may be necessary to use a puller to remove the flange from the output shaft.

> **NOTE —**
> • A deep thin-wall 30 mm socket is needed to remove the output shaft collar nut. A thin-wall socket can be made by grinding the outer surface of an ordinary socket. See Fig. 5-8.
>
> • The collar nut is assembled using Loctite®. Considerable effort may be required to remove it.

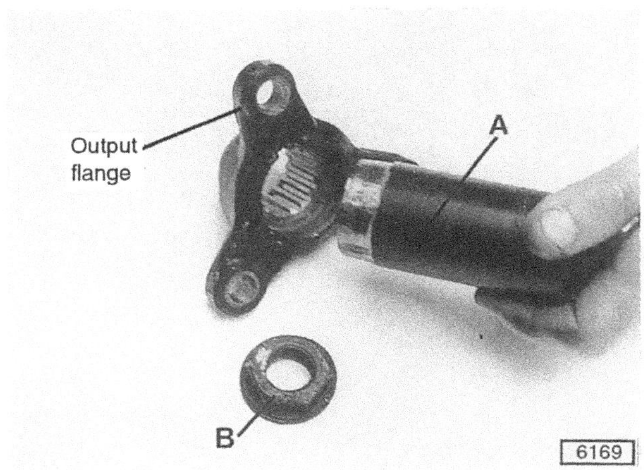

Fig. 5-8. Thin-wall deep 30 mm socket (**A**) needed to remove output flange nut (**B**).

12. Engage 2nd gear by rotating the selector shaft counterclockwise to the stop and then sliding the shaft forward.

> **NOTE —**
> It may be helpful to use a screwdriver or drift inserted in the selector shaft hole when engaging 2nd gear.

13. Remove the rear to intermediate case mounting bolts and drive out the case guide pin. See Fig. 5-9.

14. Remove the rear case by pressing out the output shaft. See Fig. 5-10.

Fig. 5-9. Rear case mounting bolts and nuts (arrows) and case guide pin **A**. Different length bolts are used.

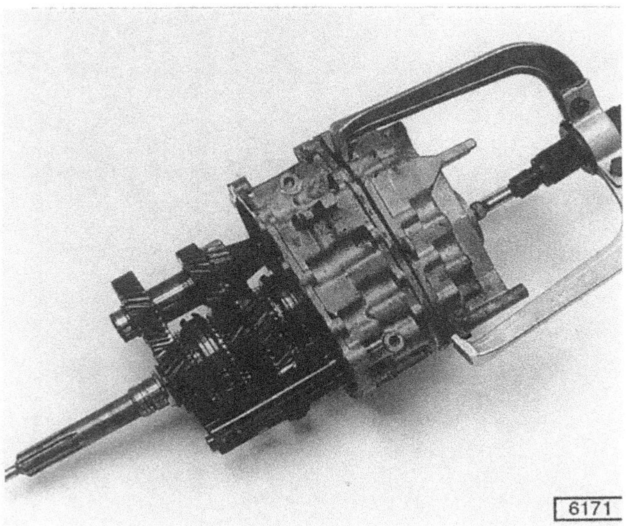

Fig. 5-10. Output shaft being pressed out of rear case.

NOTE —
- BMW special tools (BMW tool no. 23 1 350 and BMW tool no. 33 1 301) or an equivalent press should be used to press off the rear case.

- Check that the rear case separates evenly on all sides. If it doesn't, 2nd gear may not be engaged properly.

15. Remove the small roller bearings from the end of the selector shaft.

To install front and rear cases:

1. Inspect all bearings, bearing races and seals in the transmission cases. Replace any that are damaged as described below under **6. Inspecting Transmission Components.**

2. Apply Loctite® 573 or an equivalent to the mating surfaces of the intermediate case and front case. Install the front case. Drive the guide pin in and tighten the mounting bolts.

Tightening torque
• front case to intermediate case (bolts) . . . . 25 Nm (18 ft-lb)

3. Install the reverse light switch. Install the selector shaft lockpin so that the roller is horizontal with the case. See Fig. 5-11. Install the lockpin end cap.

Tightening torque
• reverse light switch. . . . . 20 Nm max. (15 ft-lb max.)

4. Heat the input shaft inner race and slide it onto the input shaft. Make sure the race is firmly seated.

Installation temperature
• input shaft bearing inner race. . . . . 176°F (80°C)

5. Install the spacer on the input shaft. Then using a feeler gauge, check the axial clearance between the circlip and its groove. See Fig. 5-12. There should be little or no clearance. If the clearance is too big, use a new circlip with a different thickness to achieve the proper clearance. Selective size circlips are listed in **Table f**.

Clearance
• input shaft to bearing (measured at circlip groove) . 0 to 0.09 mm (0 to 0.0035 in)

NOTE —
Circlips can be ordered from an authorized BMW dealer parts department.

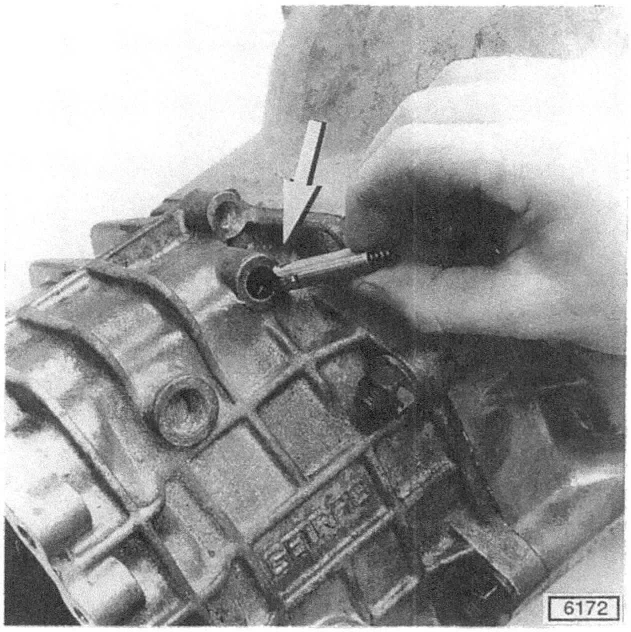

Fig. 5-11. Lockpin being installed in transmission case. Roller (arrow) on lockpin must be horizontal to case.

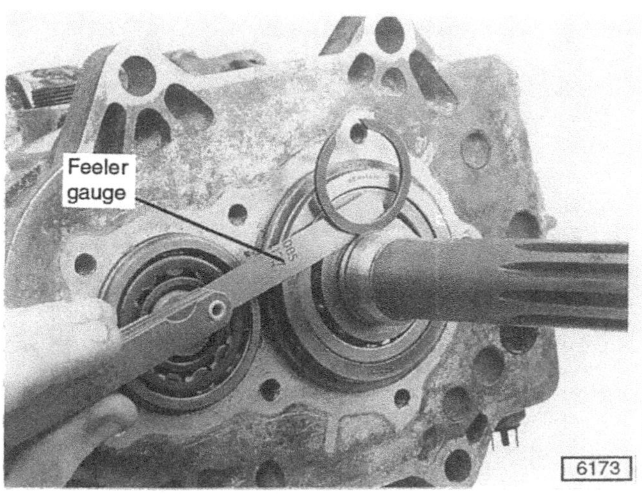

Fig. 5-12. Input shaft axial clearance being checked using old circlip.

Table f. Selective Circlips and Spacers for Input Shaft Bearing, Layshaft Bearing and Clutch Guide Sleeve (Getrag 265)

Location	Dimension and thickness in mm
Input shaft bearing to input shaft	35x2.0 (circlip) 35x2.1 (circlip) 35x2.2 (circlip) 35x2.3 (circlip)
Clutch guide sleeve to input shaft bearing	78x63x0.3 (spacer) 78x63x0.4 (spacer) 78x63x0.5 (spacer)
Layshaft bearing	50x62x0.3 (spacer)

6. Install the grooved washer and snap ring on input shaft. The grooved side of the washer goes toward the circlip.

7. Measure the depth of the flange on the clutch guide sleeve. Then measure the input shaft bearing protrusion height. See Fig. 5-13. Subtract the bearing height measurement from the guide sleeve depth. Then select the correct spacer(s) from **Table f** to bring the clearance to within specification.

Clearance
• input shaft bearing to clutch guide sleeve . . . . 0 to 0.09 mm (0 to 0.0035 in)

8. Repeat the same procedure outlined in step 7 for the layshaft bearing to guide sleeve clearance. Only one size shim is available for the layshaft to guide sleeve position. See **Table f.**

13

Clearance
• layshaft bearing to clutch guide sleeve . . . . 0 to 0.09 mm (0 to 0.0035 in)

9. Hold the shims in the clutch guide sleeve with a small amount of grease. Coat the surface of the sleeve and the mounting bolts with Loctite® 573 or equivalent and install the sleeve.

NOTE —
The clutch guide sleeve bolts are different lengths and have different torque values. The longer bolts go in the top three holes. See Fig. 5-3 above. Be sure to clean any old sealer from the bolt threads before installing them.

Fig. 5-13. Clutch guide sleeve cover flange depth (**A**) and input shaft bearing height (**B**).

Tightening torque	
• clutch release-bearing guide sleeve bolts	
(M8x22)	18 Nm (13 ft-lb)
(M8x30)	25 Nm (18 ft-lb)

10. Working at the rear of transmission, engage 2nd gear, then align the top of the 1st/2nd shift rod to same height as 5th/reverse shift rod cutout. See Fig. 5-14.

11. Place the four small roller bearings on the selector shaft using a little grease to hold them in place. See Fig. 5-15.

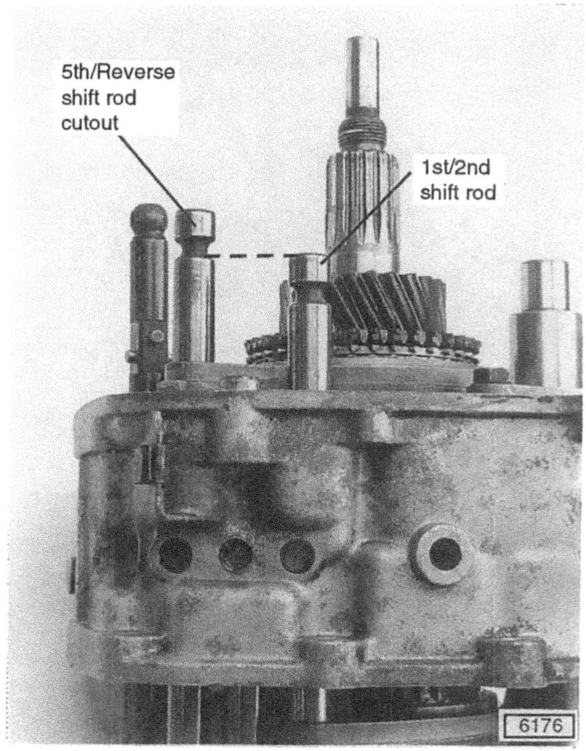

Fig. 5-14. Height of 5th/reverse shift rod cutout should be the same as end of 1st/2nd shift rod (dashed line).

Fig. 5-15. Selector shaft roller bearing being installed using grease to hold it in place.

12. Remove the oil seal from the rear of the transmission case. Pull out the output shaft bearing inner race and the spacer (or speedometer pinion) from the case. Pry out the lower end cap as shown in Fig. 5-16.

Fig. 5-16. Rear case end cap being removed. Use a thin punch to knock cap on its side and then use needle nose pliers to remove.

13. Coat the mating surface of the rear case with Loctite® 573 or an equivalent. Coat the selector shaft oil seal with oil. Slowly slide the rear case on while aligning the selector shaft and the shift rods. Use a small screwdriver to engage the lockout pin into the cutout in the 5th/reverse gear shift rod to fully seat the case. See Fig. 5-17.

NOTE —
Check that the selector shaft spring engages the lock in the case as the case is lowered into position. See Fig. 5-18.

14. Install the rear case bolts and nuts and drive in the case guide pin.

CAUTION —
The rear case uses different length bolts. See Fig. 5-9 for bolt locations.

Tightening torque
• rear case to intermediate case . . . . 25 Nm (18 ft-lb)

Fig. 5-17. Lockout pin being pressed into cutout (arrow) in 5th/reverse gear shift rod as rear case is lowered into position.

Fig. 5-18. Spring must engage lock when installing rear case.

15. Heat the output shaft inner bearing race and slide it onto the output shaft. Slide the spacer (or speedometer pinion) onto the shaft. Coat the oil seal with oil and drive it into the rear case. Where applicable, install spacer(s) on the output shaft.

Installation temperature
• output shaft inner bearing race . . . . 176°F (80°C)

16. Apply Loctite® 573 to the end cap and install the cap into the rear case.

NOTE —
End caps should be replaced whenever they are removed to prevent possible oil leaks.

17. Apply a sealer (Loctite® 270 or Hylogrip) to the output shaft threads and the bearing surface of the output shaft collar nut. Install the output flange onto the output shaft. Torque the nut and install a new lockplate.

NOTE —
Tighten the nut in two stages. First, tighten the nut to fully seat the flange, spacer and inner race. Then loosen the nut and retighten it to the specification listed below.

Tightening torque
• output flange nut, 1st stage tightening . . 170 Nm (125 ft-lb)
• output flange nut, 2nd stage tightening (after loosening) . . . . 120 Nm (89 ft-lb)

18. Tighten the drain plug, fill the transmission with oil and tighten the fill plug. Install the bellhousing and tighten the nuts.

Tightening torque
• drain and fill plugs . . . . 40 to 60 Nm (30 to 44 ft-lb)
• bellhousing nuts . . . . 72 to 80 Nm (53 to 59 ft-lb)

5.2 Removing and Installing Transmission Shafts (Getrag 265)

Removing the transmission shafts from the intermediate case requires that 5th gear be pressed off the layshaft. This pressing on and off operation requires considerable force (in excess of 9 tons). It is recommended that 5th gear be pressed off by an authorized BMW dealer or a qualified machine shop with the appropriate equipment. The remainder of the procedure can be accomplished using standard pullers.

WARNING —
When pressing or pulling is required, always use tools having a higher capacity rating than that required to remove the part. Power tools, such as air impact tools, should not be used on press equipment.

Also covered under this heading is the removal of the shift forks, shafts and shift rods. Installation of the transmission shafts into the transmission assembly requires precision measuring equipment (micrometer and vernier caliper), new rolled pins, and possibly a new shim to correctly set the axial play of the 5th/reverse gear guide sleeve (synchronizer hub).

To remove transmission shafts:

1. Remove the front and rear transmission cases from the intermediate case as described above under **5.1 Removing and Installing Front and Rear Cases (Getrag 265).**

2. Press the 5th gear off the layshaft. The bearing inner race will come off with the gear.

WARNING —
An extremely high force is required to press 5th gear off the layshaft. BMW special tool (BMW tool no. 23 0 080) and a 10-ton hydraulic bottle jack or an equivalent hydraulic press with the correct press bed adaptors must be used to remove the gear.

Pressing-off force
• 5th gear on layshaft . . . . 9 to 10 tons

3. Remove the output shaft bearing inner race from the front of 5th gear. Start the race off using pry bars, then remove the race using a suitable puller. See Fig. 5-19.

NOTE —
BMW special tools (BMW tool no. 23 1 100 and BMW tool no. 00 7 500) can be used to remove the output shaft bearing inner race.

4. Remove the spacer and 4 mm ball from in front of 5th gear.

5. Remove the 5th gear, synchro ring, and split bearings from the output shaft.

6. Drive out the rolled pin securing the selector shaft while supporting the underside of the shaft. See Fig. 5-20. Slide the shaft out toward the rear and remove the selector arm.

CAUTION —
Do not reuse rolled pins.

Fig. 5-19. Output shaft bearing inner race being removed. Using pry bars, lift race enough to get jaws of puller behind race.

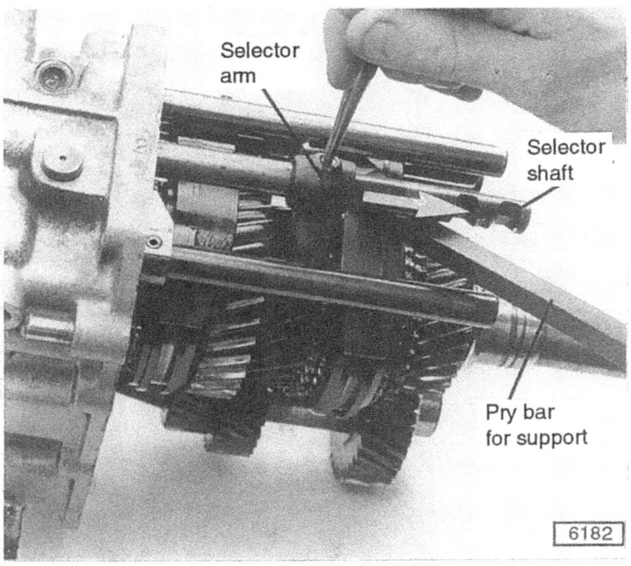

Fig. 5-20. Selector arm lever rolled pin being driven out of selector shaft. Remove lever and slide shaft out of the transmission case in direction of arrow.

NOTE —
As an aid to reassembly, temporarily reassemble the selector arm to the shaft and install the rolled pin.

7. Engage 2nd gear. Drive out the rolled pin for the shift rod turning lock while supporting the underside of the rod. Remove the turning lock and the reversing lever. Pull the shift rod from the case. See Fig. 5-21.

NOTE —
Engaging 2nd gear will help gain access to the rolled pin for the turning lock.

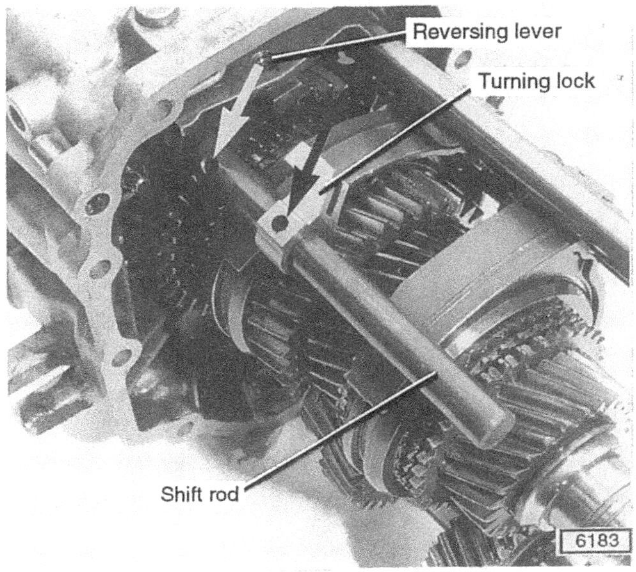

Fig. 5-21. Shift rod with turning lock. Rolled pin has been removed. Arrows show alignment holes.

8. Disengage 2nd gear.

9. Slowly remove the 5th/reverse gear shift rod with fork and operating sleeve just enough to remove the sleeve from the shift fork. Then turn the shaft while removing it from the case to prevent the shaft cutouts from engaging the detent balls in the case. See Fig. 5-22. Remove the two 8 mm detent balls from the case.

NOTE —
- When removing the 5th/reverse gear operating sleeve from the guide sleeve, be prepared to catch the three 6.5 mm balls, springs, and dogs.

- As an aid to reassembly, mark a "5" and an "R" on each side of the fork corresponding to the appropriate gears.

Fig. 5-22. 5th/reverse gear shift rod being removed. Turn rod (arrow) when removing to prevent detent balls from engaging cutouts in rod.

10. Remove the reverse gear idler plate mounting bolts. Then, working from the front side of the case, remove the idler shaft bolt while counter holding the shaft at the rear. See Fig. 5-23. Tap out the reverse idler shaft and remove the cutout washer.

11. Pull off the 5th gear bearing sleeve together with the 5th/reverse gear guide sleeve. See Fig. 5-24.

NOTE —
BMW recommends using special tools (BMW tool no. 23 1 420 and BMW tool no. 33 1 301) to remove the 5th/reverse gear guide sleeve and bearing sleeve due to the limited space behind the guide sleeve. Use care when using a standard puller, as the guide sleeve can break if the force is not equally distributed.

12. Remove the reverse gear and needle bearing from the output shaft.

13. Engage 3rd gear, then drive down the rolled pin securing 3rd/4th shift fork just until it contacts the operating sleeve below. Support the rod while driving the pin down. Slide the shift rod out through the rear and remove the shift fork. Remove the two 8 mm detent balls from the case.

CAUTION —
Do not reuse rolled pins.

Fig. 5-23. Reverse gear idler shaft bolt being removed. To prevent shaft from spinning, hold shaft with wrench.

Fig. 5-24. 5th/reverse gear guide sleeve and bearing sleeve being removed.

NOTE —

• Engaging 3rd gear allows the rolled pin to be driven down onto the gear so that the shift rod can be removed.

• As an aid to reassembly, scratch the gear numbers (3rd and 4th) on each side of the fork.

14. Remove the bearing sleeve from the front of the intermediate case on the output shaft.

NOTE —

Special tools (BMW tool no. 23 1 060 and BMW tool no. 33 1 301) are needed to remove the bearing sleeve from the output shaft. The special tools clamp onto the outside of the sleeve and then the output shaft is pressed out while pulling on the bearing sleeve. As an alternative, the output shaft can be carefully pressed out of the intermediate case using a standard puller. See Fig. 5-25.

Fig. 5-25. Output shaft with bearing sleeve being pressed out of case. Use wood between the jaws of the puller and the case sealing surface.

15. Remove and label the shim from the output shaft (in front of bearing sleeve). Slide the entire gear train assembly from the intermediate case. Remove the 1st/2nd gear shift rod detent ball (8 mm) from the case.

CAUTION —

Be sure to hold all shafts together as the case is removed.

NOTE —

The 1st/2nd selector rod is removed together with the gear train. Remove the rolled pin from the shift fork once they are removed from the case. Scratch the gear numbers (1st and 2nd) on each side of the fork.

16. Separate the input shaft from the front of the output shaft. Remove the 4th gear synchronizer ring and the roller bearing.

NOTE —

If the synchronizer rings are going to be reused, they should be installed in their original positions. All five synchronizer rings in the Getrag 265 have the same part number.

To install transmission shafts:

1. Check the position of the small oil plate in the intermediate case. The plate should be vertical when positioned correctly. See Fig. 5-26.

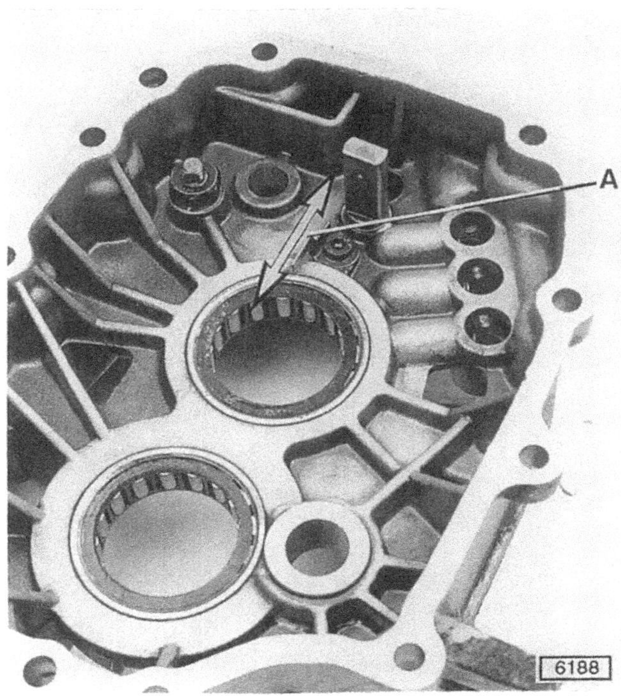

Fig. 5-26. Oil plate (**A**) in intermediate case correctly positioned straight up and down (arrow).

2. Install the reverse idler shaft thrust washer with cutout, in the case. Install the reverse idler gear, bearings, and idler shaft through the case. Apply Loctite® 270 to the shaft bolt threads and install and tighten the bolt. Install the reverse gear holder plate.

NOTE —
The collar on the reverse idler gear faces the rear of the transmission.

Tightening torques	
• reverse gear idler shaft bolt (with Loctite® 270)	49 Nm (36 ft-lb)
• reverse idler gear holder plate bolts	25 Nm (18 ft-lb)

3. Place the input shaft with its 4th gear synchro ring and roller bearing on the output shaft. Install the layshaft to the output shaft. While holding the assembly together, slide the intermediate case into position. See Fig. 5-27.

Fig. 5-27. Intermediate case being installed onto output shaft and layshaft assembly.

4. Install the intermediate case bearing inner race shim on top of the inner race. Then using a vernier caliper, measure and record the distance between the shim and the top of the 5th/reverse gear guide sleeve splines. See Fig. 5-28.

5. Place the 5th/reverse gear guide sleeve on top of the bearing sleeve and measure and record the total height of the two pieces. See Fig. 5-29.

6. Subtract the first distance (measured in step 4) from the second distance (measured in step 5). If there is any difference (play), a new shim should be selected until the clearance is eliminated.

NOTE —
Shims can be ordered from an authorized BMW dealer parts department. This shim is only available in one thickness (35 x 45 x 0.2 mm).

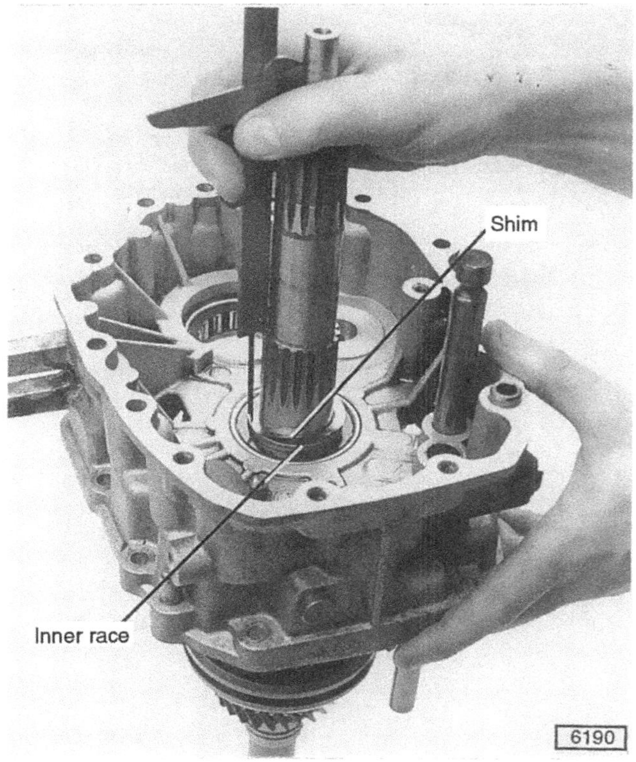

Fig. 5-28. Distance from bearing inner race to end of 5th/reverse gear guide sleeve splines being measured.

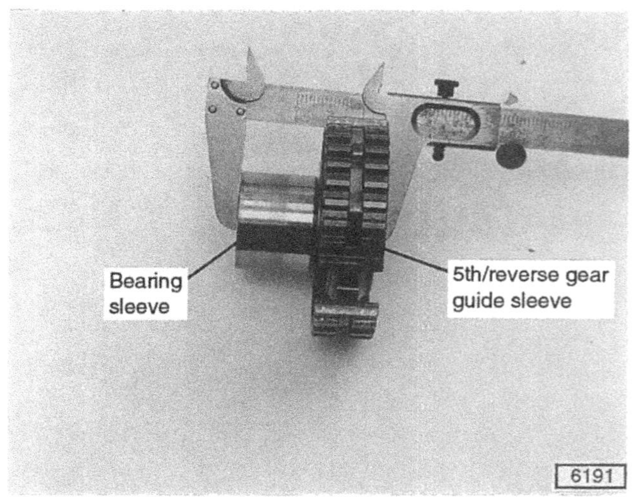

Fig. 5-29. 5th/reverse gear guide sleeve and bearing sleeve total height being measured.

7. Heat the bearing sleeve and install it on the output shaft with the collar facing the shim. Be sure the sleeve is fully seated.

Installation temperature	
• bearing sleeve	176°F (80°C)

NOTE —
Allow the bearing inner sleeve to cool before installing the reverse gear needle bearing.

8. Install the reverse gear needle bearing, reverse gear and synchronizer ring. See Fig. 5-30.

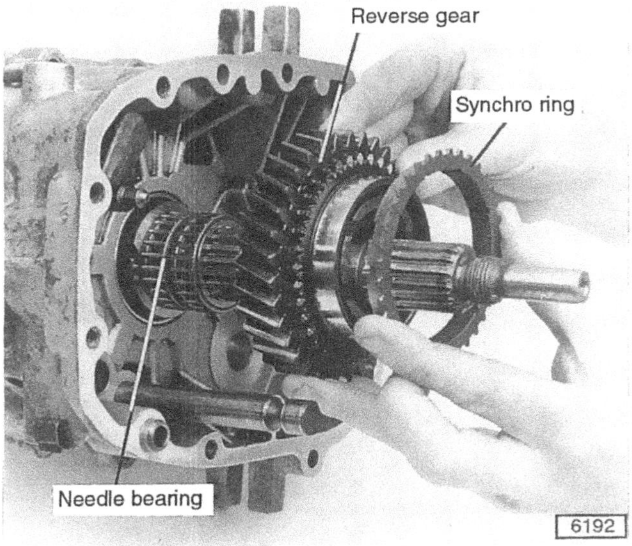

Fig. 5-30. Reverse gear, needle bearing, and synchro ring being installed on output shaft.

9. Install the 1st/2nd gear shift rod detent ball into the case. Install 1st/2nd gear shift fork on its operating sleeve using the identifying marks made earlier. Slide the 1st/2nd gear shift rod through the shift fork into case while pressing down on the detent ball and spring.

NOTE —
- Apply a small amount of grease to the detent ball to hold it in place.
- Do not confuse the 8 mm detent balls with other size balls used in the transmission.

10. While supporting the 1st/2nd gear shift rod, drive in a new rolled pin (6x32 mm). Position the rod in its neutral position. See Fig. 5-31.

CAUTION —
Do not reuse rolled pins.

11. Install an 8 mm detent ball in the 1st/2nd gear shift rod cutout. Then install another 8 mm ball on the spring for the 3rd/4th gear shift rod. See Fig. 5-32. Install 3rd/4th gear shift fork on its operating sleeve. Slide the shift rod through shift fork and into the case while pressing down on the detent ball and spring.

Fig. 5-31. 1st/2nd gear shift fork rolled pin being driven in.

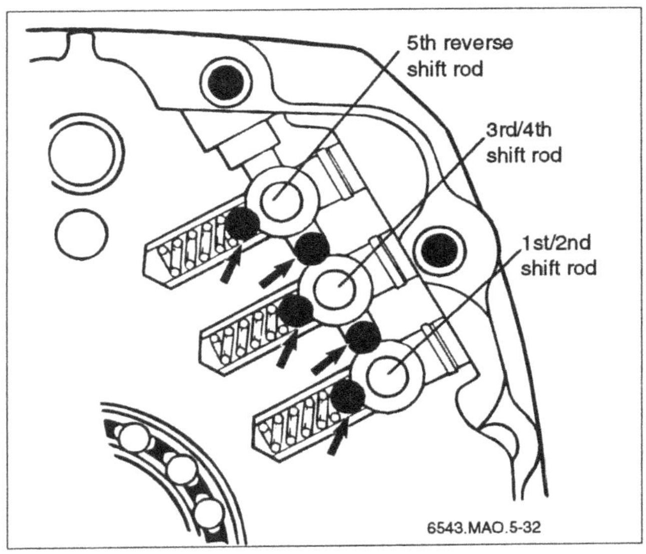

Fig. 5-32. Shift rod detent ball locations (arrows).

12. While supporting the 3rd/4th gear shift rod, drive in a new rolled pin (6x26 mm).

13. Assemble the 5th/reverse gear synchronizer guide sleeve to the operating sleeve. Make sure the stepped side of the operating sleeve is on the opposite side of the centering pin on the guide sleeve. See Fig. 5-33.

13

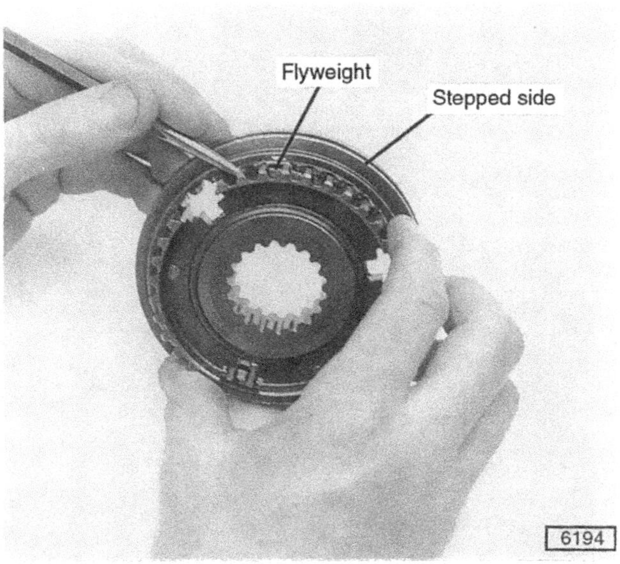

Fig. 5-33. 5th/reverse gear operating sleeve being installed to guide sleeve. Centering pin not shown (on opposite side of sleeve).

NOTE —
- When installing the guide sleeve to the operating sleeve, make sure the beveled edge of the three dogs is facing up (towards the operating sleeve).

- On some early transmissions, a special 5th/reverse gear operating sleeve with flyweights was installed. When reinstalling this guide sleeve, make sure the ends of the flyweights align with cutouts in the operating sleeve. In addition, when installing the guide sleeve to the operating sleeve, make sure the side of the operating sleeve with the cutouts is on the opposite side of the guide sleeve centering pin. See Fig. 5-34.

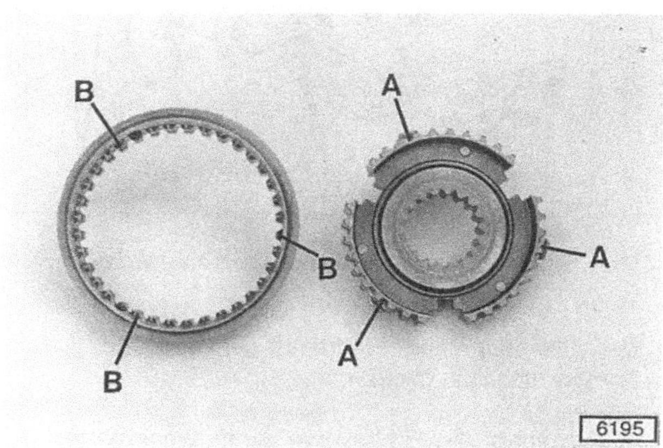

Fig. 5-34. 5th/reverse guide sleeve with flyweights (**A**) used on some early style transmissions. Align end of flyweights with cutouts in operating sleeve (**B**). Cutout side of operating sleeve must be on opposite side of centering pin in guide sleeve.

14. Install the reverse gear synchronizer ring and the 5th/reverse gear guide sleeve assembly on the output shaft with the centering pin facing 5th gear.

15. Position the 5th/reverse shift fork on its operating sleeve, insert the two remaining 8 mm detent balls. See Fig. 5-32 above. Install the 5th/reverse gear shift rod into the case.

NOTE —
- When inserting the 5th/reverse gear shift rod in the case, make sure the two cutouts in the rod are always opposite the detent balls. See Fig. 5-35. If a ball gets stuck in the cutout, twist the rod before pushing it in.

- Do not confuse the 8 mm detent balls with other size balls used in the transmission.

Fig. 5-35. 5th/reverse gear shift rod being installed. When sliding rod through case, make sure cutout (black arrow) is opposite detent balls. Additional cutout (not shown) should also be positioned away from detent balls.

16. Carefully slide the shift fork, shift rod, and operating sleeve assembly toward the rear just until a new rolled pin (6x26 mm) can be driven in. Install the new rolled pin and then position the shift rod in its neutral position.

17. Heat the 5th gear bearing sleeve and install it on the output shaft so that the cutout on the bearing sleeve aligns with the pin in the guide sleeve. Make sure the bearing sleeve is fully seated.

Installation temperature	
• 5th gear bearing sleeve	176°F (80°C)

CAUTION —
Allow the bearing sleeve to cool before installing the 5th gear needle bearing halves.

18. Install the reversing lever and turning lock. See Fig. 5-36. Slide the shift rod through the turning lock and drive in a new rolled pin (6x26 mm). If necessary, engage 2nd gear.

NOTE —
Be sure to support the rod when driving in the pin.

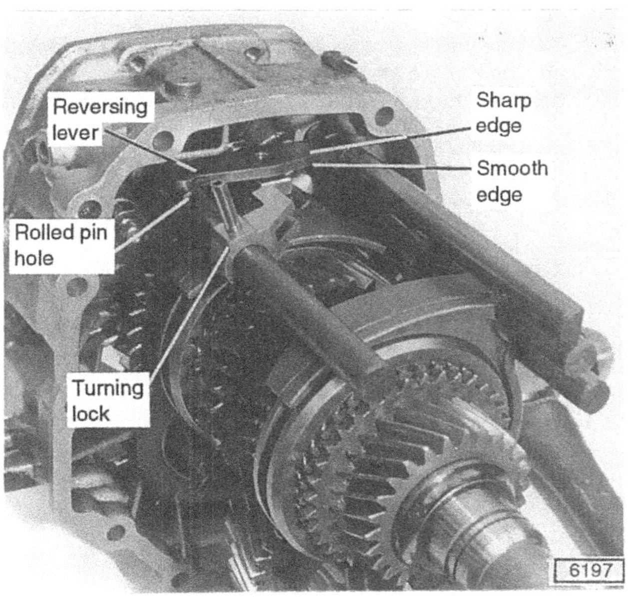

Fig. 5-36. Reversing lever and turning lock being positioned in case. Install reversing lever so that smooth edge is down.

19. Install the 5th gear needle bearing halves, the 5th gear synchronizer ring and 5th gear.

20. Heat the layshaft 5th gear to the temperature listed below. Lubricate the layshaft, then press the 5th gear on the layshaft. Pull up on and mesh 5th gear on the output shaft while pressing on 5th gear on the layshaft.

Installation temperature	
• layshaft 5th gear	350°F (180°C)

Pressing-on force	
• 5th gear to layshaft	5 to 6 tons

WARNING —
- *Use extreme care when handling 5th gear once it has been heated. Wear appropriate hand protection.*
- *Be sure a non-flammable lubricating oil is used for the layshaft.*

CAUTION —
While pressing on 5th gear, support the underneath of the layshaft. Do not allow pressure to bear on the output shaft or any of its gears, in particular 3rd gear and its synchronizer.

21. Heat the inner race for the layshaft bearing and install it on the end of the layshaft with the flange end of the race facing 5th gear.

Installation temperature	
• layshaft bearing inner race	176°F (80°C)

NOTE —
BMW special tool (BMW tool no. 23 1 030) or an equivalent can be used to drive on the inner race.

22. Install the 4 mm ball and the slotted washer to the output shaft. See Fig. 5-37. Use small amount of grease to hold the ball in position. Heat the inner race and install on the output shaft so the small cutout in the rear of the race aligns with the ball.

Installation temperature	
• output shaft rear bearing inner race	176°F (80°C)

23. Install the selector arm into position (longer side engages 3rd/4th gear shift rod cutouts). Install the selector shaft through the selector arm. Install the shaft so the cutouts in the rear of the shaft face away from the shift rod. See Fig. 5-38. While supporting the shaft drive in a new rolled pin (6x32 mm).

CAUTION —
Do not reuse rolled pins.

24. Install the front and rear transmission cases as described in **5.1 Removing and Installing Front and Rear Cases.**

13

Fig. 5-37. Inner race being installed on output shaft. 4 mm ball is used to locate race (arrow). Rear of race has cutout for ball.

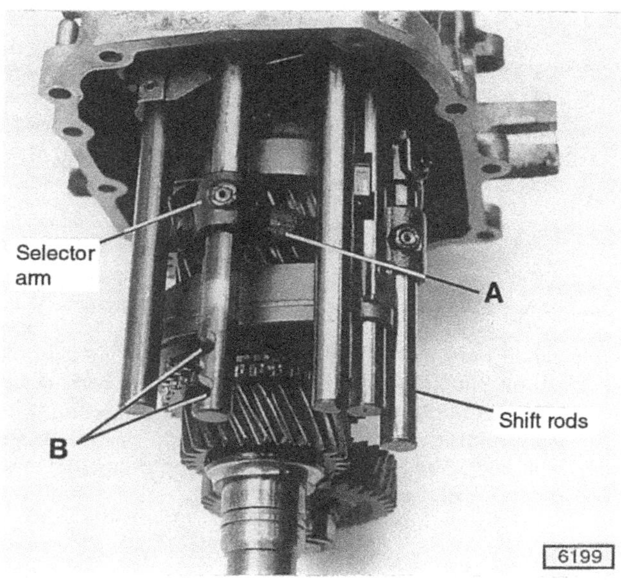

Fig. 5-38. When selector shaft is installed, long end of selector arm faces 3rd/4th gear shift rod cutout (**A**). Cutouts (**B**) at rear of shaft face away from shift rods.

5.3 Disassembling and Assembling Layshaft and Output Shaft (Getrag 265)

Disassembly of the gear train requires a hydraulic press or other suitable pullers. Special press bed adaptors are available from an authorized BMW dealer to adapt the press bed to the Getrag gear train.

Removing third gear and fourth gear from the layshaft requires considerable force (in excess of 5 tons). It is recommended that these gears be pressed off by an authorized BMW dealer or a qualified machine shop with the appropriate pressing equipment.

> ***WARNING —***
> *When pressing or pulling is required, always use tools having a higher capacity rating than that required to remove the part. Power tools, such as air impact tools, should not be used on press equipment.*

To disassemble output shaft:

1. Remove the snap ring and spacer from front of the output shaft. Pull off the 3rd/4th gear guide sleeve, 3rd gear synchro ring, and 3rd gear with needle bearing.

> ***CAUTION —***
> *Do not reuse old snap rings.*

2. Press off 1st gear together with the inner bearing sleeve. Remove the 1st/2nd gear synchro assembly and synchro rings. Remove 2nd gear and needle bearing. See Fig. 5-39.

To assemble output shaft:

1. Install the 2nd gear needle bearing, 2nd gear, and the 2nd gear synchro ring on output shaft.

> **NOTE —**
> Make sure the tabs in the synchro ring are correctly aligned with the cutouts in the operating sleeve.

2. Assemble the guide sleeve with springs, dogs, and 6.5 mm balls to the operating sleeve. The convex surface of the dogs face out towards the operating sleeve. Install the assembly on the shaft.

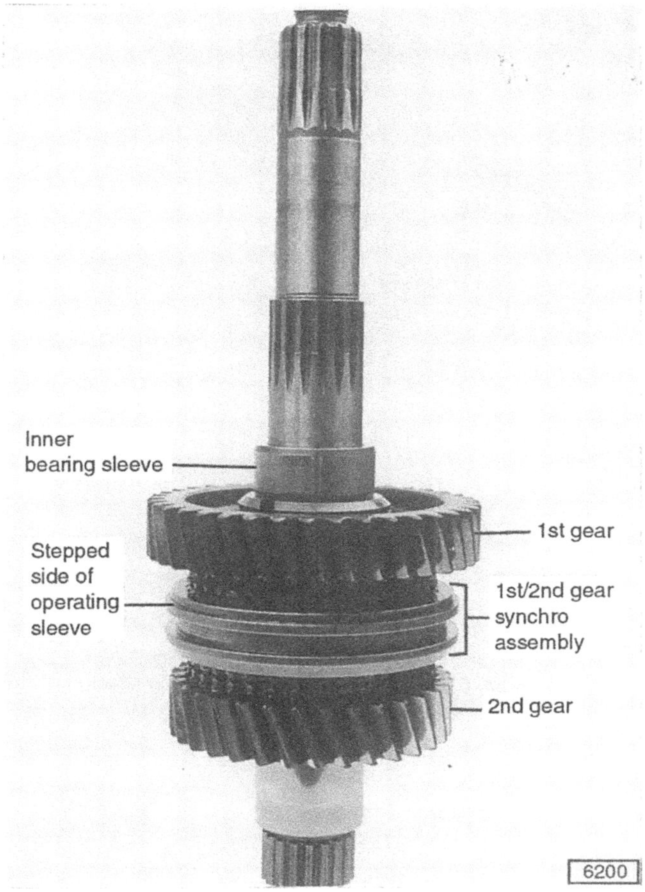

Fig. 5-39. 1st gear, 2nd gear and related components to be pressed off output shaft.

NOTE —
- **Do not confuse the 6.5 mm balls with other size balls used in the transmission.**
- **The small offset grooves machined in the guide sleeve face 1st gear. See Fig. 5-40.**
- **The step side of the operating sleeve should face 1st gear, the beveled side faces 2nd gear. Failure to orient the stepped side of the sleeve correctly can cause the transmission to jump out of gear.**

3. Install the synchro ring, 1st gear, and bearing on the output shaft.

4. Heat the inner bearing sleeve and install it on the output shaft with the flange end facing the output end of shaft.

Fig. 5-40. Small offset grooves (arrows) in the guide sleeve face 1st gear. Step side of operating sleeve **A** should also face 1st gear.

NOTE —
If the bearing sleeve is not correctly installed before cooling, it will have to be pressed off, reheated, and reinstalled. Do not drive the bearing sleeve on once it has cooled.

Installation temperature
• inner bearing sleeve . . . . 176°F (80°C)

5. Heat the inner bearing sleeve and install on it on the layshaft with the flange end facing 4th gear.

Installation temperature
• inner bearing sleeve . . . . 176°F (80°C)

6. Working at the input end of the output shaft, install the needle bearing, 3rd gear, 3rd/4th guide and operating sleeve with synchro rings, and finally the shim.

7. Position the circlip in its groove and measure the clearance between the 3rd/4th guide sleeve and the circlip using a feeler gauge. See Fig. 5-41. **Table g** lists various thickness shims available if the play is incorrect.

CAUTION —
Do not reuse circlips.

Axial play
• 3rd/4th gear guide sleeve (measured at circlip groove) . . . . 0.0 to 0.09 mm (0.0 to 0.0035 in.)

13

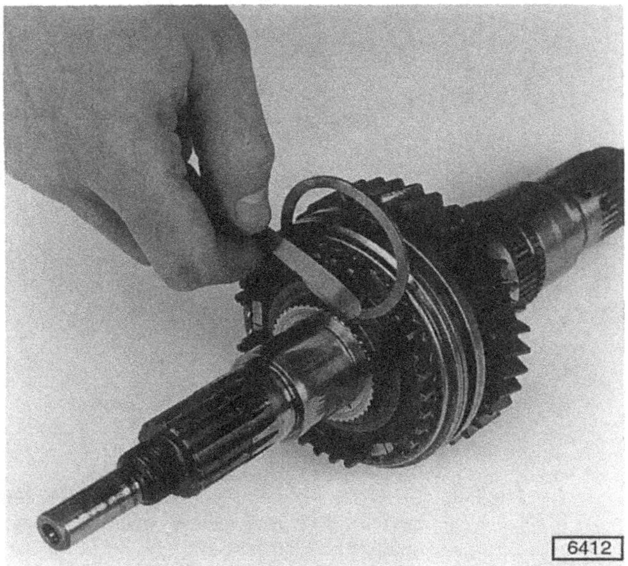

Fig. 5-41. Synchronizer guide sleeve play being measured using feeler gauge. (5th/reverse gear guide sleeve shown).

NOTE —
See an authorized BMW dealer for various thickness shims.

Table g. Selective Shims for 3rd/4th Guide Sleeve on Output Shaft (Getrag 265)

Location	Dimension and thickness in mm
Output shaft	28x37x3.7 28x37x3.8 28x37x3.9 28x37x4.0

Disassembling and Assembling Layshaft

1. Working at end of the layshaft, pry the layshaft bearing inner race away from 4th gear using a pair of pry bars. See Fig. 5-42.

2. Press 4th gear off the layshaft.

Pressing-off force
• layshaft 4th gear . . . . 5 to 7 tons

3. Remove the circlip from above 3rd gear, then press 3rd gear off the layshaft.

Pressing-off force
• layshaft 3rd gear . . . . 5 to 7 tons

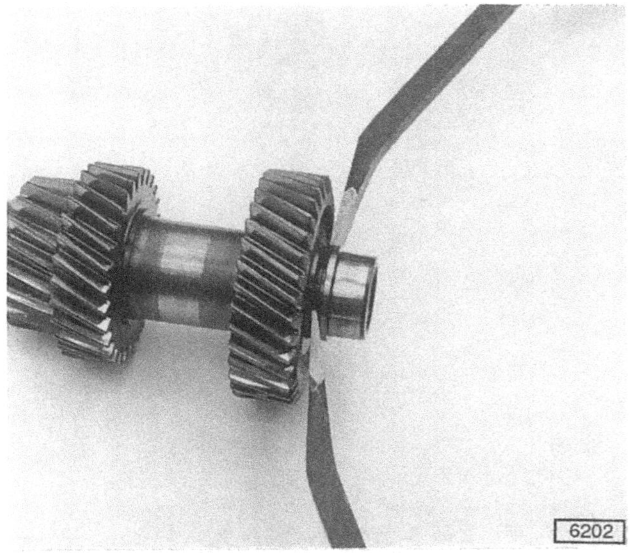

Fig. 5-42. Layshaft bearing inner race being removed.

Assembly is the reverse of removal. All of the layshaft gears must be heated prior to being pressed on the layshaft. Use lubricating oil on the layshaft when installing.

Pressing-on force
• layshaft 3rd gear and 4th gear . . . . 5 to 5.5 tons

Installation temperature
• layshaft gears . . . . 350°F (180°C)

WARNING —
- *Use extreme care when handling heated gears. Wear appropriate hand protection.*

- *Be sure a non-flammable lubricating oil is used for the layshaft.*

5.4 Transmission Oil Seals (Getrag 265)

Replacing the transmission oil seals can be done without any major disassembly of the transmission. In addition, two of the three seals can be replaced with the unit installed in the car.

Low gear oil level due to a faulty oil seal may cause problems such as hard shifting, jumping out of gear, and transmission noise. Oil at the bottom of the bellhousing may also be due to a leaking rear crankshaft seal.

Before assuming that the seals are at fault, check the transmission vent on the top of the case. A clogged or damaged vent can cause the pressure inside the case to increase to a point where the lubricant is forced past the seals.

To replace output shaft oil seal (transmission installed):

1. Remove the exhaust system as described in **EXHAUST SYSTEM.**

2. Disconnect the driveshaft and, where applicable, the vibration damper from the transmission. See **DRIVESHAFT AND FINAL DRIVE.**

3. Bend back the lockplate tabs shown in Fig. 5-43, and remove the lockplate. Hold the flange steady and remove the collar nut using a deep 30 mm thin-walled socket.

NOTE —
A thin-wall socket can be made by grinding the outer surface of an ordinary socket.

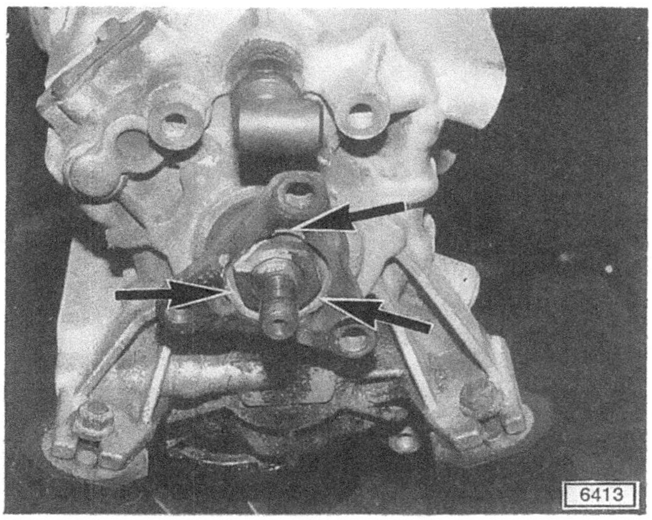

Fig. 5-43. Lockplate tabs (arrows). Bend tabs out of groove to remove lockplate and collar nut.

4. Remove the output flange from the output shaft. If necessary use a puller to remove the flange.

5. Using a seal remover or a small screwdriver, pry out the old seal. Lubricate the new seal with oil and carefully drive it into place.

CAUTION —
Be careful not to damage the transmission case housing when prying out the old seal. A damaged case can result in oil leaks.

6. Install the output flange. Coat the bearing surface of the collar nut with sealer and install the nut, torquing it to the specifications listed below.

NOTE —
• BMW recommends the use of a sealer such as Loctite® 270, Hylogrip, or Curil K2 when installing the collar nut to prevent transmission oil leaks. Such a sealant is available from your BMW dealer.

• Tighten the nut in two stages. First, tighten the nut to fully seat the flange. Then loosen the nut and retighten it. Install a new lockplate over the nut and bend the tabs into the groove.

Tightening torque
• output flange nut, 1st stage tightening . . 170 Nm (125 ft-lb)
• output flange nut, 2nd stage tightening (after loosening) . . . . 120 Nm (89 ft-lb)

7. Reinstall the driveshaft and exhaust system.

To replace selector shaft oil seal (transmission installed):

1. Remove the exhaust system as described in **EXHAUST SYSTEM.**

2. Remove the driveshaft or disconnect the flexible coupling as described in **DRIVESHAFT AND FINAL DRIVE.**

3. Bend back the lockplate tabs shown above in Fig. 5-43, and remove the lockplate. Hold the flange steady and remove the collar nut.

4. Remove the output flange from the output shaft. If necessary use a puller to remove the flange.

5. Put the transmission in 3rd gear. Disconnect the selector rod from the selector shaft by pulling back on the spring sleeve and driving out the pin.

6. Using a pick or small screwdriver, pry out the seal as shown in Fig. 5-44. Lubricate the new seal with oil and drive it into place.

CAUTION —
Be careful not to mar the housing when removing the seal.

7. Reinstall the selector rod, the driveshaft flange and driveshaft, and the exhaust system.

CAUTION —
The tightening torque on the drive flange collar nut is critical. Tighten the collar nut as described above under the procedure for replacing the output shaft oil seal.

13

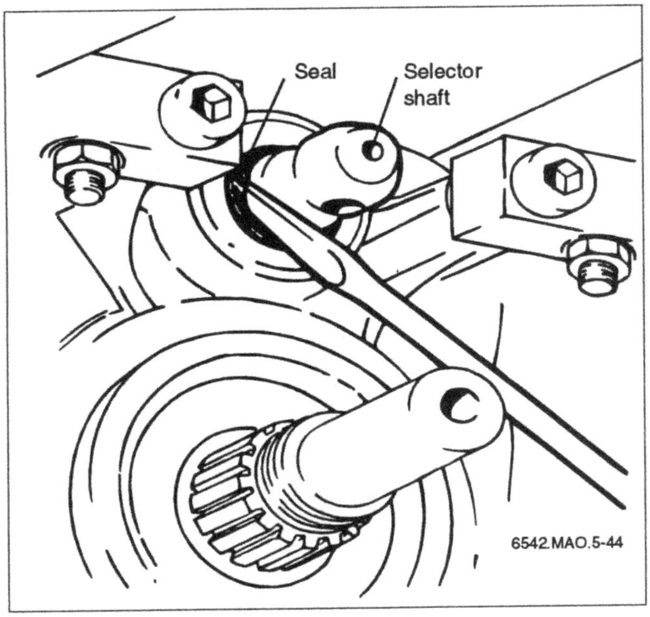

Fig. 5-44. Selector shaft oil seal being pried out with screw-driver. Take care not to scratch selector shaft.

Fig. 5-45. Clutch release bearing guide sleeve bolts. Getrag 265 transmission shown.

To replace input shaft oil seal (transmission removed):

1. Remove the transmission as described in **MANUAL TRANSMISSION AND CLUTCH**.

2. Remove the clutch release bearing and release lever from the bellhousing.

3. Remove the bolts for the clutch release-bearing guide sleeve noting their lengths. See Fig. 5-45. Remove the sleeve and any spacers under it.

4. Pry out the old seal, lubricate the new seal with oil, and drive it into place.

CAUTION —
Be careful not to mar the housing when removing the seal.

5. Clean the mounting bolts and sealing surfaces of the guide sleeve. Clean the bolt threads in the case. Apply sealer (Loctite® 573 or equivalent) to the guide sleeve sealing surface and bolts. Reinstall the guide sleeve and spacer(s). Tighten the bolts.

NOTE —
The cover bolts are different lengths and have different torque values. The longer bolts go in the top three holes.

Tightening torque	
• release bearing guide tube cover bolts	
(M8x22)	18 Nm (13 ft-lb)
(M8x30)	25 Nm (18 ft-lb)

NOTE —
A different spacer may be required if either the bearing or the guide sleeve are replaced. See **5.1 Removing and Installing Front and Rear Cases** for more information.

6. Reinstall the clutch release lever, spring, and release bearing. Reinstall the transmission.

6. INSPECTING TRANSMISSION COMPONENTS

The inspection information listed under this heading covers the transmission components that are most likely to wear. Some of the inspection procedures require precision measuring equipment such as a dial indicator setup and a vernier caliper. If this equipment is not available, a local machine shop should be able to do the work quickly and at a reasonable cost.

In addition, some of the steps below can be done prior to complete disassembly of the transmission. For example, checking for worn synchronizer rings and checking the axial play of the guide sleeves on the output shaft.

Gears and Synchronizer Rings

Gears should be inspected for wear and broken or missing teeth. Because the gears are constantly meshing with the gears on the layshaft, they are not normally a cause of transmission trouble. Check the gear's surface that the needle bearing rides on, the needle bearing itself, and the gear's journal on the output shaft. Replace any components that are found to be faulty.

The contact surface of each synchronizer ring is coated with a rough friction material that can wear away over time. Visually inspect the cone surface of the rings for damaged or missing teeth. Check that the ring's friction surface is not shiny and smooth. If no visible faults are found, check each ring with its matching gear as shown in Fig. 6-1. **Table h** lists synchronizer ring wear specifications. It is recommended that all synchro rings are replaced if the transmission is disassembled.

6204

Fig. 6-1. Synchronizer ring being checked for wear using feeler gauge. Measure in the area of the tabs (arrow).

Table h. Synchronizer Ring Wear Specifications

Gear	New	Wear Limit
Forward gears (1st–5th)	1.0–1.3 mm (0.039–0.051 in.)	0.8 mm (0.031 in.)
Reverse	0.5–0.6 mm (0.020–0.024 in.)	0.4 mm (0.016 in.)

Shift Forks and Shift Rods

Two types of shift forks were used in the transmissions covered by this manual. The early transmissions used a bronze shift fork while the later versions used aluminum forks. The bronze shift forks had better wear characteristics but were heavier.

Inspect the shift fork guides that ride in the grooves of the steel synchronizer operating sleeves. There should be little or no play. Measure each shift fork at the area shown in Fig. 6-2. **Table i** lists shift fork wear specifications. Replace any shift fork that exceeds the wear limits.

Table i. Shift Fork Guide Wear Specifications

Shift fork	Wear limit
1st/2nd, 3rd/4th	4.8 mm (0.189 in.)

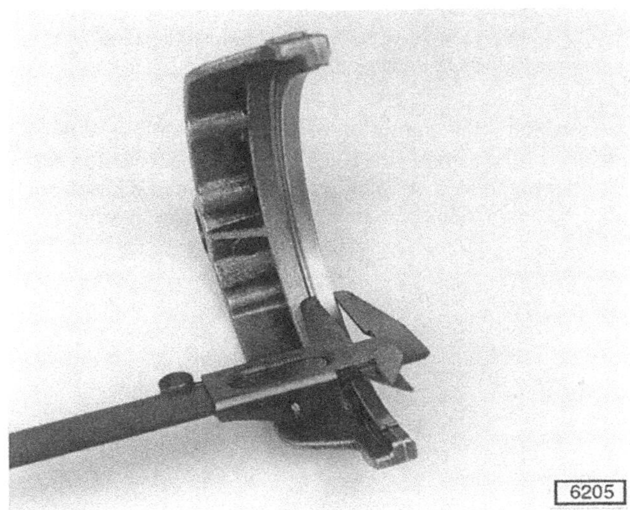

6205

Fig. 6-2. Shift fork being checked for wear.

NOTE —
Wear specifications for the 5th/reverse gear shift fork are not given by BMW.

13

The shift rods should not be bent or otherwise damaged. Make sure that the shift rods slide smoothly in their bores and that they are not worn.

Transmission Shafts

Visually inspect the output shaft and the input shaft for wear. Check for distorted or damaged splines or journals. The output shaft can be inspected for radial runout (bending) using a dial indicator setup.

Radial wear limit
• output shaft runout (measured at journals) .... 0.07 mm (0.0028 in.)

Transmission Case Bearings

Ball bearing races should turn smoothly, without roughness or tight spots. There should be no visible damage or heat discoloration. Look for galled or flattened areas on the roller bearings and the inner races. Inspect the needle bearings to see that none are flattened or heat damaged. Check that the bearing cage is not damaged. Replace any bearing that is damaged or worn.

> **NOTE —**
> Always replace the bearing races when installing new bearings.

To replace the output shaft bearing on Getrag 265 transmissions, first remove the bolts holding the horseshoe shaped retainer to the rear case then heat the case and press out the bearing. Heat the rear case and drive the new bearing into the case. Install the horseshoe shaped retainer and tighten the bolts.

Tightening torque
• Getrag 265 rear case bearing retainer (bolts) .... 10 Nm (89 in-lb)

Installation temperature
• rear case bearing removal/installation ...... 176°F (80°C)

When installing a new output shaft bearing, first measure the height of the new bearing. Then measure the bearing depth in the case (measuring from the top of the raised bolt lands on the case). Subtract the first measurement from the later. The difference in the two measurements is the thickness of the shim needed to be installed between the bearing and the rear case. Install the shim.

If the layshaft axial play is excessive, the bearings and their races may need to be replaced. All case bearings need to be pressed out and in. When installing the layshaft bearings, check protrusion height to the front case. The front case layshaft bearing is installed with the collar facing in.

Specification
• layshaft bearing protrusion height (front case).................. 0.5 mm (0.197 in)

On the Getrag 265, intermediate case bearings need to be pressed into the case flush with the case surface.

> **NOTE —**
> • Gears are brittle and can be easily damaged if dropped or hammered. Do not allow brittle gears to fall on a hard surface.
>
> • Only soft faced hammers should be used on gears and components in the transmission.

7. TECHNICAL DATA

I. Tightening Torques (Getrag 260)

Back-up light switch to transmission case	20 Nm max. (15 ft-lb max.)
Clutch release-bearing guide sleeve to front transmission case	10 Nm (89 in-lb)
Detent ball and spring locking plate to transmission case (bolt)	10 Nm (89 in-lb)
Front transmission case to rear case (bolt)	25 Nm (18 ft-lb)
Reverse gear shaft to transmission case (bolt)	25 Nm (18 ft-lb)
Reverse gear shaft retaining bracket to transmission case (bolt)	25 Nm (18 ft-lb)
Transmission drain plug/fill plug	40–60 Nm (30–44 ft-lb)
Transmission output flange collar nut	
initial	170 Nm (125 ft-lb)
final (after loosening)	120 Nm (89 ft-lb)

II. Transmission Tolerances, Wear Limits and Settings (Getrag 260)

Shift fork guide wear limit	4.8 mm (0.189 in.)
Output shaft axial play, maximum	0–0.09 mm (0–0.0035 in.)
Input shaft axial play, maximum	0–0.09 mm (0–0.0035 in.)
Output shaft radial runout, maximum	0.07 mm (0.0028 in.)
Synchronizer ring specifications (measured between ring and gear)	
Forward gears	
new	1.0–1.3 mm (0.039–0.051 in.)
wear limit	0.8 mm (0.031 in.)
Reverse	
new	0.5–0.6 mm (0.020–0.024 in.)
wear limit	0.4 mm (0.016 in.)
Transmission case bearings installation temperature	176°F (80°C)

III. Tightening Torques (Getrag 265)

Back-up light switch to transmission case	20 Nm max (15 ft-lb max.)
Bellhousing to front transmission case (nuts)	72 to 80 Nm (53 to 59 ft-lb)
Front case to intermediate case (bolts)	25 Nm (18 ft-lb)
Output shaft bearing retainer to rear transmission case (bolts)	10 Nm (89 in-lb)
Rear case to intermediate case (bolts)	25 Nm (18 ft-lb)
Release bearing guide sleeve to front case (bolts)	
(M8x22)	18 Nm (13 ft-lb)
(M8x30)	25 Nm (18 ft-lb)
Reverse gear shaft to intermediate case (bolt)	49 Nm (36 ft-lb)
Reverse gear shaft retaining bracket to intermediate case (bolt)	25 Nm (18 ft-lb)
Transmission drain plug/fill plug	40–60 Nm (30–44 ft-lb)
Transmission output flange collar nut	
initial	170 Nm (125 ft-lb)
final (after loosening)	120 Nm (89 ft-lb)

IV. Transmission Tolerances, Wear Limits and Settings (Getrag 265)

Input shaft axial play, maximum	0–0.09 mm (0–0.0035 in.)
Layshaft gears installing temperature	350°F (180°C)
Output shaft axial play, maximum	0–0.09 mm (0–0.0035 in.)
Output shaft radial runout, maximum	0.07 mm (0.0028 in.)
Shift fork guide wear limit	4.8 mm (0.189 in.)
Synchronizer ring specifications (measured between ring and gear)	
Forward gears	
new	1.0–1.3 mm (0.039–0.051 in.)
wear limit	0.8 mm (0.031 in.)
Reverse	
new	0.5–0.6 mm (0.020–0.024 in.)
wear limit	0.4 mm (0.016 in.)
Transmission case bearings installation temperature	176°F (80°C)

Section 14

AUTOMATIC TRANSMISSION

Contents

INTRODUCTION

Three different automatic transmissions are used on the various cars covered by this manual, all manufactured by ZF. Early models are equipped with a 3 HP 22 three speed. Later models are equipped with a 4 HP 22 four speed. 535i models from 1986 on are equipped with the 4 HP 22 EH, a refined four speed automatic transmission with electronic shifting control. Though the automatic transmission is entirely different from the manual transmission used in these cars, the driveshaft and final drive components are virtually identical. See **DRIVESHAFT AND FINAL DRIVE**.

With normal use and regular maintenance, the automatic transmission is very reliable and does not require internal repairs. Repairs to the internal sections of the transmission require special knowledge and equipment, and as such are beyond the scope of this manual. In the event that internal repairs or overhaul are required, it may be possible to save some of the expense of professional repair by removing and installing the transmission yourself, using the procedures described in this section.

We recommend that the outside of the transmission be thoroughly cleaned and taken to the shop fully assembled. It is important to realize that a partially disassembled transmission is a mechanic's nightmare. We strongly advise against disassembling the transmission to begin any repair you cannot properly finish.

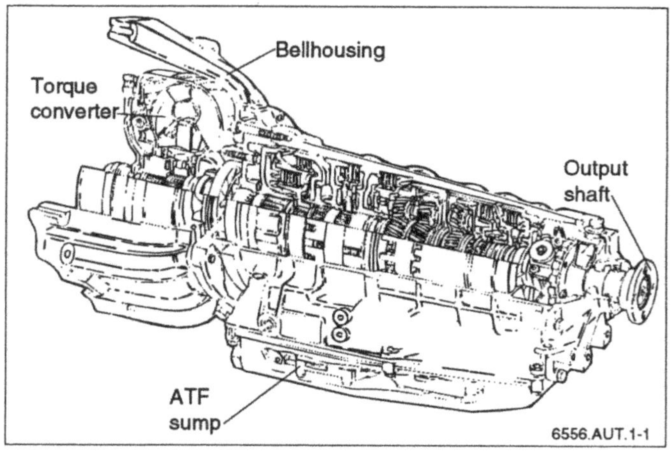

Fig. 1-1. Phantom view of 4-speed ZF 4 HP 22 automatic transmission.

CAUTION —
• BMW recommends that cars with automatic transmission be towed with the rear wheels on the ground for no more than 30 miles (50 km), at no more than 30 mph (50 km/h).

• If the distance will be greater than 30 miles (50 km), either remove the driveshaft or add 1.05 quarts (1 liter) of ATF to the transmission. Reduce the fluid level to normal before driving the car.

1. GENERAL DESCRIPTION

The 4 HP 22 automatic transmission is shown in Fig. 1-1 and Fig. 1-2. At the front end of the transmission is the bellhousing, which is bolted to the transmission and houses the torque converter. The bellhousing is bolted to the engine block. At the rear end of the transmission is the extension housing, which is also bolted to the main transmission case and contains the governor, the parking pawl and spring, and the transmission vent.

The transmission case contains a supply of automatic transmission fluid (ATF), the ATF pump, the hydraulic controls, and the planetary gear system. The planetary gear system is lubricated solely by ATF, circulated through the transmission by the ATF pump. The ATF does not circulate unless the engine is running, so the transmission parts are only partially lubricated when the engine is turned off.

CAUTION —
• Towing a BMW with an automatic transmission while the rear wheels are on the ground can cause damage due to lack of lubrication.

1.1 Automatic Transmission

The automatic transmissions used on the BMWs covered by this manual are hydraulically controlled with either three or four forward speeds. The 4 HP 22 EH transmission features additional electronic control of the hydraulic system.

The automatic transmission can best be understood by dividing it into subsystems. These are the torque converter, the ATF pump, the planetary gear system, and the hydraulic and electronic controls. All cars with automatic transmission also have an ATF cooler integrated into the radiator for the engine cooling system.

Torque Converter

The torque converter is a doughnut-shaped assembly located between the engine and the transmission inside the bellhousing. The torque converter transmits engine output to the transmission, and also multiplies engine torque at low speeds. The torque converter is a fluid coupling. Power is transmitted as internal vanes driven by the engine create a flow of ATF to drive other vanes.

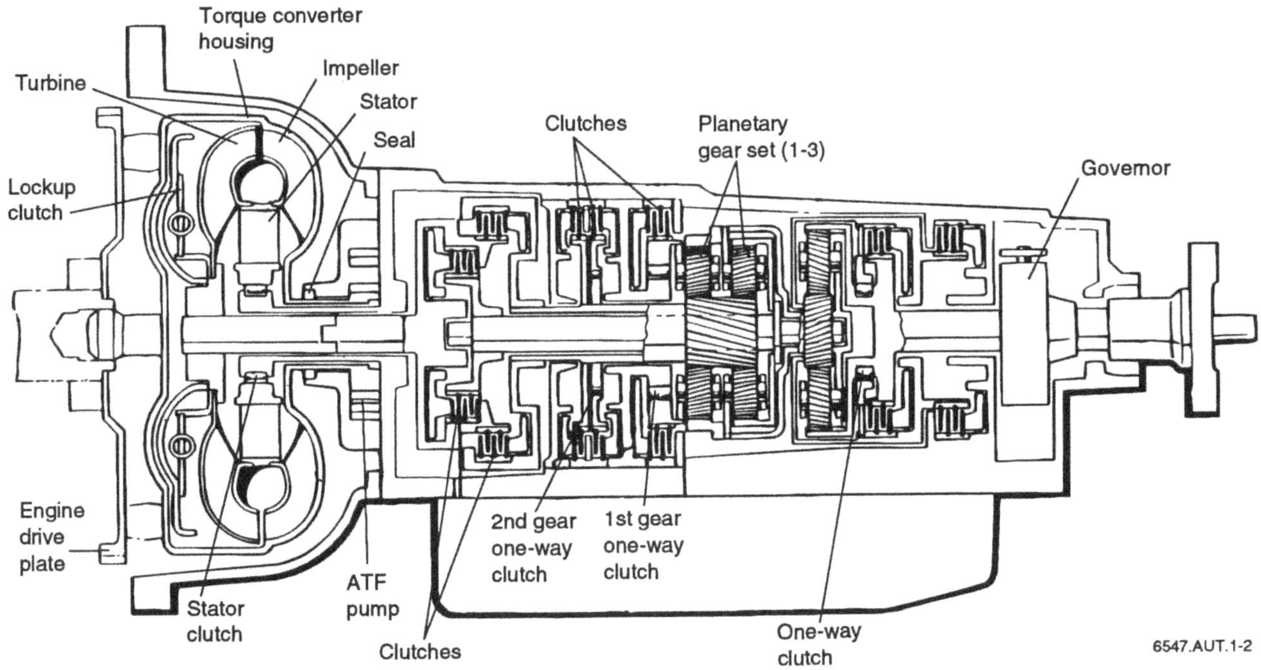

Fig. 1-2. Schematic view of 4-speed ZF 4 HP 22 automatic transmission. 3-speed ZF 3 HP 22 automatic transmission is similar.

The first set of vanes is called the impeller. The impeller is part of the converter housing. Since the housing is bolted to the engine drive plate, the impeller vanes are always driven at engine speed. As the converter spins, the curved impeller vanes set up a flow of ATF that drives the second set of vanes opposite it, called the turbine. The turbine is directly connected to the transmission gear system.

A third set of vanes, called the stator, is between the impeller and turbine. At low speeds, a one-way clutch keeps the stator stationary so that its curved vanes will redirect the flow of ATF and multiply torque. At higher speeds, when there is no longer a speed differential between the impeller vanes and the turbine, the stator freewheels and torque multiplication ceases.

Lockup Torque Converter

Because a normal torque converter is only a fluid coupling between the engine and transmission, there is always some loss of efficiency due to slippage. To ensure maximum efficiency and power transmission, some 3 HP 22 and all 4 HP 22 H and EH transmissions are equipped with a lockup torque converter. Above an ATF temperature of 68°F (20°C) and a road speed of 53 mph (85 km/h), and depending on throttle position, an additional clutch in the torque converter locks the turbine to the impeller. This provides a direct mechanical link between the transmission and the engine.

ATF Pump

Automatic transmission fluid (ATF) is circulated through the transmission under pressure by the pump. It is located at the front of the transmission case, between the torque converter and the transmission. The torque converter shaft engages the ATF pump and drives the pump when the engine is running. ATF is drawn from the ATF sump (also referred to as the transmission oil pan) through a filter screen, and fills the valve body before circulating to fill the torque converter and transmission.

14

Planetary Gear System

A torque converter alone cannot provide the torque multiplication needed for all driving conditions. The transmission therefore contains a planetary gear system which can operate at different drive ratios. The gears are driven directly by the torque converter turbine.

The planetary gear system contains a number of one-way and hydraulically-operated clutches along with the planetary gearsets. The gearsets—one gearset for the three-speed transmission, and two gearsets for the four-speed transmission—drive the transmission output shaft.

The rotational speed of the output shaft depends on which gears are rotating and which are stationary. Gear rotation is controlled by the clutches. By varying the application of hydraulic pressure to the clutches, one or more of the gears can be held stationary. This changes the gear ratio, and also controls reverse.

The clutches also affect the freewheeling of the transmission during coasting. For example, when the selector lever is at **1**, the normally freewheeling 1st gear is locked in order to provide additional engine braking.

Hydraulic Controls

The hydraulic control system directs and regulates hydraulic pressure from the ATF pump to control shifting of the planetary gear system. Shifts are produced by ring-shaped clutch pistons or brake bands, based on the flow of ATF from the pump through the valve body. Three primary control devices regulate the movement of the control valves in the valve body.

The first control device is the manual valve, which is connected to the selector lever by a rod and linkage. Moving the lever changes the setting of the valve and produces the necessary application of hydraulic pressure for the drive range selected.

The second control device, the throttle valve, is linked by cable to the position of the accelerator pedal and makes the transmission responsive to engine speed and load.

The third control device, the governor, controls ATF pressure relative to the output shaft rotational speed, making the transmission responsive to rear wheel speed.

Electronic Controls

The 4 HP 22 EH transmission is a further refinement of the 4 HP 22 H transmission, with electronic shift control. The transmission range is still selected by the selector lever and the planetary-gear clutches are actuated hydraulically, but the control valves in the valve body are operated by five electric solenoids. See Fig. 1-3. The solenoids are controlled by an electronic control unit based on inputs such as program selector switch, throttle kickdown, throttle position, engine speed, and transmission speed.

The transmission range switch, located at the base of the shift lever, sends voltage to the control unit inputs when the shift lever is in the neutral, drive, third, second, and first positions. The transmission control unit is located behind the driver's side kick panel, below the headlight switch.

ATF Cooler

Transmission performance and reliability life depend almost entirely on the condition of the transmission fluid. To ensure that the transmission fluid does not overheat, reducing its ef-

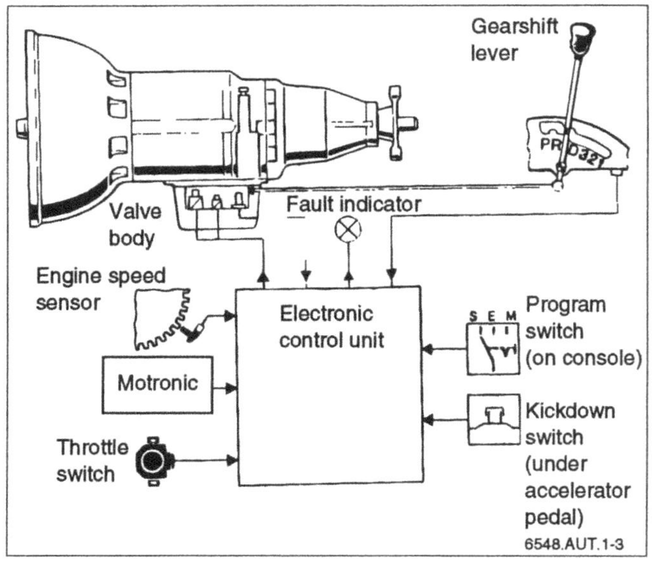

Fig. 1-3. Schematic of 4 HP 22 EH automatic transmission showing electronic control system. Courtesy Robert Bosch GMBH.

fectiveness, an ATF cooler is integrated with the engine cooling system radiator.

The ATF cooler is in the side of the radiator and is linked to the transmission by metal pipes. The ATF is circulated through the cooler by the ATF pump, and heat is removed by the engine coolant.

1.2 Identification Codes and Specifications

Due to different power characteristics and performance requirements of various engine/model combinations, there are minor variations of the basic automatic transmission, including the torque converter. The different versions are identified by code letters.

The transmission code letters and type number are located on a data plate on the left side of the transmission housing, just behind the manual valve lever. See Fig. 1-4.

Table a lists automatic transmission and torque converter combinations for all cars covered by this manual. When replacing the transmission or the torque converter, the replacement part's code letters should correspond to those of the original part.

Table a. Automatic Transmission Specifications

Model	528e (1982-1983)	528e (1984-1988)	533i (1983)	533i (1984)	535i (1985-1988)	535i (1986-1988)
Type	3 HP 22	4 HP 22	3 HP 22 H	4 HP 22 H	4 HP 22 H	4 HP 22 EH
Torque converter code	N1	R2	N1	R2	R2	R2
Gear ratios						
1st gear	2.73	2.48	2.73	2.48	2.48	2.48
2nd gear	1.56	1.48	1.56	1.48	1.48	1.48
3rd gear	1.0	1.0	1.0	1.0	1.0	1.0
4th gear	N/A	0.73	N/A	0.73	0.73	0.73
Reverse	2.09	2.09	2.09	2.09	2.09	2.09

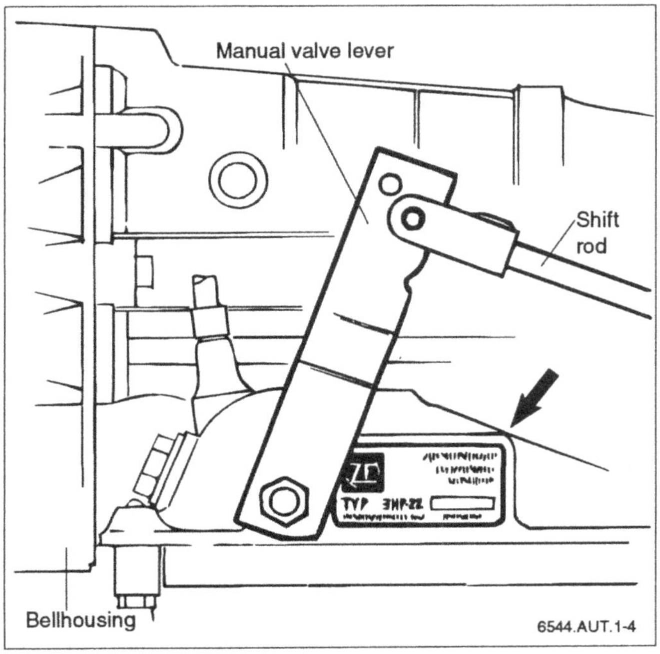

Fig. 1-4. Location of automatic transmission data plate.

2. MAINTENANCE

BMW specifies the maintenance steps below to be carried out at particular time or mileage intervals for proper maintenance of the automatic transmission. Information on automatic transmission maintenance and on the prescribed maintenance intervals can be found in **LUBRICATION AND MAINTENANCE**.

1. Checking ATF level and adding ATF
2. Changing ATF
3. Checking all transmission mating surfaces and ATF cooler lines for leaks
4. Cleaning or replacing the ATF filter screen

3. TROUBLESHOOTING

This troubleshooting section applies to problems affecting the transmission, including the torque converter. The basic function is to transfer engine power to the rear wheels. To do this, the automatic transmission must circulate clean ATF under pressure and provide the correct drive ratio in response to both the hydraulic and mechanical or electronic controls.

Most external adjustments to the transmission, as well as removal and installation, are covered in this manual, but some of the troubleshooting information in this section describes problems which can only be remedied by disassembly and internal repair. Internal transmission repairs require specialized knowledge and equipment and are, therefore, beyond the scope of this manual. The publisher recommends that such repairs be left to an authorized BMW dealer or other qualified automatic transmission repair shop.

14

NOTE —
- Automatic transmissions are available from an authorized BMW dealer as an exchange unit. Check with your dealer for the latest parts information.
- On transmissions manufactured before February 1985, transmission fluid leaks at the ATF filler tube or breather vent valve under heavy engine loads or during cruising may be caused by the transmission being overfilled. This mistake is common due to the design of the dipstick used on these earlier models. See **LUBRICATION AND MAINTENANCE** for more information on correctly filling the transmission and on retrofitting the later dipstick.
- To pinpoint the location of leaks, have the undercarriage and transmission degreased. Remember that air turbulence from driving can cause fluids to appear to be leaking from other locations.

NOTE —
- If the car has been in a collision, it is possible that the magnet located in the oil sump has been dislodged and is blocking the oil pickup tube. Remove the oil pan and check for the proper location of the magnet. See **LUBRICATION AND MAINTENANCE**.

3.1 Basic Principles

In order for automatic transmission troubleshooting to provide meaningful results, the engine must be in good mechanical condition and properly tuned. See **ENGINE—SERVICE AND REPAIR** for troubleshooting engine problems.

Inspect the transmission for external damage, loose or missing fasteners, and for any obvious leaks. Many automatic transmission problems can be traced to an incorrect ATF level, incorrect ATF type, contaminated ATF, or to misadjusted transmission controls.

Check to see if the fluid is dirty or has a burned odor. A burned odor indicates overheated fluid. This may be accompanied by burned clutches, as well as friction material which may be clogging the valve body passages.

Minor automatic transmission problems can often be corrected merely by correcting the ATF level, by draining and refilling the transmission as described in **LUBRICATION AND MAINTENANCE**, or by inspecting the adjustment and operation of the controls as described in **4.2 External Transmission Adjustments**.

Table b lists symptoms of automatic transmission problems, their probable causes, and recommended corrective actions. Numbers in bold type in the corrective action column indicate the heading in this section where the appropriate test and repair procedures can be found. Probable causes and corrective actions that refer to specific internal transmission faults are also provided for reference.

NOTE —
On 4 HP 22 EH transmissions with electronic control, a misaligned valve body may cause shorting out of the lockup clutch solenoid. This will cause the Transmission Fault Indicator to stay on after the car is started, and will allow the car only to be driven in the Emergency program. This fault may also cause engine rpm to drop below normal when braking to a stop.

Table b. Automatic Transmission Troubleshooting

Symptom	Probable cause	Corrective action
1. Park does not hold	**a.** Selector lever linkage misadjusted **b.** Parking pawl or spring broken	**a.** Adjust selector linkage. **4.2** **b.** Repair pawl or spring. To remove and install transmission, see **5.1**
2. No Drive (car will not move)	**a.** ATF level low **b.** Drive plate or torque converter incorrectly installed **c.** Selector lever linkage misadjusted **d.** Sticking pressure valves, faulty ATF pump, or broken gear or shaft—no pressure	**a.** Check and if necessary correct ATF level. See **LUBRICATION AND MAINTENANCE** **b.** Check and correct drive plate installation. See **ENGINE—RECONDITIONING**. Check and correct torque converter installation. **5.1, 5.2** **c.** Adjust linkage. **4.2** **d.** Test ATF pressure. **3.2** Overhaul or replace transmission. To remove and install transmission, see **5.1**
3. Irregular drive in all forward gears and reverse	**a.** ATF level low **b.** ATF pump filter screen partially clogged **c.** Damaged clutch pistons or brake bands	**a.** Check and if necessary correct ATF level. See **LUBRICATION AND MAINTENANCE** **b.** Remove ATF sump. Clean or replace filter screen. See **LUBRICATION AND MAINTENANCE** **c.** Overhaul or replace transmission. To remove and install transmission, see **5.1**
4. Kickdown fails to operate	**a.** Accelerator cable misadjusted **b.** Kickdown switch misadjusted or faulty (4 HP 22 EH only) **c.** Kickdown function faulty	**a.** Adjust cable. **4.2** **b.** Adjust or replace switch. **4.2** **c.** Overhaul or replace transmission. To remove and install transmission, see **5.1**
5. Poor acceleration, poor high-speed performance	**a.** Engine out of tune **b.** Transmission accelerator cable misadjusted	**a.** Tune engine. See **LUBRICATION AND MAINTENANCE** **b.** Adjust cable. **4.2**

continued on next page

Table b. Automatic Transmission Troubleshooting (continued)

Symptom	Probable cause	Corrective action
6. Transmissions lips during hard cornering	**a.** ATF level low **b.** Clogged ATF filter screen	**a.** Check and if necessary correct ATF level. See **LUBRICATION AND MAINTENANCE** **b.** Remove ATF sump. Clean or replace filter screen. See **LUBRICATION AND MAINTENANCE**
7. ATF appears dirty, smells burned	**a.** Contaminated ATF **b.** Damaged clutch friction linings	**a.** Drain and replace ATF. Remove and clean ATF filter screen. See **LUBRICATION AND MAINTENANCE** **b.** Overhaul or replace transmission. To remove and install transmission, see **5.1**
8. Selector lever will not move	**a.** Selector lever linkage binding **b.** Internal transmission failure	**a.** Check and adjust linkage. **4.2** **b.** Overhaul or replace transmission. To remove and install transmission, see **5.1**
9. Hard jolt when moving selector lever from **N** to **D** or **R**	**a.** Idle speed too high **b.** ATF level low **c.** Transmission accelerator cable misadjusted **d.** Transmission electronics faulty (4 HP 22 EH only) **e.** Internal transmission fault	**a.** Adjust idle. See **FUEL INJECTION** **b.** Check and correct ATF level. See **LUBRICATION AND MAINTENANCE** **c.** Adjust cable. **4.2** **d.** Check transmission electronics **e.** Overhaul or replace transmission. To remove and install transmission, see **5.1**
10. Hard jolt on upshift or downshift	**a.** Transmission accelerator cable misadjusted **b.** Faulty clutch **c.** Valve body faulty **d.** Transmission electronics faulty (4 HP 22 EH only)	**a.** Adjust cable. **4.2** **b.** Overhaul or replace transmission. To remove and install transmission, see **5.1** **c.** Replace valve body **d.** Check transmission electronics
11. Vibration when accelerating quickly from a stop	**a.** Driveshaft vibration **b.** Faulty clutch	**a.** See **DRIVESHAFT AND FINAL DRIVE** **b.** Overhaul or replace transmission. To remove and install transmission, see **5.1**
12. Vibration or slipping when accelerating in **R**	**a.** Internal transmission fault	**a.** Overhaul or replace transmission. To remove and install transmission, see **5.1**
13. Car cannot be started in **N** or **P**	**a.** Faulty transmission switch	**a.** To test switch see **ELECTRICAL SYSTEM**. To replace switch see **4.1**
14. No upshift into 2nd, 3rd or 4th gear	**a.** Governor bushing seized **b.** Valve body faulty **c.** Transmission electronics faulty (4 HP 22 EH only) **d.** Valve body solenoids faulty (4 HP 22 EH only)	**a.** Replace governor **b.** Replace valve body **c.** Check transmission electronics **d.** Repair or replace valve body

continued on next page

Table b. Automatic Transmission Troubleshooting (continued)

Symptom	Probable cause	Corrective action
15. Upshift delayed in all gears	**a.** ATF level low **b.** Transmission accelerator cable misadjusted **c.** Valve body contaminated or plugged **d.** Internal pressure valves sticking or faulty **e.** Governor dirty or seized **f.** Transmission electronics faulty (4 HP 22 EH only)	**a.** Check and correct ATF level. See **LUBRICATION AND MAINTENANCE** **b.** Adjust cable. **4.2** **c.** Remove valve body and clean or replace **d.** Overhaul or replace transmission. To remove and install transmission, see **5.1** **e.** Remove governor and clean or replace **f.** Check transmission electronics
16. Downshift delayed or no downshift	**a.** Transmission accelerator cable misadjusted **b.** Governor dirty **c.** Shift valves sticking **d.** Clutches slipping **e.** Transmission electronics faulty (4 HP 22 EH only) **f.** Valve body solenoids faulty (4 HP 22 EH only)	**a.** Adjust cable. **4.2** **b.** Remove governor and clean or replace **c.** Remove valve body and clean or replace **d.** Overhaul or replace transmission. To remove and install transmission, see **5.1** **e.** Check transmission electronics **f.** Repair or replace valve body

3.2 Diagnostic Tests

Although most internal transmission problems cannot be fully diagnosed unless the unit is removed and disassembled, these tests may help further diagnose transmission trouble prior to its removal. To obtain accurate test results, the ATF must be clean, of the proper type, in good condition (not previously overheated or burned), and at the proper level.

Stall Speed Test

This test is used to check for faults in the torque converter and planetary gear system when there is no other apparent cause for poor performance and acceleration. The term is somewhat misleading as the car is not actually stalled during the test. The test results are meaningless if the engine is not running properly. A precise tachometer must be used for the rpm measurements, as dashboard instruments are not sufficiently accurate.

> ***CAUTION —***
> *The stall speed test should be as short as possible, and should never extend beyond 10 seconds maximum. Prolonging the test may overheat the transmission and damage the seals or internal components.*

To test stall speed:

1. Drive the car to warm the engine and transmission to normal operating temperature. Stop the engine.
2. Connect a tachometer according to the instrument manufacturer's instructions, so that it can be read from the driver's seat. Start the engine.
3. Set the parking brake and depress the foot brake firmly to hold the car stationary.
4. Select the transmission drive range. On 3 HP 22 transmissions, place the selector lever in **R** or **1**. On 4 HP 22 H/EH transmissions, place the selector lever in **D**.
5. While holding the car stationary with the brakes, floor the accelerator for no more than 10 seconds. Note the tachometer readings.

> ***WARNING —***
> *Never stand in front of or behind the car when the engine is running and tests are being made.*

> **NOTE —**
> • Engine rpm should increase, and then hold steady. Maximum rpm achieved during this test is the stall speed. See **Table c** for stall speed specifications.
>
> • It is normal for the stall speed to be 125 rpm lower than specified for each 1000 meters (3200 ft.) above sea level. High ambient air temperature will also cause a slight drop in stall speed.

Table c. Automatic Transmission Stall Speed Specifications

Transmission	Model	Converter code	Stall speed (rpm)
3 HP 22	528e	N1	1900–2000
	533i	N1	1870–1970
4 HP 22 H	528e	R2	1900–2050
	533i, 535i	R2	1970–2120
4 HP 22 EH	535i	R2	1980–2140

Stall speed that is below the specified range by a few hundred rpm is probably due to reduced engine performance. Stall speed that is below the specified range by 400 rpm or more indicates a faulty torque converter stator or ATF pump. Check the pump pressure as described below.

Stall speed above the specified range is caused either by low ATF level, or by slippage in the hydraulic or one-way clutches.

Pressure Tests

A main pressure test will reveal internal leaks, sticking control valves, or other troubles in the hydraulic controls. The pressure gauge should have a range of at least 0 to 20 bar (0 to 300 psi), and a hose long enough to allow it to be read from the passenger compartment. Engine idle speed must be correctly adjusted.

The main pressure tap is shown in Fig. 3-1. Because the location of the tap makes it difficult to fit a thick tester hose, an additional thin extension pipe, bent at a right angle and with pressure fittings on either end, may be needed between the tap and the pressure tester hose. The test procedures are slightly different for the 3 HP 22 and 4 HP 22 transmissions.

To test pressure (3 HP 22):

1. Remove the main pressure tap plug and connect the pressure gauge to the transmission. Route the hose so that the gauge can be read from inside the car.

2. Drive the car to warm the engine and transmission to normal operating temperature. Stop the car and firmly set the parking brake and block the wheels. Have an assistant firmly push the brake pedal and be prepared to shut off the engine.

> ***WARNING —***
> *Never stand in front of or behind the car when the engine is running and tests are being made.*

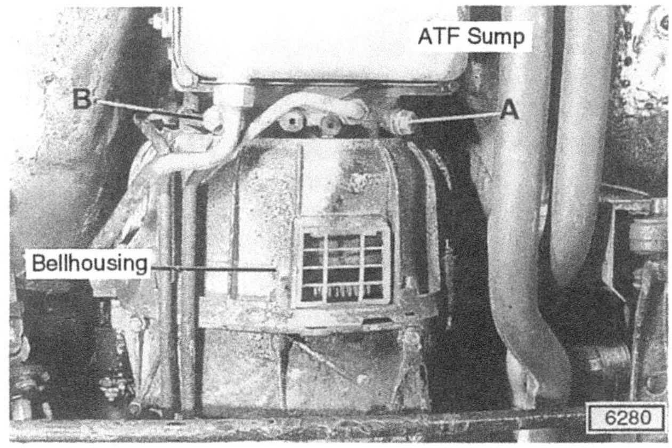

Fig. 3-1. Main pressure tap on transmission case for 3 HP 22 transmissions (**A**), and 4 HP 22 transmissions (**B**). Tap is located below torque converter bellhousing.

3. Disconnect the transmission accelerator cable from the throttle lever. See Fig. 3-2. Manually hold the throttle so that the idle speed is at 1300 rpm.

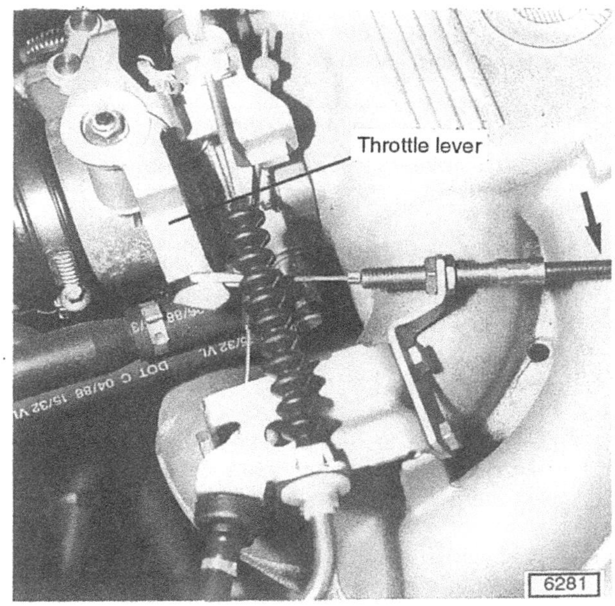

Fig. 3-2. Transmission accelerator cable (arrow).

4. With the selector lever in **R**, the pressure should be as shown in **Table d**. When the transmission accelerator cable is pulled fully out to the kickdown position, the pressure should increase. Release the transmission accelerator cable when done.

14

5. Move the selector lever to any of the other drive ranges, including **P**. See **Table d** for test pressures. When the transmission accelerator cable is pulled fully out to the kickdown position, the pressure should increase.

Table d. Automatic Transmission Pressure Tests (3 HP 22)

Selector lever position	Transmission accelerator cable	Test pressure @ engine rpm
R (reverse)	Released	12.5 to 14.5 bar (181 to 210 psi) @ 1300
	Kickdown position (cable pulled out)	17.3 to 19.4 bar (251 to 281 psi) @ 1300
P (park) D (drive) 3,2,1	Released	5.5 to 6.4 bar (80 to 93 psi) @ 1300
	Kickdown position (cable pulled out)	7.6 to 8.5 bar (110 to 123 psi) @ 1300

6. When finished, stop the engine. Reconnect the transmission cable to the throttle lever, and adjust it as described in **4.2 External Transmission Adjustments.** Remove the gauge and replace the pressure tap plug.

Tightening torque
• pressure tap plug . . . . 40 to 46 Nm (30 to 34 ft-lb)

To test pressure (4 HP 22/4 HP 22 EH):

1. Remove the main pressure tap plug and connect the pressure gauge to the transmission See Fig. 3-1 above. Route the hose so that the gauge can be read from inside the car.

2. Drive the car to warm the engine and transmission to normal operating temperature. Stop the car and firmly set the parking brake and block the wheels. Have an assistant firmly push the brake pedal and be prepared to shut off the engine.

3. With the selector lever in **D**, let the engine idle. Pressure should be as shown in **Table e**. Shift the transmission into **R**, pressure should increase as shown in the table.

4. Shift back to **D** and road test the car. Test the pressure at approximately 4000 rpm and in any gear range from **2** to **4**. Pressure should be as shown in **Table e**.

5. When finished, stop the engine, remove the gauge and replace the pressure tap plug.

Pressure that is higher or lower than specification usually indicates a malfunctioning valve body and valves, probably due

Table e. Automatic Transmission Pressure Tests (4 HP 22, 4 HP 22 EH)

Selector lever position	Test pressure @ engine rpm
D (drive)	6.0 to 7.5 bar (87 to 109 psi) at idle
R (reverse)	11 to 13 bar (160 to 189 psi) at idle
D (drive)	4.6 to 5.8 bar (67 to 84 psi) at 4000

Tightening torque
• pressure tap plug . . . . 40 to 46 Nm (30 to 34 ft-lb)

to contamination. The valve body can be removed, cleaned, and reinstalled with the transmission in place by removing the ATF sump and filter screen.

> ***CAUTION —***
> *The valve body contains many precision parts which must be reassembled in their exact locations. Because of the complexity of the valve body assembly, we recommend that these repairs be left to an authorized BMW dealer or other qualified repair shop.*

Low pressure may also indicate a worn ATF pump or internal ATF pump leaks past seals, gaskets, and metal mating surfaces. These repairs require that the transmission be removed and disassembled.

4. CONTROLS

The automatic transmission controls include the selector lever, the selector lever linkage, and the transmission accelerator cable. The selector lever and linkage allows the transmission to be manually shifted from inside the passenger compartment. The transmission accelerator cable makes the transmission responsive to throttle position.

The selector lever linkage activates the manual valve on the transmission. When the lever is moved out of park, the parking pawl is released and the manual valve is correctly positioned according to the gear selected. On 3 HP 22 transmissions a lever linkage is used. On 4 HP 22 transmissions a cable linkage is used.

The transmission accelerator cable is shown in Fig. 4-1. It modifies transmission shift points based on engine load. The cable connects the throttle operating lever to the transmission. Inside the transmission, the cable is connected to a cam on the valve body. Movement of the accelerator pedal and throttle lever also moves the transmission accelerator cable. This changes pressures in the valve body to adjust shift points.

On all except 4 HP 22 EH transmissions, the transmission accelerator cable also activates the transmission kickdown feature when the accelerator pedal is depressed to the kickdown position.

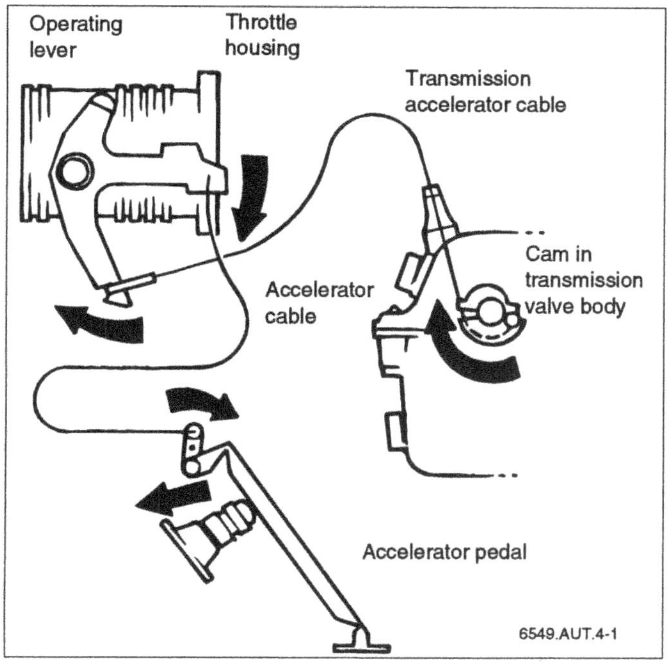

Fig. 4-1. Accelerator cable and transmission accelerator cable layout. Arrows show movement of operating lever and transmission cable when accelerator pedal is depressed.

On 4 HP 22 EH transmissions, kickdown is activated by a switch underneath the accelerator pedal. To replace the switch, disconnect the wires, loosen the locknut and unscrew the switch. Using an ohmmeter, adjust the kickdown switch for zero ohms with the connector disconnected and the accelerator pedal floored.

The transmission accelerator cable is easily adjusted. Replacement, however, requires removal of the valve body. Because of the complexity of the valve body assembly, we recommend that replacement of the cable be left to an authorized BMW dealer or other qualified repair shop.

4.1 Removing and Installing Selector Lever and Linkage

The selector lever linkage and related parts for the 3 HP 22 are shown in Fig. 4-2. The 4 HP 22 transmissions use a cable in place of the lower selector lever and selector rod. The cable is easily replaced after removing the selector lever. The neutral/park/reverse light switch prevents the car from starting when the selector lever is in gear. The switch also actuates the reverse lights when the selector lever is in reverse. For information on testing the switch, see **ELECTRICAL SYSTEM**.

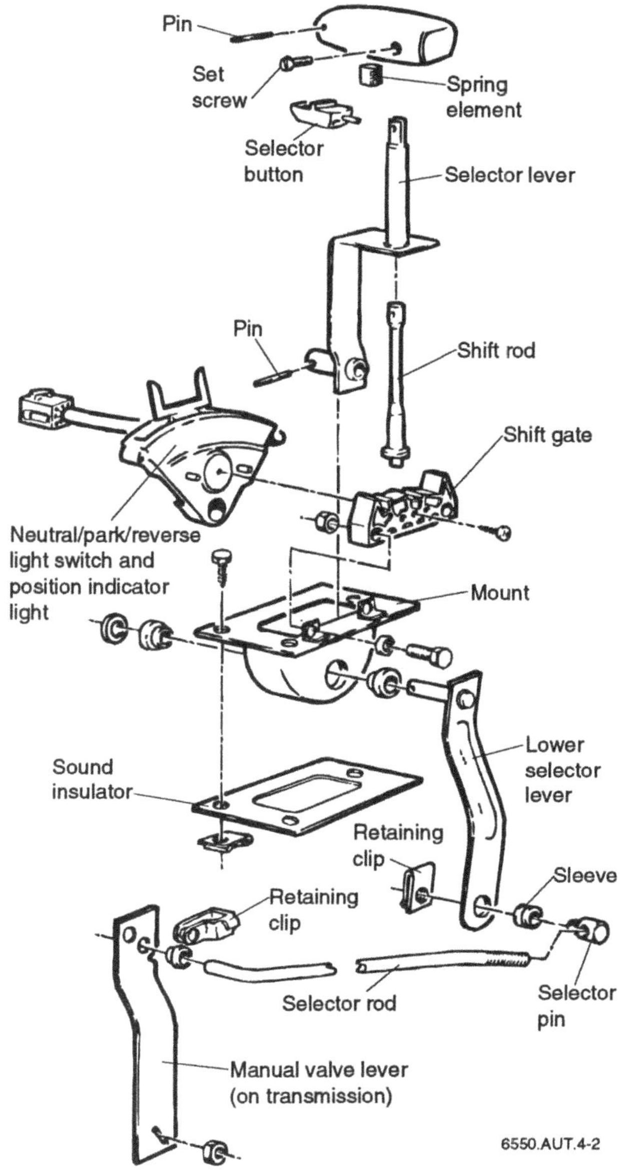

Fig. 4-2. Exploded view of 3 HP 22 automatic transmission selector lever and linkage.

To remove selector lever:

1. At the shift console inside the car, remove the console trim. Pry out the console plate, and then remove the screws that hold the trim to the console. See Fig. 4-3.

2. On 3 HP 22 transmissions, disconnect the selector rod from the bottom of the lower selector lever. See Fig. 4-4. It may be necessary to remove the exhaust system and heat shield to reach the linkage.

3. On 4 HP 22 transmissions, work inside the passenger compartment to disconnect the cable from the bottom of the selector lever.

Fig. 4-3. Console trim mounting screws (arrows).

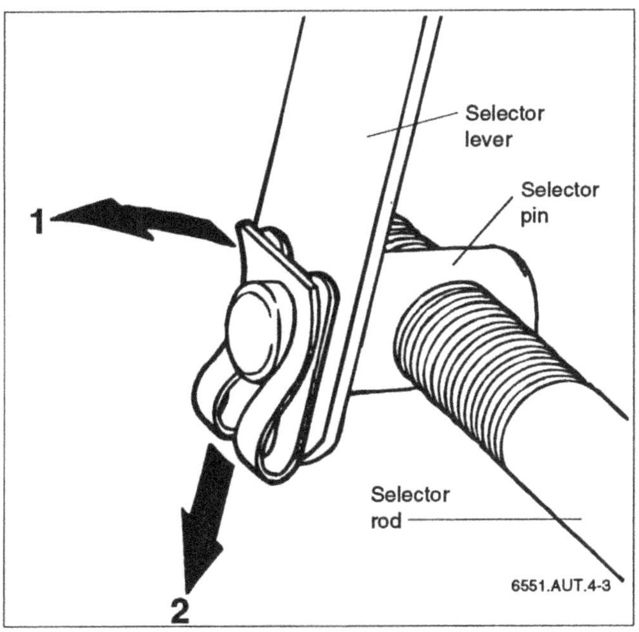

Fig. 4-4. Selector rod being disconnected from bottom of lower selector lever on 3 HP 22 models. Lift tab of retaining clip (direction **1**) and slide clip off (direction **2**). Models with cable use a similar retaining clip.

4. Remove the shift lever assembly by disconnecting the electrical switch and removing the hex-head screws holding the lever assembly to the frame. See Fig. 4-5.

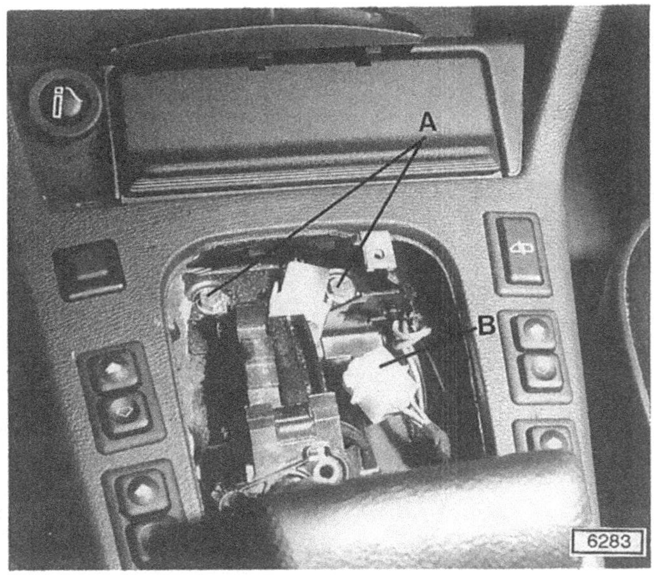

Fig. 4-5. Selector lever assembly showing screws (**A**) and wiring harness (**B**).

5. When reassembling the selector lever assembly, make sure that the sound insulator is between the assembly and the transmission tunnel, and adjust the linkage as described in **4.2 External Transmission Adjustments**.

If the lever is removed from the manual valve on the transmission, torque the nut when reinstalling the lever.

Tightening torque
• manual valve lever nut . . . . 8 to 10 Nm (71 to 89 in-lb)

On 3 HP 22 transmissions, make sure the selector rod is installed in the correct hole as shown in Fig. 4-6.

4.2 External Transmission Adjustments

External adjustments to the automatic transmission include adjustments of the selector cable or linkage and the transmission accelerator cable. A helper will be required to hold the accelerator pedal at full throttle while the cable adjustment is made. This adjustment is always made with the engine off.

Adjusting Selector Lever Linkage

This external adjustment may correct imprecise shifting without the need for removal of the transmission or costly overhaul work. On some models, it may be necessary to remove the exhaust assembly and heat shield to reach the bottom of the selector lever.

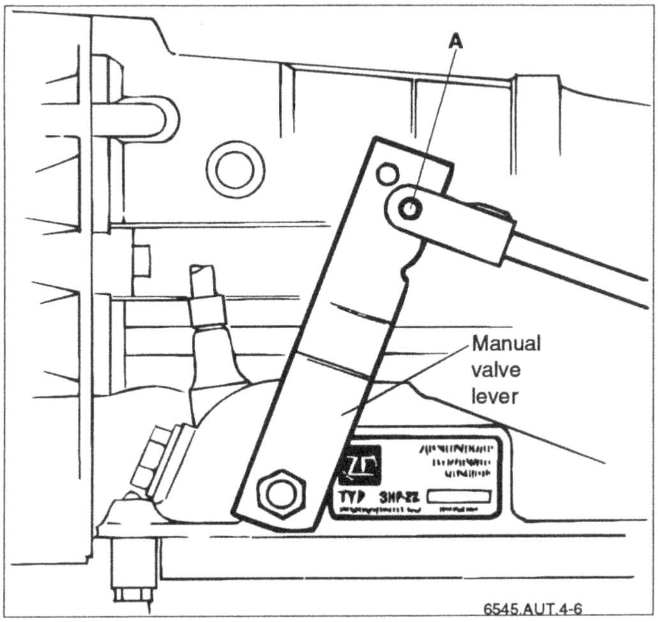

Fig. 4-6. Manual valve lever for 3 HP 22 transmissions. Selector rod should be installed in hole **A**.

To adjust linkage (3 HP 22):

1. Remove the selector rod from the bottom of the selector lever by removing the retaining clip shown in Fig. 4-3 above and pulling out the selector pin.

2. Pull the manual valve lever on the transmission all the way to the rear, then push it forward two clicks. The transmission should now be in **N**.

3. In the passenger compartment, move the selector lever to **N**, and then push it forward against the Neutral position stop in the shift gate.

4. While holding the selector lever against the gate, turn the selector pin on the threaded selector rod until the pin lines up with the hole on the lower selector lever. Then shorten the linkage by turning the pin an additional one to two turns clockwise.

5. Reinstall the selector pin and check that the linkage is correctly adjusted by starting the engine and moving the selector lever through all shift positions.

To adjust linkage (4 HP 22):

1. Position the shift lever in the P (Park) position.

2. Loosen the nut securing the cable rod to the manual valve lever.

3. Push the manual valve lever forward toward the engine (Park position) while pulling the selector cable rod backward.

4. Clamp the cable rod at the manual valve by torquing the nut.

Tightening torque
• cable rod securing nut . . . . . . 10 to 12 Nm (89 to 106 in-lb)

Adjusting Transmission Accelerator Cable and Kickdown

Proper throttle control and transmission operation depend on accurate adjustment of the accelerator cable. Before performing the adjustment, make sure the full-throttle adjustment for the accelerator cable and throttle linkage is correct, as described in **FUEL INJECTION**.

To adjust transmission accelerator cable and kickdown:

1. With the throttle at idle position, use a feeler gauge to check the distance between the lead seal and the end of the cable sleeve, as shown in Fig. 4-7. **Table f** gives the correct specification. If the distance is incorrect, loosen the locknuts and adjust the position of the cable sleeve, then tighten the nuts.

NOTE —
Always verify that the lead seal has not become loose and slipped on the cable. If the lead seal is loose, replace the cable.

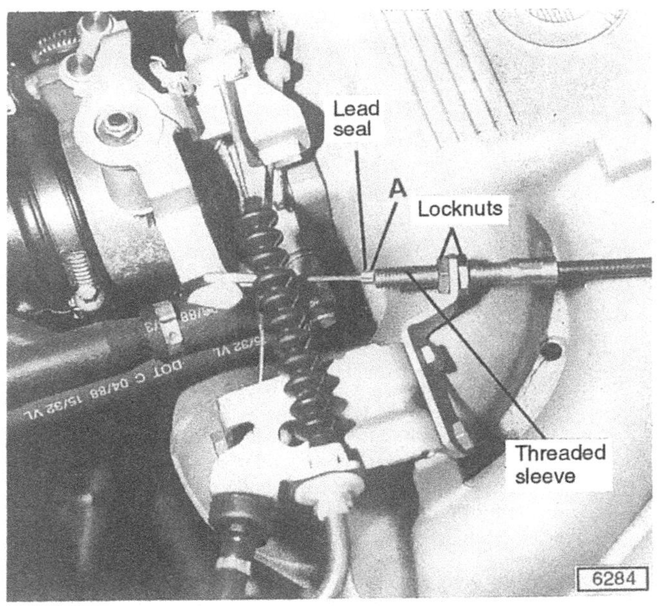

Fig. 4-7. Transmission accelerator cable adjustment point. Check distance **A** at idle and at kickdown position.

14

Table f. Transmission Accelerator Cable Adjustment Specifications

Measured between lead seal on cable and end of cable housing
at idle . . . . 0.50±0.25 mm (0.02±0.01 in.)
with accelerator pedal
at kickdown position. . . . . at least 44 mm (1.73 in.)

2. Working under the accelerator pedal, loosen the kickdown stop locknut, shown in Fig. 4-8 and screw the kickdown stop in.

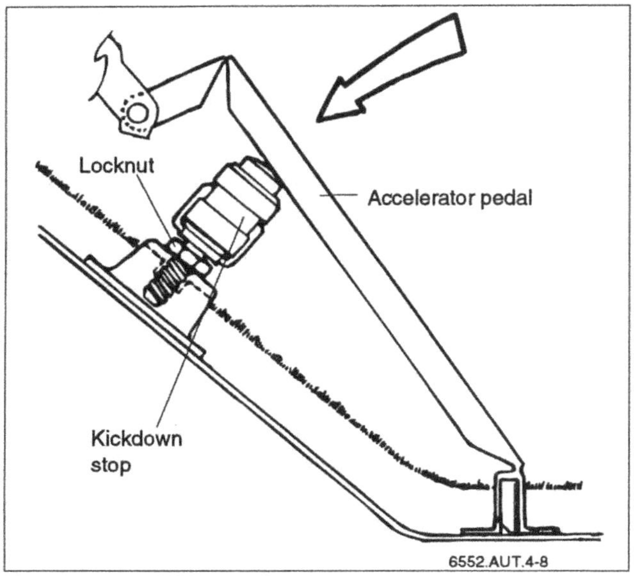

Fig. 4-8. Kickdown stop and accelerator pedal.

3. Depress the accelerator just until the point where transmission resistance is felt. Hold the pedal in this position and screw out the kickdown stop so that it just contacts the pedal. Then tighten the locknut.

4. Check the adjustment by depressing the accelerator pedal to the kickdown position and noting the position of the lead seal. The correct specification is listed in **Table f** above.

NOTE —
If the transmission accelerator cable cannot be adjusted to the correct specification, check the full-throttle adjustment of the throttle. Adjust if necessary, then recheck the transmission accelerator cable adjustment.

5. TRANSMISSION ASSEMBLY

Repair of internal transmission components requires that the transmission be removed for disassembly. Remanufactured or exchange transmissions are available through authorized BMW dealers, and these offer some savings over the purchase of a new unit.

To correct fluid leaks, the output shaft seal and the manual valve seal are replaceable with the transmission installed in the car. The transmission must be removed to replace the torque converter seal.

5.1 Removing and Installing Automatic Transmission

Using this procedure, the transmission assembly is separated from the engine, supported on a floor jack, and taken out from below. Removal of the transmission requires a transmission jack or a floor jack with transmission adaptor, and jack stands to support the car. Use extreme caution when working beneath the car and lowering the transmission. Special Torx® sockets are needed to remove the bellhousing bolts on some models.

To remove:

1. Disconnect the negative (–) and positive (+) battery cables from the battery, in that order.

2. Disconnect the transmission accelerator cable from its bracket on the intake manifold and then remove it from the throttle operating lever. Refer to Fig. 4-7 above.

3. Remove the exhaust system as described in **EXHAUST SYSTEM.** Also remove the heat shield from the transmission tunnel, where applicable.

NOTE —
It is possible to disconnect only the exhaust downpipe from the exhaust manifold, and the exhaust hanger from the transmission support, and to then push the exhaust assembly to the side and out of the way. The transmission removal procedure is somewhat easier, though, if the exhaust is removed completely.

4. Disconnect the selector rod from the bottom of the lower selector lever as described in **4.1 Removing and Installing Selector Lever and Linkage.**

5. Remove the driveshaft as described in **DRIVESHAFT AND FINAL DRIVE**. Leave the flexible coupling attached to the driveshaft. Be sure to mark the parts for correct reassembly.

6. Drain the transmission by removing the ATF sump drain plug, then remove the ATF filler tube and ATF cooler lines shown in Fig. 5-1. Plug the holes to prevent dirt from entering the transmission.

Fig. 5-1. ATF cooler lines for 4 HP 22 transmissions are at **1** and **2**. ATF filler tube is at **3**. Cooler lines on 3 HP transmission are similar.

7. On the 3 HP 22 remove the transmission cover at the bottom of the bellhousing.

8. Working through the opening in the bellhousing, remove the four bolts that hold the torque converter to the drive plate. Some models may have three bolts. See Fig. 5-2.

Fig. 5-2. Torque converter mounting bolt (arrow). Rotate torque converter for access to all bolts. Replace with new bolts when installing.

CAUTION —
Be sure to mark and use the correct bolts and washers securing the torque converter to the drive plate. Incorrect length bolts can damage the transmission.

NOTE —
Use a socket wrench on the crankshaft vibration damper and turn the crankshaft clockwise to move the bolts into position for removal.

9. On cars with the Motronic engine rpm sensor and the reference sensor mounted in the bellhousing, remove the sensors as described in **IGNITION**.

10. On cars with 4 HP 22 EH transmission, disconnect the control unit plug from the left side of the transmission, just above the ATF sump. Mark aligning marks for reinstallation.

11. Support the transmission from below. Remove the rear transmission support and, where applicable, the reinforcement bar, by removing the bolts that hold them to the body. See Fig. 5-3. Lower the engine and transmission onto the front axle carrier.

CAUTION —
At no time should the weight of the transmission be supported by the torque converter shaft.

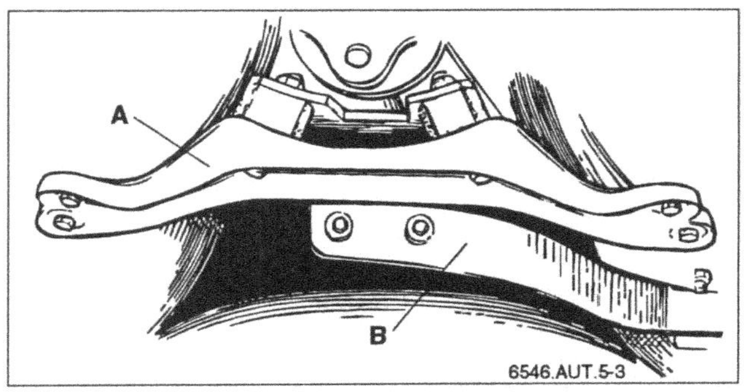

Fig. 5-3. Rear transmission support (**A**) and reinforcement bar (**B**) to be removed.

12. Remove the bolts that hold the transmission bellhousing to the engine.

13. Remove the inspection grill from the side of the bellhousing for access to the torque converter. While using a lever to make sure that the torque converter stays firmly mounted to the transmission, pull the transmission off of the engine. See Fig. 5-4.

WARNING —
Make sure the car is stable and well supported at all times during the removal procedure. Use jack stands which are designed for the purpose. A floor jack alone is not adequate support.

14

NOTE —
If the engine drive plate is cracked, it must be replaced. Use only new stretch bolts when installing the new drive plate, and coat them with thread lock. Clean the bolt holes thoroughly and torque the bolts.

Tightening torque	
• crankshaft to drive plate	
M 12	113 to 130 Nm (83 to 96 ft-lb)

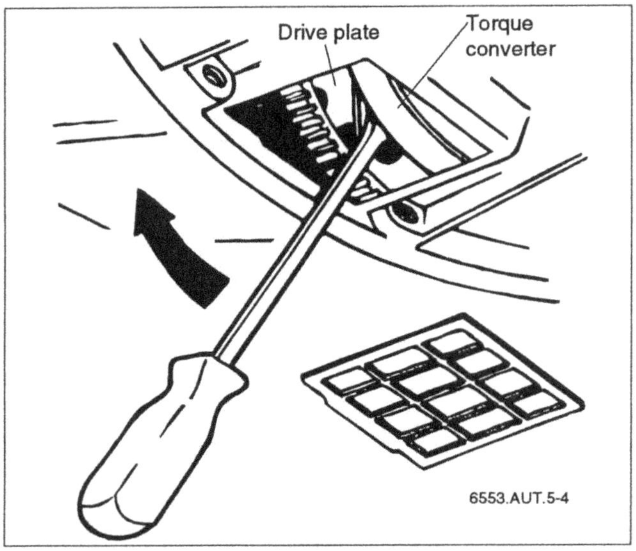

Fig. 5-4. Torque converter being held against transmission. Use screwdriver or lever to apply pressure in direction of arrow as transmission is pulled off of engine.

Installation is the reverse of removal. To avoid serious damage when installing the transmission, turn the converter back and forth to check that it has engaged its splines. The depth from the edge of the bellhousing to the flat surface where the threaded holes of the torque converter are located should be measured to verify that the converter is properly seated.

Depth: bellhousing edge to torque converter flat surface	
• 3 HP 22	30.5 mm (1.20 in.)
• 4 HP 22	below edge of case

Use a new gasket on the ATF sump drain plug, and a new O-ring on the ATF filler tube connection to the ATF sump. Adjust the transmission accelerator cable and selector lever linkage as described above in **4.2 External Transmission Adjustments**. Fastener torque specifications are listed below in **Table g**.

When reinstalling the drive shaft always replace the nuts securing the flexible coupling with new ones. Torque the nuts while holding the bolt stationary. Never torque the bolt head as the coupling will be destroyed. See **DRIVESHAFT AND FINAL DRIVE** for more information.

CAUTION —
- *When installing a new or rebuilt transmission, the ATF cooler and lines must be blown clean with compressed air and then thoroughly flushed twice with clean ATF. This will clean out any dirt or friction lining particles that could clog the passages of the new transmission.*
- *Refill the transmission with ATF as described in **LUBRICATION AND MAINTENANCE** before starting the engine or moving the car.*
- *Do not reuse drained ATF.*

NOTE —
- Torx® bolts must be used with washers. Bolts without washers may increase the effort needed to remove them.
- When reconnecting the control unit plug on 4 HP 22 EH transmissions, make sure that the marking lines are aligned.

Table g. Transmission Installation Torques

• transmission to engine (hex-head)	
M 8	24 Nm (18 ft-lb)
M 10	45 Nm (33 ft-lb)
M 12	78 to 86 Nm (58 to 63 ft-lb)
• transmission to engine (Torx®-head)	
M 8	21 Nm (15 ft-lb)
M 12	63 Nm (46 ft-lb)
• rear transmission support (to body)	22 to 24 Nm (16 to 17 ft-lb)
• torque converter to drive plate	
M 8	25 to 27 Nm (18 to 20 ft-lb)
M 10	47 to 51 Nm (35 to 38 ft-lb)
• transmission reinforcement plate	22 to 24 Nm (16 to 17 ft-lb)
• ATF cooler line coupling nuts	18 +3/-0 Nm (13 +2/-0 ft-lb)
• ATF cooler line hollow bolt	
M16	35 +3/-0 Nm (26 +2/-0 ft-lb)
• ATF cooler line adaptor coupling	
M16 to M14	35 +3/-0 Nm (26 +2/-0 ft-lb)
• ATF filler tube to ATF sump	
3 HP 22	100 to 110 Nm (74 to 81 ft-lb)
4 HP 22 H/EH	98 Nm (72 ft-lb)
• ATF sump drain plug	
M 10	5 to 17 Nm (11 to 13 ft-lb)

5.2 Seals

Three seals in the transmission are potential areas for ATF leaks. The first is the torque converter seal around the torque converter shaft. ATF leaking from this seal will usually be seen at the bottom of the bellhousing. The second seal is for the manual valve. ATF leaks here will be seen at the side of the transmission case and on the side of the ATF sump. The third seal is for the transmission output shaft, and ATF leaking here will drip down onto the extension housing.

NOTE —
- To pinpoint the location of leaks, have the undercarriage and transmission degreased. Remember that air turbulence from driving can cause fluids to appear to be leaking from other locations.

- A clogged or inoperative transmission vent can cause fluid leaks at other seals in the transmission. If in doubt, replace the vent valve located on the extension housing.

All of the seals can be replaced without extensive disassembly. It is necessary to remove the transmission to replace the torque converter seal. See **5.1 Removing and Installing Automatic Transmission.** It is also necessary to disconnect the driveshaft from the transmission to replace the output shaft seal. See **DRIVESHAFT AND FINAL DRIVE**.

Torque Converter Seal

A leaking torque converter seal is often caused by a worn bushing in the torque converter hub. The bushing should always be checked when the seal is replaced. A worn bushing will promote rapid wear of the new seal.

Fig. 5-5 shows the bearing surface that should be checked. Remove sharp edges and burrs with fine emery cloth. If the hub is deeply scored, the torque converter should be replaced.

To replace the seal, pull the converter off of its mounting on the transmission. To facilitate removal, install two long bolts halfway into the converter mounting holes and pull evenly on both of the bolts. Using a hooked seal tool or screwdriver, pry the old seal out of the transmission case. Lubricate the new seal with ATF and drive it into position.

CAUTION —
Be careful not to mar the surface of the housing when removing the seal.

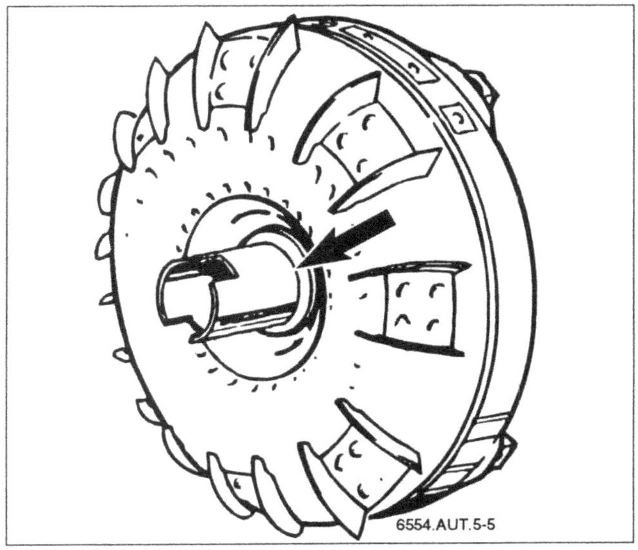

Fig. 5-5. Torque converter removed from engine showing bearing surface (arrow) that should be checked for scoring or wear.

Output Shaft Seal

The output shaft seal can be replaced without removing the transmission from the car, but a thin-walled 30 mm deep-well socket will be needed to remove the output flange collar nut. Also, installation of the collar nut requires a new lockplate, and application of a sealant to prevent transmission fluid leakage. BMW specifies Curil® K2 or Loctite® 572, which should be available from your BMW dealer.

To replace the output shaft seal, bend back the locking tab on the lockplate, shown in Fig. 5-6. Hold the output flange stationary and remove the collar nut, then use a puller to pull off the flange. To hold the output flange stationary put the shift lever in park. Pry the old seal out of the transmission, lubricate the new seal with ATF, and drive it into place.

When installing the collar nut, lightly coat the side that presses against the output flange with the sealer and torque the nut. Install the new lockplate and bend the tab into the slot.

Tightening torque	
• output flange collar nut	100 Nm (74 ft-lb)

CAUTION —
Be careful not to mar the surface of the housing when removing the seal.

14

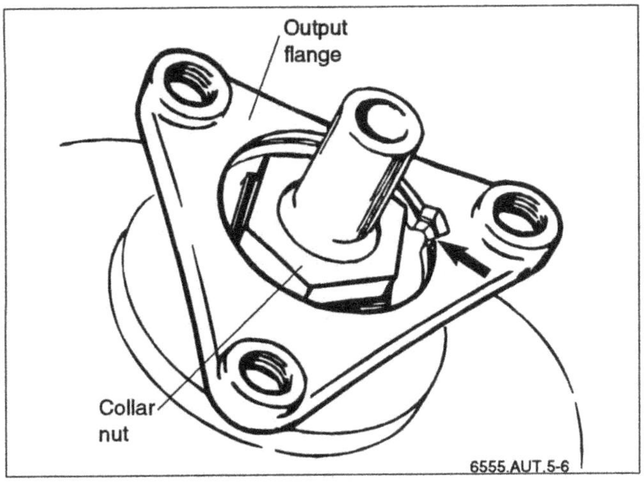

Fig. 5-6. Locking tab for output flange collar nut (arrow). Bend tab out of groove to remove nut.

Manual Valve Seal

To replace the manual valve seal, remove the manual valve lever from the transmission, shown in Fig. 5-7. Pry out the old seal using a small screwdriver or hooked tool. Lubricate the new seal with ATF, and drive it into place. When installing the manual lever, torque the nut.

Tightening torque
• manual lever nut . . . . .8 to 10 Nm (71 to 89 in-lb)

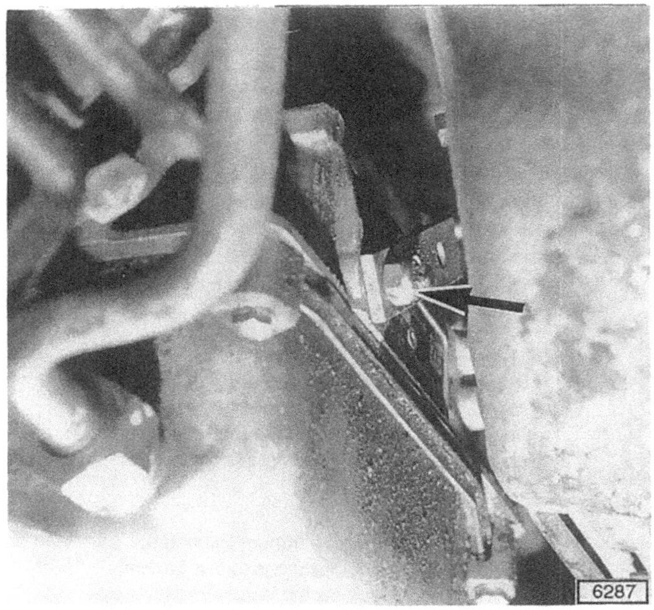

Fig. 5-7. Manual valve lever and nut (arrow) on transmission to be removed for replacement of seal.

If the linkage was not disconnected from the manual valve lever, it should not be necessary to adjust the shift linkage.

CAUTION —
Be careful not to mar the surface of the housing when removing the seal.

Section 15

DRIVESHAFT AND FINAL DRIVE

Contents

INTRODUCTION

The driveshaft transmits power from the engine and transmission to the final drive. From the final drive, power is transmitted through the drive axles and constant velocity joints to the drive wheels.

Service and repair of the drive axles and the rear wheel bearings is covered in **SUSPENSION–REAR**. Repairs to the rear brakes are covered in **BRAKES**. Checking or changing the oil in the final drive is covered in **LUBRICATION AND MAINTENANCE**.

1. GENERAL DESCRIPTION

Fig. 1-1 shows a driveshaft and final drive removed from the car. Together, the two are often referred to as the driveline. The installed position of the driveshaft and final drive in relation to the rear axle carrier and rear suspension is shown in **SUSPENSION–REAR**.

1.1 Driveshaft

The driveshaft is made of two sections. On most models, the driveshaft halves are joined by a sliding splined coupling. This coupling compensates for the slight front-and-back movement of the engine and transmission. The rear of the driveshaft is bolted to the final drive. The front is connected to the transmission by either a flexible rubber coupling, a universal joint, or a constant velocity joint.

The rubber coupling (sometimes known as the guibo or flex-disc) isolates the driveshaft and final drive from the sudden torque forces of the engine. On some models, a vibration damper is mounted between the flexible coupling and the transmission output flange. The universal joints and the constant velocity joint on the front and rear sections compensate for differences in the angle of the driveshaft as it rotates.

The middle of the driveshaft is supported by the center bearing. The bearing is rubber mounted to isolate driveshaft vibrations. The bearing housing is bolted to the car body.

1.2 Final Drive

The final drive, also known as a differential, consists of the drive pinion, the ring gear, the differential, and the output flanges. The drive pinion is also known as the final drive input shaft. It is connected to the driveshaft by the input flange, and drives the ring gear. Together, the drive pinion and ring gear are known as the crown wheel set. The final drive is lubricated by hypoid gear oil.

The differential is bolted to the ring gear and drives the rear wheels through the output flanges and drive axles. The differential allows the rear wheels to turn at different speeds, as is necessary when making turns and the outside wheel must travel farther than the inside wheel. The final drive assembly at the top and sides is bolted to the rear-axle carrier, and at the rear is connected to the car body by flexible rubber bushings.

Limited-Slip Differential

Some models have a limited-slip differential in the final drive. In a limited-slip differential, the two output flanges are connected by a number of clutch plates. The plates create a solid connection so that both wheels are always driven, even if one is slipping. In cornering, when the rear wheels move at different speeds, the clutch plates allow a limited amount of slip between the rotational speed of the two axles.

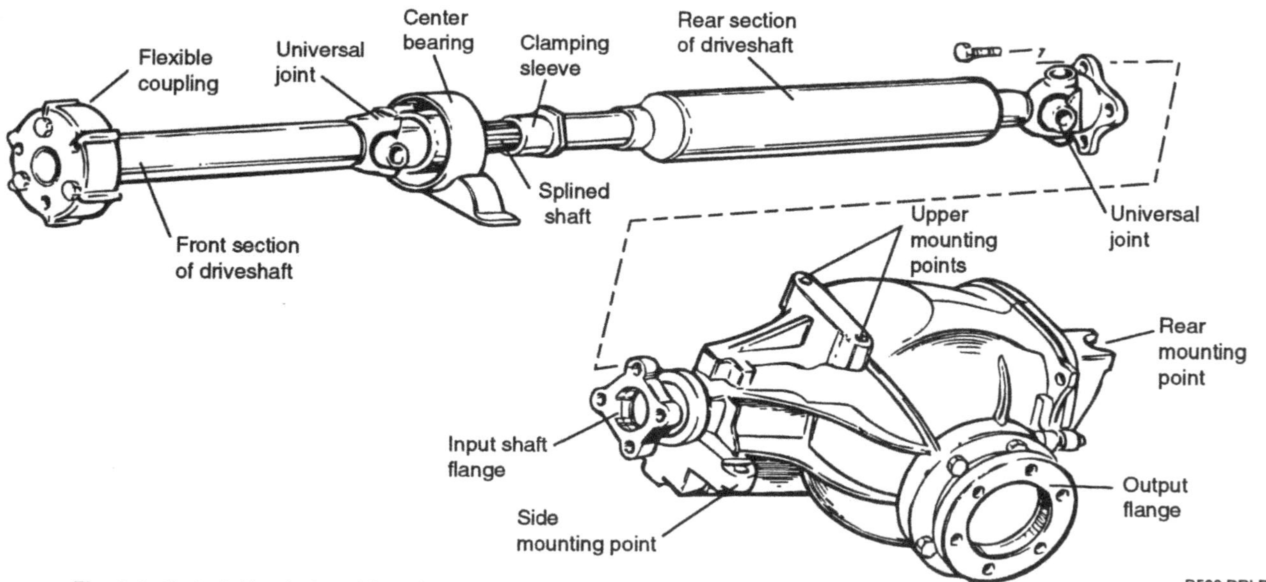

Fig. 1-1. Typical driveshaft and final drive assemblies. Version with flexible coupling shown.

NOTE —
Limited-slip differentials and non-slip differentials may require different lubricating oils. For more information on differential lubrication requirements, see **LUBRICATION AND MAINTENANCE.**

1.3 Applications–Identifying Features

Final drives are identified by a code that gives drive ratio and manufacturing number. The code is stamped on a metal tag attached to the final drive. See Fig. 1-2. The code on the tag depends on whether the final drive is the original one installed in the car or whether it is a replacement final drive.

For original final drives, the code gives the drive ratio first and then the manufacturing number. So code **3.25 4002** identifies an original equipment final drive with a ratio of 3.25:1. Limited-slip new final drives are identified with the prefix **S** (S 3.25 4002).

Fig. 1-2. Final drive showing location of identification tag (arrow).

For replacement final drives, the code also has the drive ratio and manufacturing number. However, the ratio is given indirectly, by the number of teeth on the ring and pinion. For example, **14 41 A 405** is an exchange final drive with a ratio of 2.93:1 (41 ring teeth divided by 14 pinion teeth). The **A** is an additional indicator that it is a replacement final drive. Limited-slip replacement final drives are identified by the suffix **S** (14 41 A 405 S).

2. MAINTENANCE

BMW specifies the maintenance steps below to be carried out at particular time or mileage intervals for proper maintenance of the final drive. Information on performing these maintenance steps, and on the prescribed maintenance intervals, can be found in **LUBRICATION AND MAINTENANCE**.

1. Checking final drive oil level
2. Changing final drive oil

3. TROUBLESHOOTING

This troubleshooting section applies to problems affecting the driveshaft and final drive. The basic function of these components is to smoothly transmit engine power to the drive axles and rear wheels.

The source of driveline vibrations and noise can be difficult to pinpoint. Engine, transmission, rear axle, or wheel vibrations can be transmitted through the driveshaft to the car body. Noises from the rear of the car may be caused by final drive problems, or by faulty wheel bearings or constant velocity (CV) joints. Also check that the tires are correctly inflated and are not excessively worn.

WARNING —
Never stand in front of or behind the vehicle when it is operating.

To isolate a vibration or noise problem, speed up the engine in the stopped car to the rpm range where the problem occurs. This eliminates the influence of the rotating driveshaft and will help indicate if the problem is caused by an engine condition.

Drive the car at the speed where the problem occurs, then shift to different gear ranges to see if the problem changes with road-speed or engine-speed. Road-speed dependent problems usually indicate trouble in the driveline.

Detach the axle shafts at the final drive and then operate the engine at the speed and gear/selector-lever position where the problem occurs. If the problem disappears, then it may be in the axle CV joints or wheel bearings.

CAUTION —
- *Damage to the differential and axle shaft joints can result if the car is run with the wheels off the ground, for example, on a lift. The wheels hang below their normal height putting stress on driveline components. Compress the suspension to normal height with jackstands before running the car.*

- *Never operate the car with one drive wheel turning at a different speed than the other. Damage to the driveline components can result.*

NOTE —
No-load tests with the car stopped or the drive axles disconnected will only give a general idea of the car's performance under the load of normal operation.

15

For more information on troubleshooting engine or transmission problems, see **ENGINE–SERVICE AND REPAIR**, **MANUAL TRANSMISSION AND CLUTCH**, **MANUAL TRANSMISSION OVERHAUL** or **AUTOMATIC TRANSMISSION**. For information on troubleshooting rear axle, wheel bearing or constant velocity joint problems, see **SUSPENSION–REAR**.

3.1 Basic Troubleshooting Principles

Smooth operation of the driveshaft and final drive depends on the condition of the rubber isolation components and center bearing, the lubricant level and type in the final drive, and properly tightened fasteners. Any symptom of vibration or noise may be caused by worn or damaged components.

Aside from inspection for worn or broken parts, troubleshooting should also consider the installed angle of the driveshaft. If applicable, check the splined coupling for free movement. See **4.1 Driveshaft Noise and Vibration** for more information.

Table a lists symptoms of driveshaft and final drive problems and their probable causes, and suggests corrective actions. The numbers in bold type in the corrective action column refer to headings in this section where the repairs are described.

Table a. Driveshaft and Final Drive Troubleshooting

Symptom	Probable cause	Corrective action
1. Vibration when moving off (forward or reverse)	**a.** Incorrect preload of center bearing	**a.** Check preload of center bearing. Readjust if necessary. **4.2**
	b. Center bearing rubber deteriorated or torn	**b.** Inspect center bearing. Replace if necessary. **4.4**
	c. Flexible coupling damaged or worn	**c.** Inspect flexible coupling. Replace if necessary. **4.3**
	d. Engine or transmission mounts faulty	**d.** Inspect engine and transmission mounts. Align or replace, if necessary
	e. Front centering guide worn, or driveshaft mounting flanges out of round	**e.** Check front centering guide and replace if necessary. **4.5**. Check runout of driveshaft flanges. **4.1**
	f. Universal joints worn or seized	**f.** Check universal joint play and movement. **4.1**. Replace driveshaft if necessary. **4.2**
	g. Sliding coupling seized	**g.** Remove driveshaft and check movement of sliding coupling. Clean coupling splines and replace parts as necessary. **4.2**
	h. Driveshaft misaligned	**h.** Check driveshaft alignment. **4.1**
2. Vibration at 25 to 30 mph (40 to 50 km/h)	**a.** Front centering guide worn, or driveshaft mounting flanges out of round or damaged	**a.** Check front centering guide and replace if necessary. **4.5**. Check runout of driveshaft flanges. **4.1**
	b. Universal joints worn or seized	**b.** Check universal joint play and movement. **4.1.** Replace driveshaft if necessary. **4.2**
	c. Flexible coupling damaged or worn	**c.** Inspect flexible coupling. Replace if necessary. **4.3**
	d. Center bearing rubber deteriorated or torn	**d.** Inspect center bearing. Replace if necessary. **4.4**
	e. Sliding coupling seized	**e.** Remove driveshaft and check movement of sliding coupling. Clean coupling splines and replace parts as necessary. **4.2**
	f. Driveshaft misaligned	**f.** Check driveshaft alignment. **4.1**
3. Vibration, audible rumble over 35 mph (60 km/h)	**a.** Front centering guide worn, or driveshaft mounting flanges out of round or damaged	**a.** Check front centering guide and replace if necessary. **4.5**. Check runout of driveshaft mounting flanges. **4.1**
	b. Mounting flange bolts loose or holes worn	**b.** Remove driveshaft and check transmission output flange and final drive input flange. Replace if necessary. **4.1**
	c. Driveshaft unbalanced	**c.** Check driveshaft for loose or missing balance plates. Have driveshaft rebalanced or replace if necessary. **4.1**
	d. Universal joints worn or seized	**d.** Check universal joint play and movement. **4.1**. Replace driveshaft if necessary. **4.2**
	e. Sliding coupling seized	**e.** Remove driveshaft and check movement of sliding coupling. Clean coupling splines or replace parts as necessary. **4.2**

continued on next page

Table a. Driveshaft and Final Drive Troubleshooting (continued)

Symptom	Probable cause	Corrective action
3. Vibration, audible rumble over 35 mph (cont'd)	f. Incorrect preload of center bearing g. Center bearing faulty h. Final drive rubber mount faulty i. Driveshaft misaligned	f. Check preload of center bearing. Readjust if necessary. **4.2** g. Replace center bearing. **4.4** h. Inspect final drive rubber mount and replace if necessary. **5.2** i. Check driveshaft alignment. **4.1**
4. Noise during on/off throttle or when engaging clutch	a. Final drive components worn or damaged (excessive pinion-to-ring-gear clearance) b. Drive axle or CV joint faulty c. Sliding coupling seized	a. Remove final drive and repair. **5.1** b. Inspect drive axles and CV joints. Repair or replace as necessary. See **SUSPENSION–REAR** c. Remove driveshaft and check movement of sliding coupling. Clean coupling splines and replace parts as necessary. **4.2**
5. Rattling, clicking, or groaning	a. Universal joints worn or seized b. Final drive lubricant level low c. Faulty CV joint or wheel bearing d. Loose flexible coupling mounting bolts	a. Check universal joint play and movement. **4.1.** Replace driveshaft if necessary. **4.2** b. Check for final drive leaks. Replace oil seals as necessary. **5.3.** Check and correct gear oil level if needed. See **LUBRICATION AND MAINTENANCE** c. Inspect CV joints and wheel bearings. Replace as necessary. See **SUSPENSION–REAR** d. Inspect flexible coupling and tighten bolts as necessary. **4.3**
6. Grinding noise when turning corner	a. Faulty wheel bearing b. Limited-slip differential faulty	a. Inspect wheel bearings, replace as necessary. See **SUSPENSION–REAR** b. Remove final drive and repair. **5.1**
7. Continual drumming or humming noise from rear end (goes away when accelerator pedal is released)	a. Final drive components worn or damaged (excessive pinion-to-ring-gear clearance)	a. Remove final drive and repair. **5.1**

4. DRIVESHAFT

Most components of the driveshaft are easily replaced once the driveshaft is removed from the car. Repair kits for the universal joints are available, but it is not common practice to repair the universal joints on BMW driveshafts, and there are no BMW-recommended repair procedures. Worn or damaged universal joints usually require replacement of that section of the driveshaft.

The front of the driveshaft is aligned with the transmission by a centering guide. The guide is recessed into the front of the driveshaft. It engages the end of the transmission output shaft when the driveshaft is installed.

The driveshaft is balanced to very close tolerances. Whenever it is to be removed or disassembled, the mounting flanges and driveshaft sections should be marked with paint or a punch before proceeding with work. This will ensure that the driveshaft can be reassembled or installed in exactly the same orientation.

4.1 Driveshaft Noise and Vibration

The causes of driveshaft noise and vibration can be difficult to pinpoint. When troubleshooting driveshaft problems, begin with a close visual inspection. Check the driveshaft for broken or missing balance weights. The weights are welded to the driveshaft, so broken welds will indicate where a weight may have fallen off.

Check the torque of the fasteners at the flange connections, check the rubber of the flexible coupling and center bearing for deterioration or tearing, and check that the preload for the center bearing is correct. See **4.2 Removing and Installing Driveshaft** for more information.

Check the universal joints for play. With the driveshaft installed, pull and twist the driveshaft while watching the joint. The BMW specification for play is very small, so almost any noticeable play could indicate a problem.

Universal joint play
• maximum allowable . . . . 0.15 mm (0.006 in.)

Further inspection requires removal of the driveshaft. Check the front centering guide for correct installation as described in **4.5 Replacing Front Centering Guide**. Also check runout at the transmission output flange and output shaft, and at the final drive input flange. Check the bolt hole bores in the flange for wear.

Driveshaft flange runout

- transmission output flange
  axial play . . . . .0.10 mm (0.004 in.) maximum
  radial play . . . . .0.07 mm (0.003 in.) maximum
- final drive input flange
  radial play (measured at driveshaft
  centering lip) . . . . .0.07 mm (0.003 in.) maximum

Check for smooth operation of the center bearing. There should be slight resistance, but no binding or grittiness. If applicable, check that the splines of the sliding coupling are properly lubricated and free from corrosion. Retorque the clamping sleeve to the proper torque as described in **4.2 Removing and Installing Driveshaft** and check that the coupling slides freely. If it doesn't, clean the splines and lubricate them with molybdenum disulfide grease (Molykote Longterm 2® or equivalent). Replace the clamping bushing if necessary, then recheck movement.

Check that the universal joints move freely without binding. If they are difficult to move or feel gritty, the driveshaft may have to be replaced.

NOTE —
With the driveshaft installed, the actual amount that the universal joints pivot is limited. For the most accurate test, check universal joints only in their normal range of movement.

If inspection reveals nothing wrong with the driveshaft, it may need to be rebalanced. This can be done by any repair shop with the right equipment. Also, check the alignment of the driveshaft as described below.

NOTE —
Minor driveshaft vibrations can often be corrected simply by disconnecting the driveshaft at the final drive, and repositioning the driveshaft 90°, 180° or 270° in relation to the final drive input flange.

Aligning Driveshaft

The alignment of the driveshaft does not normally need to be checked unless the engine/transmission or the final drive have been removed and installed. If, however, all other parts of the driveshaft have been inspected and found to be okay, but there is still noise or vibration, it's probably a good idea to check driveshaft alignment.

There are two important driveshaft alignment checks. The first is to make sure that the driveshaft runs straight from the transmission to the final drive, without any variation from side-to-side caused by misalignment of the engine/transmission in its mounts. Make a basic check by sighting along the driveshaft from back to front. Any misalignment should be apparent from the center bearing forward. To adjust the side-to-side alignment, loosen the transmission or engine mounts to reposition them, then retighten the mounts. The driveshaft is centered when it is positioned as shown in Fig. 4-1.

Fig. 4-1. Driveshaft side-to-side alignment. Driveshaft should be centered between the edges of the driveshaft tunnel as shown.

The second important driveshaft alignment check is more complicated. It checks the amount the driveshaft is angled vertically at the joints. This angle is known as driveshaft deflection.

In general, there should be little deflection in the driveshaft between the engine, the center bearing, and the final drive. Precise checks require the use of a large protractor or some other means of measuring the angle of the engine and the final drive and comparing these angles to the angle of the driveshaft sections.

To change the angle, shims are placed between the center bearing and the body or between the transmission and its rear support. When using shims to change a deflection angle, keep in mind that the angle of adjacent joints will also change. Deflection angles should be as small as possible. **Table b** lists the correct specifications.

CAUTION —
The maximum allowable change in height of the center bearing or transmission support using shims is 3 mm (.118 in.).

Table b. Driveshaft Deflection Angle Specifications

Deflection angle being checked	528e		533i/535i	
	manual transmission	automatic transmission	manual transmission	automatic transmission
Flexible coupling (difference between engine/transmission angle and front driveshaft angle)	–20' to +40'	–16' to +44'	50' to 1°58'	1° 03' to 2° 03'
Center bearing (difference between front driveshaft angle and rear driveshaft angle)	–36' to +24'	–38' to +22'	–02' to +58'	–07' to +53'
Rear universal joint (difference between rear driveshaft angle and final drive angle)	–18' to +42'	–18' to +42'	–46' to +14'	–46' to +14'

4.2 Removing and Installing Driveshaft

The driveshaft must be removed for complete inspection, for replacement of worn or broken driveshaft components, or when removing other components such as the engine or transmission. The procedure differs depending on whether there is a flexible coupling or a CV joint at the front of the driveshaft.

The procedure below describes the complete removal of the driveshaft from the car. It is possible to disconnect only one end of the driveshaft and leave the other end connected, for instance when removing the transmission, but the center bearing must always be unbolted and the driveshaft securely supported. To prevent damage to the driveshaft and to make working under the car easier, the publisher recommends always removing the complete driveshaft.

> ***WARNING —***
> *Removal of the driveshaft will disconnect the transmission from the final drive. Do not rely on engagement of the transmission to prevent the car from rolling. Chock all wheels that are on the ground. Use extreme caution when working beneath the car and lowering the driveshaft. Firmly set the parking brake.*

> ***CAUTION —***
> *• Do not let the driveshaft hang unsupported with only one end connected. Damage to the universal joints could result. If only one end of the driveshaft is disconnected, use stiff wire to suspend the driveshaft out of the way, in as close to the installed position as possible.*
>
> *• The driveshaft is mounted to the transmission and final drive with self-locking nuts. These nuts are designed to be used only once and should be replaced whenever they are removed.*

To remove driveshaft:

1. Remove the exhaust system as described in **EXHAUST SYSTEM.**

2. Remove the exhaust heat shield.

> ***CAUTION —***
> *On some models, the front exhaust bracket mounting bolts also secure the holder for the oxygen sensor connector. Take care not to damage the sensor connector or its wires.*

3. If applicable, loosen the sliding coupling clamping sleeve by several turns, but do not remove it. See Fig. 4-2.

> **NOTE —**
> BMW special tool 261 040 or equivalent can be used to loosen the clamping sleeve.

Fig. 4-2. Clamping sleeve for sliding coupling (arrow) to be loosened.

4. Using paint or a punch, make matching marks on the driveshaft and its attaching flanges on the final drive and transmission.

15

5. On models with a front universal joint or a flexible coupling, remove the nuts and bolts holding the coupling or universal joint to the transmission flange. See Fig. 4-3. Do not pull the driveshaft off the flange.

Fig. 4-3. Front driveshaft to transmission mounting bolts being removed.

NOTE —
Removal of the bolts may be made easier by placing a large hose clamp around the flexible coupling and tightening the clamp slightly to compress the coupling

6. On models with a front CV joint, use a suitable transmission jack to support the transmission, then loosen the rear transmission support mounting nuts and bolts and slide the mount toward the rear. See Fig. 4-4. Then remove the six CV joint to transmission mounting nuts and bolts.

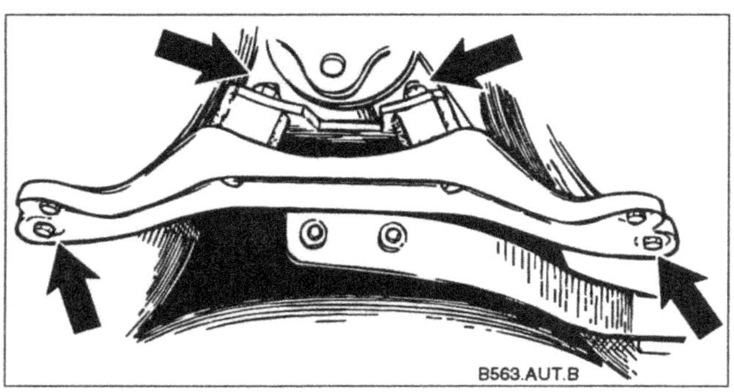

Fig. 4-4. Transmission mounting bracket mounting bolts (arrows). Loosen bolts and slide bracket to rear.

7. Remove the bolts that hold the rear driveshaft section to the final drive flange. See Fig. 4-5.

Fig. 4-5. Rear driveshaft to final drive mounting bolt being removed.

8. Support the driveshaft and remove the center bearing mounting bolts.

CAUTION —
Do not let the rear section rest on the fuel tank connection line.

9. Remove the driveshaft. Bend it down at the center bearing, and then pull it off at the front. It may be necessary to slide the driveshaft together at the splined coupling to make enough room. Do not pull the two sections apart.

NOTE —
On models with a vibration damper, the damper can be removed from the transmission flange by turning it about 60° and pulling it over the flange. See Fig. 4-6.

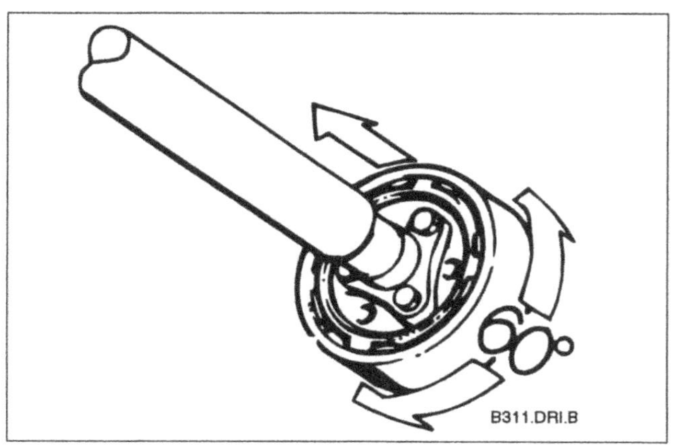

Fig. 4-6. On models with vibration damper, turn damper 60° in either direction to remove.

If the splined coupling is to be cleaned or if the rubber bushing is being replaced, mark the two sections of the driveshaft before pulling the coupling apart. Lubricate the splines with molybdenum disulfide grease (Molykote Longterm 2® or equivalent) before reassembly. Inspect the flange bolt mounting holes and replace any worn components.

NOTE —
If the two sections are pulled apart without the driveshaft being marked, reassemble the two halves so that the universal joints are on the same plane as shown in Fig. 4-7. There is still the possibility of the driveshaft being reassembled wrong by 180° and causing vibration. In this case remove the driveshaft and turn one section 180°, and reassemble.

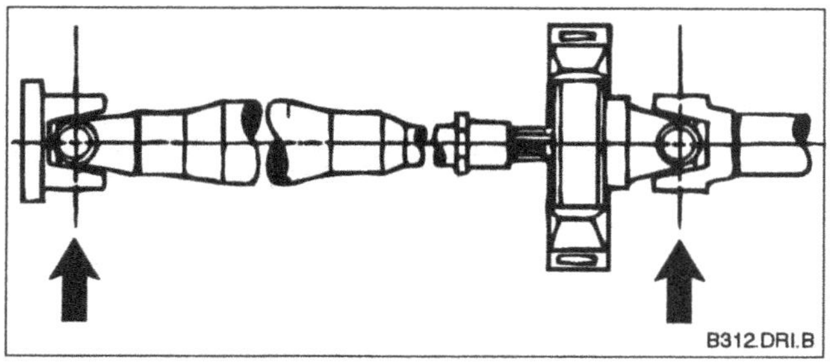

Fig. 4-7. If driveshaft sections are separated, make sure U-joints are on same plane as shown when driveshaft is reassembled.

To install driveshaft:

1. Lightly lubricate the end of the transmission output shaft with molybdenum disulfide grease (Molykote Longterm 2® or equivalent).

2. Reinstall the driveshaft by first positioning it against the final drive input flange and then sliding it onto the transmission output shaft and flange. Position the center bearing up against the body and loosely install the center mount bolts.

3. Using new self-locking nuts, reconnect the rear drive shaft section to the final drive. Align the marks made during removal. Torque the nuts while holding the bolts.

Tightening torque
• driveshaft to final drive flange . . . . 72 Nm (53 ft-lb)

CAUTION —
Do not reuse self-locking nuts. These nuts are designed to be used only once and should be replaced whenever they are removed.

4. Preload the center bearing mount toward front of car. Torque the mounting bolts. See Fig. 4-8.

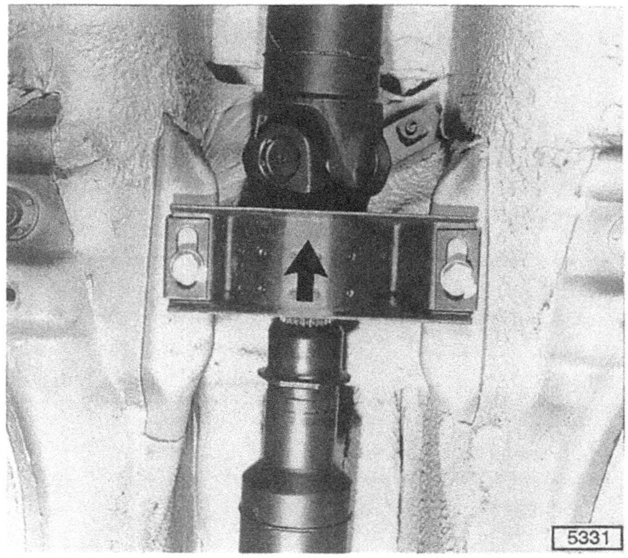

Fig. 4-8. Before tightening bolts, push center bearing towards the front (arrow) to preload it.

Center bearing mount preload	
• driveshaft with front U-joint or flex coupling	4 to 6 mm (5/32 to 1/4 in.)
• driveshaft with front CV joint	4 to 5 mm (5/32 to 7/32 in.)

Tightening torque
• driveshaft center mount to body . . . . 22 Nm (16 ft-lb)

5. Using new self-locking nuts, reconnect the front driveshaft to the transmission. Align the marks made during removal. While holding the bolts, torque the nuts.

NOTE —
• Installation of the bolts may be made easier by first placing a large hose clamp around the flexible coupling, and then tightening the clamp slightly to compress the coupling.

• The grade of the bolt is marked on the head. When replacing bolts, only use bolts of the same strength and hardness as the originals installed.

Tightening torques	
• driveshaft-to-transmission fasteners	
M10 8.8 grade	46 Nm (34 ft-lb)
M10 10.9 grade	72 Nm (53 ft-lb)
M12	123 Nm (90 ft-lb)

15

CAUTION —
Avoid stressing the flexible coupling when torquing the bolts. Do this by holding the bolts steady and turning the nuts on the flange side.

6. On models with CV joints, slide the rear transmission bracket forward and tighten the mounting bolts. Remove the transmission jack.

Tightening torques	
• rear transmission bracket to transmission and body	
M10	43 to 48 Nm (32 to 35 ft-lb)
M8	22 to 24 Nm (16 to 17 ft-lb)

7. If applicable, tighten the clamping sleeve on the splined coupling and torque it using BMW special tool 261 040 or equivalent.

Tightening torque	
• clamping sleeve	17 Nm (13 ft-lb)

8. Reinstall the heat shield, and the holder for oxygen sensor, where applicable.

9. Reinstall the exhaust system.

4.3 Replacing Flexible Coupling

The flexible coupling between the front section of the driveshaft and the transmission is made of rubber. Although BMW specifies no maintenance interval for the flexible coupling it should be inspected regularly. Check the coupling for cracks, tears, missing pieces, or distortion. Always check for worn bolt hole bores in the flange. When replacing the flexible coupling, use new self-locking nuts.

To replace flexible coupling:

1. Remove the driveshaft as described above in **4.2 Removing and Installing Driveshaft**.

CAUTION —
It is possible to only partially remove the driveshaft, leaving it connected to the final drive. Use extreme care when using this method. Use stiff wire to suspend the driveshaft in as close to the installed position as possible. If the driveshaft hangs from the final drive unsupported, the rear universal joint may be damaged. The entire rear driveshaft section will then need to be replaced.

2. Remove the flexible coupling from the driveshaft.

NOTE —
Removal and installation of the bolts may be made easier by placing a large hose clamp around the flexible coupling, and tightening the clamp slightly to compress the coupling.

3. Install the new flexible coupling using new self-locking nuts. If the coupling has arrows on it, they should face the flange arms, as shown in Fig. 4-9. While holding the bolts, torque the nuts.

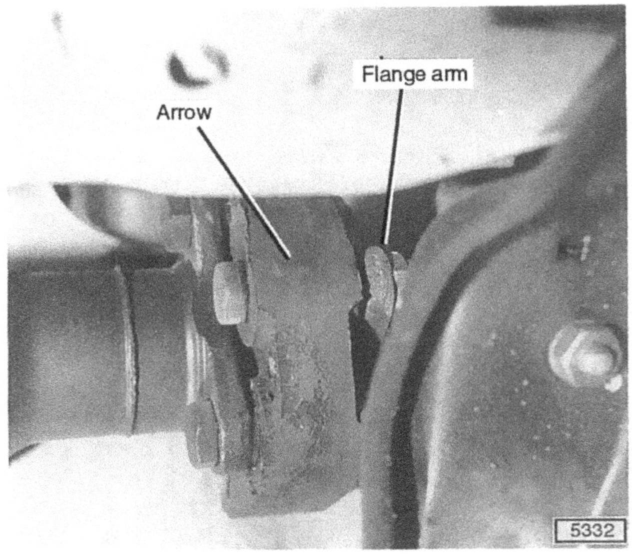

Fig. 4-9. When attaching flexible coupling to driveshaft or to transmission flange, arrows must point toward flange arms as shown.

Tightening torques	
• flexible coupling to driveshaft or transmission flange:	
M10 8.8 grade	46 Nm (34 ft-lb)
M10 10.9 grade	72 Nm (53 ft-lb)
M12	123 Nm (90 ft-lb)

CAUTION —
• Do not reuse self-locking nuts. These nuts are designed to be used only once and should be replaced whenever they are removed.

• Avoid stressing the flexible coupling when torquing the bolts. Do this by holding the bolt steady and turning the nut on the flange side.

NOTE —
• Remove the hose clamp if one was used during installation or if the flexible coupling came supplied with one.

• The bolt grade is marked on the head. When replacing bolts, only use bolts of the same strength and hardness as the originals installed.

4. Reinstall the driveshaft.

4.4 Replacing Center Bearing

The center bearing consists of a grooved ball bearing assembly press-fit in a rubber mount. The bearing assembly is pressed onto the front section of the driveshaft and secured by a circlip or a bolt. See Fig. 4-10.

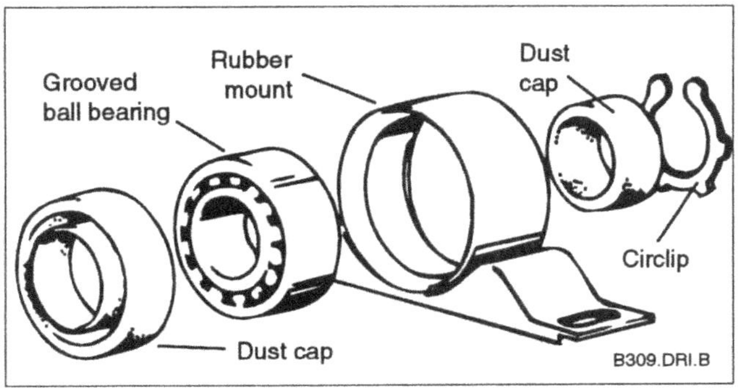

Fig. 4-10. Exploded view of center bearing components. Model with circlip shown. Some early models use bolt.

The center bearing is protected by dust caps on either side, but water and dirt may contaminate the bearing and cause it to fail. Driveshaft vibration may also contribute to bearing failure. BMW does not specify any periodic maintenance or lubrication procedures for the center bearing.

To properly inspect the bearing, the driveshaft must be removed from the car. Check the bearing for smooth movement, and check the condition of the mounting rubber. If the bearing is difficult to turn or if there is a gritty feeling, only the bearing needs to be replaced. If the rubber is cracked or deteriorated, the entire assembly should be replaced.

To replace center bearing:

1. Remove the driveshaft as described in **4.2 Removing and Installing Driveshaft**.

2. Make matching marks on the front and rear sections of the driveshaft for reassembly as shown in Fig. 4-11.

3. If applicable, unscrew the clamping sleeve all the way and then pull the two sections of the driveshaft apart. Remove the rubber bushing, washer, and clamping sleeve from the front section of the driveshaft.

4. If applicable, inspect the condition of the rubber bushing for the splined coupling. It should be replaced if it is cracked or torn.

5. Remove the center bearing circlip or mounting bolt, and dust guard if applicable. See Fig. 4-12. Then use a puller to pull the bearing off of the driveshaft.

Fig. 4-11. Before pulling apart driveshaft sections, make matching marks as shown.

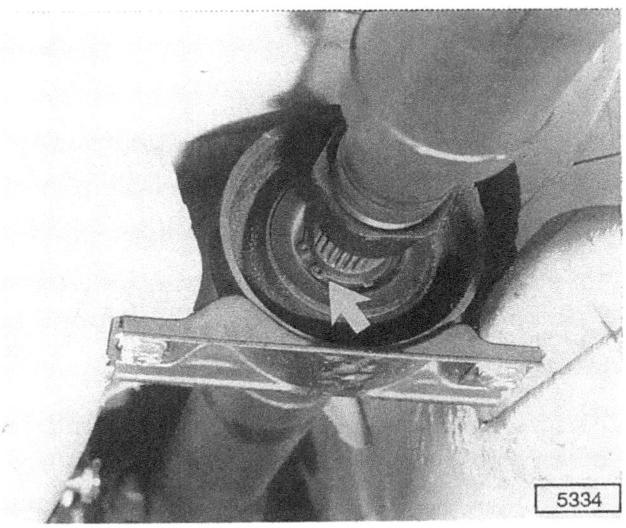

Fig. 4-12. Center bearing circlip (arrow) to be removed. Driveshaft shown installed.

NOTE —
- Install the puller so that it pulls on the inner hub of the bearing. Pulling on the outer ring of the mount may tear the rubber, and the entire bearing assembly will need to be replaced.

- On models without a sliding driveshaft, it may be necessary to grind a 19-mm box-end wrench to reach and loosen the driveshaft yoke mounting bolt.

15

6. If the bearing is being replaced, press it out of the bearing mount and press a new bearing in.

7. Make sure that the dust guard is on the driveshaft, and then press the center mount onto the driveshaft so that it is flush with the dust guard.

8. If applicable, place the clamping sleeve, washer, and rubber bushing on the front driveshaft section. Lubricate the splines with molybdenum disulfide grease (Molykote Longterm 2® or equivalent) and then reassemble the driveshaft. Make sure the matching marks align. Do not retighten the clamping sleeve until the driveshaft is installed.

9. Reinstall the driveshaft.

10. On models without a sliding driveshaft, install the driveshaft yoke mounting bolt with Loctite®.

Tightening torque
• driveshaft yoke to driveshaft (install with Loctite®) . . . . 100 Nm (74 ft-lb)

11. On models with a sliding driveshaft, torque the clamping sleeve using BMW special tool 261 040 or equivalent.

Tightening torque
• clamping sleeve . . . . 17 Nm (13 ft-lb)

4.5 Replacing Front Centering Guide

The front centering guide precisely centers the driveshaft in relation to the transmission. The guide is press-fit into a cavity in the front of the driveshaft and slides onto the end of the protruding transmission output shaft.

A worn centering guide will allow the front of the driveshaft to wobble and cause vibration. It is necessary to remove the driveshaft to inspect and replace the centering guide. No specifications are given for wear of the guide, but generally the guide should fit snugly on the transmission output shaft.

NOTE —
Some driveshafts have a dust cap installed on the end of the driveshaft, over the centering guide. The dust cap may become bent or distorted when the driveshaft is removed or installed. Damage to the dust cap should not affect the centering guide and should not be mistaken for guide wear.

To replace front centering guide:

1. Remove the driveshaft as described in **4.2 Removing and Installing Driveshaft.**

2. Pack the cavity behind the centering guide with a heavy grease until the grease is flush with the bottom edge of the guide.

3. Insert a 14 mm (approximately 1/2 in.) diameter mandrel or metal rod into the guide, then strike it with a hammer. The pressure of the mandrel against the grease should force the centering guide out of the driveshaft.

NOTE —
The mandrel needs to fit snugly in the centering guide so that the grease cannot escape around the sides of the mandrel.

4. Remove the old grease from driveshaft, lubricate the new centering guide with molybdenum disulfide grease (Molykote Longterm 2® or equivalent) and then drive it into the driveshaft. The sealing lip of the guide should face outward and it should be driven into the driveshaft to a specified depth. See Fig. 4-13.

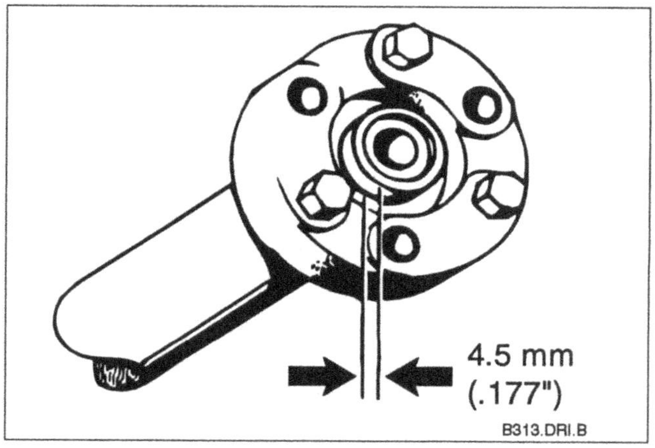

Fig. 4-13. When installing new driveshaft centering guide, drive guide in until its protrusion depth is as shown.

5. Reinstall the driveshaft.

4.6 Replacing Driveshaft Constant Velocity Joint

The constant velocity (CV) joint on the front of the driveshaft bolts directly to the transmission output flange. The CV joint can be replaced after removing the driveshaft. Always replace clamps, circlips, boots, gaskets, and covers. Have these items on hand before starting the job.

Capacity	
• CV joint grease	approximately 60 grams

To remove CV joint:

1. Remove the driveshaft as described in **4.2 Removing and Installing Driveshaft.**
2. Remove the boot clamp at the back of the CV joint.
3. At the front of the joint, remove the circlip securing the joint to the shaft. Pull the joint off the shaft using a suitable puller. Remove the bolts from the joint.

 NOTE —
 The CV joint is cemented (Loctite®) to the shaft and will require a puller to break the joint free. The CV joint mounting bolt shoulders are knurled and need to be driven out of the joint.

To install CV joint:

1. Pack the new joint with 60 grams of CV joint grease, working the grease down into the cage. Install a new gasket and dust cover.

 NOTE —
 Do not swivel the CV joint inner race when packing it with grease. The balls can fall out of the cage.

2. Tap the bolts and retainers through the joint holes. Clean the splines on the driveshaft and apply a small amount of bolt cement to the splines.
3. Place the boot on the shaft. Install the joint onto the shaft and tap it on using a wooden dowel.
4. Install a new circlip on the shaft, and a new boot clamp.
5. Reinstall the driveshaft.

Tightening torque	
• CV joint bolts to transmission flange (M8)	32 Nm (24 ft-lb)

5. FINAL DRIVE

This heading covers removal and installation of the final drive, as well as those repair operations that do not require complicated disassembly of the final drive. Disassembly of the final drive requires special tools and knowledge, and is beyond the scope of this manual.

All final drive work requires some method of raising the car and supporting it securely while the work is performed. Jack stands and a floor jack can easily be used, but use extreme caution when working beneath the car and lowering the final drive. For more information on raising the car see **FUNDAMENTALS.**

5.1 Removing and Installing Final Drive

The final drive is removed or installed using ordinary hand tools. New self-locking nuts are required when reattaching the driveshaft to the final drive.

If the final drive is being replaced with a remanufactured unit, remove the speedometer pulse sender from the old unit and install it into the new unit. The sender is not normally supplied on replacement final drives. Remanufactured final drives are essentially new, and therefore require break-in. For more information see **LUBRICATION AND MAINTENANCE.**

To remove final drive:

1. Raise the rear of the car and support it securely. Drain the oil from the final drive unit. See **LUBRICATION AND MAINTENANCE.**
2. Make matching marks on the driveshaft and final drive input flange, and then remove the bolts and nuts that hold the driveshaft to the final drive. Push the driveshaft forward slightly and suspend it from the body with stiff wire.

 CAUTION —
 • Do not let the driveshaft hang unsupported from the center bearing. Damage to the center bearing could result. If only one end of the driveshaft is disconnected, use stiff wire to suspend the driveshaft out of the way, in as close to the installed position as possible.

 • Do not rest the driveshaft on the fuel lines or brake lines.

3. Disconnect the drive axles from the final drive as described in **SUSPENSION–REAR.**

 CAUTION —
 Suspend the detached drive axle from the car body with a stiff wire hook to prevent damage to the outer constant velocity joint.

4. Disconnect the wires for the speedometer pulse sender. See Fig. 5-1.

Fig. 5-1. Wires removed from speedometer pulse sender (arrow).

NOTE —
If the final drive is being replaced, remove the two bolts that hold the speedometer pulse sender, and remove the sender. Using a new O-ring, install the sender on the new final drive. Torque the bolts.

Tightening torque
• speedometer sender to final drive. . . . . 10 Nm (89 in-lb)

5. Support the final drive with the floor jack. Then remove the two upper mounting bolts and the two side mounting bolts. See Fig. 5-2.

6. Remove the rearmost final-drive mounting bolt from the rubber mount. See Fig. 5-3. Lower the final drive, and remove it from under the car.

NOTE —
On some 1982 and 1983 528e models, the rubber mount bolt may have an additional locknut on top of the rubber mount. The top locknut must be counter-held when removing the mounting bolt. The locknut can be reached through the access plug in the trunk floor.

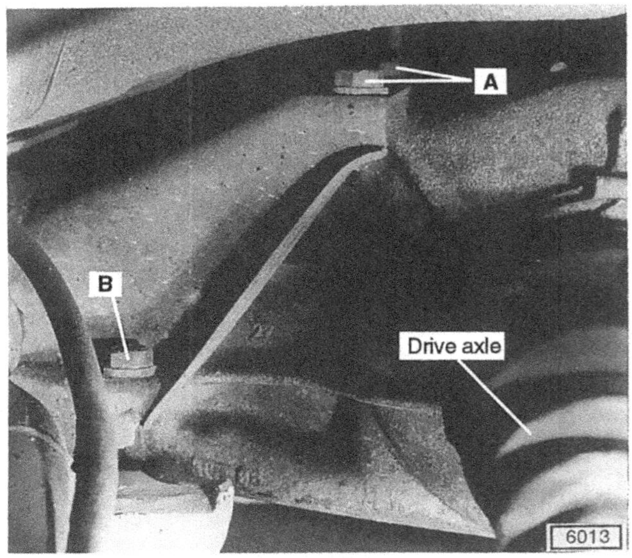

Fig. 5-2. Final drive upper (**A**) and side (**B**) mounting bolts. Also remove side bolt from other side (not shown).

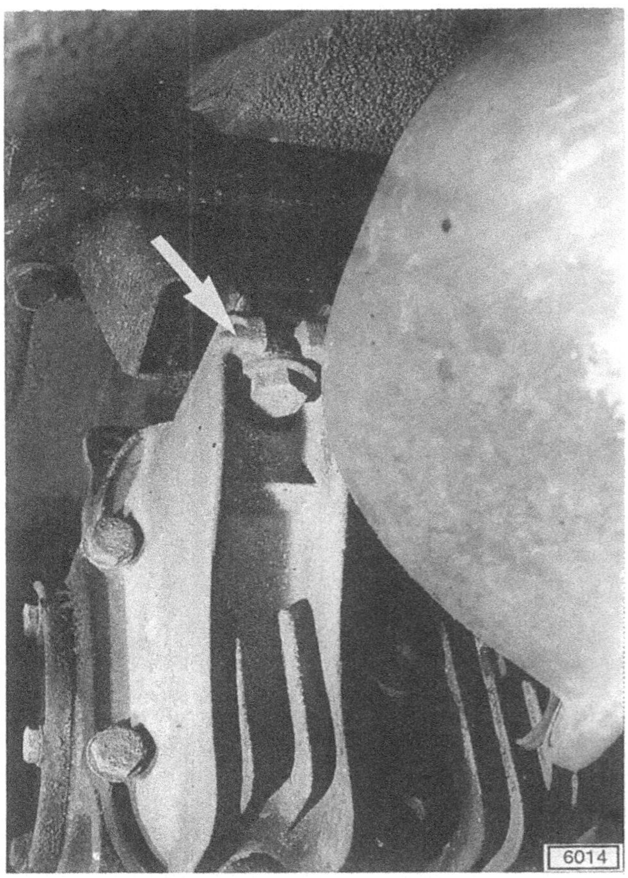

Fig. 5-3. Rear final drive mounting bolt (arrow) for rubber mount.

NOTE —
Oil leaks from the differential vent pipe on 528e models can be fixed by replacing the vent pipe in the cover. The installed depth of the new vent pipe in the cover is 27mm (1.063 in.) and the pipe should be pressed using Loctite® 271 (red). The vent bore, when installed, should point to the rear.

To install final drive:

1. Position the final drive in place against its mounts and loosely install the two upper bolts first. Then install the two side bolts and the bolt for the rubber mount. Fastener torque specifications are listed below.

CAUTION —
Do not reuse self-locking nuts. These nuts are designed to be used only once and should be replaced whenever they are removed.

NOTE —
Where applicable, do not forget to install the additional locknut on top of the rubber mount.

Tightening torques	
• final drive to rear axle carrier	80 Nm (59 ft-lb)
• final drive to rubber mount	87 Nm (64 ft-lb)
• oil drain and fill plug	55 Nm (41 ft-lb)
• drive axle to final drive	58 Nm (43 ft-lb)

2. Reattach the drive axles as described in **SUSPENSION–REAR.**

3. Using new self-locking nuts, reconnect the rear driveshaft section to the final drive. Align marks made during removal. Torque the nuts, while holding the bolts.

Tightening torque	
• driveshaft to final drive flange	72 Nm (53 ft-lb)

CAUTION —
Do not reuse self-locking nuts. These nuts are designed to be used only once and should be replaced whenever they are removed.

4. Reconnect the speedometer pulse sender wires.

5. Check the final drive oil level and add or fill as necessary. See **LUBRICATION AND MAINTENANCE.**

5.2 Replacing Final Drive Rubber Mount

The final drive rubber mounting bushing can be replaced without removing the final drive or the drive axles.

To replace final drive rubber mount:

1. Remove the wires for the speedometer pulse sender.

2. Support the final drive from below and remove the front and side final drive mounting bolts. See Fig. 5-2 above.

3. Remove the mounting bolt from the rubber bushing and lower the final drive just until the four rubber mount to body mounting bolts are accessible.

NOTE —
On some 1982 and 1983 528e models, the rubber mount may have an additional locknut on top of the rubber mount. The top locknut must be counter-held when removing the long mounting bolt. The locknut can be reached through the access plug in the trunk floor.

4. Remove the old bushing and install the new bushing. Install the four mounting bolts using Loctite® 270 or an equivalent thread locking compound.

Tightening torque	
• rubber mount to body (install with Loctite® 270 or equivalent)	48 Nm (35 ft-lb)

WARNING —
On some 1982 and 1983 528e models, the captive nut on the top of the rubber mount may have been over-tightened at the factory. Carefully inspect the nut for small radial cracks. If the nut is cracked, the entire mount should be replaced. The old style mount has been superseded with an improved part. If the nut is not cracked, BMW recommends replacing the old mounting bolt with a longer bolt (BMW Part No. 07 11 9 914 660) and adding a self-locking nut (BMW Part No. 07 12 9 922 734) on top of the new bolt.

NOTE —
BMW has initiated a recall campaign (NHTSA assigned no. 82V-108) for this possible defect on 1982 and 1983 528e models. If this recall has been performed, there will be a white paint dot on the trunk insulation under the carpeting. If this recall has not be performed, check with an authorized BMW dealer for more information.

15

5. Raise the final drive up against its mounts and install the two upper bolts first, then install the two side bolts and the bolt for the rubber mount.

NOTE —
Where applicable, do not forget to install the additional locknut on top of the rubber mount.

Tightening torques	
• final drive to rear axle carrier	80 Nm (59 ft-lb)
• final drive to rubber mount	87 Nm (64 ft-lb)

5.3 Replacing Final Drive Oil Seals

Low oil level caused by leakage due to faulty final drive oil seals may be the cause of noisy operation or limited-slip chatter. The drive flange oil seals can be replaced while the final drive is installed in the car, but the final drive must be removed as described in **5.1 Removing and Installing Final Drive** to replace the input shaft oil seal. Read the procedure to determine what other new parts are required before beginning repairs. Do not mistake leaking CV joints for flange seal leaks. It may be helpful to degrease the final drive to pinpoint the source of the leak prior to replacing seals.

To replace drive flange oil seals:

1. Detach the drive axle from the final drive as described in **SUSPENSION–REAR**.

CAUTION —
Suspend the detached drive axle from the car body with a stiff wire hook to prevent damage to the outer constant velocity joint.

2. Pry the drive flange from the final drive as shown in Fig. 5-4. Then remove the flange snap ring shown in Fig. 5-5. Inspect the flange and replace it if there is a groove worn where it contacts the oil seal.

3. Pry the faulty seal from its recess using a hooked seal removal tool, or a large screwdriver.

CAUTION —
Be very careful not to mar the housing when removing the seal.

4. Dip the new seal in final drive lubricant and drive the new seal into place until it is fully seated.

5. Install a new snap ring in the groove of the differential housing. Make sure both ends of the ring are fully seated in the groove.

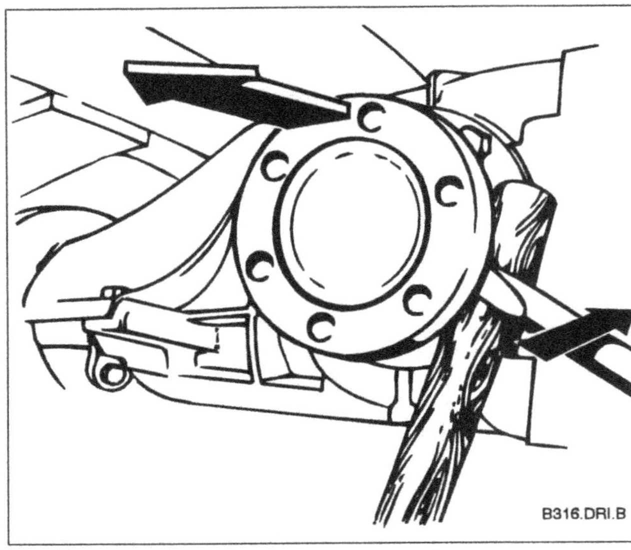

Fig. 5-4. Drive flange being pried off. Use extreme care to avoid damaging final drive. For leverage, use wooden dowels as shown.

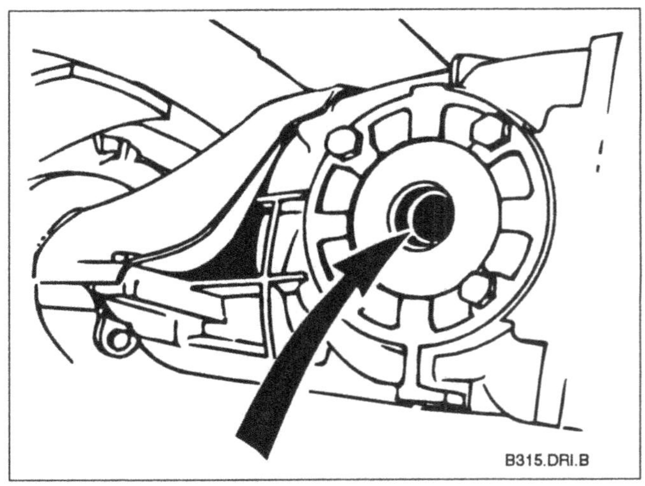

Fig. 5-5. Drive flange snap ring (arrow) to be removed from differential housing. Always use a new snap ring.

6. Install the drive flange by pressing it in by hand until the snap ring engages. It may be necessary to turn the drive flange slightly while pushing.

7. Attach the drive axle and torque the bolts.

Tightening torque	
• drive axle to final drive flange	58 Nm (43 ft-lb)

To replace final drive input shaft oil seal:

1. Remove the final drive as described in **5.1 Removing and Installing Final Drive.**

2. Drain the oil from the final drive.

3. Make matching marks on the input shaft and input shaft collar nut as shown in Fig. 5-6.

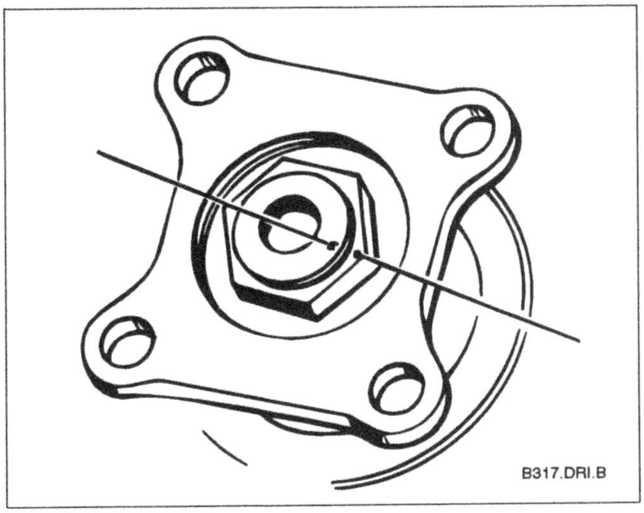

Fig. 5-6. Final drive input shaft, collar nut, and flange, showing matching marks made with punch.

4. Remove the lockplate, then hold the input flange and remove the collar nut. Using a puller, remove the input flange.

5. Pry the faulty seal from its recess using a hooked seal removal tool or a large screwdriver. Dip the new seal in final drive lubricant and drive the new seal into position.

CAUTION —
Be very careful not to mar the housing when removing the seal.

6. Lightly lubricate the input shaft and press the input flange back on, but do not bottom it out. Install the collar nut and slowly tighten it until the matching marks line up, coming as close as possible to the specified torque.

Tightening torque	
• final drive input shaft collar nut	
M20 pinion thread diameter	150 Nm (111 ft-lb) or until matching marks line up
M22 pinion thread diameter	310 Nm (229 ft-lb) or until matching marks line up

7. Install a new lockplate.

CAUTION —
Do not tighten the collar nut past the matching marks to reach the specified torque, and then move the nut back. This may over-compress the pinion shaft bushing and require disassembly of the final drive for replacement.

6. TECHNICAL DATA

I. Tightening Torques

Driveshaft to final drive flange (bolts and self-locking nuts)	72 Nm (53 ft-lb)
Center bearing to body (bolts)	22 Nm (16 ft-lb)
Flexible coupling to driveshaft or transmission (bolts and self-locking nuts)	
M10 8.8 grade	46 Nm (34 ft-lb)
M10 10.9 grade	72 Nm (53 ft-lb)
M12	123 Nm (90 ft-lb)
Driveshaft clamping sleeve	17 Nm (13 ft-lb)
Driveshaft yoke to driveshaft (install with Loctite®)	100 Nm (74 ft-lb)
CV joint to transmission (M8 bolts)	32 Nm (24 ft-lb)
Speedometer sender to final drive (bolts)	10 Nm (89 in-lb)
Final drive to rear axle carrier	80 Nm (59 ft-lb)
Final drive to rubber mount	87 Nm (64 ft-lb)
Final drive rubber mount to body (install with Loctite® 270)	48 Nm (35 ft-lb)
Oil drain and fill plugs	55 Nm (41 ft-lb)
Drive axle to final drive (bolts)	58 Nm (43 ft-lb)
Final drive input (pinion) shaft flange to input shaft (collar nut)	
M20 pinion thread diameter	150 Nm (111 ft-lb)
M22 pinion thread diameter	310 Nm (229 ft-lb)

II. Driveshaft and Final Drive Specifications

Final drive oil type	SAE 90 GL-5 Hypoid gear oil*
Final drive oil capacity	
528e, 533i models	1.8 liters (1.9 qt.)
535i/is models	1.9 liters (2.0 qt.)
Driveshaft center mount preload	
with front U-joint or flex coupling	4 to 6 mm (5/32 to 1/4 in.)
with front CV joint	4 to 5 mm (5/32 to 7/32 in.)

*On models with limited-slip final drives, final drive oil type may vary. See **LUBRICATION AND MAINTENANCE** for more information.

15

Section 16

SUSPENSION—FRONT

Contents

TABLES

INTRODUCTION

BMW 5-Series cars are equipped with a strut-type independent front suspension. The double-pivot front suspension struts are integral spring and shock absorber units. This section covers only the front suspension. For information on steering, see **STEERING AND WHEEL ALIGNMENT**.

1. GENERAL DESCRIPTION

The BMW 5-series front suspension is engineered to provide a sophisticated compromise between taut, responsive handling and ride comfort. Fig. 1-1 shows the components of the front suspension.

The control arm and the thrust arm connect the steering arm at the lower end of each suspension strut to mounting points on the subframe and chassis. The two-point mountings of the arms control the front-to-rear and side-to-side movements of the suspension strut.

Each front suspension strut assembly includes a shock absorber cartridge inside a tubular strut housing, and a concentrically-mounted coil spring. The lower end of each strut includes the stub axle for the front wheel bearing and hub.

Suspension travel is limited by rubber bump stops on the top of each strut assembly. A stabilizer bar mounted to the suspension struts and chassis helps to reduce body roll when cornering.

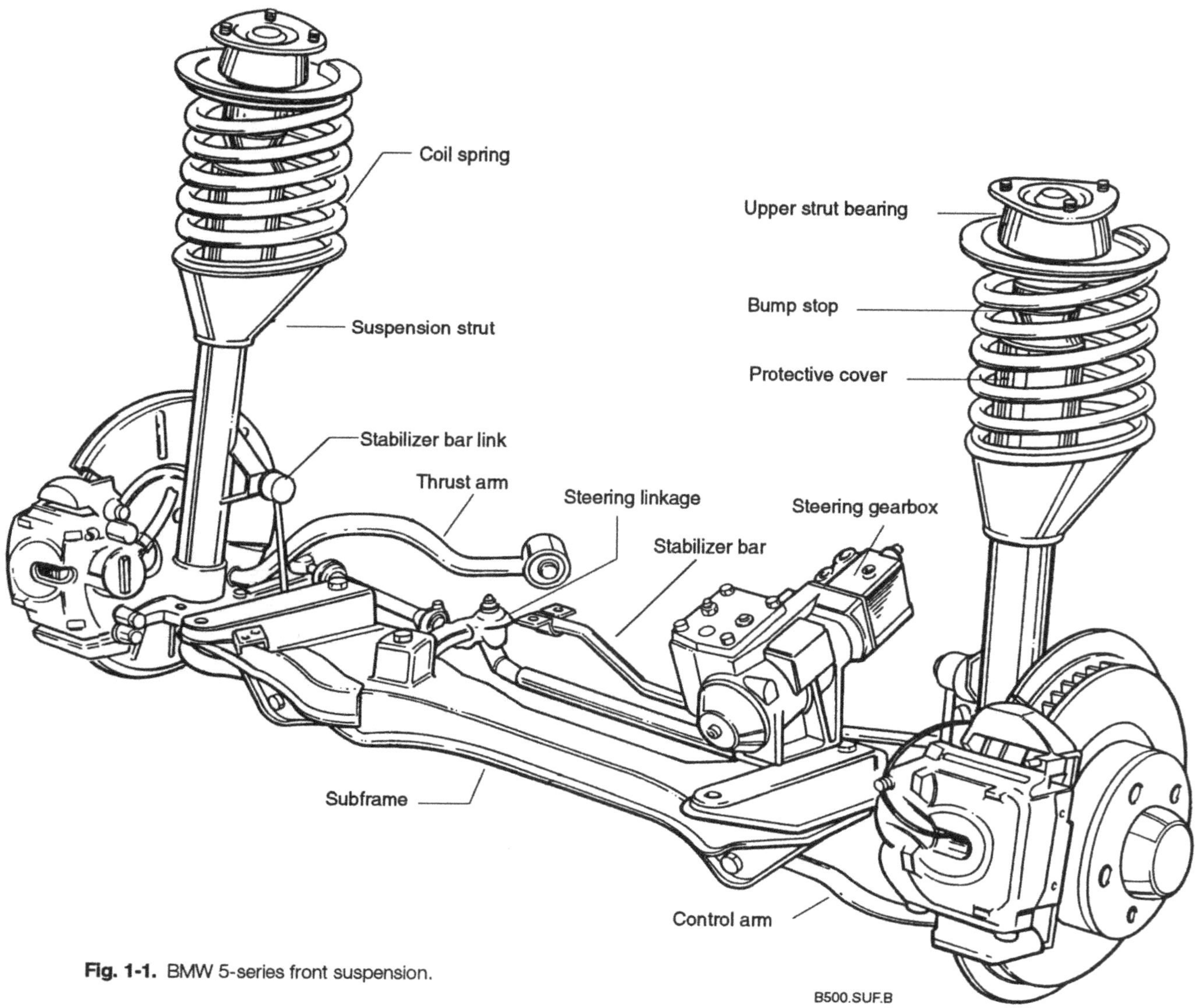

Fig. 1-1. BMW 5-series front suspension.

2. MAINTENANCE

Scheduled maintenance includes periodic inspection of the front suspension components for wear and damage.

BMW recommends that the following items be routinely checked at periodic intervals as described in **LUBRICATION AND MAINTENANCE**. Detailed descriptions of these maintenance operations can be found either in this section of the manual under the boldface numbered headings shown, or in **LUBRICATION AND MAINTENANCE**.

1. Inspecting front suspension components, including control arm and thrust arm joints. **4.2**
2. Checking wheel bearings. **4.3**
3. Checking tightness of anchor bolts at front suspension subframe. **4.5** (Additional periodic maintenance recommended by the publisher of this manual.)

3. TROUBLESHOOTING

This troubleshooting section applies to problems with the front suspension that affect ride comfort, handling and stability. Similar problems may be caused by faults in the brakes or steering system. For example, a problem such as consistently pulling to one side may be caused by a bent control arm or by faulty brakes. For brake system troubleshooting, see **BRAKES**. For more basic help in determining the appropriate section for troubleshooting a particular symptom, see **FUNDAMENTALS** at the front of the manual.

3.1 Basic Troubleshooting Principles

Stable handling and ride comfort both depend on the integrity of the suspension components. These systems must precisely position the wheels so that the car is stable and controllable, but also allow movement so that the wheels can steer and react to bumps. Any symptom of instability or imprecise road feel may be caused by worn or damaged suspension components.

In addition to inspecting for worn parts, troubleshooting must also consider the condition of tires, wheels, and their alignment. Tire wear and incorrect inflation pressures can dramatically affect handling. Subtle irregularities in wheel alignment angles also affect stability. Mixing different types or sizes of tires, particularly on the same end of the car, can affect alignment and may unbalance a car's handling. See **STEERING AND WHEEL ALIGNMENT** for more information.

Table a lists symptoms of suspension problems and their probable causes, and suggests corrective actions. The boldface numbers in the corrective action column refer to headings where the repairs are described.

Table a. Front Suspension Troubleshooting

Symptom	Probable cause	Corrective action
1. Pull to one side, wandering	a. Incorrect tire pressure b. Defective/unevenly worn tire c. Incorrect wheel alignment d. Faulty brakes (pulls only when braking)	a. Check and correct tire pressures. See **LUBRICATION AND MAINTENANCE** b. Inspect tires and replace as needed. See **STEERING AND WHEEL ALIGNMENT** c. Check and adjust wheel alignment. See **STEERING AND WHEEL ALIGNMENT** d. See **BRAKES**
2. Steering heavy, poor return-to-center	a. Worn upper strut bearing(s) b. Incorrect tire pressure c. Power steering system faulty	a. Replace strut bearings. **4.1** b. Check and correct tire pressures. See **LUBRICATION AND MAINTENANCE** c. Check power steering system. See **STEERING AND WHEEL ALIGNMENT**
3. Front-end vibration or shimmy	a. Severely worn strut shock absorber cartridges b. Worn control arm or thrust arm ball joints or rubber bushings c. Unbalanced or bent wheels/tires d. Loose wheel lug bolts	a. Replace strut cartridges. **4.1** b. Replace thrust or control arm(s) or replace bushing. **4.2** c. Balance tires. Check tires for uneven wear patterns. Check wheels for damage. See **STEERING AND WHEEL ALIGNMENT** d. Tighten lug bolts to proper torque. See **STEERING AND WHEEL ALIGNMENT**

continued on next page

Table a. Front Suspension Troubleshooting (continued)

Symptom	Probable cause	Corrective action
4. Poor stability, repeated bouncing after bumps, suspension bottoms out easily	a. Worn (weak) strut shock absorber cartridges	a. Replace strut cartridges. **4.1**
5. Suspension noise (especially over bumps)	a. Worn front upper strut bearings(s) b. Loose front strut shock absorber cartridge c. Faulty rubber shock absorber damper (bump stop) d. Worn control arm or thrust arm ball joints or rubber bushings e. Loose or worn stabilizer bar rubber mounts f. Loose front suspension subframe	a. Replace upper strut bearings. **4.1** b. Disassemble strut, check for damage. **4.1** c. Replace rubber damper. **4.1** d. Replace thrust or control arm(s) or replace bushing. **4.2** e. Retorque stabilizer bar rubber mounts. Replace rubber mounts as necessary. **5.4** f. Check for subframe damage. Tighten mounting bolts. **4.5**
6. Uneven ride height	a. Incorrect coil springs b. Bent or damaged suspension components c. Sagging coil springs	a. Measure ride height and replace springs as required. **4.1** b. Inspect and repair/replace as necessary c. Replace front springs in pairs. **5.1**
7. Front wheel noise, continuous growling, may be more noticeable when turning	a. Worn front wheel bearing	a. Replace front wheel bearing and hub. **4.3**

3.2 Diagnostic Inspection and Testing

Front suspension problems can usually be isolated and at least partially diagnosed by careful observation of the symptoms and inspection of the components that are the most likely cause.

Tire Wear

Tire wear can be a good indicator of front suspension problems. For more information on tire wear, see **STEERING AND WHEEL ALIGNMENT**. For information on maintaining proper tire inflation pressures, as well as other general tire maintenance, see **LUBRICATION AND MAINTENANCE**.

Isolating Pulling Symptoms

Consistent pulling to one side in a car driven straight ahead on a level road may be caused either by worn or damaged (bent) suspension components or a faulty tire. In more unusual cases, a brake problem may be the cause.

To help decide whether the problem is caused by the tires or the suspension, temporarily swap the front wheels (and tires), and then road-test the car. If a tire problem is the cause of the pulling symptom, the problem should switch to the other side of the car when the wheels are switched, and it should now pull to the other side. If the symptom persists, then the problem is probably caused by faulty suspension components.

> **NOTE —**
> If no suspension components are found to be faulty, the car may only need a front wheel alignment. See **STEERING AND WHEEL ALIGNMENT** for more information.

Vibration

Abnormal vibration, if not caused by a mechanical problem, will most often be caused by wheels and tires that are out of balance. Other, less likely causes of vibration are bent wheels or a tire that has become distorted and out-of-round. These and other tire conditions can be checked by any reputable tire professional.

Before spending the money for balancing, look for an obvious cause of imbalance. Caked-on mud, ice, or snow can dramatically affect wheel balance. Clean the wheels and tires and road-test the car before investigating for more serious causes.

4. FRONT SUSPENSION

The BMW 5-series cars covered by this manual feature a relatively simple front suspension design. The limited number of parts that wear can be easily replaced. Some of these repairs do, however, require special tools and press equipment. To avoid starting a job that may be difficult to complete, please read the entire procedure before beginning.

When doing any front suspension maintenance or repair, please observe the following general cautions:

> ***CAUTION —***
> - *Do not install bolts and nuts coated with undercoating wax, as correct tightening torque cannot be assured. Always clean the threads with solvent before installation, or install new parts.*
> - *Do not attempt to weld or straighten the suspension struts, the control arms, thrust arms, or the subframe. Replace damaged parts.*
> - *Do not reuse self-locking nuts or cotter pins. These parts are designed to be used only once.*

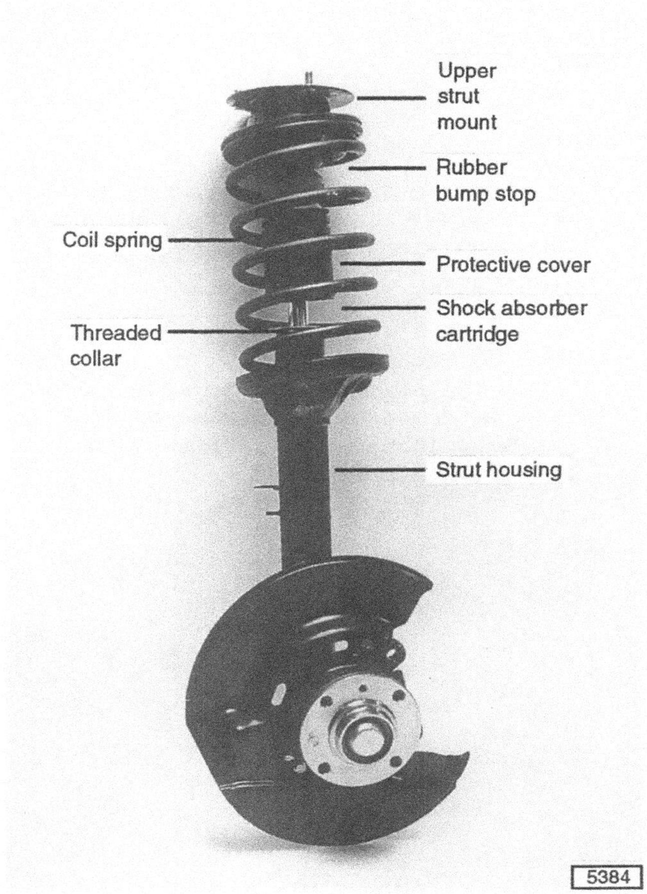

Fig. 4-1. Front suspension strut assembly.

4.1 Front Suspension Struts

Fig. 4-1 shows the main components of the front strut assembly. These are the strut housing, the shock absorber cartridge, the upper strut bearing, and the coil spring. Each of these main components can be replaced separately as required. The strut housing itself need only be replaced if it is damaged.

The procedures that follow, beginning with **Removing and Installing Front Suspension Struts**, describe in order the steps that should be followed to completely disassemble the strut.

Differences from side to side will affect handling and stability. It is strongly recommended that springs, shock absorber cartridges, or strut bearings be replaced in pairs.

Checking Shock Absorbers

Springs are what support the weight of the car and allow the suspension to travel smoothly over bumps and other road irregularities. A spring that is compressed by a bump simply rebounds, springing in the other direction with nearly the same force that compressed it.

The function of a shock absorber, or damper, is to moderate the spring action. It slows the bounce and helps the spring return to its normal position. Shock absorbers require no routine maintenance and cannot be serviced.

The best evidence of failing shock absorbers is their behavior in normal driving. Worn shock absorbers will allow extra skittishness over bumps, and a less-controlled and wallowing feel after bumps and in corners. When seriously worn, the shock absorbers present little resistance to spring oscillations. Because they so easily reach the limits of their travel, they may knock when going over bumps.

The most common–though not entirely accurate–test of shock absorber function involves vigorously bouncing each end or corner of the car, and then releasing and observing how quickly the bouncing stops. More than one bounce usually suggests that the shock absorbers are not properly damping the spring action and need to be replaced.

Removing and Installing Front Suspension Struts

The struts can be easily removed for replacement of components. It is not necessary to remove any ball joints or tie rod ends when removing the strut assemblies.

16

The exact mounting of the struts affects wheel alignment. A wheel alignment is highly recommended any time the struts have been removed.

To remove front struts:

1. Loosen the wheel lug bolts, then raise the front of the car and support it securely on jack stands that are designed for the purpose.

2. Remove the wheel.

3. Remove the wires attached to the strut housing. On the left side only, disconnect the electrical connector for the brake pad wear sensor. On cars equipped with anti-lock brakes (ABS), remove the wheel speed sensor from its mounting. See Fig. 4-2.

Fig. 4-2. Brake pad wear sensor connector (**A**) and ABS wheel speed sensor (**B**). Wheel speed sensor is secured with 5 mm internal-hex-head screw.

CAUTION —
Make sure that the ignition is off when removing ABS sensors.

4. Without disconnecting the flexible brake hose, remove the brake caliper mounting bolts shown in Fig. 4-3. Slide the caliper off the brake rotor and strut housing.

CAUTION —
Do not step on the brake pedal or operate the brakes while the caliper is removed. Doing this will damage the brake caliper.

NOTE —
To avoid kinking or damaging the hose, use stiff wire to suspend the caliper from the underbody.

Fig. 4-3. Front brake caliper mounting bolts (arrows).

5. Disconnect the stabilizer bar connecting link from the strut assembly. See Fig. 4-4.

Fig. 4-4. *Nut (**A**) securing stabilizer bar connecting link to strut. Hold slotted part of connecting link joint (**B**) with wrench when removing nut.*

6. Remove the three steering arm (knuckle) mounting bolts. See Fig. 4-5.

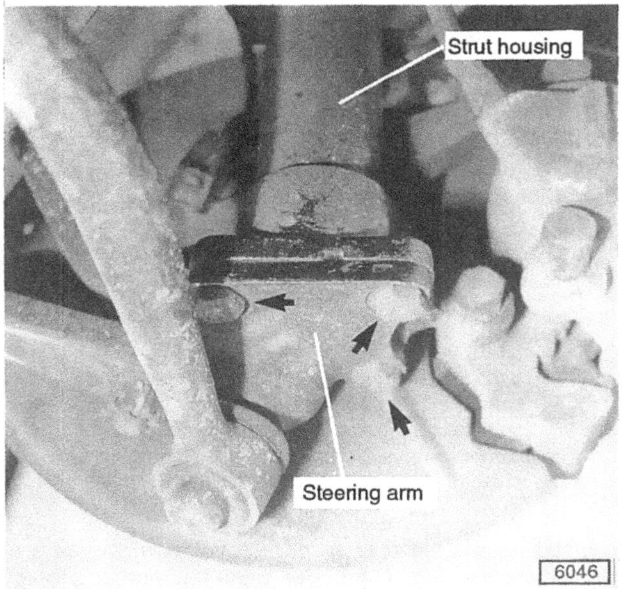

Fig. 4-5. Bolts (arrows) securing steering arm to strut housing.

7. Pull the bottom of the strut out away from the car, far enough to clear the steering arm.

8. Working in the engine compartment, remove the three mounting nuts at the top of the strut tower and remove the strut. See Fig. 4-6.

> ***CAUTION —***
> *Support the weight of the strut while removing the bolts.*

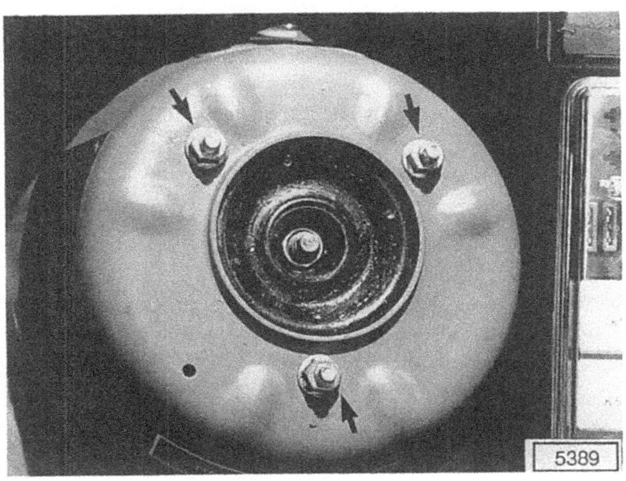

Fig. 4-6. Strut mounting nuts (arrows).

> ***WARNING —***
> *Further disassembly of the strut requires special tools and extreme caution to avoid serious personal injury. Do not proceed without reading the procedure below.*

Installation is the reverse of removal. Be sure the steering arm mounting bolts, the threaded holes for the bolts, and the mating surfaces are clean and free of grease. When installing the steering arm to the strut, be sure the mating notches are aligned before tightening the bolts. See Fig. 4-7. Install the steering arm mounting bolts using a bolt locking compound (Loctite® 270 or equivalent).

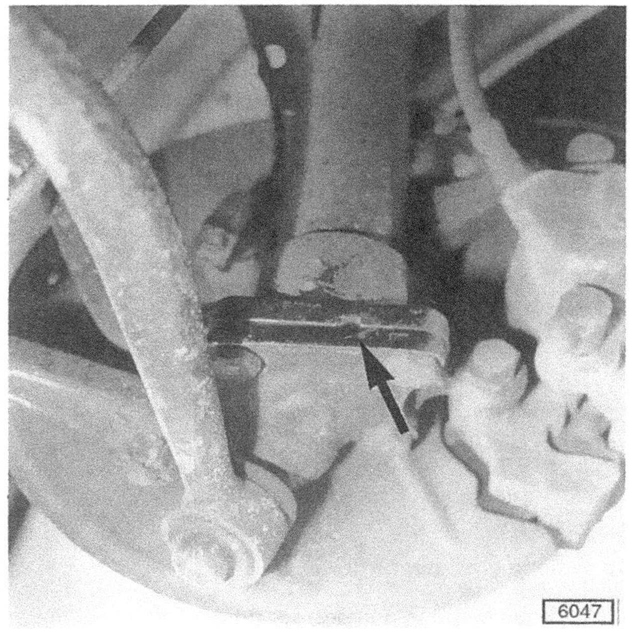

Fig. 4-7. Steering arm and strut housing mating notch (arrow).

16

Tightening torques

- brake caliper mounting bolts . . . . 110 to 123 Nm (81 to 91 ft-lb)
- steering arm to strut housing (install with bolt locking compound (Loctite® 270 or equivalent)) . . . . 65 Nm (48 ft-lb)
- stabilizer bar link to strut housing (self-locking nut)
  - yellow chrome . . . . 20 Nm (15 ft-lb)
  - white chrome . . . . 33 Nm (24 ft-lb)
  - white . . . . 59 Nm (43 ft-lb)
- strut bearing (upper mount) to wheel housing . . . . 22 Nm (16 ft-lb)
- wheel lug bolts . . . . 100±10 Nm (74±7 ft-lb)

CAUTION —
Shock absorbers should be stored upright, imitating their normal installed position. If they are stored horizontally or upside-down for an extended period, they should be placed upright at room temperature with shock absorber rods fully extended for at least 24 hours prior to installation.

Removing and Installing Strut Bearing (Upper Mount)

The best way to check for a worn strut bearing is to remove the strut assembly from the car and relieve the coil spring pressure on the bearing. Replacement of the strut bearing requires partial disassembly of the strut. Even when the strut is removed from the car, the spring is preloaded, exerting considerable force on the ends of the strut and bearing. Attempting to disassemble the strut without first compressing the spring is extremely dangerous. Read the procedure carefully.

WARNING —
Do not attempt to disassemble the struts without a spring compressor designed specifically for this particular job.

CAUTION —
Do not attempt to disassemble a strut by removing the upper mounting nut while the strut is still installed in the car. To avoid damage to other steering and suspension components, the strut must be completely removed from the car before being disassembled.

To remove strut bearing:

1. Install a spring compressor on the removed strut and relieve the tension on the upper spring retainer.

WARNING —
Work slowly while watching the spring compressor. Make sure the spring compressor stays securely in place.

2. Pry the protective cap off the top nut. Hold the shock absorber shaft so that it cannot turn and slowly loosen the top self-locking nut. Remove the nut and washer.

WARNING —
Do not remove the top nut from the shock absorber rod until the coil spring is compressed and there is no spring force against the upper bearing.

NOTE —
BMW Special Tool 31 3 115 is an extension bar that temporarily threads onto the end of the shock absorber shaft to make disassembly and reassembly easier and safer.

3. Lift off the strut bearing, the insulator, and the large washer.

4. To remove the coil spring, lift off the upper spring retainer and the rubber ring at the top of the spring.

5. Release the spring compressor carefully, letting the spring slowly expand to its free length.

NOTE —
If the strut is to be disassembled for only a short time, it may be convenient to leave the spring compressor attached. If so, make certain the compressor is secure and handle the compressed spring with extreme caution.

Installation is the reverse of removal. Always use a new self-locking top nut. The larger diameter washer should be placed between the bearing and cartridge shaft and the smaller diameter washer should be placed between the bearing and the mounting nut. Make sure that the inside convex surface of the insulator faces the bearing. Check the shock absorber protective tube and rubber bumper and replace them if they are worn or damaged.

CAUTION —
Do not reuse self-locking nuts or cotter pins. These parts are designed to be used only once and should be replaced whenever they are removed.

Align the ends of the coil spring to mate properly with the shoulder of the rubber ring and with the spring retainers, as shown in Fig. 4-8. With the spring compressed so that there is no pressure on the upper bearing, tighten the top nut while holding the shock absorber shaft stationary.

Tightening torque
• upper strut bearing (self-locking) mounting nut. . . . . 65 Nm (48 ft-lb)

NOTE —
Springs of different specifications are identified by a number stamped on the top coil, and may or may not be marked with a red stripe. Springs are also matched to one of two different size rubber rings. All of these characteristics should match when installing replacement parts. See **Checking and Correcting Ride Height** for more information.

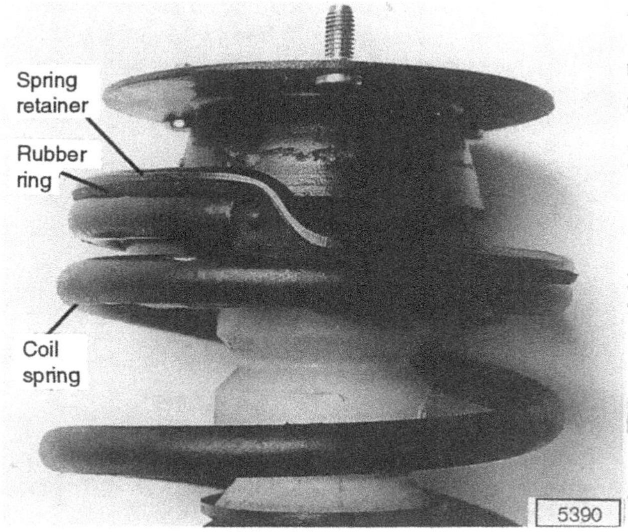

Fig. 4-8. Correct assembly of coil spring, rubber ring, and spring retainer.

Replacing Front Strut Shock Absorber Cartridges

To replace the cartridges, the struts must be removed from the car and partially disassembled as described in **Removing and Installing Strut Bearing (Upper Mount)** above. With the strut bearing and the coil spring removed, loosen and remove the threaded collar. See Fig. 4-9. The shock absorber cartridge can then be pulled out of the strut housing.

Fig. 4-9. Typical threaded collar (arrow) on top of strut housing retains shock absorber cartridge.

To install shock absorber cartridges:

1. If applicable, pour the old oil from the strut housing.

2. On conventional hydraulic type shock absorbers, pour the specified amount of engine oil into strut housing.

Strut housing oil capacity
• oil added to strut housing prior to installing cartridge . . . . 30 to 35 ml (about 1 fl. oz.).

CAUTION —
In struts fitted with gas-pressure shock absorber cartridges, the strut housing must NOT be filled with oil when installing the cartridge.

NOTE —
In struts fitted with conventional hydraulic shock absorber cartridges, the oil helps cool the shock absorber by transferring heat to the strut housing. The grade of engine oil is not critical.

3. Slide the shock absorber cartridge into the strut housing and install and tighten the threaded collar.

Tightening torque
• strut housing threaded collar . . . . 130 Nm (96 ft-lb)

4. Reassemble the strut as described above in **Removing and Installing Strut Bearing (Upper Mount).**

Checking and Correcting Ride Height

Ride height is measured from the lower edge of the wheel arch to the center of the bottom edge of the wheel rim. See Fig. 4-10.

Fig. 4-10. Ride height measurement (**A**) is taken from center of wheel arch to lower edge of wheel rim. Measure with wheels on ground and car loaded. See text below.

16

Table b lists front suspension ride height specifications for the cars covered by this manual. These specifications apply to a car loaded as follows:

Suspension loading requirements:

- each front seat—68 kg (150 lb.)
- center of rear seat—68 kg (150 lb.)
- trunk—21 kg (46 lb.)
- fuel tank full

Table b. Front Ride Height Specifications

Wheel (rim) installed	Ride Height*
14 in. rims	563±10 mm (22±3/8 in.)
15 in. rims	576±10 mm (22 1/2±3/8 in.)
165 TR 390 (TRX) rims	577±10 mm (22 1/2±3/8 in.)

*On models with wheel arch trim, subtract 3 mm (1/8 in.) from the ride height specifications listed.

If the ride height is outside the specification listed, new springs can be selected to correct it. Be sure to have the old spring code number on hand when ordering new springs.

NOTE —
The springs are identified by a code number found on the top coil of the spring.

The code numbers marked on the springs are not BMW part numbers. To get the BMW part number for a spring with a red stripe, add 1 to the last digit of the spring number. For springs without a red stripe, add 2. For example, if the spring is number 1 125 332 and has a red stripe, it is BMW Part No. 1 125 333. If the spring is number 1 125 332 and does not have a red stripe, it is BMW Part No. 1 125 334.

CAUTION —
BMW part numbers are provided for reference only and are subject to change. Always rely on an authorized BMW dealer parts department for the most up-to-date and accurate information.

To change springs, remove the front suspension struts from the car, as described in **Removing and Installing Front Suspension Struts**. Then remove the upper strut bearing and the old springs as described in **Removing and Installing Strut Bearing (Upper Mount)**.

All springs are installed with a rubber ring on either end of the spring. These rubber rings are available in two thicknesses. The correct size depends on whether or not the spring is marked with a red stripe.

Springs marked with a red stripe use a thicker rubber ring with a shoulder approximately 9 mm (3/8 in.) thick. Springs without a red stripe use the thinner rubber ring with a shoulder approximately 3 mm (1/8 in.) thick.

NOTE —
Minor ride height adjustment can be made by exchanging the rubber rings on the ends of the springs. Replacing thinner rings with thick ones will result in a slightly increased ride height. Replacing thick rings with thinner ones will result in a slightly reduced ride height.

4.2 Control Arms and Thrust Arms

The control arms link the front suspension subframe to the steering arm at the bottom of the suspension strut. The thrust arms link the car's chassis to the steering arm. There are two mounting points on each arm, a rubber bushing and a ball joint. See Fig. 4-11. The stiff rubber bushings allow the outer end of the control and thrust arms to pivot up and down. The ball joints connect the control and thrust arms to the suspension strut and allow the steering arm to pivot.

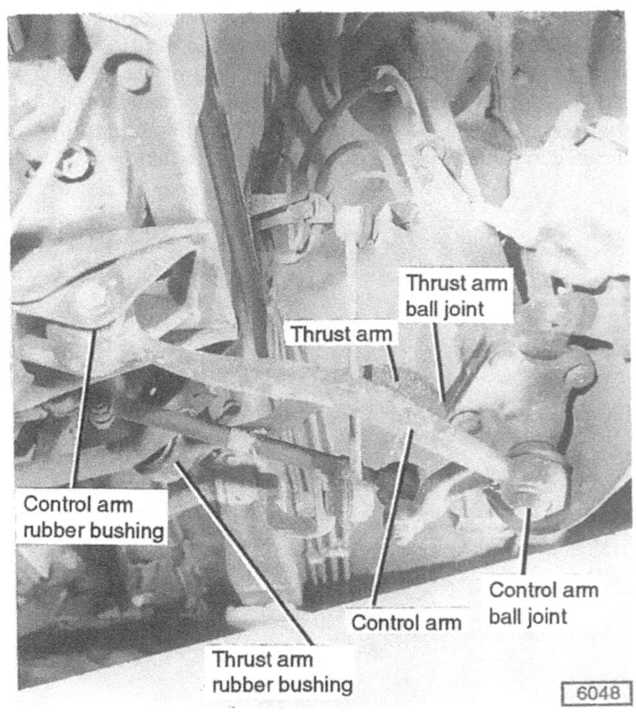

Fig. 4-11. Front suspension control arm and thrust arm as viewed from below.

The rubber bushings are replaceable. The ball joints are permanently lubricated and surrounded by a protective rubber boot. The ball joints cannot be replaced separately. The only recommended routine maintenance is a periodic inspection for damage and wear. If the ball joints are worn, the entire control arm or thrust arm must be replaced as a unit.

Inspecting Control Arm and Thrust Arm Mounts

Ball joints and bushings can be checked for wear, however, BMW gives exact specifications only for checking thrust arm busing wear.

To inspect the ball joints for freeplay, lift the front end of the car and support it securely on jack stands. Check for movement between the joint and the steering arm. Even a worn joint may be stiff. Grip the top and bottom of the joint with a large pliers and squeeze to check for play. See Fig. 4-12.

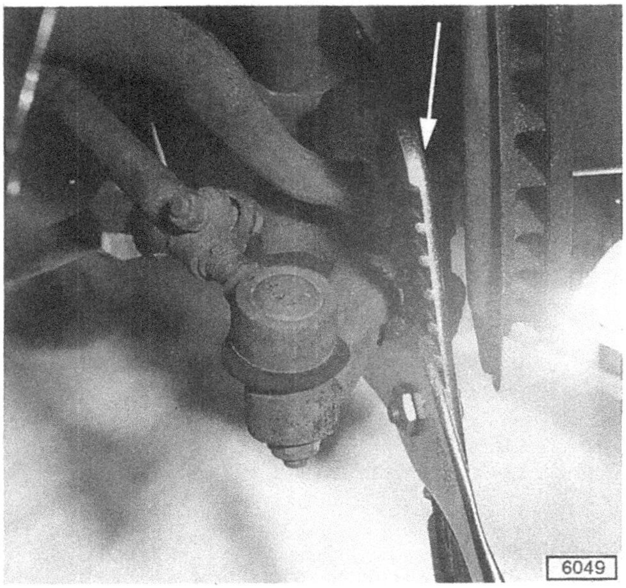

Fig. 4-12. Control arm ball joint being checked for wear using water pump pliers.

Thrust arm and control arm joint wear
• maximum axial travel (with testing force of 265 lbs.). . . . . 1.4 mm (0.06 in.)

To check the thrust arm rubber bushing, the car must be in its normal loaded position on the ground as described below. Using a feeler gauge, measure the gap in the bushing as shown in Fig. 4-13. If the gap is smaller or larger than that specified, the bushing should be replaced as described in **Replacing Control Arm and Thrust Arm Rubber Bushings**.

Suspension loading requirements:

- each front seat—68 kg (150 lb.)
- center of rear seat—68 kg (150 lb.)
- trunk—21 kg (46 lb.)
- fuel tank full

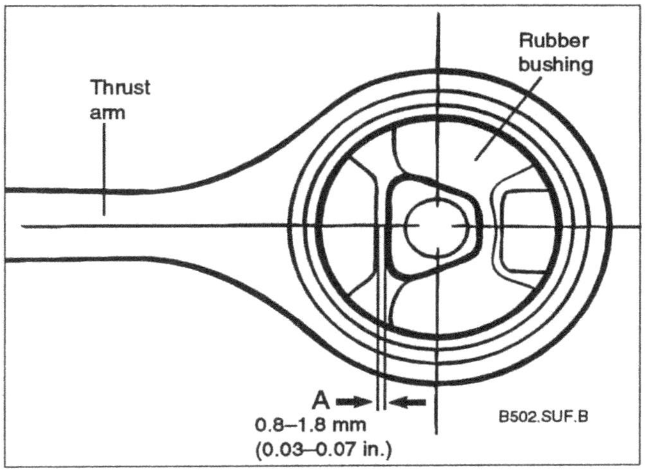

Fig. 4-13. Thrust arm rubber bushing and thrust arm. Measure gap (**A**) with wheels on ground and car loaded. See text above.

Removing and Installing Control Arms and Thrust Arms

Replacement of the arm is the only way to correct for worn or damaged ball joints. Installation of a new control arm also requires a new rubber bushing, as the old bushing cannot be reused once it is removed. See **Replacing Control Arm and Thrust Arm Rubber Bushings** below.

> ***CAUTION —***
> *Do not try to straighten a bent thrust arm or control arm.*

In the procedure below, the suspension strut remains attached to the body at its top mounting point.

To remove control arms or thrust arms:

1. Loosen the wheel lug bolts, then raise the front of the car and support it on jack stands that are designed for the purpose. Remove the wheel.

> **NOTE —**
> The front wheels should be off the ground so that the suspension is not bearing any weight.

2. If the control arm is being removed, remove the three bolts from the steering arm and push the strut assembly away from the arm. See Fig. 4-14.

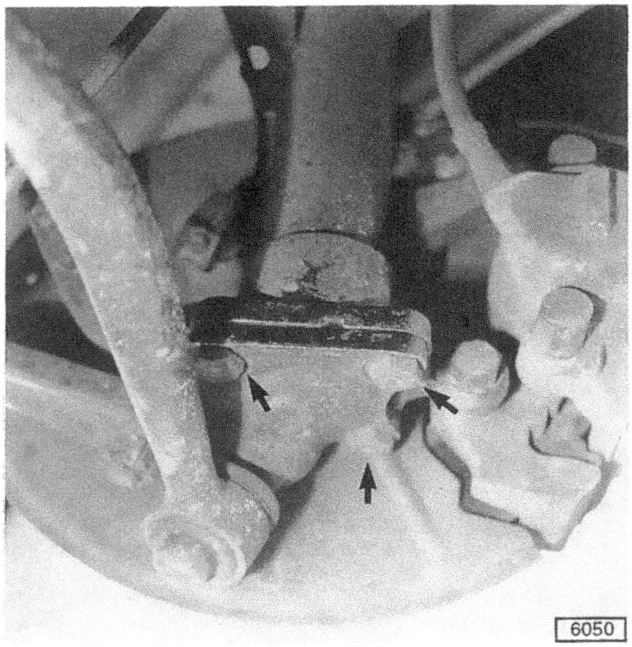

Fig. 4-14. Steering arm mounting bolts (arrows).

3. Remove the self-locking nut and the through-bolt that fastens the rear mount of the control arm or the thrust arm. See Fig. 4-15.

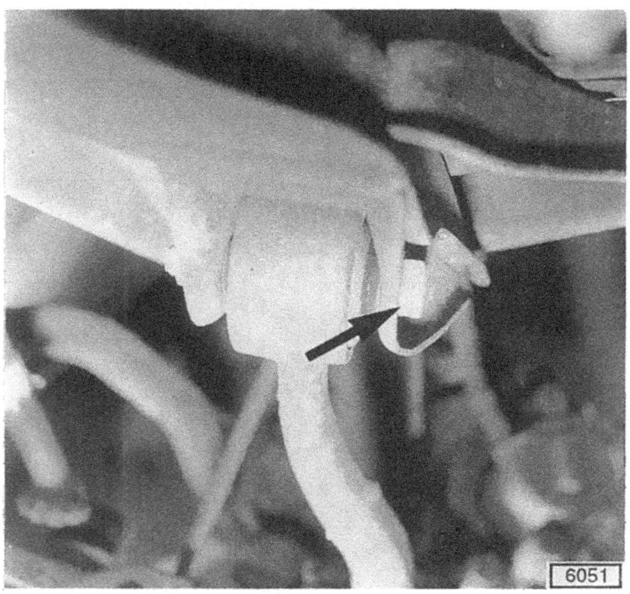

Fig. 4-15. Control arm through-bolt (arrow). Thrust arm is similar.

4. Remove the cotter pin and nut from the ball joint. See Fig. 4-16. Support the steering arm and press or knock the ball joint out of the steering arm.

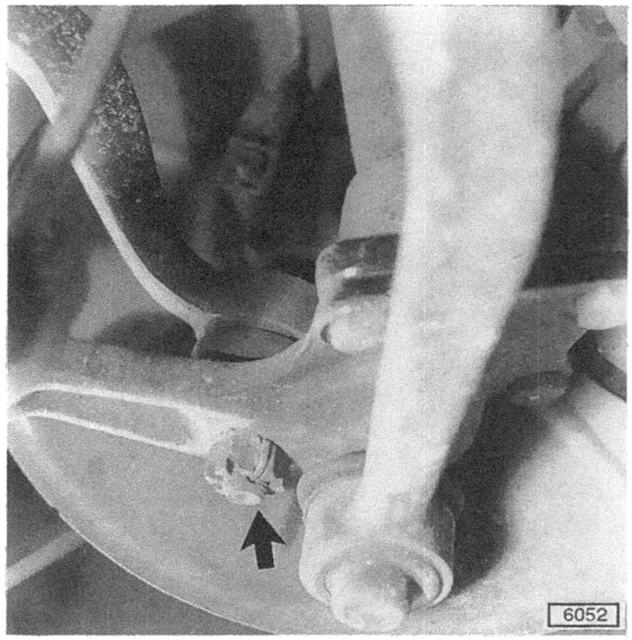

Fig. 4-16. Thrust arm joint mounting nut and cotter pin (arrow). Control arm is similar.

NOTE —
BMW special tool no. 31 1 110 is available to press ball joints out of the steering arm.

Installation is the reverse of removal. Use a new cotter pin on the ball joint nut. Use a new self-locking nut on the bushing through-bolt and use washers on both sides of the through-bolt. The mating surfaces should be kept clean and free of grease. If applicable, install the steering arm mounting bolts using a bolt locking compound (Loctite® 270 or equivalent). Before tightening the bushing through-bolts to the final torque, the car should be on the ground and loaded to its normal position as described below.

CAUTION —
Do not reuse self-locking nuts or cotter pins. These parts are designed to be used only once and should be replaced whenever they are removed.

NOTE —
Thrust arms are marked with an L for the left (driver's) side and a R for the right (passenger's) side. Be sure to check the marking before installing a replacement arm.

Suspension Loading Requirements

- each front seat—68 kg (150 lb.)
- center of rear seat—68 kg (150 lb.)
- trunk—21 kg (46 lb.)
- fuel tank full

Tightening torques

- steering arm to suspension strut (install with bolt locking compound (Loctite® 270 or equivalent)) . . . 65 Nm (48 ft-lb)
- control arm ball joint to steering arm. . . . . . 85 Nm (63 ft-lb)
- thrust arm ball joint to steering arm. . . . . . . 85 Nm (63 ft-lb)
- control arm through-bolt (with car in normal loaded position) . . . . . . 78 Nm (57 ft-lb)
- thrust arm through-bolt (with car in normal loaded position) . . . . . 130 Nm (96 ft-lb)

Replacing Control Arm and Thrust Arm Rubber Bushings

The control arm and thrust arm rubber bushings can be replaced separately once the arm is removed as described above in **Removing and Installing Control Arms and Thrust Arms**. The bushings are tightly fitted to the arms and must be pressed out. BMW's method calls for special tools for both removal and installation of the bushings.

NOTE —
The bushings can be replaced with the arms installed in the car by removing the bushing through-bolt and swinging the arm down. This procedure requires the BMW special tools. If the special tools are not available, it is recommended that the arm be removed from the car. At this point, the old bushing can be pressed out and the new bushing pressed in.

The rubber bushings should always be replaced in pairs, on each side. Both bushings should have the same markings, indicating that they come from the same manufacturer.

When installing the new bushing into the arm, make sure the bore in the arm is clean and free of grease. Press the bushing in working from the tapered side of the bore. Make sure there is an equal amount of bushing overhang on either side of the arm.

When replacing the thrust arm bushing, special installation instructions apply. Fig. 4-17 shows the correct orientation of the thrust arm bushing.

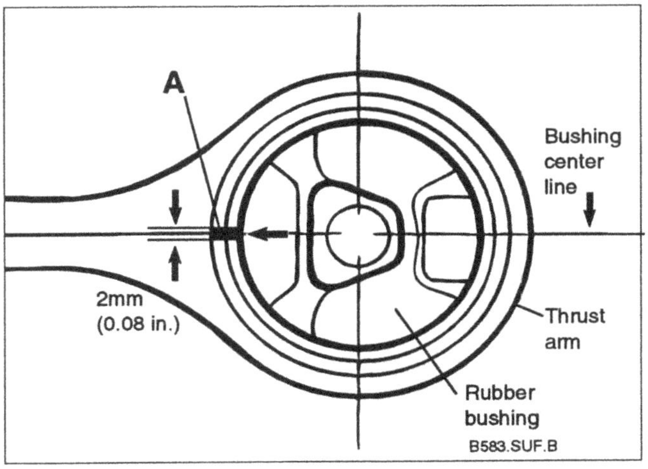

Fig. 4-17. Rubber bushing correctly installed in thrust arm. Arrow on rubber bushing is aligned with mark on arm (**A**). Bushing must be centered in the arm bore and must be no more than 1 mm (0.04 in.) above or below the bushing's centerline.

4.3 Front Wheel Bearings

The front wheel bearings are permanently sealed and lubricated. They do not require any routine maintenance. The sealed bearing and the front hub are one unit and are removed and installed together. The bearing cannot be replaced separately. The bearing inner race is press-fit onto the stub axle and the bearing/hub assembly is held in place with a large collar nut.

The bearing and hub will normally be damaged as it is removed, and therefore cannot be reinstalled. In addition to a new bearing and hub, a new collar nut and a new grease cap are required when replacing a wheel bearing.

Removing the bearing and hub requires a puller, and installation requires special tools. Read the procedure before starting work. If the necessary tools are not available, this work may be best left to an authorized BMW dealer or other qualified repair shop.

Check the wheel bearing with the wheel off the ground and the car properly supported. Spin the wheel and listen for any roughness. Alternately apply pressure to either side of the wheel while checking for play. If any faults are found and no suspension ball joints are found to be faulty, the wheel bearing should be replaced.

To remove front wheel bearings:

1. Loosen the wheel lug bolts, then raise the front of the car and support it securely on jack stands. Remove the wheel.

2. Using a flat-blade screwdriver or drift, remove the dust cap in the center of the front wheel hub.

3. Using a chisel, unlock the collar nut by breaking away the portion that engages the slot in the stub axle. See Fig. 4-18.

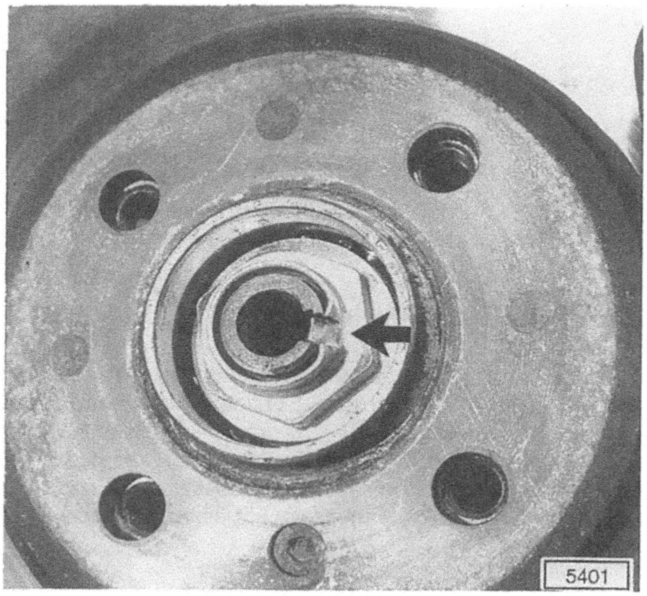

Fig. 4-18. Collar nut locked into front stub axle. Unlock nut by breaking locked portion (arrow) with chisel.

4. Remount the wheel, install and tighten the lug bolts, and lower the car to the ground.

5. Pry off the cap from the wheel center and loosen the collar nut but do not remove it.

> ***WARNING —***
> *Always loosen the collar (axle) nut with the wheel installed and the car on the ground. The leverage required to loosen the nut could cause the car to topple off a lift or jack stand.*

> **NOTE —**
> The collar nut requires a 46 mm socket.

6. Raise the front end of the car, support it securely on jack stands, and remove the wheel.

7. Remove the brake caliper and the brake rotor as described in **BRAKES**. Keep the brake line attached to the caliper. Fig. 4-19 shows the bearing assembly.

Fig. 4-19. Front brake caliper and rotor removed from wheel bearing assembly.

> ***CAUTION —***
> *Do not step on the brake pedal or operate the brakes while the caliper is removed. Doing this will damage the brake caliper.*

> **NOTE —**
> To avoid kinking or damaging the hose, use stiff wire to suspend the caliper from the underbody.

8. Remove the collar nut and, using a puller, remove the hub and bearing assembly from the stub axle.

9. If the bearing inner race stays on the stub axle, remove the dust shield behind the bearing. Then use a puller to remove the inner race.

To install front wheel bearings:

1. Install a new dust shield and position the bearing assembly on the stub axle.

2. Applying force only to the bearing's inner race, press the hub and bearing onto the stub axle.

> **NOTE —**
> BMW special tool 31 2 110 should be used to press the bearing onto the stub axle. Driving the bearing on or using incorrect tools may damage the bearing.

3. Install a new collar nut and tighten it slightly. Do not try to tighten it to its final torque value.

4. Mount the disc brake rotor with its countersunk screw.

5. Mount the wheel and lower the car to the ground. Tighten the collar nut and bend the collar to lock it to the stub axle.

Tightening torque
• wheel bearing collar nut . . . . 290 Nm (214 ft-lb)

WARNING —
Always tighten the collar (axle) nut with the wheel installed and the car on the ground. The leverage required to tighten the nut could cause the car to topple off a lift or jack stand.

6. Raise the front end of the car, support it securely on jack stands, and remove the wheel.

7. Install the brake caliper as described in **BRAKES**.

8. Apply Loctite® 638 sealant or equivalent to the new grease cap, then install it using a soft-faced mallet.

NOTE —
Install only the later, smaller volume grease cap, BMW Part No. 31 21 1 130 124 This later cap will fit all cars covered by this manual. The cap is not reusable and must be replaced any time it is removed.

9. Lubricate the wheel and hub contact surface with multipurpose grease and mount the wheel. Lower the car to the ground and tighten lug bolts.

Tightening torque
• wheel lug bolts . . . . 100±10 Nm (74±7 ft-lb)

4.4 Front Stabilizer Bar

The stabilizer bar is essentially a large spring that helps to distribute cornering loads and reduce body roll. The one-piece stabilizer bar is mounted to the underbody of the car in rubber bushings that allow it to twist. The ends of the bar are mounted via short connecting links to the control arms. The rubber bushings can be replaced separately if necessary. The connecting links are also available separately.

To remove front stabilizer bar:

1. Remove the lower connecting link nut and press the link joint out of the stabilizer bar. See Fig. 4-20.

2. Remove the mounting bolts and two brackets that fasten the bar to the subframe and remove the bar from the car. See Fig. 4-21.

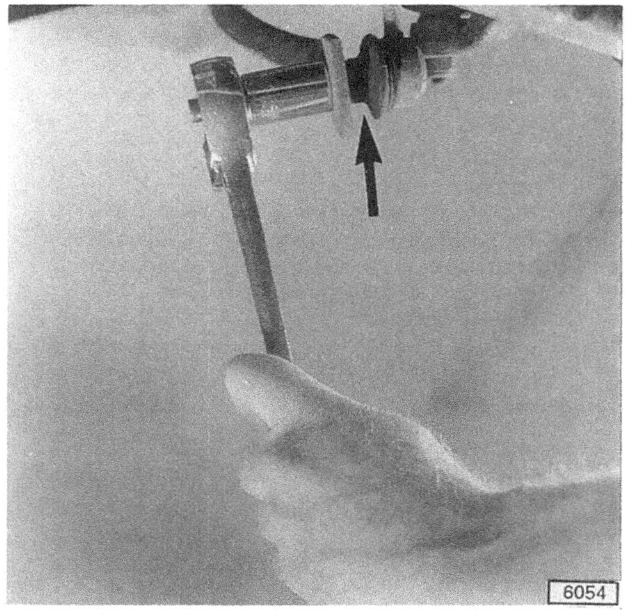

Fig. 4-20. Stabilizer bar connecting link being removed from stabilizer bar. Hold slotted part of link joint (arrow) when removing nut.

Fig. 4-21. Stabilizer bracket and rubber bushing (arrow).

3. To replace the connecting links, simply remove the nut that fastens each link to the suspension struts.

Installation is the reverse of removal. Fastener torques are listed below.

CAUTION —
Do not reuse self-locking nuts or cotter pins. These parts are designed to be used only once and should be replaced whenever they are removed.

NOTE —
Installation of the bar and proper tightening of the mountings will be easiest with the car level (front wheels at the same height), and as near to normal ride height as possible.

Tightening torques	
• stabilizer bar mounting brackets	22 Nm (16 ft-lb)
• stabilizer bar link to strut housing (locknut)	
yellow chrome	20 Nm (15 ft-lb)
white chrome	33 Nm (24 ft-lb)
white	59 Nm (43 ft-lb)

4.5 Front Suspension Subframe

Fig. 4-22 shows the subframe and related components. The subframe is designed to add rigidity to the unit-body construction and to provide mounting points for the engine mounts and other components. The subframe does not normally need to be removed and only requires attention if it is bent, or if the mounting bolts have become loose.

Tightening torques	
• crossmember to body	60 Nm (44 ft-lb)
• subframe mounting bolts	42 Nm (31 ft-lb)
• control arm through bolt (see **4.2 Control Arms and Thrust Arms**)	

5. TECHNICAL DATA

I. Tightening Torques

Control arm ball joint to steering arm (locknut)	93 Nm (69 ft-lb)
Control arm rubber bushing to subframe mount (bolt and locknut)	78 Nm (57 ft- lb)
Front brake caliper to suspension strut (bolt)	110—123 Nm (81–91 ft-lb)
Front wheel bearing collar (axle) nut	290 Nm (214 ft-lb)
Stabilizer bar connecting link to suspension strut or stabilizer bar (locknut)	
yellow chrome	20 Nm (15 ft-lb)
white chrome	33 Nm (24 ft-lb)
white	59 Nm (43 ft-lb)
Stabilizer bar mounting brackets to subframe	22 Nm (16 ft-lb)
Steering arm to strut housing (bolts installed with locking compound)	65 Nm (48 ft-lb)
Strut bearing to shock absorber cartridge (locknut)	65 Nm (47 ft-lb)
Strut housing threaded collar	130 Nm (94 ft-lb)
Subframe anchor bolts (M10)	42 Nm (31 ft-lb)
Subframe crossmember to body	60 nm (44 ft-lb)
Suspension strut bearing (upper mount) to wheel housing (nuts)	22 Nm (16 ft-lb)
Thrust arm ball joint to steering arm (locknut)	85 Nm (63 ft- lb)
Thrust arm rubber bushing to chassis mount (bolt and locknut)	130 Nm (96 ft-lb)
Wheel lug bolts	100±10 Nm (74±7 ft-lb)

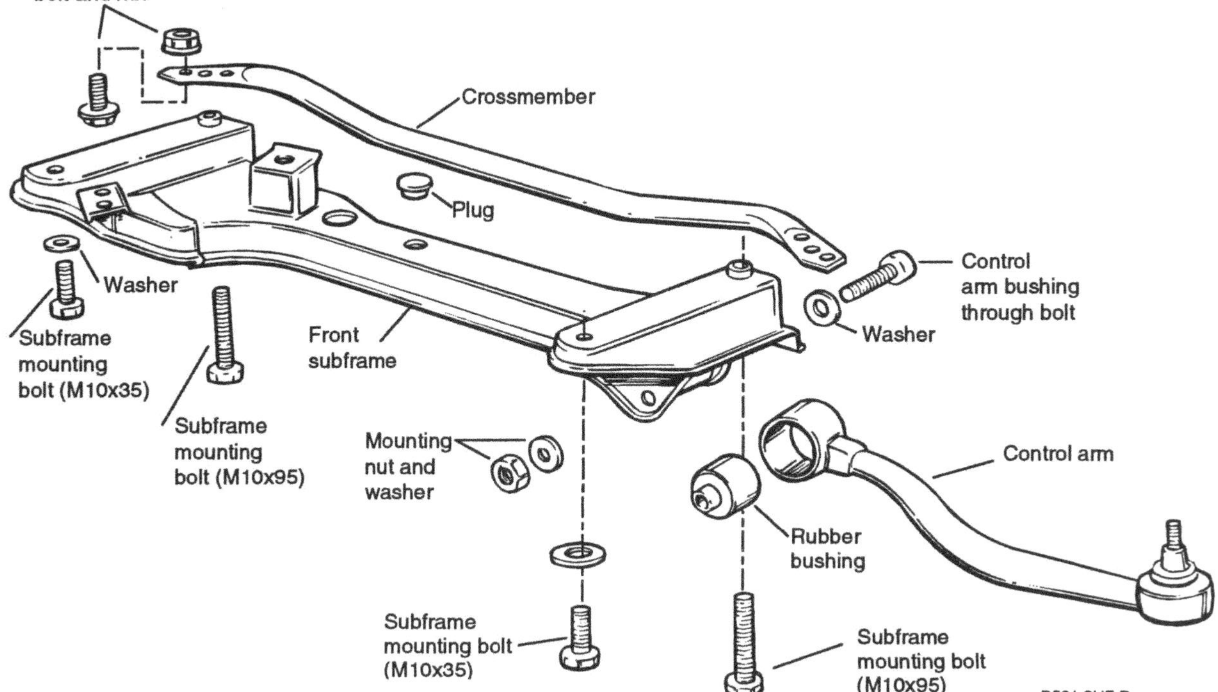

Fig. 4-22. Subframe and related components.

Section 17

SUSPENSION — REAR

Contents

INTRODUCTION

BMW 5-Series cars are equipped with a strut-type independent rear suspension with semi-trailing arms. The suspension struts are integral spring and shock absorber units. This section of the manual covers servicing of the rear suspension and the drive axles. Information on the final drive and its mounting is in **DRIVESHAFT AND FINAL DRIVE**.

The rear axle carrier spans the width of the car and is the main mounting for the final drive housing and the rear suspension components. Trailing arms, mounted to pivot points on the rear axle carrier, are the main suspension components. The trailing arms also house the rear wheel bearings. Suspension loads are taken up by integral coil spring and shock absorber units mounted between the trailing arms and the body.

1. GENERAL DESCRIPTION

The BMW 5-Series rear suspension system is engineered to provide a sophisticated compromise between taut, responsive handling and ride comfort. Although the front and rear suspension systems are independent subsystems, and are of completely different designs, they work together to achieve BMW's overall combination of precise handling and comfort.

Rear Suspension

The BMW 5-Series rear suspension is shown in Fig. 1-1. The final drive housing is mounted to the rear axle carrier and the chassis. The independent suspension allows each rear wheel to react independently to bumps and other suspension loads. The rear suspension is designed with anti-squat geometry. It reduces the normal tendency for the rear of the car to "squat" during acceleration.

2. MAINTENANCE

Scheduled maintenance of the rear axle includes only periodic inspections of the suspension components for wear and damage.

BMW recommends that the following items be routinely checked at periodic intervals as described in **LUBRICATION AND MAINTENANCE.** Detailed descriptions of these maintenance operations can be found either in this section of the manual under the boldface numbered headings shown, or in **LUBRICATION AND MAINTENANCE**.

1. Inspecting suspension components. **4.**
2. Inspecting drive axle constant velocity (CV) joint boots. **4.2**
3. Checking wheel bearings. **4.4**

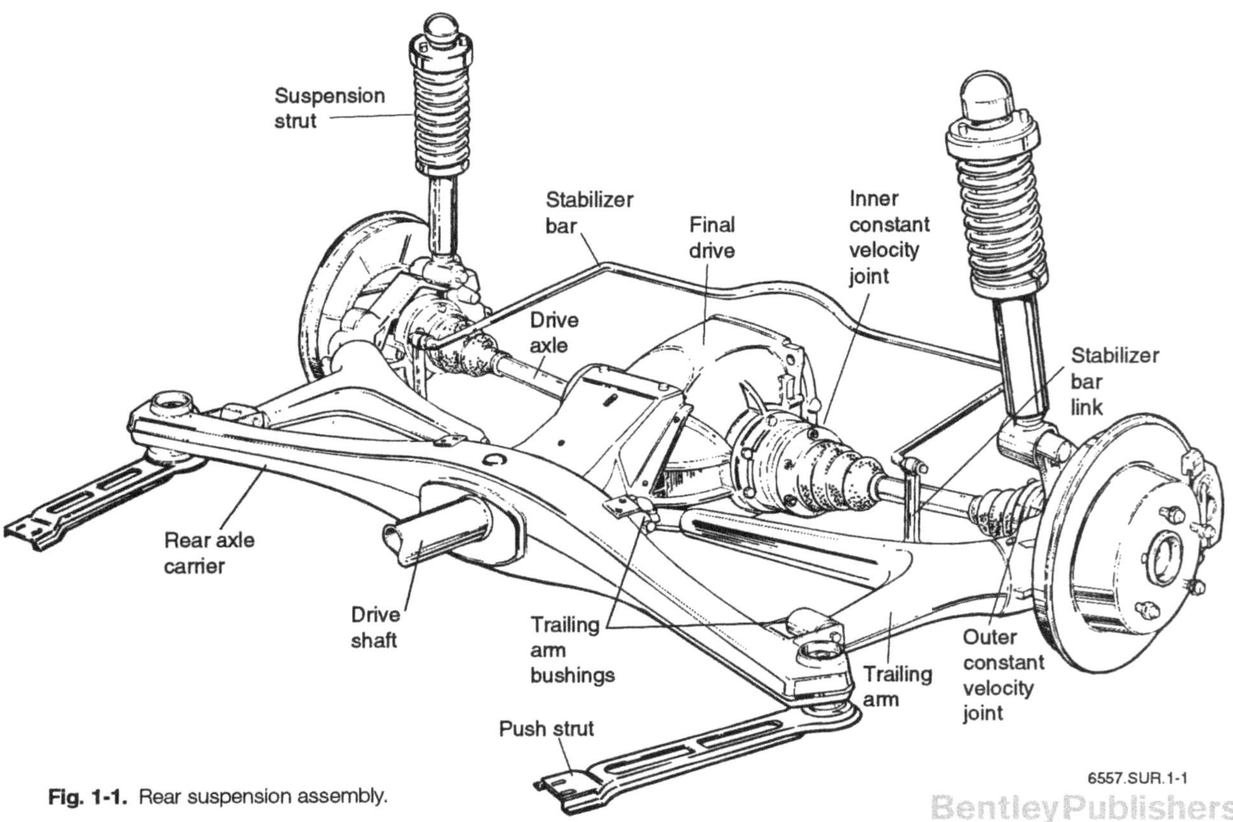

Fig. 1-1. Rear suspension assembly.

3. TROUBLESHOOTING

This troubleshooting section applies to rear suspension problems affecting ride comfort, handling and stability. Such problems are usually caused by faults in the suspension, but a problem such as consistently pulling to one side may also be caused by faulty brakes. For brake system troubleshooting, see **BRAKES**. For more basic help in determining the appropriate section for troubleshooting a particular symptom, see **FUNDAMENTALS** at the front of the manual.

3.1 Basic Troubleshooting Principles

Stable handling and ride comfort depends on the integrity of the suspension components. The suspension system must precisely position the wheels so that the car is stable and controllable, but also allow movement so the wheels can react to bumps.

In addition to inspecting for worn parts, troubleshooting must consider the condition of tires, wheels, and their alignment. Tire wear and incorrect inflation pressures can dramatically affect handling. Subtle irregularities in wheel alignment angles also affect stability. Mixing different types or sizes of tires, particularly on the same end of the car, can affect alignment and may unbalance a car's handling. See **STEERING AND WHEEL ALIGNMENT** for more troubleshooting information.

Table a lists symptoms of rear suspension problems and their probable causes, and suggests corrective actions. The boldface numbers in the corrective action column refer to headings where the repairs are described.

3.2 Diagnostic Inspection and Testing

Suspension problems can usually be isolated and at least partially diagnosed by careful observation of the symptoms and inspection of the components that are the most likely cause.

Tire Wear

Tire wear can be a good indicator of rear suspension problems. For more information on tire wear, see **STEERING AND WHEEL ALIGNMENT**. For information on proper tire inflation pressures, as well as other general tire maintenance, see **LUBRICATION AND MAINTENANCE.**

Table a. Rear Suspension Troubleshooting

Symptom	Probable cause	Corrective action
1. Rear-end vibration or shimmy	a. Incorrect tire pressure	a. Check and correct tire pressures. See **LUBRICATION AND MAINTENANCE**
	b. Unbalanced wheels/tires or misaligned rear wheels	b. Balance tires. Check for uneven wear patterns. Align front and rear end. See **STEERING AND WHEEL ALIGNMENT**
	c. Loose wheel lug bolts	c. Tighten lug bolts to proper torque
	d. Bent wheel rim (radial or lateral runout)	d. Inspect wheels and replace as necessary. See **STEERING AND WHEEL ALIGNMENT**
	e. Tire(s) out-of-round (radial runout)	e. Measure tire radial runout and remount or replace as necessary. See **STEERING AND WHEEL ALIGNMENT**
	f. Severely worn shock absorbers	f. Replace strut shock absorber. **4.1**
	g. Worn or defective trailing arm rubber bushings	g. Replace rubber bushings. **4.3**
2. Uneven ride height	a. Incorrect coil springs	a. Measure ride height and replace springs as required. **4.1**
	b. Bent or damaged suspension components	b. Inspect and repair/replace as necessary
	c. Sagging coil springs	c. Replace front and/or rear springs in pairs. **4.1**
3. Rear wheel noise scraping grinding	a. Worn rear wheel bearing	a. Replace bearings. **4.4**
	b. Faulty rear brakes	b. See **BRAKES**
4. Knocking or rattling noise from trunk or rear of car	a. Faulty shock absorber lower or upper mount	a. Replace rubber mounts. **4.1**
	b. Faulty trailing arm rubber bushings	b. Replace rubber bushings. **4.3**

17

Isolating Pulling Symptoms

Consistent pulling to one side in a car driven straight ahead on a level road may be caused either by worn or damaged (bent) suspension components or a faulty tire/wheel. In more unusual cases, a brake problem may be the cause.

To help decide whether the problem is caused by the tires or the suspension, temporarily swap the rear wheels (and tires), and then road test the car. If a tire problem is the cause of the pulling symptom, the problem should switch to the other side of the car when the wheels are switched, and it should now pull to the other side. If the symptom persists, then the problem is probably caused by faulty suspension components.

NOTE —
If no suspension components are found to be faulty, the car may only need a wheel alignment. See **STEERING AND WHEEL ALIGNMENT** for more information.

Vibration

Abnormal vibration, if not caused by a mechanical problem, will most often be caused by wheels and tires that are out of balance. Another, less likely cause of vibration is a distorted or out-of-round tire. These and other tire and wheel conditions can be checked by any reputable tire professional.

Before spending the money for balancing, look for an obvious cause of imbalance. Caked-on mud, ice, or snow can dramatically affect wheel balance. Clean the wheels and tires and road-test the car before investigating more serious causes.

4. REAR SUSPENSION

The main rear suspension components are the trailing arms and the suspension struts. The trailing arms are mounted in rubber bushings at pivot points on the rear axle carrier. The trailing arms house the wheel bearings and serve as lower mounting points for the struts.

The limited number of rear suspension parts that normally wear can be replaced. Some of these repairs require special tools and equipment. To avoid starting a job that may be difficult to complete, please read the entire procedure before beginning each repair.

4.1 Shock Absorbers and Springs

Fig. 4-1 shows the main components of the rear strut assembly. These are the shock absorber, the upper mount, and the coil spring. Each of these main components can be replaced separately as required.

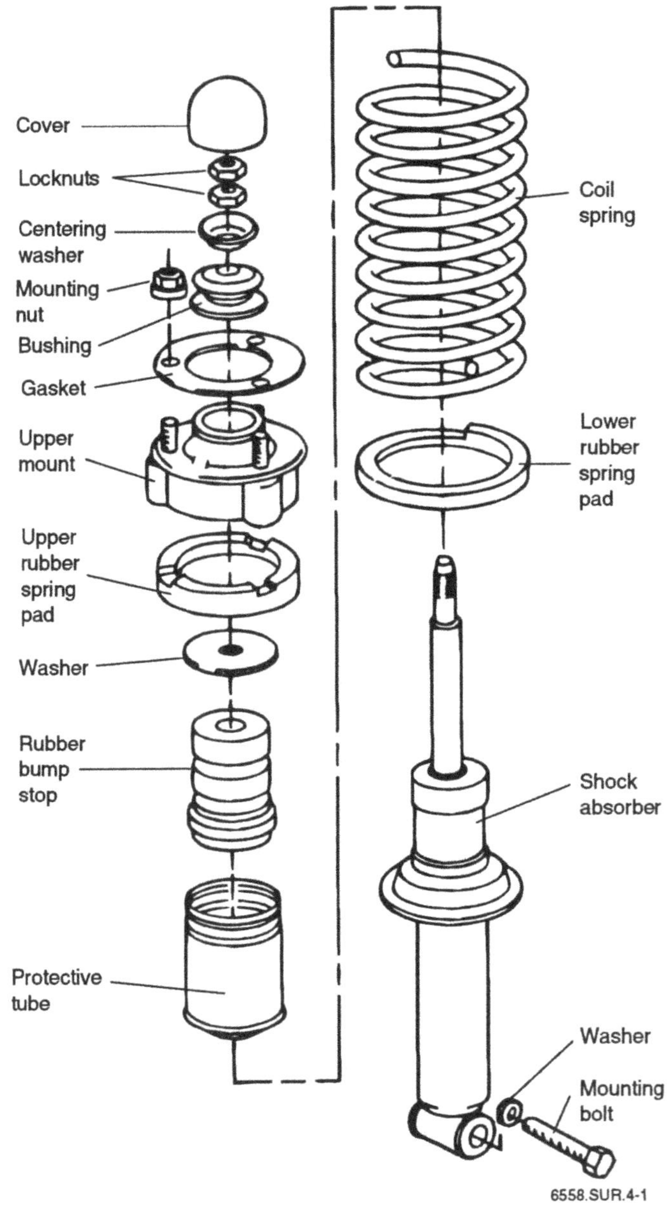

Fig. 4-1. Exploded view of rear strut assembly.

The shock absorbers and the springs are separate components. If the shock absorbers need replacing, special tools will be required to compress the spring to remove it from the shock absorber.

NOTE —
Differences from side to side will affect handling and stability. It is strongly recommended that springs and shock absorbers be replaced in pairs.

Checking Shock Absorbers

Springs are what support the weight of the car and allow the suspension to travel smoothly over bumps and other road irregularities. A spring that is compressed by a bump simply rebounds, springing in the other direction with nearly the same force that compressed it.

The function of a shock absorber, or damper, is to moderate the spring action. It slows the bounce and helps the spring return to its normal position. Shock absorbers require no routine maintenance and cannot be serviced.

The best evidence of failing shock absorbers is their behavior in normal driving. Worn shock absorbers will allow extra skittishness over bumps, and a less-controlled and wallowing feel in corners and after bumps. When seriously worn, the shock absorbers present little resistance to spring oscillations. Because they so easily reach the limits of their travel, they may knock when going over bumps.

The most common test of shock absorber condition involves vigorously bouncing each end or corner of the car, and then releasing and observing how quickly the bouncing stops. More than one bounce usually suggests that the shock absorbers are not properly damping the spring action and need to be replaced.

Removing and Installing Rear Suspension Struts

The struts are easily removed using ordinary hand tools. Note that at full extension, the rear suspension struts also limit rear suspension travel. This helps prevent damage to the drive axle constant velocity (CV) joints.

> ***CAUTION —***
> *To avoid damaging the CV joints when removing the rear struts, the suspension must remain approximately within its normal range of travel. It should be supported either by a stand under the trailing arm or by the wheel in contact with the ground before removing the strut.*

To remove rear suspension struts:

1. Remove the lower strut mounting bolt at the trailing arm. See Fig. 4-2.

2. In the trunk, partially remove the trim to expose the upper shock absorber mounts on each side.

3. While supporting the trailing arm, remove the strut mounting nuts, the strut assembly and its gasket. See Fig. 4-3.

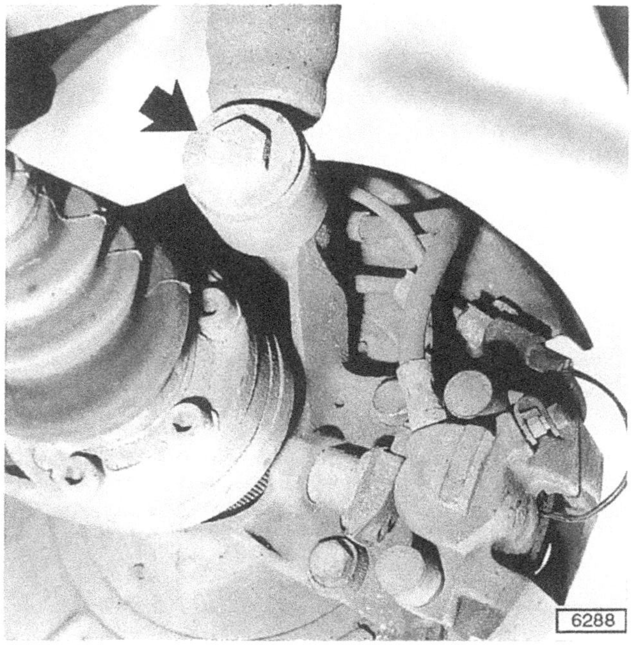

Fig. 4-2. Rear strut lower mounting bolt (arrow).

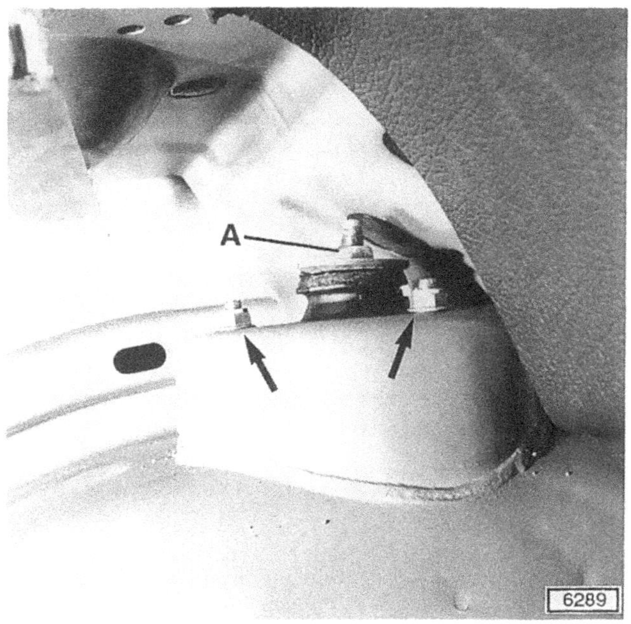

Fig. 4-3. Rear strut upper mounting nuts (arrows) in trunk. Third nut not visible in picture. Do not remove center nut (**A**). Gasket is between upper mount and body.

Installation is the reverse of removal. Make sure there is a gasket between the upper mount and the body. Tighten the lower mounting bolt to its final tightening torque with the car on the ground and loaded as specified below.

17

Suspension loading requirements:

- each front seat—68 kg (150 lb.)
- center of rear seat—68 kg (150 lb.)
- trunk—21 kg (46 lb.)
- fuel tank full

Tightening torques
• strut lower mount. . . . . 125 to 143 Nm (92 to 105 ft-lb)
• strut upper mount . . . . 22 to 24 Nm (16 to 18 ft-lb)

CAUTION —
Shock absorbers should be stored upright, imitating their normal installed position. If they are stored horizontally or upside-down for an extended period, they should be placed upright at room temperature with shock absorber rods fully extended for at least 24 hours prior to installation.

Disassembling Rear Strut Assembly

To replace either the coil spring or the shock absorber, the strut assembly must be disassembled using a spring compressor. The top center locknut(s) must not be removed from the shock until the spring tension is relieved.

To disassemble rear struts:

1. Install a spring compressor on the removed strut and relieve the spring tension.

   WARNING —
   Work slowly while watching the spring compressor. Make sure the spring compressor stays securely in place.

2. Pry the protective cap off the top nut. Holding the shock absorber shaft so that it cannot turn, slowly loosen the top locknut(s). Remove the nut(s) and the centering washer.

   WARNING —
   Do not remove the top locknuts from the shock absorber rod until the coil spring is compressed and there is no spring force against the upper mount.

3. Lift off the upper mount, the large washer, and the rubber spring pad.

4. Release the spring compressor carefully, letting the spring slowly expand to its free length.

NOTE —
If the strut is to be disassembled for only a short time, it may be convenient to leave the spring compressor attached. If so, make certain the compressor is secure and handle the compressed spring with extreme caution.

Installation is the reverse of removal. Inspect all rubber components, including the upper mounts, the spring pads, the bump stop, and the protective tube. Replace any damaged components. Use Fig. 4-1 above as a guide during assembly of the strut. With the spring compressed, tighten the top nuts while holding the shock absorber shaft stationary.

Tightening torque
• upper spring mount to shock absorber (locknut) . . . . . 22 to 25 Nm (16 to 18 ft-lb)

The rear springs should only be replaced in pairs, and should always be replaced with springs of the same number and color code as those that were removed. Note that different models are equipped with different springs. They are not necessarily interchangeable.

Checking and Correcting Ride Height

Ride height is measured from the lower edge of the wheel arch to the center of the bottom edge of the wheel rim. See Fig. 4-4. Measure the height with the car on the ground and loaded as specified. **Table b** lists rear ride height specifications.

Fig. 4-4. Ride height measurement (**A**) is taken from center of wheel arch to lower edge of wheel rim. (Measure with car on ground and loaded.)

Suspension loading requirements:

- each front seat—68 kg (150 lb.)
- center of rear seat—68 kg (150 lb.)
- trunk—21 kg (46 lb.)
- fuel tank full

Table b. Rear Ride Height Specifications

Wheel (rim) Installed	Ride Height (Dimension A)*
14 in. rims	501±10mm (19 3/4±3/8 in.)
165 TR 390 (TRX) rims	514±10 (20 1/4±3/8 in.)

*On models with wheel arch trim, subtract 3 mm (1/8 in.) from the ride height specifications listed.

If the ride height is outside the specification listed, new springs can be selected to correct the ride height.

4.2 Drive Axles and Constant Velocity (CV) Joints

Since the 5-series BMW models covered by this manual feature independent rear suspension and a final drive mounted to the underbody, the drive axles must be able to move with the suspension while delivering power to the wheels. To accomplish this, a constant velocity (CV) joint is located at each end of each drive axle.

A CV joint is similar in function to the more familiar universal joint. The CV joint allows for the smooth flow of power to the drive wheels, even though the final drive, axles and wheels are not precisely straight. The advantage of the more complicated CV joint is that it operates more smoothly and can accept more radical drive axle angles than a universal joint. A drive axle in its installed position is shown in Fig. 4-5.

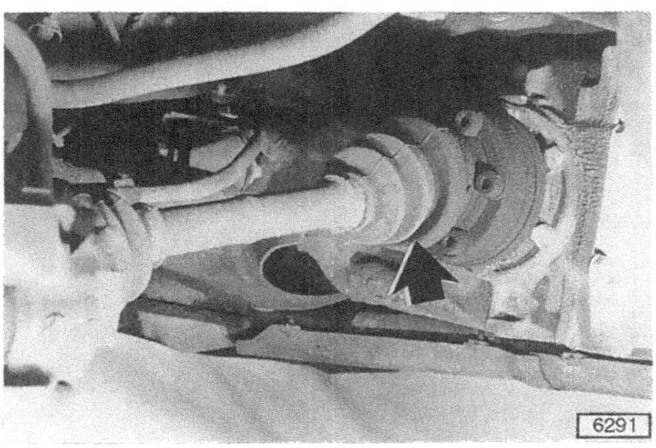

Fig. 4-5. Drive axle and inner CV joint. Rubber boot (arrow) retains lubricant and protects joint from dirt.

The inner and outer CV joints are bolted to the drive flanges of the final drive and stub axle. Each joint is packed with a special lubricant and sealed by a rubber boot.

The boots should be inspected periodically to see that they are in good condition. A damaged boot will let in dirt, which will very quickly destroy the joint. The CV joints are not rebuildable. The inner and outer joints are the same part and can be replaced separately. If the drive axle itself is damaged, only the complete axle assembly, including joints and boots, is available from BMW.

Removing and Installing Drive Axles

The CV joints are mounted to the drive flanges with internal-hex-head bolts that require a special wrench. In addition, the CV joints are tightly fitted to the drive axle ends and should be pressed off if either the boot or the joint needs to be replaced.

To remove drive axle:

1. Firmly set the parking brake.

2. Clean around the inner CV joint and remove the six internal-hex-head bolts that hold it to the drive flange. See Fig. 4-6.

> ***CAUTION —***
> *To prevent damage to the outer CV joint, do not allow the inner end of the axle to hang free. Use stiff wire to suspend the inner end of the drive axle from a convenient point on the underbody.*

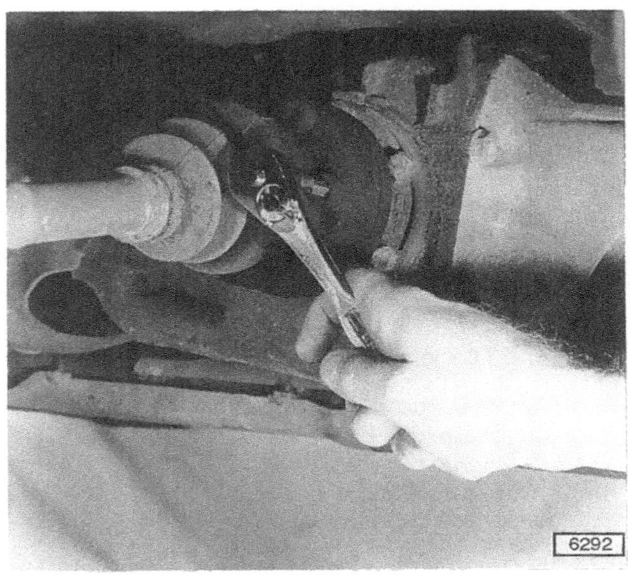

Fig. 4-6. Inner CV joint mounting bolt being removed.

17

3. Clean around the outer CV joint and remove the six internal-hex-head bolts that hold it to the drive flange. See Fig. 4-7.

Fig. 4-7. Outer CV joint mounting bolt being removed.

4. Remove the drive axle from the car.

NOTE —
If the joints are stuck to the flange, use a soft faced hammer to knock them loose.

Installation is the reverse of removal. The drive axle CV joints are identical and can be installed in either orientation.

Tightening torque
• CV joint to drive flange . . . . 58 to 63 Nm (43 to 46 ft-lb)

Replacing CV Joints and Protective Boots

The CV joints and boots are serviced by pressing the joints off the drive axle. Begin the procedure by removing the drive axle from the car.

To replace CV joints or boots:

1. Lift off the outer dust cover from the joint and remove the circlip. See Fig. 4-8.

2. While supporting the inner hub of the joint, press the drive axle out of the joint. See Fig. 4-9.

3. Clean all the old lubricant off the shaft splines and the joint's inner splines.

NOTE —
Inspect the drive axle and the splines for wear or damage. If the shaft is damaged, the complete assembly will have to be replaced.

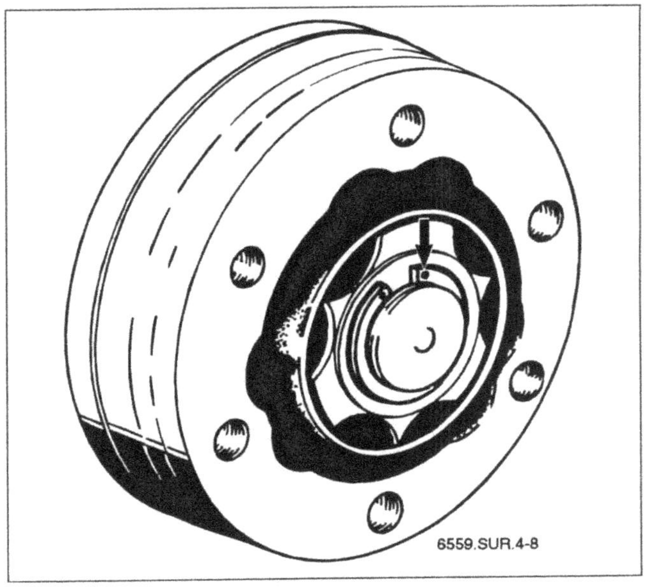

Fig. 4-8. Circlip (arrow) securing CV joint to axle shaft. Always replace.

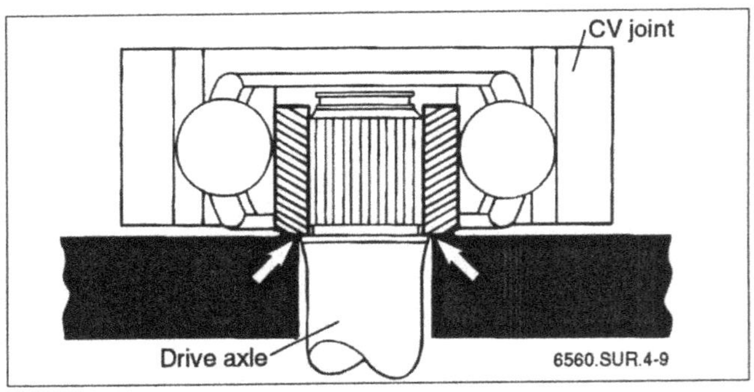

Fig. 4-9. Support CV joint on inner hub only (arrows) when pressing drive axle out of joint.

4. Place the protective boot and the inner joint cover over the drive axle.

NOTE —
When replacing the CV joint boot, use a complete CV joint boot repair kit. The kit will include a new boot, clamps, lubricant, and a new inner CV joint circlip. The kit is available from an authorized BMW dealer parts department.

5. Apply Loctite®270 or an equivalent locking compound to the drive axle splines. Position the new CV joint on the shaft so that the hub with the raised or taller side is facing the shaft. See Fig. 4-9 above.

CAUTION —
Do not let the locking compound contact the balls in the joint. Apply only a thin coat to cover the splines.

6. Press the shaft into the joint while supporting the inner hub of the joint. Install the new circlip.

CAUTION —
• Always install a new circlip.

• Do not let the ball hub pivot more than 20° in the outer ring of the joint. The balls will fall out if the hub is pivoted too far.

7. Pack each CV joint and rubber boot with the specified amount of the lubricant supplied in the boot kit. Apply adhesive to the large end of the boot and mount it on the joint. Secure the boot with the clamps. Then apply sealer to the joint cover(s) and install them.

Capacity	
• CV joint grease (per joint)	120 gm (4.2 oz.)

NOTE —
BMW recommends Bostik 1513 or Epple 4851 adhesive, and Epple 39 or Curil K sealer. These materials or suitable equivalents are available from your BMW dealer.

Inspecting CV Joints

The components of each CV joint are precisely matched during manufacture and cannot be serviced or replaced individually. To inspect a CV joint, clean away the grease and look for galling, pitting and other signs of wear or physical damage. Any of these is cause for replacement. Discoloration due to overheating indicates lack of lubrication. A joint that is otherwise in good condition may be reinstalled if thoroughly cleaned and repacked with the proper amount of new grease.

NOTE —
Polished interior surfaces or visible ball tracks alone are not necessarily cause for replacement.

The balls and grooves allow the hub to move, but the parts should fit snugly and move only with some effort. A joint with obvious freeplay between inner hub, balls and outer cage should be replaced.

Disassembly of the CV joints is not normally recommended. If the joint must be disassembled for cleaning, mark the relative positions of the inner hub, the ball cage, and the outer housing. Remove the balls one at a time, and keep track of them so that each can be reinstalled in its original position in the hub and cage.

CAUTION —
• The cage, housing, and balls are precisely matched. When disassembling more than one joint, do not intermix components.

• The CV joint should go together firmly but smoothly. Heavy force should not be required. If in doubt, start over and recheck the alignment. A joint that is forced together may lock and not come apart again.

4.3 Rear Suspension Trailing Arms

The trailing arms and their mounting bushings control the position of the rear wheels. A damaged trailing arm or worn mounting bushings will change rear wheel alignment and may adversely affect handling and stability.

CAUTION —
Do not attempt to straighten a damaged trailing arm. Bending or heating may weaken the original part. Replace damaged parts.

The rear brake lines must be disconnected to remove a trailing arm. This means the brakes will have to be refilled with fluid and bled when the trailing arm is reinstalled.

To remove rear trailing arms:

1. Raise the rear end of the car and support it securely on jack stands. Remove the wheel.

2. Remove the outer CV joint from the drive flange as described in **4.2 Drive Axles and Constant Velocity (CV) Joints**.

3. Disconnect the rear brake lines at the union shown in Fig. 4-10. Drain the brake fluid or plug the line.

4. Disconnect the parking brake cable from the parking brake actuator, as described in **BRAKES**. Unclip the parking brake cable form the trailing arm.

5. If applicable, remove the ABS wheel speed sensor from the trailing arm and unclip the sensor's wiring from the trailing arm.

NOTE —
Set the sensor aside so it will not be damaged when the trailing arm is removed and installed.

6. If removing the right trailing arm, separate the brake pad wear sensor connector.

7. Disconnect the stabilizer bar connecting link from the trailing arm. See Fig. 4-11.

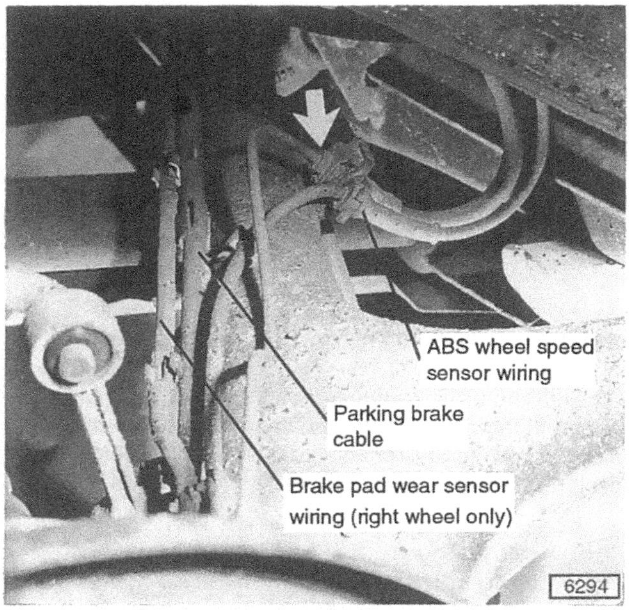

Fig. 4-10. Rear brake line union (arrow) at rear suspension trailing arm.

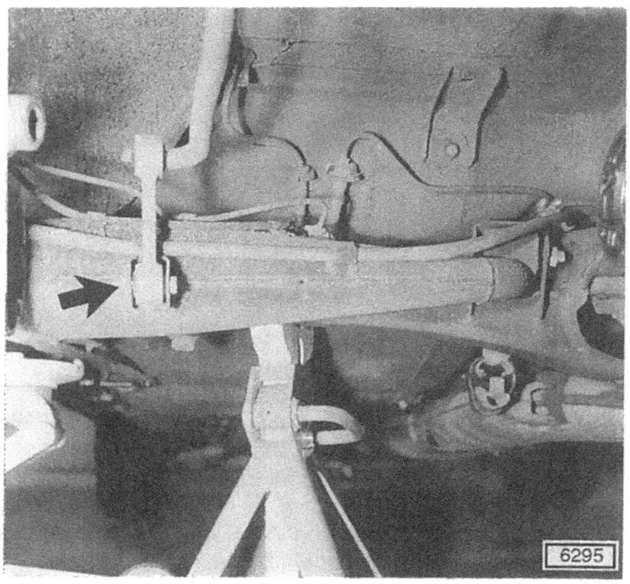

Fig. 4-11. Rear stabilizer bar connecting link mounting point (arrow).

8. On 1983 and later models, remove one of the mounting bolts from the trailing arm to rear axle carrier connecting link. See Fig. 4-12.

9. Disconnect the rear strut lower mounting bolt. See Fig. 4-13. Pull the strut away from the trailing arm.

10. Remove the two trailing arm through-bolts and nuts and remove the trailing arm from the car.

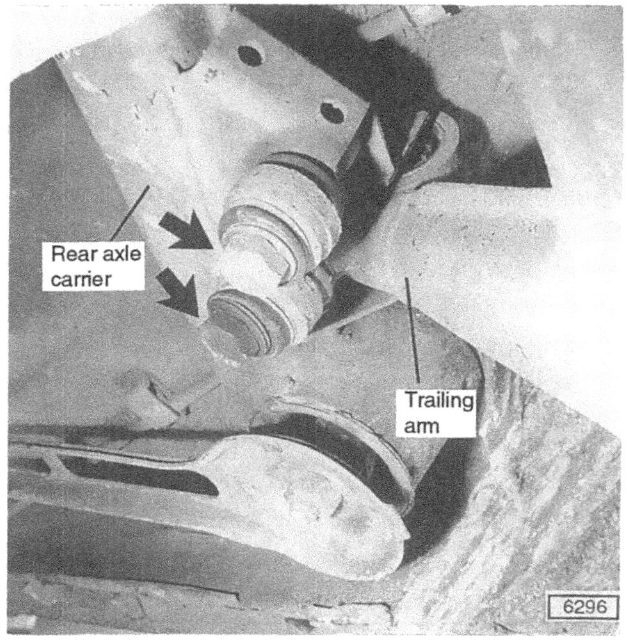

Fig. 4-12. Rear trailing arm to axle carrier connecting link used on 1983 and later models. Arrows indicate mounting bolts.

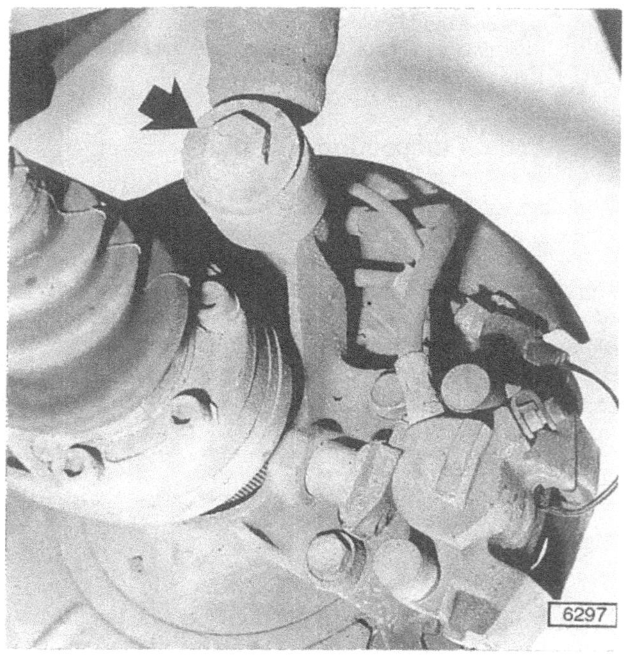

Fig. 4-13. Rear strut lower mounting bolt (arrow).

Installation is the reverse of removal. When installing the trailing arm mounting bolts, begin with the inner bolt. Fastener torque specifications are given below. Tighten the mounting bolts to the final tightening torque with the car on the ground and loaded as specified below. Refill and bleed the brake system as described in **BRAKES**.

Suspension loading requirements:

- each front seat—68 kg (150 lb.)
- center of rear seat—68 kg (150 lb.)
- trunk—21 kg (46 lb.)
- fuel tank full

Tightening torques
• trailing arm to rear axle carrier (rubber bushing through-bolt and nut) . . . . 67 Nm (49 ft-lb)
• trailing arm to axle carrier connecting link (1983 and later models) . . . . 127 Nm (93 ft-lb)
• shock absorber lower mount. . . . . 125 to 143 Nm (92 to 105 ft-lb)
• CV joint to drive flange . . . . 58 to 63 Nm (43 to 46 ft-lb)
• stabilizer bar connecting link to trailing arm . . . . 22 to 24 Nm (16 to 17 ft-lb)

Removing and Installing Trailing Arm Bushings

With the trailing arm removed as described above, the rubber bushings can be removed for replacement due to wear.

The rubber bushings are pressed into the trailing arms. BMW specifies special tools to remove them and to install replacements. With the trailing arm removed, most machine shops should have the equipment to press the old bushings out and install the new ones. When replacing the bushings, note that the bushing inner sleeve is longer on one side. When installing a new bushing, the longer side of the inner sleeve should always face the center of the car.

4.4 Rear Wheel Bearings

Each wheel bearing is pressed into the trailing arm wheel bearing housing and secured with a large circlip. The stub axle is in turn pressed into the bearing inner race and held in place by the drive flange and its nut. See Fig. 4-14.

The wheel bearings are permanently sealed and lubricated. They do not require any routine maintenance. Bearings are serviced by replacement only, as they are usually damaged when removed.

Check the wheel bearing with the wheel off the ground and the car properly supported. Spin the wheel and listen for any roughness. Alternately apply pressure to either side of the wheel while checking for play. If any faults are found, the wheel bearing should be replaced.

The large nut securing the stub axle to the drive flange is very tight, and the stub axle is press-fit into the wheel bearing inner race. Removal and installation of the drive flange, the stub axle, and the wheel bearing, and torquing the nut properly requires some special tools, including a puller and a press. If proper tools are not available, this work should be left to an authorized BMW dealer or other qualified repair shop. Read the procedure carefully before starting work.

To remove rear wheel bearings:

1. Remove the outer CV joint from the drive flange as described in **4.2 Drive Axles and Constant Velocity (CV) Joints**.

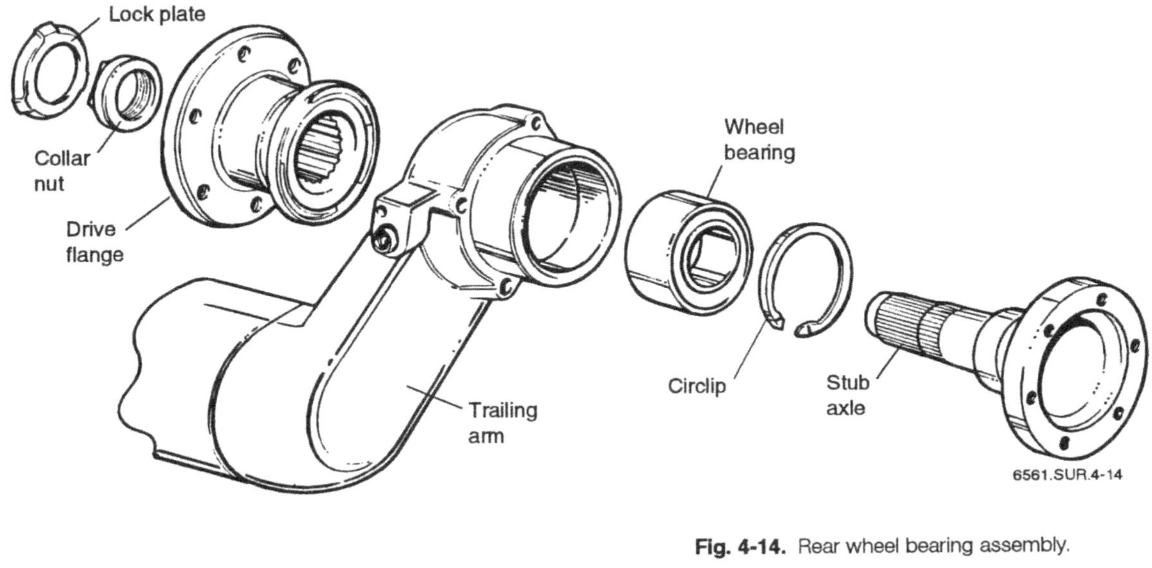

Fig. 4-14. Rear wheel bearing assembly.

2. Pry out the lockplate around the drive flange mounting nut. See Fig. 4-15. With the wheel on the ground and the parking brake firmly set, loosen (do not remove) the nut.

> ***WARNING —***
> *Always loosen and tighten axle nuts only while the car is on the ground. The leverage required could topple the car from a lift or jack stands.*

Fig. 4-15. Drive flange axle nut and lockplate (arrow) with CV joint removed.

3. Raise the rear end of the car and support it securely on jack stands. Remove the wheel.

4. Dismount the brake caliper, leaving the brake hose connected. Remove the brake rotor. See **BRAKES**.

> ***CAUTION —***
> *Do not allow the caliper to hang by the brake line. Doing this could damage the line and cause later brake failure. Use stiff wire to suspend the caliper from the underbody.*

> **NOTE —**
> Leaving the brake line attached to the caliper eliminates the need to bleed the brake system after reassembly.

5. Remove the drive flange mounting nut. Then using a suitable puller, remove the drive flange. See Fig. 4-16.

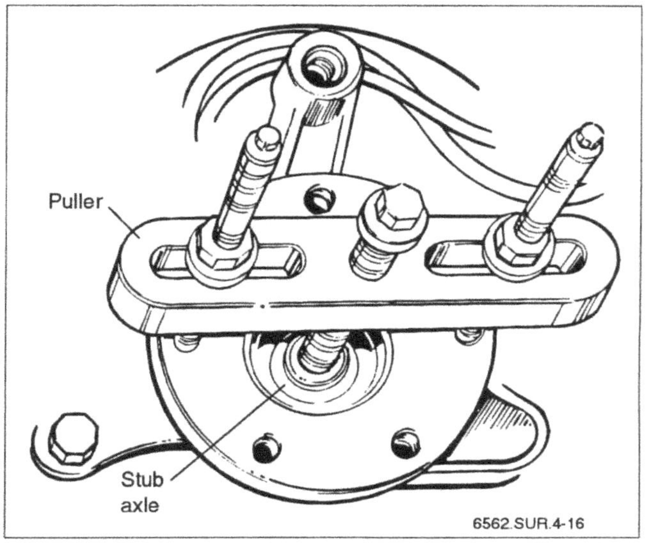

Fig. 4-16. Drive flange being remove from stub axle.

6. Thread the drive flange nut onto the stub axle until it is flush with the end of the axle. Using a soft-faced hammer, drive the stub axle out of the bearing.

> **NOTE —**
> If the bearing inner race remains attached to the stub axle when the axle is removed, as it probably will, it will be necessary to use a puller to remove the bearing race from the hub.

7. Remove the large circlip that holds the wheel bearing in the wheel bearing housing, then drive the bearing out from behind.

To install rear wheel bearings:

1. Applying force to the bearing on the outer race only, press the bearing into position in the trailing arm wheel bearing housing and fit a new circlip.

> ***CAUTION —***
> *Make sure that the bearing is pressed in far enough to contact the shoulder at the backside of the housing, and that the circlip is fully seated in its groove.*

2. Install the stub axle into the bearing.

> **NOTE —**
> BMW specifies special tools (Part No. 23 1 300, No. 33 4 080, and No. 33 4 020) that pull the stub axle into the bearing. If using an alternate method to pull or press the axle into position, support or apply force to the bearing only on the inner race.

3. Reinstall the drive flange. Install and tighten the drive flange mounting nut. Install the lockplate over the nut.

> ***WARNING —***
> *Take care not to topple the car from jack stands when tightening the axle collar nut.*

> **NOTE —**
> For safety, the wheel should be installed and the car lowered to the ground when tightening the nut. However, loading the bearing without the nut tightened sufficiently may damage the bearing. First snug down the nut as much as possible. Then install the brake rotor and caliper. Firmly set the parking brake and carefully tighten the nut further. Temporarily install the wheel, lower the car to the ground, then tighten the nut to its final torque.

Tightening torque	
• drive flange axle nut	
M22	175 to 210 Nm (129 to 155 ft-lb)
M27	234 to 260 Nm (173 to 192 ft-lb)

4. Reinstall the brake rotor and remount the brake caliper as described in **BRAKES**.

5. Reinstall the CV joint as described above under **4.2 Drive Axles and Constant Velocity (CV) Joints**.

6. Lubricate the wheel and hub contact surface with a thin film of multi-purpose grease, and mount the wheel. Lower the car to the ground, and torque the wheel lug bolts.

Tightening torque	
• wheel lug bolts	100±10 Nm (74±7 ft-lb).

5. TECHNICAL DATA

I. Tightening Torques

Connecting link between trailing arm and rear axle carrier (1983 and later models)	127 Nm (93 ft-lb)
CV joint to drive flange (bolts)	58–63 Nm (42–46 ft-lb)
Drive flange to stub axle (axle nut)	
M22	175–210 Nm (129–155 ft-lb)
M27	234–260 Nm (173–192 ft-lb)
Shock absorber lower mount to trailing arm (bolt)	125–143 Nm (92–105 ft-lb)
Shock absorber upper mount to body (nuts)	22–24 Nm (16–17 ft-lb)
Strut upper spring mount to shock absorber (locknuts)	22–25 Nm (16–18 ft-lb)
Trailing arm to rear axle carrier (through-bolt and nut)	67 Nm (49 ft-lb)
Wheel lug bolts	100±10 Nm (74±7 ft-lb)

17

Section 18

STEERING AND WHEEL ALIGNMENT

Contents

TABLES

INTRODUCTION

This section covers service and repair of the steering system as well as alignment information. For information relating to the front suspension, see **SUSPENSION—FRONT**. Much of the work described in this section requires special tools and experience. If you lack the skills, the tools, or a suitable workplace for steering or alignment work, we suggest you leave such repairs to an authorized BMW dealer or other qualified repair shop.

1. GENERAL DESCRIPTION

All models covered by this manual feature a power-assisted steering system, where hydraulic pressure, supplied by an engine-driven pump, reduces steering effort.

The two main components of the system are the pressure pump mounted in the engine compartment and driven by a V-belt, and the steering gearbox, mounted on top of the front suspension subframe. Tie rods connect the gearbox to the steering arms, which are mounted to the bottom of each front suspension strut. The steering column is connected to the steering gearbox by a universal joint.

The power-assisted steering provides varying amounts of assist, depending on engine speed. The fluid flow rate from the pressure pump is highest at low engine speeds, to provide more assist for low speed maneuvers such as parking. It is reduced at higher engine speeds, to prevent over-assist and to maximize road feel at highway speed.

Neither the steering gearbox nor the tie rod ends require any routine lubrication service during normal service life. The steering gearbox can be adjusted to compensate for small amounts of wear.

2. MAINTENANCE

Scheduled maintenance includes checking the fluid level for the power-assisted steering, and periodic inspections of the steering components for wear and damage.

Another periodic requirement is precise adjustment of wheel alignment. This can be done by an authorized BMW dealer service department or other qualified alignment shop. See **4. Wheels, Tires, and Alignment**.

BMW recommends that the following items be routinely checked at periodic intervals as described in **LUBRICATION AND MAINTENANCE**. Detailed descriptions of these maintenance operations can be found either in this section of the manual under the boldface numbered headings shown, or in **LUBRICATION AND MAINTENANCE**.

1. Checking fluid level and checking power steering for leaks
2. Checking steering for play and adjusting if necessary. **5.2**
3. Inspecting steering gearbox, steering coupling, and steering linkage. **5.2**

3. TROUBLESHOOTING

This troubleshooting section applies to problems affecting steering and wheel alignment. Such problems are usually caused by faults in the steering system, but a problem such as consistently pulling to one side may also be caused by faulty brakes or damaged suspension components. For brake system troubleshooting, see **BRAKES**. For suspension troubleshooting, see **SUSPENSION—FRONT** or **SUSPENSION—REAR**. For more basic help in determining the appropriate section for troubleshooting a particular symptom, see **FUNDAMENTALS** at the front of the manual.

3.1 Basic Troubleshooting Principles

The steering system must precisely position the wheels so that the car is stable and controllable. Any symptom of instability or imprecise road feel may be caused by worn or damaged steering components or incorrectly aligned wheels.

In addition to inspecting for worn parts, troubleshooting must also consider the condition of tires, wheels, and their alignment. Tire wear and incorrect inflation pressures can dramatically affect handling. Subtle irregularities in wheel alignment angles also affect stability. Mixing different types or sizes of tires, particularly on the same end of the car, can affect alignment and may unbalance a car's handling.

Table a lists symptoms of steering and wheel alignment problems and their probable causes, and suggests corrective actions. The boldface numbers in the corrective action column refer to headings where the repairs are described.

3.2 Diagnostic Inspection and Testing

Steering and wheel alignment problems can usually be isolated and at least partially diagnosed by careful observation of the symptoms and inspection of the components that are the most likely cause.

Tire Wear

Tire wear is a good indicator of steering and wheel alignment problems. Proper tread wear is difficult to notice, so tires are made with wear-indicator bands that indicate when the tire is nearly worn-out. On an evenly worn tire, these wear-indicator bars will eventually appear as evenly spaced bald "stripes" about 1/2 in. wide running across the tread surface, as shown in Fig. 3-1. The appearance of these tread-wear bands on only one part of the tread indicates uneven wear.

Table a. Steering and Wheel Alignment Troubleshooting

Symptom	Probable cause	Corrective action
1. Pull to one side, wandering	**a.** Incorrect tire pressure	**a.** Check and correct tire pressures. See **LUBRICATION AND MAINTENANCE**
	b. Defective/unevenly worn tire	**b.** Inspect tires and replace as needed. **4.1**
	c. Incorrect wheel alignment	**c.** Check and adjust wheel alignment. **4.2**
	d. Faulty brakes (pull under braking)	**d.** See **BRAKES**
2. Steering heavy, noisy, poor return-to-center	**a.** Incorrect tire pressure	**a.** Check and correct tire pressures. See **LUBRICATION AND MAINTENANCE**
	b. Low power steering fluid	**b.** Check power steering fluid and add as required. See **LUBRICATION AND MAINTENANCE**
	c. Loose or broken power steering pump V-belt	**c.** Inspect V-belt. Tighten or replace as necessary. See **LUBRICATION AND MAINTENANCE**
	d. Worn front suspension strut bearing(s)	**d.** Replace strut bearings. See **SUSPENSION—FRONT**
	e. Faulty power steering pump	**e.** Test and, if necessary, replace pump. **5.3**
	f. Air in power steering fluid	**f.** Repair air leak and, if necessary, add fluid. **5.3**
3. Front-end vibration or shimmy	**a.** Incorrect tire pressure	**a.** Check and correct tire pressures. See **LUBRICATION AND MAINTENANCE**
	b. Unbalanced wheels/tires or front wheels out of alignment	**b.** Balance tires. Align front wheels. Check tires for uneven wear patterns. **4.**
	c. Loose wheel lug bolts	**c.** Tighten lug bolts to proper torque. **4.**
	d. Bent wheel rim (radial or lateral runout)	**d.** Inspect wheels and replace as necessary. **4.**
	e. Tire(s) out-of-round (radial runout)	**e.** Measure tire radial runout and remount or replace as necessary. **4.**
	f. Severely worn shock absorbers or suspension components	**f.** Replace worn parts. See **SUSPENSION—FRONT**
	g. Worn control arm or thrust arm bushings	**g.** Replace bushings. See **SUSPENSION—FRONT**
4. Steering loose, imprecise	**a.** Incorrect tire pressure	**a.** Check and correct tire pressures. See **LUBRICATION AND MAINTENANCE**
	b. Loose steering gearbox mounting bolt(s)	**b.** Inspect and tighten bolts. **5.2**
	c. Worn tie rod end(s)	**c.** Replace tie rod(s) and align wheels. **5.2**
	d. Faulty front wheel bearing	**d.** Replace wheel bearing. See **SUSPENSION—FRONT**
	e. Worn or damaged steering gear box	**e.** Adjust or replace steering gearbox. **5.2**

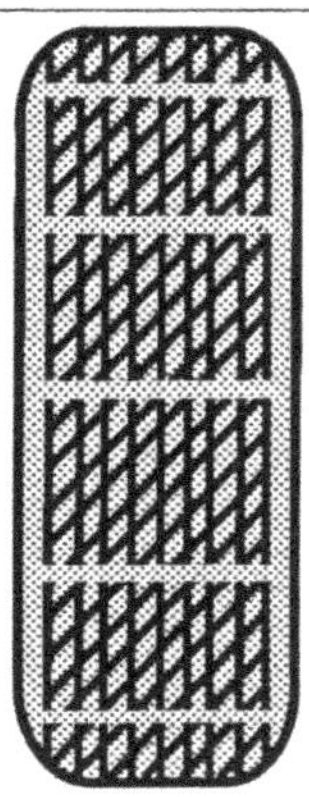

Fig. 3-1. Tire tread wear-indicator bands showing on worn tire indicate that replacement is necessary.

Uneven tire wear is usually caused by improper tire inflation pressures or misalignment. Fig. 3-2 illustrates the influence of inflation pressures on tire wear. Fig. 3-3 illustrates uneven tread wear resulting from prolonged over- or underinflation. For more information on maintaining proper tire inflation pressures, as well as other general tire maintenance, see **LUBRICATION AND MAINTENANCE**.

Other kinds of uneven tread wear may be the result of faulty components or improper alignment. See Fig. 3-4. Tire wear that is uneven across the tread—one side worn more than the other, or unusual wear of individual tread ribs—indicates an alignment problem, and perhaps worn suspension or steering components. Cupping or scalloping—wear that is uneven around the circumference of the tire—is a telltale sign of an unbalanced tire or a worn-out shock absorber.

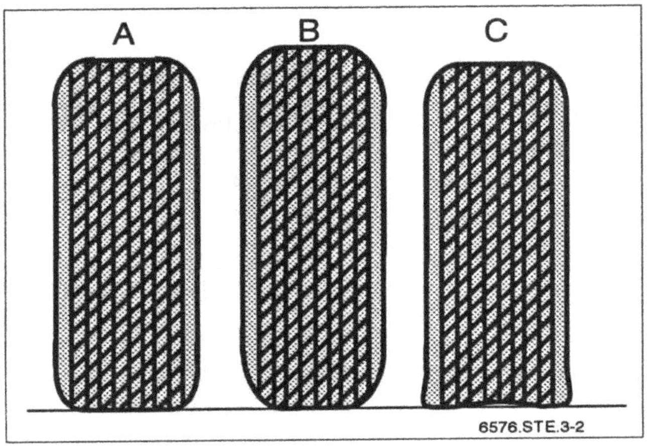

Fig. 3-2. Effect of tire inflation pressure on wear pattern. Condition **A** is normal. Overinflation (**B**) causes increased wear in center of tire tread. Underinflation (**C**) causes increased wear on outer edges of tire tread.

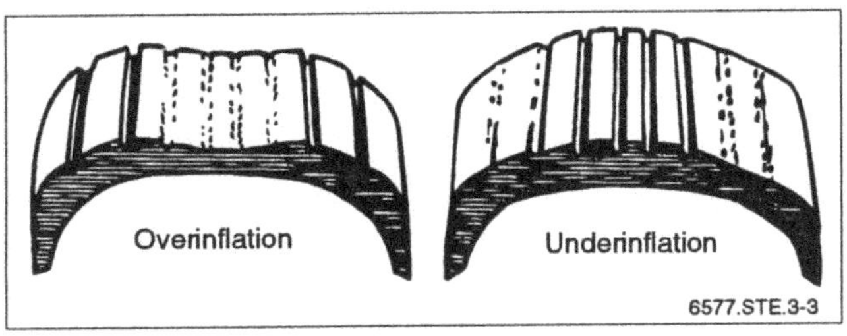

Fig. 3-3. Examples of uneven tread wear caused by overinflated and underinflated tires. Drawings courtesy of Goodyear Tire and Rubber Co.

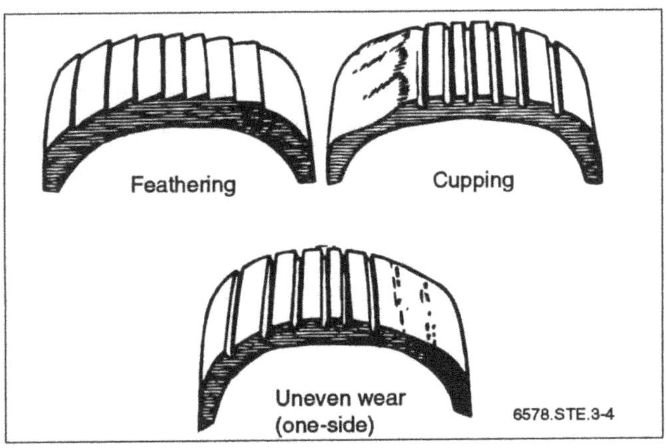

Fig. 3-4. Examples of uneven tread wear caused by misalignment and/or worn or damaged components. Feathering suggests incorrect alignment. Uneven wear on one side suggests incorrect alignment or repeated hard cornering. Cupping results from vibration, often caused by unbalanced tire or faulty shock absorber. Drawings courtesy of Goodyear Tire and Rubber Co.

Isolating Pulling Symptoms

Consistent pulling to one side in a car driven straight ahead on a level road may be caused either by wheel misalignment or by a faulty tire. In more unusual cases, a brake problem may be the cause.

To help decide whether tires or alignment are at fault, temporarily swap the front wheels (and tires), and then road-test the car. If a tire problem is the cause of the pulling symptom, the problem should switch to the other side of the car when the wheels are switched, and it should now pull to the other side. If the symptom persists, then the problem is probably caused by faulty alignment.

Vibration

Abnormal vibration, if not caused by a mechanical problem, will most often be caused by wheels and tires that are out of balance. Another, less likely cause of vibration is a tire that has become distorted and out-of-round. These and other tire conditions can be checked by any reputable tire professional.

Before spending the money for balancing, look for an obvious cause of imbalance. Caked-on mud, ice, or snow can dramatically affect wheel balance. Clean the wheels and tires and road-test the car before investigating more serious causes.

4. WHEELS, TIRES, AND ALIGNMENT

For stability and control, wheels and tires must be in good condition and be properly aligned. Tire inflation pressures, tire wear and wheel alignment will all influence how the car feels and responds on the road.

Precise wheel alignment can only be accomplished when the tires, the steering, and the suspension are in good condition. Uneven tire wear, different size tires, or worn suspension and steering parts all affect wheel alignment. Reputable wheel alignment technicians will always inspect the suspension and steering for worn parts before an alignment, and will recommend that any necessary repairs be made first. Aligning the wheels of a car with worn suspension and steering parts is a complete waste of time and money.

4.1 Wheels and Tires

Wheels and tires are subject to many stresses. They will perform as intended only if undamaged, properly inflated, and correctly balanced. If properly maintained, tires of the correct size will provide long service with comfort and safety. See **LUBRICATION AND MAINTENANCE** for information on wheel size, tire size, recommended inflation pressures, and other routine tire maintenance.

NOTE —
When installing the wheels, check that the wheel-contact area of the wheel hub is lightly coated with grease. If a lubricant is not used, corrosion between the wheel and the hub can occur, making the wheel difficult to remove. Plastilube® (BMW Part No. 81 22 9 400 207) is recommended by BMW for this application.

Wheels

The BMW wheels supplied as original equipment are designed for tubeless, radial-ply tires of a specific size. Replacement tires of non-standard size or construction should be installed only if the tire manufacturer specifies them for your specific make and model.

The wheels are sized according to diameter, width, and offset. Offset is the distance between the wheel's mounting surface and the rim's true centerline. When fitting a non-standard wheel with different dimensions, there is danger of wheels or tires interfering with the body or with suspension parts. Even if there is no interference, incorrectly sized wheels may place additional loading on wheel bearings or adversely affect steering geometry. For best results, rely on the advice of your BMW dealer or reputable wheel and tire professionals.

Wheel lug bolts should always be tightened using a torque wrench. Too little torque is obviously dangerous as the wheel may come loose, but excessive torque is also a problem. Over-tight wheel lug bolts can warp the hub, causing vibration and other problems affecting stability and braking. The proper torque value for wheel lug bolts is listed in **Table b**.

Damaged wheels may also be a source of vibration problems. Alloy wheels are constructed of a soft material and can be easily damaged. Check for a bent wheel by measuring lateral or radial runout at the part of the wheel where the tire bead meets the rim. Maximum allowable runout values are listed in **Table b**.

Table b. Wheel Bolt-Torque and Runout Specifications

Wheel lug bolt torque	100±10 Nm (74±7 ft-lb)	
Model year	**Lateral Runout (maximum)**	**Radial Runout (maximum)**
1982 (all)	1.0 mm (0.039 in.)	1.0 mm (0.039 in.)
1983–1987	0.5 mm (0.020 in.)	0.5 mm (0.020 in.)
1988 (w/alloy wheels)	0.6 mm (0.024 in.)	0.6 mm (0.024 in.)
1988 (w/steel wheels)	1.0 mm (0.039 in.)	1.0 mm (0.039 in.)

Tires

Radial-ply tires are installed as standard equipment on all cars covered by this manual. To retain the car's excellent handling characteristics, it is recommended that the tires only be replaced with those of the same (radial-ply) construction, size and speed rating.

Winter Tires

Though inferior to regular tires for dry-road wear and handling, winter (mud and snow) tires offer a big improvement on snow-covered winter roads. Studded winter tires improve traction on icy surfaces, but may be more dangerous than non-studded tires on dry roads. They should be used only if icy conditions predominate. Check local laws. Studded tires are restricted or completely prohibited in many states.

WARNING —
If winter tires are to be installed on only two of the four wheels, they should be installed on the rear driving wheels. To avoid creating dangerous handling characteristics, they must be of the same type (radial-ply construction) as those on the front of the car.

4.2 Alignment

Wheel alignment is the precise adjustment of the steering and suspension components to ensure that all wheels are oriented correctly, compared to the other wheels, the chassis, and the direction of travel. Small changes can have a big effect on how the car drives. Proper alignment provides the best compromise between responsiveness, stability, and tire wear.

Alignment specifications differ from model to model. It is important to use the correct alignment specifications for the specific car. Wheel alignment can also be fine-tuned for specific driving conditions. If, for example, the car normally carries only one person or is often heavily loaded, this information may help the alignment technician. Like tire pressure, wheel alignment also has some influence on fuel economy, because of its potential to affect the tires' rolling resistance. The important alignment angles are camber, caster, and toe.

Camber is the angle at which the wheels tilt away from vertical, as viewed from the front or rear. Camber is illustrated in Fig. 4-1. Wheels tilting in at the top display negative (–) camber. Wheels tilting out at the top display positive (+) camber.

Camber influences cornering and directional stability. A difference in camber between two front or two rear wheels may cause the car to pull to one side. Misadjusted camber will cause uneven tire wear.

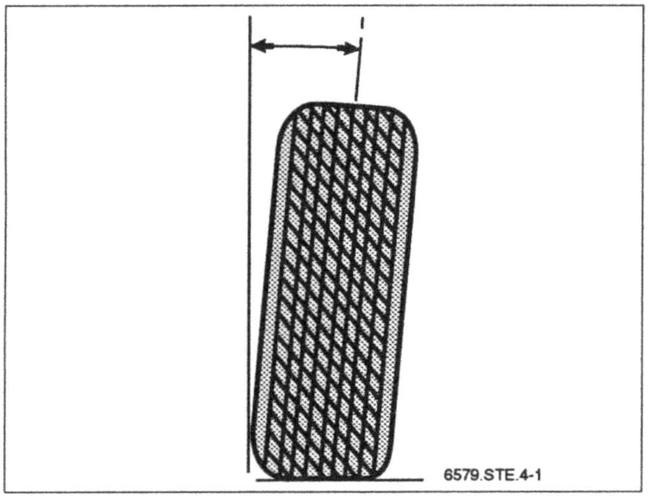

Fig. 4-1. Camber angle viewed from front or rear.

On the cars covered by this manual, camber is slightly negative by design and is not normally adjustable. Some camber correction can be achieved on the front wheels with the installation of alternate parts, available from BMW. See **Front Wheel Camber Correction** below.

Caster is the angle at which the steering axis deviates from vertical, illustrated in Fig. 4-2. The steering axis is an imaginary line about which the front wheels turn. Most cars are designed with a steering axis that is inclined toward the rear at the top (positive caster), giving them directional stability and self-centering steering.

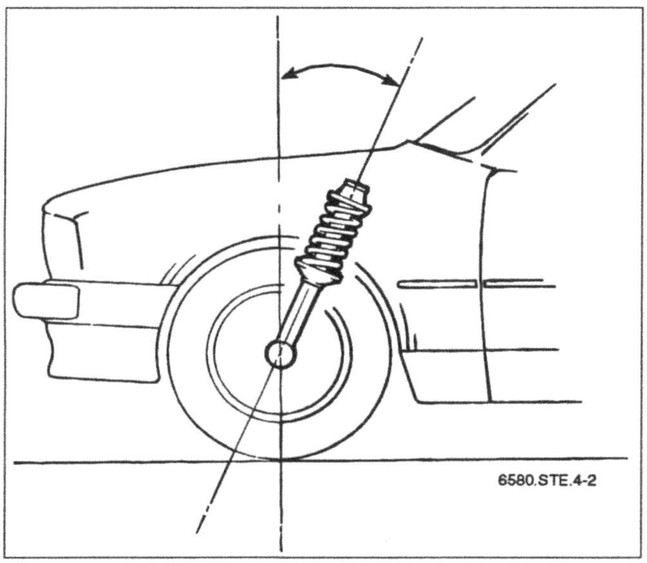

Fig. 4-2. Caster: angle of steering axis inclination from vertical (arrow). Caster is non-adjustable on cars covered by this manual.

Caster should be checked as part of any alignment. It is not, however, an adjustable angle on cars covered by this manual. Caster that is out of specification suggests worn or damaged components. Uneven front wheel caster will cause the car to pull to one side. Too little caster reduces directional stability. Too much caster increases turning effort.

Toe is a measurement of the amount that two wheels on the same axle point toward each other (toe-in) or away from each other (toe-out) in their direction of travel. Toe-in is illustrated in Fig. 4-3. Toe is the primary alignment adjustment for cars covered by this manual.

Toe affects directional stability and tire wear, and also has some effect on how the car responds to steering inputs. A small amount of toe-in is usually needed to offset the forces that tend to spread the tires outward when the car is traveling forward.

Too much toe will cause the tires to "scrub" and to wear unevenly and more quickly. Too little toe—too near zero—may cause the car to be less stable and wander slightly at highway speeds. See **Front Wheel Toe Adjustment** for information on adjusting toe.

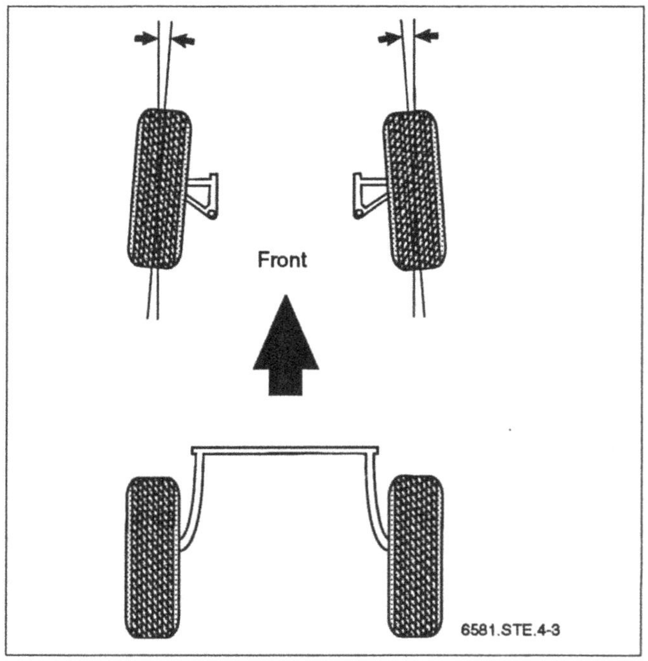

Fig. 4-3. Illustration of toe-in of front wheels. Toe angle is the measurement of how far the wheels point away from the direction of travel. The total toe specification is the sum of the toe angles of both wheels.

Modern computerized equipment, using special optics or laser beams for precise measurement, helps the qualified technician do the job more quickly and more accurately than is possible for the do-it-yourselfer. For this reason, only the specific adjustment methods are given, and not the basic procedures for measuring wheel alignment angles. The reasonable cost of a professional wheel alignment is money well spent.

Four-wheel Alignment

A four-wheel alignment considers not just the individual wheels or pairs of wheels, but also the alignment of the wheels relative to the car's centerline and the relationships between the front and rear wheels. Because of the independent rear suspension design on cars covered by this manual, it is especially important to have all four wheels properly aligned.

Front Wheel Toe Adjustment

Toe at the front wheels is adjusted by changing the length of the steering tie rods. Because camber also affects toe, any camber correction as described in **Front Wheel Camber Correction** below should be made before adjusting toe.

First, center the steering wheel so that the front wheels are pointing straight ahead. Check that the mark on the steering shaft is aligned with the mark on the gearbox housing. See Fig. 4-4. To adjust toe on each side, loosen the inner and outer tie rod clamping bolts and turn the rod to change its length. See Fig. 4-5.

Fig. 4-4. Marks on steering gearbox housing (arrow) align as shown when gearbox is centered and front wheels are in straight-ahead position.

NOTE —
Each tie rod has one left-hand-threaded ball joint end and one right-hand-threaded ball joint end. This facilitates increasing and decreasing the rod length without having to disconnect the ends. Shorten the tie rod to increase toe-in; lengthen it to reduce toe-in.

Fig. 4-5. Front wheel toe is adjusted by shortening or lengthening tie rod. Loosen tie rod clamping bolts (outer bolt shown) and turn tie rod.

Tightening torque	
• tie rod clamping bolt	14 Nm (10 ft-lb)

Front Wheel Camber Correction

Although there is no provision for routine camber adjustment on the BMW 5-series models covered by this manual, some correction is possible for those cases where the measured camber falls outside the specification.

18

The original three-cornered upper strut mount, shown in Fig. 4-6, can be replaced with an eccentric version that repositions the top of the strut. Depending on the installed position, the new strut mounts add or subtract 30' (1/2°) of camber.

WARNING —
Eccentric strut mounts should not be installed to try to correct camber measurements that are out of specification due to crash damage.

Fig. 4-6. Three mounting points (arrows) of upper strut mount. Available replacement mount moves strut off-center to alter front wheel camber.

For more information on removing and installing the front suspension struts and replacing the upper mounts, see **SUSPENSION—FRONT.**

> **NOTE —**
> Always rely on an authorized BMW dealer parts department for the most up-to-date and accurate parts information.

Rear wheel camber is not adjustable. Rear wheel camber that does not meet specifications suggests damaged components. See **SUSPENSION—REAR** for more information.

Rear Wheel Toe Correction

There is no provision for routine rear wheel toe adjustment on the cars covered by this manual, some correction is possible in those cases where rear wheel toe falls outside the limits of the toe specification.

The correction is made by removing the trailing arm and replacing the outer rubber trailing arm pivot bushing with an eccentric version. The eccentric bushing repositions the trailing arm slightly, resulting in a toe change.

The eccentric bushings are marked, indicating increments of change, and can be installed in any position to produce a specific amount of toe correction. In extreme cases, two eccentric bushings can be installed in both pivot points of a single trailing arm for twice as much correction. For information on removing and installing the trailing arms, see **SUSPENSION—REAR**.

> ***WARNING —***
> *Eccentric bushings should not be installed to try to correct toe measurements that are out of specification due to crash damage.*

4.3 Alignment Specifications

Alignment specifications for all models covered by this manual are listed in **Table c** and **Table d**. These specifications apply only under the following conditions:

1. Correct wheels and tires are installed, in good condition, at the correct inflation pressures.
2. Wheel bearings are in good condition.
3. Ride height is in accordance with specifications.
4. Car is loaded as follows: 68 kg (150 lb.) on each front seat, 68 kg (150 lb.) in the middle of the rear seat, 21 kg (46 lb.) in the trunk, and a full tank of fuel.

> **NOTE —**
> Detailed alignment specifications are furnished for reference only. Many of these measurements are beyond the capabilities of anyone but an experienced alignment technician using sophisticated equipment.

Table c. Front Wheel Alignment Specifications*

	1982 (all models)	1983–1986 (all except TRX tires)	1983–1986 (all with TRX tires)	1987–1988 (all except TRX tires)	1987–1988 (all with TRX tires)
Toe angle (total)	18'±05'	18'±05'	18'±05'	18'±05'	18'±05'
corresponding linear toe measurement	2.0±0.6 mm (0.079±0.024 in.)	2.0±0.6 mm (0.079±0.024 in.)	2.1±0.6 mm (0.083±0.024 in.)	2.0±0.6 mm (0.079±0.024 in.)	2.1±0.6 mm (0.083±0.024 in.)
toe difference angle (inside wheel @ 20° lock)	–1° 50'±30'	–1° 50'±30'	–1° 50'±30'	–1° 50'±30'	–1° 50'±30'
Camber					
wheels straight ahead	–20'±30'	–20'±30'	–20'±30'	–20'±30'	–20'±30'
Caster					
wheels straight ahead	8° 17'±30'	NA	NA	NA	NA
@ 10° wheel lock	NA	8°±30'	8°±30'	8°±30'	8°±30'
@ 20° wheel lock	NA	8° 15'±30'	8° 15'±30'	8° 20'±30'	8° 20'±30'
maximum permissible difference between sides	30'	30'	30'	30'	30'
Kingpin Inclination					
wheels straight ahead	12° 8'±30	12°±30	12°±30	NA	NA
@ 10° wheel lock	NA	NA	NA	11° 57'±30	11° 57'±30
@ 20° wheel lock	NA	NA	NA	12°±30	12°±30
Front Wheel Displacement	±15'	±15'	±15'	±15'	±15'
Full lock					
inside wheel (approximate)	43.7°	43.7°	43.7°	43.7°	43.7°
outside wheel (approximate)	33.1°	33.1°	33.1°	33.1°	33.1°
Turning circle diameter	10.9 mm (35.7 ft.)	10.9 mm (35.7 ft.)	10.9 mm (35.7 ft.)	10.9 mm (35.7 ft.)	10.9 mm (35.7 ft.)

*special measurement conditions apply. See **4.3** above.

Table d. Rear Wheel Alignment Specifications*

	1982 (all models)	1983–1988 (all except TRX tires)	1983–1988 (all with TRX tires)
Toe angle (total)	18'±07'	18'±07'	18'±07'
corresponding linear toe measurement	2.0±0.8 mm (0.079±0.031 in.)	2.0±0.8 mm (0.079±0.031 in.)	2.1±0.8 mm (0.083±0.031 in.)
maximum permissible deviation from driving direction	15'	15'	15'
Camber			
each side	–1° 48'±30'	–2° 20'±30'	–2° 20'±30'
maximum permissible difference between sides	NA	30'	30'

*special measurement conditions apply. See **4.3** above.

18

5. STEERING

The steering wheel and steering column are connected to the steering gearbox by a universal joint assembly. Turning the steering wheel turns the steering gear pinion which in turn moves the pitman arm and steering linkage. The steering gearbox and steering linkage is shown in Fig. 5-1.

The steering gearbox is power-assisted. Hydraulic pressure boosts the response of the gearbox and reduces steering effort. An engine-driven pump delivers power steering fluid to the steering gearbox under pressure.

Wear and excessive play or clearance anywhere in the system will cause sloppy, loose-feeling, and imprecise steering. On cars with high mileage, the tie rod end joints and the steering column universal joint are particularly prone to wear.

There are no provisions for periodic lubrication of the gearbox or the steering linkage. In general, the steering system is serviced mainly by the replacement of worn parts. Checking and filling power steering fluid is covered in **LUBRICATION AND MAINTENANCE**.

If steering requires too much effort or lacks precise feel, the steering mechanism may not be to blame. Faulty front strut bearings (upper mounts) or other worn front suspension parts may also cause or contribute to the symptoms. Before checking for any steering problems, check the tires, the tire inflation pressures, and the front suspension components. For more information, see **3. Troubleshooting** and **SUSPENSION—FRONT**.

Supplemental Restraint System (SRS)

Some of the cars covered by this manual may be equipped with a Supplemental Restraint System (SRS), which automatically deploys an airbag to protect the driver in the event of a frontal impact. Crash sensors in the front of the car detect frontal impact and electrically ignite a powerful gas generator to inflate the airbag, located in the steering wheel.

The airbag unit is a pyrotechnic device. Handled improperly or without adequate safeguards, the system can be very dangerous. Both BMW and the publisher of this manual strongly recommend that the system be serviced only by professional, BMW-trained technicians. Procedures for servicing the SRS system and the components related to the SRS system are, therefore, not included in this manual.

On cars equipped with the Supplemental Restraint System, the following precautions must be observed, to prevent injury due to unwanted activation of the system:

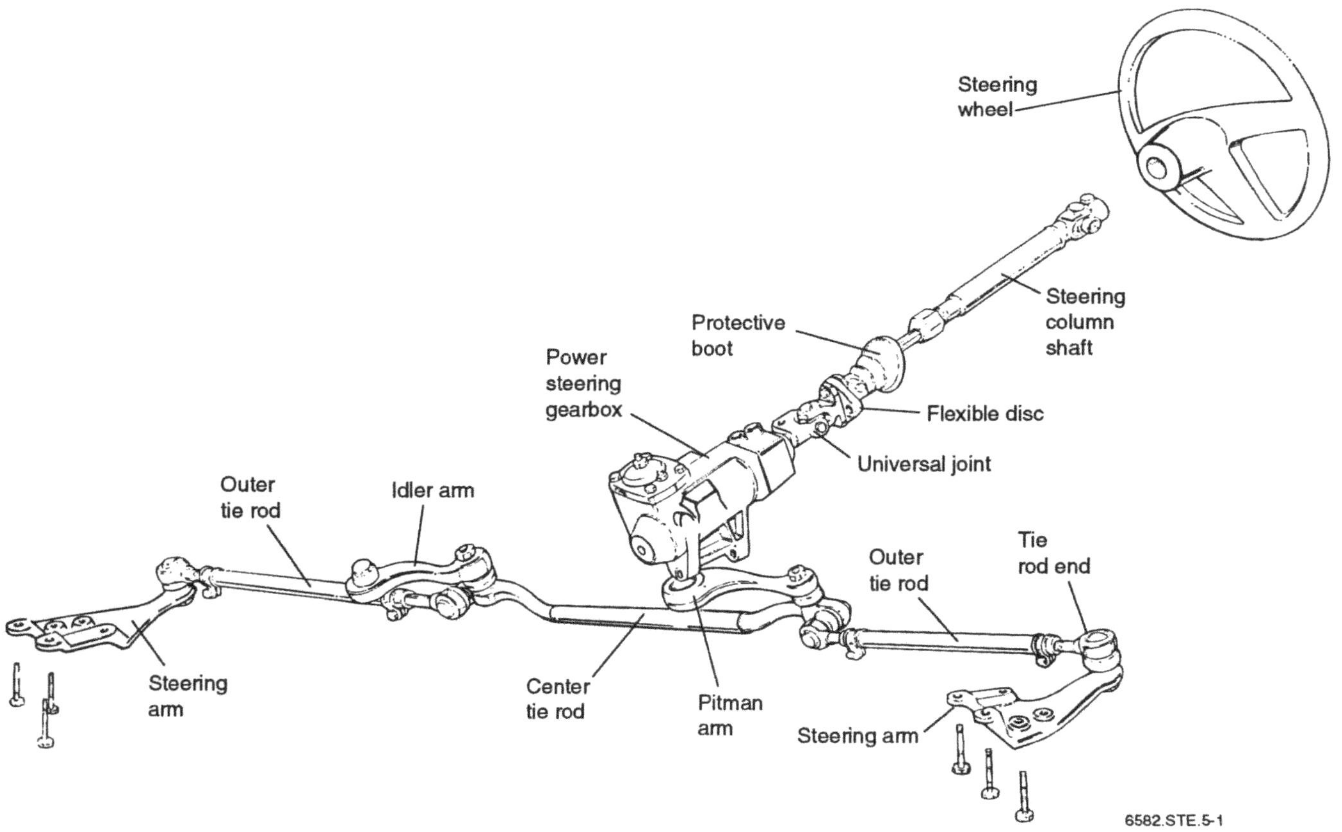

Fig. 5-1. Steering gearbox and linkage.

WARNING —
• Tests and installation/removal should only be performed by personnel with qualified training in BMW service.

• The supplemental restraint system can only be checked electrically, in the car, and only with the testers specified by BMW.

• Always disconnect the battery, cover the negative terminal, disconnect both plugs from the front crash sensors in the engine compartment, and disconnect the orange SRS plug in the steering column trim below the steering wheel to be sure that the power supply to the gas generator is interrupted prior to any work on the SRS, any body straightening, or any electric welding.

• Never treat the airbag unit with cleaning solutions or grease.

• Never subject an airbag unit to temperatures above 212°F (100°C).

• Never install airbag units, crash sensors, or SRS electronic units that have been dropped from a height of 1/2 meter (1 1/2 ft.) or more.

• If an airbag has been activated, all components must be replaced. The wiring harness, if it is undamaged, may be reused.

• Cars that are to be scrapped must have the gas generator rendered unusable prior to scrapping. Consult an authorized BMW dealer service department for the proper procedure.

5.1 Steering Column

The steering column connects the steering wheel to the steering gearbox. The column is supported by the steering column housing that is fastened to the instrument panel (by shear-bolts) and to the firewall. Several accessory switches are also mounted on the steering column housing.

Removal and installation of the steering column switches, including the windshield wipers and washers, the turn signals, and the headlight high beam switch, are covered here. For information on the electrical functions and troubleshooting of these switches, see **ELECTRICAL SYSTEM**.

Removing and Installing Steering Wheel

The steering wheel must be removed for access to the steering column switches and the ignition switch. The steering wheel is held in place with a large self-locking nut that should always be replaced.

WARNING —
On models equipped with SRS, removing and installing the steering wheel requires special knowledge and special tools. It is recommended that this work be left to an authorized BMW dealer. Incorrect removal or installation procedures could set off the airbag charge, causing serious personal injury.

To remove steering wheel:

1. Disconnect the negative (–) battery cable.
2. Use a small screwdriver to pry up the center cover (the BMW emblem). Center the wheel and mark its position on the steering column shaft. See Fig. 5-2.

Fig. 5-2. Large nut secures steering wheel to steering column. Mark wheel and steering column shaft prior to removal to ensure proper alignment for reinstallation (arrows).

3. While holding the steering wheel, remove the large nut and washer.
4. On 1986 and later models, insert the ignition key into the ignition switch. Turn the key to the first position to unlock the ignition lock. Do not switch on the ignition.
5. Pull the steering wheel straight off the steering column.

Installation is the reverse of removal. If the steering linkage was repositioned while the steering wheel was removed, the gearbox will have to be centered before the wheel is installed. See **Steering Wheel Alignment** below.

Tightening torque
• steering wheel (self-locking nut). . . . . 80 Nm (59 ft-lb)

18

Steering Wheel Alignment

If the steering was disturbed while the steering wheel was removed, or if a new unmarked steering wheel is being installed, the steering gear will have to be centered prior to installing the wheel in the straight ahead position.

To center the steering gear, temporarily position the steering wheel on the column and turn the wheel until the wheels are straight ahead, and the mark on the steering gear housing is aligned with the mark on the steering column universal joint as viewed from beneath the car. See Fig. 5-3. When the gear is correctly positioned, install the wheel in the straight ahead position.

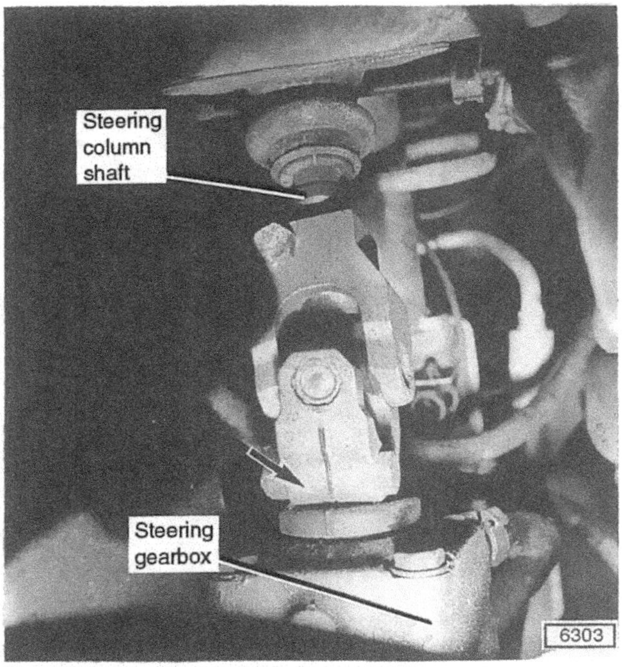

Fig. 5-3. Mark on steering gearbox housing mark aligned with mark on universal joint as viewed from beneath car (arrow). With marks aligned, wheels are in straight ahead position.

Fig. 5-4. Lower steering column trim. Arrows indicate fastener locations.

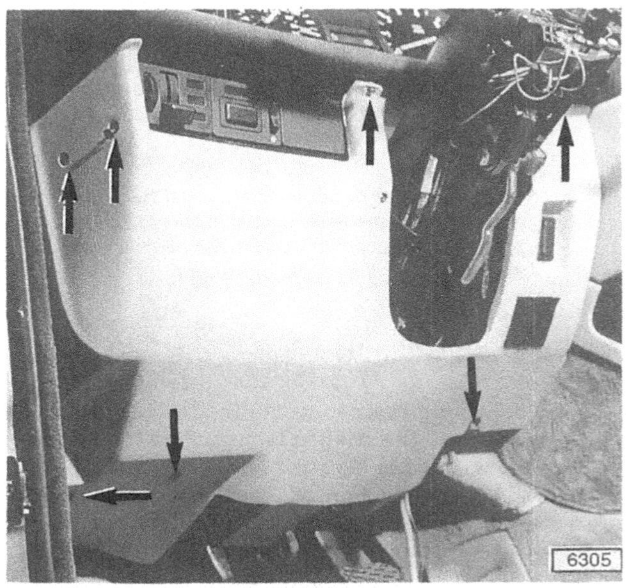

Fig. 5-5. Lower instrument panel trim. Arrows indicate fastener locations.

Removing and Installing Steering Column Switches

Before removing any switches from the steering column, remove the negative (–) battery cable. For access to the turn signal/headlight flasher switch, the windshield wiper/washer switch, or the cruise control switch, first remove the steering wheel as described above. Then remove the lower instrument panel trim and the lower steering column trim. See Fig. 5-4 and Fig. 5-5.

Each switch is fastened to the steering column with screws, as shown in Fig. 5-6. The lower screws are also ground wire connections. To remove a switch, remove the two screws on that side of the steering column and disconnect the appropriate wiring harness connectors. Fig. 5-7 shows the main harness connectors for the steering column wiring.

Installation of the switches is reverse of removal. When installing the cruise control switch, make sure the two locating pins engage the holes on the front of the steering column. See Fig. 5-8.

Fig. 5-6. Screws fasten steering column switches (arrows).

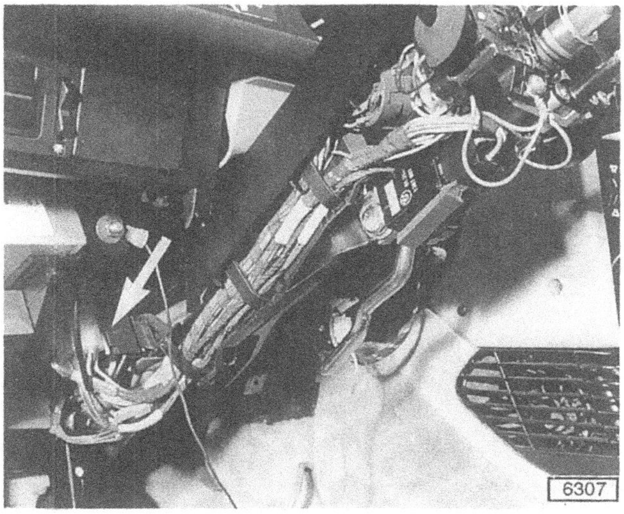

Fig. 5-7. Main harness connectors for steering column wiring. (1986 model shown). Separate harnesses can be disconnected at lower connectors (arrow).

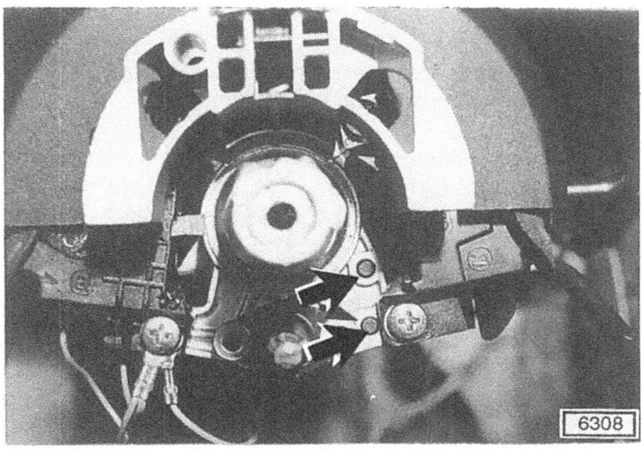

Fig. 5-8. Cruise control switch locating pins (arrows).

Replacing Ignition Switch

The ignition switch is mounted in the rear of the steering lock housing and is easily removed and tested. For information on testing the ignition switch, see **ELECTRICAL SYSTEM.**

To access the ignition switch, first remove the negative (–) battery cable. Remove the steering wheel as described above. Then remove the lower instrument panel trim and the lower steering column trim. See Fig. 5-4, above.

On 1982 though 1985 models, the ignition switch and switch harness can be removed from the back of the ignition lock by loosening a small screw. See Fig. 5-9. When installing the screw, seal it in place using a small amount of clear lacquer or paint.

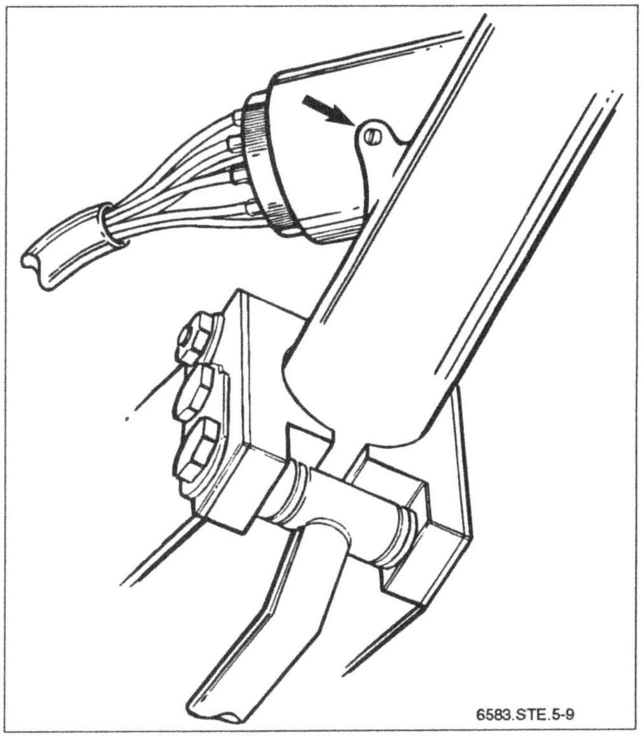

Fig. 5-9. Ignition switch retaining screw (arrow) on 1982 through 1985 models. Install and tighten screw using lacquer or paint to seal.

On 1986 and later models, the ignition switch and switch harness can be removed from the back of the ignition lock by using a small screwdriver to release the spring catch. See Fig. 5-10.

18

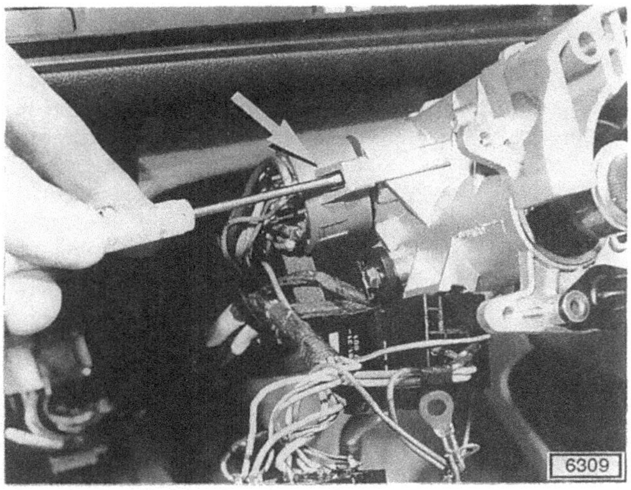

Fig. 5-10. Ignition switch retaining clip (arrow) on 1986 and later models being released.

Replacing Steering Lock

To remove the ignition/steering lock cylinder, remove the steering wheel and steering column trim as described above. Insert the key and work through the small hole to release the lock. Turn the key until it stops and then pull the cylinder out. See Fig. 5-11.

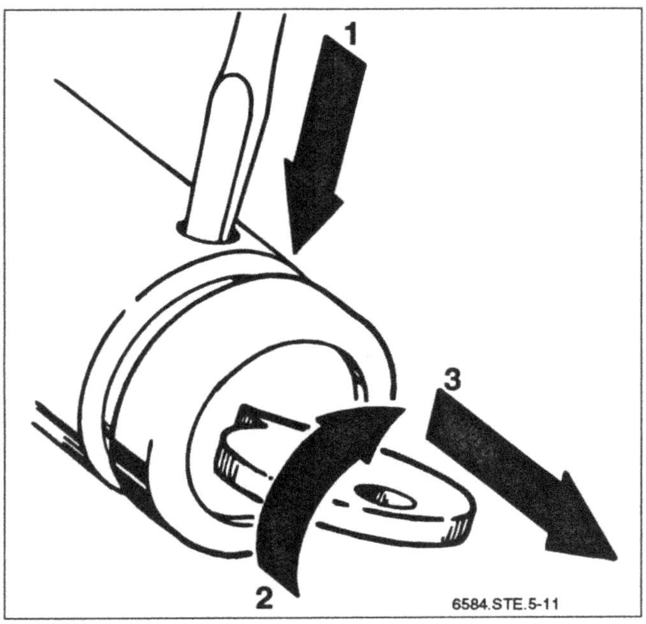

Fig. 5-11. Inserting screwdriver into small hole (**1**) while turning key (**2**) will release ignition/steering lock (**3**).

Removal of the steering lock mechanism from the steering column is more involved. The steering lock assembly is mounted using special shear bolts that must be chiseled off. During installation, the bolts are tightened until the heads actually shear off.

To remove steering lock mechanism (1982 through 1985 models):

1. Disconnect the negative (–) battery cable.
2. Remove the trim and steering column switches as described above under **Removing and Installing Steering Column Switches.**
3. Disconnect the emergency flasher and remove its mounting bracket from the steering column.
4. Chisel off the heads of the four shear bolts that hold the steering column to the dashboard. Chisel off the shear bolt that holds the steering lock to the steering column. See Fig. 5-12.

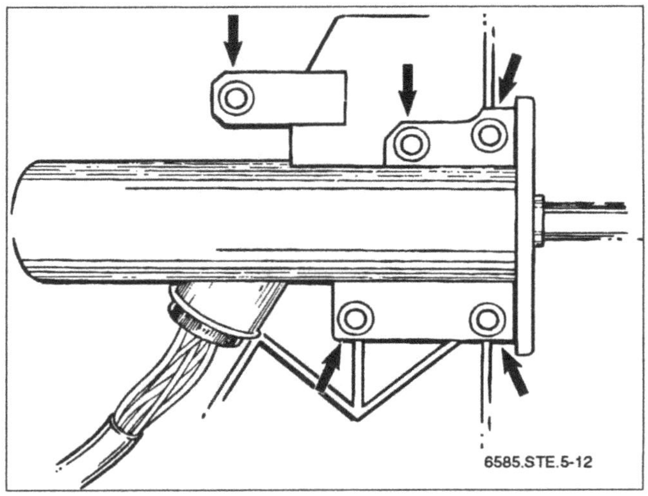

Fig. 5-12. Steering column and steering lock shear bolts (arrows)

5. Remove the ignition switch from the rear of the lock assembly.
6. Apply downward pressure to the steering column and withdraw the steering lock assembly.

Installation is the reverse of removal. If the lock mechanism is being replaced, remove the carbon contact from the old mechanism by drilling out the rivets, and install it on the new mechanism. Screws, washers, and nuts for this purpose are available from your authorized BMW dealer. When installing the ignition switch, seal the screw using a small amount of lacquer or paint.

Tightening torque

- steering lock and steering column shear bolts . . . . tighten until bolt head shears

To remove steering lock mechanism (1986 through 1988 models):

1. Disconnect the negative (–) battery cable.
2. Remove the trim and steering column switches as described above under **Removing and Installing Steering Column Switches.**
3. Disconnect the emergency flasher and remove its mounting bracket from the steering column. Disconnect the electrical connector from the rear of the horn contact.
4. Remove the steering column collar, snap ring, washer, spring, the plastic sleeve, in that order, from the end of the steering column. See Fig. 5-13.

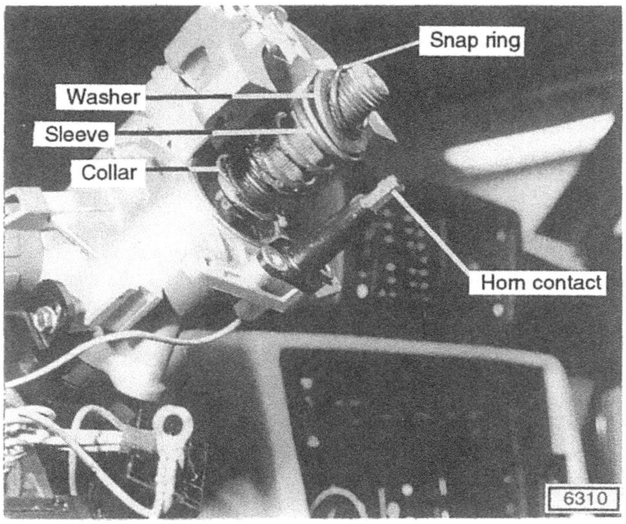

Fig. 5-13. Steering lock assembly mounting parts on 1986 though 1988 models.

5. Use two screwdrivers to pry up the upper steering column bearing. See Fig. 5-14.
6. Remove the ignition switch from the steering lock as described above under **Replacing Ignition Switch**.
7. Chisel off the head of the shear bolt (shown above in Fig. 5-12). Then slide the steering lock mechanism off the steering column.

Installation is the reverse of removal. If the lock mechanism is being replaced, remove the carbon contact from the old mechanism by drilling out the rivets, and install it on the new mechanism. Screws, washers, and nuts for this purpose are available from your authorized BMW dealer.

Tightening torque	
• steering lock and steering column shear bolts	tighten until bolt head shears

Fig. 5-14. Steering column bearing being removed using two screwdrivers.

5.2 Steering Gearbox and Steering Linkage

Gaskets and seals are the only internal gearbox replacement parts available from BMW. If any gearbox components are found to be faulty, the complete unit must be replaced.

In the recirculating-ball steering gearbox, the steering pinion translates rotating motion of the steering wheel into the side-to-side motion of the steering linkage. Tie rods connect the pitman arm of the gearbox to the steering arms on the front suspension struts.

The tie rod ends are ball-jointed to combine precise steering with the flexibility necessary to accommodate suspension movement. Rubber boots protect the joints and retain lubricant. Other than periodic inspections for wear or damage, the joints require no maintenance or additional lubrication. Adjustment or replacement of tie rods will always affect wheel alignment. A wheel alignment should be included as the final step in any such repair work.

CAUTION —
Parts of the steering mechanism are fastened with self-locking nuts. Do not reuse self-locking nuts. These nuts are designed to be used only once and should be replaced whenever they are removed.

NOTE —
Re-alignment is not necessary as long as tie rod length is unchanged.

Inspecting Steering Linkage

Fig. 5-15 shows the components of the steering linkage. When inspecting tie rods, look for wear and excess play in the tie rod end ball joints, look for damage to the boots that will allow dirt to get in and cause them to wear faster, and look for any damage to the rod section itself. Inspect the rubber bushing in the idler arm for wear.

Remove the steering wheel and lower steering column trim to inspect the steering column flexible disc. The flexible disc is located in approximately the middle of the steering column. Check the torque on the disc mounting bolts. Check for any loose mounting bolts. If any are loose, use new self-locking nuts and tighten them to the correct torque.

Tightening torque
• flexible disc (self-locking nut) . . . . 24 Nm (18 ft-lb)

To inspect the undercar steering linkage, raise the car and support it securely on jack stands. Grasp the tire firmly and check for play by rocking the tire back and forth.

There should be little or no play in any of the ball joints. There should be no visible damage to any of the tie rods, the steering arms, or the idler arm and its bushing. All turning motion of the wheel should be translated directly to the steering gearbox with no play in between. Try forcing the mating parts in opposite directions and looking for freeplay. See Fig. 5-16. Replace any worn tie rod end as described in **Replacing and Adjusting Tie Rods**.

Replacing and Adjusting Tie Rods

To replace a worn tie rod end, first measure and record the overall length of the tie rod. If applicable, remove the cotter pin and the mounting nut from the worn joint. Loosen the tie rod end clamping bolt. See Fig. 5-17. Remove the joint from its mounting and thread the end out of the rod. Install the new end, adjusting it until the rod length is the same as the measurement made earlier.

Fig. 5-16. Inspecting tie rod ends. There should be no noticeable play (arrows) between mating parts.

CAUTION —
If the tie rod is only being removed and installed, use a puller to avoid damaging the joint's threads and the rubber boot. Forked-type ball joint and tie rod separators may damage the rubber boot. Hammering on the top of the joint may damage the threads. A damaged joint should always be replaced.

NOTE —
A complete tie rod assembly consists of one left-hand thread end and one right-hand thread end. When replacing a tie rod end, make sure the replacement part has the same thread direction as the part being replaced.

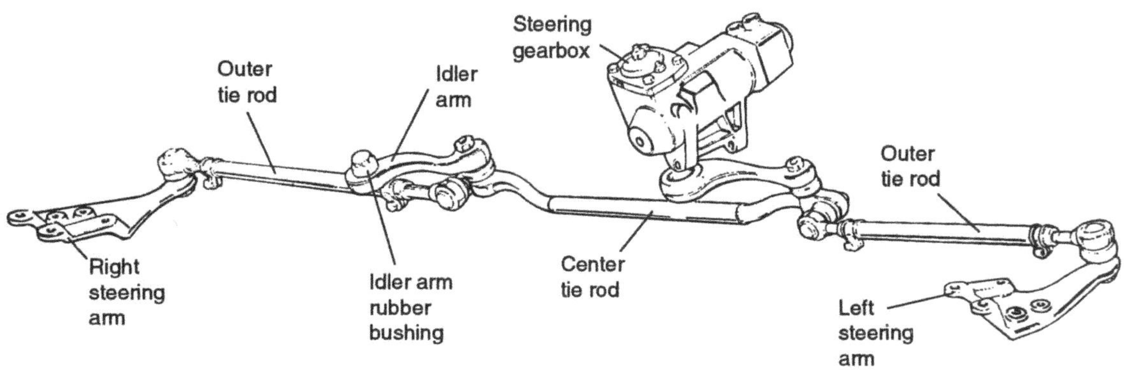

Fig. 5-15. Steering linkage assembly.

Fig. 5-17. Tie rod end clamping bolt and ball joint self-locking nut (arrows).

Installation is the reverse of removal. Use a new self-locking nut to fasten the outer tie rod end ball joint to the steering arm. Recheck toe as part of a wheel alignment after the job is complete. See **4.2 Alignment**.

> ***CAUTION —***
> *Early style tie rod ends mounted with cotter pins and castellated nuts are no longer available as replacement parts. The new style replacement end with a self-locking nut can be used on all cars covered by this manual. The self-locking nut is only designed to be used with the new style joint and must not be used to replace the castellated nut on the early style tie rod end.*

Tightening torques	
• tie rod end clamping bolt	14 Nm (10 ft-lb)
• tie rod end to steering arm	36.5±3.5 Nm (27±2.6 ft-lb)

Removing and Installing Steering Gearbox

If the steering gearbox requires repair or replacement, it is most easily removed as a unit. Internal seals and gaskets, available as a set, are the only repair parts available from BMW. To guard against hydraulic leaks when servicing the hydraulic fluid lines, the sealing washers should always be replaced. Removing the gearbox requires special tools to press the pitman arm off the steering gear.

To remove steering gearbox:

1. Depress the brake pedal approximately 20 times to discharge the hydraulic system.

2. Using a large syringe, a hand pump, or other suitable method, empty the power steering fluid reservoir in the engine compartment.

3. Remove the four mounting bolts from the subframe crossmember bar and remove the bar from the car. See **SUSPENSION—FRONT.**

4. Remove the two nuts and bolts that secure the universal joint to the steering gearbox. See Fig. 5-18. Slide the joint up and off the gearbox shaft.

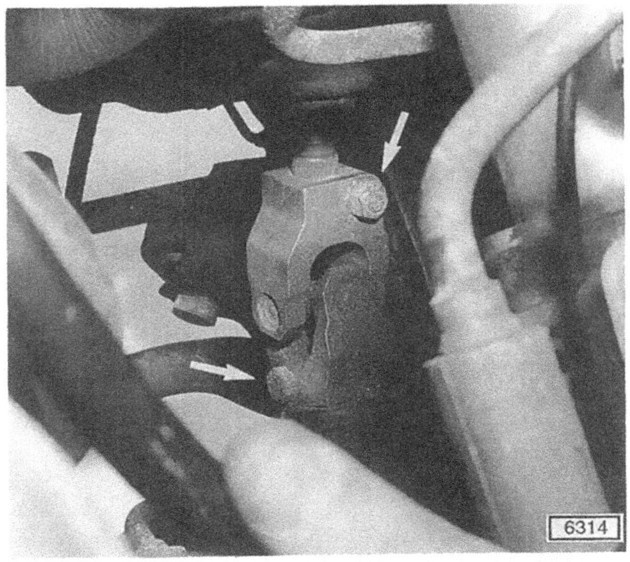

Fig. 5-18. Steering universal joint mounting nuts and bolts (arrows). Bolts fit in slot on steering gearbox pinion shaft and must be completely removed.

> **NOTE —**
> If the universal joint is worn or moves with difficulty, the joint assembly should be replaced.

18

5. Disconnect the hydraulic pressure line and the return line from the gearbox. See Fig. 5-19.

> ***CAUTION —***
> *Cover or cap the open lines to protect from dirt. Even a small particle of dirt can damage the system.*

6. Press the tie rod end out of the pitman arm, as described in **Replacing and Adjusting Tie Rods.**

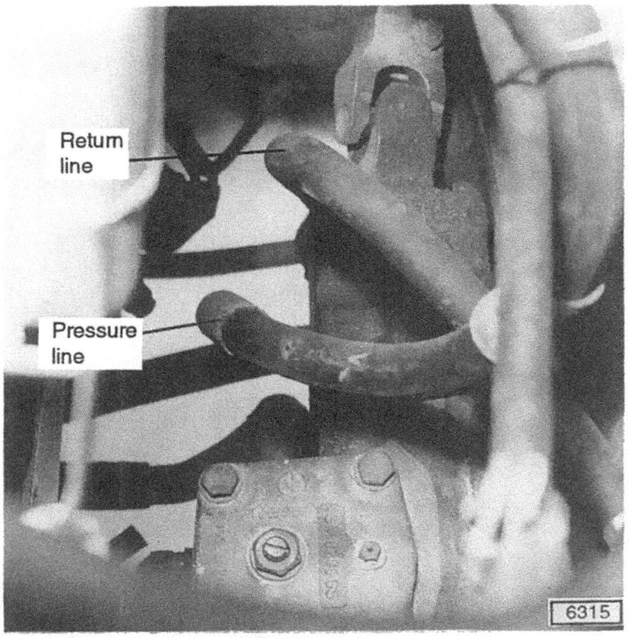

Fig. 5-19. Power steering gearbox hydraulic lines.

7. Remove the steering gearbox mounting bolts and nuts. See Fig. 5-20. Remove the steering gearbox from below.

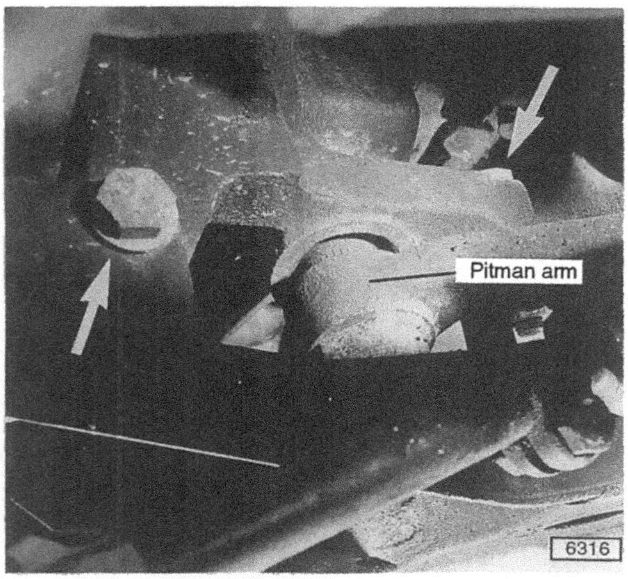

Fig. 5-20. Steering gearbox mounting bolts (arrows) and self-locking nuts. Always replace self-locking nuts.

Installation is the reverse of removal. Make sure the steering gearbox is centered, as shown in Fig. 5-3 above. Use new self-locking nuts on the universal joint, the tie rod end, the steering gearbox mounting bolts, and the crossmember bar mounting bolts. If applicable, use a new cotter pin. Use new sealing washers on the power steering pressure line connection to the steering gearbox. If the gearbox is being replaced, transfer the pitman arm to the gearbox.

> ***CAUTION —***
> *Do not reuse self-locking nuts. These nuts are designed to be used only once and should be replaced whenever they are removed.*

Fastener torque specifications are listed below. Refill the power steering system with new fluid and bleed the air from the system as described in **5.3 Power Steering Pressure System**.

> ***CAUTION —***
> *Do not reuse power steering fluid that has been drained from the system.*

Tightening torques	
• hydraulic line (pressure) connection	45 Nm (33 ft-lb)
• steering gearbox to front suspension subframe	42 Nm (31 ft-lb)
• tie rod end self-locking nut	36.5±3.5 Nm (27±2.6 ft-lb)
• universal joint clamping bolts	22 Nm (16 ft-lb)
• crossmember bar to body	60 Nm (44 ft-lb)
• pitman arm to gearbox (torque may be exceeded to align locking plate)	140 (103 ft-lb)

> **NOTE —**
> If hydraulic fluid leaks from the screw on the top of the gearbox, it may be due to loose-fitting threads on the screw. To correct the problem, carefully remove the top nut while counter- holding the screw. **The position of the screw must not be disturbed.** Wrap the threaded part of the screw four to six times with teflon tape and install a new nut (BMW Part No. 32 13 1 130 398). Tighten the nut while holding the screw stationary.

Tightening torque	
• steering gearbox adjusting screw locknut	30 Nm (22 ft-lb)

5.3 Power Steering Pressure System

This heading covers the hydraulic pressure portion of the power steering system. Information on adjusting or replacing the power steering pump V-belt is found in **LUBRICATION AND MAINTENANCE**.

Filling and Bleeding

Checking and correcting the fluid level in the fluid reservoir is covered in **LUBRICATION AND MAINTENANCE**. The layout of the system ensures that, under normal conditions, any trapped air escapes from the fluid in operation. If, however, the system is being refilled after being drained or to compensate for severe leaks, it may be necessary to deliberately bleed unwanted air from the system. Air in the system may cause noisy and inconsistent operation.

To bleed the system, make sure the fluid level in the reservoir is at the correct level, as shown in **LUBRICATION AND MAINTENANCE**, and start the engine. With a helper, turn the steering wheel back and forth from full left to full right, gradually adding fluid to the reservoir as the level drops. On 533i and 535i models, pump the brake pedal approximately five times before turning the steering wheel to the opposite side. When the fluid level remains constant, continue turning the steering wheel until no more air bubbles appear in the reservoir. Add fluid until the fluid level is correct. Replace the cap gasket if it is damaged.

> **NOTE —**
> Total fluid volume in the power steering hydraulic system is approximately 1200 ml (40 fl. oz.) on 528e models and 2000 ml (68 fl. oz.) on 533i and 535i models.

Pressure Testing

The power steering pump is tested by measuring its output pressure, using a pressure gauge (0-2000 psi) installed inline. See Fig. 5-21.

> **NOTE —**
> Accurate test results can only be achieved if the power steering pump V-belt is correctly tensioned and in good condition as described in **LUBRICATION AND MAINTENANCE.**

Connect the gauge between the pump and the pressure hose. Use new sealing washers to avoid leaks and to ensure an accurate measurement. Tighten the connections. Center the steering. Start the engine and let it idle.

Tightening torque
• hydraulic line at power steering pump . . . . 45 Nm (33 ft-lb)

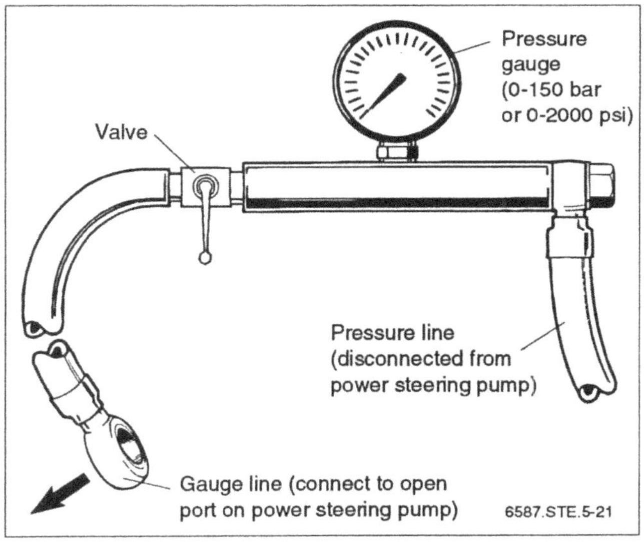

Fig. 5-21. Pressure gauge set-up for measuring power steering pump pressure. Disconnect original pressure line from power steering pump and connect to gauge as shown. Connect gauge line to open pressure line connection on power steering pump.

To test the pump, close the valve for no more than ten seconds, quickly read the pressure, then open the valve. The pressure should rise to at least 100 bar (1470 psi) and should not exceed 130 bar (1885 psi). If not, the pump should be replaced.

> ***CAUTION —***
> *The valve should be closed for no more than 10 seconds. A longer test risks damaging the steering gearbox, the pump, and/or the V-belt.*

To test system pressure, have a helper turn the steering wheel to 1/2 turn before the full right and full left stop. Hold the wheel in each position for about 5 seconds.

> **NOTE —**
> BMW recommends stopping the linkage (1/2 turn before full lock) from below using a block of wood and then applying a force of 100 N (22 lb.) at the steering wheel, measured with a spring-force gauge.

At each wheel lock position, the pressure should equal the pump pressure measured above. Lower pressure during this test indicates internal leaks, and the steering gearbox should be overhauled or replaced. Higher pressure indicates a faulty power steering pump.

18

Power Steering Pump

The power steering pump is mounted on a bracket attached to the engine block. The pump, mounting brackets, and belt-tensioner are shown in Fig. 5-22 and Fig. 5-23. For information on V-belt adjustment or replacement, see **LUBRICATION AND MAINTENANCE**.

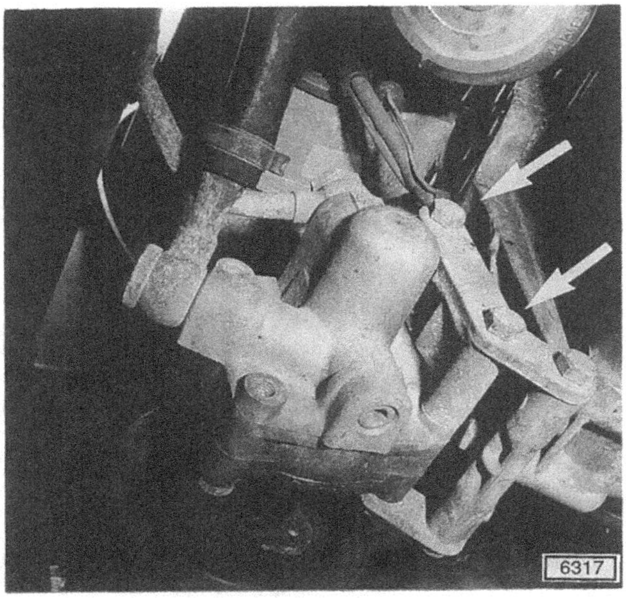

Fig. 5-22. Power steering pump mounting bolts (arrows).

Fig. 5-23. Power steering pump belt tension adjuster (arrow).

NOTE —
On 533i and 535i models, always discharge the hydraulic system by depressing the brake pedal approximately 20 times before loosening any hydraulic line fittings.

Tightening torques

- power steering pump to bracket. . . . . 22 Nm (16 ft-lb)
- power steering pump bracket to engine block . . . . 22 Nm (16 ft-lb)
- power steering pump hydraulic fittings (banjo bolts) . . . . 45 Nm (33 ft-lb)
- V-belt toothed adjusting nut. . . see **LUBRICATION AND MAINTENANCE**

6. TECHNICAL DATA

I. Tightening Torques

Hydraulic line fittings at steering gear box and power steering pump	45 Nm (33 ft-lb)
Power steering pump to pump bracket (bolt)	22 Nm (16 ft-lb)
Power steering pump bracket to engine block (bolt)	22 Nm (16 ft-lb)
Power steering pump V-belt toothed adjusting nut	see **LUBRICATION AND MAINTENANCE**
Steering wheel to steering column (self-locking nut)	80 Nm (59 ft-lb)
Steering column flexible disc to steering wheel and steering column (self-locking nut)	24 Nm (18 ft-lb)
Steering gearbox to front suspension subframe (self-locking nut)	42 Nm (31 ft-lb)
Steering gearbox adjusting screw locknut	30 Nm (22 ft-lb)
Steering gearbox pitman arm to gearbox (nut with locking plate) torque may be exceeded to align locking plate	140 (103 ft-lb)
Suspension crossmember bar to body (self-locking nut)	60 Nm (44 ft-lb)
Tie rod end to tie rod (clamping bolt)	14 Nm (10 ft-lb).
Tie rod end to steering arm (self-locking nut)	36.5±3.5 Nm (27±2.6 ft-lb).
Universal joint to steering column (self-locking nuts)	22 Nm (16 ft-lb)
Wheel to wheel flange (lug bolt)	100±10 Nm (74±7 ft-lb)

Section 19

BRAKES

Contents

TABLES

19

INTRODUCTION

Properly functioning brakes are essential to safe driving. If either the brake warning indicator or the ABS warning indicator comes on while driving, it is imperative that the system be given a thorough check, even if braking action still seems satisfactory. The brakes should be regularly inspected, and all brake work must be done with careful attention to cleanliness, correct specifications and proper working procedures.

This section contains information needed to perform routine maintenance and service of the brakes, although some of the information may only be useful to the professional mechanic.

1. GENERAL DESCRIPTION

All cars covered by this manual are equipped with disc brakes on all four wheels. The brake system is a diagonal dual-circuit design, hydraulically actuated by the master cylinder. A brake power booster system is used to reduce pedal effort. 1985 535i models and all 1986 and later 535i and 528e models are equipped with a Bosch anti-lock braking system (ABS). This electronically-controlled system maintains stability and control during emergency braking by preventing wheel lock-up.

Fig. 1-1 is a schematic diagram of the dual-circuit hydraulic brake system. The master cylinder, operated by the brake pedal, creates pressure in the hydraulic system. At the wheels, the hydraulic pressure acts on the calipers, which in turn mechanically apply the brakes. The use of hydraulics makes it possible for the driver to generate high braking forces with a comparatively small amount of effort.

> ***WARNING —***
> - *Brake friction materials such as brake linings or brake pads may contain asbestos fibers. Do not create dust by grinding, sanding, or cleaning the pads with compressed air. Avoid breathing any asbestos fibers or dust. Breathing asbestos can cause serious diseases such as asbestosis or cancer, and may result in death.*
> - *Brake fluid is poisonous. Wear safety glasses when working with brake fluid, and wear rubber gloves to prevent brake fluid from entering the bloodstream through cuts or scratches. Do not siphon brake fluid with your mouth.*

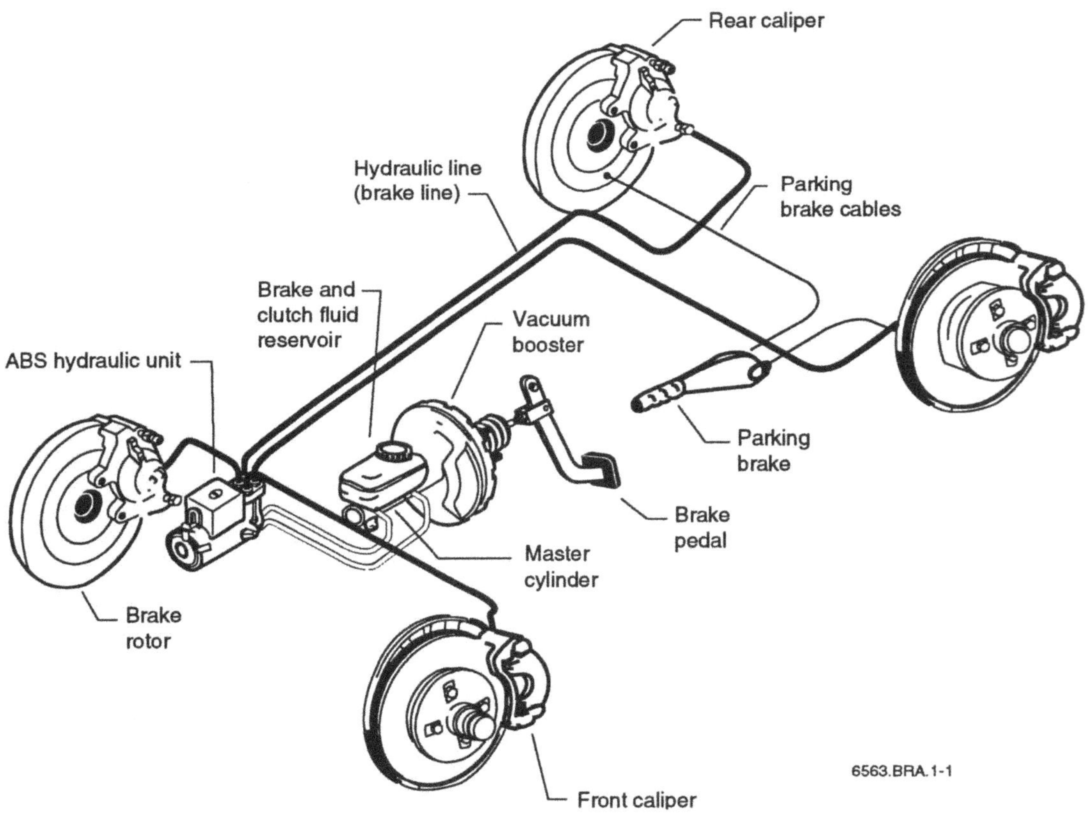

Fig. 1-1. Schematic view of dual-circuit brakes used on 528e models with ABS. Dual-circuit system used on other models is similar.

CAUTION —
Brake fluid is very damaging to paint. Any brake fluid that spills on the car should be cleaned off immediately.

Master Cylinder and Brake Lines

The master cylinder has separate chambers for the two hydraulic circuits. The front chamber operates the brakes on the front left wheel and the rear right wheel, while the rear chamber operates the brakes on the front right wheel and the rear left wheel. In the event of a loss of pressure in one circuit, the other will still supply a portion of the normal braking force.

The rigid brake lines transmit hydraulic pressure to the wheels and resist expansion and pressure loss. Flexible hoses joining the rigid lines to the brakes accommodate wheel movement due to steering and suspension action.

Brake Booster System

On 528e models, a vacuum booster supplied with engine vacuum assists braking effort. On 533i and 535i models, a hydraulic brake booster supplied with hydraulic pressure from the power steering pump assists braking effort.

The pedal pushrod is connected to the master cylinder so that a failure of the vacuum booster or the hydraulic brake unit, will not result in total brake failure, although it will greatly increase braking effort.

Disc Brakes

Each disc brake uses a caliper with a single hydraulic cylinder to clamp the rotor (disc) between two brake pads. Front and rear rotors are mounted to the wheel hubs. Disc brakes automatically adjust for brake pad and rotor wear.

Anti-Lock Braking System (ABS)

The anti-lock braking system (ABS) provides optimum deceleration and stability during emergency braking. It automatically adjusts the hydraulic pressure at each wheel to prevent wheel lock.

The system's main components are the wheel speed sensors, the electronic control unit, and the hydraulic control unit. See Fig. 1-2. The wheel speed sensors continuously send wheel speed signals to the control unit. The control unit compares these signals to determine, in fractions of a second, whether any of the wheels are about to lock. If any wheel is nearing a lock-up condition, the control unit signals the hydraulic unit to maintain or reduce hydraulic pressure at the appropriate wheel(s). Pressure is modulated by electrically-operated solenoid valves in the hydraulic unit.

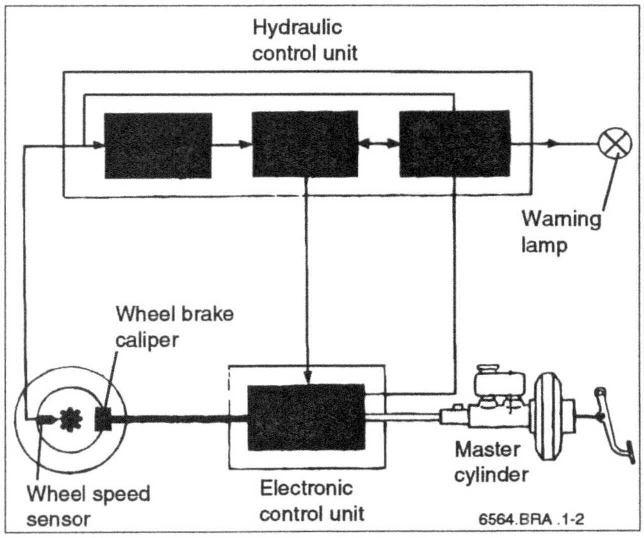

Fig. 1-2. Schematic representation of the ABS system used on cars covered by this manual.

Parking Brakes

The parking brake is mechanically operated. The parking brake system is independent of the rear disc brake system. The drum-style parking brake shoes act on a small drum, which is integral with the rotor.

2. MAINTENANCE

BMW specifies the maintenance steps below to be carried out at particular time or mileage intervals for proper maintenance of the brakes. A number in bold type indicates that the procedure is covered in this section, under that numbered heading. Information on other brake system maintenance and on the prescribed maintenance intervals can be found in **LUBRICATION AND MAINTENANCE**.

1. Checking brake fluid level
2. Inspecting brake hoses, lines, and wheel calipers for leaks
3. Checking brake pad and/or lining wear
4. Replacing brake fluid. **4.2**

3. TROUBLESHOOTING

This heading describes symptoms of trouble with the brakes. The sole function of the brakes is to generate friction to slow or stop the car. Brake problems are usually obvious because they affect the way the car slows and stops. Noise or problems with the car's handling may be caused by the brakes, but may also be caused by faults in the suspension or steering systems. For help in selecting the appropriate repair section, see the discussion of troubleshooting in **FUNDAMENTALS**.

3.1 Basic Troubleshooting Principles

Reliable braking depends on creating and applying hydraulic pressure. Brake performance is mainly affected by three things: the level and condition of the brake fluid, the system's ability to create and maintain pressure, and the condition of the friction components. Because air is compressible and brake fluid is not, air in the fluid will make the brake pedal feel spongy during braking, or will increase the force required on the brake pedal to stop the car. Fluid contaminated by moisture or dirt will increase internal corrosion and wear.

Seals are used throughout the system to maintain hydraulic pressure. Faulty seals or wear and corrosion on the sealing surfaces will reduce braking efficiency. A symptom of this condition is the need to pump the pedal to get good braking. Simple leaks at the brake line or hose unions may cause the same problems.

On cars so-equipped, a fault in the anti-lock braking system may affect brake performance during emergency stops.

NOTE —
If an ABS system malfunction is detected by the electronic control unit, the ABS warning lamp will light and the ABS system will automatically turn off. Once the ABS system is turned off, the brakes will operate in the normal mode (i.e., the wheels may lock during emergency braking).

Worn or contaminated brake pads will cause poor braking performance. The friction material is slowly consumed by braking and must be periodically replaced. Also, pads oil-contaminated or glazed due to overheating cannot produce as much friction, and stopping distances will increase.

For ABS-equipped cars, most of this same troubleshooting information applies. ABS has no effect on normal braking. ABS activates automatically only when the brakes start to lock during hard braking or on a slippery road surface. When the ABS is activated and is modulating brake pressure to one or more wheels to prevent lock-up, there will be some pulsation in the system that may be felt through the brake pedal. This condition is normal.

Table a lists symptoms of problems commonly associated with the brakes, their probable causes, and suggested corrective actions. The numbers in bold type in the corrective action column refer to numbered headings in this section where the suggested repairs are described.

Table a. Brake Troubleshooting

Symptom	Probable cause	Corrective action
1. Brake squeak or squeal	**a.** Normal condition	**a.** See **3.3**
	b. Incorrectly installed brake pads or parking brake shoes	**b.** Check installation. **5.1, 6.1**
	c. Brake pad carriers dirty or corroded	**c.** Clean calipers. **3.3, 5.1, 6.1**
	d. Brake pad anti-rattle springs faulty or missing	**d.** Install/replace anti-rattle springs. **5.1, 6.1**
	e. Brake pads heat-glazed or oil-soaked	**e.** Replace brake pads. Clean rotors. Replace leaking calipers as required. **5., 6.**
	f. Wheel bearings worn	**f.** See **FRONT SUSPENSION** or **REAR SUSPENSION**
2. Pedal goes to floor when braking	**a.** Brake fluid level low	**a.** Check fluid level. See **LUBRICATION AND MAINTENANCE** Check for leaks. Fill and bleed system. **4.2**
	b. Master cylinder faulty	**b.** Replace master cylinder. **4.3**
3. Low pedal even after brake adjustment and bleeding	**a.** Master cylinder faulty	**a.** Replace master cylinder. **4.3**
4. Pedal feels spongy or brakes work only after pedal is pumped	**a.** Brake fluid level low	**a.** Check fluid level. See **LUBRICATION AND MAINTENANCE** Check for leaks. Fill and bleed system. **4.2**
	b. Air in brake fluid	**b.** Bleed system. **4.2**
	c. Master cylinder internal return spring weak	**c.** Replace master cylinder spring. **4.3**
	d. Leaking line or hose unions	**d.** Repair or replace lines and hoses. Bleed system. **4.2**

continued on next page

Table a. Brake Troubleshooting (continued)

Symptom	Probable cause	Corrective action
5. Excessive braking effort with little effect	**a.** Brake pads wet	**a.** Use light pedal pressure to dry pads while driving
	b. Brake pads heat-glazed or oil-soaked	**b.** Replace brake pads. Clean rotors. Replace leaking calipers as required. **5.**, **6.**
	c. Vacuum booster or vacuum connections faulty (528e only)	**c.** Inspect vacuum lines. Test vacuum booster and replace as required. **4.4**
	d. Hydraulic brake booster faulty (533i and 535i only)	**d.** Test hydraulic booster and replace as required. **4.4**
6. Brakes pulsate, chatter, or grab	**a.** Warped disc brake rotors	**a.** Resurface or replace rotors. **5.1**, **6.1**
	b. Brake pads worn	**b.** Recondition brakes. **5.1**, **6.1**
	c. Brake pads heat-glazed or oil-soaked	**c.** Replace brake pads. Clean rotors. Replace leaking calipers as required. **5.**, **6.**
7. Uneven braking, car pulls to one side, rear brakes lock	**a.** Incorrect tire pressures or worn tires	**a.** Inspect tire condition. Check and correct tire pressures. See **LUBRICATION AND MAINTENANCE**, or **STEERING AND WHEEL ALIGNMENT**
	b. Brake pads on one side of car heat-glazed or oil-soaked	**b.** Replace brake pads. Clean rotors. Replace leaking calipers as required. **5.**, **6.1**
	c. Caliper or brake pads binding	**c.** Clean and recondition brakes. **5.**, **6.**
	d. Brake fluid contaminated	**d.** Drain, flush, refill, and bleed brakes. **4.2**
	e. Worn suspension components	**e.** See **SUSPENSION**, or **STEERING AND WHEEL ALIGNMENT**
8. Brakes drag, bind, or overheat	**a.** Brake caliper or brake pad binding	**a.** Clean and recondition brakes. **5.**, **6.**
	b. Master cylinder faulty	**b.** Replace master cylinder. **4.3**

3.2 Diagnostic Checks

Component inspection and some general brake system checks can help isolate problems.

Check the brake booster first by pumping the brake pedal approximately 10 times (vacuum-assist systems) or 20 times (hydraulic-assist systems) with the engine off. While holding the pedal down start the engine. The pedal should fall slightly. If not, check for any visible faults before suspecting a faulty brake booster.

On vacuum-assist systems, check for leaks in the vacuum line. A faulty check valve or a faulty O-ring between the master cylinder and the vacuum booster will decrease the vacuum assist. Low engine vacuum or a clogged vacuum booster filter could also decrease vacuum assist. Test the check valve by removing it from the booster vacuum line and blowing through it. Air should pass through in the direction of the arrow, but not pass through the opposite way. See **4.4 Brake Booster System** for more information.

On hydraulic-assist systems, check that the fluid level in the power steering fluid reservoir is correct. Check for any leaks or damaged hoses and lines. If no faults are found, special pressure testing equipment is required to further test the system. See **4.4 Brake Booster System**.

NOTE —
Be sure the power steering pump is operating correctly before testing the hydraulic brake booster system. See **STEERING AND WHEEL ALIGNMENT** for more information on the power steering pump.

On all models, check the master cylinder by holding the pedal down hard with the car stopped and the engine running. The pedal should feel solid and stay solid. If the pedal slowly falls to the floor, either the master cylinder is leaking internally, or fluid is escaping from the system. If no leaks can be found, the master cylinder is faulty and should be replaced.

3.3 Brake Noise

Occasional groaning or squealing sounds from the disc brakes are usually caused by vibration being transmitted through the brake pads. Brake friction materials that contain little or no asbestos may contribute to brake noise. These noises are normal and rarely indicate a problem.

Although there is no good solution, proper maintenance and repair can help minimize brake noise. Disc brake caliper assemblies include anti-rattle springs to minimize vibration and noise. Brake pads and calipers should be kept clean and free of corrosion. Always resurface or replace brake rotors when changing brake pads.

19

4. BRAKE SERVICE

This section includes important information about brake fluid, methods for bleeding the brakes, and service and repair procedures for the master cylinder, the brake booster system, and the pressure regulator.

4.1 Brake Fluid

Brake fluid is the heart of the hydraulic system. It transmits the mechanical braking force at the pedal to the brake mechanisms at the wheels. Any contamination of the fluid will directly affect braking performance and reliability. Dirt in the system will clog small passages and increase wear of rubber parts. Moisture in the fluid can cause a spongy pedal and poor braking performance.

Brake fluid is hygroscopic, meaning it absorbs moisture. Moisture in the fluid lowers its boiling point. If the brakes generate enough heat, the fluid will boil and create air bubbles. Since air is compressible, braking effectiveness will be reduced. Moisture in the system also causes corrosion.

When adding brake fluid, use only new fluid from unopened containers. Do not reuse brake fluid that has been bled from the system, even if it is brand new. See **LUBRICATION AND MAINTENANCE** for brake fluid specifications. For best performance and system reliability, use only BMW-approved brake fluid.

Brake fluid should be changed annually because of its moisture absorbing properties, regardless of the number of miles driven. The system should also be bled of air after any repairs in which brake lines are disconnected. See **4.2 Bleeding Brakes** for more information.

> ***WARNING —***
> *Brake fluid is poisonous. Wear safety glasses when working with brake fluid, and wear rubber gloves to prevent brake fluid from entering the bloodstream through cuts or scratches. Do not siphon brake fluid with your mouth.*

> ***CAUTION —***
> *Brake fluid is very damaging to paint. Immediately wipe up any brake fluid that spills on painted surfaces and wash the surface with water.*

4.2 Bleeding Brakes

Bleeding brakes is the process of removing trapped air or contaminated brake fluid from the system and replacing it with new, clean fluid. Bleeding brakes is essential if brake lines or components have been disconnected, or if air enters the system. A spongy brake pedal or excessive pedal travel may indicate the need to bleed air from the system.

There are three widely used methods of bleeding brakes. Each uses some means of forcing new fluid through the system and out of the caliper bleeder valves. The system can be bled using a pressure bleeder, using a hand-held vacuum bleeder, or manually using the brake pedal and a helper. Each method accomplishes the same result. Pressure bleeding, if available, is fastest. Manual bleeding is simplest, since it requires no special tools.

For the best possible results, bleeding should start at the wheel farthest from the master cylinder and end at the wheel closest to the master cylinder. This bleeding sequence is as follows:

1. **Right rear** caliper
2. **Left rear** caliper
3. **Right front** caliper
4. **Left front** caliper

Although it is possible to bleed the brakes with the car on the ground, for best access to the bleeder valves, the car should be raised and firmly supported on jack stands. The wheels should be removed and the engine should not be running.

Use a box wrench to open and close the bleeder valve. Use a clear container to catch the expelled fluid and a piece of clear, flexible tubing (about 4 mm or 5/32 in. inside diameter) connected to the bleeder valve, as shown in Fig. 4-1, so that outgoing air bubbles are visible.

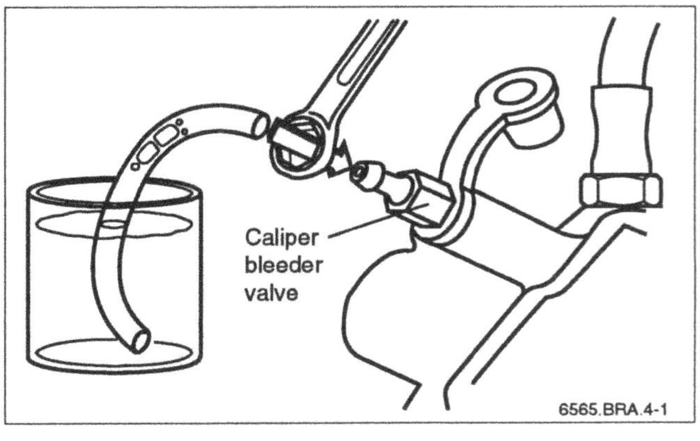

Fig. 4-1. Container (partially filled with brake fluid) connected to caliper bleeder valve with clear tubing for bleeding brakes.

Tightening torques	
• 7-mm bleeder valves	3.5 to 5.0 Nm (31 to 44 in-lb)
• 9-mm bleeder valves	4.0 to 6.0 Nm (35 to 53 in-lb)

Pressure Bleeding

Pressure bleeding, using special equipment and compressed air, is the method used by most repair shops. It is the quickest and the best way to prevent contaminants and air from entering the system. The pressure bleeder is connected to the brake fluid reservoir and fills the system with fluid under pressure. Follow the instructions supplied by the equipment manufacturer. A pressure of 0.8 to 1.0 bar (11.6 to 14.5 psi) should be used to bleed the brakes.

> ***CAUTION —***
> *Do not exceed a pressure of 2 bar (28 psi) when pressure bleeding the brake system. Excessive pressure will damage the brake fluid reservoir.*

> **NOTE —**
> On cars with ABS, slowly pump the brake pedal a minimum of 12 times each time the bleeder valve is opened to help expel air trapped in the hydraulic unit.

Vacuum Bleeding

Vacuum bleeding is a practical, do-it-yourself alternative to pressure bleeding, is faster than manual bleeding, and can be accomplished by only one person. Vacuum bleeding uses an inexpensive hand pump connected at the bleeder valve to draw the fluid through the system. Pumps of this general type are available from auto parts and supply outlets.

Follow the pump manufacturer's instructions. Make sure that the hose connection to the brake bleeder valve is secure and air-tight. Check the brake fluid level in the master cylinder often to guard against emptying the reservoir.

> **NOTE —**
> On cars with ABS, slowly pump the brake pedal a minimum of 12 times each time the bleeder valve is opened to help expel air trapped in the hydraulic unit.

Manual Bleeding

Manual bleeding uses pressure created by pumping the brake pedal. It is the most economical method, but a helper is needed during the procedure.

When manually bleeding the brakes, first fill the brake fluid reservoir on top of the master cylinder until the level is well above the MAX mark. Replace the cap to prevent contamination of the fluid.

Starting with the right rear caliper, clean the area around the bleeder valve and remove the dust cap. Fit the box wrench to the bleeder valve, then slip one end of the hose onto the valve and submerge the other end in clean brake fluid in a clear container.

Have a helper slowly pump the brake pedal about three times (12 times or more on cars with ABS). The last time, hold the pedal down. Slowly open the bleeder valve approximately one-half turn. Close the bleeder valve when brake fluid stops flowing from it, then release the brake pedal.

> ***CAUTION —***
> - *Be sure that the bleeder valve is fully closed before releasing the brake pedal to avoid drawing air back into the system.*
>
> - *On a car with high mileage, it is safest to avoid using the full stroke of the pedal while bleeding. This will prevent overrunning the master cylinder's normal stroke, which may damage the master cylinder piston seals if the inside of the master cylinder is worn or corroded beyond the normal piston stroke.*

> **NOTE —**
> On cars with ABS, pump the brake pedal a minimum of 12 times before opening the bleeder valve. This will help to expel air trapped in the hydraulic unit.

Repeat the bleeding procedure until the escaping fluid is free of air bubbles. Then tighten the valve and replace the dust cap. Using the sequence described above, bleed the other three wheels. Check the fluid level in the master cylinder frequently, and add more to keep the level at the MAX mark.

Replacing Brake Fluid

BMW recommends that the brake fluid be replaced at least once per year. This short replacement interval is due to the fact that brake fluid readily absorbs moisture. Moisture in the brake fluid can adversely affect braking performance and may also damage the system, leading to costly repairs.

Replace the brake fluid using one of the procedures described above to expel the old fluid. Remove the filter/strainer from the brake fluid reservoir and clean it in new, unused brake fluid. Using new, unused brake fluid, pump at least 1 pint (500 cc) of brake fluid through each caliper to completely flush the system and expel the old fluid. Then refill the reservoir and bleed the brakes as described above. See **LUBRICATION AND MAINTENANCE** for brake fluid specifications. Use only BMW-recommended brake fluid.

> **NOTE —**
> On cars with manual transmission, the brake fluid reservoir also supplies the hydraulic clutch master cylinder. It is a good idea to also flush the clutch slave cylinder when replacing the brake fluid. See **MANUAL TRANSMISSION AND CLUTCH.**

Flushing Brake System

Do not rely on flushing alone to clean a system contaminated with dirt or corrosion. The flushing procedure may actually force dirt in the lines into the calipers. To do the job thoroughly, the system must be disassembled and the parts individually cleaned. Use only brake fluid to flush the lines. Alcohol must not be used since it will encourage the accumulation of water in the system.

Replacing Brake Lines

Only straight brakes lines are available as replacements. To replace a brake line, first remove the old line from the car and measure its length. Using a new straight line of the same length, bend it to match the pattern of the old line. Each bend radius should be no smaller than about 15 mm (9/16 in.). Check that the protective coating on the brake line is not damaged after bending.

Make sure the new brake line has sufficient clearance from the body and suspension when installed, to allow for suspension bottoming and steering travel.

Tightening torques	
• brake line unions	10 to 15 Nm (7 to 11 ft-lb)
• brake hose unions	13 to 16 Nm (10 to 12 ft-lb)

CAUTION —
The use of a tube bender is highly recommended for shaping replacement brake lines. Bending brake lines by hand can cause kinks that weaken the line and restrict fluid flow. Inexpensive tube benders are available at most auto parts stores. See Fig. 4-2.

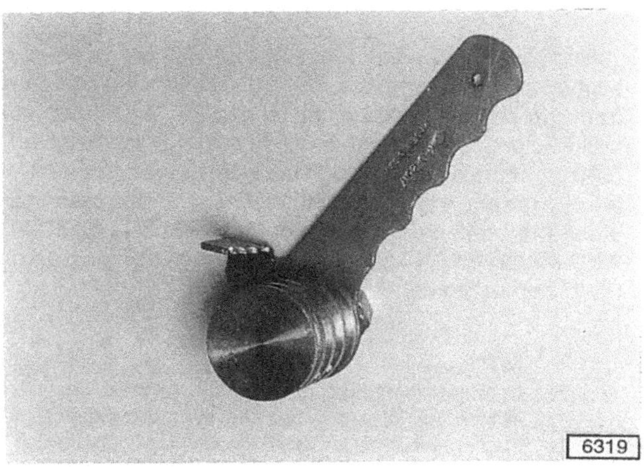

Fig. 4-2. Typical tube bender used to shape replacement brake lines.

4.3 Master Cylinder

The master cylinder is mounted to the front of the brake booster. The brake pedal operates the master cylinder via a pushrod that passes through the firewall. The master cylinder piston acts on the brake fluid to create pressure in the system. The brake fluid reservoir is mounted on top of the master cylinder. Fig. 4-3 shows the master cylinder and related parts.

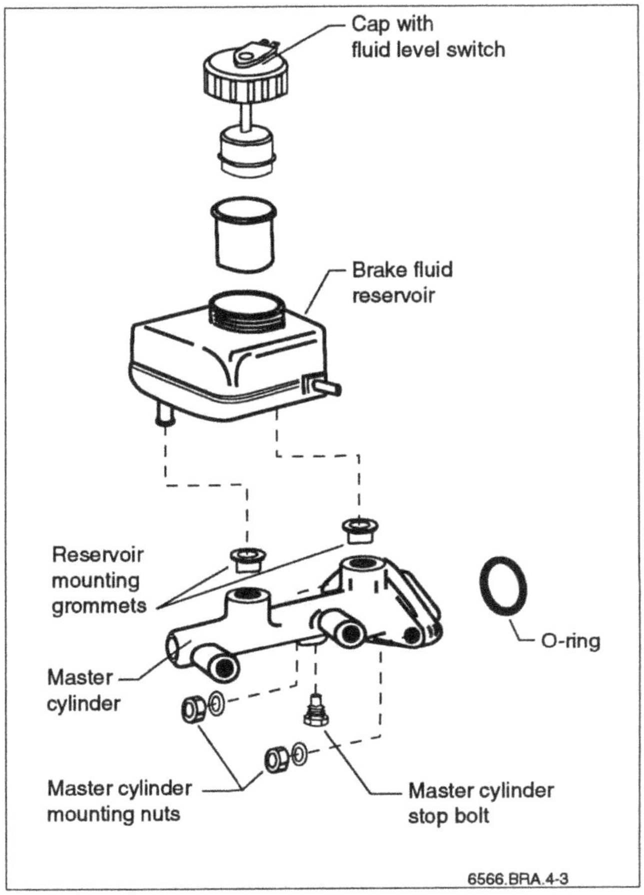

Fig. 4-3. Master cylinder and related parts.

Removing and Installing Master Cylinder

The master cylinder can be removed and installed with ordinary hand tools. Be sure to have an adequate supply of new, unopened brake fluid on hand, as bleeding the brakes will be necessary after installation. See **4.2 Bleeding Brakes.**

Although master cylinder parts and rebuild kits are available for most models, replacing the master cylinder as a complete unit may be preferable. It will take more time to rebuild the master cylinder than to replace it. Also, whether or not it can be successfully rebuilt depends on its internal condition, which can only be determined after it is disassembled.

NOTE —
On 1984 533i models with chassis numbers 1190516 to 1191101, incorrect machining of the surface between the master cylinder and the hydraulic booster may result in a warped or cracked master cylinder flange and possible brake system malfunction. A BMW recall campaign (84V-006) applies to these models. A white paint dot on the hydraulic brake booster indicates that the inspection and possible repair has been done by an authorized BMW dealer. If there is no dot, contact an authorized BMW dealer for more information on this inspection.

To remove master cylinder:

1. On 533i and 535i models, discharge the hydraulic accumulator by operating the brake pedal at least 20 times with the engine stopped.

2. Disconnect the negative (–) battery cable.

3. Loosen and remove the brake fluid reservoir cap.

4. Using a clean syringe or some equivalent, empty the brake fluid reservoir.

WARNING —
Brake fluid is poisonous. Do not siphon brake fluid with your mouth. Wear safety glasses when working with brake fluid, and wear rubber gloves to prevent brake fluid from entering the bloodstream through cuts or scratches.

5. Disconnect the brake lines from the master cylinder. On models with manual transmission, disconnect the hydraulic clutch supply hose.

6. If the master cylinder is being replaced, remove the brake fluid reservoir from the top of the master cylinder. Install the reservoir on the new master cylinder.

CAUTION —
Use care when removing the reservoir from the master cylinder to avoid damaging the plastic fittings. Gently rock the reservoir while pulling it from the master cylinder.

7. Remove the master cylinder by removing the two nuts or bolts holding it to the brake booster. On 528e models, remove the O-ring from the front of the master cylinder.

NOTE —
On 528e models, the O-ring between the master cylinder and the vacuum booster should always be replaced. A faulty O-ring can be the source of a vacuum leak, causing a reduction in braking performance or an erratic idle speed.

To install master cylinder:

1. Install the new master cylinder using new self-locking nuts or bolts and a new O-ring, if applicable.

Tightening torques	
• master cylinder to brake booster	
528e	24 to 29 Nm (18 to 21 ft-lb)
533i, 535i	26 to 32 Nm (19 to 24 ft-lb)

CAUTION —
Do not exceed the recommended mounting torque on master cylinder fasteners. Over tightening may damage the brake booster.

2. Connect the brake lines and torque the unions. If applicable, connect the hydraulic clutch supply hose using a new hose clamp.

Tightening torque	
• brake lines to master cylinder	10 to 15 Nm (7 to 11 ft-lb)

3. Fill the reservoir with new brake fluid and bleed the system as described in **4.2 Bleeding Brakes**.

4. Reconnect the negative (–) battery cable.

5. Road test the car and bleed the brakes again, if necessary.

4.4 Brake Booster System

Two types of brake booster systems are used on the cars covered by this manual. On 528e models, a vacuum-assist system is used. On 533i and 535i models, a hydraulic-assist system is used.

Vacuum Brake Booster
(528e models)

Fig. 4-4 shows the vacuum booster and related components used on 528e models. The vacuum booster uses engine vacuum to help actuate the master cylinder and reduce pedal effort when the engine is running. A check valve in the vacuum hose holds vacuum in the booster when the engine is stopped. The pedal pushrod is connected directly to the master cylinder, so failure of the vacuum booster does not normally result in total brake failure.

19

The vacuum booster is not repairable, and if faulty must be replaced. It can be removed and installed with ordinary hand tools.

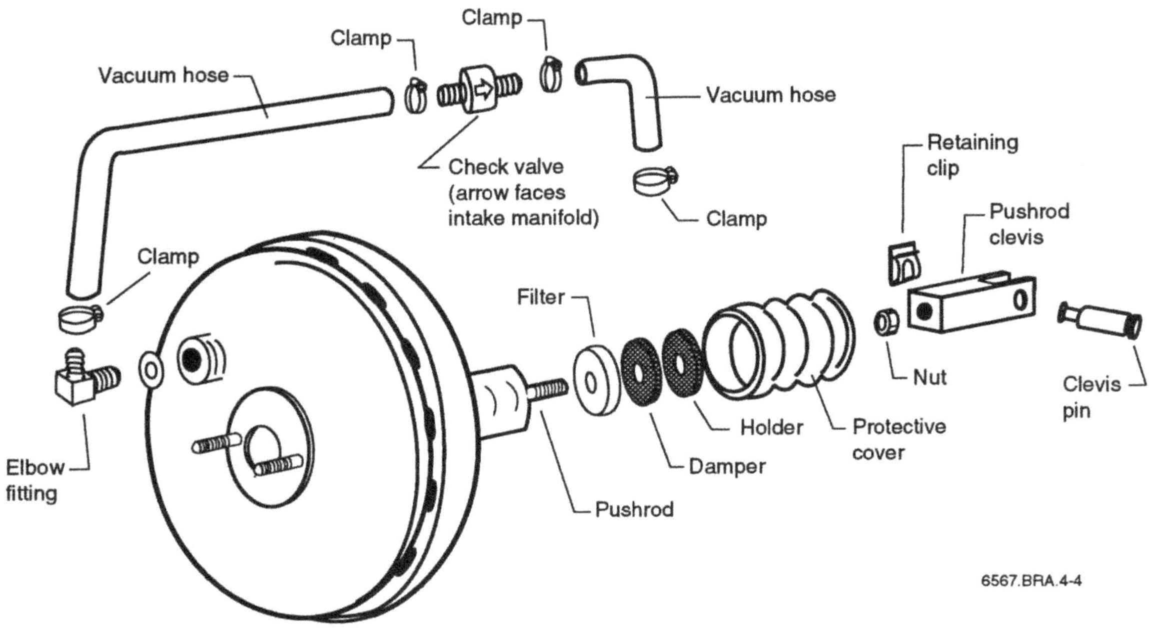

Fig. 4-4. Vacuum brake booster assembly used on 528e models.

To remove vacuum booster:

1. Remove the master cylinder as described in **4.3 Master Cylinder**.
2. Disconnect the vacuum hose from the booster.
3. Working from inside the passenger compartment, remove the lower left trim panels above the brake pedal.
4. Disconnect the pushrod from the brake pedal by disconnecting the brake pedal return spring and prying off the retaining clip. See Fig. 4-5. Slide the clevis pin out.
5. Remove the four mounting nuts as shown in Fig. 4-6. Working from the engine compartment side, pull the vacuum booster from the pedal base.
6. Inspect the small foam filter beneath the rubber boot on the pushrod. See Fig. 4-7. If the filter is clogged, it could affect the booster's performance. The filter can be cleaned using a mild soap. If it cannot be sufficiently cleaned it should be replaced.

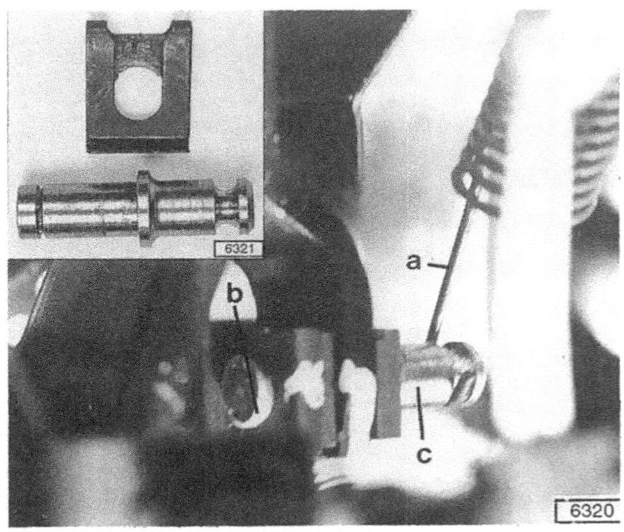

Fig. 4-5. Brake pedal return spring (**a**), retaining clip (**b**) and clevis pin (**c**) as viewed from inside passenger compartment. Inset shows removed clevis pin and clip.

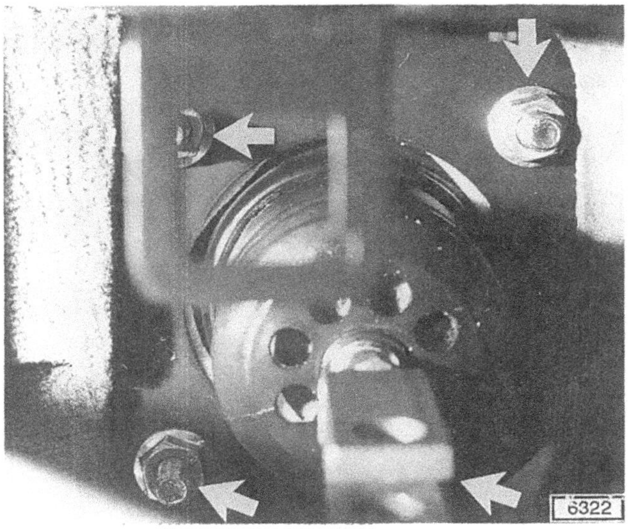

Fig. 4-6. Vacuum booster mounting nuts (arrows).

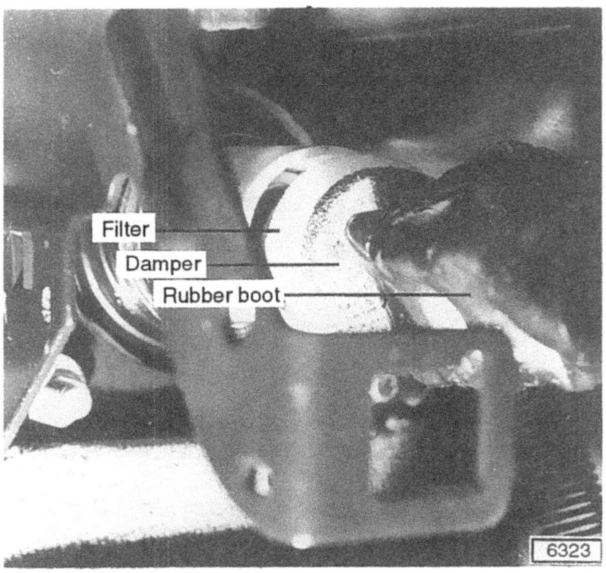

Fig. 4-7. Vacuum booster foam filter and damper as viewed from inside passenger compartment.

Installation of the vacuum booster is the reverse of removal. When reinstalling the filter, be sure the notches in the filter offset the notches in the damper by 180°. Adjust the brake pedal travel and the brake light switch as described below under **4.5 Brake Pedal Adjustments.**

Tightening torque	
• vacuum booster to pedal base	22 to 24 Nm (16 to 18 ft-lb)

Hydraulic Brake Booster
(533i and 535i models)

Fig. 4-8 shows the hydraulic brake booster system used on 533i and 535i models. When the engine is running, the power steering pump supplies hydraulic fluid under pressure to the power flow regulator and accumulator. The regulator and accumulator assembly stores and regulates the pressure to the brake booster. When the brake pedal is depressed, the pressure in the brake booster helps actuate the master cylinder to reduce pedal effort.

The pedal pushrod is connected to the brake booster, which is connected directly to the master cylinder, so failure of the booster does not normally result in total brake failure. The hydraulic brake booster is not repairable, and if faulty must be replaced. Special high-pressure measuring equipment is needed to test the system.

Before testing the system, check the fluid level in the power steering/brake booster reservoir. With the engine off, remove the cover from the reservoir. Depress the brake pedal until the fluid level stops rising (approximately 20 times). See Fig 4-9. Add fluid if the level is lower than 3/8 in. (10 mm) from the top rim of the reservoir. See **LUBRICATION AND MAINTENANCE** for fluid specifications.

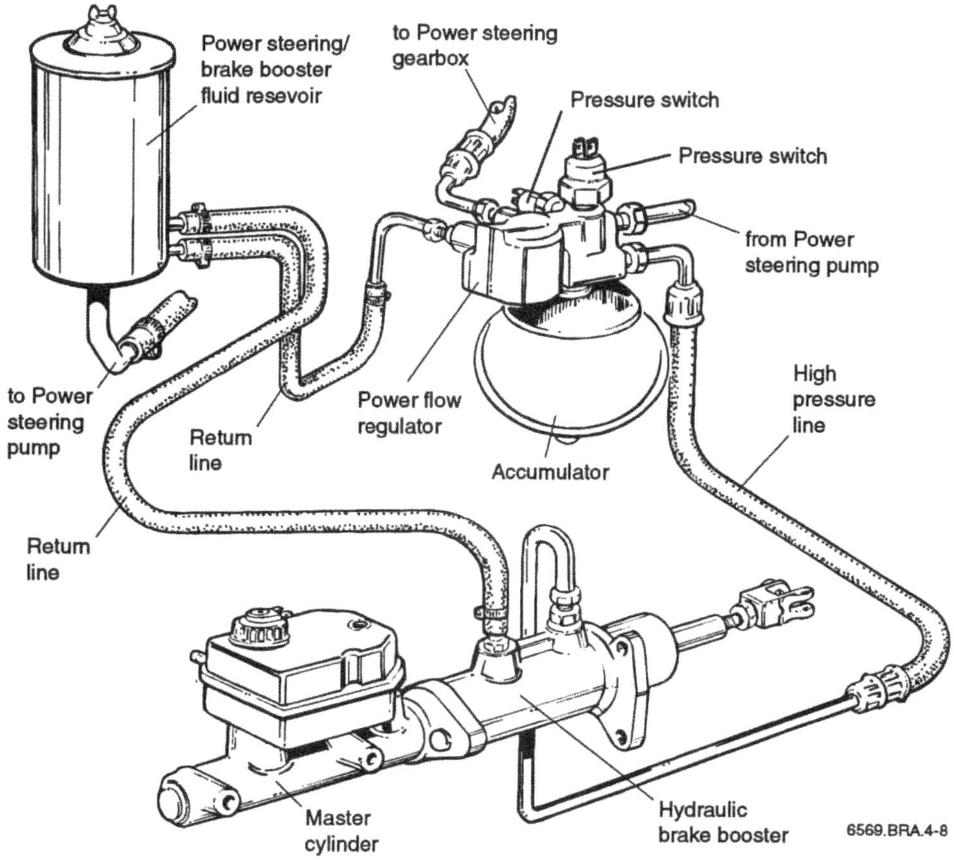

Fig. 4-8. Hydraulic brake booster system used on 533i and 535i models.

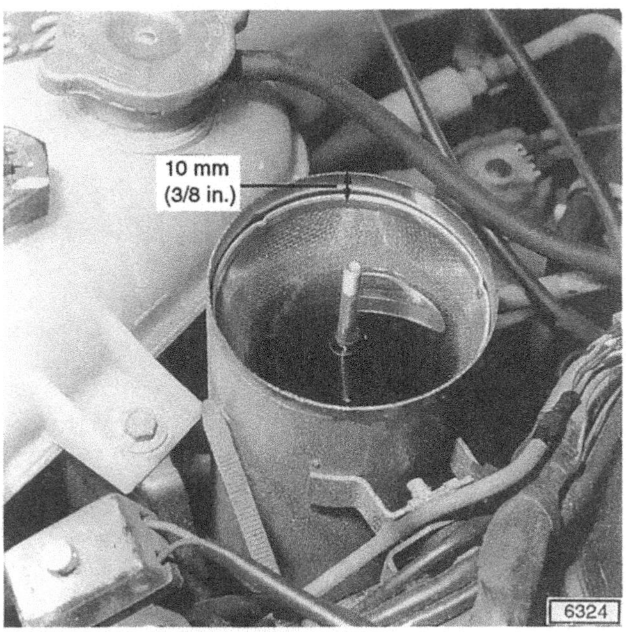

Fig. 4-9. Hydraulic brake booster fluid reservoir. Correct fluid level is 10 mm (3/8 in) below top rim of reservoir.

To test hydraulic booster system:

1. Discharge the hydraulic accumulator by operating the brake pedal at least 20 times with the engine stopped.

2. Thoroughly clean the area around the hydraulic pressure switch on the side of the power flow regulator and remove the switch. See Fig. 4-10.

> ***CAUTION —***
> *Extreme cleanliness is essential when working on the hydraulic brake booster system. Even a small piece of dirt could damage the brake booster system or the power steering system.*

3. Connect a pressure gauge in place of the pressure switch.

> **NOTE —**
> The gauge should have a range of 0 to 60 bar (0 to 1000 psi).

Fig. 4-10. Hydraulic pressure switch in power flow regulator (arrow).

4. Start the engine and check for leaks. With the engine running, the pressure should be at the upper switching specification listed in **Table b**.

Table b. Hydraulic Brake Booster Test Specifications

Upper switching pressure	52 to 57 bar (754 to 827 psi)
Lower switching pressure	36 to 41 bar (522 to 595 psi)
Maximum allowable pressure drop after 5 minutes	5 bar (73 psi)

5. Apply the brake pedal until the pressure drops to the lower switching specification. Check that the pressure quickly returns to the upper switching specification.

NOTE —
If any faults are found up to this point, the power flow regulator is faulty and should be replaced.

6. Turn the engine off. After five minutes, the pressure drop should be no greater than the specification listed in **Table b**.

NOTE —
If the pressure drop exceeds the limits, proceed with step 7 below. If the pressure drop is within limits, proceed with step 9.

7. Remove the reservoir return line from power flow regulator to check for leaking fluid. See Fig. 4-11.

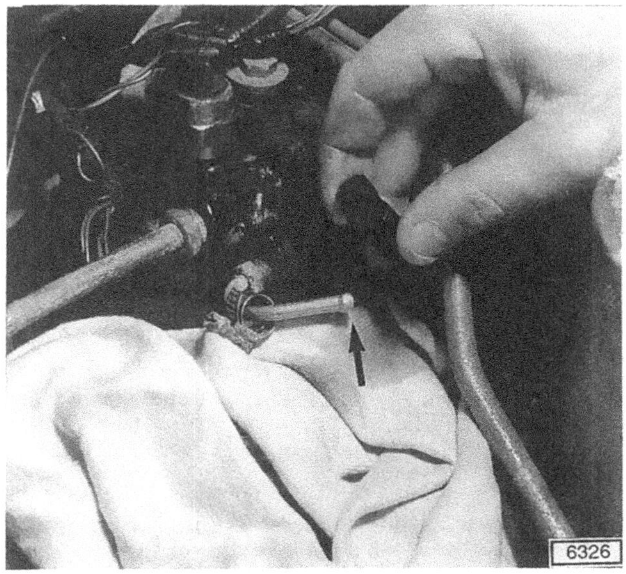

Fig. 4-11. Return line removed from power flow regulator (arrow).

NOTE —
If fluid comes out of the power flow regulator, the regulator is faulty and should be replaced. If no fluid comes out, the brake booster is leaking internally and should be replaced.

8. Install the return line.

9. Discharge the hydraulic accumulator as described in step 1. Start the engine. If the pressure does not immediately rise to the specifications listed in **Table c**, the hydraulic accumulator is faulty and should be replaced.

Table c. Hydraulic Accumulator Pressure Specifications

Temperature	Pressure
50°F (10°C)	9.75 to 25.0 bar (141 to 363 psi)
68°F (20°C)	10.0 to 26.0 bar (145 to 377 psi)
86°F (30°C)	10.25 to 27.75 bar (149 to 402 psi)

NOTE —
The hydraulic accumulator, which is threaded into to the bottom of the power flow regulator, can be replaced separately as described below.

10. Remove the test equipment and install and tighten the return line and the pressure switch.

Tightening torque	
• hydraulic pressure switch to power flow regulator	15 to 18 Nm (11 to 13 ft-lb)

To remove hydraulic brake booster:

1. Discharge the hydraulic accumulator by operating the brake pedal at least 20 times with the engine stopped.

2. Remove the master cylinder as described above under **4.3 Master Cylinder.**

3. Clean the supply line and return line fittings on the brake booster. Remove the two lines from the booster. See Fig. 4-12.

CAUTION —
Extreme cleanliness is essential when working on the hydraulic brake booster system. Even a small piece of dirt could damage the brake booster system or the power steering system.

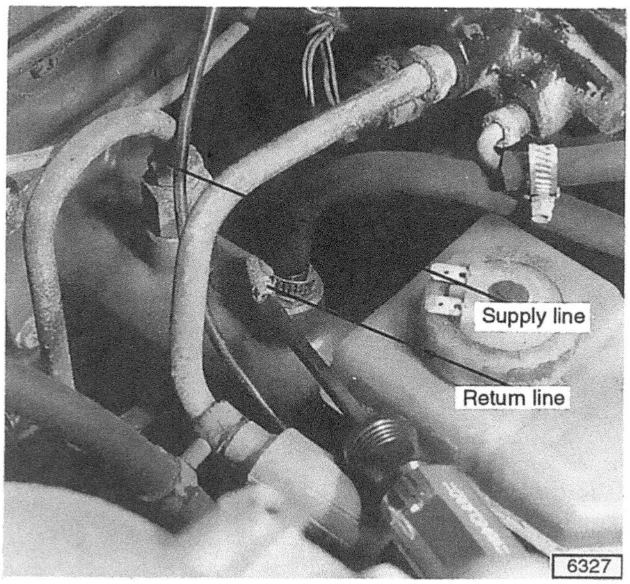

Fig. 4-12. Brake booster return line being removed.

4. Working from inside the passenger compartment, remove the lower left trim panels above the brake pedal.

5. Disconnect the pushrod from the brake pedal by disconnecting the brake pedal return spring and prying off the retaining clip. Slide the clevis pin out. See Fig. 4-13.

6. Remove the four mounting nuts as shown in Fig. 4-14. Working from the engine compartment side, pull the brake booster from the pedal base.

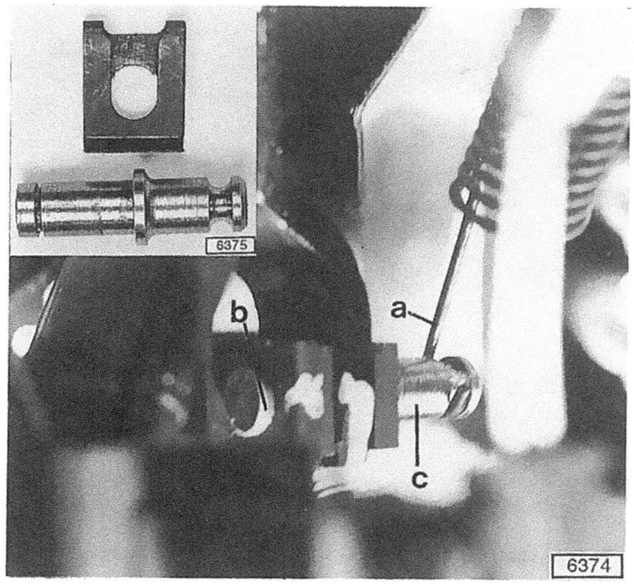

Fig. 4-13. Brake pedal return spring (**a**), retaining clip (**b**) and clevis pin (**c**) as viewed from inside passenger compartment. Inset shows clevis pin and clip.

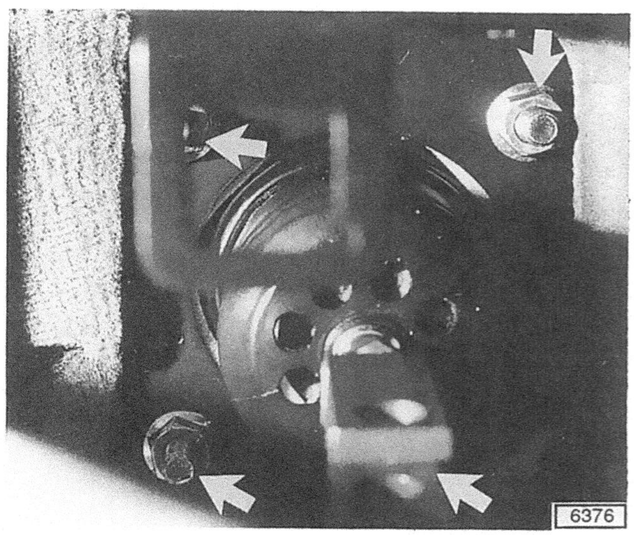

Fig. 4-14. Brake booster mounting nuts (arrows) as viewed from inside the passenger compartment.

Installation of the brake booster is the reverse of removal. Tighten all fittings to the specified torque. Adjust the brake pedal travel and the brake light switch as described below under **4.5 Brake Pedal Adjustments.**

Tightening torque	
• hydraulic lines to brake booster	25 to 35 Nm (18 to 26 ft-lb)

CAUTION —
Hydraulic fittings should be installed using a crowfoot flare-nut wrench and an accurate torque wrench. Failure to tighten the fittings to the specified torque with the correct wrench could result in damage to the component or the fitting. Damaged fittings or components will result in leaks, which will affect system operation.

To remove power flow regulator:

1. Discharge the hydraulic accumulator by operating the brake pedal at least 20 times with the engine stopped.

2. Thoroughly clean the power flow regulator and its fittings.

CAUTION —
Extreme cleanliness is essential when working on the hydraulic brake booster system. Even a small piece of dirt could damage the brake booster system or the power steering system.

3. Remove and label the electrical wires from the two pressure switches on the regulator.

4. Remove the 4 hydraulic lines from the power flow regulator. See Fig. 4-15.

Fig. 4-15. Power flow regulator hydraulic lines to be removed (arrows).

5. Remove the regulator by removing the two mounting bolts shown in Fig. 4-16.

Fig. 4-16. Power flow regulator mounting bolts to be removed.

Installation of the power flow regulator is the reverse of removal. If the power flow regulator is being replaced, install the old switches and the accumulator to the new regulator. To remove the accumulator from the regulator, clamp the regulator in a soft-jaw vise and thread the accumulator off. Tighten all fittings to the specified torque. See Fig. 4-17 for reference. When installation is complete, bleed the power steering system as described in **STEERING AND WHEEL ALIGNMENT**.

CAUTION —
Hydraulic fittings should be installed using a crowfoot flare-nut wrench and an accurate torque wrench. Failure to tighten the fittings to the specified torque with the correct wrench could result in damage to the component or the fitting. Damaged fittings or components will result in leaks, which will affect system operation.

NOTE —
If the accumulator is difficult to remove from the regulator, use a strap wrench or a large oil filter wrench on the accumulator while the regulator is in the vise.

19

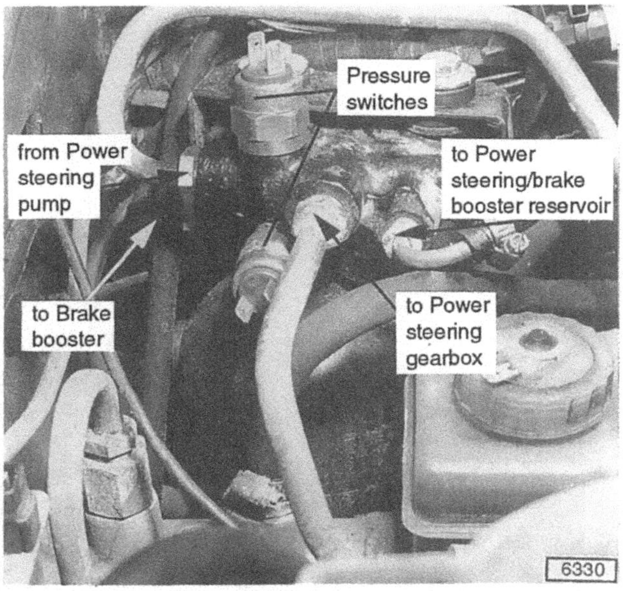

Fig. 4-17. Power flow regulator hydraulic lines.

Tightening torques	
• line from power flow regulator to fluid reservoir	15 to 18 Nm (11 to 13 ft-lb)
• line from power flow regulator to steering gearbox	45 to 50 Nm (33 to 36 ft-lb)
• line from power flow regulator to power steering pump	30 to 40 Nm (22 to 29 ft-lb)
• line from power flow regulator to brake booster	25 to 35 Nm (18 to 25 ft-lb)
• pressure switches to power flow regulator	15 to 18 Nm (11 to 13 ft-lb)
• accumulator to power flow regulator	43 to 50 Nm (31 to 36 ft-lb)

4.5 Brake Pedal Adjustments

Brake pedal height and brake light switch adjustment should be checked any time the master cylinder or the brake booster are removed or replaced.

Correct brake pedal height and brake light switch positions are shown in Fig. 4-18. To adjust pedal height, loosen the locknut on the pushrod clevis and rotate the pushrod until the correct dimension is obtained. For brake light switch adjustment, remove the electrical connectors from the switch, loosen the locknut, and rotate the switch. Be sure to retighten the locknuts.

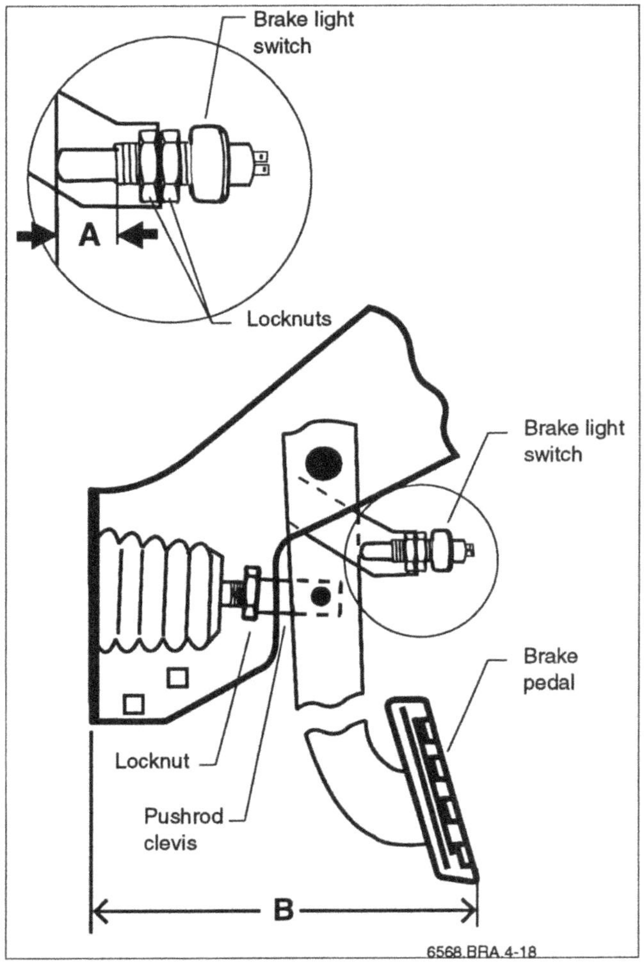

Fig. 4-18. Brake pedal assembly showing brake light switch plunger height (dimension **A**) and brake pedal height (dimension **B**).

Brake pedal adjustment specifications	
• dimension A	5 mm (1/5 in.)
• dimension B	231 mm (9 1/10 in.)

4.6 Brake Pad Wear Warning System

The brake pad wear warning system warns the driver with a light on the instrument panel when the pads are worn. This warning system should not be ignored. Failure to replace the pads soon after the light comes on may cause caliper or rotor damage that will add to the cost of repairs.

On 1982 through most 1985 models, the brake pad wear warning system includes an early warning lamp that lights only when the brake pedal is depressed. This light notifies the driver in advance that the pads will need replacement. On all models, the pads should be replaced when the red instrument panel lamp comes on continuously.

All models have one sensor in the front left wheel brake pad and one in the rear right wheel brake pad. In either case, a sensor will turn the instrument panel light on if an open circuit exists or if the sensor's terminals are grounded through the rotor. For brake pad sensor replacement, see **5.1 Reconditioning Front Disc Brakes** or **6.1 Reconditioning Rear Disc Brakes**.

4.7 Anti-Lock Brake System (ABS)

Fig. 4-19 shows the main components of the ABS. The wheel sensors supply wheel speed information to the electronic control unit in the form of electrical impulse signals. The control unit continuously compares these signals to determine whether any of the wheels is about to lock. The hydraulic unit modulates hydraulic pressure to prevent wheel lock.

CAUTION —

- *Voltage spikes can damage the ABS control unit. If electrical welding is to be carried out on the car, disconnect the electrical connector from the ABS electronic control unit.*

- *Extreme heat can damage the ABS control unit. If the car is to be subjected to sustained temperature above 185°F (85°C), the control unit should be removed from the car. The ABS control unit is located beneath the instrument panel to the left of the steering column.*

NOTE —

The ABS warning indicator should come on briefly when the car is started, then go out. If it comes on while the car is running, the ABS system has detected a fault. Once a fault has been detected by the built-in safety circuit, the system is automatically turned off.

The hydraulic unit consists of a three-way electrically operated solenoid valve for each wheel, a brake fluid reservoir for each brake circuit, a return pump, and two relays. During normal braking, the solenoid valves allow brake fluid to pass freely from the master cylinder to the wheels. If a wheel sensor detects that a wheel is about to lock, the control unit signals the appropriate solenoid valve to hold the brake pressure to that wheel constant.

If holding constant pressure does not prevent the wheel from locking, the valve is repositioned and a small amount of brake fluid is allowed to pass from the brake line into the hydraulic unit reservoir to reduce the hydraulic pressure at that wheel. During this regulating period, the return pump is activated and any brake fluid in the hydraulic unit reservoir is returned to the master cylinder reservoir. The regulating cycle continues as necessary until the car has reached a safe speed (approximately 4 mph, or until the brake pedal is released).

The ABS system is designed to be maintenance free. There are no adjustments that can be made to the system. Repair and troubleshooting of the ABS system requires special test equipment and knowledge, and is therefore outside the scope of this manual.

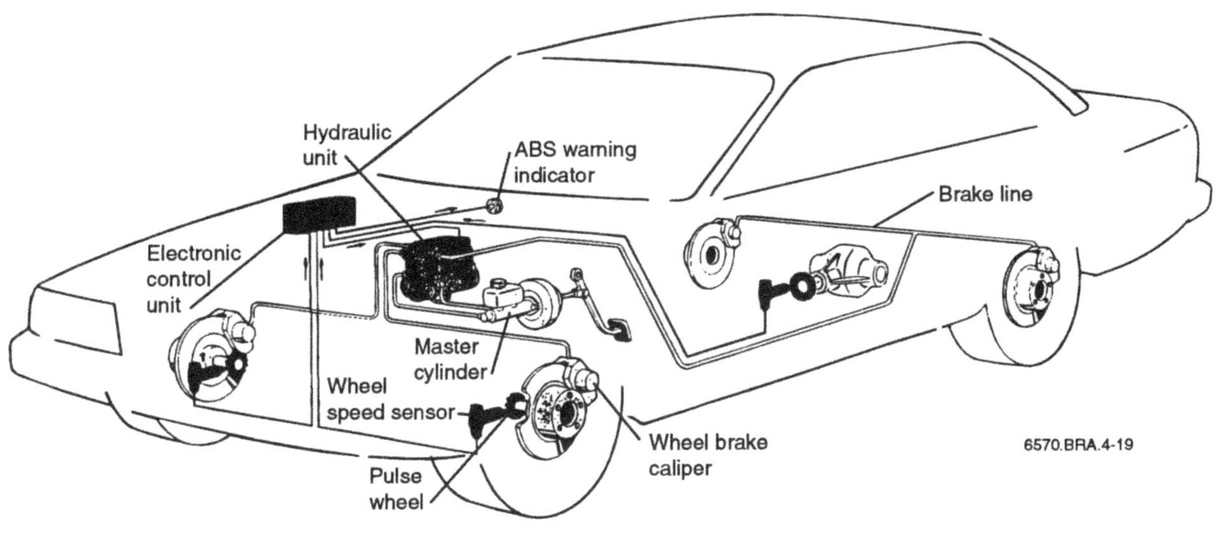

Fig. 4-19. Schematic representation of typical anti-lock braking system (ABS). Exact location and configuration of components may vary.

19

Inspecting ABS System

If the ABS warning indicator comes on, indicating a problem, a visual inspection of the system may help to locate the fault. Carefully inspect the wiring harness. Check especially the ABS harness and connections near each wheel for chafing or damage due to incorrectly routed wires. Remove the speed sensor from its mounting and check it for damage. See Fig. 4-20. If the wheel sensor or its harness is damaged, the wheel sensor with its integral harness should be replaced. Apply a small amount of grease to the sensor's O-ring to aid in reinstalling the sensor.

WARNING —
If the main ABS wiring harness is damaged in any way, it must be replaced. Do not attempt to repair the wiring harness. The ABS system is sensitive to very small changes in resistance. Repairing the wiring harness could alter resistance values and cause the system to malfunction.

CAUTION —
The ignition must be off before connecting or disconnecting any electrical connections.

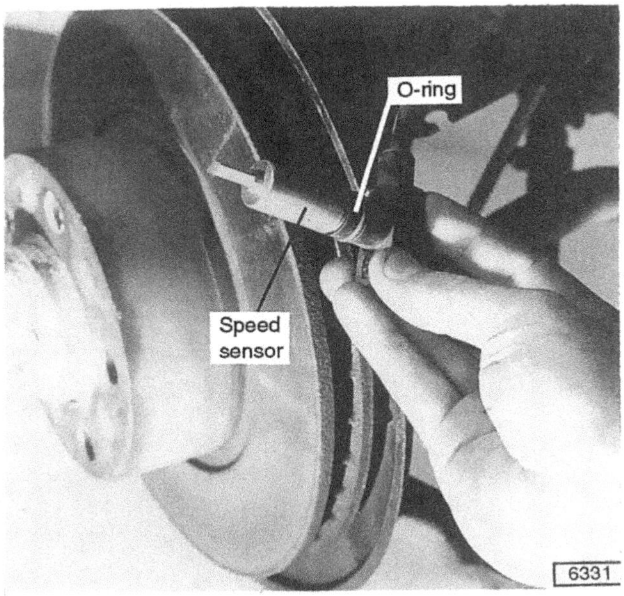

Fig. 4-20. ABS wheel speed sensor.

The pulse wheels themselves may be responsible for faulty speed signals reaching the control unit. Inspect the pulse wheels for damage such as cracks or missing teeth. Check to see that a crack has not made the pulse wheel loose. Spin the tire to check to see if the pulse wheel is out-of-round. Check for dirt or worn wheel bearings that could reduce the clearance between the sensor and the wheel. Even accumulated metallic brake-pad dust has been known to cause problems. A faulty pulse wheel should be replaced. See Fig. 4-21.

NOTE —
Front pulse wheels are replaced as part of the front wheel bearing assembly as described in **SUSPENSION—FRONT**. Rear pulse wheels are replaced as part of the wheel bearing assembly as described in **SUSPENSION—REAR**.

Fig. 4-21. Front left ABS pulse wheel.

5. FRONT DISC BRAKES

Fig. 5-1 is an exploded view of a front disc brake. When the brakes are applied, the caliper pushes the inside brake pad against the rotor. This moves the caliper body at the same time, pulling the outside pad against the rotor.

Because the front brakes do more of the work of stopping the car, they are likely to wear faster and may require service more frequently than the rear brakes. The brake pads and rotors are subjected to the greatest wear, and are the components most often needing attention.

The brake pads are designed to be routinely replaced as they wear out. While it is possible to restore the brakes by replacing only the brake pads, the rotors should be resurfaced or replaced at the same time to achieve full braking performance and maximize pad life.

WARNING —
Brake pad friction materials may contain asbestos fibers. Do not create dust by grinding, sanding, or cleaning the pads with compressed air. Avoid breathing any asbestos fibers or dust. Breathing asbestos can cause serious diseases such as asbestosis or cancer, and may result in death.

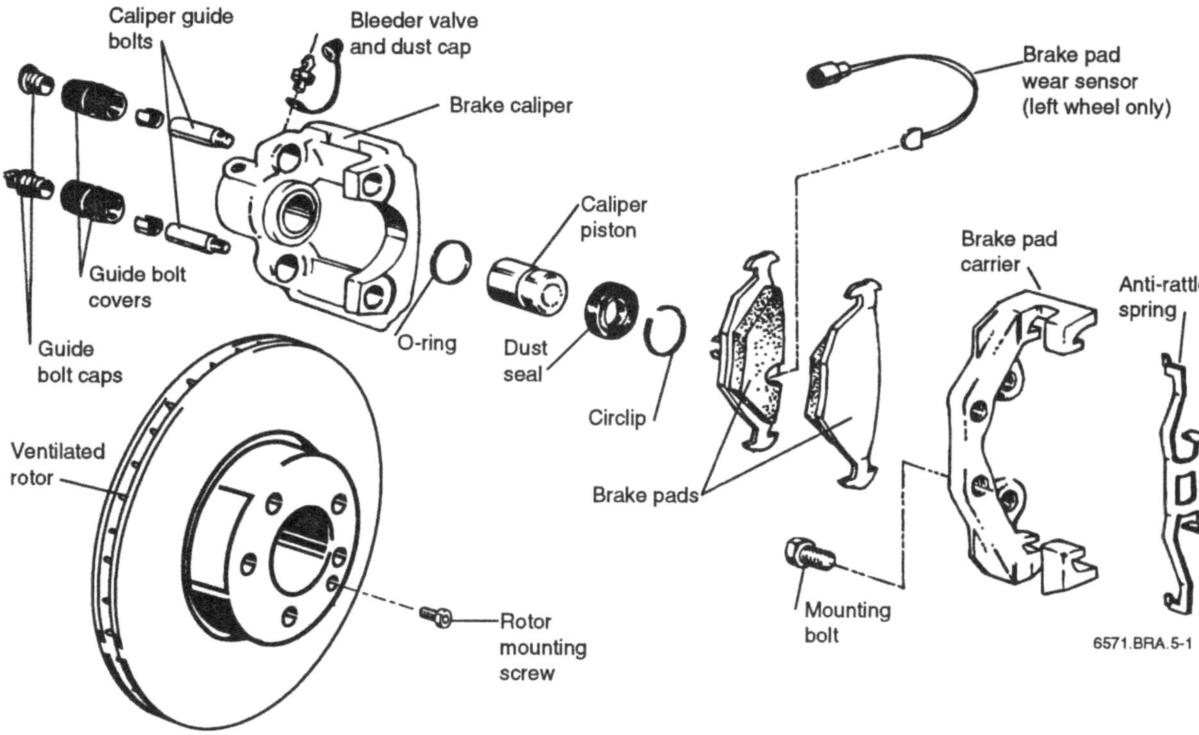

Fig. 5-1. Exploded view of front disc brake assembly.

Checking Brake Pad Wear

The inspection procedure and specifications for checking the brake pads are found in **LUBRICATION AND MAINTENANCE**. For more complete inspection of pad condition, the pads must be removed as described in **5.1 Reconditioning Front Disc Brakes**.

5.1 Reconditioning Front Disc Brakes

Reconditioning the front disc brakes typically includes replacing the brake pads and resurfacing or replacing the brake rotors. Unless the calipers and hoses are damaged, visibly worn, or leaking fluid, more extensive front brake repairs are not normally required.

If the brake pads are soaked with oil, grease, or brake fluid, the cause of the contamination must be found and corrected before new pads are installed. Brake rotors must always be replaced when the resurfacing thickness exceeds the specified limits. See **Removing, Reconditioning, and Installing Brake Rotors** for rotor thickness and runout (out-of-round) specifications.

Removing and Installing Front Brake Pads

Replacing the brake pads can be done without disconnecting the flexible brake hose. Keeping the hydraulic system sealed eliminates the need to bleed the brakes afterward.

Brake pads and the surfaces of the rotors wear slightly differently. Always replace brake pads in complete sets. If the old pads are to be reinstalled, such as after inspecting them, always make sure they are refitted in their original locations.

To remove front brake pads:

1. Raise the front of the car and support it securely on jack stands. Remove the front wheels.

2. Pry out the two plastic caps from the guide bolt covers. Remove the two internal-hex-head guide bolts. See Fig. 5-2.

3. If applicable, separate the brake pad wear sensor connector and remove the wiring from the brake bleeder dust cap. See Fig. 5-3.

19

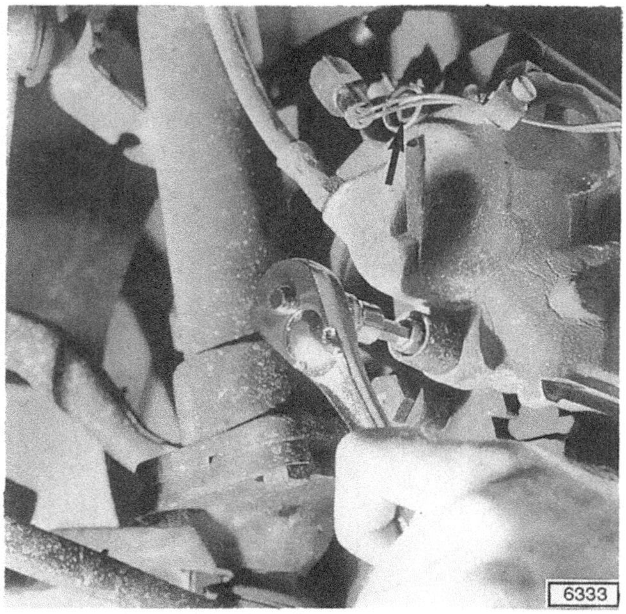

Fig. 5-2. Lower caliper guide bolt being removed. Upper guide bolt shown at arrow.

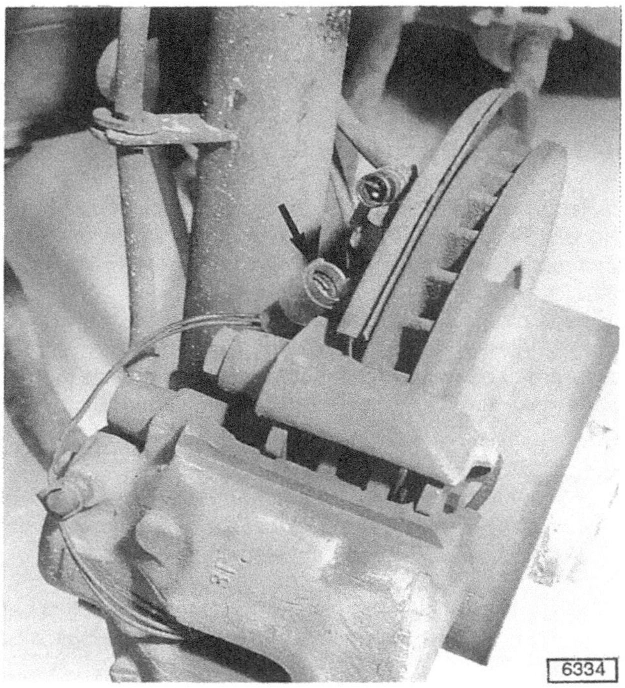

Fig. 5-3. Brake pad wear sensor connector (arrow) on front left wheel.

4. Remove the anti-rattle spring. See Fig. 5-4.

Fig. 5-4. Front caliper anti-rattle spring being removed. Unhook spring from caliper (arrow).

5. Pull the caliper straight off. See Fig. 5-5. Remove the brake pads from the caliper. Rest the caliper on the pad carrier.

NOTE —
The inner pad has a spring clip that holds the pad to the caliper piston.

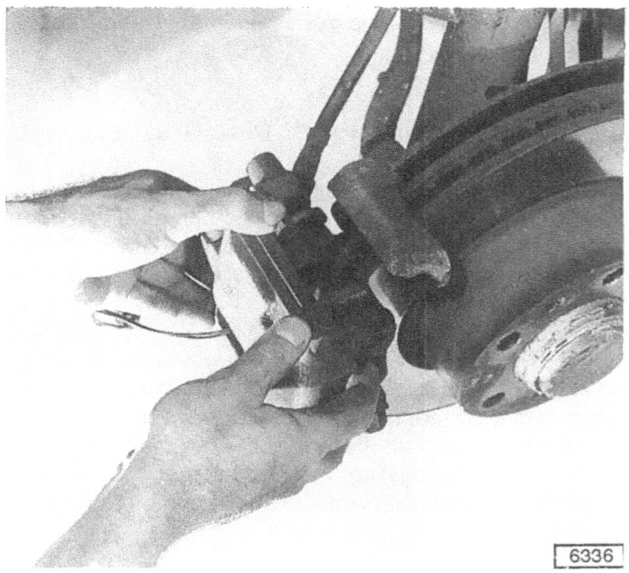

Fig. 5-5. Front brake caliper with brake pads being removed.

CAUTION —
Do not let the brake hose support the weight of the caliper. Suspend the caliper and pad carrier assembly from the suspension or body. Avoid stretching or kinking the hose.

NOTE —
If the piston creeps out of the caliper due to pressure in the brake lines, place a strong rubber band around the piston and caliper to hold the piston in.

6. If the pads are being replaced, carefully pry the pad wear sensor straight off the left-wheel inner pad. See Fig. 5-6.

NOTE —
• Inspect the plastic part of the brake pad wear sensor. If the plastic part of the sensor is worn through to the wires, the sensor should be replaced.

• If the pads are being replaced because the brake pad wear indicator light on the instrument panel came on, the sensor should be replaced.

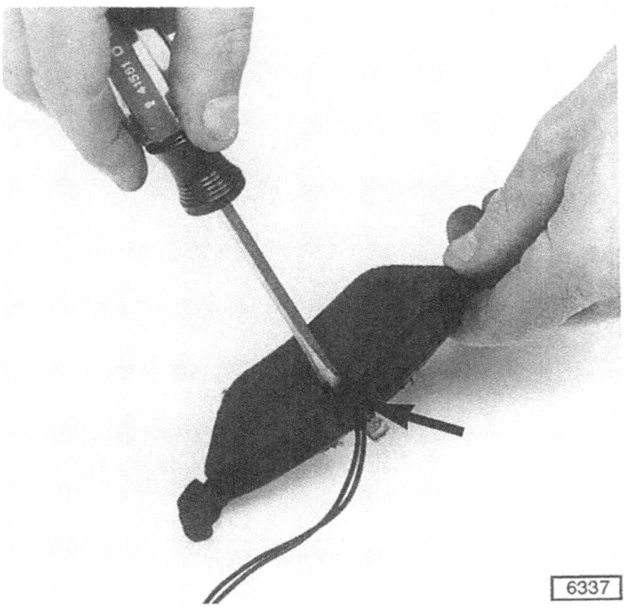

Fig. 5-6. Front brake pad wear sensor (arrow) being removed from brake pad.

To install front brake pads:

1. Insert the brake pad wear sensor into the cutout in the new pad.

2. Place the outer brake pad onto the caliper. Place the inner brake pad with its spring onto the brake caliper piston. Route the pad wear sensor wiring through the caliper opening and bleeder dust cap.

CAUTION —
If the brake pad wear sensor wiring is incorrectly routed, it could be damaged by wearing against the tire.

3. Slowly push the piston back into the caliper to provide clearance for the thicker new brake pads. Be careful not to push the piston past the outer edge of the piston dust seal. See Fig. 5-7.

CAUTION —
Use a shop rag or a thin piece of wood to protect the caliper piston from being marred by the jaws of the water pump pliers.

NOTE —
Pushing in the caliper piston will cause brake fluid to overflow a full master cylinder fluid reservoir. To prevent this, use a clean syringe or some equivalent to first remove some fluid from the reservoir.

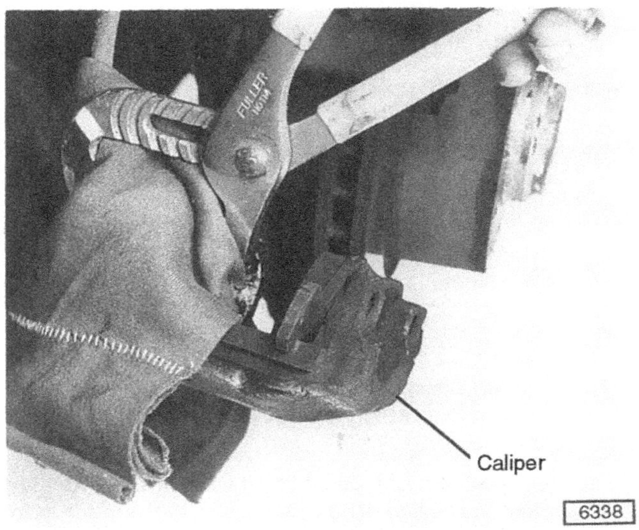

Fig. 5-7. Water pump pliers being used to push piston into caliper.

4. Install the caliper onto the pad carrier. Install and tighten the guide bolts. See Fig. 5-8. Install the two plastic guide bolt caps.

Tightening torque
• front brake caliper to pad carrier (guide bolts) . . . . 30 to 35 Nm (22 to 26 ft-lb)

CAUTION —
• Carefully inspect the guide bolts for damage. If the bolts are in any way damaged, they should be replaced.

• When installing the caliper, make sure the dust seal is positioned so that it does not catch the edge of the piston when the pedal is depressed. Installing it incorrectly will damage the dust seal.

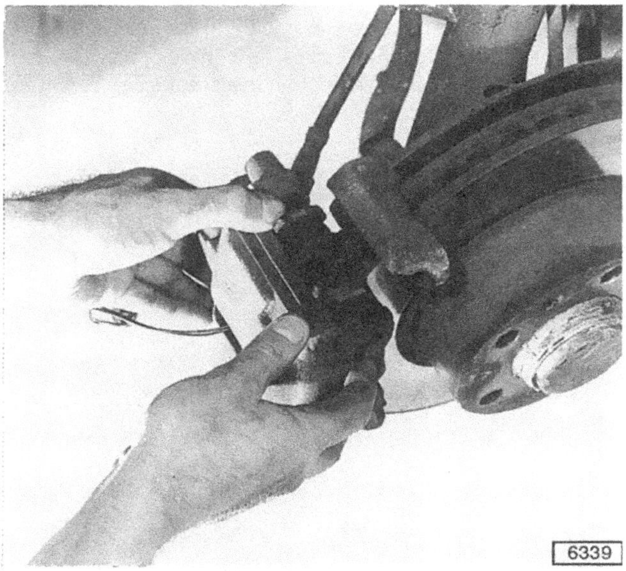

Fig. 5-8. Caliper being installed.

5. Mount the wheel and loosely install the lug bolts. Lower the car and then torque the lug bolts.

Tightening torque
• wheel lug bolts .... 100±10 Nm (74±7 ft-lb).

NOTE —
Lightly lubricate the wheel hub center with multipurpose grease before installing the wheel.

6. Depress the brake pedal using several short strokes to adjust the caliper and brake pads to the rotor before moving or running the car.

7. Check the level of brake fluid in the reservoir. If necessary, add new brake fluid to fill the reservoir to the MAX mark.

WARNING —
New brake pads require some break-in. Allow for slightly longer stopping distances for the first 100 to 150 miles of city driving, and avoid hard stops.

Removing, Reconditioning, and Installing Brake Rotors

To remove the brake rotors, remove the caliper mounting bolts and remove the caliper. See Fig. 5-9. Using a hex wrench, remove the rotor mounting screw. See Fig. 5-10. Pull the rotor off the hub. Use a soft-faced mallet to free a stuck rotor.

CAUTION —
• Do not remove the rotor without first removing the caliper assembly. Excessive force may damage the rotor or the wheel bearing housing.

• Do not let the caliper hang by its hose. Suspend the caliper from the body using stiff wire.

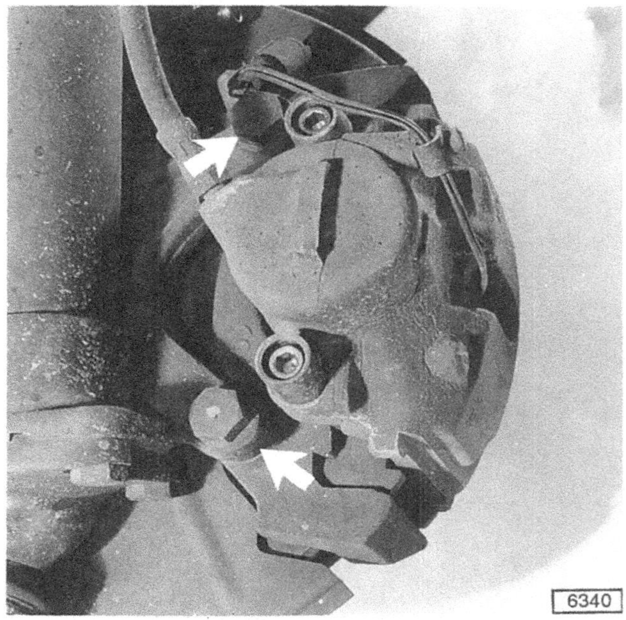

Fig. 5-9. Front caliper mounting bolts (arrows).

Installation is the reverse of removal. Make sure that the rotor and wheel hub mounting surfaces mate properly and are free of dirt and corrosion.

Tightening torque
• front brake caliper to steering knuckle .... 110 to 123 Nm (81 to 91 ft-lb)

NOTE —
New replacement brake rotors should be cleaned with a grease-free solvent, such as a commercially available brake cleaner, before installing the caliper and brake pads.

Check the brake rotor for wear anytime the brakes are serviced. Rotors that are scored with sharp ridges, warped, worn irregularly or cracked should be replaced.

Fig. 5-10. Front rotor mounting screw (arrow).

Fig. 5-11. Front rotor minimum dimension stamped on rotor hub (arrow).

Brake rotors should always be resurfaced in pairs, with an equal amount of material removed from both sides of each rotor. **Table d** lists the dimensions for resurfacing rotors. Rotors which fail to meet these requirements should be replaced.

NOTE —
On original equipment rotors, the minimum thickness dimension can be found stamped into the rotor's hub. See Fig. 5-11.

Table d. Front Rotor Reconditioning Specifications

	528e, 533i models	535i models
Minimum thickness after machining	20.4 mm (0.803 in.)	23.4 mm (0.921 in.)
Wear limit (minimum permissible thickness)	20.0 mm (0.787 in.)	23 mm (0.905 in.)
Thickness tolerance (difference between any two measurements)	0.02 mm (0.0008 in.)	0.02 mm (0.0008 in.)
Axial runout (maximum permissible)		
rotor removed	0.05 mm (0.002 in.)	0.05 mm (0.002 in.)
rotor installed	0.20 mm (0.008 in.)	0.20 mm (0.008 in.)

Use a micrometer to measure the rotor thickness at eight to ten positions along the rotor's braking surface. Rotor runout should be measured using a dial indicator setup. If a low speed shimmy goes away when the brakes are released, excessive rotor runout is probably the cause of the shimmy. If the rotor's runout is within limits, and the shimmy still persists, check the front wheel bearings as described in **SUSPENSION—FRONT**. Brake rotors can be resurfaced by most local automotive machine shops.

5.2 Front Calipers

Fig. 5-12 is an exploded view of the front disc brake caliper. The piston seal holds the hydraulic pressure in the system as the piston moves, and also contributes to the caliper's self-adjusting action.

Brake fluid seeping or leaking from around the caliper piston is the result of a failed or damaged piston seal. A damaged piston seal is usually caused by corrosion, scoring, or pitting of the piston or caliper bore. The seal can be replaced separately, but a damaged piston will quickly damage the new seal. To remedy a leaking caliper piston seal and avoid future problems, complete replacement of the caliper is recommended.

To remove front calipers:

1. Raise the front of the car and support it securely on jack stands. Remove the front wheels.

2. If applicable, separate the brake pad wear sensor connector as shown above in Fig. 5-3.

3. Loosen the hose fitting on the caliper. Remove the two mounting bolts from the steering knuckle. See Fig. 5-13.

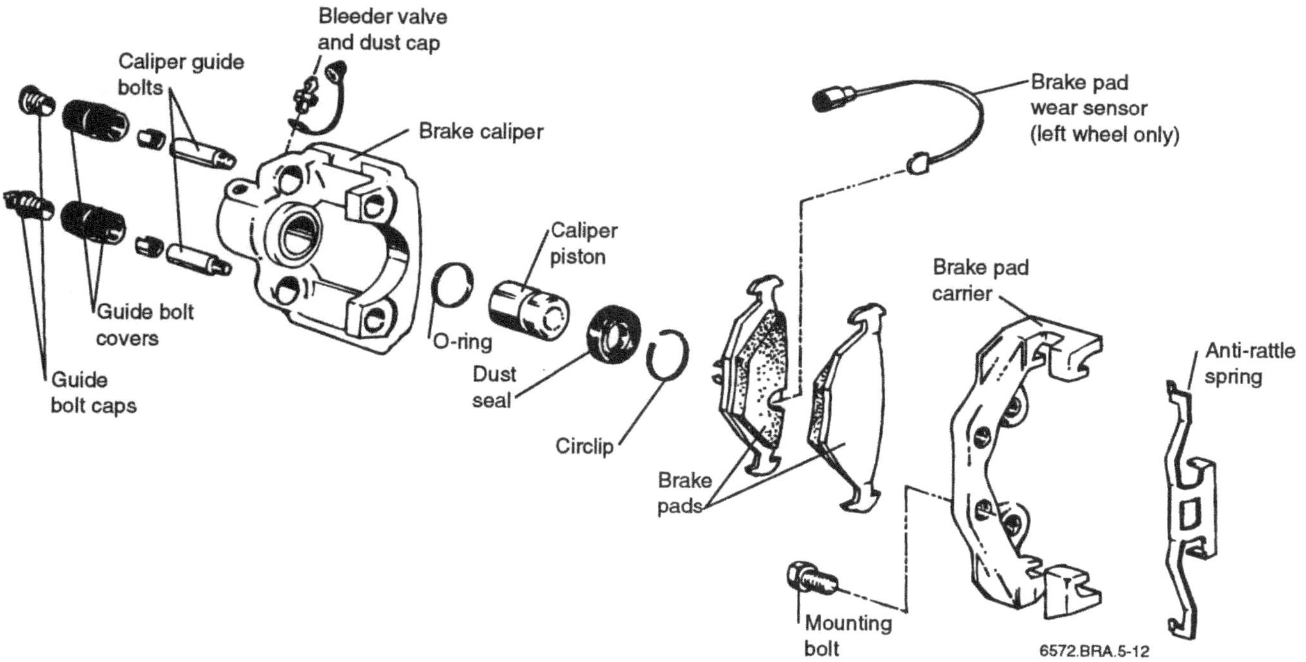

Fig. 5-12. Exploded view of front disc brake caliper.

Fig. 5-13. Front brake caliper mounting bolts (arrows).

4. Disconnect the brake hose from the caliper by spinning the caliper off the fitting. Drain the brake fluid into a container.

> ***CAUTION —***
> *Do not let brake fluid spill onto the brake pads or brake rotor surface. Cap the end of the brake line to prevent dirt and moisture from entering the brake system.*

Installation is the reverse of removal. Start all brake line fittings by hand to avoid cross-threading. Bleed the brakes as described in **4.2 Bleeding Brakes.**

Tightening torques
• brake hose to caliper . . . . 13 to 16 Nm (10 to 12 ft-lb).
• front brake caliper to steering knuckle . . . . 110 to 123 Nm (81 to 91 ft-lb).

6. REAR DISC BRAKES

Fig. 6-1 is an exploded view of the rear disc brake. The rear disc brakes operate in much the same way as the front disc brakes, except that the rear rotor doubles as the brake drum for the parking brake. The rotor is easily removed without having to remove the rear wheel bearings.

Because the rear brakes do less of the work of stopping the car, they are not likely to wear as quickly or require service as frequently as the front brakes. The brake pads and rotors are subjected to the greatest wear, and are the components most often needing attention.

The brake pads are designed to be routinely replaced as they wear out. While it is possible to restore the brakes by replacing only the brake pads, the rotors should be resurfaced or replaced at the same time to achieve full braking performance and maximize life of the new pads.

> ***WARNING —***
> *Brake pad friction materials may contain asbestos fibers. Do not create dust by grinding, sanding, or cleaning the pads with compressed air. Avoid breathing any asbestos fibers or dust. Breathing asbestos can cause serious diseases such as asbestosis or cancer, and may result in death.*

Checking Brake Pad Wear

The inspection procedure and specifications for checking the brake pads are found in **LUBRICATION AND MAINTENANCE**. For more complete brake pad inspection, the pads must be removed, as described below.

6.1 Reconditioning Rear Disc Brakes

Reconditioning the rear disc brakes typically includes replacing the brake pads and resurfacing or replacing the brake rotors. Unless the calipers and hoses are damaged, visibly worn, or leaking fluid, more extensive rear brake repairs are not normally required.

If the brake pads are soaked with oil, grease, or brake fluid, the cause of the contamination must be found and corrected before new pads are installed. Brake rotors must always be replaced when the resurfacing thickness exceeds the specified limits. See **Removing, Reconditioning, and Installing Rear Brake Rotors** for rotor thickness specifications.

Removing and Installing Rear Brake Pads

Replacing the brake pads can be done without disconnecting the flexible brake hose. Keeping the hydraulic system sealed eliminates the need to bleed the brakes afterward.

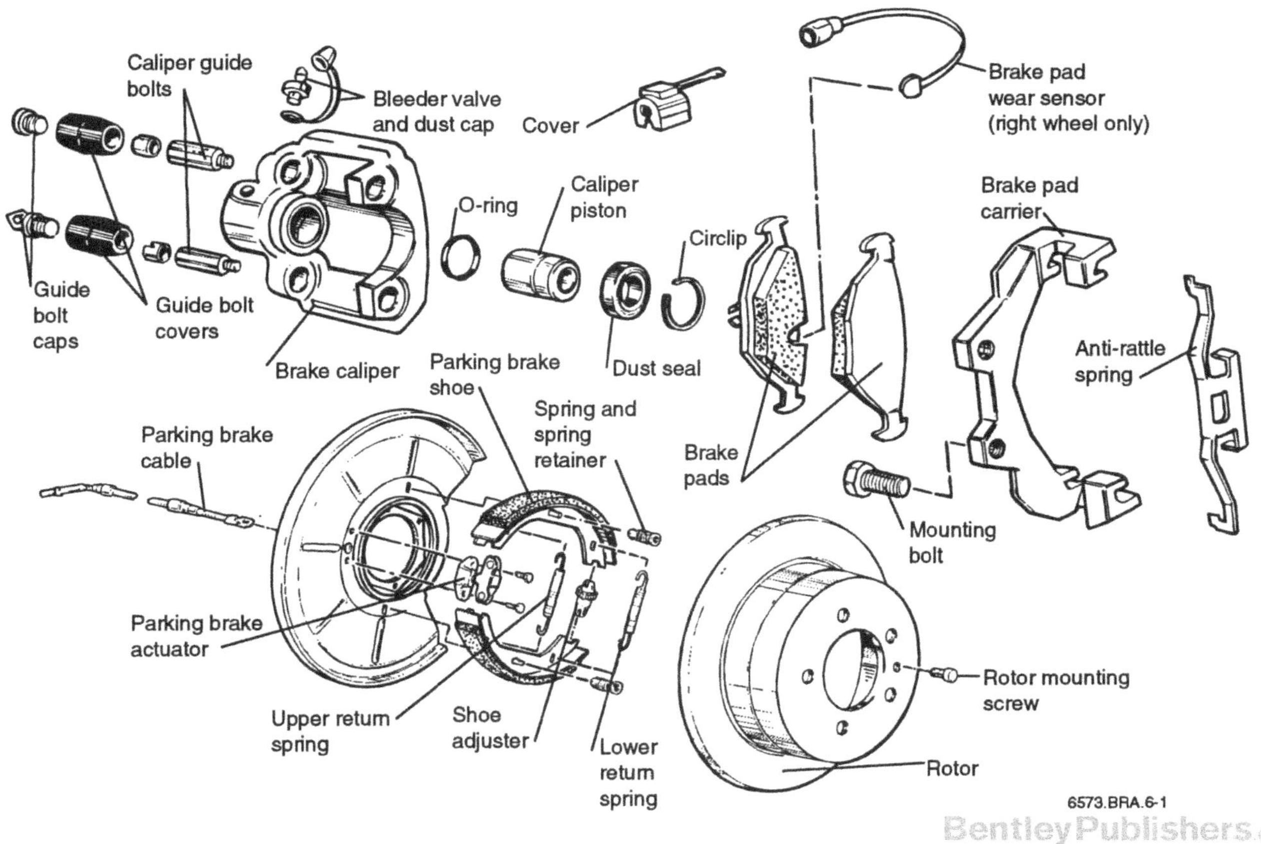

Fig. 6-1. Exploded view of rear disc brake assembly.

Brake pads and the surfaces of the rotors wear slightly differently. Always replace brake pads in complete sets. If old pads are to be reinstalled, after inspecting them for example, always make sure they are refitted in their original locations.

To remove rear brake pads:

1. Raise the rear of the car and support it securely on jack stands. Release the parking brake and remove the rear wheels.

2. Pry out the two plastic caps from the guide bolt covers. Remove the two internal-hex-head guide bolts. See Fig. 6-2.

Fig. 6-2. Rear caliper guide bolts (arrows).

3. If applicable, separate the brake pad wear sensor connector.

4. Remove the anti-rattle spring. See Fig. 6-3.

5. Pull the caliper straight off. See Fig. 6-4.

6. Remove the inner pad from the caliper piston and the outer pad from the caliper. Rest the caliper on the pad carrier.

> ***CAUTION —***
> *Do not let the brake hose support the weight of the caliper. Avoid stretching or kinking the hose.*

> **NOTE —**
> If the piston creeps out of the caliper due to pressure in the brake lines, place a strong rubber band around the piston and caliper to hold the piston in.

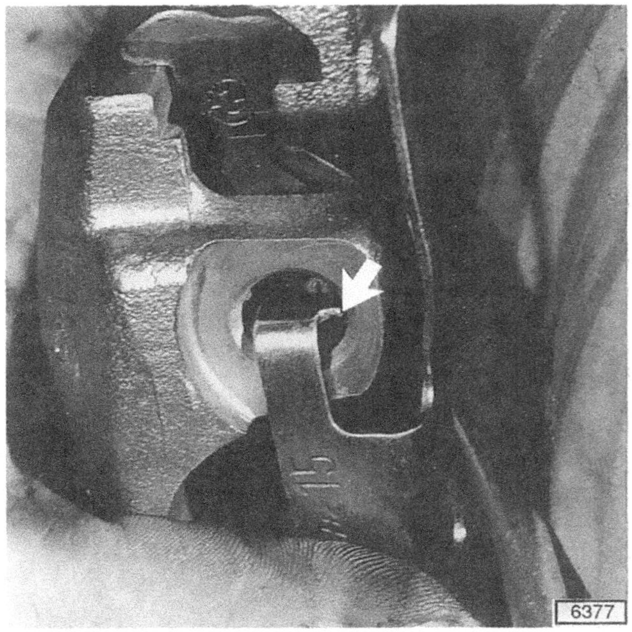

Fig. 6-3. Anti-rattle spring being removed. Unhook spring from caliper (arrow).

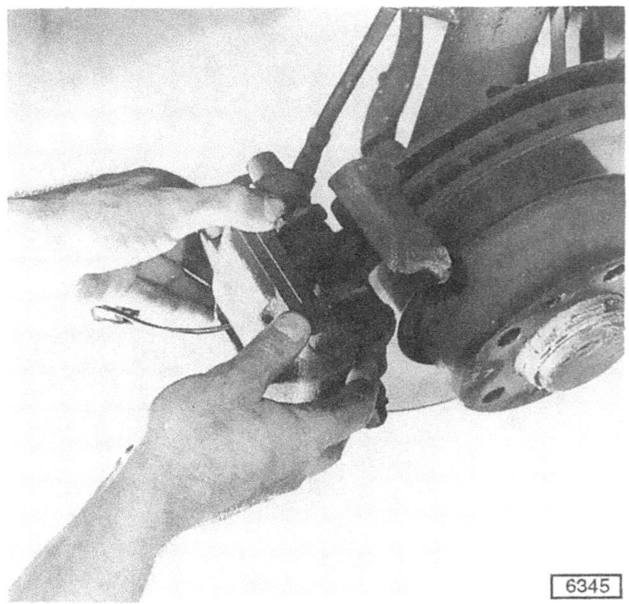

Fig. 6-4. Brake caliper being removed.

7. If the pads are being replaced, carefully pry out the pad wear sensor from the pad. See Fig. 6-5.

NOTE —
• Inspect the plastic part of the brake pad wear sensor. If the plastic part of the sensor is worn through to the wires, the sensor should be replaced.

• If the pads are being replaced because the brake pad wear indicator light on the instrument panel came on, the sensor should be replaced.

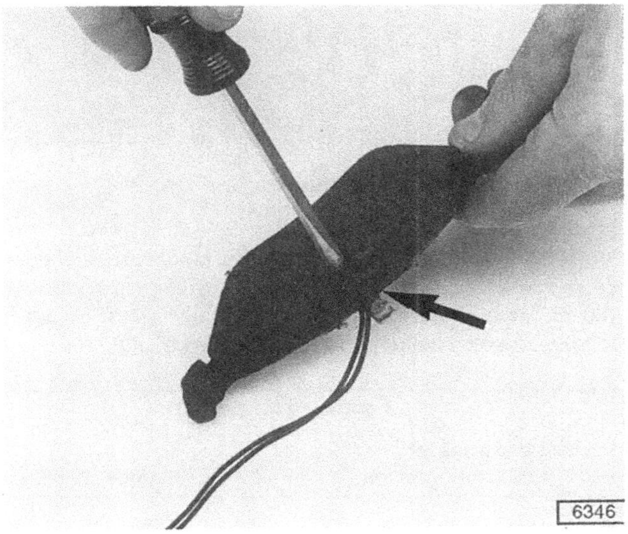

Fig. 6-5. Rear brake pad wear sensor (arrow) being removed.

To install rear brake pads:

1. If applicable, install the pad wear sensor into the cutout in the new pad.

2. Slowly push the piston back into the caliper to provide clearance for the thicker new brake pads. See Fig. 6-6. Be careful not to push the piston past the outer edge of the piston dust seal.

CAUTION —
Use a shop rag or a thin piece of wood to protect the caliper piston from being marred by the jaws of the water pump pliers.

NOTE —
Pushing in the caliper piston in will cause brake fluid to overflow a full master cylinder fluid reservoir. To prevent this, use a clean syringe or some equivalent to first remove some fluid from the reservoir.

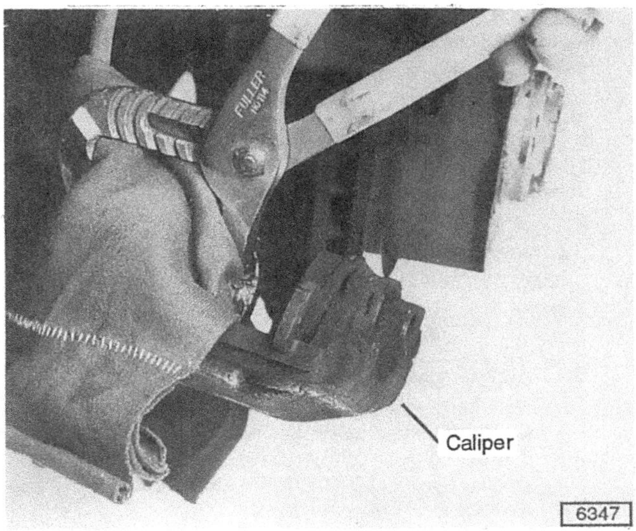

Fig. 6-6. Water pump pliers being used to push piston into caliper.

3. Place the outer brake pad onto the caliper. Place the inner brake pad with its spring onto the brake caliper piston. Route the pad wear sensor wiring through the caliper opening.

CAUTION —
If the brake pad wear sensor wiring is incorrectly routed, it could be damaged by wearing against the tire.

4. Install the caliper onto the pad carrier. Install and tighten the guide bolts. Install the two plastic guide bolt caps. If applicable reconnect the pad wear sensor connector.

Tightening torque	
• rear brake caliper to pad carrier (guide bolts)	30 to 35 Nm (22 to 26 ft-lb)

CAUTION —
• Carefully inspect the guide bolts for damage. If the bolts are in any way damaged, they should be replaced.

• When installing the caliper, make sure the dust seal is positioned so that it does not catch the edge of the piston when the pedal is depressed. Installing it incorrectly will damage the dust seal.

5. Mount the wheel and loosely install the lug bolts. Lower the car and then torque the lug bolts.

Tightening torque	
• wheel lug bolts	100±10 Nm (74±7 ft-lb).

19

NOTE —
Lightly lubricate the wheel hub center with multipurpose grease before installing the wheel.

6. Depress the brake pedal using several short strokes to adjust the caliper and brake pads to the rotor before moving the car.

7. Check the level of brake fluid in the reservoir. If necessary, add new brake fluid to fill the reservoir to the MAX mark.

WARNING —
*New brake pads require some break-in. Allow for slightly longer stopping distances for the first 100 to 150 miles of city driving, and avoid hard stops. See **LUBRICATION AND MAINTENANCE** for more information.*

Removing, Reconditioning, and Installing Rear Brake Rotors

To remove the brake rotors, remove the caliper mounting bolts and remove the caliper. See Fig. 6-7. Using a hex wrench, remove the rotor mounting screw and pull the rotor off the hub. See Fig. 6-8. Use a soft-faced mallet to free a stuck rotor.

CAUTION —
- *Do not remove the rotor without first removing the caliper assembly. Excessive force may damage the rotor or the wheel bearing housing.*
- *Do not let the caliper hang by its hose. Suspend the caliper from the body using a stiff wire.*

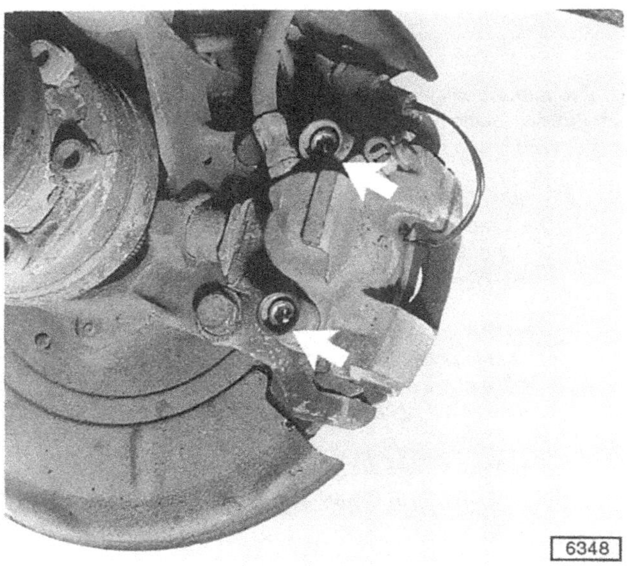

Fig. 6-7. Rear caliper mounting bolts (arrows).

Fig. 6-8. Rear rotor mounting screw (arrow).

Installation is the reverse of removal. Make sure that the rotor and wheel hub mounting surfaces mate properly and are free of dirt and corrosion. After replacing rotors, adjust the parking brake as described in **6.3 Parking Brake.**

Tightening torque
• rear brake caliper to trailing arm . . . . 60 to 67 Nm (44 to 49 ft-lb).

NOTE —
New replacement brake rotors should be cleaned with a grease-free solvent, such as a commercially available brake cleaner, before installing the caliper and brake pads.

Check the rotors for wear anytime the brakes are serviced. Rotors that are scored with sharp ridges, warped, worn irregularly or cracked should be replaced.

Brake rotors should always be resurfaced in pairs, with an equal amount of material removed from both sides of each rotor. **Table e** lists the dimensions for resurfacing rotors. Rotors which fail to meet these requirements should be replaced. Use a micrometer to measure the rotor thickness at eight to ten positions along the rotor's braking surface. Rotor runout should be measured using a dial indicator setup. Brake rotors can be resurfaced by most local automotive machine shops.

NOTE —
On original equipment rotors, the minimum thickness dimension can be found stamped into the rotor's hub. See Fig. 6-9.

Fig. 6-9. Rear rotor minimum dimension stamped on rotor hub (arrow).

Table e. Rear Rotor Reconditioning Specifications

Minimum thickness	
(after machining)	8.4 mm (0.331 in.)
Wear limit	
(minimum permissible thickness)	8.0 mm (0.315 in.)
Thickness tolerance	
(difference between	
two measurements)	0.02 mm (0.0008 in.)
Axial runout	
(maximum permissible)	
rotor removed	0.05 mm (0.002 in.)
rotor installed	0.20 mm (0.008 in.)

6.2 Calipers

Fig. 6-10 is an exploded view of the rear disc brake caliper. Brake fluid seepage or leaks around the brake caliper piston are usually the result of a damaged piston seal. A damaged piston seal is usually caused by corrosion, scoring, or pitting of the piston or caliper bore. The seal can be replaced separately, but a damaged piston will promptly destroy the new seal. To remedy a leaking caliper piston seal and avoid future problems, complete replacement of the caliper is recommended.

To remove rear caliper:

1. Raise the rear of the car and support it securely on jack stands. Remove the rear wheels.
2. If applicable, separate the brake pad wear sensor connector.
3. Loosen the hose fitting on the caliper. Remove the two caliper mounting bolts from the trailing arm. See Fig. 6-11.
4. Disconnect the brake hose from the caliper by spinning the caliper off the fitting. Drain the brake fluid into a container.

> ***CAUTION —***
> *Do not let brake fluid spill onto the brake pads or brake rotor surface. Cap the end of the brake line to prevent dirt and moisture from entering the brake system.*

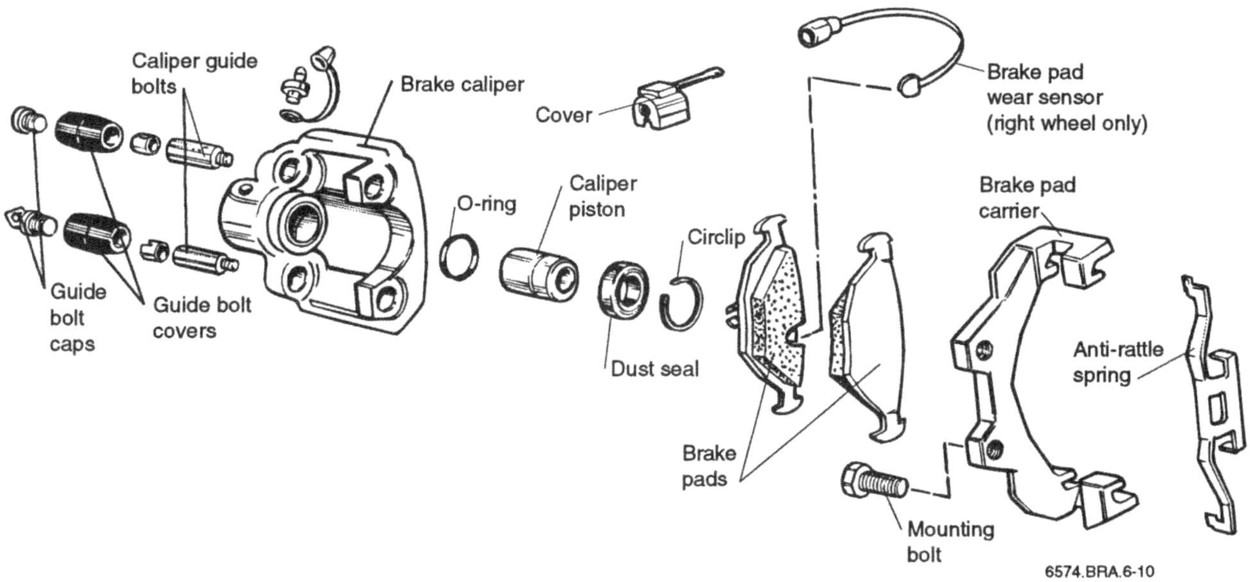

Fig. 6-10. Exploded view of rear brake caliper.

Fig. 6-11. Rear brake caliper mounting bolts (arrows).

Installation is the reverse of removal. Start all brake line fittings by hand to avoid cross-threading. Bleed the brakes as described in **4.2 Bleeding Brakes.**

Tightening torques	
• brake hose to caliper	13 to 16 Nm (10 to 12 ft-lb).
• rear brake caliper to trailing arm	60 to 67 Nm (44 to 49 ft-lb)

> ***WARNING —***
> *New brake pads require some break-in. Allow for slightly longer stopping distances for the first 100 to 150 miles of city driving, and avoid hard stops.*

6.3 Parking Brake

The parking brake is a small brake drum inside the rotor hub. The parking brake shoe assembly is fitted in the drum. See Fig. 6-1 above. The mechanical parking brake operates only on the rear wheels and is independent of the main hydraulic brake system.

Each of the two parking brake cables is connected to an actuator lever on each of the brake shoe assemblies. When the parking brake is applied, the brake shoes are forced out against the brake drum.

Adjusting Parking Brake

There is no automatic adjusting mechanism for the parking brake. The parking brake needs periodic adjustment to compensate for wear. The parking brake should also be adjusted if the cable, the brake rotor, or the parking brake shoes are replaced. Check to see that the parking brake functions properly whenever the cables have been detached for rear brake service.

To quickly check the parking brake adjustment, apply the parking brake while counting the number of clicks at the lever. If the parking brake lever can be pulled up further than the eighth detent (click), the parking brake should be adjusted.

To adjust parking brake:

1. Working from inside the passenger compartment, remove the parking brake lever trim and loosen the parking brake cable adjusting nuts. See Fig. 6-12.

Fig. 6-12. Parking brake cable adjusting nuts.

2. Raise the rear of the car and support it securely on jack stands. Remove one lug bolt from each rear wheel. Spin the wheel until the adjuster (star wheel) is visible. See Fig. 6-13.

3. Turn the adjuster until the brake shoes just contact the brake drum. Then back off the brake shoes so the wheel spins freely. (three to four teeth on the adjuster). See Fig. 6-14.

Fig. 6-13. Parking brake adjuster (arrow). Wheel and rotor removed for clarity.

Fig. 6-14. Screwdriver being used to turn parking brake adjuster.

NOTE —
- If the adjuster is difficult to move, the wheel and brake rotor may have to be removed. This will allow for lubrication and better leverage on the adjuster.
- Turn adjuster clockwise to move shoes out; turn adjuster counterclockwise to move shoes in.

4. With the rotor installed, set the parking brake three times to stretch and seat the cables. Then pull the lever up to the fifth detent (click) from the bottom.

5. Tighten the parking brake cable adjusting nuts by equal amounts until the rear brake shoes contact the brake drum. Check that both wheels rotate with the same amount of friction.

6. Release the parking brake and check that both rear wheels rotate freely. Readjust if necessary.

7. Loosely install the lug bolt. Lower the car and torque the lug bolts.

Tightening torque
• wheel lug bolts . . . . 100±10 Nm (74±7 ft-lb).

Replacing Parking Brake Shoes

Always replace parking brake shoes in complete sets. Make sure the parking brake is released before beginning. Remove the rear rotor as described in **6.1 Reconditioning Rear Disc Brakes**.

Using brake spring pliers, disconnect the lower return spring from the forward brake shoe. Push the spring retainer in and rotate it 90° to remove it. See Fig. 6-15. Repeat the step for the other retainer. Spread the shoes apart at the top and lift the shoes out from below.

Fig. 6-15. Rear parking brake shoe retainer (arrow) being removed.

Installation is the reverse of removal. Inspect the return springs and replace any that are damaged. Adjust the parking brake as described above in **Adjusting Parking Brake.**

19

NOTE —
New parking brake linings require a break-in procedure. Begin by making five full stops from a speed of 30 mph using the parking brake lever. Allow the brakes to cool, then repeat the procedure. This break-in procedure ensures proper seating of the parking brakes.

Replacing Parking Brake Actuator

Remove parking brake shoes as described above. Remove the parking brake actuator by pulling the outer cam from the actuator off towards the rear. See Fig. 6-16. Push out the pivot pin. Disconnect the inner cam from the parking brake cable

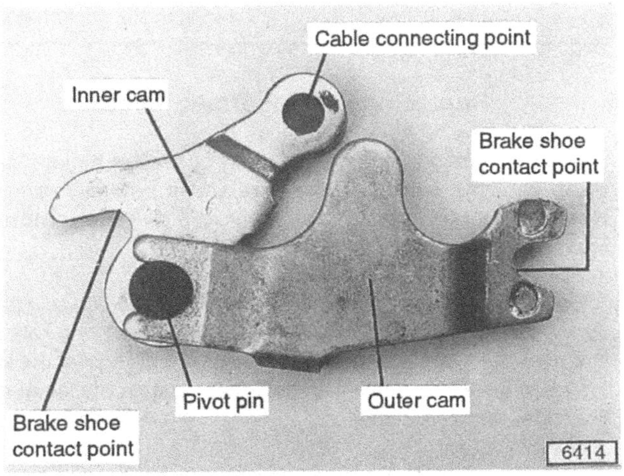

Fig. 6-16. Parking brake actuator components used on cars with rear disc brakes.

Installation is the reverse of removal. Lightly coat the moving parts and the pins with molybdenum disulfide grease (Molykote G paste or equivalent). Adjust the parking brake as described above in **Adjusting Parking Brake**.

CAUTION —
Apply grease sparingly. Do not allow grease to contact the brake linings.

Replacing Parking Brake Cable

The cables can be replaced separately.

To replace parking brake cables:

1. Remove the adjusting nut that holds the cable to the parking brake lever.

2. Remove the parking brake shoes and the parking brake actuator as described above in **Replacing Parking Brake Shoes** and **Replacing Parking Brake Actuator**.

3. Unhook the cable from the guides and hangers that secure it to the trailing arm and the body. Pull the old cable out of the body, working from the wheel end of the cable.

Installation is the reverse of removal. Lubricate the new cable with multipurpose grease before installing it. Adjust the parking brake as described under **Adjusting Parking Brake**.

7. TECHNICAL DATA

I. Tolerances, Wear Limits, and Settings

Brake rotor, front	
528e, 533i models	
thickness after machining (minimum)	20.4 mm (0.803 in.)
wear limit (minimum thickness)	20.0 mm (0.787 in.)
axial runout (maximum permissible)	
rotor installed	0.20 mm (0.008 in.)
rotor removed	0.05 mm (0.002 in.)
thickness tolerance (maximum permissible)	0.02 mm (0.0008 in.)
535i models	
thickness after machining (minimum)	23.4 mm (0.921 in.)
wear limit (minimum thickness)	23 mm (0.905 in.)
axial runout (maximum permissible)	
rotor installed	0.20 mm (0.008 in.)
rotor removed	0.05 mm (0.002 in.)
thickness tolerance (maximum permissible)	0.02 mm (0.0008 in.)
Brake rotor, rear	
thickness after machining (minimum)	8.4 mm (0.331 in.)
wear limit (minimum thickness)	8.0 mm (0.315 in.)
axial runout (maximum permissible)	
rotor installed	0.20 mm (0.008 in.)
rotor removed	0.05 mm (0.002 in.)
thickness tolerance (maximum permissible)	0.02 mm (0.0008 in.)

II. Tightening Torques

Front brake caliper to steering knuckle (bolt)	110–123 Nm (81–91 ft-lb)
Front brake caliper to brake pad carrier (guide bolt)	30–35 Nm (22–26 ft-lb)
Rear brake caliper to brake pad carrier (guide bolt)	30–35 Nm (22–26 ft-lb)
Rear brake pad carrier to trailing arm (bolt)	60–67 Nm (44–49 ft-lb)
Bleeder valve to caliper	
7-mm	3.5–5.0 Nm (31–44 in-lb.)
9-mm	4.0–6.0 Nm (35–53 in-lb.)
Brake line unions (union nuts)	10–15 Nm (7–11 ft-lb)
Brake hose unions (union nuts)	13–16 Nm (10–12 ft-lb)
Master cylinder to brake booster	
528e models (nut)	24–29 Nm (18–21 ft-lb)
533i, 535i models	26-32 Nm (19–24 ft-lb)
Brake booster to pedal base	22–24 Nm (16–18 ft-lb)
Wheel to rotor (lug bolt)	100±10 Nm (74±7 ft-lb)
Brake booster hydraulic line fittings (533i, 535i models)	
power flow regulator to fluid reservoir	15–18 Nm (11–13 ft-lb)
power flow regulator to steering gearbox	45–50 Nm (33–37 ft-lb)
power flow regulator to power steering pump	30–40 Nm (22–30 ft-lb)
power flow regulator to brake booster	25–35 Nm (18–26 ft-lb)
Brake booster pressure switches to power flow regulator (533i, 535i models)	15–18 Nm (11–13 ft-lb)
Brake booster accumulator to power flow regulator (533i and 535i models)	43–50 Nm (32–37 ft-lb)

Section 20

BODY AND INTERIOR

Contents

TABLES

INTRODUCTION

The BMW 528e, 533i, and 535i have unit construction steel bodies that are exceptionally strong and light. Because very few screws and bolts are used in assembling the body, fewer rattles are likely to develop. The ride is quieted further by the application of sound-dampening material to the floor plates and the body panels.

During manufacture, the various body panels, subassemblies, and a number of smaller pressed-steel panels and plates are joined by electric welding. Although all body panels are available as replacement parts, many of these replacement panels must be butt-welded to the body after the damaged panels have been cut away. This work should be left to an experienced body repair technician.

The front fenders, however, are bolted to the main body structure and can be easily replaced. The hood, the grille, the doors, and the trunk lid are also removable. These bolt-on components are easily replaced even if you have little or no knowledge of auto body repair.

Repairs to the heating and air conditioning system are covered in **HEATING AND AIR CONDITIONING**. Electrical repairs to the instrument cluster, including the gauges and lights, are covered in **ELECTRICAL SYSTEM**. Care of the body, trim, upholstery, and windows is described in **LUBRICATION AND MAINTENANCE**.

Some of the cars covered by this manual may be equipped with a Supplemental Restraint System (SRS) that automatically deploys an airbag. The airbag unit uses a pyrotechnical device to electrically ignite a powerful gas. On cars so equipped, any work involving the steering wheel should only be performed by an authorized BMW dealer. Performing repairs without disarming the SRS may cause serious personal injury. SRS components should not be subjected to temperatures in excess of 212° F (100°C) or excessive vibration.

1. GENERAL DESCRIPTION

Body

The body is of welded, unitized construction, meaning it does not have a separate frame. This design forms a very rigid passenger compartment, with large crumple zones in the front and rear for energy absorption in the event of a collision.

For corrosion protection, all steel is treated with a multi-layer finish. The body seams are then sealed using a PVC compound. The front fenders are flanged to allow trapped moisture to evaporate. The body is undercoated and all interior cavities are flooded with a rust preventative sealant. In areas subject to corrosion, galvanized metal is used.

WARNING —
Welding and grinding galvanized steel produces a poisonous gas containing zinc oxide. This type of work should always be carried out by experienced body technicians having the proper respiratory equipment and work area.

Seats and Interior

The front seats are mounted to the floor. Rear seats are bolted to the body and are easily removed for access to the rear seat belt mountings. The seat belts are typical three-point belts that cross the hips and the shoulder. All interior trim is easily removed using ordinary hand tools.

Instrument Cluster, Center Console, and Instrument Panel

The padded instrument panel houses the instrument cluster and the ventilation and heating system. It is fastened to the body and can be removed using ordinary hand tools. The instrument cluster is removable as a unit without removing the instrument panel.

2. MAINTENANCE

BMW specifies the maintenance steps below to be carried out at particular time or mileage intervals for proper maintenance of body and interior components. A number in bold type indicates that the procedure is covered in this section, under that numbered heading. Information on other body and interior maintenance and on the prescribed maintenance intervals can be found in **LUBRICATION AND MAINTENANCE**.

1. Lubricating door hinges and striker plates
2. Checking headlight and driving light aim. **6.1**
3. Cleaning and lubricating sunroof slide rails. **7**
4. Checking seat belts. **4.5**
5. Lubricating accelerator and throttle linkage
6. Have an authorized BMW dealer perform the annual anti-corrosion inspection

3. TROUBLESHOOTING

Because the components and assemblies covered in this section vary widely, specific troubleshooting is covered with the repair information in this section. For electrical troubleshooting and electrical tests, see **ELECTRICAL SYSTEM**. For a more general description of troubleshooting, see **FUNDAMENTALS** at the beginning of the manual.

4. INTERIOR

This section covers the removal and replacement of interior components. Repairs to interior electrical components are described in **ELECTRICAL SYSTEM**. Information on interior door trim, door locks, window controls, and other mechanical parts of the door are covered below under **5. Doors**.

4.1 Instrument Cluster, Center Console, and Instrument Panel

This heading covers the removal and installation of instrument panel components and related trim.

Removing and Installing Instrument Cluster and Instrument Panel Trim

The instrument cluster and other instrument panel trim can be removed using ordinary hand tools and without removing the instrument panel.

> ***WARNING —***
> *Some of the cars covered by this manual may be equipped with a Supplemental Restraint System (SRS) that automatically deploys an airbag. The airbag unit uses a pyrotechnical device to electrically ignite a powerful gas. On cars so equipped, any work involving the steering wheel should only be performed by an authorized BMW dealer. Performing repairs without disarming the SRS may cause serious personal injury.*

To remove instrument cluster:

1. Disconnect the negative (–) battery cable.

2. Remove the upper instrument cluster retaining screws. See Fig. 4-1.

3. Carefully pull the cluster away from the instrument panel just enough to reach the electrical connectors, then disconnect them. Release the three large connectors (23-point) by sliding the connector lock out. See Fig. 4-2.

Installation is the reverse of removal. Be certain to test all instrument cluster functions after reassembly to be sure all connectors have been reconnected.

To remove center console:

1. Disconnect the negative (–) battery cable.

2. Remove the screws securing the lower trim panel below the steering column and remove the panel.

3. Remove the ashtray and cigar lighter.

Fig. 4-1. Instrument cluster retaining screws (arrows) to be removed from instrument cluster.

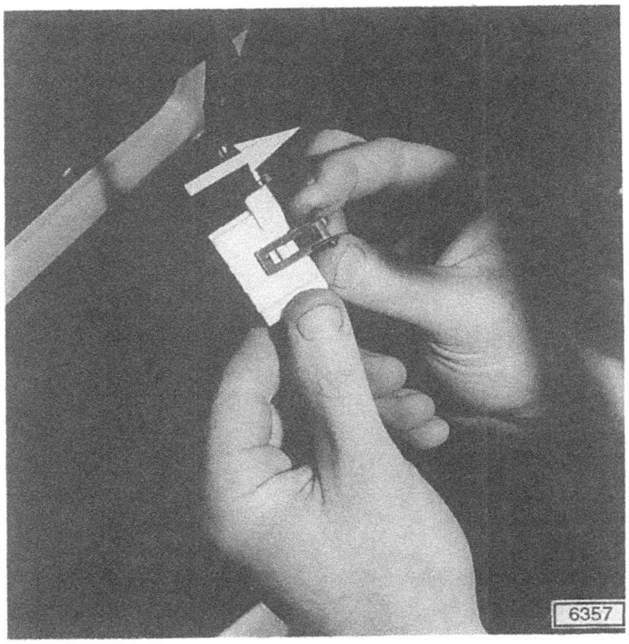

Fig. 4-2. Large instrument cluster harness connectors (23-point) are locked in place. Slide out lock (arrow) to remove.

4. On cars with manual transmission, carefully pry off the shifter boot.

5. On cars with automatic transmission, pry out the selector lever trim plate, remove the two screws shown in Fig. 4-3, and then remove the shift lever trim panel.

Fig. 4-3. Automatic transmission shift lever trim retaining screws (arrows).

6. Remove the screws on each side of the center console. See Fig. 4-4.

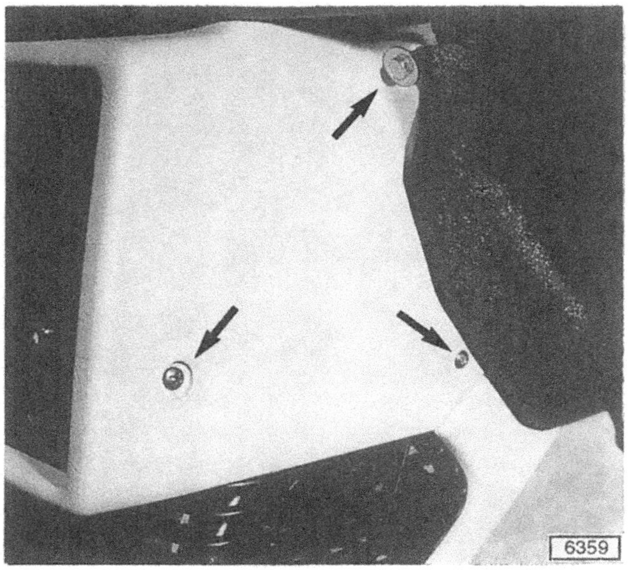

Fig. 4-4. Screws securing right side of center console (arrows). Remove screws from both sides.

7. Carefully pry up all of the window switches and disconnect their harness connectors.

8. Remove the other console mounting screws. There are two below dummy plastic covers behind the handbrake lever, one under the rear of the handbrake lever brush, and two in the ashtray opening. Remove the handbrake trim by sliding it to the rear and lifting it out. Disconnect any electrical connectors. Remove the nut directly below the shift lever and unplug the wires to the cigar lighter and cigar light. See Fig. 4-5.

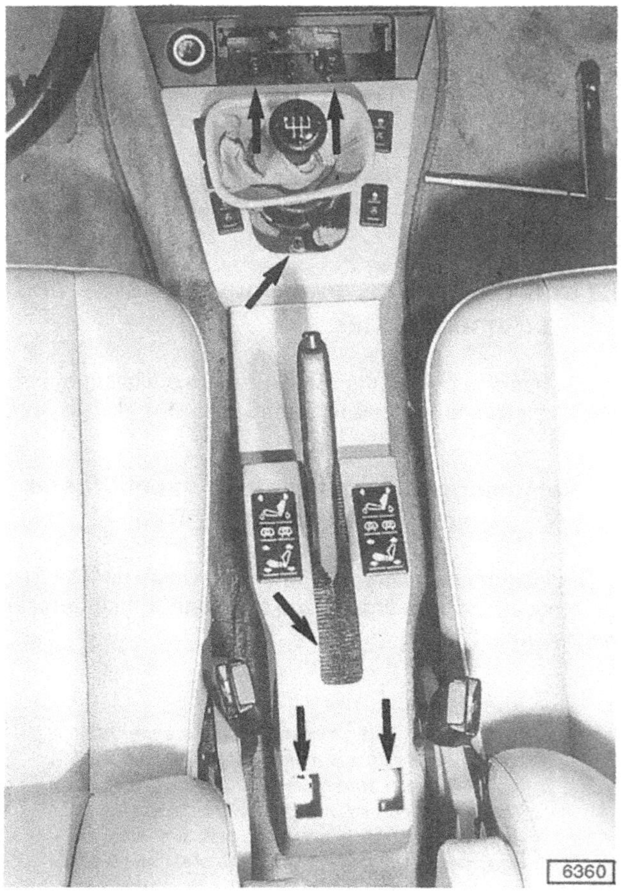

Fig. 4-5. Arrows show screws to be removed through access cutouts, under handbrake brush and under ashtray. Also, remove nut behind the gear shift.

9. Slide back the console to reveal two screws above the radio. Remove these screws and carefully push the radio through the opening.

10. Carefully slide the console out of the way. Mark and disconnect any wires or cables.

Installation is the reverse of removal. Be certain to test all console functions after reassembly to be sure all connectors have been reconnected.

To remove glove box:

1. Open the glove box and disconnect the retaining straps by pushing out the securing pins. See Fig. 4-6.

2. Close the glove box. Working from below the glove box, remove the screws that hold the hinge to the body.

3. Lower the glove box slowly while disconnecting the light wiring from the side of the glove box.

Fig. 4-6. Glove box retaining strap with pin partially removed (arrow). Disconnect straps from both sides.

Installation is the reverse of removal. It may be necessary to loosen the striker plate screws and adjust the plate up or down so that the glove box latches correctly.

Removing and Installing Instrument Panel

The instrument panel is held to the body by six main mounting bolts, four on either side and two underneath. It may be helpful to label all wires prior to removal to ensure correct reinstallation.

> ***WARNING —***
> *Some of the cars covered by this manual may be equipped with a Supplemental Restraint System (SRS) that automatically deploys an airbag. The airbag unit uses a pyrotechnical device to electrically ignite a powerful gas. On cars so equipped, any work involving the steering wheel should only be performed by an authorized BMW dealer. Performing repairs without disarming the SRS may cause serious personal injury.*

To remove and install instrument panel:

1. Disconnect the negative (–) battery cable.

2. Remove the instrument cluster, instrument panel trim, shift console, and glove box as described above. Additional trim at the top and left side of the passenger footwell must also be removed.

3. Disconnect the wiring from the instrument panel switches. It is necessary to pry out the switches first. Remove the radio as described in **4.2 Radio and Antenna**.

4. Remove the wiring and cables from the heater controls as described in **HEATING AND AIR CONDITIONING**, then remove the controls.

5. Remove the interior A-pillar trim above the ends of the instrument panel by prying back the edge guard.

6. Remove the steering wheel as described in **Steering and Wheel Alignment**.

7. Locate and undo all plastic ties that hold the instrument panel and radio wiring harnesses to the instrument panel.

8. Remove the instrument panel mounting bolts. Make one last check for any installed bolts or wires that may still be connected, then carefully pull out the instrument panel.

Installation is the reverse of removal. Make sure the ventilation air ducts are fully seated in the instrument panel before installation, and that the cruise control lever engages both tabs when installed on the steering column. Be certain to test all console functions after reassembly to be sure all connectors have been reconnected.

4.2 Radio and Antenna

Two types of factory-installed radios are used. One type is held in by spring clips. The special wire tools needed for removing this radio are available from an authorized BMW dealer or from a specialty car stereo shop. The other type of radio is an anti-theft unit held in by securing clips which are moved in and out by internal-hex-head screws. See Fig. 4-7. Whenever removing or installing the radio, always disconnect the negative (–) battery cable.

When troubleshooting a problem with the sound system, always check fuse no.5 (8 amp) and fuse no.12 (8 amp). When troubleshooting a problem with the power antenna, always check fuse no.4 (25 amp). All are located in the main fuse/relay panel.

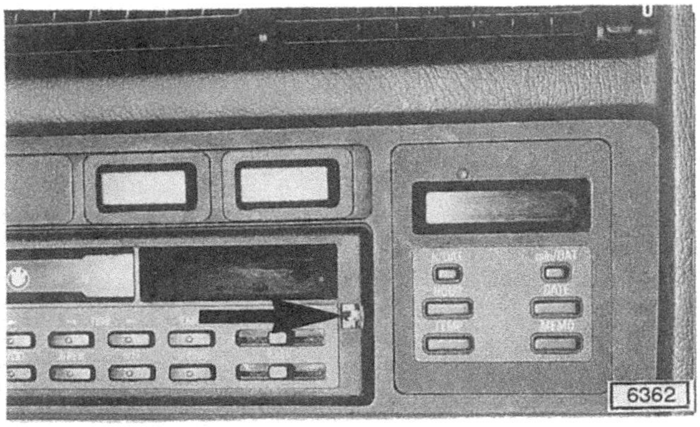

Fig. 4-7. Access hole and concealing flap (arrow) for internal-hex-head screws of BMW anti-theft radios. The screws move a retaining clip in and out.

NOTE —
Anti-theft radios have two replaceable fuses in the circuit. One is located at the back of the radio chassis and the other is located on the power amplifier in the luggage compartment. Always check these fuses as a first step when experiencing trouble.

Anti-theft radio and amplifier fuse specifications
• radio fuse ....7 amp (BMW Part No. 88 88 2 600 000)
• amplifier fuse....10 amp (BMW Part No. 61 13 1 380 468)

CAUTION —
On BMW cars equipped with anti-theft radios, make sure you know the correct radio activation code before disconnecting the battery. If the wrong code is entered into the radio when power is restored, the radio may lock up, even if the correct code is then entered. For more information, see your owner's manual.

NOTE —
If the radio does not take the anti-theft code, check that the battery is fully charged. Low battery voltage can cause the radio to not recognize the code. Once the code has been entered three times unsuccessfully, the key will have to be left in the on position for one hour after the battery is charged. This will reset the radio so that the correct code can be entered again.

Power Antenna and Mast

Failure of the mast to operate may be due to an electrical fault, to dirt or corrosion in the power assembly, or to a faulty, dirty or bent mast. Replacement power antenna masts are available from an authorized BMW dealer. The entire power antenna assembly does not have to be replaced due to a damaged mast. On 1984 and later antennas, the printed circuit board (control unit) is also available as a replacement part.

BMW recommends monthly cleaning of the antenna mast. Clean and lubricate the mast in its up position.

NOTE —
• The power antenna mast can be cleaned using special antenna mast cleaning pads available from an authorized BMW dealer parts department (BMW Part No. 65 12 9 056 514). As an alternative, a lubricating product such as WD-40® will also work well.

• Two different replacement masts are available, depending on the power antenna assembly installed. On power antennas with a plastic housing, use replacement mast BMW Part No. 88 88 0 825 948. On antennas with a metal housing and a plastic cover, use replacement mast BMW Part No. 88 88 0 882 090. The power antenna assembly is located in the trunk. See Fig. 4-8.

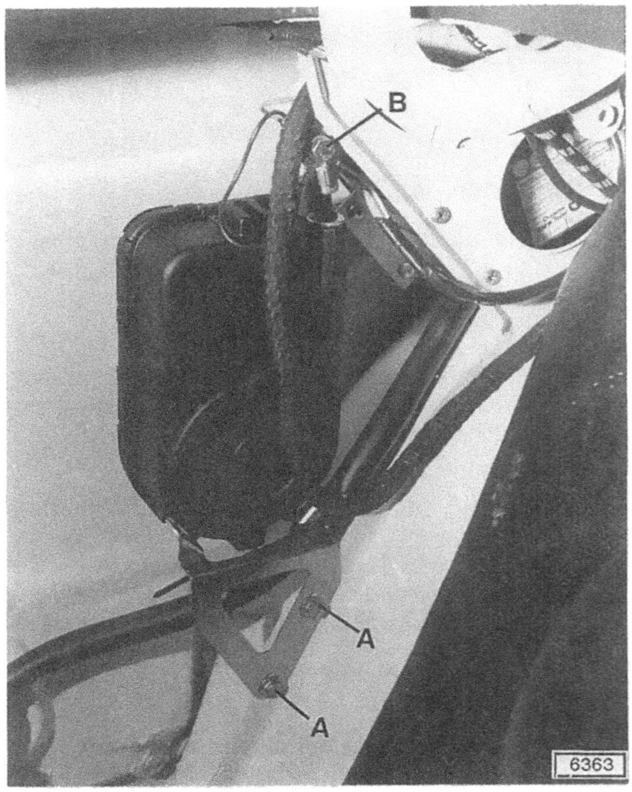

Fig. 4-8. Power antenna with plastic housing in trunk. Antenna with metal housing is similar. Housing mounting bolts are at **A**, ground strap is at **B**.

To replace power antenna mast:

1. Turn on the radio to extend the mast as far as possible, then loosen the mast clamping nut shown in Fig. 4-9.

2. Firmly pull on the antenna mast until the mast and the plastic actuating cable are free of the guide sleeve.

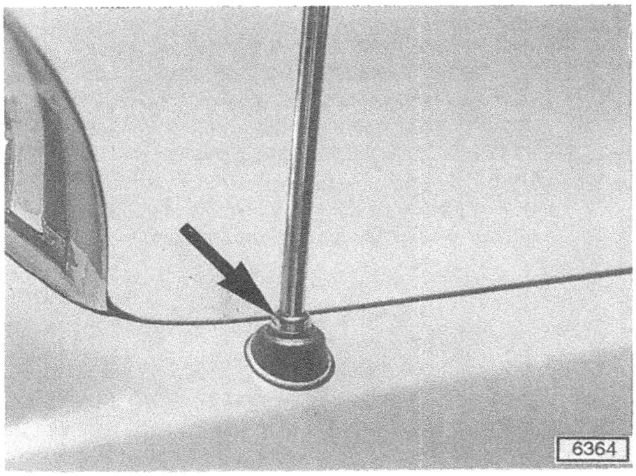

Fig. 4-9. Antenna mast clamping nut (arrow).

3. Slide the plastic tip of the new antenna mast into the guide sleeve until it is felt to contact the motor guide roller.

4. Turn off the radio so that the plastic cable is retracted by the motor, and guide the mast into the sleeve at the same time. Tighten the clamping nut when finished and recheck antenna operation. Finally, lubricate the mast.

4.3 Interior Lights

A light failure may be caused by a blown fuse, especially if more than one bulb is out. Check and, if necessary, replace fuses as described in **ELECTRICAL SYSTEM**. Dashboard light replacement is also covered there. **Table a** lists interior light bulb specifications.

Table a. Interior Light Bulb Specifications

Light	Bulb
Interior dome	12 volt, 10 watt
Luggage compartment	12 volt, 5 watt
Glove compartment	12 volt, 4 watt
Center stop light	12 volt, 21 watt

Replacing Interior Light Bulbs

Always turn off the ignition before replacing a bulb. Wipe off all finger prints after installing a bulb. Fingerprints can cause hot spots or evaporate when the glass gets hot and dim the light or reflector.

To replace the interior dome light bulbs, carefully pry the assembly away and withdraw the assembly. See Fig. 4-10. Press the assembly back into position after installing the new bulbs.

Fig. 4-10. Interior dome light assembly in headliner removed. Assembly contains two bulbs.

To replace the luggage compartment light bulb, use a small flat-blade screwdriver to pry the light assembly out. See Fig. 4-11. Install the bulb and press the assembly back into position.

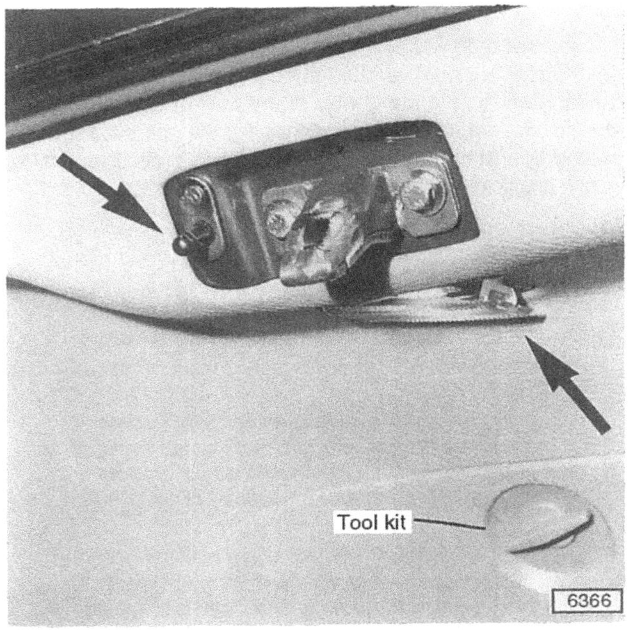

Fig. 4-11. Luggage compartment light assembly and switch (arrows).

20

To replace the glove box bulb, open the glove box door and locate the bulb directly above the light switch. The bulb is held in a spring loaded bayonet socket. Push and turn the bulb to remove.

For 1986 and later cars, the center stop light is reached from inside the car. Pull the assembly cover up at an angle, then compress the tabs on either side of the bulb holder and pull out the holder. Replace the bulb and press the holder back into place until the tabs lock. See Fig. 4-12.

Fig. 4-12. Center stop light for 1986 and later models is reached from inside passenger compartment.

4.4 Seat Belts

The seat belts installed on the cars covered by this manual are combination lap-shoulder belts (three-point), except for the lap seat belt in the center of the rear seat. The seat belts should be periodically inspected for webbing defects and proper operation. If the seats belts are removed for any reason, it is extremely important to reinstall them correctly.

WARNING —
- *For maximum protection from injury, seat belts should be replaced as a set (including all hardware) if they are subjected to occupant loading in a collision.*

- *Do not install seat belts or seat belt hardware purchased from an auto-recycler or salvage yard. These parts may have been subjected to occupant loading from a collision.*

- *Seat belt assemblies must remain together as a matched set. Improper latching of the seat belt may occur if the components have been mixed from other sets or other cars.*

- *For maximum protection from injury, do not interchange buckle and retractor assemblies with those designated for other seating positions or other models.*

- *Do not bleach or dye seat belt webbing. Webbing that is severely faded or redyed will not meet the strength requirements and must be replaced.*

WARNING —
- *Only BMW original equipment mounting bolts and hardware should be used with seat belt assemblies. Never substitute after-market mounting hardware. Check with an authorized BMW dealer for the latest parts information.*

- *Always consult the BMW owner's manual for proper seat belt usage and applications.*

Inspecting Seat Belts

When inspecting belt webbing, replace belts with broken or pulled threads, cut loops at the belt edge, bowed, creased, or melted webbing, faded areas, or cuts. Pull the belt out fully and let it retract. If it does not move smoothly in either direction, check for dirt, grease, or gum on the webbing. If the belt cannot be cleaned using only a mild soap solution recommended for cleaning upholstery or carpets, replace the belt.

Replace the belt if the buckle cover is cracked, if the push button is loose, or if the buckle does not lock securely. Check all mounting points. The mounting bolts must be tight, yet allow the hardware to swivel freely. Clean any corrosion away from the anchoring points, and replace any corroded hardware.

The belt mechanism should lock up when the belt is tugged out quickly, or when the car is stopped quickly from a speed approximately twice that of walking speed. Replace any belt that does not lock up as described.

Installing or Replacing Seat Belts

All seat belt mounting bolts should be installed and correctly torqued.

Tightening torque
• seat belt mounting bolt. . . . . . . 43 to 48 Nm (32 to 35 ft-lb)

Front seat belt buckles and lower belt anchors are bolted to the seat. The lower belt is attached to a rail that is bolted to the body. The belt retractor is bolted to the body behind the B-pillar (center pillar) trim.

To remove the B-pillar trim panel, first remove the fasteners shown in Fig. 4-13 and then pull the panel out. Guide the belt through the trim panel opening.

NOTE —
The height of the upper anchor point for front seat belts can be changed. A special kit is available from BMW that permits the anchor point to be moved lower or higher. This kit should only be installed by an authorized BMW dealer.

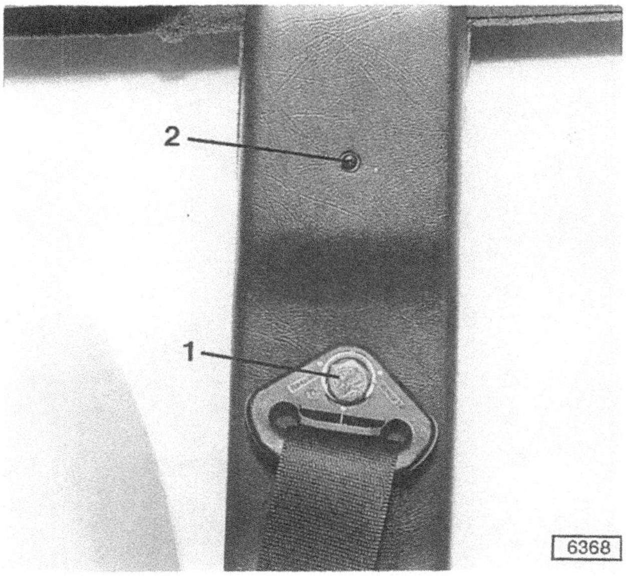

Fig. 4-13. B-pillar trim panel showing upper seat-belt anchor bolt (**1**) and panel retaining screw (**2**).

Rear seat belt lower anchor points are underneath the rear seat, as shown in Fig. 4-14. The belt retractor is mounted in the trunk. For access, first remove the rear seat and backrest as described in **4.6 Rear Seats.**

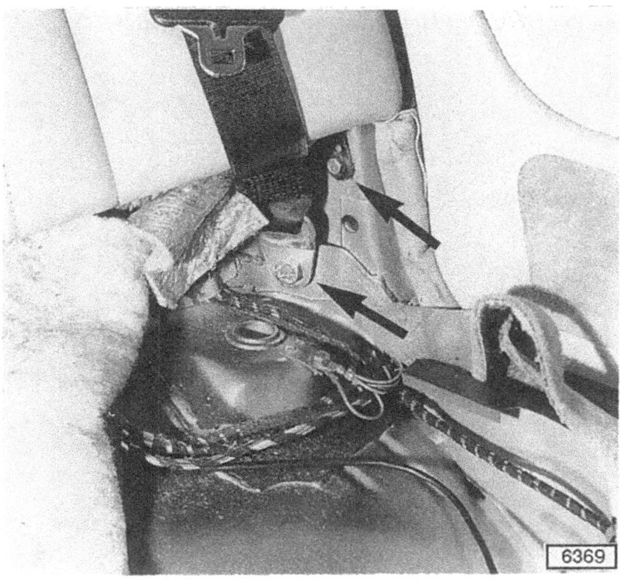

Fig. 4-14. Rear seat backrest and rear seat belt anchor bolts (arrows).

4.5 Front Seats

Front seats are held to the floor by four bolts, one at each corner. To remove a seat, pry off the seat belt trim cover and remove the seat belt mounting bolt from the side of the seat. Move the seat all the way to the rear and remove the front bolts, then move the seat forward and remove the two rear mounting bolts. See Fig. 4-15. On power seats with memory, remove the trim cover from the front lower part of the seat. Then on all power seats, disconnect the main harness connector (white) and the smaller two-point connector. Disconnect any other wiring connectors and remove the seat from the car. Installation is the reverse of removal.

Tightening torque
• front seat rails to floor (M10) . . . . 47 Nm (35 ft-lb)

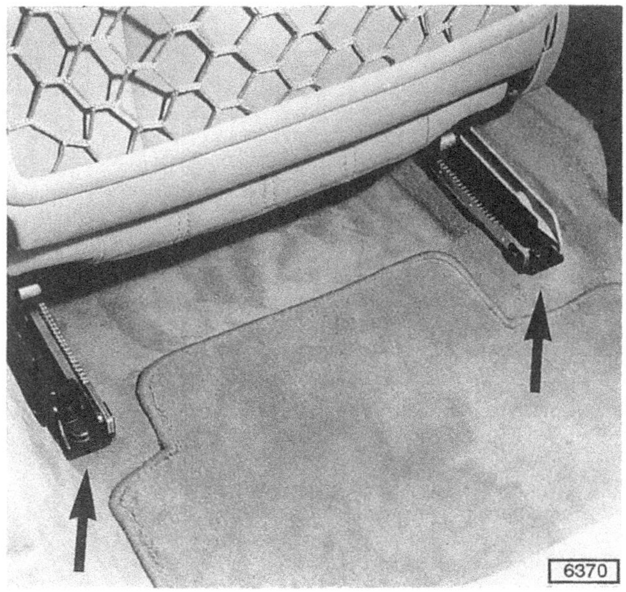

Fig. 4-15. Front-seat mounting bolts at rear of seat rails (arrows). Also remove two bolts at front of seat rails.

NOTE —
On models with power seats with memory, do not mix up the color-coded electrical connectors for the power motors. If necessary, use Fig. 4-16 as a guide during installation.

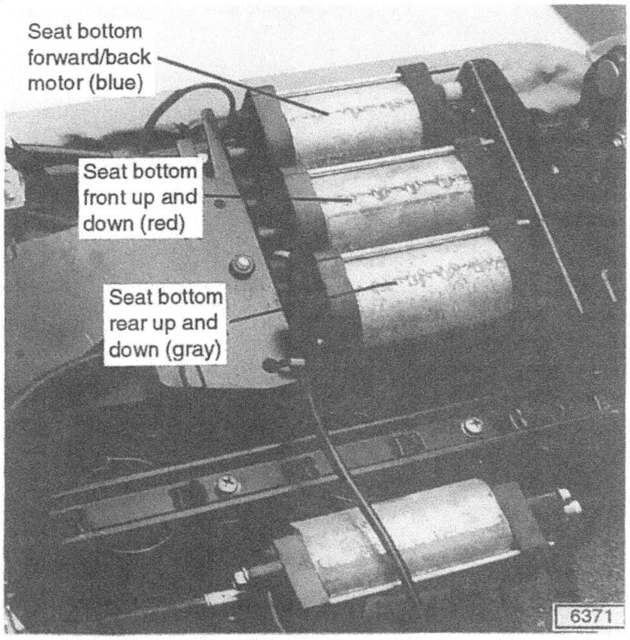

Fig. 4-16. Power seat with memory electrical motors and color-coded harness connectors.

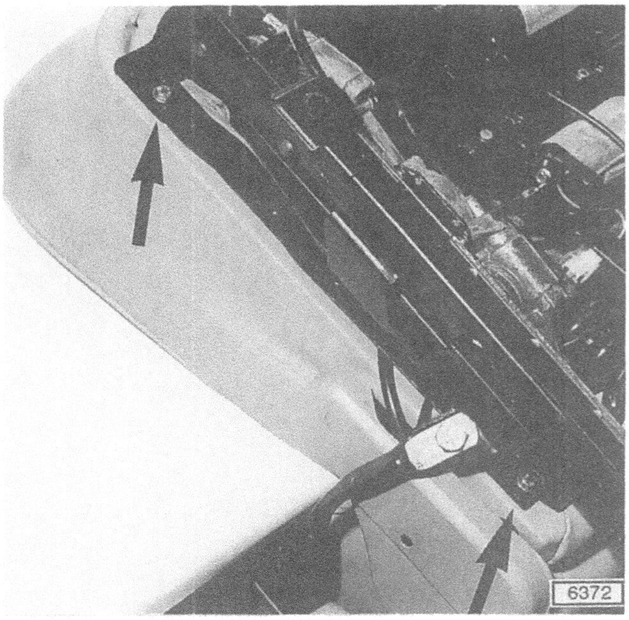

Fig. 4-17. Front seat to seat rail mounting bolts (arrows). Remove these bolts if power seat cannot be moved to reach floor bolts.

WARNING —

- *Only BMW original equipment mounting bolts and hardware should be used with seat assemblies. Never substitute after-market mounting bolts or other pieces. Check with an authorized BMW dealer for the latest correct parts information.*

- *For maximum protection from injury, front seats should be carefully inspected for structural damage (including all hardware) if they have been subjected to occupant loading in a collision.*

If the seat mounting bolts cannot be accessed owing to an electrical failure of the power seat, the seat can be removed from the seat rails so that the seat motors and mechanical components can be repaired.

To remove the seat from the seat rails, remove the four mounting bolts shown in Fig. 4-17. Disconnect the wiring and lift the seat out of the car.

4.6 Rear Seats

The rear seat cushion is held in with a bolt located in the center above the transmission tunnel. See Fig. 4-18.

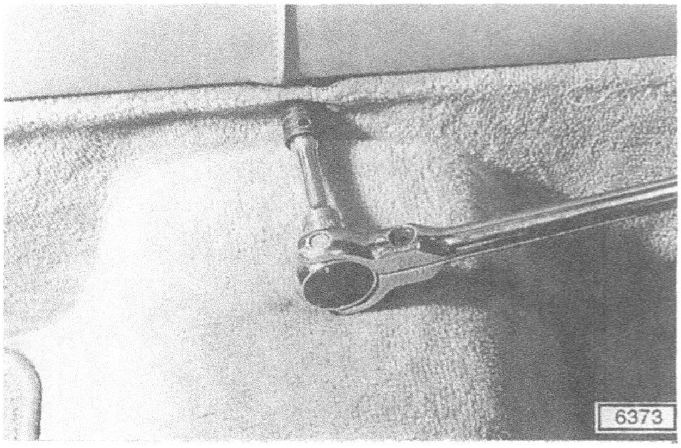

Fig. 4-18. Rear seat cushion mounting bolt (arrow) being removed.

The rear seat backrest is held in place by two bolts and clips. Remove the bolts, then pull the backrest straight up and off the clips. See Fig. 4-14 above.

5. DOORS

This heading covers removal and alignment of the doors, as well as removal and installation of internal door components and door trim.

Removing and Installing Doors

Doors are mounted to the body with hinges. The front half of each hinge is welded to the car body, while the rear half is bolted to the inside of the door. Fig. 5-1 shows a front door with the trim panel removed.

The door can be easily removed once the door trim panel is removed. Before loosening the hinge bolts, scribe a mark showing the outline of the bolt on the hinge so that door can be realigned in its original position.

When installing a new replacement door, the door parts from the old door will have to be transferred to the new door. Replacement doors will also probably require new door shims to align the door.

NOTE —
- If a sharp cracking noise is heard when operating the door, the problem is most likely due to a worn check rod or check rod pin. Inspect the check rod components and replace any faulty parts. The pin is available as a separate replacement part.

- If a scraping noise is heard when operating the door, spray the door check rod with a motorcycle chain lubricant. Spray in both directions and then open and close the door several times.

- The body is painted at the factory after assembly. Realignment of body panels may expose unpainted metal. To prevent rust, paint all exposed metal as soon as possible.

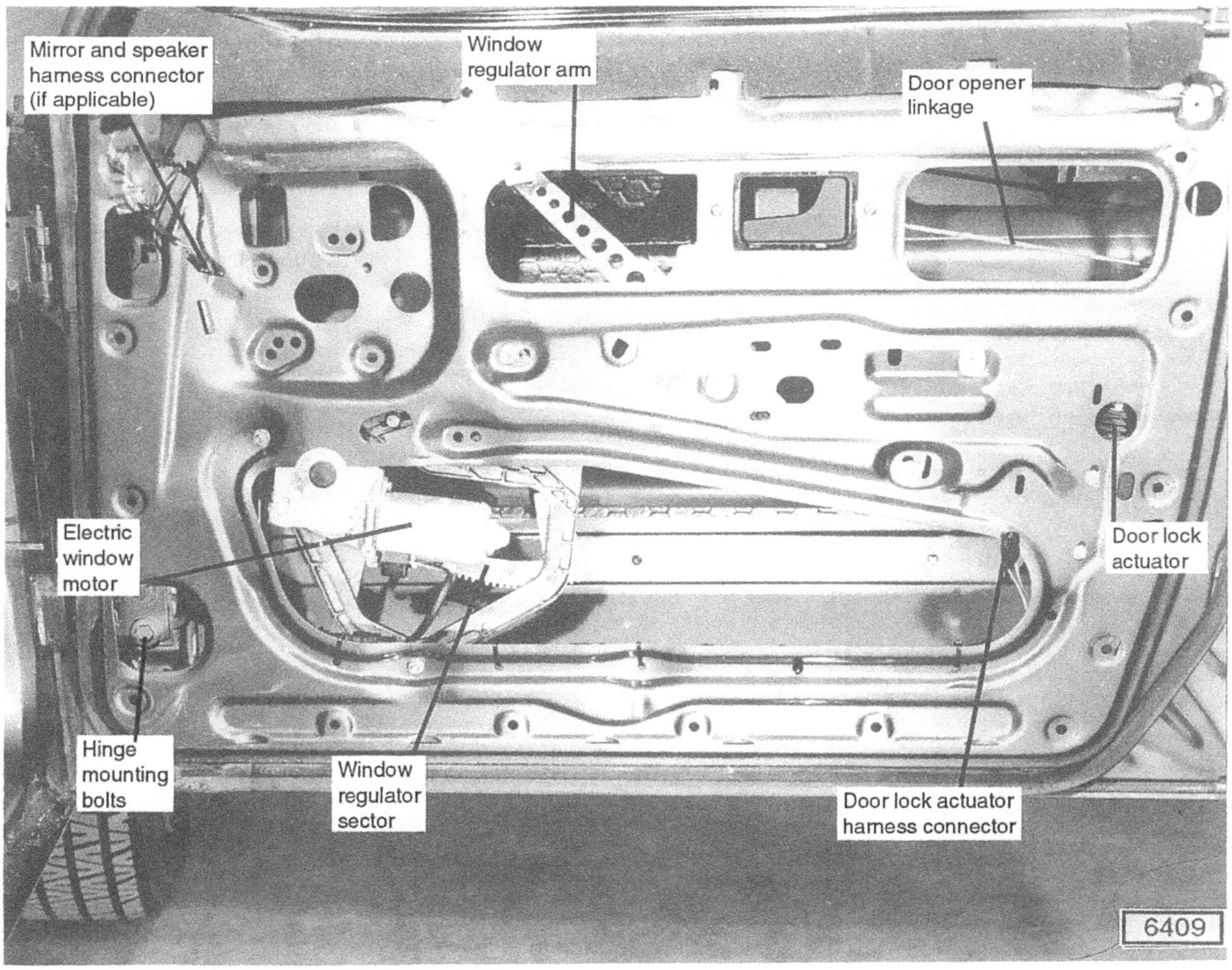

Fig. 5-1. Front door with trim panel removed.

20

To remove doors:

1. Remove the interior trim panel. See **5.1 Door Assembly**.

2. Pull back the rubber protector on the door check rod and the door wiring. Disconnect the check rod by removing the retaining circlip and driving the pin up. See Fig. 5-2.

Fig. 5-2. Door check rod disconnected from body.

3. Working inside the door, disconnect the following door wiring harness connectors: door lock actuator, driver's door lock microswitch, power window motor, mirror and speaker. Carefully pull the wire harnesses out of the door.

4. Unscrew and remove the upper and lower hinge bolts while supporting the door. See Fig. 5-1 above.

5. Slide the door off the hinges, being careful not to scratch the paint.

NOTE —
An easy way to remove, install, and adjust doors when help is not available is to use a floor jack to support the door. Be sure to put a protective covering on the jack cradle to prevent damaging the door's finish.

Installation is the reverse of removal. If necessary, adjust the door as described below.

Adjusting Doors

Misaligned doors can cause wind noise and paint damage due to abrasion by road debris.

Check the door's depth compared to the adjacent body panels, and the door's fore-aft and height position compared to the door opening of the car body. Adjustments can be made at both ends of the door. Some adjustments may require shims between the hinge and the door.

NOTE —
For adjustment of the door, the door's interior trim panel must first be removed to access the hinge bolts. See **5.1 Door Assembly**.

If the door's fore-aft position requires adjustment, or if the door is crooked in its opening, the door can be easily realigned by simply loosening the hinge bolts and repositioning the door. If the front edge depth requires alignment, a combination of shims will be needed. See Fig. 5-3. By varying the number of shims between the hinge and the door, door alignment is changed.

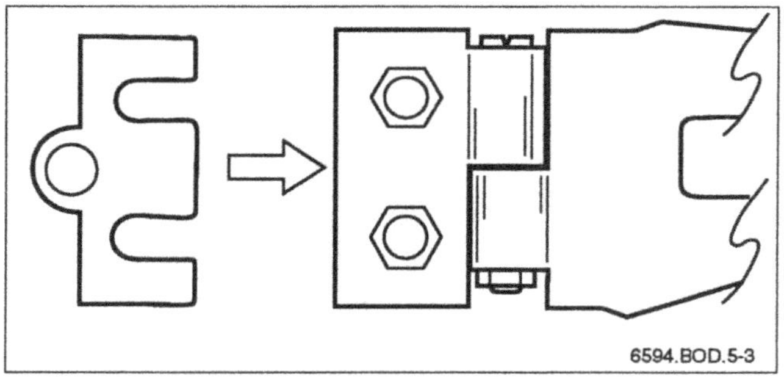

Fig. 5-3. Door hinge showing location of shims between hinge and door.

NOTE —
Adjustment shims are available only in 1.0 mm (0.04 in.) thicknesses through an authorized BMW dealer (BMW Part No. 41 51 1 873 064).

Adjust the door's rear edge depth by changing the position of the door striker plate. See Fig. 5-4. It may be necessary to use an impact driver to loosen the screws. Make small adjustments until the alignment is correct. Make sure that the striker plate is adjusted vertically so that the door lock engages the striker equally on the top and bottom when the door is closed.

NOTE —
- The fore-aft position of the door is correct when there is an equal amount of space between the door's edges and its adjacent panels. The door height is correct when the stamped creases in the upper and lower part of the door are aligned with the same creases in the adjacent panels.

- Door depth is correct when the door's front edge is slightly deeper than its adjacent panel and its rear edge is slightly higher than its adjacent door or panel.

Fig. 5-4. Door striker plate. Loosen bolts and adjust to align closed position of door.

5.1 Door Assembly

The interior door trim, the glass, door handles and locks, side mirrors, and internal door components are all removable with the door installed on the car. The interior trim must be removed for access to the internal door components.

Interior Trim

Although the procedures below apply to the front doors, they can be used as a general guide when removing the interior trim from the rear doors.

To remove interior trim panel:

1. Unscrew and remove the lock button. Fully lower the window.
2. Remove the door handle trim by sliding it to the rear and then pulling it off. See Fig. 5-5.

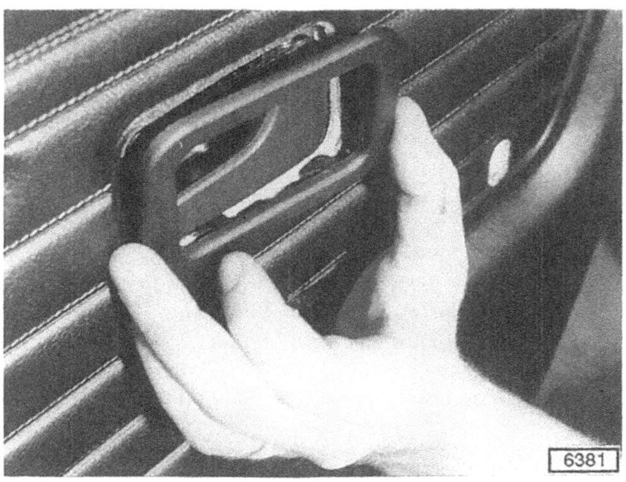

Fig. 5-5. Door handle trim being removed. Slide trim to rear then pull off.

3. Remove the screws that hold the armrest to the door. Rotate the armrest up about 45° to disengage the locking tab from the door. See Fig. 5-6.

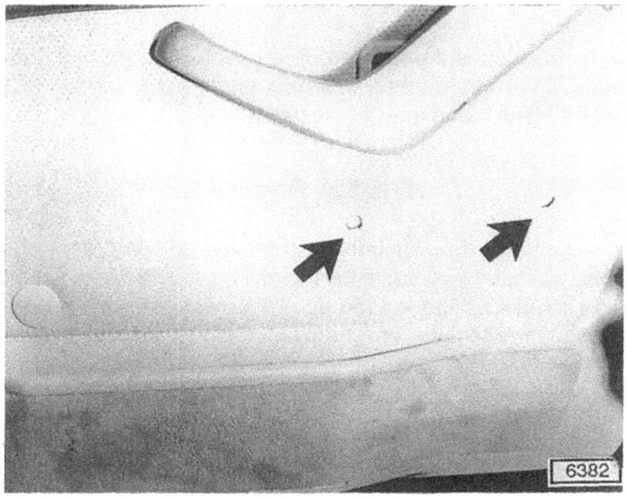

Fig. 5-6. Front armrest being disengaged from door panel. Rotate armrest up 45° and pull out. Arrows show screw locations.

4. On rear doors, pry out the window switch and disconnect the wiring.
5. Beginning with one of the lower corners of the doors, work to unclip the trim panel from the door. See Fig. 5-7. Work around to the top, and work slowly to prevent breaking the clips.

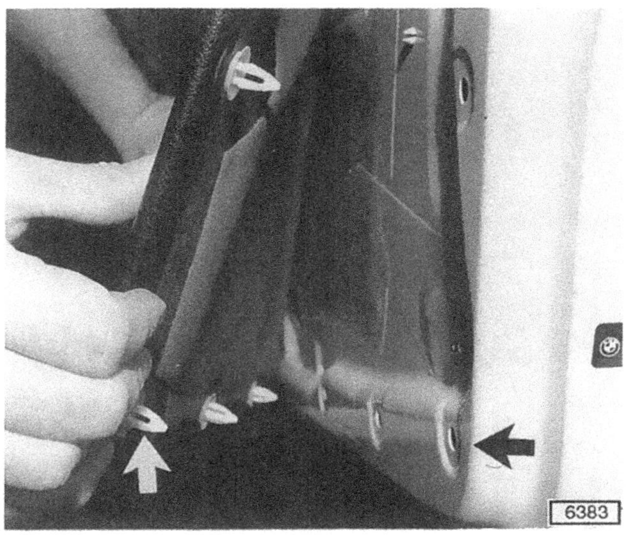

Fig. 5-7. Door interior trim panel being removed. Pull panel retaining clips out of holes (arrows).

20

6. Pry out the power mirror switch from the upper trim panel. Working from the lower corners of the upper trim piece (pad), remove the mounting screws and lift off the panel. Peel back and remove the plastic door moisture seal, taking care not to tear it.

Installation is the reverse of removal. Replace any broken clips. If the window seal was removed with the trim panel, install it at the same time as the panel.

Window Regulator

Whenever the regulator is removed and installed, check that the window contacts the top of the window frame squarely as it is raised. If the window does not meet the frame squarely, loosen the regulator or window mounting bolts and reposition the window.

> **NOTE —**
> • Check the power window electrical circuit and switches as described in **ELECTRICAL SYSTEM** before assuming that the power regulator is faulty.
>
> • Mark the location of the window regulator mounting bolts and nuts before loosening them. This will make realignment easier.

To remove front door power window regulator:

1. Remove the interior trim panel as described above.

2. Raise the window enough for access to the window securing nuts shown in Fig. 5-8. While supporting the window, remove the nuts.

Fig. 5-8. Nuts (arrows) that secure window glass to window regulator.

3. Swing the front of the window out of the regulator guide roller and lower the window into the door.

4. Remove the window regulator mounting bolts and nuts. See Fig. 5-9.

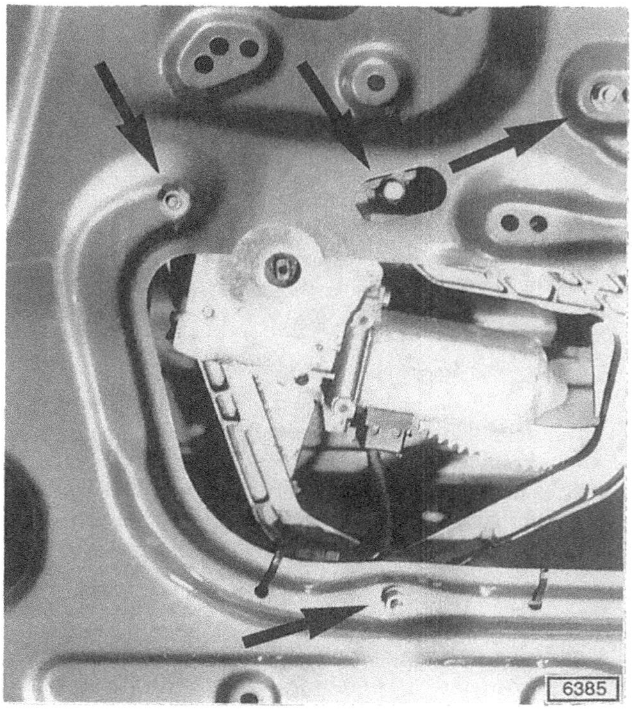

Fig. 5-9. Front window regulator mounting fasteners (arrows).

5. Disconnect the regulator electrical connector. Tilt the regulator up slightly and remove the regulator from the door.

Installation is the reverse of removal. Be certain to test all functions before reinstalling the trim panel.

To remove rear power window regulator:

1. Remove the rear interior trim panel as described above.

2. Loosen the window guide bolt at the bottom rear of the door. See Fig. 5-10.

3. Lower the window enough to disconnect the regulator guide roller from the window frame. Disconnect the regulator motor electrical harness connector.

4. Raise and hold the window in its up position.

5. Remove the three bolts securing the motor/regulator to the door. See Fig. 5-11. Remove the regulator from the access hole in the bottom of the door.

Fig. 5-10. Rear door window guide mounting bolt (arrow).

Fig. 5-11. Rear door power window regulator mounting bolts (arrows).

Windows

This heading covers removal of front and rear door window glass as well as the fixed rear door window glass.

> ***WARNING —***
> *Always wear hand and eye protection when working with broken glass.*

> **NOTE —**
> If the window is broken, all of the glass bits should be vacuumed out of the door cavity. Use a blunt screwdriver to clean out any remaining glass pieces from the window guide rails.

To remove front door window:

1. Remove the interior trim panel as described above.

2. Remove the inner and outer window seals from the top of the door by carefully prying them off.

> **NOTE —**
> When removing the outer window seal, be careful not to damage the door's finish. Use a wooden wedge or similar wooden tool to pry the seal from the door. See Fig. 5-12.

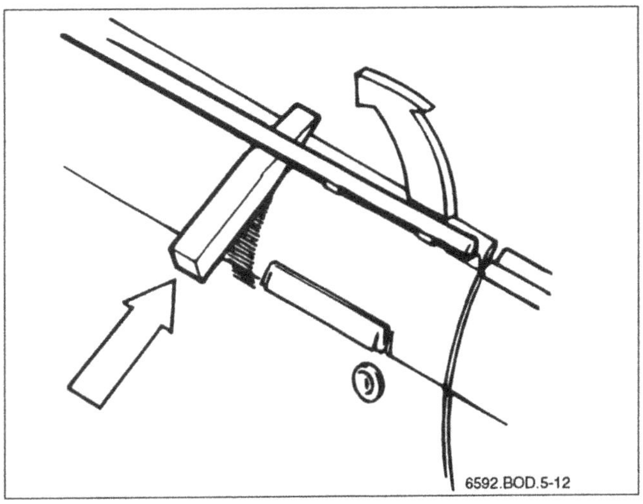

Fig. 5-12. Outer window channel seal being removed using wooden wedge. Be careful not to damage paint.

3. While supporting the glass, remove the window to regulator mounting nuts. See Fig. 5-8 above.

4. Remove the glass from above.

Installation is the reverse of removal. When installing the window make sure the window regulator rollers are inserted into the guides on the bottom of the glass before tightening the retaining nuts.

Lubricate any assemblies that were pre-lubed before removal. Adjusting bolts for the front window are shown in Fig. 5-13. Be certain to test all functions before reinstalling the trim panel.

Fig. 5-13. Loosen bolt (**A**) to adjust uniformity of gap with window open approximately one inch. Loosen bolt (**B**) to adjust lower guide so that window opens and closes smoothly.

To remove rear door windows:

1. Remove the rear interior trim panel as described above.

2. Remove the inner and outer window seals from the top of the door by carefully prying them off. See Fig. 5-12 above.

3. Raise the window until the bolts securing the glass to the regulator track can be reached. Remove the bolts. See Fig. 5-14.

4. Loosen the rear window guide mounting bolt shown in Fig. 5-10 above.

5. Disengage the window glass from the front guide roller and remove the glass from the left and right guide rails. Slide the glass toward the rear of the door and remove the glass from above.

6. Remove the fixed glass panel by removing the upper and lower guide rail bolts. Pull down on the rail while swinging it towards the front of the car. Carefully remove the glass.

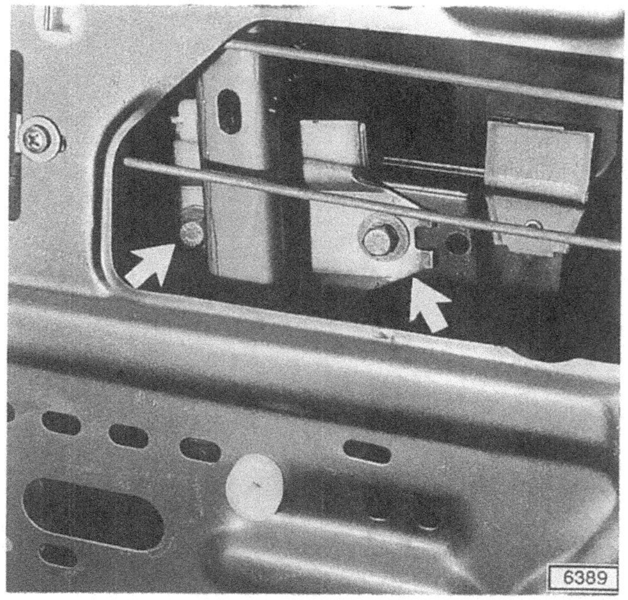

Fig. 5-14. Rear door window to regulator mounting bolts (arrows).

Installation is the reverse of removal. The use of a light soap solution or silicone spray will make installation of the fixed window and rubber guides easier.

When adjusting the window, first loosen the bolts shown in Fig. 5-14. With the window approximately one inch open, adjust the glass until there is a uniform gap, then tighten the bolts. Test all functions before reinstalling the trim panel.

Handles and Locks

Fig. 5-15 shows the basic lock linkage used on the cars covered by this manual. The lever on the top of the lock assembly is actuated by the outside door handle to open the door. It is not necessary to remove the outside door handle and the lock assembly for access to the door lock cylinder. On some models, lifting the driver's outside door handle actuates a microswitch that turns on the lock heater.

> **NOTE —**
> When repairing or adjusting locks or lock linkages it is possible—if the repair is not complete—to close and lock a door so that it cannot be opened. While performing repairs or adjustments, test all lock functions before closing the door.

To remove lock assembly and linkage:

1. Close the window. Remove the door interior trim panel as described above.

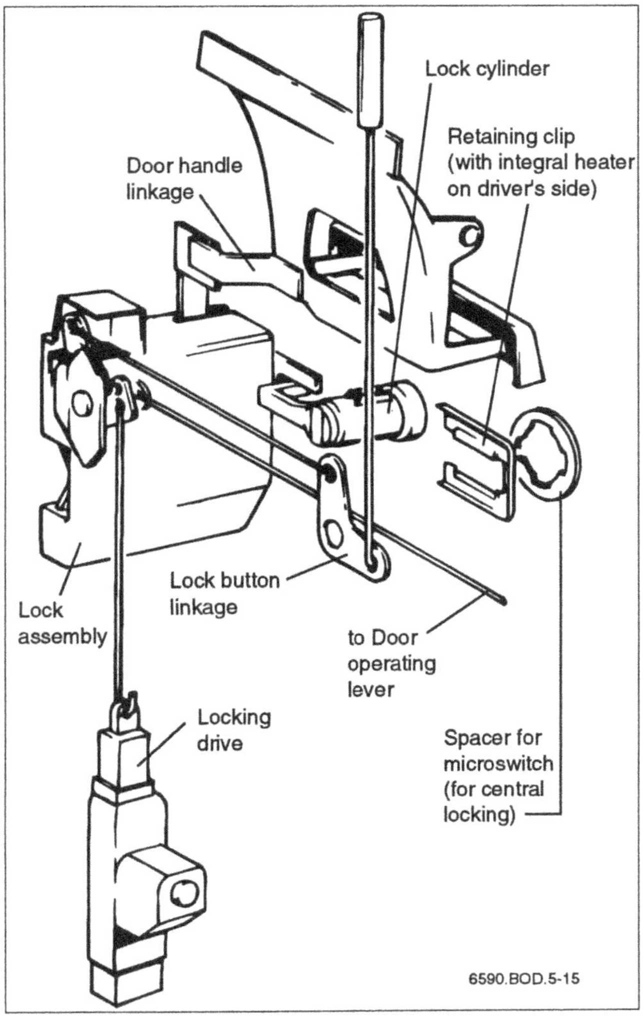

Fig. 5-15. Phantom view of lock linkage. Linkage differs slightly for rear doors.

2. Remove the screws that secure the door operating lever, shown in Fig. 5-16. Unclip the operating rod from the lever and let the lever hang down.

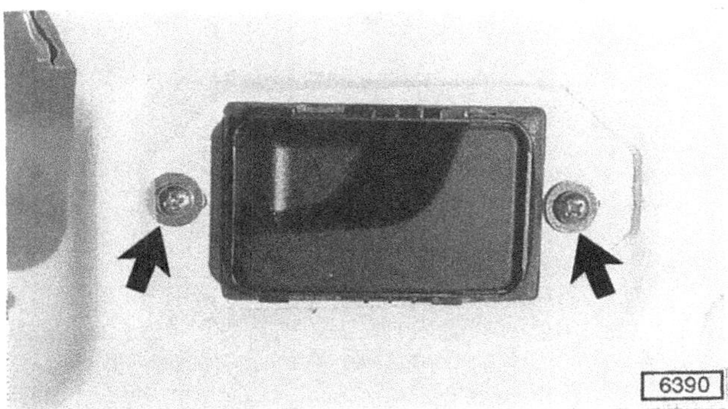

Fig. 5-16. Door operating lever and securing screws (arrows).

3. On front doors, remove the window guide rail mounting bolt and remove the rail. See Fig. 5-17.

Fig. 5-17. Guide rail mounting bolt on front door **A** and central locking drive mounting bolts **B**. Mark locations before removing.

4. On front doors, disconnect the linkage from the door lock cylinder.

5. On rear doors, disconnect the linkages from the locking rod and from the inside door handle at the lock assembly.

6. Remove the bolts that hold the locking drive to the door. Disengage the locking linkage from the lock assembly, disconnect the locking drive harness connector, and remove the drive from the door.

7. Rotate the catch mechanism to the closed door position. Remove the screws securing the lock assembly to the door. Then remove the lock assembly and linkage.

Installation is the reverse of removal. When installing the door operating lever, preload it slightly towards the front of the car before tightening the screws. On models with central locking, install and adjust the locking drive as described in **5.2 Central Locking System**. Be sure the lock mechanism is lubricated with grease.

20

NOTE —
Test all of the door functions before reinstalling the trim panel.

To remove door lock cylinder:

1. Remove the door interior trim panel as described above.
2. Working inside the door, disconnect the linkage from the cylinder to the lock mechanism. Pull the retaining clip and spacer off the lock cylinder. See Fig. 5-15 above.
3. Remove the spacer and withdraw the lock cylinder from the outside of the door.

To remove outside door handle mechanism:

1. Remove the door interior trim panel as described above.
2. Remove the outside door handle by removing the screws shown in Fig. 5-18.

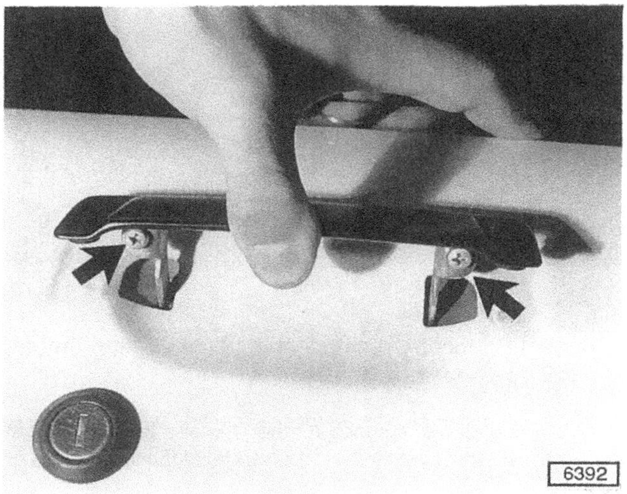

Fig. 5-18. Door handle retaining screws (arrows).

3. On models with heated door locks, disconnect the electrical connector inside the driver's side door that leads to the microswitch and control unit on the door handle mechanism.
4. Working from inside the door, remove the four nuts that secure the handle mechanism to the door panel and remove the mechanism.

When installing the door handle on models with heated door locks, the tab on the handle linkage should actuate the microswitch when the handle is pulled up. If it does not, carefully bend the actuating tab on the handle linkage so that the switch operates correctly.

NOTE —
Periodically lubricate all of the lock cylinders with BMW lock lubricant (BMW Part No. 81 22 9 407 421). This lubricant works down to -40°F and prevents moisture buildup.

5.2 Central Locking System

The central locking system consists of solenoid locking drives at each of the doors, at the rear lid, and at the gas tank filler flap. The drives are connected to the lock by a linkage rod. When the key is turned in the driver's door or trunk lock, the central locking control unit activates the drives to open or close the locks. The gas filler flap can be manually opened if the central locking system fails by pressing the locking rod located behind the luggage compartment carpeting. The central locking system control unit incorporates a safety impact switch that opens the locks in an accident. It is located above the speaker in the passenger's footwell.

This heading covers only the removal and installation of the mechanical parts. For general electrical information to aid in troubleshooting the central locking circuit, see **ELECTRICAL SYSTEM.**

Different locking drives are used depending on the application. The drives are identified by codes stamped on the housing. See Fig. 5-19 and **Table b.** When installing the locking drives, make sure the beveled "roof" of the door drives faces toward the front of the car.

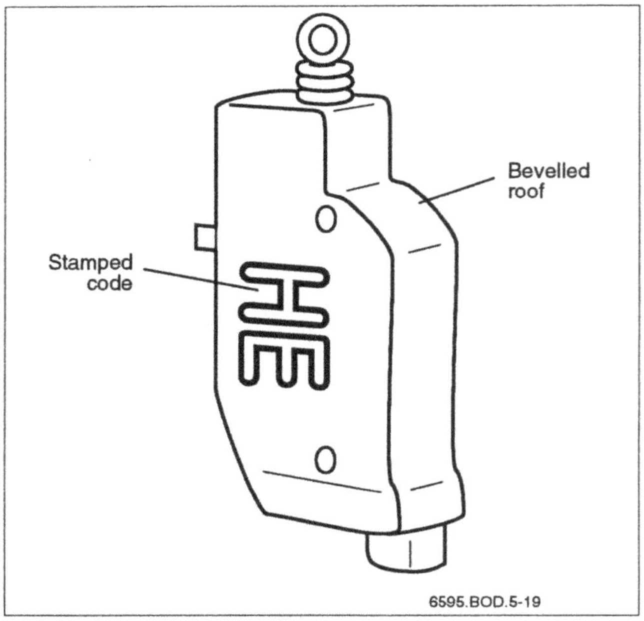

Fig. 5-19. Central locking system locking drive. See **Table b** for correct identification of drives.

Table b. Central Locking System Locking Drive Identification

Locking drive	Stamped code
Driver's door (without deadbolt)	TV
Passenger door (without deadbolt)	TV
Rear doors (without deadbolt)	TH
Driver's door (with deadbolt)	FT
Passenger door (with deadbolt)	BT
Rear doors (with deadbolt)	HT
Trunk lid	HE
Filler flap	no code

Removing, Installing, and Adjusting Locking Drives

The door locking drives are removed by first removing the interior door trim panel as described in **5.1 Door Assembly**. Remove the drive mounting bolts, disconnect the lock linkage and harness connector, and remove the drive.

The trunk locking drive is mounted behind the trunk rear trim panel. See Fig. 5-20. The gas tank flap activator is located behind the trunk right side trim panel. Remove the mounting bolts, disconnect the lock linkage and harness connector, and remove the drive.

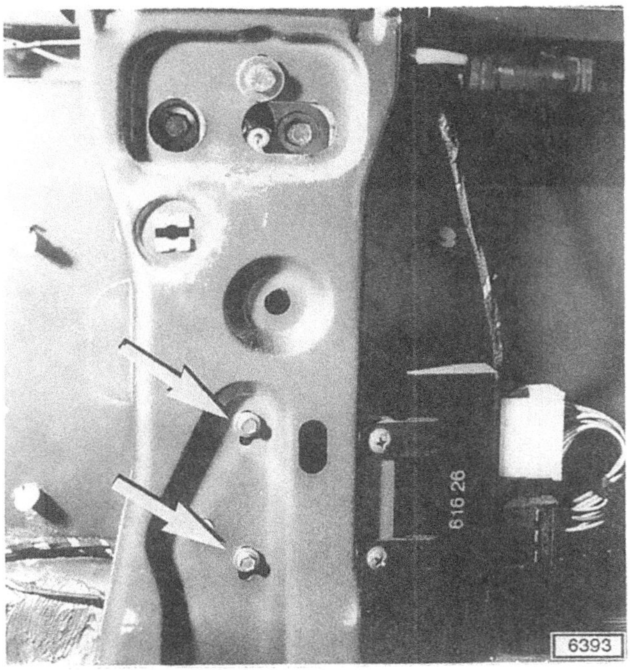

Fig. 5-20. Mounting bolts (arrows) for trunk central locking drive. Trim panel has been removed.

To install and adjust drives:

1. Place the door or trunk lock in the locked position. Make sure the locking linkage is correctly installed.
2. Engage the locking drive with the lock linkage and loosely install the bolts to hold the drive in position.
3. Pull down lightly on the entire drive assembly to take up slack in the linkage, then tighten the mounting bolts.

5.3 Side Mirrors

The mirror housing is held to the car door by two mounting bolts. The mirror glass is held in the housing by a locking ring or an adhesive-backed plate with ball and socket. The version with the locking ring has a small slot for a screwdriver at the bottom of the glass.

To remove the mirror housing, pry out the mirror trim panel or upper speaker on the inside of the door. Unclip the wiring harness connector. Remove the two bolts shown in Fig. 5-21 and remove the mirror housing.

Fig. 5-21. Side mirror housing mounting bolts (arrows), behind small trim panel or radio speaker.

To remove the mirror glass on the locking ring version, disconnect the negative (–) battery cable. Insert a screwdriver into the housing from below as shown in Fig. 5-22. Engage the tab on the locking ring with the screwdriver and push it to one side or the other until the ring is disengaged. Fig. 5-23 shows the mounting screws for the mirror motor on electric side mirrors. To install the glass, push the locking ring to its stop on one side, then mount the glass and use the screwdriver to push the tab to one side to lock the glass. 20

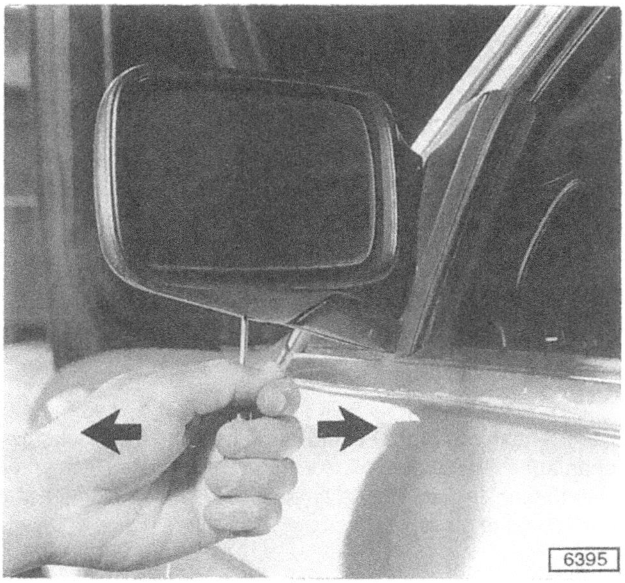

Fig. 5-22. Insert screwdriver as shown to engage tab on mirrors with locking ring, then push in either direction to unlock glass.

Fig. 5-23. Motor and mounting screws (arrows) for electric side mirrors.

To remove the mirror on the ball and socket version, disconnect the negative (–) battery cable. Insert a screwdriver or putty knife and carefully pry the mirror away.

NOTE —
Pry gently on the mirror, as it can be easily broken.

On mirrors with the integral heating element, disconnect the wires to the mirrors.

6. EXTERIOR

The exterior body parts and trim discussed under this heading are fastened with conventional fasteners, and are easily removed using ordinary hand tools. A thorough knowledge of automobile body repair is not needed to perform the procedures.

6.1 Headlights

The cars covered by this manual have dual high and low sealed beam headlights. 1985 and later models use halogen headlamps. The outer bulbs are low beam. Each headlight is aimed with two knurled knobs, and the bulb and lens is replaced as one unit.

It should not be necessary to aim the headlights after replacing a bulb if the adjustment knobs have not been touched. However, it is a good idea to check headlight aim as described below.

To replace sealed beam headlights:

1. Remove the front grille as described in **6.4 Trim, Bumpers, and Body.**

2. Remove the retaining ring screws shown in Fig. 6-1, then remove the light and disconnect the wiring.

Fig. 6-1. Sealed beam headlight and retaining ring screws (arrows).

3. When installing the new light, make sure the part of the lens marked TOP is uppermost, and that the tabs on the light engage the recesses in the headlight housing.

NOTE —
Do not interchange the bulbs for high and low beams.

Aiming Headlights

Headlights are usually aimed using special adaptors and equipment. Headlight aiming can also be done as described below without any special tools, however, the results will not be as exact.

NOTE —
Check your state laws to see if headlight adjustments must be made by a licensed shop. Some states may also have different aiming specifications from those described here.

To aim headlights without using special tools, position the car on a level surface, 7.65 meters (25 ft.) from a vertical wall. The gas tank should be full, tire pressures should be correct, and there should be a weight of approximately 155 lbs. in the back seat. Turn the headlights to low beam. With a person in the driver's seat, the low beams must be in the areas shown in Fig. 6-2. If re-aiming is necessary, open the hood, then move the vertical and horizontal headlight adjusting knobs by hand to move the headlight. See Fig. 6-3.

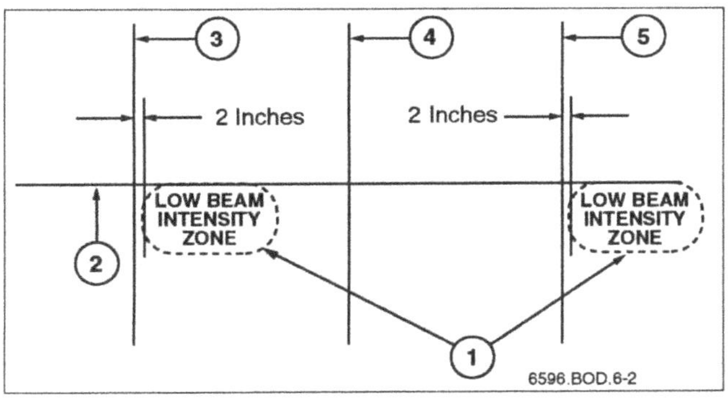

Fig. 6-2. Headlight aiming target on vertical wall. Low beam light intensity areas are at (**1**). Line (**2**) is at height of headlight centers. Line (**4**) is car centerline. Lines (**3**) and (**5**) are distance from centerline to centers of headlights.

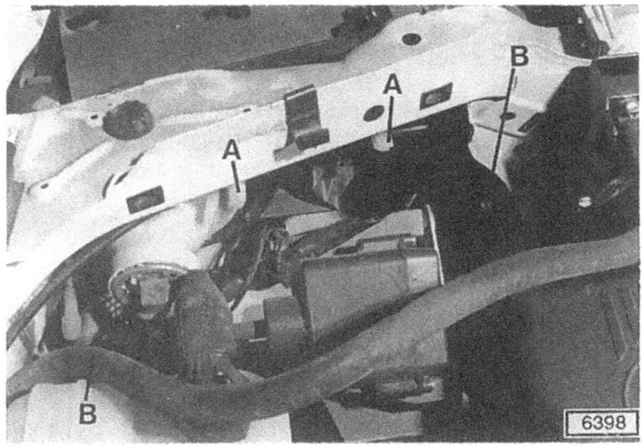

Fig. 6-3. Headlight aiming knobs for vertical adjustment (**A**) and horizontal adjustment (**B**).

6.2 Taillight Assembly

All rear bulbs and the taillight lens are replaced from the luggage compartment. Remove the plastic pieces securing the carpeting by turning them 90°. Replacement of the center stop light is covered in **4.3 Interior Lights**.

Replace a faulty taillight bulb by first removing the bulb holder. Turn the bulb holder and remove it. Push in on the bulb and twist to remove, push in and twist to install.

To replace the taillight lens, remove the six nuts that hold the lens to the body. See Fig. 6-4. From outside, pull the lens off the car with its gasket.

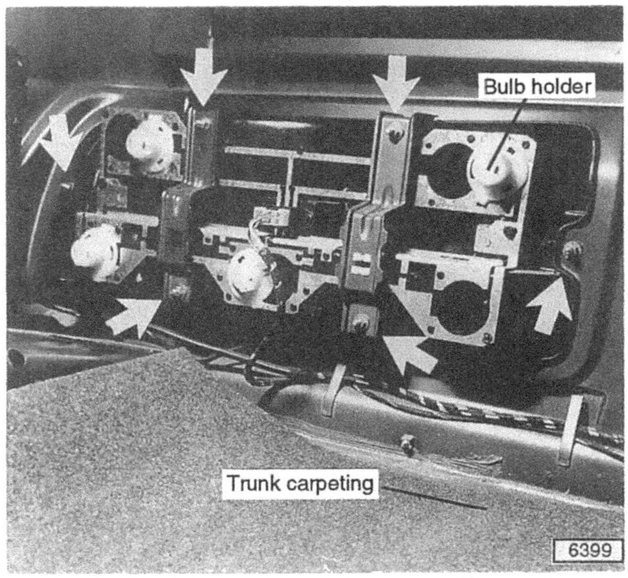

Fig. 6-4. Retaining nuts (arrows) for taillight lens. Trim carpeting has been removed.

20

6.3 Side Marker, Front Turn Signal, and Fog Lights

To replace a front turn signal bulb or lens, remove the two screws. Pry the lens assembly away from the bumper. Push and turn the bulb counterclockwise to remove it.

To replace a side marker bulb or lens, remove the screws and pry out the lens. Remove the bulb.

To replace a fog light bulb, remove the two screws securing the lens. Remove the bulb.

To replace the entire fog light lens and assembly, disconnect the harness connector, then remove the two mounting screws. The fog lights are adjusted from the front, with two screws. The screws are behind two rubber plugs.

6.4 Trim, Bumpers, and Body

This heading covers removal of components such as the fenders, hood, and trunk, as well as alignment of those components.

> ***WARNING —***
> - *Welding and grinding galvanized steel produces a poisonous gas containing zinc oxide. This type of work should always be carried out by experienced body technicians having the proper respiratory equipment and work area.*
>
> - *Pressurized strut assemblies like the hood strut rods may explode if exposed to high heat sources such as painting ovens, torches, or welders.*

Removing and Installing Fenders

For corrosion protection, the fenders are assembled with a sealant strip between the top mating surfaces to prevent moisture from becoming trapped. This strip should be replaced any time the fenders are removed.

> **NOTE —**
> - The sealant strip is installed as part of the corrosion warranty. If in doubt about warranty protection, consult an authorized BMW dealer.
>
> - The body is painted at the factory after assembly. Realignment of body panels may expose unpainted metal. To prevent rust, paint all exposed metal as soon as possible.

To replace fenders:

1. Remove the front bumper, the front grill, and where applicable the bumper side pieces, as described below.
2. Working from within the rear of the wheelhouse, remove the plastic wheel house moisture barrier by removing the four mounting bolts.
3. Remove the fender mounting bolts. There are six bolts along the top of the fender, four bolts inside the wheelhouse (near the front edge of the door), two bolts at either end of the fender bottom, and two bolts holding the front of the fender to the front valence.
4. Peel back the rubber seal along the top edge of the fender. Carefully use a heat gun and razor knife to cut through the sealant along the edge of the fender, and to cut through the PVC coating along the A-pillar where the fender meets the body.

> ***WARNING —***
> *Do not use any heat or flame sources near fuel tanks, lines, or fuel filters.*

> ***CAUTION —***
> *Use extreme care when applying heat to the body to avoid blistering the paint or discoloring the PVC body coating.*

5. Gently pry the old fender off of the body. Clean the old sealant and coating off of the body mounting surfaces.

When installing the new fender, first install a new sealant strip between the top edge of the fender and the body. The sealant strip is available from an authorized BMW dealer. Loosely install the fender mounting bolts. Position the fender so that it lines up correctly with the door and A-pillar, then tighten the bolts. Reinstall the moisture barrier and the bumper.

Removing and Installing Body Trim

Body trim includes the body side moldings, the rain gutter trim, and the front grille components. The body side moldings are held in place with clips and screws. The grille trim is held by clips.

To remove body side moldings on the front fender, remove the screw at the front of the molding. Pry the molding straight off the clips. Replace any broken clips, and press the molding onto the clips.

To remove a door panel body molding or rear side panel body molding, first remove the screw at rear of molding. See Fig. 6-5. Pry the molding straight off the clips while sliding the molding off its front clip. Replace any broken clips. When installing the molding, first slide it onto the beveled clip, as shown in Fig. 6-6, and then press the molding onto the other clips.

Fig. 6-5. Mounting nut (arrow) for front door body side molding. Rear side panel body molding is similar.

Fig. 6-6. Beveled clip (arrow) for body side moldings. Slide molding onto clip first, then press onto other clips.

To remove a rain gutter molding, carefully lift off the strip beginning at the back of the car. Twist and lift the molding as you go along. Removing the strip almost always distorts it such that it cannot be reused. Before installing the new strip, replace any clips that are deformed, and lightly lubricate the inside of the molding with silicone. To install the molding properly, use the special BMW crimping tool no. 511 060.

Removing and Installing Bumpers

The bumpers are bolted to shock absorbers that are bolted to the frame. When removing the front bumper, disconnect the turn signal and side marker light harness connectors. To remove either bumper, unscrew the four nuts (two on either side) securing the bumper to the shock absorber bracket. Pull the bumper straight off the car. When installing the bumper, be sure to correctly torque the bumper-to-bracket mounting bolts.

Tightening torque
• bumper to shock absorber bracket . . . . 22 to 25 Nm (16 to 18 ft-lb).

WARNING —
If the car has been in a collision and the bumper is collapsed or appears collapsed, the bumper shock absorbers may be under very high pressure and can cause serious injury. Any repairs should be performed by an authorized BMW dealer or other qualified repair shop.

The bumper trim pieces can be removed after the bumper is removed. The trim pieces are held to the car by screws and bolts. When removing the front trim, the side marker light will have to be removed to reach one of the side mounting bolts. When removing the rear side pieces, first remove the side marker light to disconnect its electrical connector before removing the mounting bolts.

Hood and Hood Lock

When removing the hood, first mark the hinge plate and bolt locations on the hood for reinstallation. Raise the hood and unclip hoses and wiring. Loosen the mounting bolts, then with a helper supporting the opposite side, remove the bolts. Install the original hood by aligning the matching marks made earlier.

NOTE —
The body is painted at the factory after assembly. Realignment of body panels may expose unpainted metal. To prevent rust, paint all exposed metal as soon as possible.

Hood alignment considers hood-to-body panel gap, and hood height. All adjustments except front-of-hood height are made with the mounting bolts. Front of hood height is adjusted using the threaded rubber stops at the front corners of the hood. See Fig. 6-7. Thread the rubber stops in or out until adjustment is correct. Check height with the hood closed.

The hood lock plunger is adjustable by loosening the lock nut at the plunger base. See Fig. 6-8. Thread the plunger in or out until the hood latches securely and opens without excessive effort.

Fig. 6-7. Rubber stop (arrow) used for final hood height adjustment. Screw threaded stop in and out.

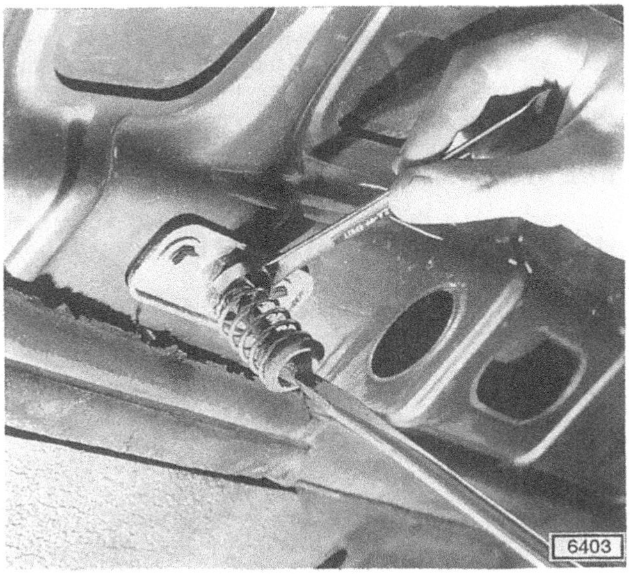

Fig. 6-8. Threaded lock plungers are adjustable. The left side is shown, the right side is similar.

The hood cables are adjustable. Loosen the cable clamping bolts and adjust the cable until all of the slack is removed. Adjust both left and right sides if necessary.

To replace hood lock cable:

1. Remove the large trim cover above the center grille.
2. Loosen the left and right cable clamping bolts and slide the cable out of the locking mechanisms.
3. Remove the entire cable assembly by unhooking the cable clips along the body.
4. Working in the passenger compartment, remove the screws holding the release handle to the body. Pull the cable through the firewall.
5. When installing the new cable, check that the firewall grommet is correctly seated for an air-tight seal. Install the release handle and cable to the body first before attaching the cable to the lock.

> **NOTE —**
> If the hood cables are inoperative, the only method of opening the hood is to release each hood lock using a long screwdriver through the front grille. Remove the front side grilles. Pry the release lever toward the driver side of the car until the latch releases. See Fig. 6-9. Repeat the procedure for the other latch.

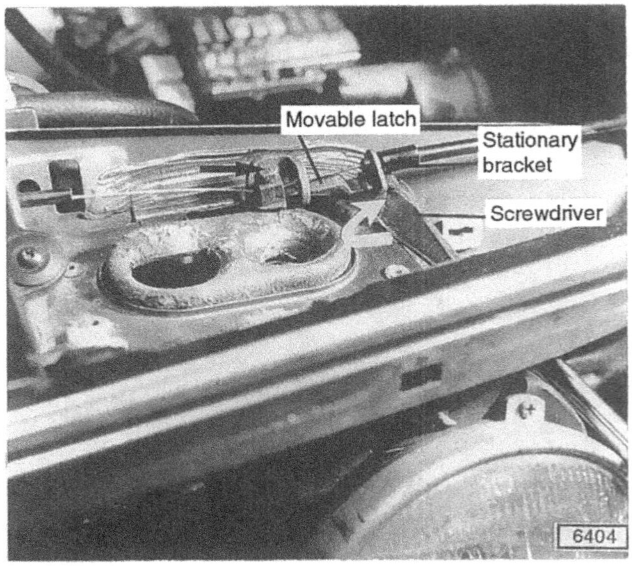

Fig. 6-9. Screwdriver being used to release hood lock. Lever screwdriver in direction of arrow to release lock. Hood is open for clarity.

Trunk Lid and Trunk Lock

When removing the trunk lid, first mark the hinge plate and bolt locations on the lid for reinstallation. Raise the trunk lid and loosen the mounting bolts, then with a helper supporting the opposite side, remove the bolts. Install the original lid by aligning the matching marks made earlier.

> **NOTE —**
> The body is painted at the factory after assembly. Realignment of body panels may expose unpainted metal. To prevent rust, paint all exposed metal as soon as possible.

To make trunk lid front height adjustments, loosen the hinge bolts shown in Fig. 6-10. Tighten the bolts when the front height is even with the adjacent panels.

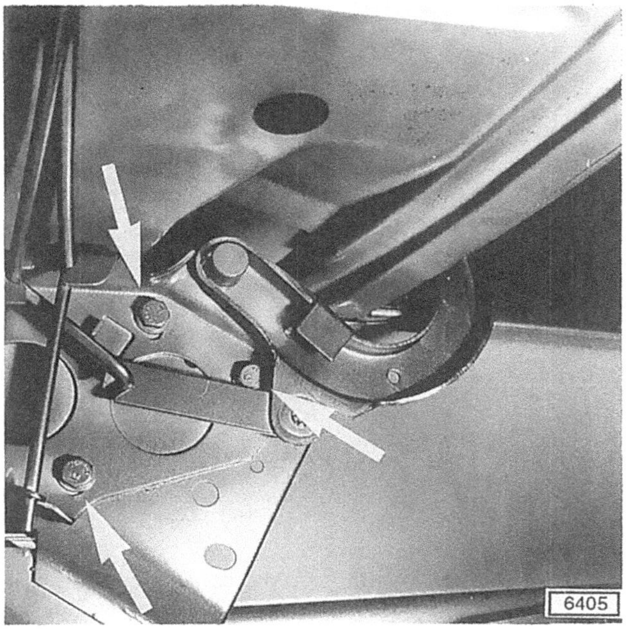

Fig. 6-10. Loosen hinge bolts (arrows) to adjust front trunk lid height.

To make rear height adjustments, remove the carpeting trim around the trunk latch. Then loosen the bolts on the lower latch assembly. Adjust the latch assembly so that the lid is at the same height as the adjacent body panels when the lid is closed. Open the lid and retighten the bolts. Make final height adjustments by unscrewing the rubber stops until they just contact the body panel.

To make side-to-side adjustments, loosen the bolts holding the lid to each hinge. The trunk lid should be adjusted until the seam between the lid and the body is uniform on both sides. Tighten the bolts.

The trunk lock is located behind the rear trunk trim panel. To remove the lock, first pull out the rear trunk trim panel. Remove the lock mounting bolts and slide the lock out from the side. The lock cylinder is held in by a circlip. On models with central locking, see **5.2 Central Locking System** for information on removing the locking drive.

7. SUNROOF

The sunroof is controlled by a set of cables that move the sunroof panel along guide rails when the motor is operated. The sunroof can be adjusted without removing it from the car. Replacement of the sunroof liner or components such as the cable assembly require that the sunroof panel be removed. If the sunroof leaks or sticks, adjust it as described in **7.2 Adjusting Sunroof Fit** before deciding that it needs to be removed from the car.

> **NOTE —**
> Leaks may be caused by clogged drain hoses. Clean the hoses with a length of flexible cable, such as an old speedometer cable or low pressure compressed air.

There are two styles of sunroofs used on the cars covered by this manual. Removal and adjustment procedures differ slightly depending on the style installed. The main difference between the two styles is that the late style model used a plastic gate. To check which style is installed, raise the rear end of the panel (two-way position) and look for a plastic gate on the side of the panel. See Fig. 7-1.

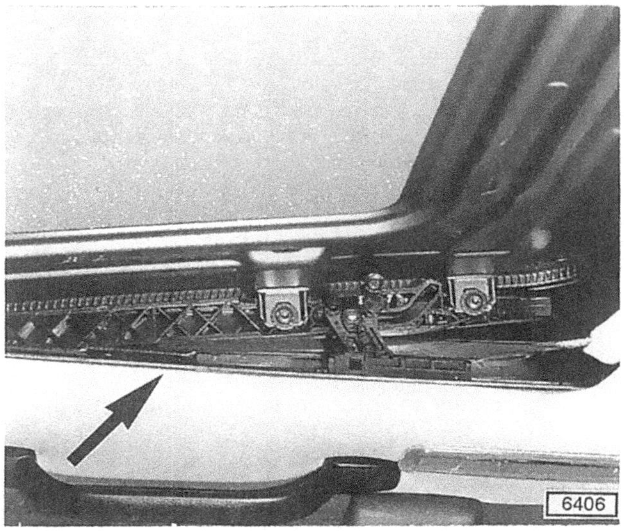

Fig. 7-1. Late style sunroof with plastic gate (arrow). Early style sunroofs do not use plastic gate.

Creaking of the sunroof when opened or closed on hot and humid days may be due to grease and dirt build-up on the slide rails. To correct this problem, use a grease-free solvent to clean the slide rails, cables, and any other mechanical components that come in contact with the plastic shoes of the sunroof. The guide rails can be lubricated with Vaseline.

20

7.1 Removing and Installing Sunroof Panel

On models with sunroofs using the plastic gate, the sunroof panel fasteners are coated with a special adhesive to prevent loosening due to vibration. These screws should always be replaced anytime they are removed. These screws can usually be identified by the colored adhesive on the threads.

If the sunroof seal is being replaced, it must be installed as shown in Fig. 7-2.

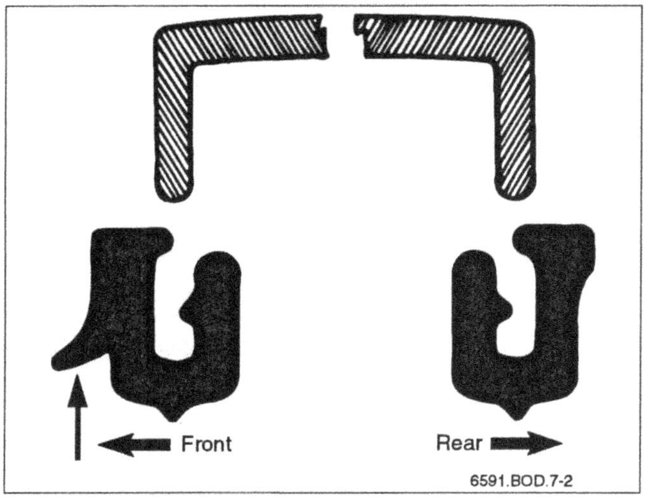

Fig. 7-2. Cross-section of sunroof seal showing correct installation of seal to sunroof panel. Note sealing lip (arrow) on forward portion.

To remove and install sunroof panel:

1. Open the sunroof approximately a quarter of the way. Using a wooden or plastic wedge, unclip the trim panel retaining clips at the front edge of the sunroof. See Fig. 7-3. Pry as close as possible to each of the clips. Close the sunroof and push the trim panel back into the roof.

Fig. 7-3. Clips (arrows) that hold trim panel to sunroof.

2. On models with a plastic gate, remove the sunroof mounting screws from the left and right side. See Fig. 7-4.

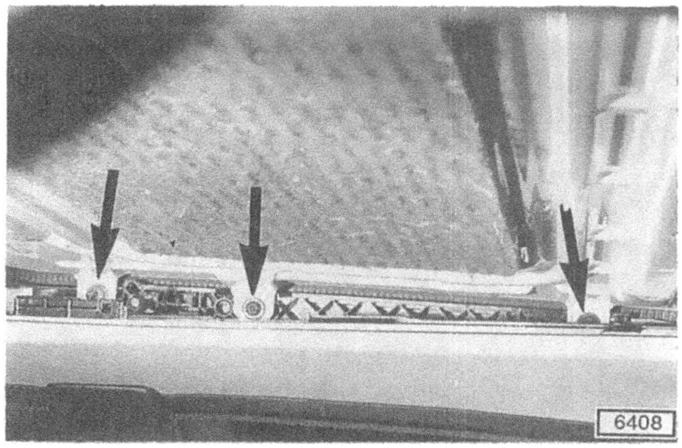

Fig. 7-4. Sunroof left-side mounting screws (arrows) on models with plastic gate. Remove screws from both right and left sides. Replace with new screws when installing.

3. On models without a plastic gate, open the sunroof. Working from above, remove the five screws and loosen the end screw as shown in Fig. 7-5. Remove the guide rail covers.

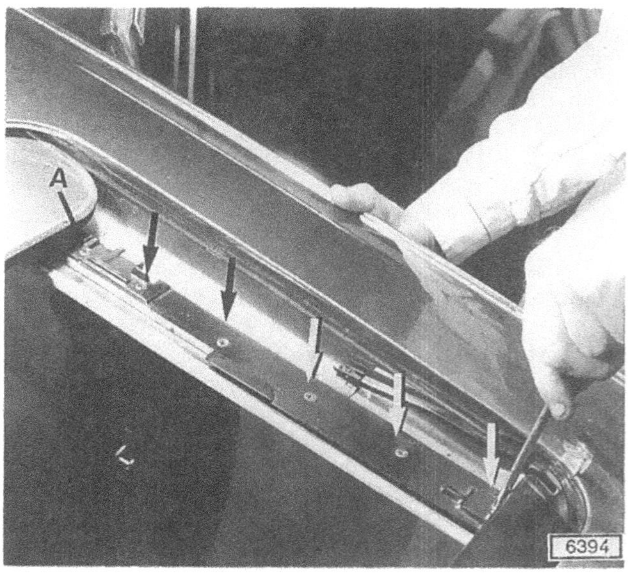

Fig. 7-5. Sunroof panel guide rail cover mounting screws (arrows) on models without plastic gate. Also loosen screw at **A**.

4. On models without a plastic gate, close the sunroof and remove the three nuts from the triangular holders on either side of the panel. See Fig. 7-6.

5. Carefully push the sunroof panel out of the car from below.

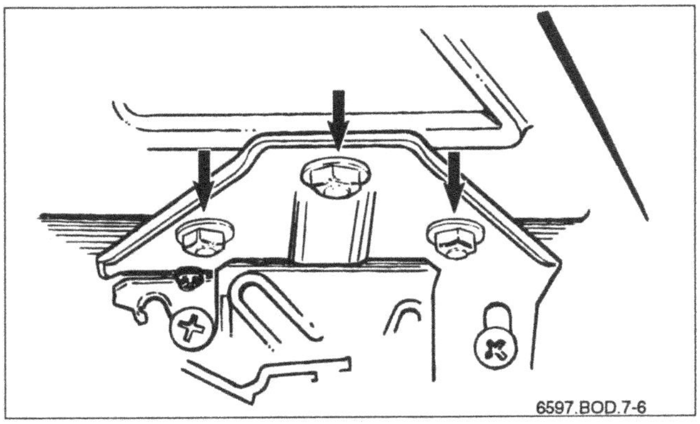

Fig. 7-6. Sunroof panel mounting nuts (arrows) on models without plastic gate. Remove nuts from both brackets.

6. Install the sunroof panel into the opening. Loosely install the screws or the nuts.

> **NOTE —**
> On models with plastic gate, always use new pre-coated screws.

7. On models without a plastic gate, slowly push the sunroof open with the help of the motor. Then install the guide rail covers and the mounting screws.

8. Adjust the sunroof panel as described in **7.2 Adjusting Sunroof Fit**.

9. Pull the sunroof trim panel forward. While holding the rear of the trim panel up, open the sunroof slightly, then press in the clips securing the front liner to the frame.

> **NOTE —**
> If the sunroof panel does not want to come forward all the way, the channels in the trim panel are not correctly engaging the rollers on the drive cable levers. These guides allow the sunroof panel to pull the trim panel up when the rear of the sunroof is raised. Try pushing up slightly on the rear of the panel as it is drawn forward to guide the channels into the rollers.

7.2 Adjusting Sunroof Fit

The sunroof should be adjusted whenever the top of the closed sunroof does not lie flush with the roof of the car, if it does not close squarely, if there are wind noises at high speed, if there are water leaks, or if the sunroof has been removed.

To adjust (with plastic gate):

1. Open the sunroof approximately a quarter of the way. Using a wooden or plastic wedge, unclip the trim panel retaining clips at the front edge of the sunroof. See Fig. 7-3 above. Close the sunroof and push the trim panel back into the roof.

2. Loosen the mounting screws on both sides that hold the sunroof to the guides. See Fig. 7-4 above. Then lock the sunroof guides in place to ensure that the panel is squarely positioned in the opening. Using a 4 mm internal-hex-head wrench or equivalent on both sides, insert the wrench through the rear guide and into the lift channel. See Fig. 7-7.

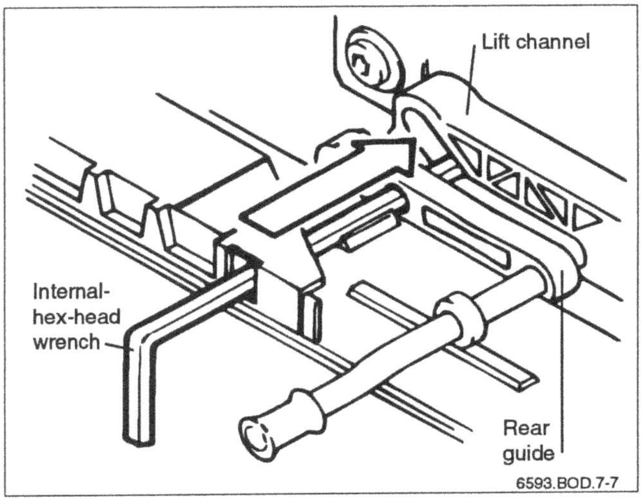

Fig. 7-7. Internal-hex-head wrench inserted through rear guides to lock guides in place on sunroof with plastic gate. Lock both sides.

3. Adjust the sunroof by tilting it as necessary until the front edge is lower than the roof by 1 mm (0.04 in.), and the rear edge is higher than the roof by 1 mm (0.04 in.). When the adjustment is correct, torque the sunroof mounting screws.

Tightening torque
• sunroof mounting screw . . . . 4 to 5 Nm (35 to 44 in-lb)

To adjust (without plastic gate):

1. Open the sunroof approximately a quarter of the way. Using a wooden or plastic wedge, unclip the trim panel retaining clips at the front edge of the sunroof. See Fig. 7-3 above. Close the sunroof and push the trim panel back into the roof.

2. To adjust the front height, loosen the front bracket mounting screws. Then turn the adjusting screw until the front edge of the panel is approximately 1 mm (0.04 in.) lower than the roof edge. See Fig. 7-8. When the adjustment is correct, tighten the mounting screws.

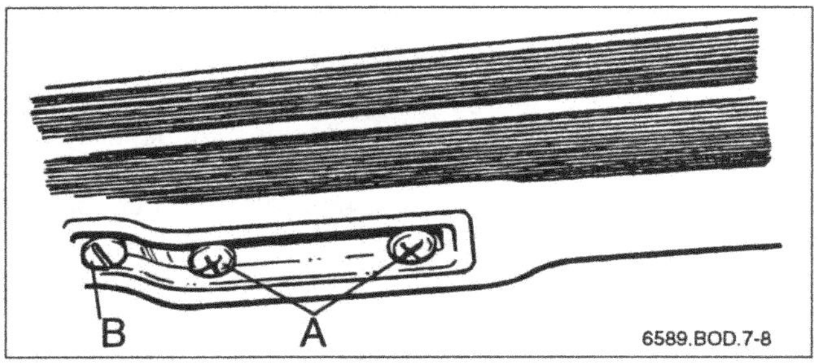

Fig. 7-8. Sunroof panel front height adjustment screws. Loosen clamping screws at **A** and turn screw at **B** until height is correct. Tighten clamping screws when adjustment is complete.

3. To adjust the rear height, loosen the mounting screws shown in Fig. 7-9 and adjust the rear panel until the rear edge is higher than the roof by 1 mm (0.04 in.). When the adjustment is correct, tighten the mounting screws.

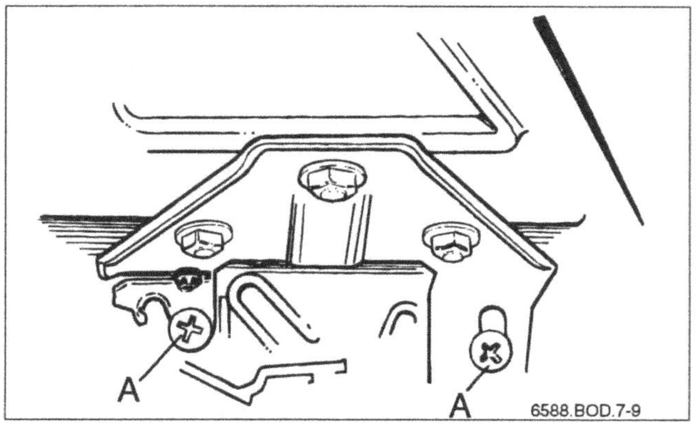

Fig. 7-9. Sunroof panel rear height adjustment screws. Loosen screws at **A** and adjust panel until height is correct.

Section 21

HEATING AND AIR CONDITIONING

Contents

TABLES

INTRODUCTION

All of the 5-series cars covered by this manual are equipped with an electronically-controlled heating and ventilation system. This climate-control system automatically regulates the passenger compartment temperature based on the temperature selected at the temperature regulating knob.

1. GENERAL DESCRIPTION

Fresh air enters the passenger compartment at the base of the windshield and passes through the heater box. The heater box is mounted behind the dashboard. Fresh air can be directed past the heater core through the lower or upper vents or directly into the passenger compartment through the center vents. Air exits through slots near the rear side windows.

Heating is controlled by regulating coolant flow through the heater core via the electric solenoid-type heater valve. See Fig. 1-1. Control of the heater valve is electronic. Depending on the position of the temperature regulating knob compared with the passenger compartment temperature and the heater core temperature, the temperature regulating control unit automatically turns the heater valve on and off to maintain a constant temperature.

The air conditioning (A/C) evaporator and blower assembly are mounted in front of the heater box. The A/C system is separate from the heating and ventilation system and uses its own blower motor. With the A/C on, the air is cooled and dehumidified by the evaporator to maintain the desired temperature. The A/C blower always runs on low speed when the A/C switch is on, even if the blower switch is off.

Air conditioning is electronically controlled in much the same way as the heating and ventilation system. The temperature regulating control unit uses the inputs from the interior temperature sensor and the evaporator temperature sensor to cycle the A/C compressor on and off via the evaporator regulator.

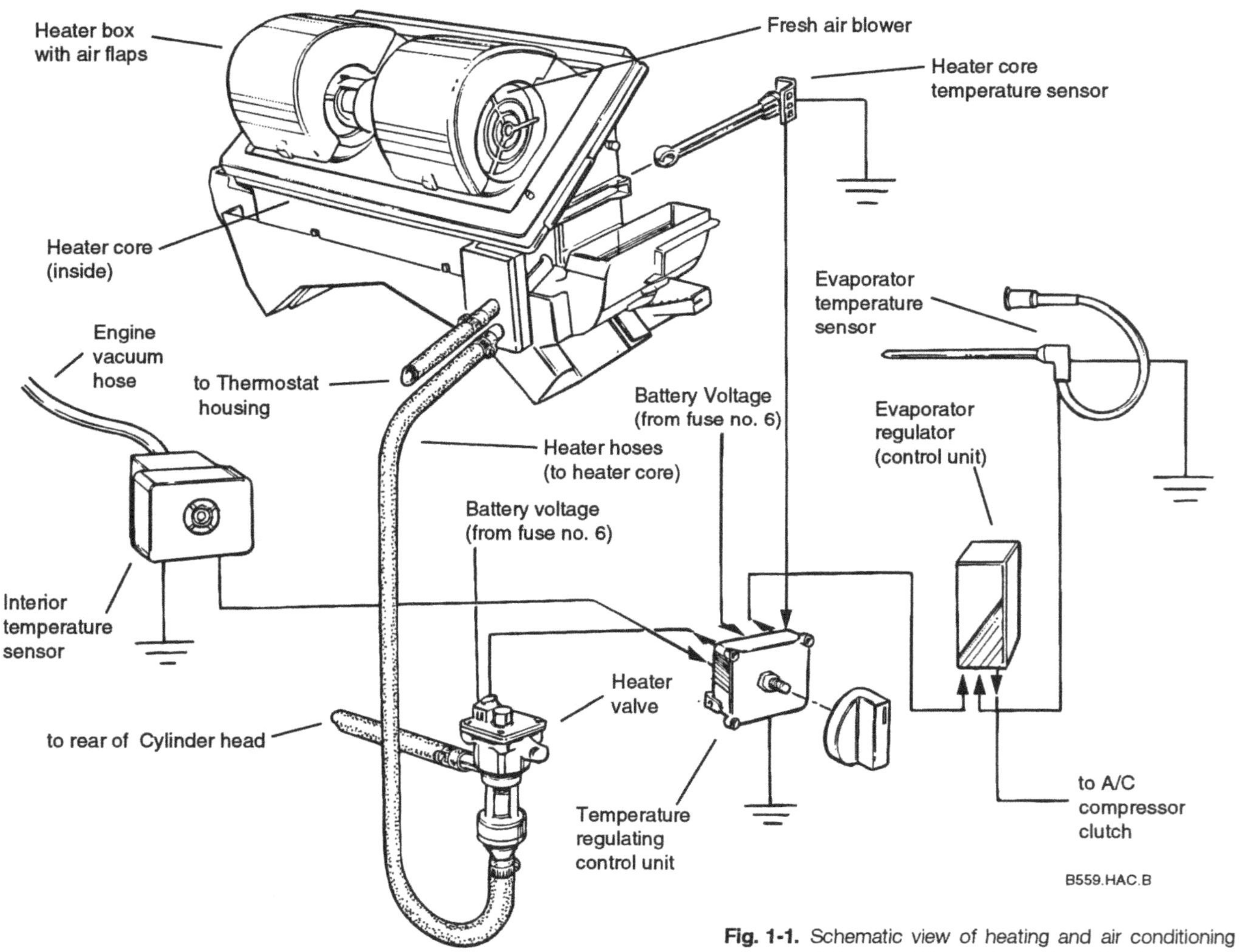

Fig. 1-1. *Schematic view of heating and air conditioning system.*

2. MAINTENANCE

There is no regularly scheduled maintenance of the heating and air conditioning system other than visual inspection and verifying system operation. Cooling system repair is covered in **COOLING SYSTEM.** All other cooling system maintenance including the following items are covered in **LUBRICATION AND MAINTENANCE.**

1. Checking coolant level and anti-freeze concentration
2. Inspecting coolant pump V-belt tension and condition
3. Inspecting coolant hoses for leaks; tightness of hose clamps
4. Replacing engine coolant

WARNING —
• Adding refrigerant or servicing the air conditioning system without the proper tools, equipment, or knowledge may cause severe personal injury as well as damage to A/C components.

• Wear eye protection when inspecting the system. R-12 at normal atmospheric pressures can evaporate and freeze anything it contacts.

• At normal operating temperature the heating system is pressurized with hot engine coolant. Allow the system to cool as long as possible before opening. Wait a minimum of an hour. Then remove the coolant expansion cap very slowly to allow safe release of pressure.

• Releasing cooling system pressure lowers the coolant's boiling point, and the coolant may boil suddenly. Use heavy gloves and wear eye and face protection to guard against scalding.

CAUTION —
Avoid adding cold water to the coolant while the engine is hot or overheated. If it is absolutely necessary to add coolant to a hot system, do so only with the engine running and coolant pump turning.

3. TROUBLESHOOTING

Problems with the heating or air conditioning system can have many causes. The first step in diagnosing any problem is to understand how the system works. For all practical purposes, the heating and ventilation system is separate from the A/C system. Therefore, problems should fall within one of two categories: heating faults or air conditioning faults. Each system has its own causes and corrective actions.

Insufficient heater output may be caused by a faulty heater valve, a cooling system fault, or a faulty electronic component. Coolant leaking visibly into the passenger compartment is a sign of a faulty heater core. A sweet, anti-freeze odor in the car's interior, or a constantly fogged windshield may also indicate a faulty core. Inspect the carpet and the area near the footwell vents for any moisture or coolant. Water appearing on the carpets or near the footwells could be a clogged A/C evaporator condensation drain hose.

To quickly check heater valve operation, turn the key on with the engine cold. Slowly turn the knob clockwise starting from its coldest setting. When the temperature setting on the knob reaches the car's interior temperature, the valve should close, making a sound like a light thump. To check the automatic function of the valve, turn the knob to a slightly warmer setting and listen for the valve to pulse on and off. The pulse period is approximately every 4 seconds.

NOTE —
If the valve functions as specified, but heater output is inadequate, the heater valve screen may be plugged or the valve diaphragm may be damaged. See **4.1 Heater Core and Heater Valve** for additional information.

Table a. Heating and Air Conditioning System Troubleshooting

Symptom	Probable cause	Corrective action
1. Heater output inadequate (temperature gauge reading normal)	**a.** Heater hose restricted **b.** Heater core or heater valve clogged or faulty **c.** Heater controls broken or out of adjustment	**a.** Replace hose. See **COOLING SYSTEM** **b.** Clean or replace heater valve. Have heater core cleaned. **4.1** **c.** Adjust or replace controls. **4.3, 4.4, 4.5**
2. Cooling output inadequate	**a.** System charge (refrigerant) low **b.** Evaporator temperature sensor faulty **c.** Evaporator regulator faulty	**a.** Check system charge at sight glass. Have system recharged if necessary. Check for leaks. **5.2** **b.** Test sensor and replace with new updated part if necessary. **5.4** **c.** Test regulator and replace with new updated part if necessary. **5.4**

4. HEATER AND CONTROLS

Temperature and air flow are controlled by instrument panel levers, a knob connected to a temperature control unit, and a blower switch. The upper and lower levers control variable-temperature air flow. The center lever controls only fresh air flow. On models up to 1986, the flaps that admit fresh air are controlled by a cable connected to the lever. On 1986 and later models, the flaps are controlled by a motor. The motor is also controlled by the center lever. The temperature regulating control unit controls the solenoid-type heater valve.

4.1 Heater Core and Heater Valve

The temperature regulating control unit regulates the opening and closing of the solenoid heater valve electrically. The valve is closed when voltage is applied and open when voltage is removed. The valve is either fully open or fully closed. There is no variable position.

If the temperature regulating knob is turned to its full hot position, automatic regulation is bypassed and the valve remains open. When the knob is turned to its cold setting, voltage is applied to the valve, closing it and preventing coolant flow through the heater core.

The replacement of the heater core is not covered in this manual. Heater core replacement involves complete removal of the center console, heater box and the A/C evaporator assembly. In order to do this the A/C system must be discharged and special equipment is needed to remove moisture from the system before it can be closed back up and recharged.

When replacing the heater valve, be prepared to catch any coolant that may spill when the coolant hose connections are opened. Also have some pre-mixed coolant on hand to top up the cooling system when finished.

To replace heater valve:

1. Disconnect the negative (–) cable from the battery.

2. Drain the coolant as described in **COOLING SYSTEM**.

    NOTE —
    It is possible to replace the valve without completely draining the cooling system. Release the coolant expansion tank cap, pinch each heater hose off near the heater valve with long nose locking pliers, then remove the heater valve.

3. Disconnect the electrical connector and the coolant hoses to the heater valve. See Fig. 4-1.

4. Remove the two mounting bolts and remove the valve.

What may seem to be a faulty heater valve may only be a clogged valve filter screen or torn diaphragm. To check for a clogged screen or damaged diaphragm, remove the four small screws from the top of the valve. Pull the diaphragm assembly from the valve. Inspect the diaphragm and clean the screen. If the screen cannot be cleaned, or the diaphragm is damaged, the diaphragm assembly should be replaced.

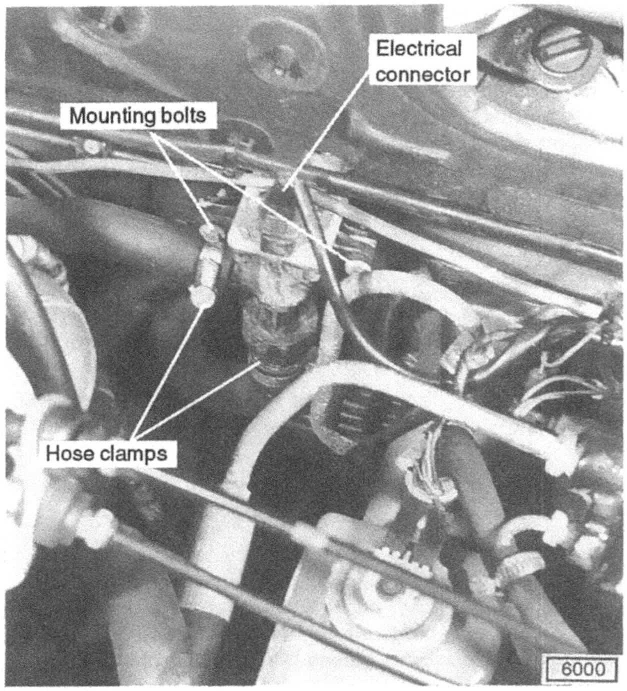

Fig. 4-1. Heater valve mounted on firewall.

NOTE —
The diaphragm assembly is available as a replacement part.

Installation is the reverse of removal. Fill and bleed the cooling system as described in **COOLING SYSTEM.**

4.2 Fresh Air Blower and Blower Switch

The fresh air blower is located behind the firewall in the engine compartment. If replacing the blower motor, check the blower motor code number on the motor housing to identify the correct replacement part.

CAUTION —
Do not remove the fan wheels or reposition them on the motor shaft. Blower motors are balanced during assembly. Changing the position of the fans can lead to noisy operation and rapid bearing failure.

The blower switch, located in the center console, controls both the fresh air blower and the A/C blower. To replace the switch, remove the center console as described in **BODY AND INTERIOR**. Pull off the knob, and remove the retaining nut from the front. Remove the switch.

Fresh Air Blower Electrical Tests

Fig. 4-2 shows the electrical circuit for the fresh air blower motor and resistor pack.

The speeds of the fresh air blower motor are controlled by routing voltage through different resistors. There are two resistors for three speeds of the motor. Direct battery voltage allows the fan to run at high speed. Maximum resistance produces the lowest speed.

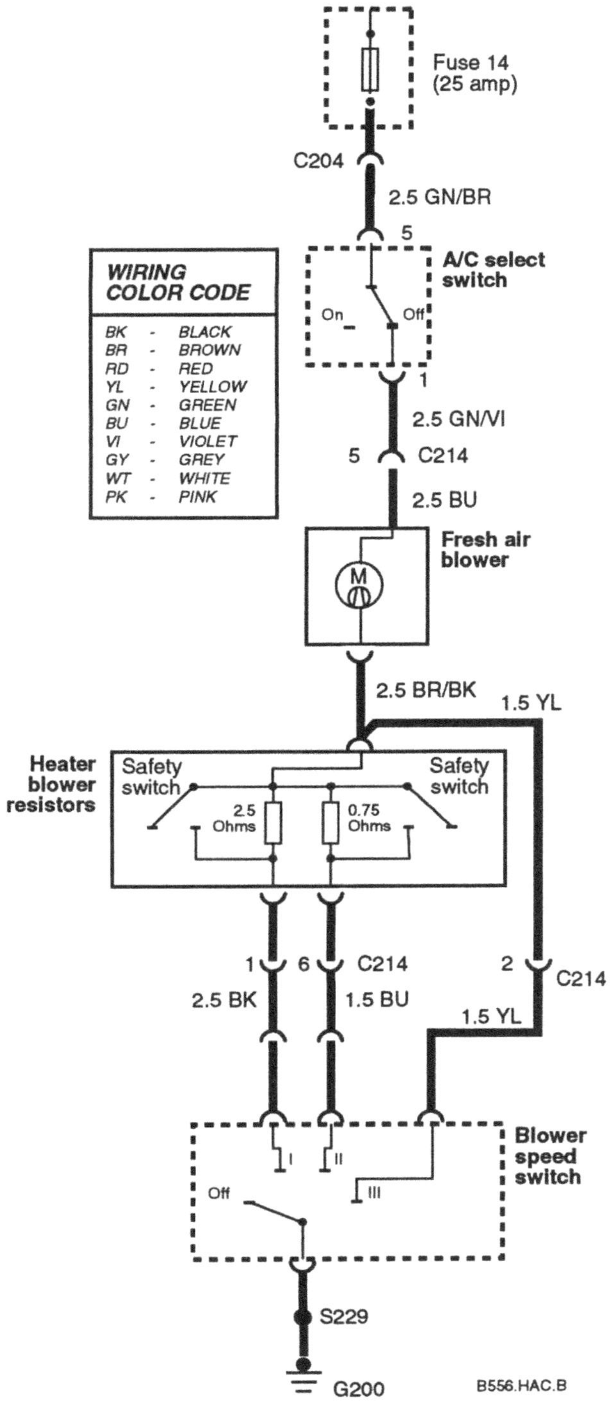

Fig. 4-2. Fresh air blower motor circuit and resistor pack.

NOTE —
A blower motor that only runs on high speed usually indicates a fault with the blower motor resistors.

To quickly check for a faulty blower motor, remove the access panel and blower motor housing covers as described above. With the ignition off, disconnect the connectors from the blower motor. Turn the ignition on and turn the blower motor speed switch to the high position. Make sure the A/C select switch is in the off position. Check for voltage between the harness connectors. If voltage is present, the blower motor is faulty and should be replaced as described above.

If there is no voltage reaching the motor, test the blower motor switch as described below under **5.4 Air Conditioning Electrical Tests**.

To remove and install fresh air blower:

1. Disconnect the negative (–) cable from the battery.

2. Remove the firewall access cover shown in Fig. 4-3. The cover is held in place by six screws. Disconnect all wiring harnesses from the cover by cutting the plastic ties.

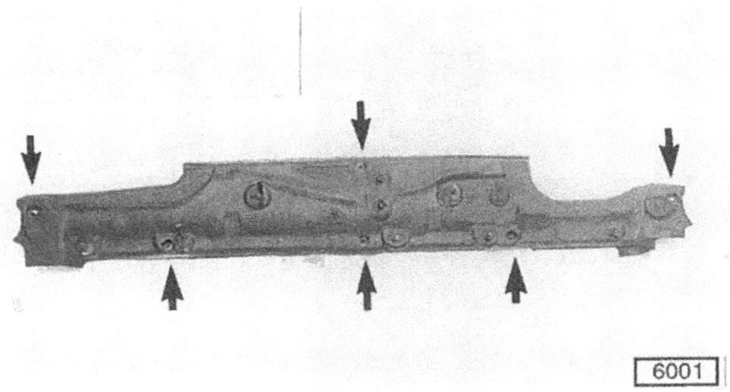

Fig. 4-3. Firewall cover mounting points (arrows).

3. Remove the blower housing shells by unhooking them at the front and rear. See Fig. 4-4.

4. Disconnect the blower wiring, then unclip the blower retaining clip shown in Fig. 4-5 and remove the blower motor.

NOTE —
The motor is located by a tab on its mounting. It may be necessary to rotate the motor around slightly in position until it is felt to engage the tab.

21

Fig. 4-4. Left blower motor housing shell being removed.

Fig. 4-5. Blower motor clip (arrow).

Installation is the reverse of removal. The fresh air blower resistor pack is located in the heater box. Replacement of the pack requires that the A/C system be discharged and the heater box be removed and is therefore not within the scope of this manual.

4.3 Heater Temperature Regulating Control Unit and Sensors

The temperature regulating knob is mounted directly to the control unit shaft. Changing the position of the knob changes the electrical setting of the control unit. The heater core temperature sensor is inserted into the left side of the heater box. The interior temperature sensor is mounted next to the hood release lever. These items are available as replacement parts. Check with an authorized BMW dealer for the latest information.

The temperature regulating control unit is also used to automatically control passenger compartment temperature for the A/C system when the A/C switch is on. For testing of other sensors and control units used in the A/C system, see **5.4 Air Conditioning Electrical Tests.**

Testing Temperature Regulating Control Unit

The control unit is solid state and cannot be repaired. The only test is to check for voltage and ground at control unit. If no fault is found with any other sensor or component, and the correct voltage and ground signals are present, the control unit is probably faulty and should be replaced.

Test voltage to the control unit by removing the harness connector from the control unit and turning the ignition on. Battery voltage should be present at terminal 2 (green/white wire) of the connector and ground. If voltage is not present check fuse no. 6 (8 amp). Test for ground at terminal 3 (brown wire from ground lug above the brake pedal).

To replace control unit:

1. Disconnect the negative (–) cable from the battery.
2. Remove the center console as described in **BODY AND INTERIOR**.
3. Pull off the temperature control knob.
4. Remove the harness connector from the back of the control unit, remove the three screws and remove the unit. See Fig. 4-6.

Installation is the reverse of removal. Make all electrical connections with the ignition off.

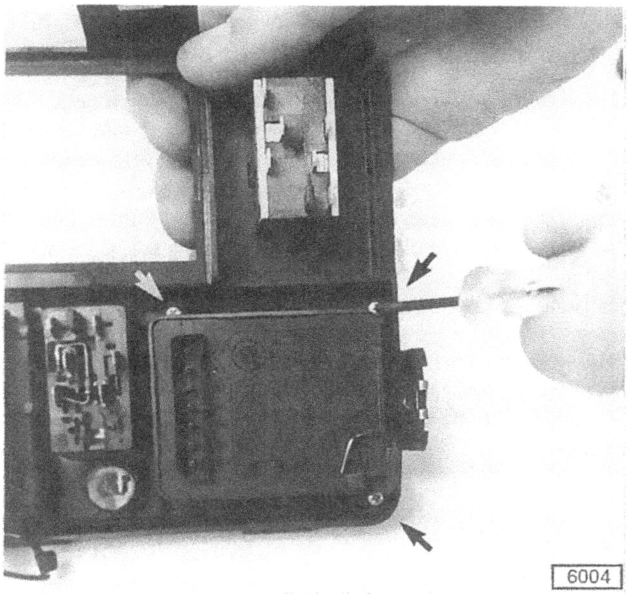

Fig. 4-6. Temperature regulating control unit being removed from trim panel. Mounting screws shown at arrows.

Testing Heater Core Temperature Sensor

The heater core temperature sensor can be tested after removing it from the heater box as described below. Fig. 4-7 shows the location of the sensor in the heater box. Use an ohmmeter at the sensor terminals for all tests listed in **Table b.**

Fig. 4-7. Heater core temperature sensor (arrow) partially removed in left side of heater box.

Table b. Heater Core Temperature Sensor Test Values

Temperature	Resistance range (ohms)
212°F (100°C)	580–770
140°F (60°C)	2190–2780
77°F (25°C)	9000–11000
68°F (20°C)	11130–13830
32°F (0°C)	28890–36400
14°F (-10°C)	48580–62080
-4°F (-20°C)	84390–109610

To replace heater core temperature sensor:

1. Disconnect the negative (–) cable from the battery.
2. Remove the screws securing the lower trim panel below the steering column and remove the panel.
3. Disconnect the harness connector from the sensor and remove sensor from heater box by pulling straight out. See Fig. 4-7 above.

Installation is the reverse of removal.

Testing Interior Temperature Sensor

The interior temperature sensor can be tested after disconnecting the harness connector from the rear of the sensor. **Table c** lists resistance values at various test temperatures. Fig. 4-8 shows the location of the interior temperature sensor on the lower left trim panel.

Table c. Interior Temperature Sensor Test Values

Temperature	Resistance range (ohms)	Terminal and wire color
77°F (25°C)	33±8.5	1 and 3 (green/yellow and brown)
104°F (40°C)	4700–5400	2 and 3 (blue and brown)
86°F (30°C)	7500–8400	
77°F (25°C)	9500–10500	
68°F (20°C)	12000–13300	
59°F (15°C)	15100–17200	
50°F (10°C)	19400–22100	
32°F (0°C)	32400–37700	

21

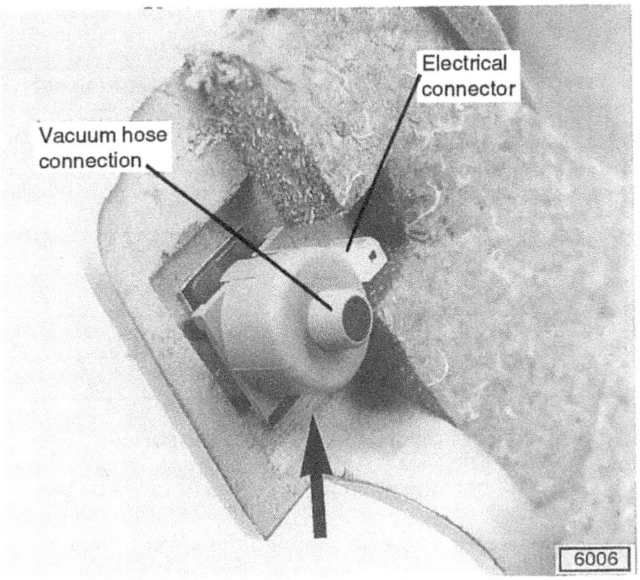

Fig. 4-8. Location of interior temperature sensor on trim panel (arrow).

To replace interior temperature sensor:

1. Disconnect the negative (–) cable from the battery.

2. Remove the screws securing the lower trim panel below the steering column and remove the panel and the sensor.

3. Disconnect the harness connector and vacuum hose from the sensor. Squeeze the two retaining tabs and remove the sensor from the bezel. See Fig. 4-9.

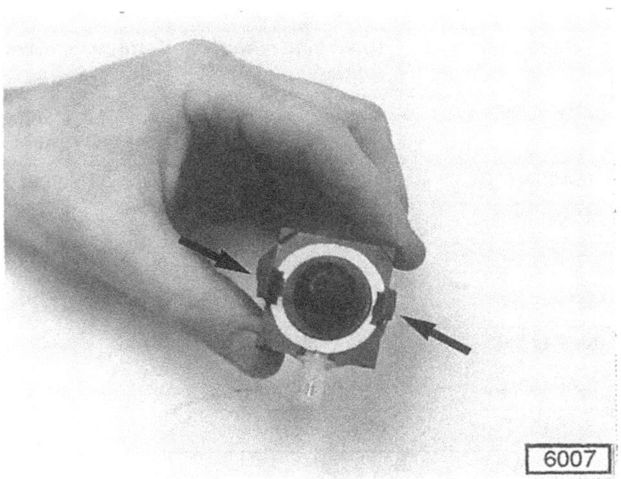

Fig. 4-9. Interior temperature sensor retaining tabs (arrows).

Installation is the reverse of removal.

4.4 Fresh Air Flap Control Unit, Actuator and Potentiometer (1986 and later models)

On 1986 and later models, fresh air entry is electronically controlled by a small motor. When the fresh air slide control is moved left or right, the electronic control unit actuates the flap motor. The control unit is powered from fuse no. 6 (8 amp). The unit is also powered from fuse no. 14 (25 amp) when the A/C switch is on to close the fresh air door if A/C is selected.

The slide lever is a potentiometer where resistance changes as the lever is moved. **Table d** lists resistance values for the potentiometer. Make all electrical tests at the connector on the rear of the panel. See Fig. 4-10. Also check that there are no sudden jumps in continuity as the lever is moved from side to side.

Table d. Fresh Air Potentiometer Test Values

Terminals	Lever position	Resistance (ohms)
1 and 2	left	approx. 10000
	right	approx. 0
1 and 3	left	approx. 0
	right	approx. 10000

Fig. 4-10. Fresh air flap potentiometer connector on 1986 and later models (arrow). Terminals are numbered left to right.

The fresh air flap motor is riveted to the right side of the heater box. To replace the motor the heater box must be removed from the car. Removal of the heater box is not covered in this manual.

To replace the fresh air flap potentiometer:

1. Disconnect the negative (–) cable from the battery.
2. Remove the center console as described in **BODY AND INTERIOR**. Remove the radio.
3. Pull the trim bezel containing the switches and levers away for easier access.
4. Detach the clips securing the cables. See Fig. 4-11. Note the color and position of each cable before detaching. Disconnect the harness connector to the center lever.

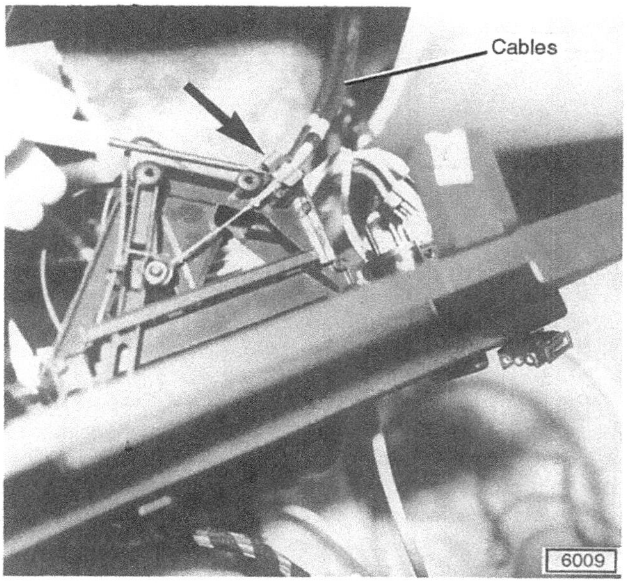

Fig. 4-11. Heater control cable being unclipped (arrow) at operating lever. Make sure threaded plastic piece on cable is directly under the clip before attaching clip.

5. Remove the screws securing the lever assembly to the trim bezel and remove the assembly.

Installation is the reverse of removal.

> ***CAUTION —***
> *Be very careful when reattaching cables to the lever assembly. The plastic posts on the assembly are delicate and will snap off easily.*

4.5 Control Cables

The heating and ventilation control cables that control air circulation cannot be easily reached without removing the heater box assembly. Removal of the cables is not covered in this manual.

The control cables are color coded and are held to the operating levers by removable clips. When reinstalling the cables to the levers, make sure that the cable is installed in its original position before reinstalling the clip. See Fig. 4-11 above.

5. AIR CONDITIONING

Air conditioning (A/C) service and repair requires special equipment and knowledge. Incorrect procedures may not only damage the system, but also may be hazardous. Pressures in excess of 300 psi are created in the system when it is operating. The refrigerant used (R-12) is not poisonous, but in its vapor form it can accumulate in areas with poor ventilation and cause suffocation. Also, in vapor or liquid form R-12 can immediately freeze anything it contacts, including eyes and skin.

To regulate passenger compartment temperature and to prevent condenser freeze-up and A/C system damage, the compressor is switched on and off by the evaporator temperature regulator. This is a control unit located to the left of the evaporator. During A/C operation, some air must flow from the outlets to prevent evaporator freeze-up.

Although tests can determine A/C efficiency, it is recommended that all service to the system be left to an authorized BMW dealer or other qualified repair shop. For information on the A/C selector switch assembly, as well as other electrical components, see **5.4. Air Conditioning Electrical Tests**. Dismounting of the A/C compressor and the condenser, without disconnecting the hoses, is covered as part of the engine removal procedure in **ENGINE—SERVICE AND REPAIR**. If any of the hoses or components are disconnected and the system is opened, special equipment will be needed to remove moisture from the system before it can be closed up.

5.1 System Description

Fig. 5-1 is a schematic view of a typical BMW A/C system. The A/C system removes heat and moisture from the passenger compartment. It accomplishes this through the application of four principles: materials absorb heat as they change from a liquid to a gas (evaporate); materials give off heat as they change from a gas to a liquid (condense); the boiling point of a liquid varies with its pressure; and, heat always flows from hot to cold. For example, the first principle is demonstrated by wetting your hand and then blowing on it. As the water evaporates, it takes some heat with it and your hand feels cooler.

In the air conditioning system, the heat from the passenger compartment boils (evaporates) the refrigerant (R-12) in the evaporator, causing heat to be absorbed by the R-12. This heat is then released into the atmosphere when the R-12 is cooled and condensed into a liquid at the condenser. Moisture is removed at the evaporator in the same way that water drops form on a cold glass. The moisture drips onto the water tray beneath the evaporator in the heater box and is routed outside via drain hoses. This is the reason a water puddle may often be seen under the car when the A/C is operating.

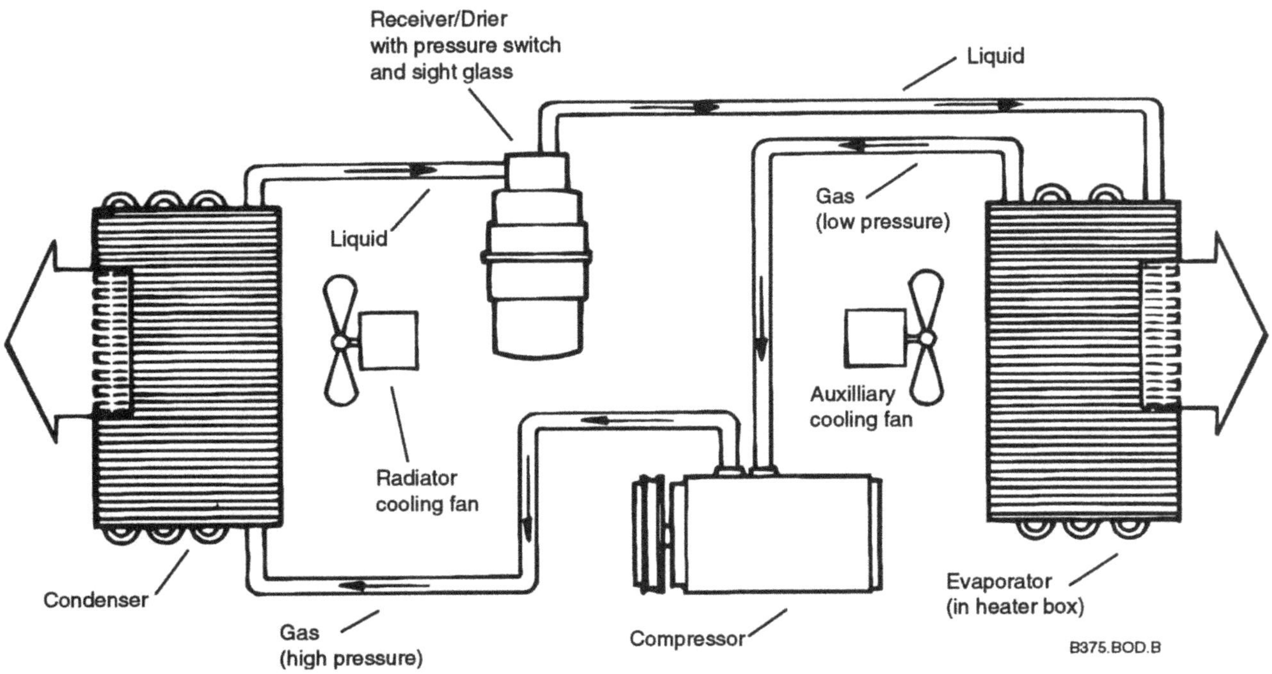

Fig. 5-1. View of typical BMW air conditioning system.

The compressor forces the R-12 through the system and at the same time pressurizes it, raising the R-12's boiling point to make it more easily condensed. The compressor is engaged by an electro-magnetic clutch that is actuated when the A/C is turned on. The temperature sensors and control unit automatically disengages the clutch when the temperature in the passenger compartment reaches the level set on the operating controls.

The condenser, which looks like a small radiator, is located in front of the engine radiator. The receiver/drier removes small amounts of moisture and dirt from the system. The evaporator regulator control unit, via the temperature sensor on the evaporator, interrupts power to the compressor clutch if the temperature of the evaporator gets too cold. 1986 and later models have an A/C oil temperature switch to shut off the compressor, and an in-line diode to the compressor.

Safety Features

To prevent system freeze-up, the A/C blower comes on at low speed whenever the A/C is turned on. The evaporator regulator turns the compressor on and off. A high-pressure switch interrupts power to the compressor clutch if the pressure in the system is excessive. Additional coolant temperature switches on the radiator increase auxiliary radiator fan speed in two steps if the coolant temperature is excessive. Depending on the model, there may be two switches or one combined switch.

5.2 Inspections and Tests

Periodic inspections will help keep the A/C operating at its peak. There are almost always small leaks in the system that will require that it eventually be recharged with R-12. The V-belt that drives the compressor is subject to wear. The condenser fins can become bent or covered with debris, reducing air flow and raising system pressure to damaging levels.

> **NOTE —**
> Run the air conditioning system for a few minutes every few weeks to keep the seals lubricated.

With the engine off, clean any debris or bugs from the front of the condenser. Straighten any bent fins using a fin comb. Check the fresh air intake for obstructions. Inspect the compressor, the hoses, and all visible components for any oil leaks. These leaks are often seen at the bottom side of the fittings and components. Inspect the wiring to the pressure switches and compressor clutch. Check the compressor mountings for tightness, and check the condition and tension of the V-belt as described in **LUBRICATION AND MAINTENANCE**. Finally, be certain that the condensation drain hoses from the passenger compartment are clear. This can be accomplished by blowing them out with low pressure compressed air.

> ***WARNING —***
> *Wear eye protection when inspecting the system. R-12 at normal atmospheric pressures can evaporate and freeze anything it contacts.*

The following tests will aid in evaluating system performance.

Checking Refrigerant Charge

Inspect the refrigerant charge of R-12 by starting the engine and turning the air conditioner on to MAX A/C. With the compressor running (clutch cycled on) view the sight glass, located in the engine compartment just behind the passenger-side headlights. See Fig. 5-2. There should be few or no bubbles visible in the glass. A constant foaming indicates that the system charge is low. Streaks on the interior of the glass may indicate that the system is totally discharged.

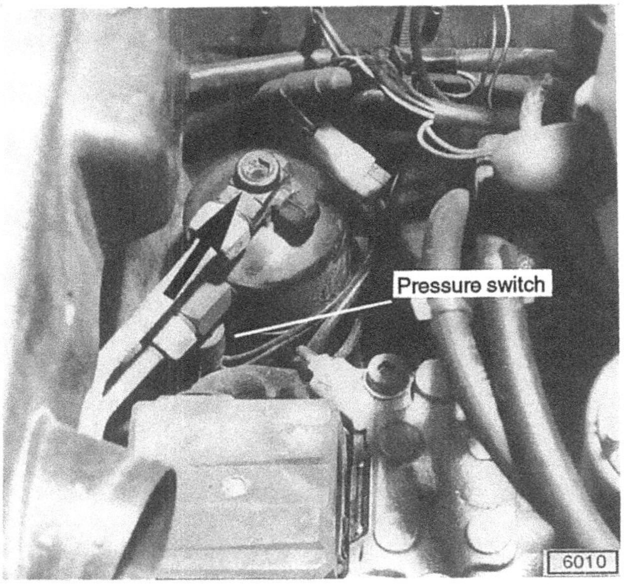

Fig. 5-2. Air conditioning sight glass (arrow) on receiver/drier.

If using an aftermarket refrigerant product to recharge the system, follow the manufacturer's directions closely. To prevent component damage, a totally discharged system must be evacuated (sometimes called pulling a vacuum) using special equipment before recharging. This removes any moisture from the lines and components, which can damage the system.

> ***WARNING —***
> *Adding refrigerant or servicing the system without the proper tools, equipment, or knowledge may cause severe personal injury as well as damage to A/C components.*

5.3 Air Conditioning Specifications

The specifications listed below in **Table e** are intended to be used by those experienced in air conditioning service.

Table e. Air Conditioning Specifications

Refrigerant capacity (R-12)	1300 grams (2.9 lbs.)
Refrigerant oil capacity (total)	170 cc
High pressure switch	
Opens above	26.6 bar (385 psi)
Closes below	21.2 bar (307 psi)
Coolant temperature switches	
Low fan speed, closes	approx. 197°F (91°C)
High fan speed, closes	approx. 210°F (99°C)

5.4 Air Conditioning Electrical Tests

Covered under this heading are tests for the A/C blower motor, the A/C compressor clutch, and the A/C selector switch. Also covered here are the electrical tests for the evaporator regulator (control unit) and the evaporator temperature sensor. Both of these components are used to automatically regulate passenger compartment temperature when the A/C system is on.

> **NOTE —**
> Tests for the temperature regulating control unit, which is also used for automatic A/C temperature control, are covered earlier under **4.3 Heater Temperature Regulating Control Unit and Sensors.**

Testing A/C Blower Motor

When checking the evaporator blower motor, first remove the center console as described in **BODY AND INTERIOR.** Disconnect the connector for the blower motor and turn the ignition and the A/C selector switch on. With the fan switch on high, check for battery voltage at the connector. If voltage is present and the blower does not operate, the motor is faulty and should be replaced. If voltage is not present, check for a faulty fuse (no. 14), damaged wires or connectors, or a faulty blower motor switch. Switch continuity tests are listed below in **Table f.** Switch terminals are identified in Fig. 5-3.

Table f. Evaporator Blower Motor Switch Tests

Test terminals	Switch position	Test results
Fresh air blower		
1 and +	high (III)	continuity
2 and +	medium (II)	continuity
3 and +	low (I)	continuity
Evaporator blower		
4 and +	low (I)	continuity
5 and +	medium (II)	continuity
6 and +	high (III)	continuity

21

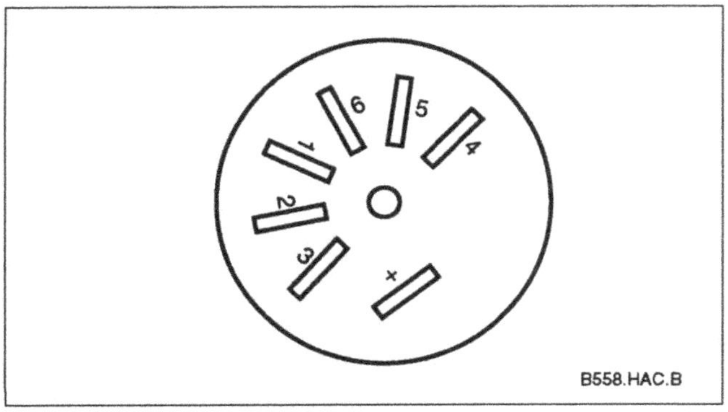

Fig. 5-3. Blower motor switch terminal identification as viewed from rear.

Testing A/C Compressor

To quickly check the operation of the compressor clutch, turn the ignition on and turn the temperature regulating knob to its full cold position. Turn the A/C switch on and listen for the compressor clutch to click on.

If no click is heard, the compressor clutch may not be receiving voltage or the clutch itself may be faulty. Check for voltage between the single-wire connector leading out of the compressor and ground. If voltage is present, the compressor clutch is faulty and should be replaced. If no voltage is present, check for a faulty fuse (fuse 14), an open high pressure switch (in fitting of receiver/drier), or damaged wiring. Also check for a faulty (open) suppression diode using an ohmmeter. The diode is mounted on the compressor.

If no faults are found up to this point, test the evaporator regulator and the evaporator temperature sensor as described below. Both of these components control the operation of the compressor clutch.

Testing A/C Selector Switch

The A/C selector switch assembly can be tested after removing it. To remove the air conditioning selector switch, carefully pry the switch away from the trim panel and unplug the connector harness. Make continuity checks at the electrical connections of the switch assembly with the switch in the ON and OFF positions. Fig. 5-4 shows a schematic of the A/C selector switch assembly.

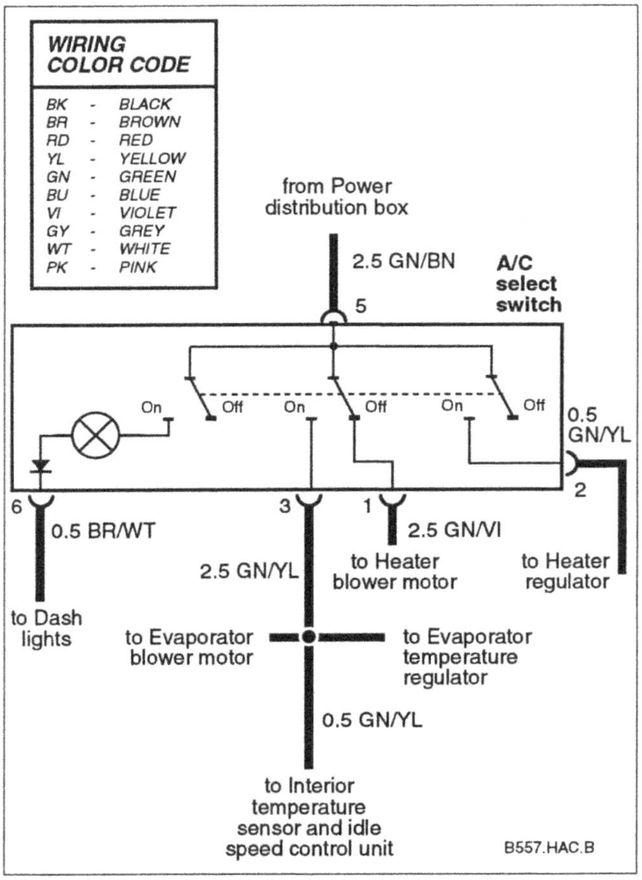

Fig. 5-4. Schematic of air conditioning selector switch assembly.

Testing Evaporator Temperature Sensor

The evaporator temperature sensor can be tested after removing it from the heater box as described below. Use an ohmmeter at the sensor terminals for all tests listed in **Table g.** Fig. 5-5 shows the location of the sensor in the evaporator assembly. Remove the center console to access the sensor as described in **BODY AND INTERIOR.**

CAUTION —
Static can permanently damage solid state modules and control units. Be sure to use the proper handling techniques and equipment.

NOTE —
Various versions of evaporator temperature sensors were used in the cars covered by this manual. Only one sensor is presently available from BMW. This updated improved sensor supersedes all other earlier sensors.

Fig. 5-5. Evaporator temperature sensor being removed (arrow).

Table g. Evaporator Temperature Sensor Test Values

Temperature	Resistance range (ohms)
95°F (35°C)	1700–1900
86°F (30°C)	2100–2300
77°F (25°C)	2600–2900
68°F (20°C)	3300–3600
59°F (15°C)	4200–4500
50°F (10°C)	5300–5600
41°F (5°C)	6800–7200
32°F (0°C)	8400–9200
23°F (-5°C)	11400–11900

Testing Evaporator Regulator

The evaporator regulator turns the compressor on and off. When the temperature of the evaporator is too cold, the regulator turns the compressor off. The regulator also cycles the compressor on and off based on the input from the temperature regulating control unit.

The evaporator regulator is an electronic component and cannot be easily tested. The only easy tests that can be made are checks for voltage and ground at the regulator.

NOTE —
Three different evaporator temperature regulators were used on the cars covered by this manual. The earliest allowed the compressor to ice up under certain conditions. Newer ones correct this problem. On regulators with code no. 1 368 753-64.5 (BMW part no. 64 50 1 368 753), check for a yellow or green paint dot(s). If no paint dots are present, the regulator should be replaced with the later version. Check with an authorized BMW dealer for replacement under the BMW warranty. Regulators with code no. 1 376 710 64.5 and all regulators with code no. 1 368 753-64.5 having paint dots should not be replaced unless found to be faulty.

To check for voltage and ground, remove the center console as described in **BODY AND INTERIOR.** Remove the connector from the regulator and check for battery voltage between terminal 15 and ground. See Fig. 5-6. Check for continuity to ground at terminal 31. If no faults are found and no other components are found to be faulty, the regulator is probably faulty and should be replaced.

Fig. 5-6. Location of evaporator regulator (arrow).

Most other testing of the A/C system, including the operation of the pressure switches, requires specialized knowledge and equipment, and is beyond the scope of this manual. Servicing by an authorized BMW dealer or other qualified air conditioning shop is highly recommended.

Section 22

ELECTRICAL SYSTEM

Contents

2 ELECTRICAL SYSTEM

TABLES

INTRODUCTION

The electrical system is an efficient means for distributing power to the electrical components of the car. It does this with the help of the alternator, which converts some of the engine's mechanical energy into electrical energy. The electrical energy is then carried through wires to the various electrical components such as motors, light bulbs, or electronic control units. The battery in the system supplies electrical power when the engine is not running, and also supplies power to start the engine.

1. GENERAL DESCRIPTION

The electrical system is based on negative (–) ground. In other words, the negative terminal of the battery is connected to the car body, and any electrical connection to the body is a connection to ground. Voltage from the battery to the various electrical components is carried by the wiring harness. Most components are then grounded through either direct mounting to the car body or by a ground wire leading to the car body.

The connecting terminals on components and wiring harness connectors can be either pins, or sockets into which pins fit. Most terminals are identified with a number stamped next to the pins on the components. On relays, the terminals are also identified by numbers on the relay. Throughout this section, sockets on the fuse/relay panel are identified by referring to the pin that fits into them. The terminal numbers and the location of all the major electrical connections can be found in the diagrams included in this section and in the BMW Electrical Troubleshooting Manual (ETM) for your specific vehicle. BMW ETMs are available through an authorized BMW dealer parts department.

All electrical circuits except those required for starting and operating the engine are protected by fuses. To prevent accidental shorts that might blow a fuse or damage wires and electrical components, the negative (–) battery cable should always be disconnected before working on the electrical system. On models equipped with anti-theft radios, the anti-theft circuitry is activated when power to the battery is removed, such as when the battery cables are removed. Be sure to have the anti-theft reactivation code on hand before removing the battery terminals or the radio fuse.

A brief description of the principal parts of the electrical system is presented here for familiarization with the system. The components are discussed in greater detail later in this section.

Voltage and Polarity

The cars covered by this manual have a 12-volt, direct current (DC), negative-ground electrical system. The voltage regulator maintains the voltage in the system at approximately 12 to 14.8 volts. All circuits are grounded by direct or indirect connection to the negative (–) terminal of the battery.

Battery

Battery capacity is determined by the amount of current needed to start the car and by the number of electrical accessories. For more information on checking the battery and on battery maintenance, see **BATTERY, STARTER, ALTERNATOR**.

Wiring, Fuses, and Relays

Nearly all parts of the wiring harness connect to components of the electrical system with keyed, push-on connectors that lock into place. Notable exceptions are the heavy battery cables and the alternator wiring. The wiring is color-coded for circuit identification.

With the exception of the battery charging system, all electrical power is routed from the ignition switch or the battery through the fuse/relay panel, located in the left side of the engine compartment. Fuses prevent excessive current from damaging components and wiring. Fuses are color coded to indicate their different current capacities.

Ceramic type fuse rating and color	
• 8 amp	white
• 16 amp	red
• 25 amp	blue

The relays are electromagnetic switches that operate on low current to switch a high-current circuit on and off. Most of the relays are mounted on the fuse/relay panel. For information concerning relay and fuse locations, see **12. Fuse/Relay Panel**.

Lights

The lighting system includes all exterior and interior lights. The headlight circuit contains separate relays for the high beams, the low beams, and the fog lights. High/low beam switching is handled at the contacts in the headlight dimmer switch, which is mounted to the steering column. The low beam check relay monitors headlight low beam operation. Information on changing bulbs and lenses can be found in **BODY AND INTERIOR**. Information on relay location can be found under **12. Fuse/Relay Panel.**

Heating and Air Conditioning

All of the cars covered by this manual are equipped with an electronically-controlled heating and ventilation system. This climate-control system automatically regulates the passenger compartment temperature based on the temperature selected at the temperature regulating knob. For information on the heating and air conditioning system, see **HEATING AND AIR CONDITIONING**.

Windshield Wipers and Washers

The blades of the two-speed windshield wiper system with intermittent operation come to rest in the "park" position automatically when the wiper switch is turned off. The wiper control unit, located in the main fuse/relay panel, controls the wiper motor when the wiper switch is in the pause and fast positions. The wiper switch includes a windshield washer control. A motor-driven pump supplies the washers with fluid. Models from 1984 on have heated spray jets.

Instruments

The dashboard instruments, including the speedometer, are all electrical. All differences in instrumentation among models are covered in the applicable wiring diagrams. Removal and installation of the instrument cluster is described in **BODY AND INTERIOR**.

2. MAINTENANCE

No routine lubrication or maintenance is required of any electrical motor or electrical system component. For more information on battery maintenance, see **BATTERY, STARTER, ALTERNATOR**, or **LUBRICATION AND MAINTENANCE**.

3. TROUBLESHOOTING

This heading describes general procedures for electrical system circuit and component troubleshooting, and provides specific troubleshooting information. For information regarding the fuses and relays mounted in the fuse/relay panel, see **12. Fuse/Relay Panel.** For information regarding splice, ground, harness connector, relay, and component locations see **11. Harness Connector, Ground, Splice and Component Locations.** Read the following cautions before testing any part of the electrical system.

CAUTION
- *Connect or disconnect multiple connectors and test equipment leads only while the ignition is off. Switch multimeter functions or measurement ranges only with the test leads disconnected.*

- *Before operating the starter without starting the engine (as when making a compression test), disable the ignition. For more information see **IGNITION.***

- *Do not connect terminal 1 of the coil to ground as a means of preventing the engine from starting (for example, when installing or servicing anti-theft devices).*

CAUTION —
- *Do not connect test instruments with a 12-volt current supply to terminal 15 (+) of the ignition coil. The voltage backflow may damage the Motronic control unit. In general, make test connections only as specified by BMW, as described in this manual, or as described by the instrument's manufacturer.*

- *Do not disconnect the battery while the engine is running.*

- *Disconnect the battery when doing any electric welding on the vehicle or charging the battery.*

- *Do not quick-charge the battery (for boost starting) for longer than one minute, and do not exceed 16.5 volts at the battery with the boosting cables attached. Wait at least one minute before boosting the battery a second time. On models equipped with on-board computer, remove the computer fuses (no. 5, no. 6, no. 9, and no. 12) prior to quick-charging to prevent damaging the computer.*

- *Do not wash the engine while it is running, or any time the ignition is switched on.*

- *Do not use a test lamp that has a normal incandescent bulb to test circuits containing electronic components. The high electrical consumption of these test lamps may damage the components.*

- *Many of the solid-state modules are static sensitive. Static discharge will permanently damage them. Always handle the modules using proper static prevention equipment and techniques.*

3.1 Basic Electrical Troubleshooting Principles

Four things are required for current to flow in any electrical circuit: a voltage source, wires or connections to transport the voltage, a consumer or device that uses the electricity, and a connection to ground. It is very important for trouble-free operation that the ground connections, including the negative battery cable and the body ground strap, remain clean and free from corrosion. Most problems can be found using only a multimeter (volt/ohm/amp meter) to check for voltage supply, for breaks in the wiring (infinite resistance/no continuity), or for a path to ground that completes the circuit.

Electric current is logical in its flow, always moving from the voltage source toward ground. Keeping this in mind, electrical faults can be located through a process of elimination. When troubleshooting a complex circuit, separate the circuit into smaller parts. The general tests outlined below may be helpful in finding electrical problems. The information is most helpful when used with the wiring diagrams found throughout this section and in the applicable BMW ETM.

Testing for Voltage and Ground

The most useful and fundamental electrical troubleshooting technique is checking for voltage and ground. A voltmeter or a simple test light should be used for this test. For example, if a parking light does not work, checking for the presence of voltage at the bulb socket will determine if the circuit is functioning correctly or if the bulb itself is faulty. If voltage and ground are found at the bulb connector, but the bulb does not illuminate, the bulb is most likely faulty.

> **NOTE —**
> Test leads with flat male connectors should be used to prevent damaging terminal connections. For basic information on using a multimeter or constructing test leads, see **FUNDAMENTALS** at the front of the manual.

To check for positive (+) battery voltage using a test light, connect the test light wire to a clean, unpainted metal part of the car or a known good ground. Use the pointed end of the light to probe the positive (+) wire. See Fig. 3-1. To check for continuity to ground, connect the test light wire to the positive (+) battery post or a battery source. Use the pointed end of the light to probe the wire leading to ground. In either case, the test light should light up.

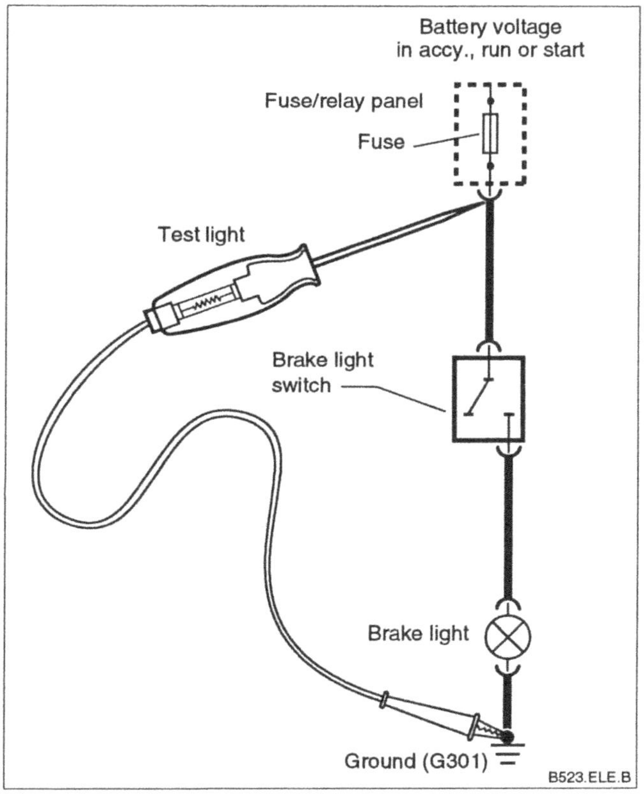

Fig. 3-1. Test light being used to check for voltage. A test light is the best tool to quickly check for voltage or ground.

> **NOTE —**
> • A test light only determines if voltage or ground is present—it does not determine how much voltage or how good the path to ground is. If the actual voltage reading is important, such as when testing a battery, use a digital voltmeter. To check the condition of the ground connection, check the voltage drop on the suspected connection as described below.
>
> • The pointed end of the test light can be used to pierce through the wire's insulation. This may help locate areas where wires are broken or connections are faulty.

To check for voltage using a voltmeter, the meter should be set to DCV and the correct scale. Connect the negative (–) test lead to the negative (–) battery terminal or known good ground. Touch the positive (+) test lead to the positive wire or connector. To check for ground, connect the positive (+) test lead to the positive (+) battery terminal or voltage source. Touch the negative (–) test lead to the wire leading to ground. In either case the meter should read battery voltage. See Fig. 3-2.

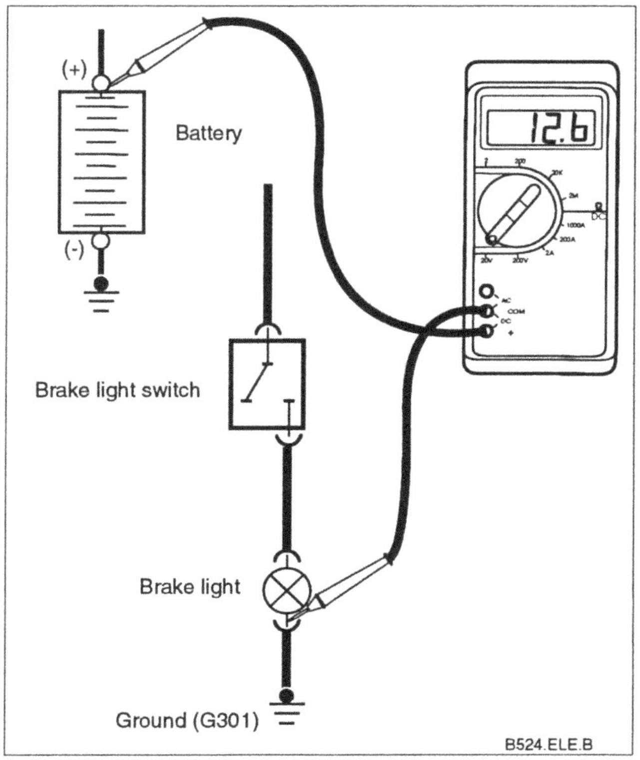

Fig. 3-2. Voltmeter being used to check for ground.

> **NOTE —**
> When using an analog (swing needle) voltmeter, be careful not to reverse the test leads. Reversing the polarity may damage the meter.

Continuity Test

The continuity test can be used to check a circuit or switch. Because most automotive circuits are designed to have little or no resistance, a circuit or part of a circuit can be easily checked for faults using an ohmmeter or a self-powered test light. An open circuit or a circuit with high resistance will not allow current to flow. A circuit with little or no resistance allows current to flow easily.

When checking continuity, the ignition should be off. On circuits that are powered at all times, the battery should be disconnected. Fig. 3-3 shows a brake light switch being tested for continuity using an ohmmeter. Using the appropriate wiring diagram, a circuit can be easily tested for faulty connections, wires, switches, relays, and engine sensors by checking for continuity.

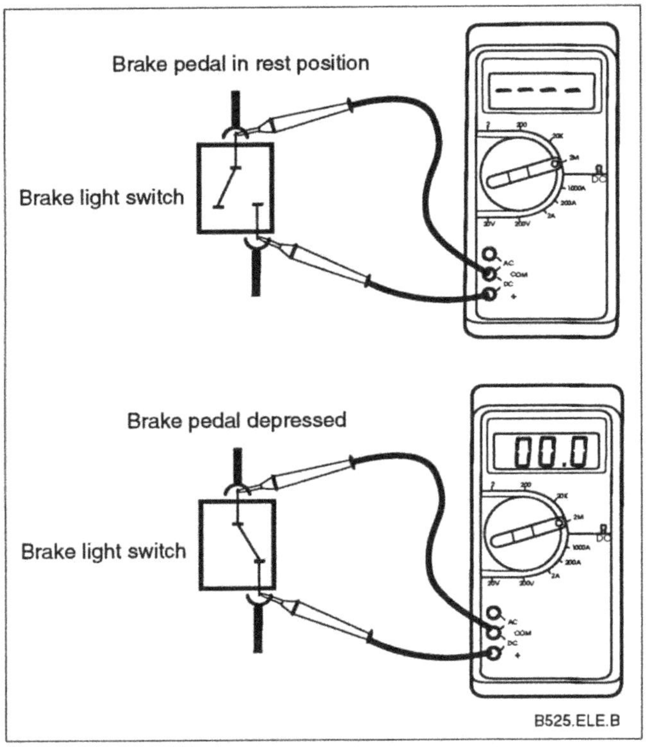

Fig. 3-3. Brake light switch being tested for continuity. With brake pedal in rest position (switch open) there is no continuity (infinite ohms). With the pedal depressed (switch closed) there is continuity (zero ohms).

CAUTION —
Do not use an analog (swing-needle) ohmmeter to check circuit resistance or continuity on any electronic (solid-state) components. The internal power source used in most analog meters can damage solid state components. Use only a high quality digital ohmmeter having high input impedance.

Short Circuit Test

A short circuit is exactly what the name implies. The circuit takes a shorter path than it was designed to take. The most common short that causes problems is a short to ground where the insulation on a positive (+) wire wears away and the metal wire is exposed. When the wire rubs against a metal part of the car or other ground source, the circuit is shorted to ground. If the exposed wire is live (positive battery voltage), a fuse will blow and the circuit may possibly be damaged.

CAUTION —
On circuits protected with large fuses (25 amp and greater), the wires or circuit components may be damaged before the fuse blows. Always check for damage before replacing fuses of this rating.

Shorts to ground can be located with a voltmeter, a test light, or an ohmmeter. Short circuits are often difficult to locate and may vary in nature, Therefore, it is important that the correct wiring diagram is available. Short circuits can be found using a logical approach based on the current path.

CAUTION —
*When replacing blown fuses, use only fuses having the correct rating. Always confirm the correct fuse rating printed on the fuse/relay panel cover. See **12. Fuse/Relay Panel** at the end of this section.*

To check for a short circuit to ground, remove the blown fuse from the circuit and disconnect the cables from the battery. Disconnect the harness connector from the circuit's load or consumer. Using a self-powered test light or an ohmmeter, connect one test lead to the load side fuse terminal (terminal leading to the circuit) and the other test lead to ground. See Fig. 3-4.

Short circuits can also be located using a test light or a voltmeter. Connect the instrument's test leads across the fuse terminals (fuse removed) and turn the circuit on. See Fig. 3-5. If necessary, check the wiring diagram to determine when the circuit is live.

Working from the wire harness nearest to the fuse/relay panel, move or wiggle the wires while observing the test light or the meter. Continue to move down the harness until the test light blinks or the meter displays a reading. This is the location of the short. Visually inspect the wire harness at this point for any faults. If no faults are visible, carefully slice open the harness cover or the wire insulation for further inspection. Repair any faults found.

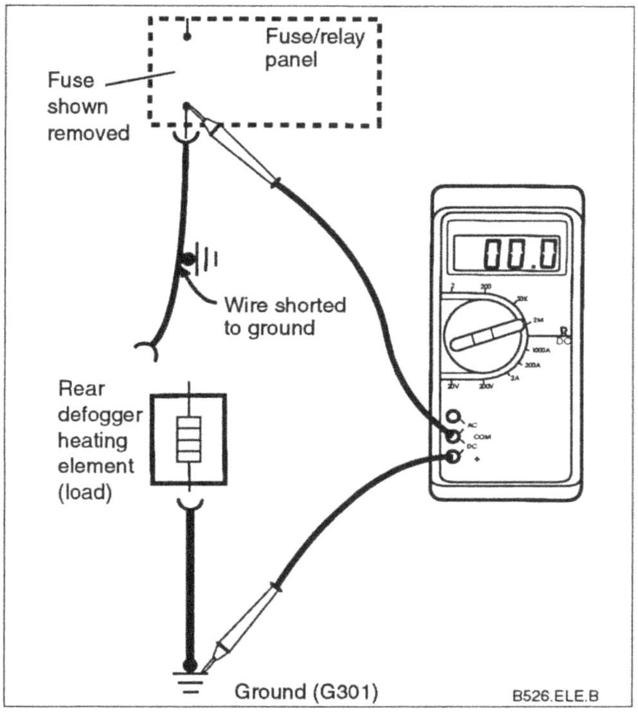

Fig. 3-4. Ohmmeter being used to check for a short circuit to ground.

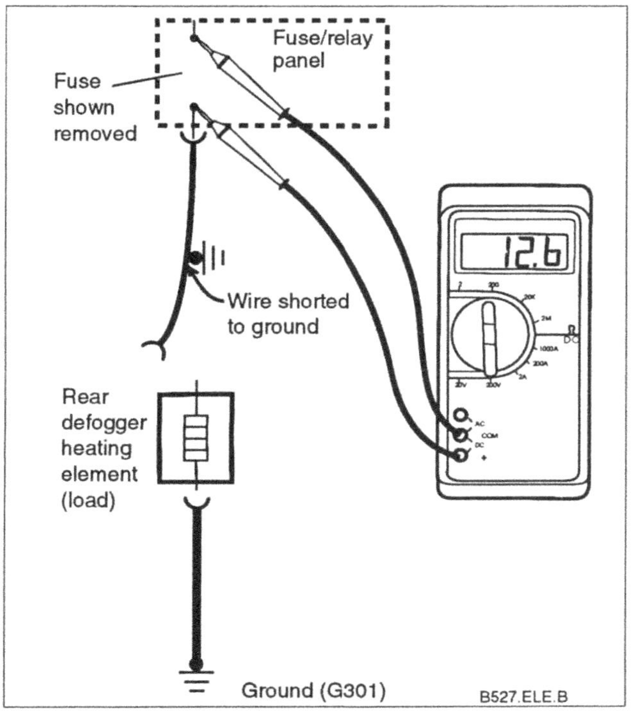

Fig. 3-5. Voltmeter being used to check for a short circuit to ground.

Voltage Drop Test

The wires, connectors, and switches that carry current are designed with very low resistance so that current flows with a minimum loss of voltage. A voltage drop is caused by higher than normal resistance in a circuit. This additional resistance actually decreases or stops the flow of current. A voltage drop can be noticed by problems ranging from dim headlights to sluggish wipers. Some common sources of voltage drops are faulty wires or switches, dirty or corroded connections or contacts, and loose or corroded ground wires and ground connections.

Voltage drop can only be checked when current is running through the circuit, such as by operating the starter motor or turning on the headlights. Making a voltage drop test requires measuring the voltage in the circuit and comparing it to what the voltage should be. Since these measurements are usually small, a digital voltmeter should be used to ensure accurate readings. If a voltage drop is suspected, turn the circuit on and measure the voltage at the circuit's load. See Fig. 3-6.

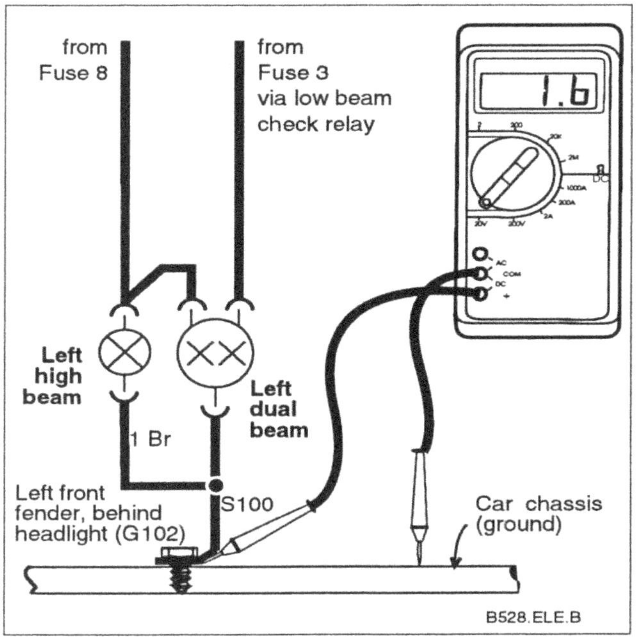

Fig. 3-6. Example of voltage drop test on dim headlights. Voltmeter showed 1.6 Volt drop between ground connector and chassis ground. After removing and cleaning headlight ground connector (G102), voltage drop returned to normal (0.2 volts) and headlights were bright.

NOTE —
- A voltage drop test is generally more accurate than a simple resistance check because the resistances involved are often too small to measure with most ohmmeters. For example, a resistance as small as 0.02 ohms results in a 3 volt drop in a typical 150 amp starter circuit. (150 amps x 0.02 ohms = 3 volts).

- Keep in mind that voltage with the key on and voltage with the engine running are not the same. With the ignition on and the engine off (battery voltage), voltage should be approximately 12.6 volts. With the engine running (charging voltage), voltage should be approximately 14.5 volts. Measure voltage at the battery with the ignition on and then with the engine running to get exact measurements.

- The maximum voltage drop in an automotive circuit, as recommended by the Society of Automotive Engineers (SAE), is as follows: 0 volt for small wire connections; 0.1 volt for high current connections; 0.2 volt for high current cables; and 0.3 volt for switch or solenoid contacts. On longer wires or cables, the drop may be slightly higher. In any case, a voltage drop of more than 1.0 volt usually indicates a problem.

To troubleshoot a circuit:

1. Check to see that all the connections in the circuit are tight and free of corrosion. Pay special attention to all ground connections.

2. Check the fuse. In the circuit example shown in Fig. 3-7 below, the circuit receives power through an 8 amp fuse in position 6 on the fuse/relay panel.

   NOTE —
   A fuse that repeatedly fails indicates an unusually high current flow, probably caused by a damaged wire, a faulty component, or a short directly to ground.

3. Check for voltage reaching the circuit. Connect a voltmeter or test light between a point in the circuit and ground (a clean, unpainted metal part of the car). In the example, the ignition must be switched on since fuse no. 6 only receives power when the ignition is on or the engine is running.

4. The blue/white wire (terminal 1) at the rear light assembly should be getting voltage when the ignition is on and the transmission is in reverse. Voltage indicated by the voltmeter (or by the test light coming on) means that the circuit is doing its job.

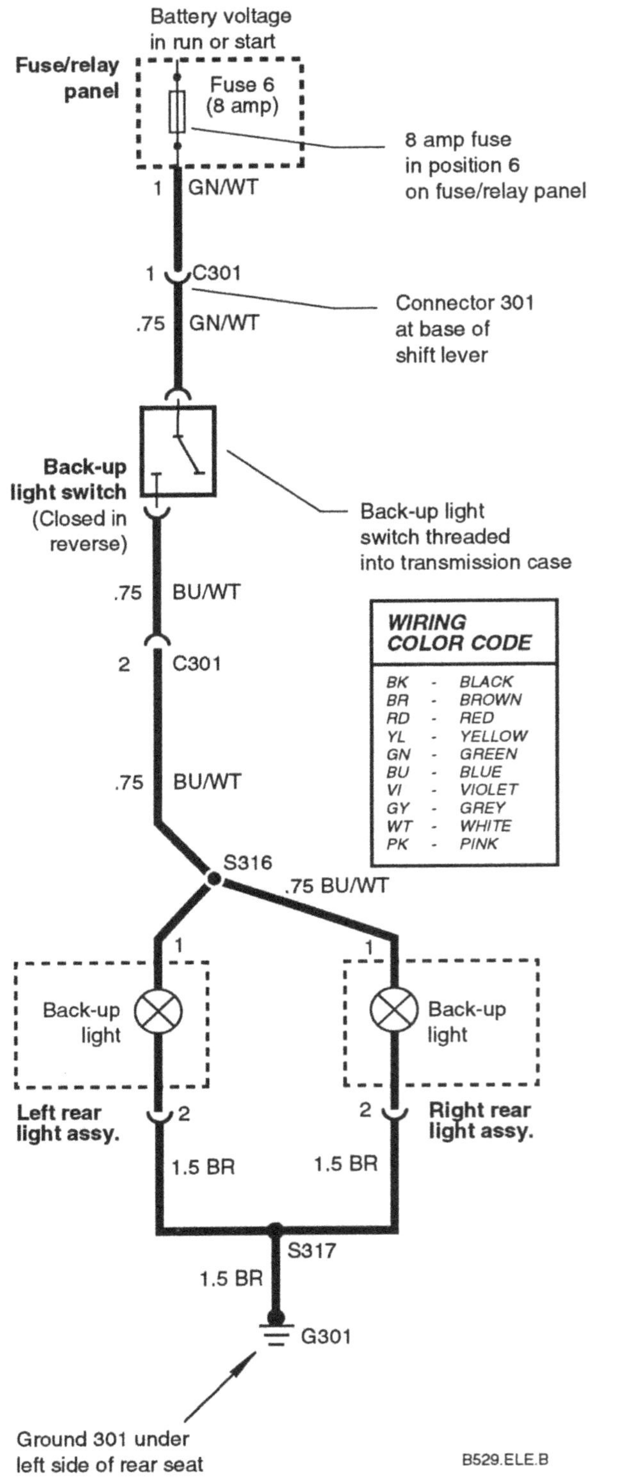

Fig. 3-7. Sample of wiring diagram of back-up lights circuit for cars with manual transmission.

5. If there is no indication of voltage, something has interrupted the circuit between the power source and the test point. Pick a new test point farther "upstream" in the current path. In the example, the green/white wire at the back-up light switch would be a good point. If there is voltage at this point, then the fault lies somewhere between the two test points.

6. Check circuit integrity between two points using an ohmmeter. Continuity (little or no resistance) between two points indicates that part of the circuit is complete and is allowing current flow. No continuity (infinite resistance) indicates that part of the circuit is interrupted and would not allow current flow. In the example, operation of the back-up light switch could be checked by disconnecting the connectors from the switch and testing for continuity across its terminals. There should be continuity only when in reverse gear—otherwise the switch is faulty and should be replaced.

7. If the back-up lights work intermittently, check voltage drop at the switch with the transmission in reverse. Connect the digital voltmeter test leads to the rear of each connector at the switch. Turn the ignition on. A voltage reading of more than 1 volt indicates a faulty switch.

3.2 How To Use The Wiring Diagrams

The wiring diagrams presented in this section of the manual are organized to indicate current flow, from positive to negative. The diagrams are also organized to show the actual routing of the wires in the car's wiring harness.

As a general rule, the diagrams show current flow from positive (+) to negative (–). The fuse, or source of voltage, is normally shown at the top of the diagram and the circuit ground is normally at the bottom of the diagram.

Fig. 3-8 is an example of a wiring diagram in this section, showing the meanings of various symbols. These are general examples. They do not show all of the symbols used. For a complete listing of the symbols, see **Wiring Diagram Symbols** below.

Wiring Diagram Symbols

Fig. 3-9 is a full listing of the wiring and component symbols appearing in the diagrams. A complete list of ground, splice, connector, and component locations, can be found under **11. Harness Connector, Ground, Splice, and Component Locations.**

Terminal and Circuit Identification

Most terminals are identified by numbers on the components and harness connectors. The terminal numbers for the major electrical connections are shown in the circuit diagrams throughout this section and in the BMW ETM.

Though many terminal numbers appear only once to identify a particular terminal, several numbers appear in numerous places throughout the electrical system and identify certain types of circuits. A letter suffix is sometimes added to the number to distinguish between two different circuits, or between two parts of the same circuit. **Table a** lists several of the most common circuit numbers, and identifies the circuit type and the wire color normally used.

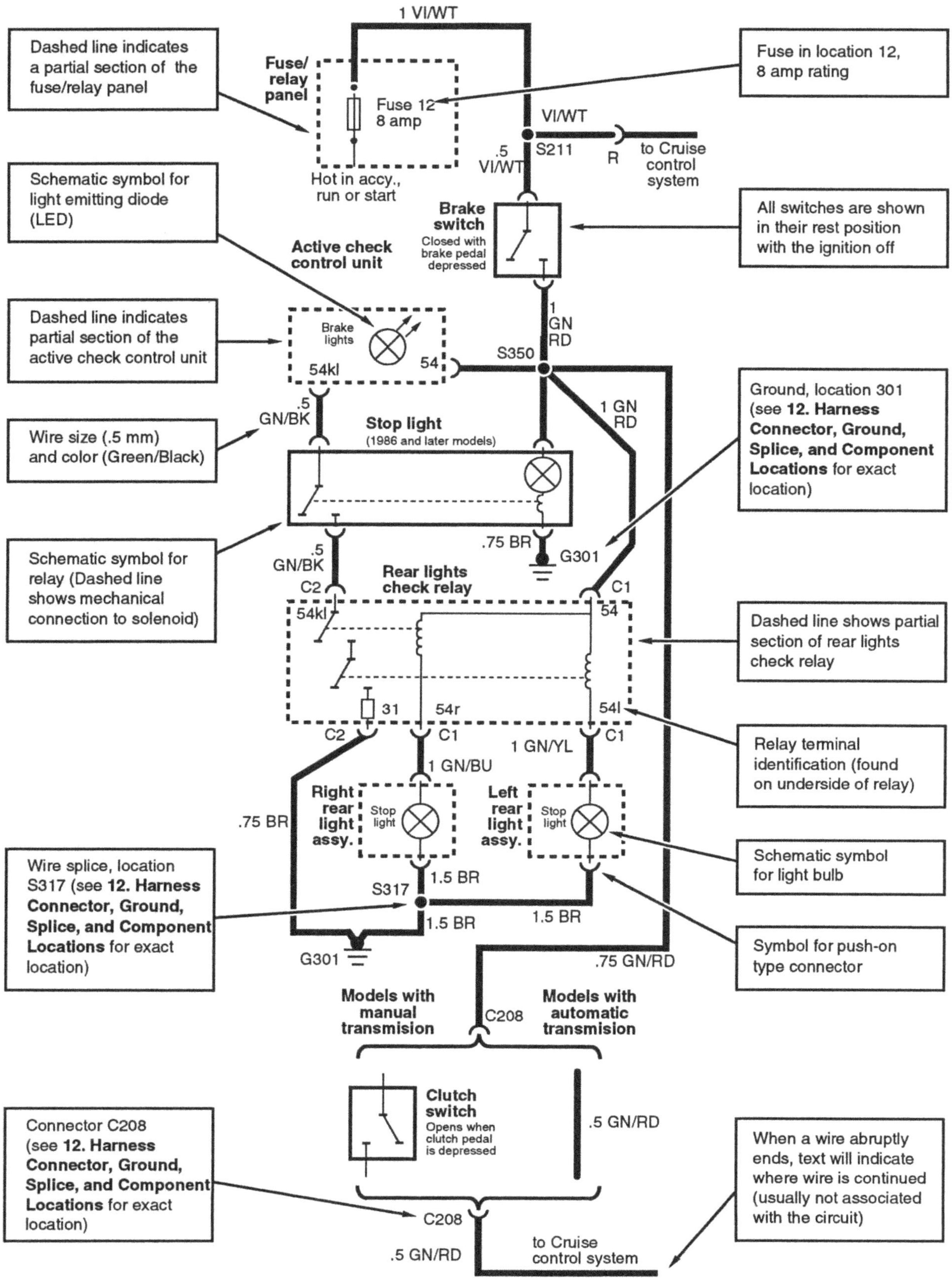

Fig. 3-8. Sample wiring diagram of stop light circuit.

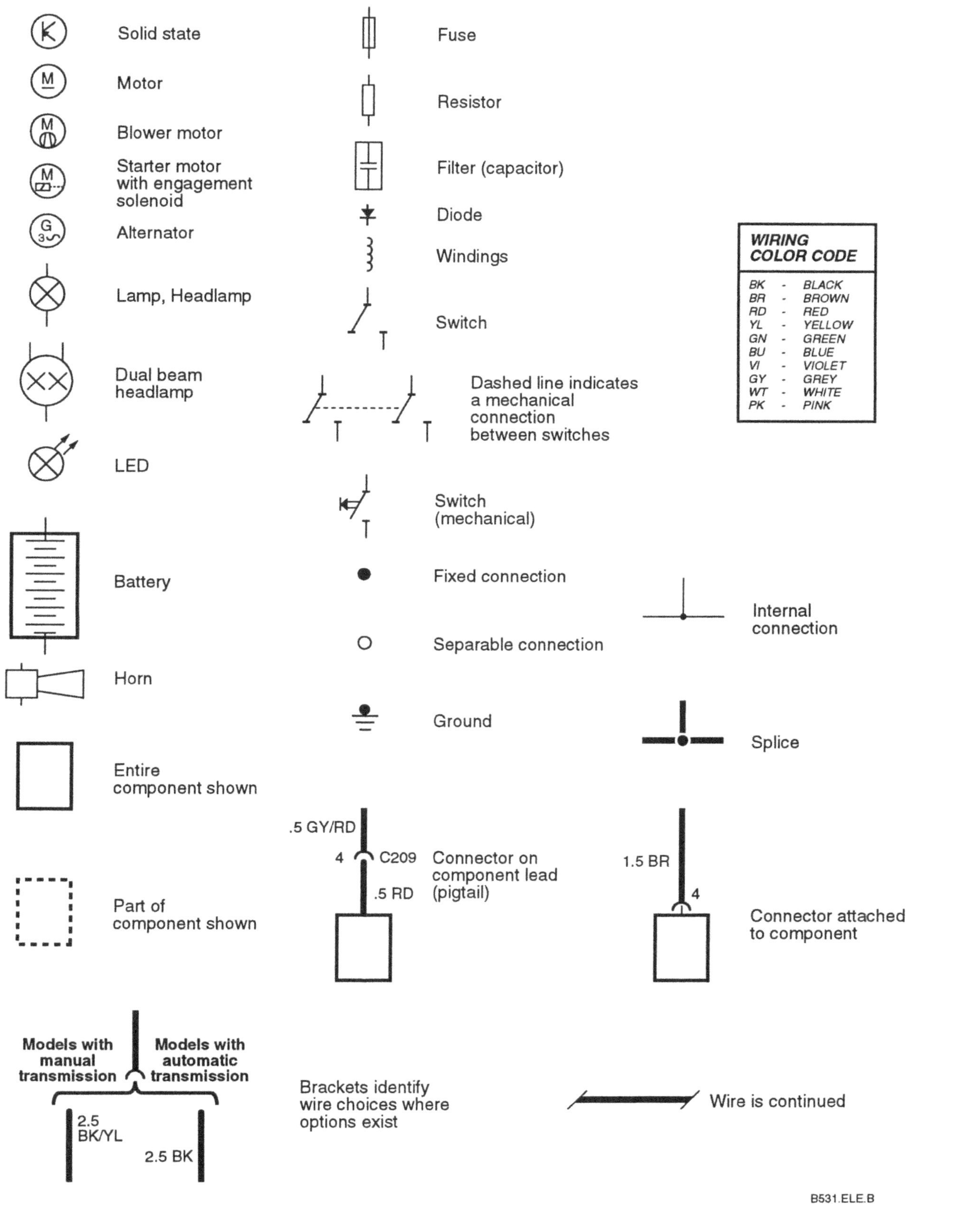

Fig. 3-9. Wiring and component symbols in the wiring diagrams in this section of the manual.

Table a. Terminal and Circuit Numbers

Number	Circuit description	Most common wire color
1	Low voltage switched (1) terminal of coil	Black
4	High voltage center terminal of coil	NA
15	Originates at ignition switch. Supplies power when ignition switch is in ON or START position	Green
30	Battery positive (+) voltage. Supplies power whenever battery is connected. (Not dependent on ignition switch position)	Red
31	Ground battery negative (–)	Brown
50	Supplies power from battery to starter solenoid when ignition switch is in START position	Black/yellow
D+	Alternator warning light and field energizing circuit	Blue
85	Ground side (–) of relay coil	Brown
86	Power-in side (+) of relay coil	NA
87	Relay change-over contact	NA

4. EXTERIOR LIGHTS

This heading covers exterior light electrical circuits. Replacement of bulbs and lenses for the driving lights, taillights, side marker lights, turn signals, brake lights, back-up lights, and headlights is covered in **BODY AND INTERIOR**.

4.1 Troubleshooting Exterior Lights

Most of the exterior lighting involves pairs of lights, and this fact is an aid to troubleshooting. If only one of a pair of lights is out—one taillight for example—then the problem is most likely due to a failed bulb or some other problem with that particular light socket or its wiring. A simple test is to exchange the bulb with its counterpart that is known to be good. If the same bulb fails to light in a new location, then the bulb is faulty and should be replaced. If the same light fails to light with the other bulb, then the problem is in the socket or wiring.

Many lighting problems are due to dirty or corroded sockets, or loosely-fitting bulb contacts or connectors. Check that voltage is reaching the bulb and that the socket has a good connection to ground.

If a pair of bulbs are both out—both taillights for example—then the problem is most likely in some part of the system that is common to both lights. Begin by checking for a failed fuse. Test switches using simple continuity checks made with an ohmmeter. Check the switch connectors for voltage and continuity to ground.

Headlights

Fig. 4-1 is a diagram for the headlight circuit. Power to the headlights comes only with the ignition on (15 circuit). When the headlight switch and the ignition switch are on, the low beam relay closes to supply power to the low beams. If the high beams are selected with the headlight dimmer switch, the high beam relay closes and the low beam relay opens.

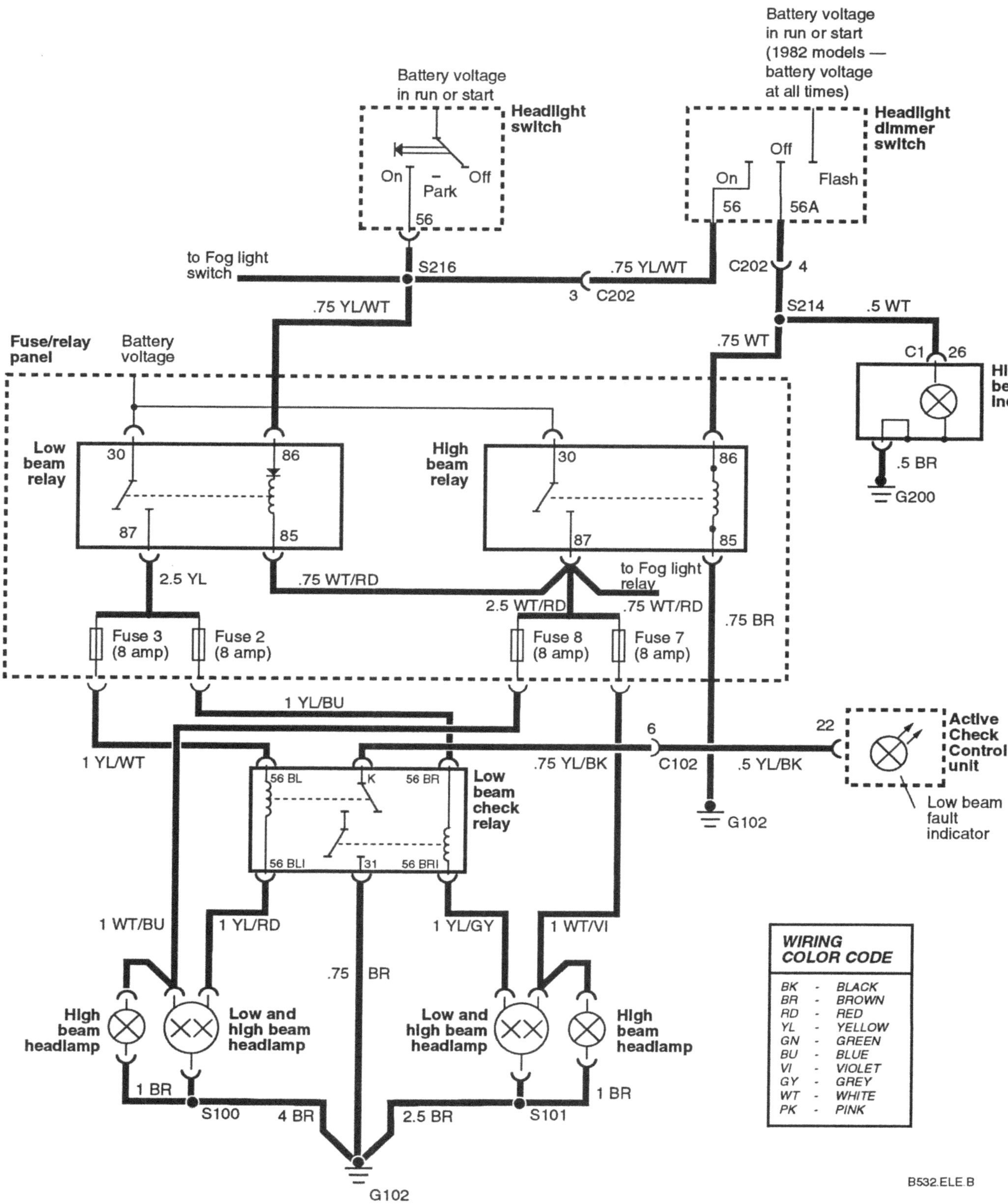

Fig. 4-1. Headlight circuit. Circuit may vary slightly on some models.

Begin troubleshooting headlight problems by checking for voltage at the faulty headlamp electrical connector with the light switch on and the ignition in the ON position. If no voltage is present, a fuse or relay is probably faulty and should be replaced. If voltage is present, the headlamp itself is faulty and should be replaced.

If either both low beams or both high beams are out and no fuses are faulty, the problem is most likely due to a faulty relay. With the ignition off, remove the low beam and high beam relays. See **12. Fuse/Relay Panel.** Check the headlight circuit using the information in **Table b**. Do all electrical checks at the sockets of the fuse/relay panel corresponding to the relay terminal numbers found on the bottom of the relay.

NOTE —
The low beam check relay, part of the check control system, is not the same as the low beam relay. The low beam check relay is mounted at the left front of the main fuse/relay panel.

Table b. Headlight Circuit Electrical Tests

Fuse/Relay panel sockets (relays removed)	Relay terminal	Test conditions	Correct test results
High beam relay socket	30 and ground	—	Battery voltage (12 VDC)
	85 and ground (G102)	—	Continuity
	86 and ground	Ignition ON headlight switch on high beam switch on	Battery voltage (12 VDC)
	86 and ground	Ignition ON headlight switch off high beam switch in flash position	Battery voltage (12 VDC)
	87 and ground	Headlight switch off high beam switch off	Continuity
Low beam relay socket	30 and ground	—	Battery voltage (12 VDC)
	85 and ground	Headlight switch off high beam switch off	Continuity
	86 and ground	Ignition ON headlight switch on	Battery voltage (12 VDC)
	87 and ground	—	Continuity

Test the headlight switch by first removing it from the instrument panel. Carefully pry the switch assembly away from the dashboard with a small screwdriver. Disconnect the electrical connector from the light switch. Using an ohmmeter, check for continuity between the switch terminals listed in **Table c.** Terminals identification can be found on rear of switch. If no faults are found, test the headlight dimmer switch as described under **8. Steering Column Switches.**

Table c. Headlight Switch Continuity Tests

Headlight switch position	Test terminals	Correct test results
Headlights on	56 (8) and 15u (9)	Continuity
Parking Lights on	58L (4) and 30 (5)	Continuity
	58R (3) and 30 (5)	Continuity
	58S (1) and 58K (2)	Continuity
Instrument Illumination (parking lights on or headlights on)	31 (7) and 31g (6)	Resistance changes as illumination knob is rotated
Switch Illumination	15u (9) and 30 (5)	Continuity

NOTE —
If the low beam indicator on the check control panel lights up when the cruise control is active and the headlights are on, the cruise control actuator may be causing interference with the low beam check relay. Exchange the entire relay and socket from the front position to one of the rear relay positions. See Fig. 4-2.

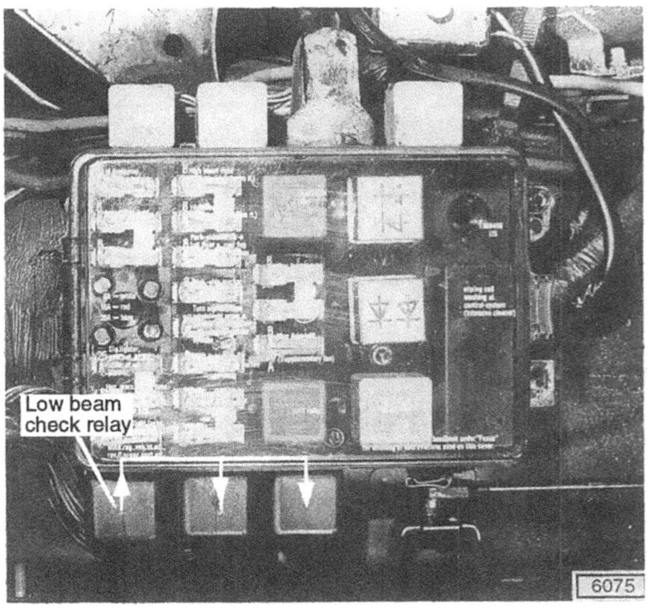

Fig. 4-2. Relay positions on main fuse/relay panel. Move low beam check relay into adjacent position (arrows) if cruise control causes check control panel light to come on.

Turn Signals and Emergency Flashers

Power for the turn signal circuit is routed through the emergency flasher switch only when the ignition is on (15 circuit). The emergency flasher circuit receives power from the battery at all times. When the emergency flashers switch is on, the turn signal switch is bypassed. Fig. 4-3 shows the location of the turn signal/flasher relay. Fig. 4-4 is a schematic of the turn signals and emergency flasher electrical circuit.

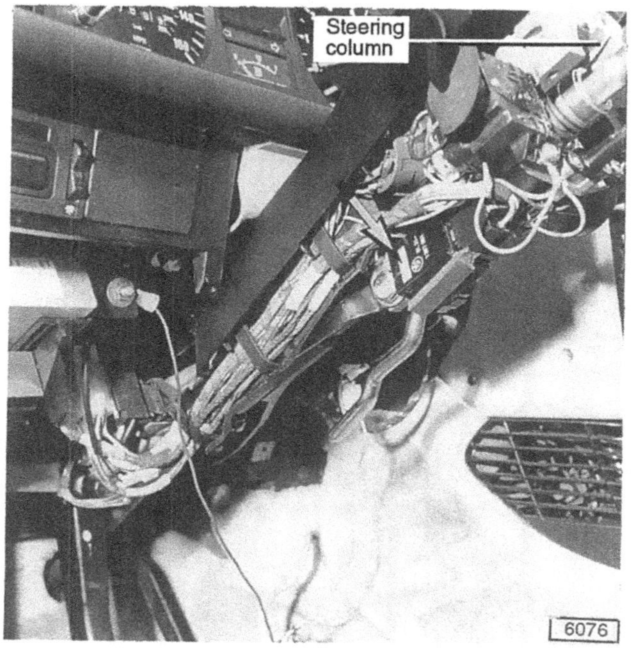

Fig. 4-3. Turn signal/emergency flasher relay (arrow) and harness connector. Flasher is located beneath steering column, below steering wheel.

The turn signals and the emergency flasher share the same flasher relay, but each circuit has its own fuse. One system working indicates that the flasher relay is good. Look for failed fuses no. 5 or no. 11.

If both the flashers and turn signals are inoperative, then it is probably either the flasher relay or the emergency flasher switch that is faulty. Check to see if power is reaching the relay.

Remove the lower steering column trim and remove the relay. Check that the emergency flasher switch is in the OFF position. With the ignition ON, check for voltage between terminal 49 (green/violet wire) of the harness connector and ground. Check for continuity between the terminal 31 (brown wire) of the connector and ground. Repair any wiring faults found. If no faults are found, the relay is probably faulty and should be replaced.

Remove the emergency flasher switch by carefully prying it away with a small knife or screwdriver. Check for continuity in the ON and OFF positions. Use **Table d** as a guide. If no faults are found, test the turn signal switch as described in **8. Steering Column Switches**.

Table d. Emergency Flasher Switch Continuity Tests

Switch Position	Terminals	Test result
Off	1 (15) and 5 (49)	Continuity
On	5 (49) and 3 (30)	Continuity
	4 (49a) and 6 (BR)	Continuity
	4 (49a) and 2 (BL)	Continuity

Parking Lights, Taillights, Side Marker Lights, and License Plate Lights

The parking, side marker, tail, and license plate lights all receive power directly from the battery through the light switch in the first or second position, independent of the ignition switch. If these lights fail together, first look for failed fuses no. 9 or no. 10. If no faults are found, test the switch as described above under **Headlights**. Fig. 4-5 is a circuit diagram for the license plate and rear side marker lights. Fig. 4-6 is a typical circuit diagram for the parking, tail, and front side marker lights.

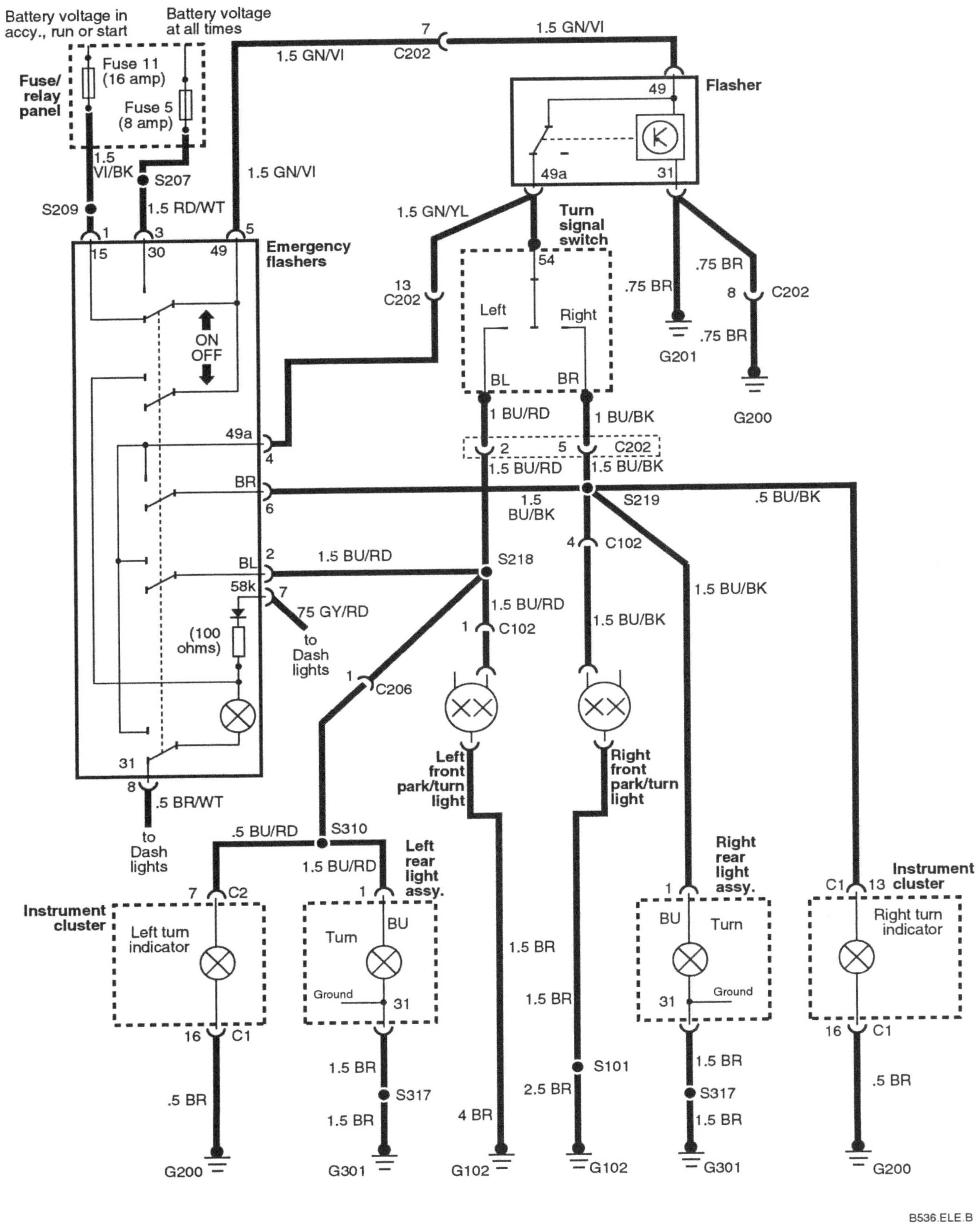

Fig. 4-4. Turn signal and emergency flasher circuit.

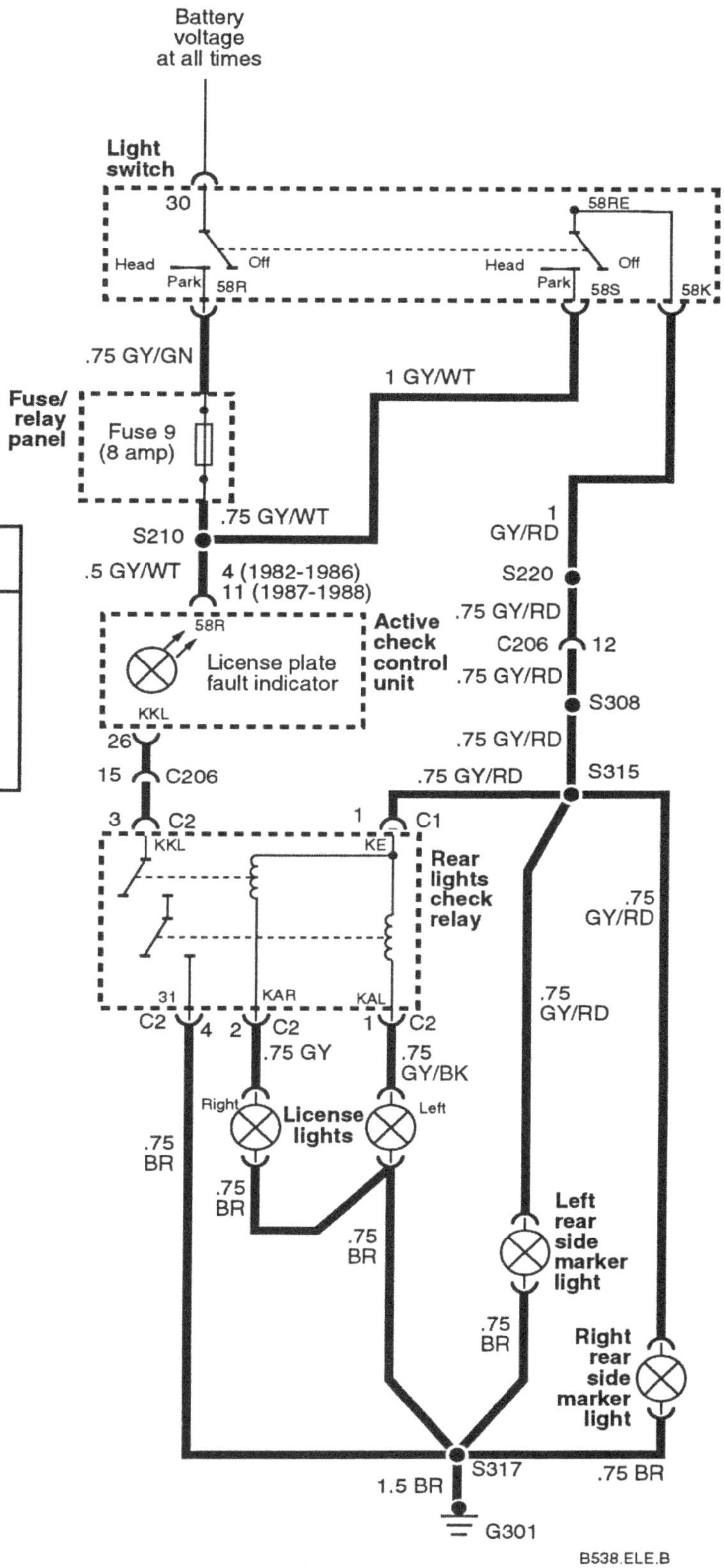

WIRING COLOR CODE		
BK	-	BLACK
BR	-	BROWN
RD	-	RED
YL	-	YELLOW
GN	-	GREEN
BU	-	BLUE
VI	-	VIOLET
GY	-	GREY
WT	-	WHITE
PK	-	PINK

Fig. 4-5. License plate lights and rear side marker lights circuit.

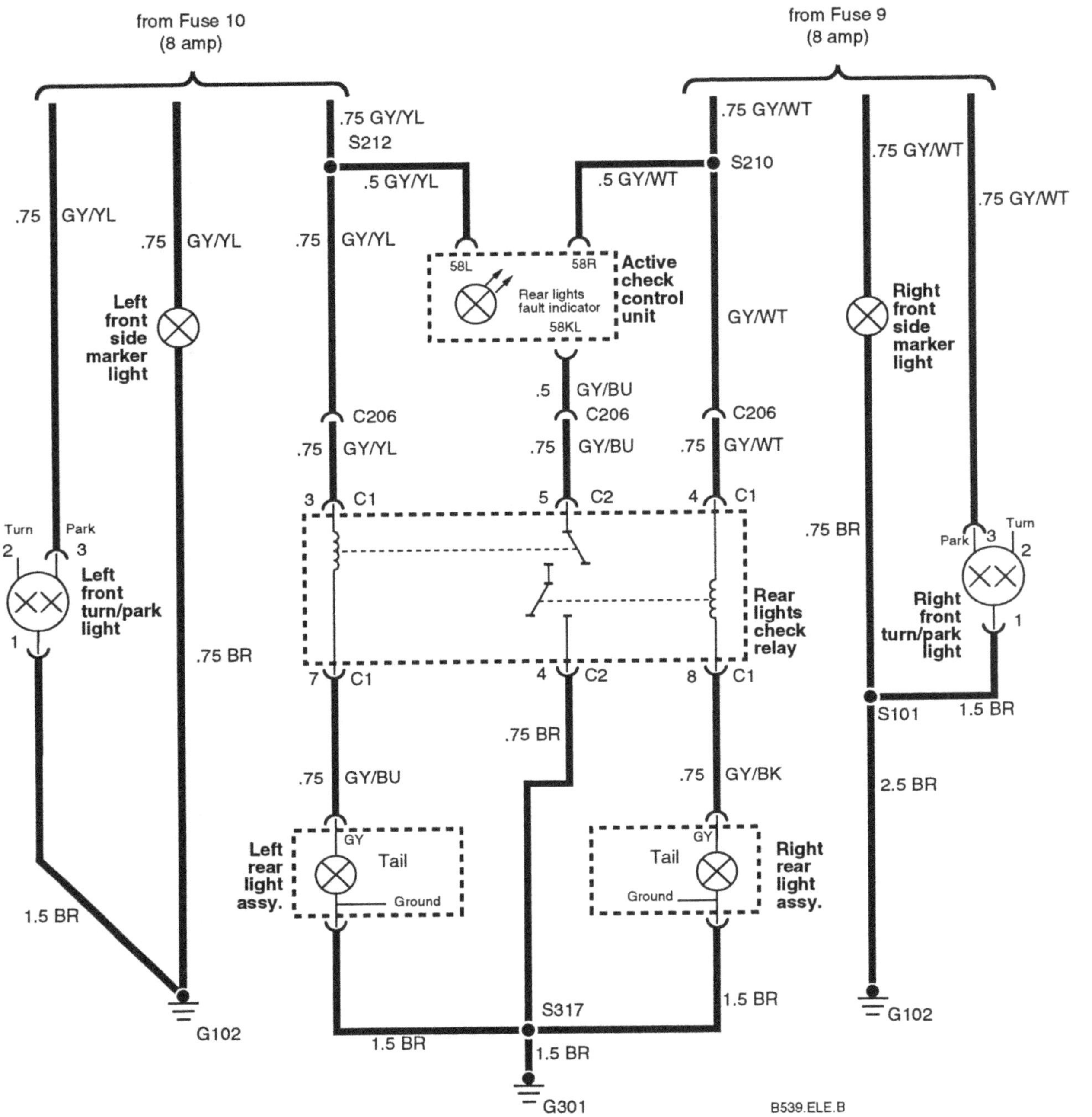

Fig. 4-6. Parking lights, front side marker lights, and taillights circuit. Circuit may vary slightly on some models.

Brake Lights

The brake lights receive their power only when the ignition switch is in accessory, run, or start. The brake light switch closes when the brake pedal is depressed to complete the circuit to the brake lights. Fig. 4-7 shows the wiring for the brake light circuit.

To quickly check if the problem is in the switch or some other part of the circuit, remove the electrical connector(s) from the brake light switch above the brake pedal. Using a fused jumper wire, bridge the wires in the connector(s) and turn the ignition ON.

If the brake lights come on, either the switch needs adjustment as described in **BRAKES** or the switch is faulty and

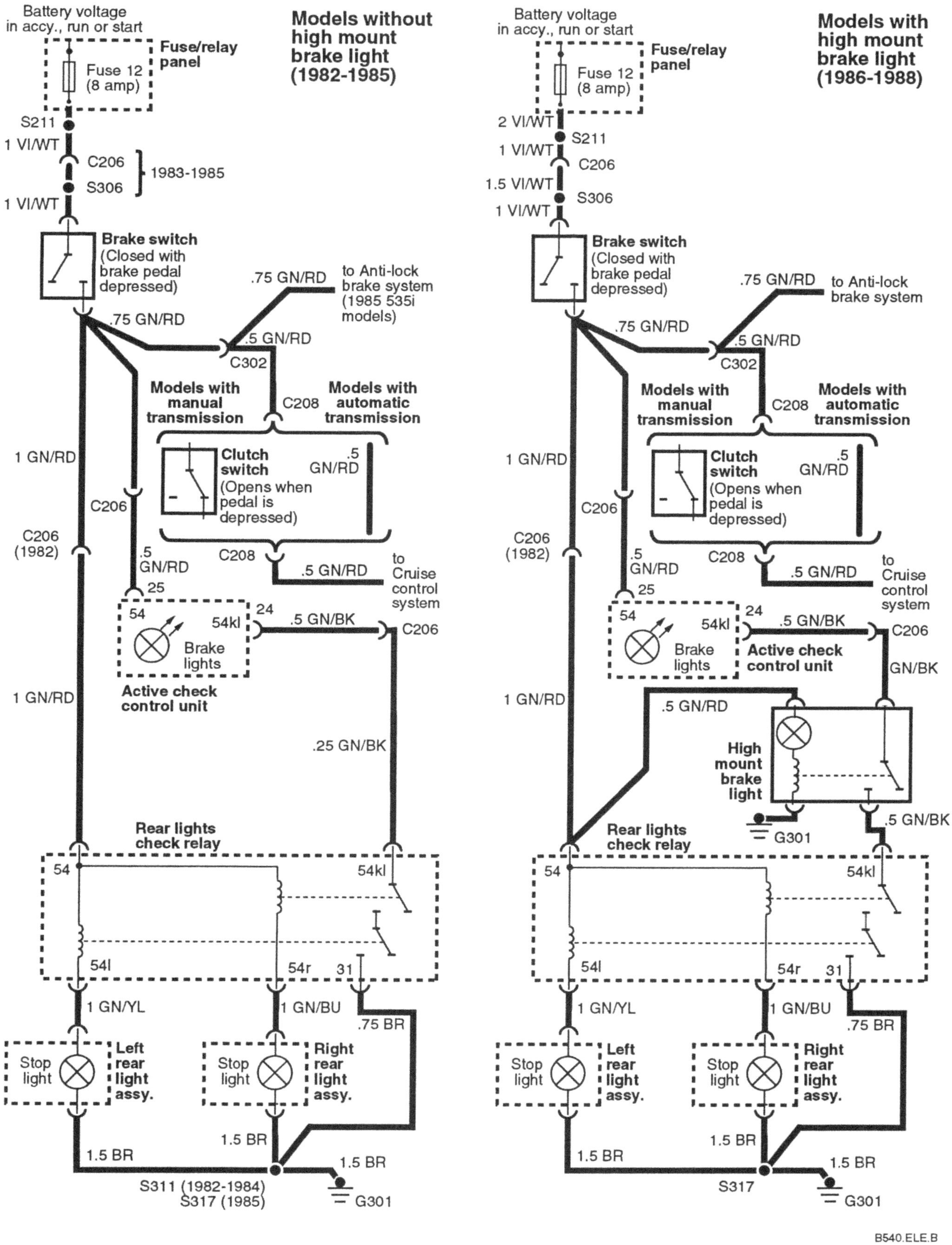

Fig. 4-7. Brake light circuit. 1987 and 1988 circuits vary slightly.

should be replaced. If the lights do not come on, check for battery voltage between the violet/white wire and ground with the ignition ON. If voltage is not present, check the wire between the fuse/relay panel (fuse no. 12) and the brake light switch. If voltage is present, check for faults in the green/red wire between the brake light switch and the brake light assembly.

> **NOTE —**
> All 1986 and 1987 cars with high mount brake light may have a defective brake light switch. The extra current consumed by the third brake light can cause the switch to overload and malfunction. BMW has initiated a recall campaign #86V-151 to replace defective brake light switches. If the new switch was not installed, contact an authorized BMW dealer for more information.

Back-up Lights

On cars with manual transmission, the back-up lights receive power through the back-up light switch when the transmission is in reverse gear. The switch is mounted on the transmission as shown in Fig. 4-8 and is accessible from beneath the car.

Fig. 4-8. Typical location of back-up light switch (arrow) on manual transmission housing.

The circuit for the back-up lights is shown in Fig. 4-9. The information below also includes electrical tests for the neutral/-park/back-up light switch.

Check the switch by removing the electrical connectors from the switch and checking for continuity across the switch terminals with the transmission in reverse. If there is no continuity, the switch is faulty and should be replaced.

Tightening torque
• back-up light switch. . . . . .6 to 10 Nm (35 to 89 in-lb)

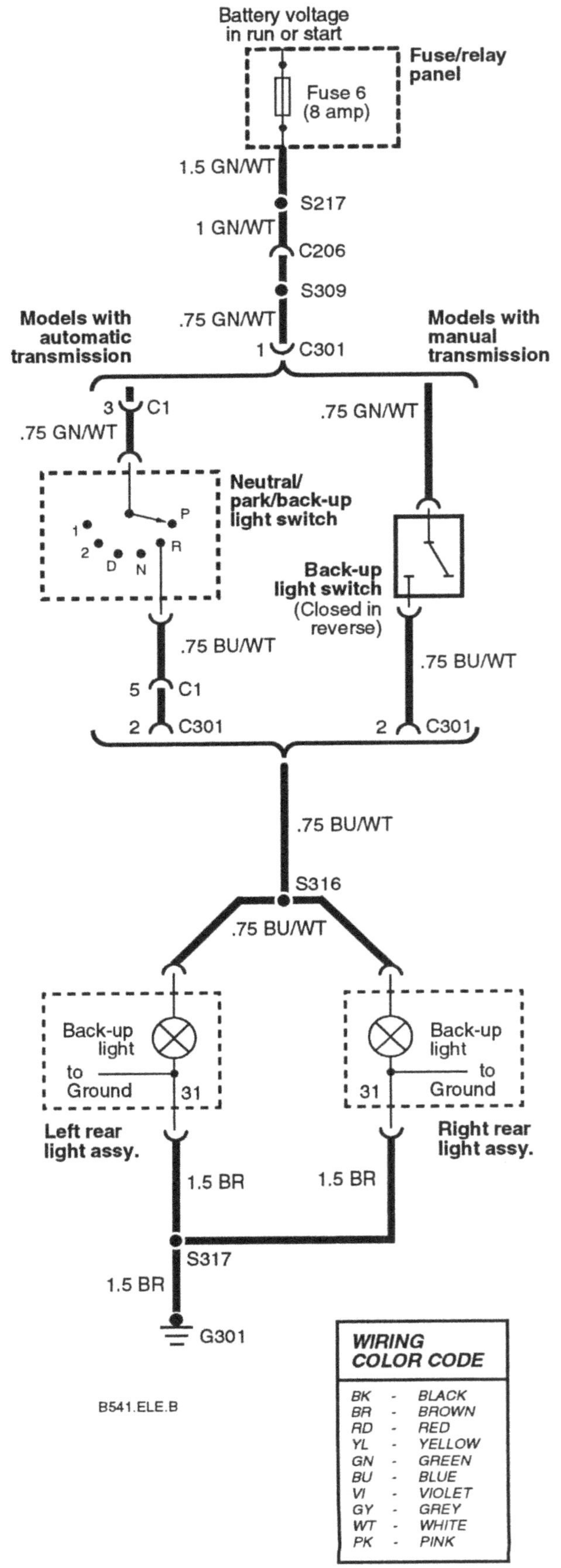

Fig. 4-9. Back-up lights circuit

On cars with automatic transmission, the back-up lights receive their power through the neutral/park/back-up light switch mounted at the base of the shift lever in the passenger compartment. The switch also prevents the car from starting if the shift lever is not in the "P" or "N" position. To test the switch, pry out the trim from around the base of the shift lever. Remove two screws and the trim bezel. Disconnect the connector shown in Fig. 4-10. Make the continuity checks listed in **Table e.** Make all tests at the switch side of the connector.

Table e. Neutral/Park/Back-up Light Switch Continuity Tests

Selector lever position	Connector and terminals	Test result
Park	C1: 3 and 4	Continuity
Reverse	C1: 3 and 5	Continuity
Neutral	C1: 3 and 6	Continuity
Drive	C1: 3 and 7	Continuity
1	C1: 3 and 9	Continuity
2	C1: 3 and 8	Continuity
3	C1: 3 and C2: 1	Continuity

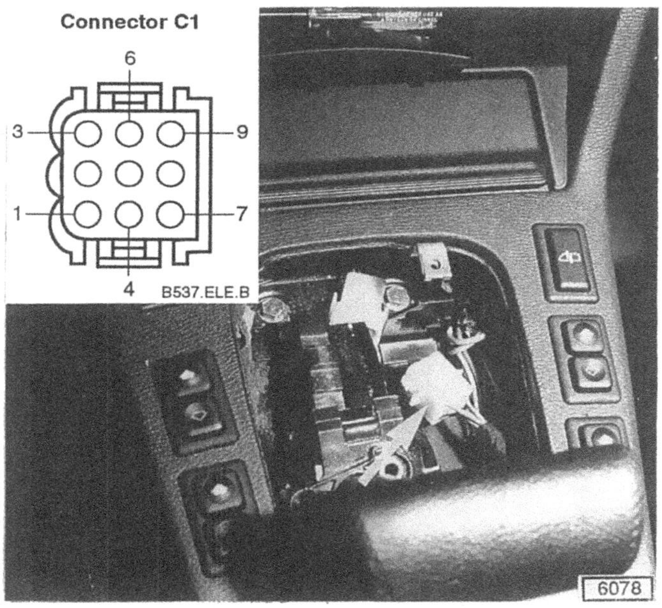

Fig. 4-10. Location of neutral/park/back-up light switch on cars with automatic transmission. Switch harness connector shown at arrow. Inset shows connector terminal locations.

5. WINDSHIELD WIPERS AND WASHERS

A control unit on the fuse/relay panel controls the operation of the wipers, and runs the wipers for several sweeps if the washer is actuated while the wiper switch is in the OFF position. On 1984 and later models, the washer spray nozzles are heated to prevent freezing in cold weather.

The wiper motor and its linkage are located at the base of the windshield and are accessible from the engine compartment after removing the rubberized firewall access panel.

If the wiper or washer motor fails to operate when the steering column lever is moved, see **5.1 Windshield Wiper and Washer Troubleshooting.** If the wiper motor runs but the wiper arms do not move, see **5.2 Windshield Wiper Motor and Linkage.**

5.1 Windshield Wiper and Washer System Troubleshooting

Fig. 5-1 is a schematic of the windshield wiper and washer system circuit. For best results, the troubleshooting tests below should be done in the sequence given. Before diagnosing windshield wiper and washer problems, check the fuse in position 11 (16 amp).

> **NOTE —**
> If the wipers operate erratically or sluggishly, the control unit contacts may be dirty or corroded. Remove the control unit and thoroughly clean the control unit and the fuse/relay panel terminals using an electrical contact cleaner.

If no visible faults are found, remove the wiper control unit from the fuse relay/panel as shown in Fig. 5-2 and make the electrical tests outlined in **Table f.** Make all tests at the fuse/relay panel sockets.

22

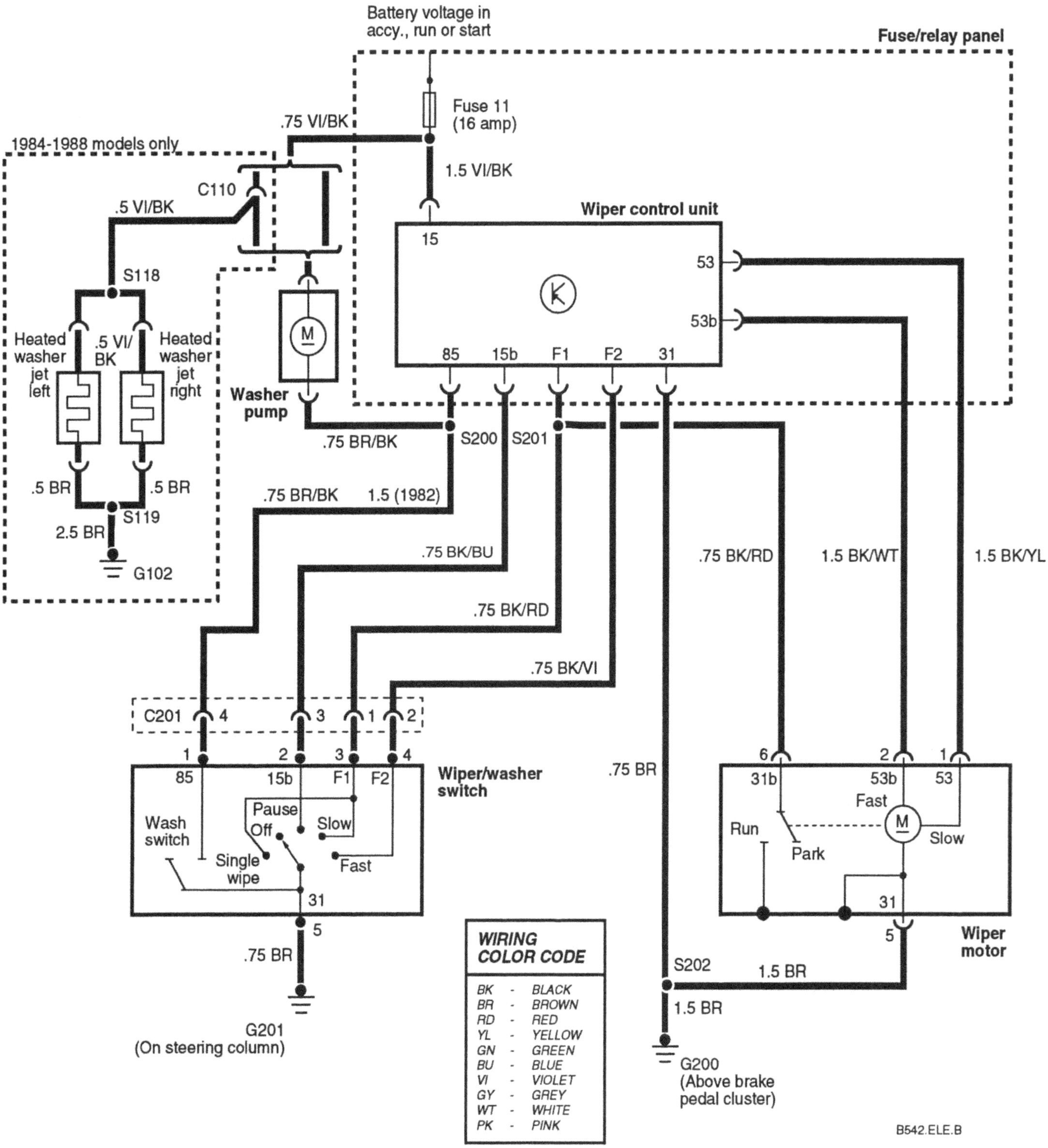

Fig. 5-1. Windshield wiper and washer system circuit.

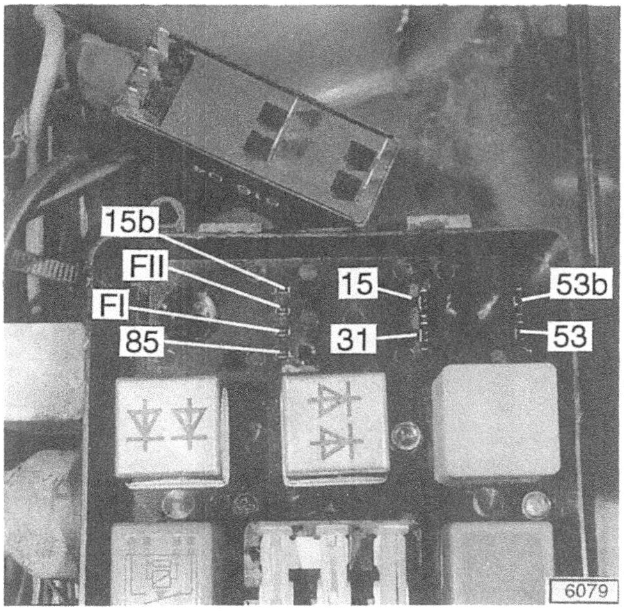

Fig. 5-2. Wiper control unit location (control unit removed) and terminal identification on fuse/relay panel.

Table f. Windshield Wiper and Washer Tests

Fuse/relay panel socket(s) (wiper control unit removed)	Test conditions	Correct test results
15 and ground	Ignition ON	Battery voltage
31 and ground	—	Continuity

If voltage is not present, the fault is in the fuse/relay panel or the connections to the panel. If there is no continuity to ground, check the main ground connection that is below the instrument panel, directly above the brake pedal. See Fig. 5-3. If the voltage and ground signals are as specified, continue the tests as outlined below.

Testing Windshield Wiper and Washer Motors

The wiper and washer motors can be quickly checked at the fuse/relay panel. Using a 1.5 mm (14 AWG) fused jumper wire with flat spade connectors on either end, make the electrical tests in **Table g**. Make all tests at the wiper control unit sockets identified in Fig. 5-2 above.

CAUTION —
Use only a fused (16 amp or less) jumper wire of the specified diameter. Make all electrical connections with the ignition OFF. When connections have been safely made, turn the ignition ON.

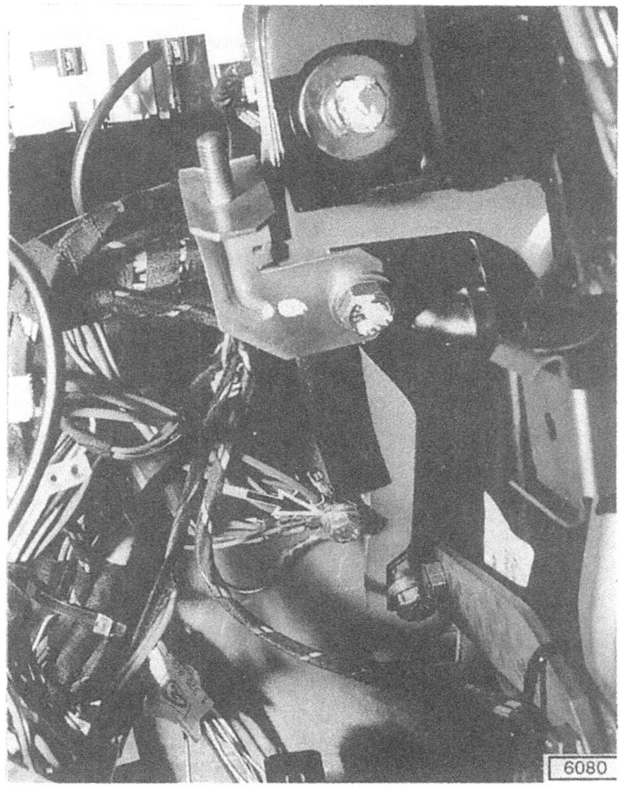

Fig. 5-3. Location of ground point (G200) for wiper motor (arrow).

Table g. Windshield Wiper and Washer Motor Tests

Fuse/relay panel socket(s) (wiper control unit removed)	Test conditions	Correct test results
15 and 53	Ignition ON	Wiper motor runs on low speed
15 and 53b	Ignition ON	Wiper motor runs on high speed
85 and ground	Ignition ON	Washer motor runs

NOTE —
If during cold weather the intermittent cycle of the wiper motor is erratic, the wiper motor may have been overfilled with grease during manufacture. To correct this problem, the wiper motor will need to be removed and disassembled. See **5.2 Windshield Wiper Motor and Linkage** for wiper motor removal.

If either the wiper motor or washer motor does not operate as specified and battery voltage is present at terminal 15 with the ignition ON, check the wires between the wiper control unit and the wiper or washer motors. If no wiring faults can be found, either the wiper motor or washer motor is faulty and should be replaced. See **5.2 Windshield Wiper Motor and Linkage.**

If the motor operates as specified, the problem lies elsewhere in the system. Continue troubleshooting by testing the wiper/washer switch as described below.

Testing Wiper/Washer Switch

The switch supplies the wiper control unit with ground inputs and routes them depending on the position of the lever. The switch can be tested using an ohmmeter. To test the switch, first remove the steering wheel as described under **STEERING AND WHEEL ALIGNMENT.** Remove the lower steering-column cover retained by four screws. Disconnect the harness connector as shown in Fig. 5-4.

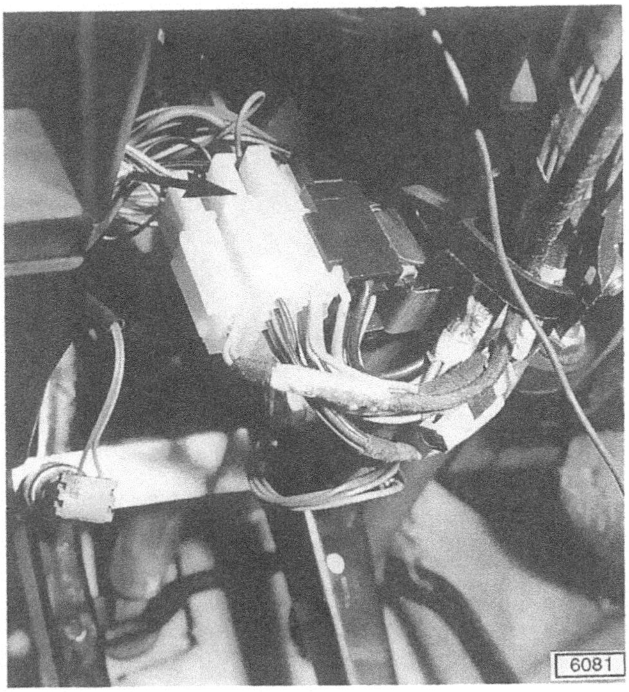

Fig. 5-4. Wiper/washer switch harness connector (arrow).

WARNING —
Some of the cars covered by this manual may be equipped with a Supplemental Restraint System (SRS) that automatically deploys an airbag. The airbag unit uses a pyrotechnical device to electrically ignite a powerful gas. On cars so equipped, any work involving the steering wheel should only be performed by an authorized BMW dealer. Performing repairs without disarming the SRS may cause serious personal injury.

Before testing the switch, check the ground wire at the front of the steering column. See Fig. 5-5. Also check the wiper motor ground wire lug below the instrument cluster, above the brake pedal cluster. (Shown earlier in Fig. 5-3.) If these wires are faulty, the wipers will not operate. Use an ohmmeter to test the switch by making continuity checks as indicated in **Table h**. If the switch fails any of these tests, it is faulty and should be replaced. See **STEERING AND WHEEL ALIGNMENT** for switch removal.

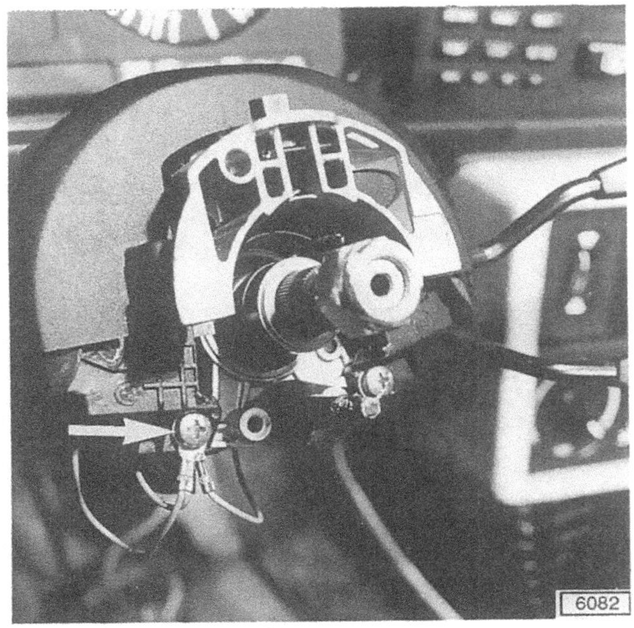

Fig. 5-5. Location of ground point—G201 for wiper/washer switch (arrow).

Table h. Wiper/Washer Switch Continuity Tests

Wiper/washer switch position	Terminal and wire color	Test result
Single wipe	F1 (black/red) and ground	Continuity
Low Speed	F1 (black/red) and ground	Continuity
High Speed	F2 (black/violet) and ground	Continuity
Intermittent	15b (black/blue) and ground	Continuity
Washer	85 (brown/black) and ground	Continuity
—	5 (brown) and ground	Continuity

Wiper Control Unit

Except for the washer motor, all wiper functions are wired through the wiper control unit. If the wiper motor and the wiper switch are functioning correctly, and there are no wiring faults, the wiper control unit is most likely faulty. Because the control unit is an electronic component, it cannot be easily tested. The only remedy for a faulty control unit is replacement.

NOTE —
Inspect the terminals of the control unit. Dirty or corroded terminals can cause erratic and sluggish wiper operation. Control unit and fuse/relay panel terminals can be cleaned using an electrical contact cleaner.

5.2 Windshield Wiper Motor and Linkage

The wiper motor and its linkage are accessible for replacement from the engine compartment. The wiper motor can be removed separately from the wiper linkage. For information on windshield wiper blades, see **LUBRICATION AND MAINTENANCE**.

Removing and Installing Windshield Wiper Motor and Linkage

To remove:

1. Disconnect the negative (–) battery cable.
2. Remove the wiper arms. Pry up the wiper arm retaining nut cover and mark the position of the wiper arm on the wiper arm shaft. Remove the wiper arm retaining nut and the wiper arm. See Fig. 5-6.

   NOTE —
   If the wiper arm is difficult to remove, gently rock the arm back and forth to loosen it on the shaft.

3. Working in the rear of the engine compartment, remove the six screws securing the rubber firewall cover surrounding the wiper motor and linkage. See Fig. 5-7.
4. Mark the position of the drive crank to the motor shaft. Remove the nut securing the wiper motor shaft to the drive crank. Remove the drive crank.
5. Disconnect the harness connector from the wiper motor. Thread out the rubber buffer mount. Remove the 3 wiper motor mounting bolts, and remove the motor. See Fig. 5-8.
6. Working from above, remove the wiper arm shaft rubber covers, the large shaft retaining nuts, and the washers.

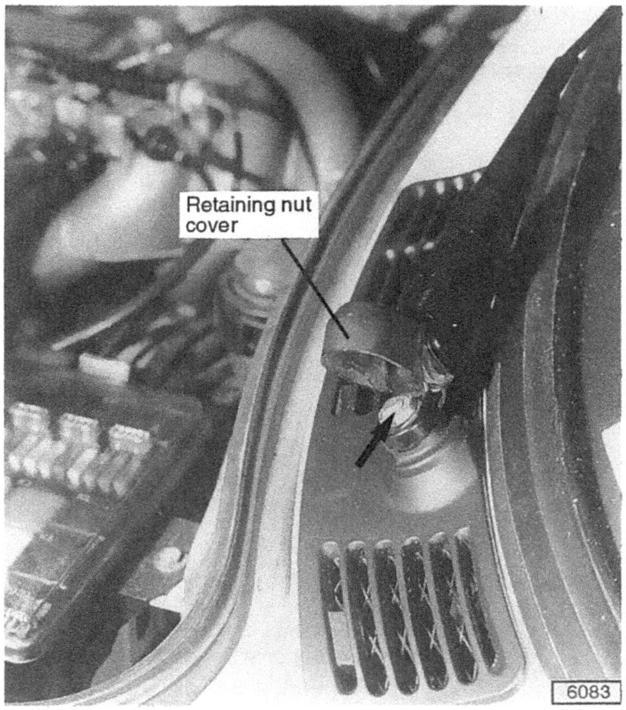

Fig. 5-6. Wiper arm retaining nut (arrow).

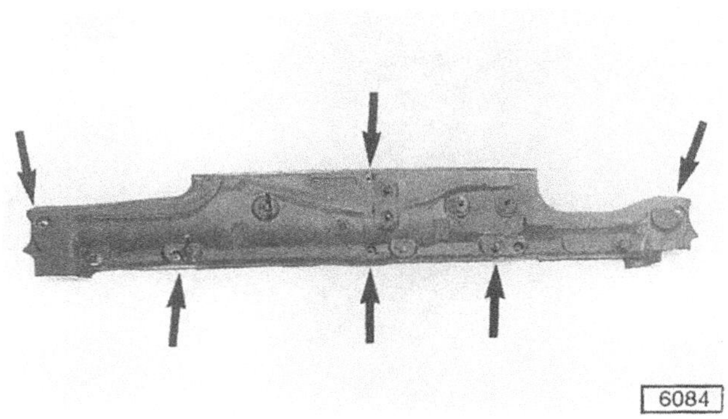

Fig. 5-7. Rubber firewall cover mounting points (arrows).

7. Unclip and remove the two blower motor covers. See Fig. 5-9. Unclip the center mount clip that retains the blower motor and remove the blower motor. See Fig. 5-10.
8. Remove the wiper linkage assembly being careful not to scratch the painted surfaces. See Fig. 5-11.

Installation is the reverse of removal. Install the linkage and wiper arms, aligning the matching marks made earlier. Position the arms in the park position. Install the new motor onto

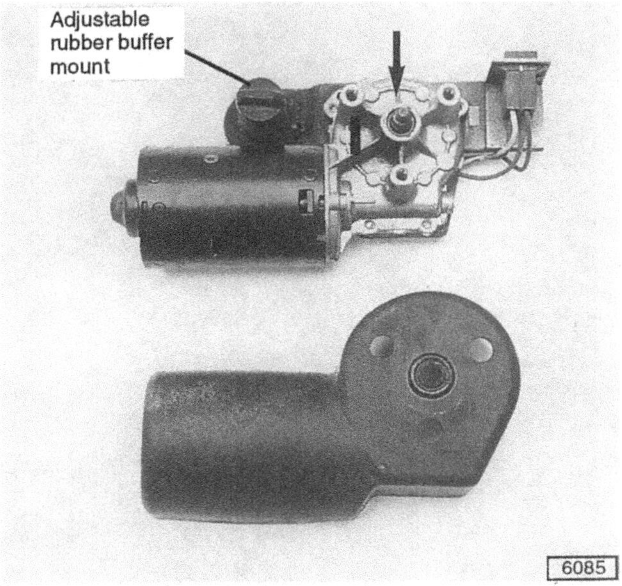

Fig. 5-8. Wiper motor and cover. Mark position of drive crank to motor shaft (arrow) before removing the crank from motor.

Fig. 5-10. Blower motor mount clip (arrow). Blower motor covers removed for clarity.

Fig. 5-9. Blower motor cover being removed.

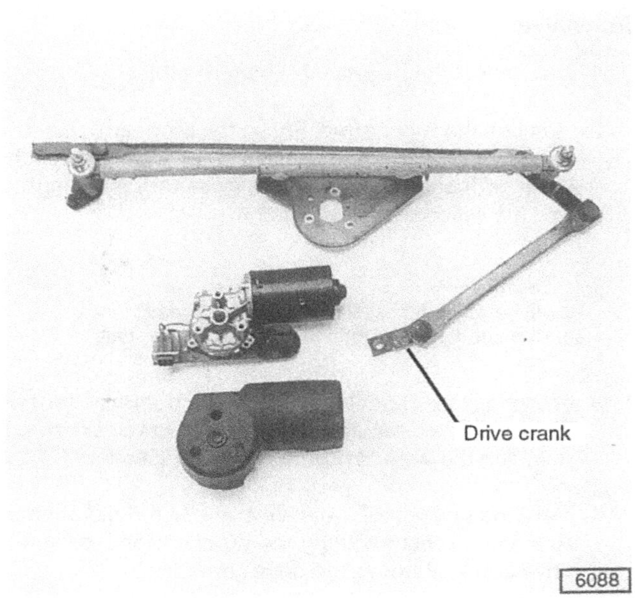

Fig. 5-11. Wiper linkage assembly and motor.

the bracket but leave the drive crank off temporarily. Connect the wiring and operate the wiper motor for 15 or 20 seconds then turn it off. The motor will stop in its park position. Now install the drive crank. It should align with the matching mark made earlier. Install the nut to the drive crank.

> **NOTE —**
> Always make sure the wiper arms are correctly positioned on the shafts before attaching the drive crank and operating the motor. If the wiper arms are improperly aligned on the wiper arm shafts, they can scratch or damage the painted surfaces of the car when the wiper motor is started.

6. HORNS

The components of the horn system are the dual horns, the horn relay, and the horn buttons on the steering wheel. A brush and slip ring assembly makes the electrical contact between the rotating steering wheel and the steering column.

The horns are connected to ground (–). Depressing the horn button with the ignition ON activates the horn relay, which in turn supplies the horns with positive (+) voltage from fuse no. 11 (16 amp). See Fig. 6-1.

6.1 Horns Troubleshooting

If the horns do not sound, check the fuse and all the wiring connections. If no faults are found, remove the horn relay and do the electrical tests in **Table i**. See **12. Fuse/Relay Panel** for relay locations.

> **NOTE —**
> The horns are mounted beneath the car directly above the fog lights. On some later models, the splash guard will have to be removed before the horns can be reached.

Table i. Horn Tests

Fuse/relay panel socket(s) (horn relay removed)	Test condition	Correct test result
Terminal 30 and ground	Ignition ON	Battery voltage
86 and ground	Ignition ON	Battery voltage
86 and 85	Ignition OFF, horn button depressed	Continuity
85 and ground	Ignition OFF, horn button depressed	Continuity

If no electrical faults are found, the relay is faulty and should be replaced. If any electrical test is not as specified, check for faulty wiring to the relay socket.

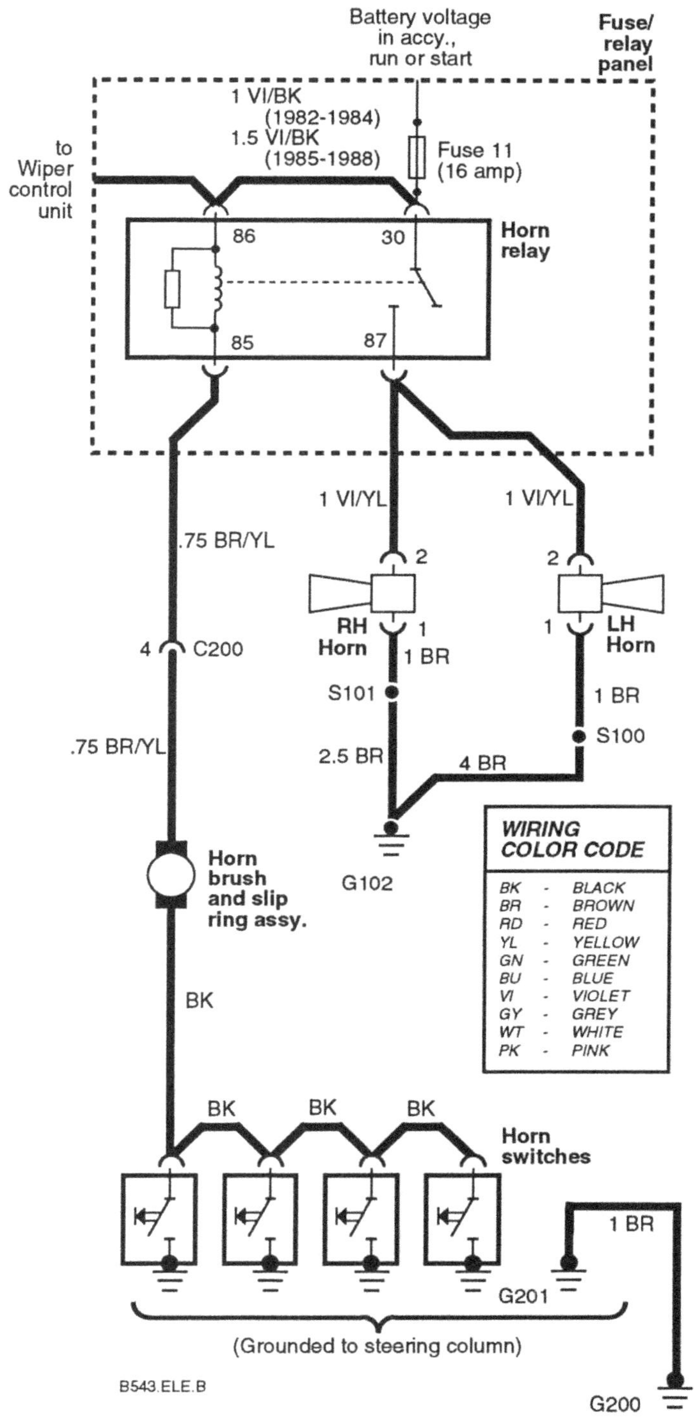

Fig. 6-1. Dual horn system circuit.

7. REAR WINDOW DEFOGGER

The rear window defogger is a series of resistance elements that heat up as electrical current passes through them. One or two wires that do not heat up suggests a break in the heating element. Repair material for the element is available from an authorized BMW dealer to repair the breaks as described below. An element that does not operate at all suggests that power is not reaching the heating element.

7.1 Rear Window Defogger Troubleshooting

The rear window defogger is powered from the unloader relay (power interrupted when starter is operated). Check for voltage to the window element (from fuse no. 13) with ignition on and the rear window defogger switch on. Check the main ground connection under the rear seat (G301). Refer to Fig. 7-1 for details of the rear window defogger circuit.

The switch can be carefully pried away from the dashboard using a small screwdriver.

To find and repair heating element breaks:

1. Wrap a small piece of aluminum foil (about the size of a thumb print) around the negative (–) probe of a voltmeter.
2. With the heating element's wiring connected, touch the positive (+) voltmeter probe to the end of the heater element closest to the incoming voltage source.
3. With the ignition and the defogger switch ON, place the foil-covered negative probe on the faulty wire near the positive side of the element. Slowly slide it toward the negative side while pressing on the foil with your thumb. The point at which the voltmeter deflects from zero volts to several volts is where the wire is broken.
4. Clean the broken wire area thoroughly using alcohol or a mixture of vinegar and water.
5. Apply a strip of masking tape to either side of the break, leaving the break exposed. Apply the repair material over the break and allow it to dry for one hour at room temperature. Remove the tape and retest the wire.

> **NOTE —**
> The rear window repair kit, BMW Part No. 81 22 9 407 066, is available from an authorized BMW dealer parts department. The kit contains enough material to repair approximately three feet of heating strip.

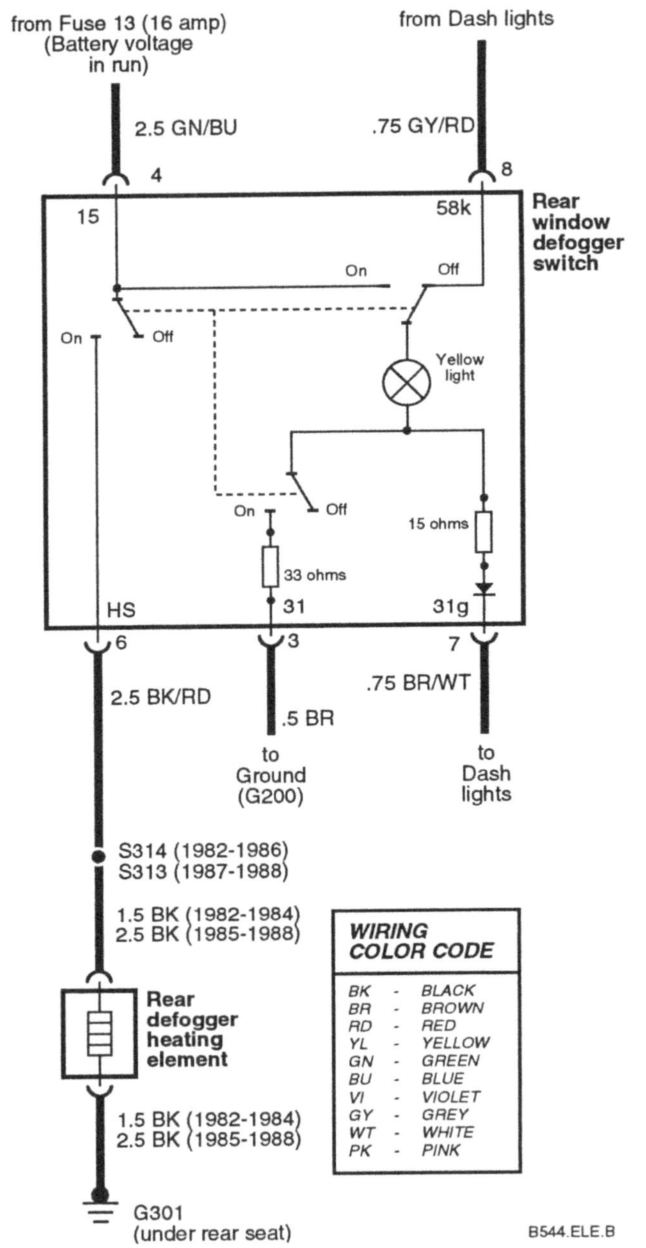

Fig. 7-1. Rear window defogger circuit.

8. STEERING COLUMN SWITCHES

The steering column switches can be tested without being removed by making electrical tests at the appropriate harness connector below the instrument panel. See Fig. 8-1. To access the connectors, remove the lower instrument panel trim retaining screws and remove the trim panel.

If a switch is found to be faulty, the steering wheel will need to be removed. Be sure to mark the position of the steering wheel to the steering column shaft before removing it. Replace faulty switches as described in **STEERING AND WHEEL ALIGNMENT**.

> ***WARNING —***
> *Some of the cars covered by this manual may be equipped with a Supplemental Restraint System (SRS) that automatically deploys an airbag from the steering wheel. The airbag unit uses a pyrotechnical device to electrically ignite a powerful gas. On cars so equipped, any work involving the steering wheel should only be performed by an authorized BMW dealer. Performing repairs without disarming the SRS may cause serious personal injury.*

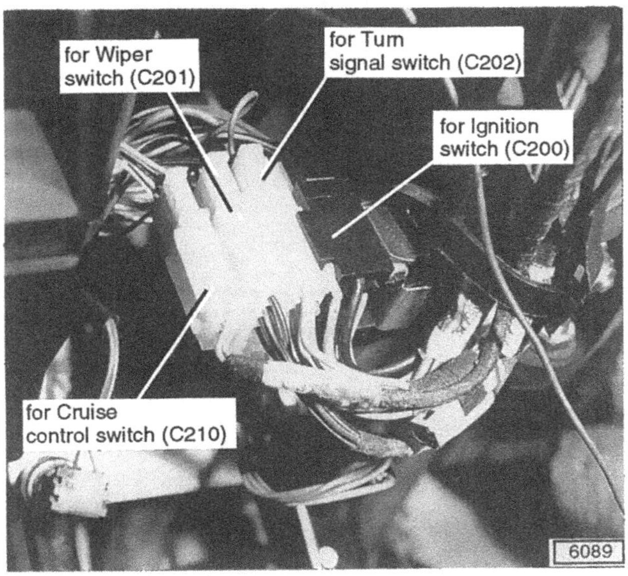

Fig. 8-1. Harness connectors for steering column switches.

To test the windshield wiper switch, see **5. Windshield Wipers and Washers.** To test the cruise control switch and the on-board computer switch (1985 and later models), see **10. Power Options and Accessories.** Tests for the ignition switch, the headlight dimmer switch, and the turn signal switch are covered below.

Ignition Switch

Fig. 8-2 is a schematic diagram of the ignition switch. Terminal 30 brings power into the ignition switch from the battery. Terminal 15, and terminal R, provide power to the ignition system and other parts of the electrical system when the ignition key is in the ON position. Terminal 50 switches power to the starter. Terminal C and terminal 31 are used to check the brake fluid level warning bulb when the ignition key is in the START position.

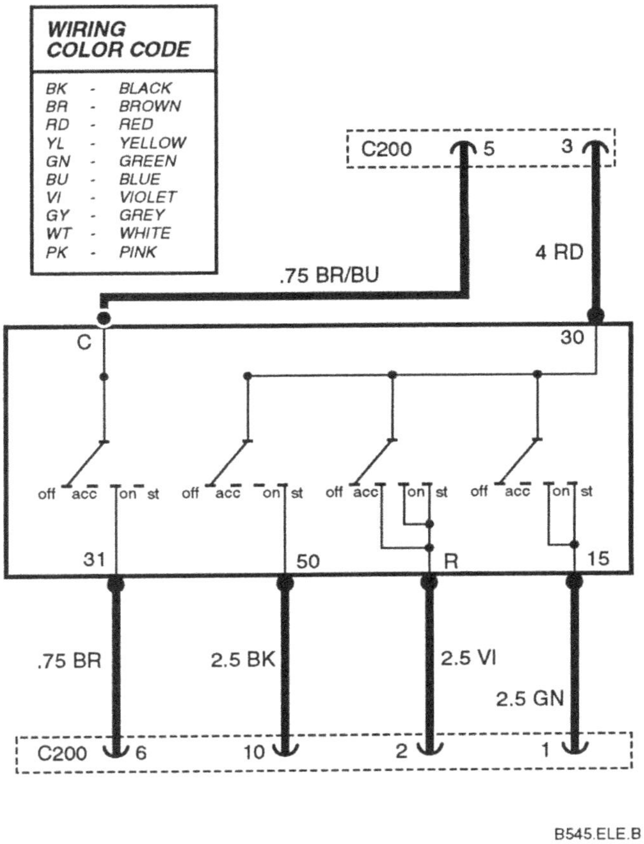

Fig. 8-2. Schematic diagram of ignition switch.

Fig. 8-3 identifies the terminals of the 10-point ignition switch harness (C200) connector beneath the steering column. **Table j** lists ignition switch continuity checks. A switch that fails any of these tests is faulty and should be replaced.

> ***CAUTION —***
> *Disconnect the negative (–) battery cable before testing the ignition switch.*

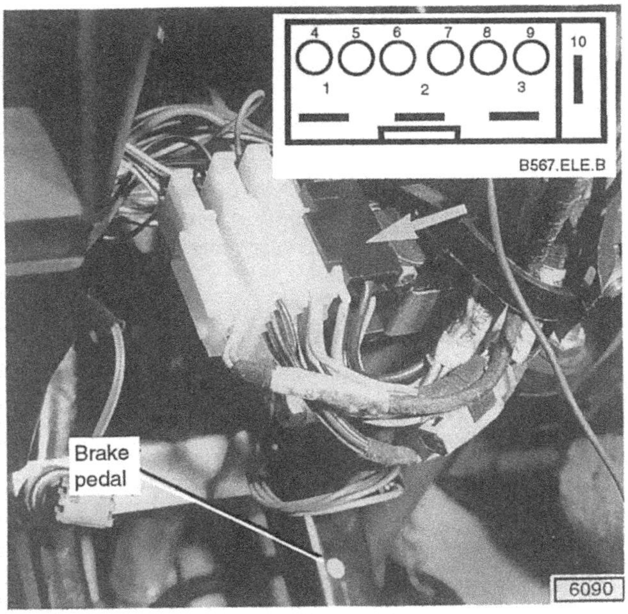

Fig. 8-3. Ignition switch harness connector (arrow). Lower steering column trim removed from steering column. Make continuity tests at switch side of connector. Inset shows connector (C200) terminal identification.

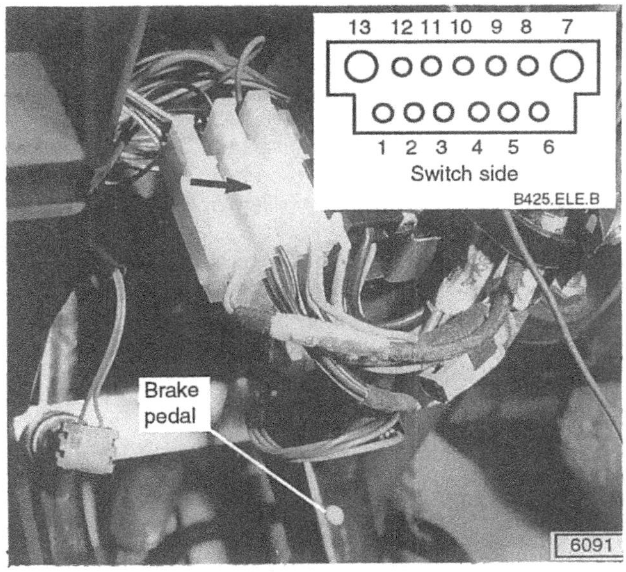

Fig. 8-4. Harness connector for headlight dimmer switch (arrow). Lower steering column trim panel removed. Inset identifies connector (C202) terminals.

Table j. Ignition Switch Continuity Tests

Ignition switch position	Test terminals (wire color)	Correct test result
Ignition ACCY, ON or START	2 (violet) and 3 (red)	Continuity
Ignition ON or START	1 (green) and 3 (red)	Continuity
START only	10 (black) and 3 (red) 5 (brown/blue) and 6 (brown)	Continuity

Headlight Dimmer Switch

Test the headlight dimmer switch at the switch side of the 13-point harness connector (C202) shown in Fig. 8-4. Using the information in **Table k**, check for continuity at the harness connector with the ignition OFF. If any faults are found, the switch is faulty and should be replaced. For more information on the headlight electrical circuit, see **4. Exterior Lights.**

Table k. Headlight Dimmer Switch Continuity Tests

Headlight dimmer switch position	Test terminals (wire color)	Correct test result
Low beam	4 (white) and 3 (yellow)	No continuity
High beam	4 (white) and 3 (yellow)	Continuity
Flash	4 (white) and 9 (green)	Continuity

Turn Signal Switch

Test the turn signal switch at the switch side of the 13-point harness connector (C202) shown in Fig. 8-4 above. Using the information in **Table l** and Fig. 8-5, check for continuity at the harness connector with the ignition OFF. If any faults are found, the switch is faulty and should be replaced.

Table l. Turn Signal Switch Continuity Tests

Turn signal switch position	Test terminals (wire color)	Correct test result
Left turn	13 (green/yellow) and 2 (blue/red)	Continuity
Right turn	13 (green/yellow) and 5 (blue/black)	Continuity

WIRING COLOR CODE	
BK	BLACK
BR	BROWN
RD	RED
YL	YELLOW
GN	GREEN
BU	BLUE
VI	VIOLET
GY	GREY
WT	WHITE
PK	PINK

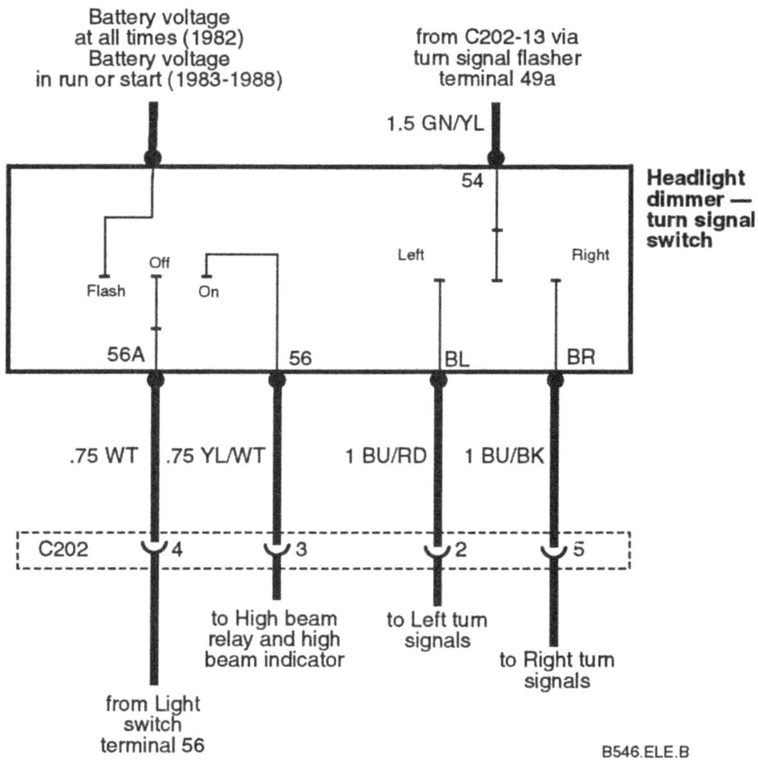

Fig. 8-5. Schematic of headlight dimmer and turn signal switch.

9. INSTRUMENTS

A printed circuit is used in place of the numerous wires and connectors that would otherwise be required to connect the instruments and switches. Three main wire harness connectors supply the power and electrical signals to the instruments. See Fig. 9-1.

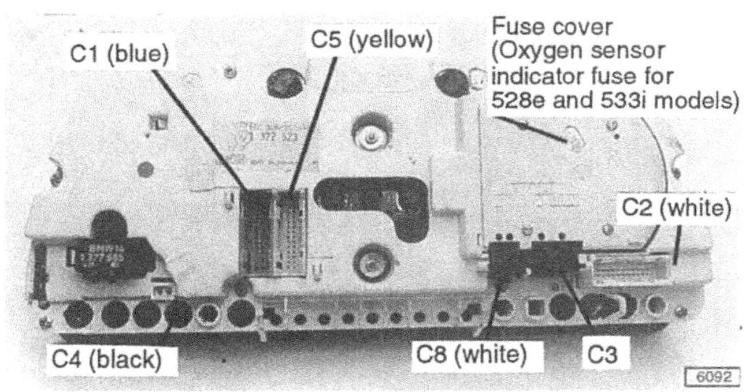

Fig. 9-1. Instrument cluster viewed from the rear.

On 1983 and later models the instrument cluster also houses an electronic control unit for the service indicator. This control unit is not repairable.

Troubleshooting or repair of the instrument cluster is easiest with the cluster removed. See **BODY AND INTERIOR** for the removal procedure. Disassembly and replacement of the separate instruments is covered under **9.2 Removing and Installing Instruments**. Cluster lamps can be serviced without disassembly of the instruments.

9.1 Troubleshooting Instrument Cluster and Gauges

Begin diagnosing instrument cluster electrical problems by checking the fuses. Make a thorough check in the engine compartment and under the instrument panel for disconnected or damaged wires or connectors.

If only one function is affected, the problem is most likely one that affects only that individual circuit. If more than one function of the instrument cluster is affected, it is logical to begin with the instrument cluster's common wiring. Check the main power supply from the ignition switch and battery and the main ground wire.

Testing Instrument Cluster Voltage and Ground

If the warning lamps or gauges do not respond with the ignition ON, it may be that either battery voltage is not reaching the instrument cluster or the cluster is not properly grounded.

With the battery negative (–) cable disconnected, remove the instrument cluster as described in **BODY AND INTERIOR**. Make sure that the blue, yellow, and white 26-point connectors at the rear of the cluster are also disconnected. Reconnect the battery cable.

CAUTION —
- *Do not allow any unprotected part of the instrument cluster electrical circuitry to touch any metal part of the car while the cluster is removed and the battery is connected.*
- *Do not touch the printed circuit boards or components. Damage may result from static discharge.*

With the ignition switched ON, use a voltmeter to check for voltage and ground at the harness connector terminals listed in **Table m**. The test terminals for the 26-point connectors are identified in Fig. 9-2.

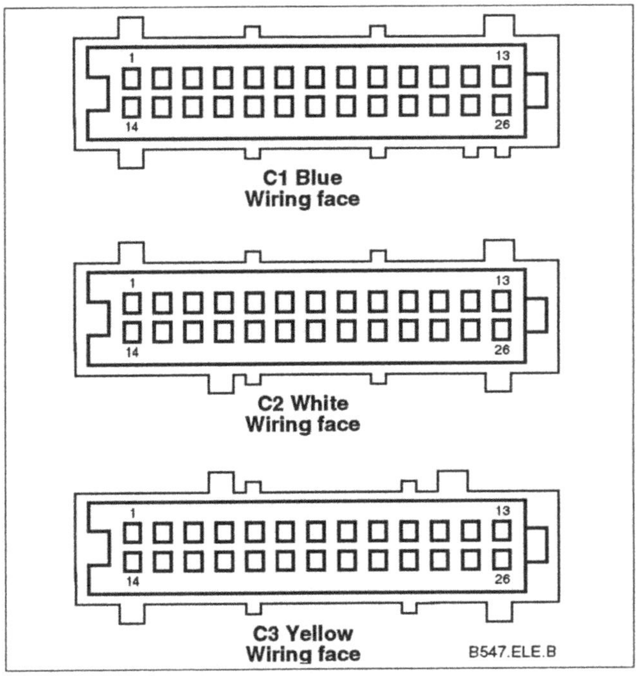

Fig. 9-2. Terminal identification for 26-point instrument cluster connectors. There are three 26-point connectors attached to the rear of the cluster. Use care not to damage connector terminals with test probes.

Table m. Instrument Cluster Electrical Tests

Circuit	26-point connector (color)	Test terminals	Test conditions	Correct test value
Voltage supply to instrument cluster	C1 (blue)	7 and ground	Ignition ON	Battery voltage (approx. 12 VDC)
Voltage supply to charge indicator lamp	C1 (blue)	22 and ground	Ignition ON	Battery voltage (approx. 12 VDC)
Instrument cluster gauges	C1 (blue)	4 and ground	Ignition OFF or ON	Battery voltage (approx. 12 VDC)
Main ground to instrument cluster	C1 (white)	16 and ground	—	Continuity

If voltage is not present in any of the above tests, check fuses no. 5 and no. 6. If no faults are found, check the wires between the fuse/relay panel and the instrument cluster connectors and also between the ignition switch and the instrument cluster connector. If there is no continuity to ground, check the wire leading from the instrument panel to the main ground connection above the brake pedal. See Fig. 9-3.

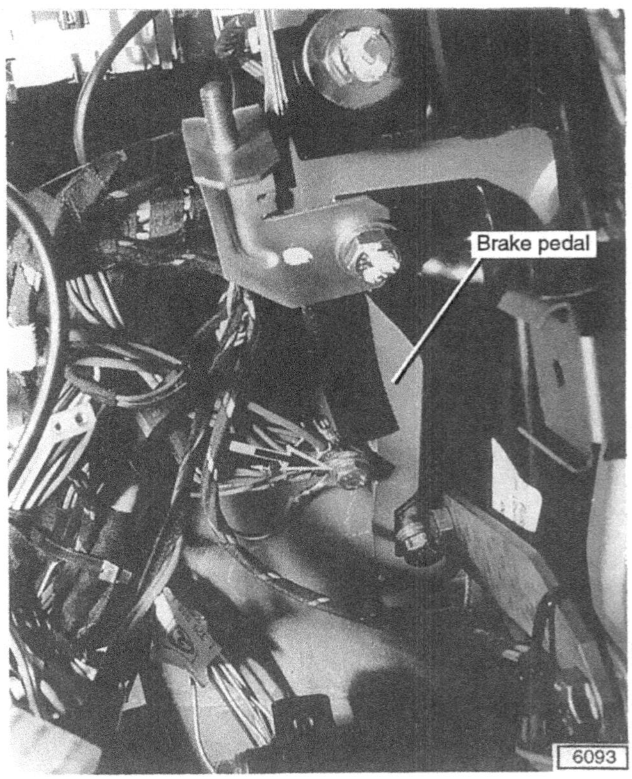

Fig. 9-3. Main ground connection (G200) for instrument cluster (arrow). Instrument trim panel removed.

Testing and Replacing Indicator Bulbs

Test indicator bulbs by first removing the instrument cluster as described in **BODY AND INTERIOR.** Remove the bulb housings from the rear of the cluster by turning the housing 90° and pulling the housing from the cluster. The ABS indicator bulb pulls straight out. Using an external 12-volt power source and jumper wires, apply voltage to the terminals of the removed bayonet socket. See Fig. 9-4.

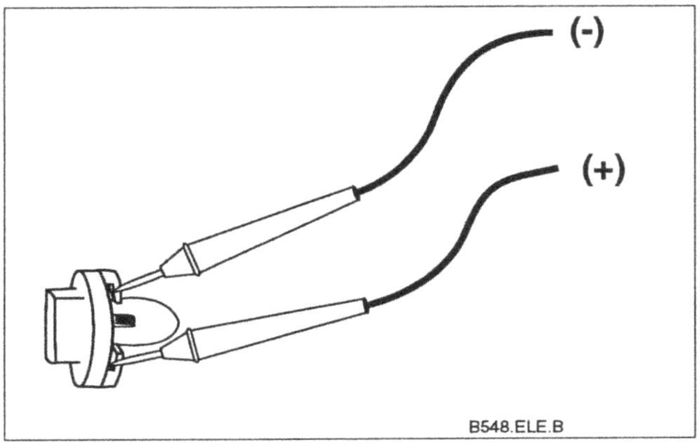

Fig. 9-4. Instrument cluster indicator bulb being tested using battery voltage (shown schematically).

If the bulb lights when tested but does not light when installed, check the printed circuit for continuity at the wire terminal housings.

Turning Out the Oxygen Sensor Indicator Lamp (528e and 533i models)

On 528e models and 533i models, an oxygen sensor indicator lamp is designed to come on at a specified mileage interval. At the specified mileage interval, a control unit in the instrument cluster short circuits and blows a small fuse (0.1 amp). When the fuse blows, the indicator lamp comes on.

During service, when the oxygen sensor is replaced, the oxygen sensor indicator bulb should be turned out. On 1982 through 1984 models, pry out the active check control unit and remove and discard the bulb from the rear of the unit. On 1985 though 1988 models, remove the lower instrument panel trim. Locate the mileage counter on the left-hand side of the steering column beneath the instrument panel and depress the reset button.

NOTE —
If the oxygen sensor lamp comes on before the 30,000 mile service interval, the 0.1 amp fuse may have blown. The fuse is under a cover at the rear of the instrument cluster, directly behind the speedometer. The cover can be unsnapped and the fuse can be easily replaced once the instrument cluster is removed. See **BODY AND INTERIOR** for instrument cluster removal.

NOTE —
On 1982 through 1984 models, see **10.3 Active Check Control** for removal of the active check control unit.

9.2 Removing and Installing Instruments

Individual replacement instruments are available from an authorized BMW dealer parts department. To remove instruments from the cluster, first remove the instrument cluster as described in **BODY AND INTERIOR.** Remove the eight mounting screws from the rear cover of the cluster. Separate any electrical connectors. Carefully remove the rear cover from the front housing. Remove the two nuts securing the fuel—temperature gauge assembly. See Fig. 9-5. Carefully remove individual instruments from the printed circuit board. Installation is the reverse of removal.

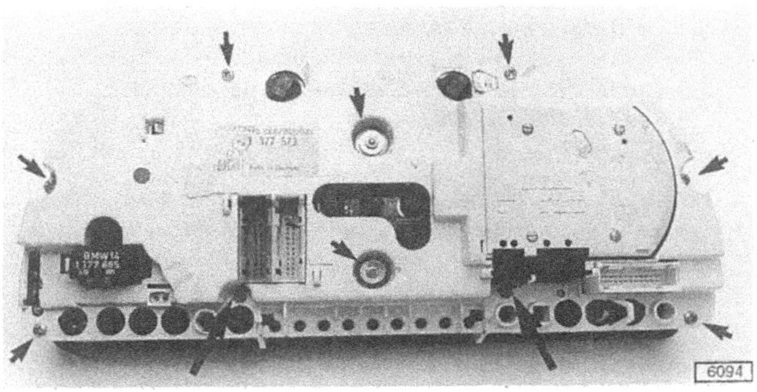

Fig. 9-5. Instrument cluster viewed from rear showing cover mounting screws and securing nuts (arrows).

Service Interval Indicator

The service interval indicator board needs voltage to maintain the board's stored memory and to keep its integral ni-cad batteries charged. This voltage is supplied to the board directly from the battery via fuse no. 5 (8 amp). If the car's battery is disconnected or becomes discharged, the ni-cad cells are designed to hold the board's memory for up to four months. Over time, these buffer batteries may become faulty or discharged if the correct voltage supply is not present (i.e., missing or blown fuse, broken or damaged wire to the instrument cluster).

Discharged ni-cad batteries can cause all of the interval lights to come on at the same time. In addition, the lights will not go out after doing a reset. If this happens, check that the board is receiving the voltage and ground at the instrument cluster harness connectors. **Table n** lists the connector test terminal and test result. See Fig. 9-1 and Fig. 9-2 above for terminal and harness identification. If no faults are found, the ni-cad batteries are probably faulty and the complete board will need to be replaced. Ni-cad batteries are not available as replacement parts from BMW.

Table n. Service Interval Indicator Voltage and Ground Tests

Circuit	26-point connector terminal (color)	Test conditions	Correct test value
Service interval indicator	C1-4 (blue) and C1-16	Ignition OFF or ON	Battery voltage (approx. 12 VDC)
	C1-7 (blue) and C1-16	Ignition ON	Battery voltage (approx. 12 VDC)

NOTE —
If the green lights go out after approximately 175 to 200 miles (especially after a memory reset), the problem is most likely caused by a faulty speedometer (gives the service interval indicator its mileage input) and not a faulty service interval indicator board. Replace the speedometer as described above.

To remove the service interval indicator board from the instrument cluster housing, first remove the instrument cluster as described above. Working from the rear of the instrument cluster housing, separate the printed circuit from the instrument panel and remove the service interval indicator board. Installation is the reverse of removal.

WARNING —
The solid-state service indicator interval board is static sensitive. Static discharge will permanently damage the board. Always handle the modules using proper static prevention equipment and techniques. See an authorized BMW dealer for the most up-to-date parts and installation information.

NOTE —
When installing a new service interval indicator board, the ni-cad buffer batteries may need to be charged. When 12-volt battery voltage is applied to the board, the ni-cad batteries begin to charge after approximately one minute and the inspection light will come on and all other lights will go out. At this point the service indicator should be reset as described in **LUBRICATION AND MAINTENANCE**. Battery voltage to the board should not be interrupted for at least three hours after the reset. The buffer batteries will be fully charged after approximately 85 hours.

10. POWER OPTIONS AND ACCESSORIES

When troubleshooting power options and accessories, always check first for blown or corroded fuses. Fuses are simple to check and are a likely cause of trouble.

10.1 Power Sunroof

Fig. 10-1 is a diagram of the sunroof circuits. Removal of the sunroof and related parts are covered in **BODY AND INTERIOR.** Since late 1986, all models have an illuminated sunroof switch.

10.2 On-Board Computer

The on-board computer was installed in all 1983 and later models.

The major components of the on-board computer are the electronic computer module with integral digital display, the on-board computer horn, the chime module, and the steering column stalk switch function (1985 and later models).

The on-board computer computes its information using inputs from the following sensors and components: the outside temperature sensor, the speedometer, the fuel gauge and the low fuel warning switch in the fuel tank, and a fuel injection pulse input.

No adjustments can be made to the system. Use the circuit diagram as shown in Fig. 10-2 to help find any faults in the wiring. If no faults can be found, the control module or the coding plug on the rear of the control module may be faulty. If the system is completely inoperative, a fuse may have blown. There are three fuses in the on-board computer circuit: nos. 5, 6, and 12.

NOTE —
If the on-board computer digital readout displays either AAAA or PPPP, the system has detected an electronic fault. If either of these fault codes appear, have the system checked by an authorized BMW dealer.

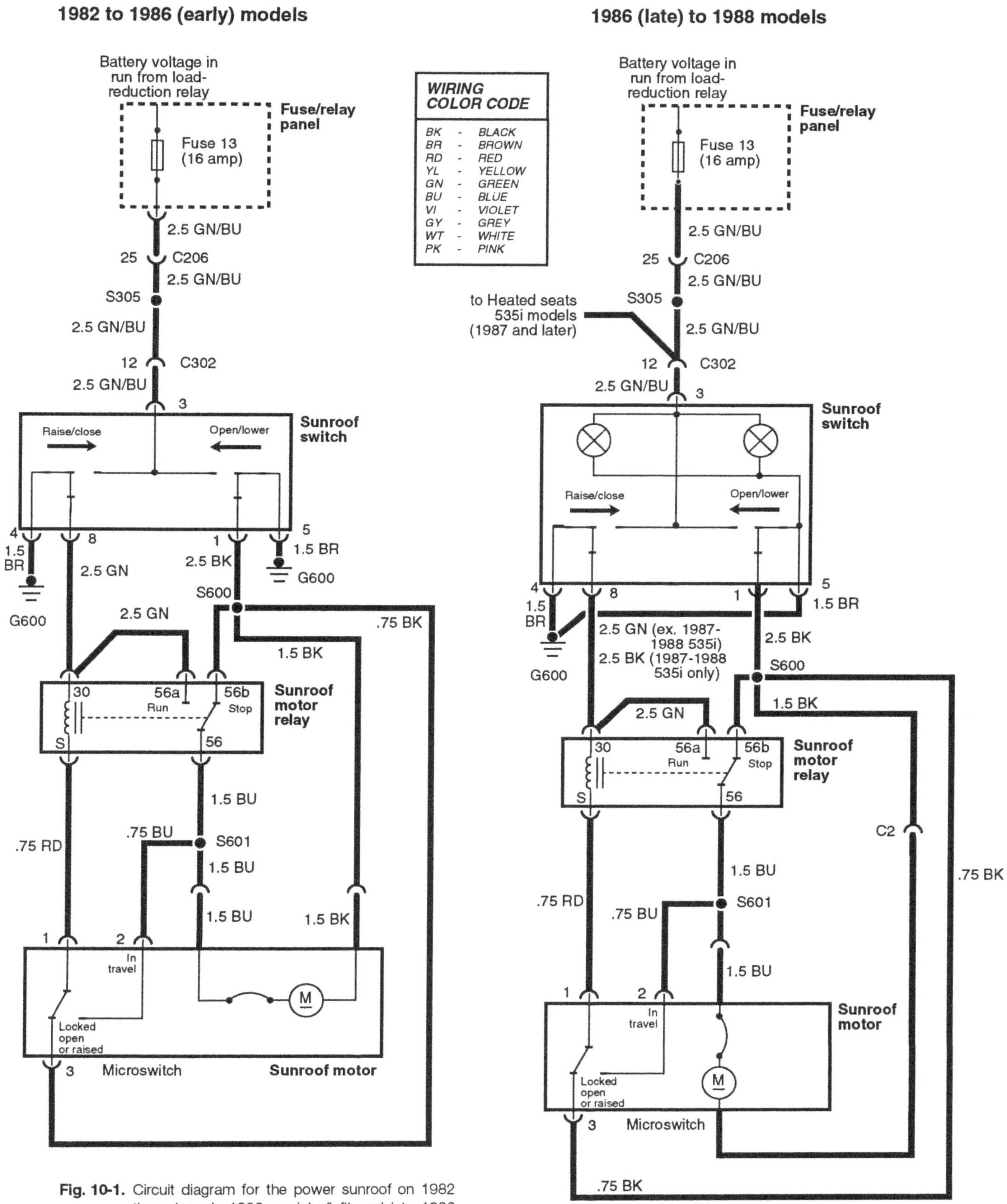

Fig. 10-1. Circuit diagram for the power sunroof on 1982 through early 1986 models (left) and late 1986 through 1988 models (right).

B549:ELE.B

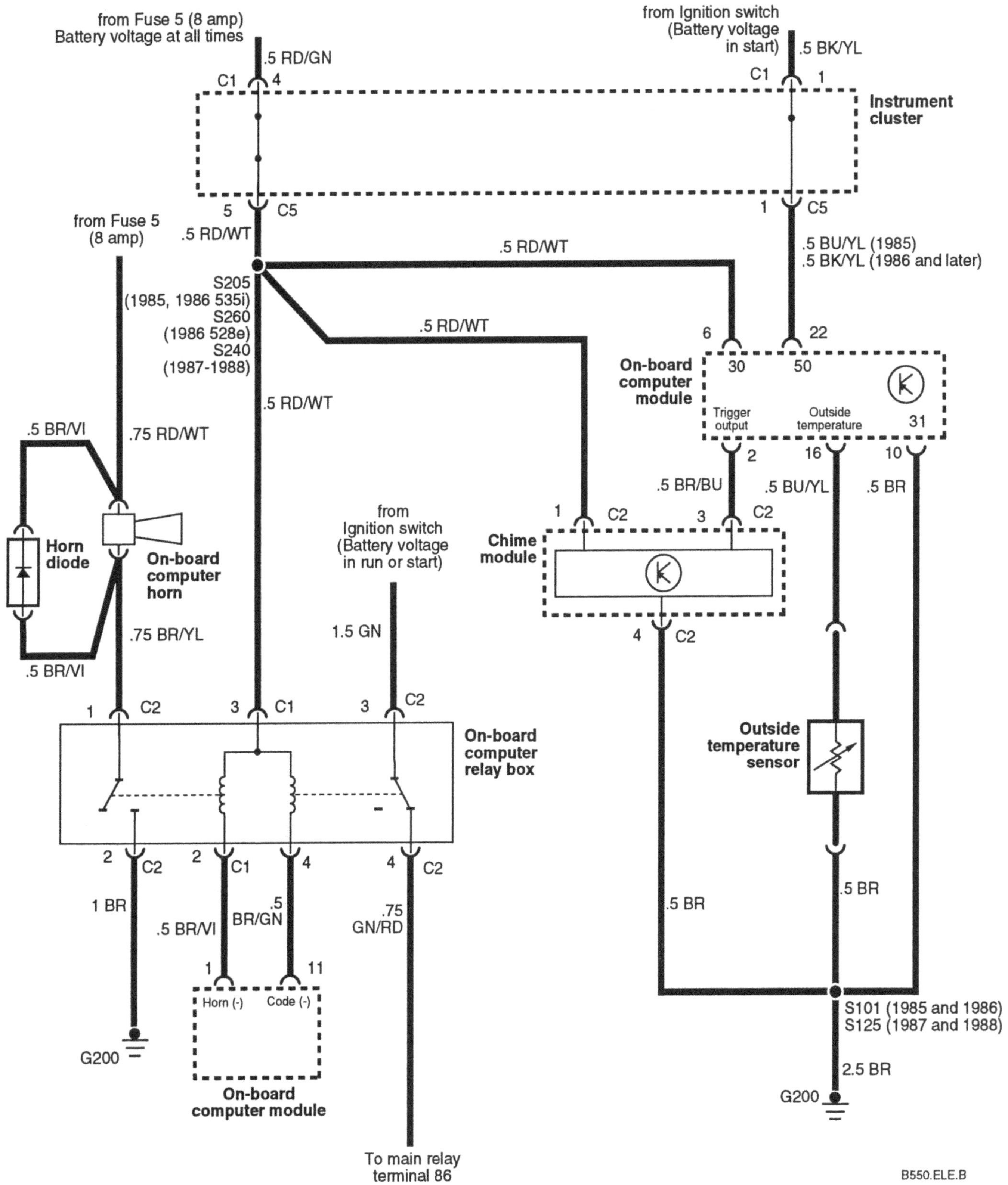

Fig. 10-2. On-board computer circuit for 1985 and later models. 1983 and 1984 models are slightly different.

continued on next page

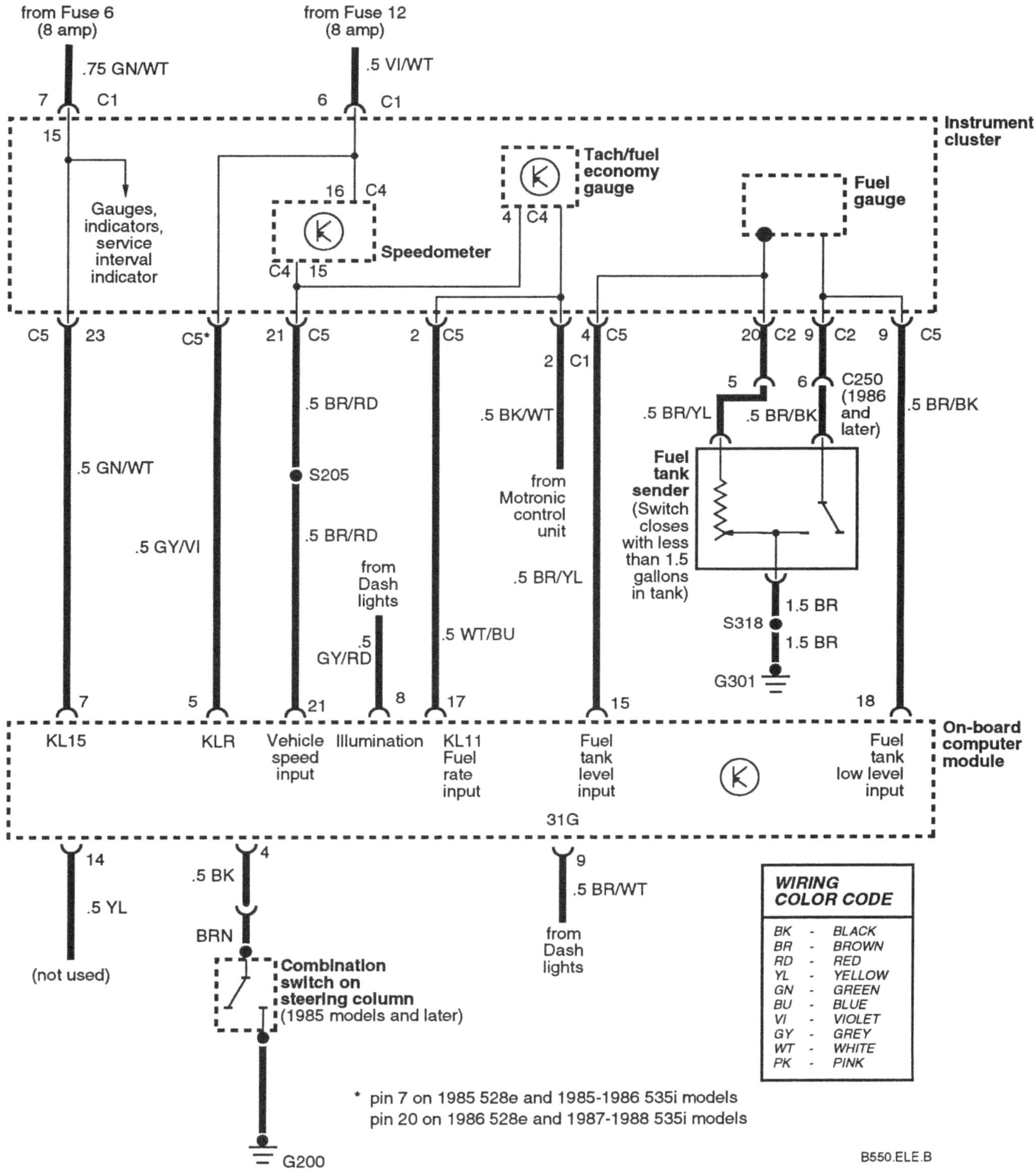

10.3 Active Check Control

Fig. 10-3 is a circuit diagram for the active check control. The active check control uses various relays, switches and an oil level sensor to monitor the lights and engine fluids. Testing procedures for the oil level sensor can be found in **ENGINE—SERVICE AND REPAIR.**

The low beam check relay is located on the side of the fuse/relay panel. The rear light check relay is located in the center of the trunk below the trunk lock motor. The check control panel, above the rear view mirror, can be easily removed by pushing it up and prying it out from the bottom.

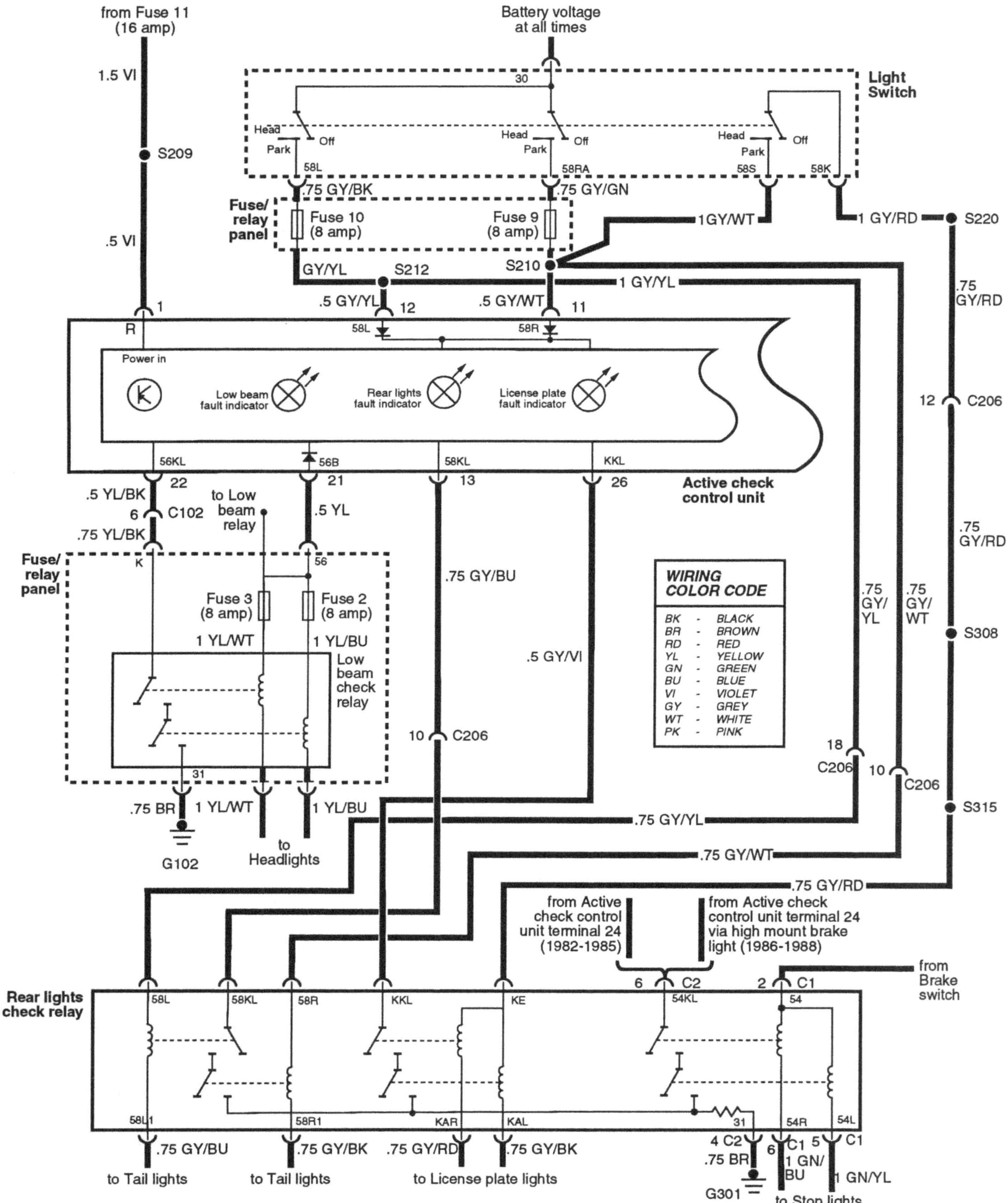

Fig. 10-3. Active check control circuit. Circuit may vary slightly on some models.

continued on next page

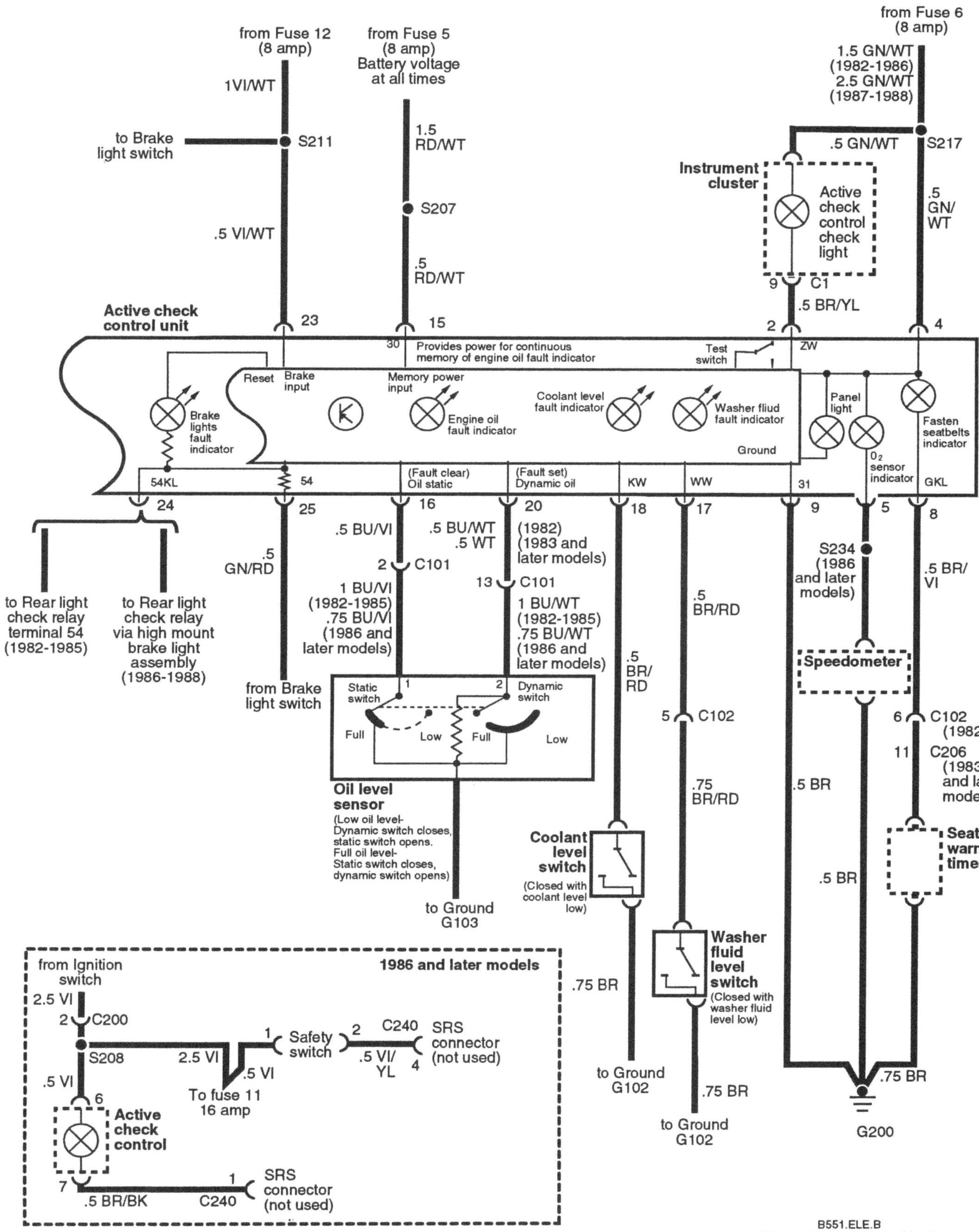

from Fuse 12 (8 amp)
1VI/WT
to Brake light switch
S211
.5 VI/WT
from Fuse 5 (8 amp) Battery voltage at all times
1.5 RD/WT
S207
.5 RD/WT
from Fuse 6 (8 amp)
1.5 GN/WT (1982-1986) 2.5 GN/WT (1987-1988)
.5 GN/WT
S217
Instrument cluster
Active check control check light
.5 GN/WT
C1
.5 BR/YL
Active check control unit
Provides power for continuous memory of engine oil fault indicator
Test switch
ZW
Reset
Brake input
Memory power input
Brake lights fault indicator
Engine oil fault indicator
Coolant level fault indicator
Washer fluid fault indicator
Panel light
Fasten seatbelts indicator
Ground
O2 sensor indicator
54KL
54
(Fault clear) Oil static
(Fault set) Dynamic oil
KW
WW
GKL
to Rear light check relay terminal 54 (1982-1985)
to Rear light check relay via high mount brake light assembly (1986-1988)
.5 GN/RD
from Brake light switch
.5 BU/VI
C101
1 BU/VI (1982-1985) .75 BU/VI (1986 and later models)
.5 BU/WT .5 WT
(1982) (1983 and later models)
1 BU/WT (1982-1985) .75 BU/WT (1986 and later models)
Static switch
Dynamic switch
Full
Low
Oil level sensor
(Low oil level- Dynamic switch closes, static switch opens. Full oil level- Static switch closes, dynamic switch opens)
to Ground G103
.5 BR/RD
Coolant level switch
(Closed with coolant level low)
.75 BR
to Ground G102
.5 BR/RD
C102
.75 BR/RD
Washer fluid level switch
(Closed with washer fluid level low)
.75 BR
to Ground G102
S234 (1986 and later models)
Speedometer
.5 BR
.5 BR/VI
C102 (1982)
C206 (1983 and later models)
Seatbelt warning timer
.75 BR
G200
from Ignition switch
1986 and later models
2.5 VI
C200
S208
2.5 VI
Safety switch
C240
SRS connector (not used)
.5 VI/YL
.5 VI
To fuse 11 16 amp
.5 VI
Active check control
.5 BR/BK
C240
SRS connector (not used)

B551.ELE.B

10.4 Power Windows

The power windows are operated by polarity-reversing motors. Depending on the position of the switches, current is routed to change the rotation direction of the motor. Fuse no. 6 and the power window circuit breaker protect the circuit. The power window relay is located on the relay panel, under the dash, by the hood release lever. The rear window lockout switch prevents operation of the rear windows by interrupting the voltage supply to the rear window switches. Fig. 10-4 shows the power window circuit.

For removal and replacement of the power window motor assemblies, see **BODY AND INTERIOR.** The console-mounted switches and the rear door switches are pressed into place and can be carefully pried out for testing.

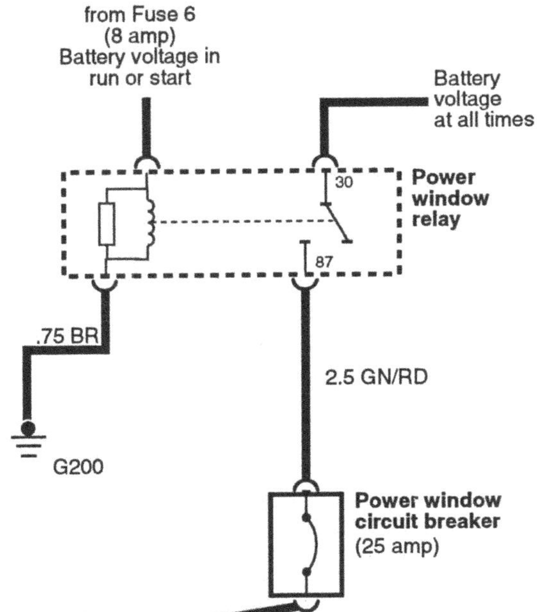

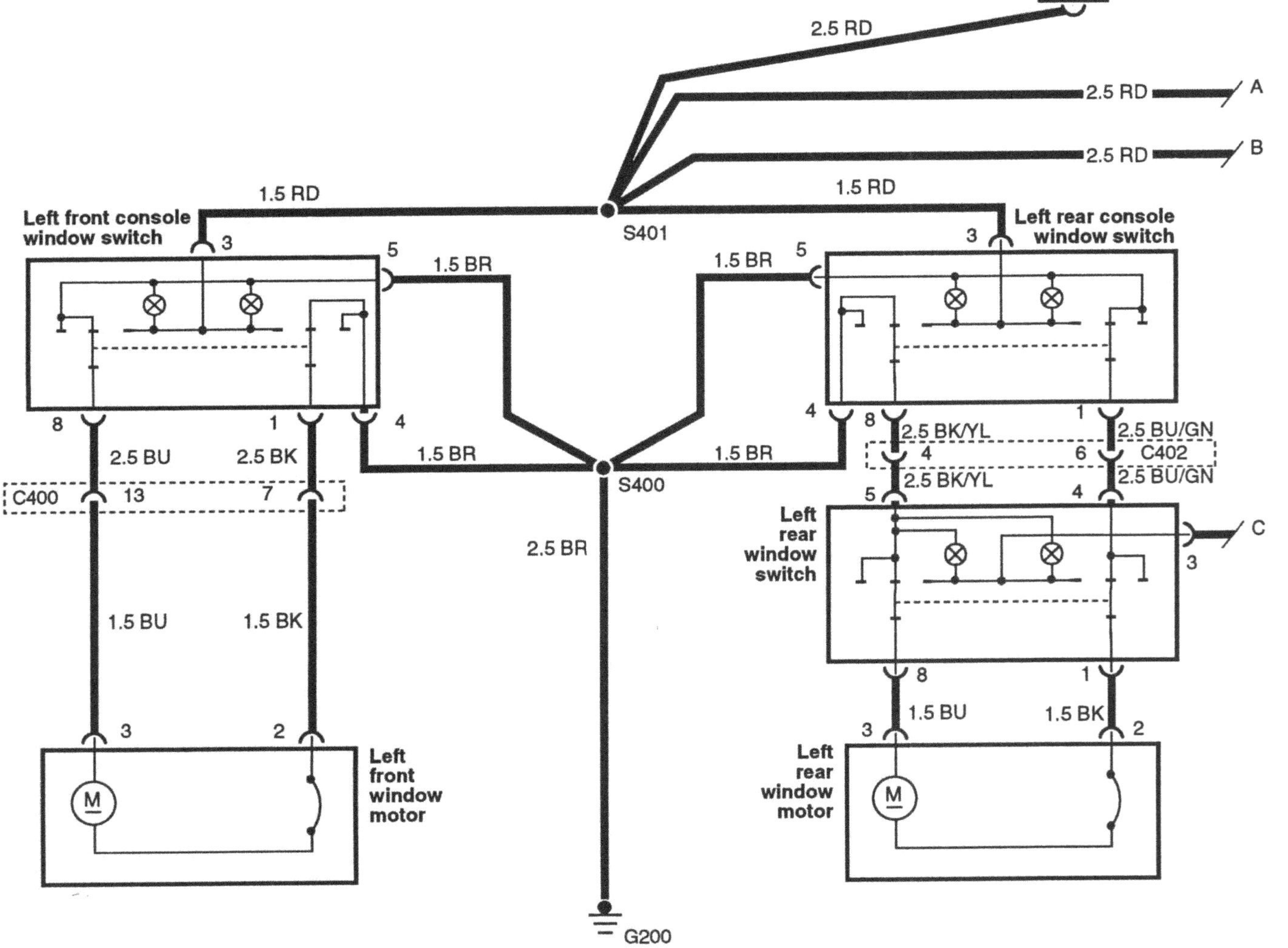

Fig. 10-4. Power window circuit shown with lighted switches.

continued on next page

A single power window that does not operate can be tested without disassembling the door. Remove the rocker switch and disconnect its harness connector. Turn the ignition ON and check for voltage at the connector. Also check ground. If voltage is reaching the switch and there is a good ground, check continuity on the switch. If no faults are found with the switch, the window motor is probably faulty and should be replaced.

10.5 Power Outside Mirrors

The mirror motors are operated by a door-mounted switch. On all models, a single motor controls the mirror's movement via a magnetic clutch. On 1983 and later models, the mirrors include a heating element that turns on at 59°F and off at 122°F. If the mirrors do not work, check for a blown fuse (no.6-8 amp). For removal and replacement of the mirror assemblies or motors, see **BODY AND INTERIOR**. The door-mounted switch is pressed into place and can be carefully pried out for testing. Fig. 10-5 is a circuit diagram for the electric mirrors.

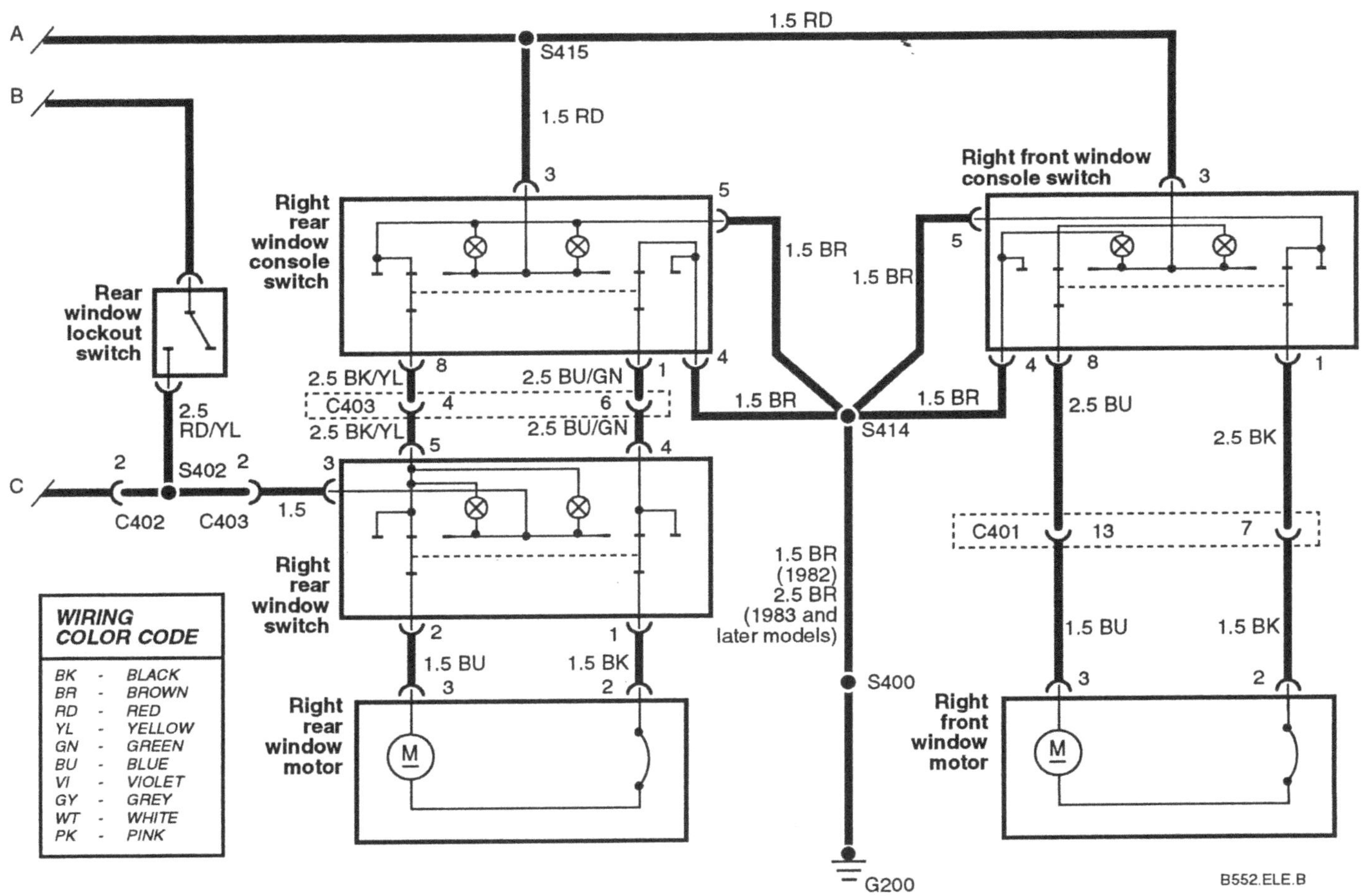

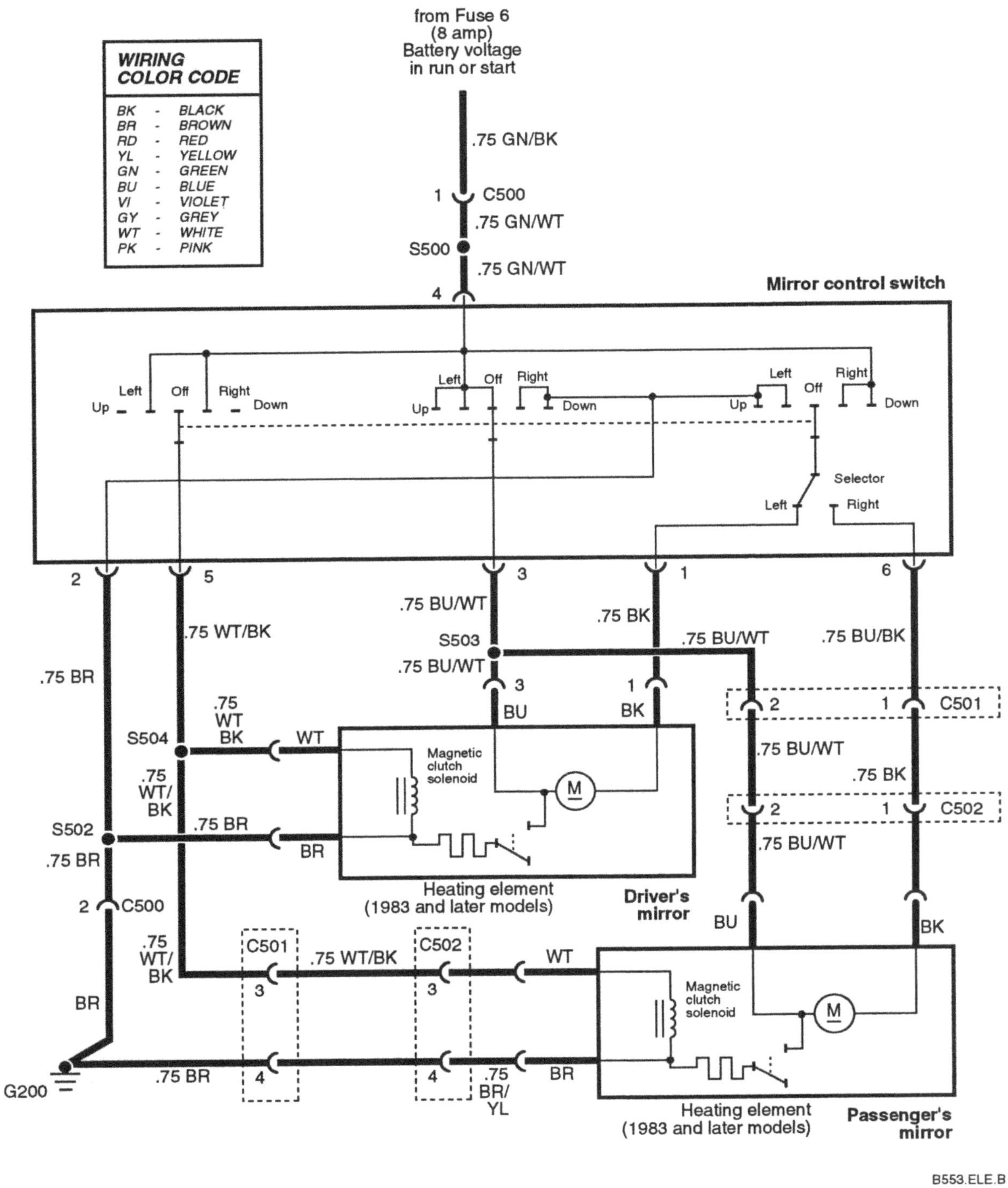

Fig. 10-5. Power mirrors circuit.

10.6 Central Locking System

The electric locking system is controlled by a central electronic control unit that is mounted above the right front speaker in the passenger side foot well. Always check fuse no. 5 (8 amp) before beginning work on the system. The system is actuated by small switches in the front doors and trunk lid lock assemblies. An additional feature allows automatic unlocking of the doors in the event of an accident.

Use Fig. 10-6 as a guide when troubleshooting the central locking system. For removal and replacement of the central locking drive assemblies, see **BODY AND INTERIOR**.

> ***CAUTION —***
> *• The control unit may be damaged if the central locking system is operated in quick succession more than eight times.*
>
> *• Remove and install the control unit only with the battery disconnected. Do not operate the system with the control unit removed from its mounting or the control unit may be damaged.*

> **NOTE —**
> As a general rule, blue wires are used for the locking side of the circuit and white wires are used for the unlocking side of the circuit.

10.7 Cruise Control

The circuit of the cruise control is shown in Fig. 10-7. The cruise control system consists of an electronic control unit, a cruise control servo, a cruise control switch on the steering column, and switches at the brake and clutch pedals.

The control unit, using the signal from the speedometer, compares the car's true speed to the cruise control speed selected by the driver. The control unit then turns on the control motor in the servo to open or close the throttle via a cable. If the driver steps on either the brake pedal or clutch pedal, the cruise control is switched off.

When troubleshooting the cruise control system, first check for a faulty fuse no. 6 (8 amp). Check that the brake light bulbs are operating correctly.

> **NOTE —**
> The cruise control system can not be engaged unless there is ground at pin 9 of the control unit. This ground comes from the chassis through the brake light bulbs. If the brake light bulbs are burned out, the cruise control system will not work.

Check the operation of the clutch switch (models with manual transmission only) and the brake light switch by making continuity checks at the connectors with the pedals depressed and released. Check that the actuating cable is not kinked or damaged.

The control unit is located beneath the dashboard to the left of the steering column. The cruise control servo motor is mounted on the forward part of the driver's side fender in the engine compartment.

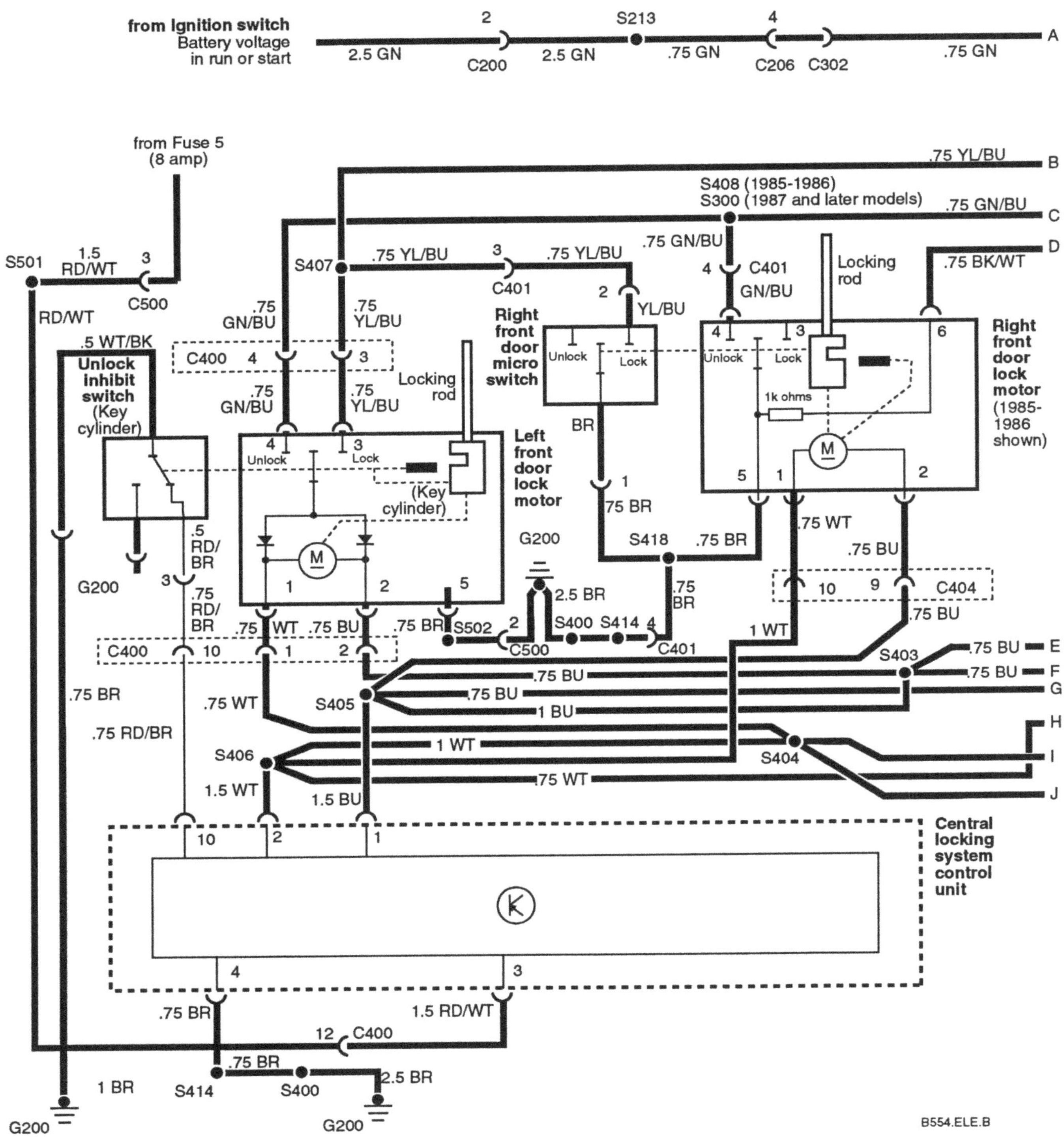

Fig. 10-6. Central locking system. 1982 through 1984 models are slightly different.

continued on next page

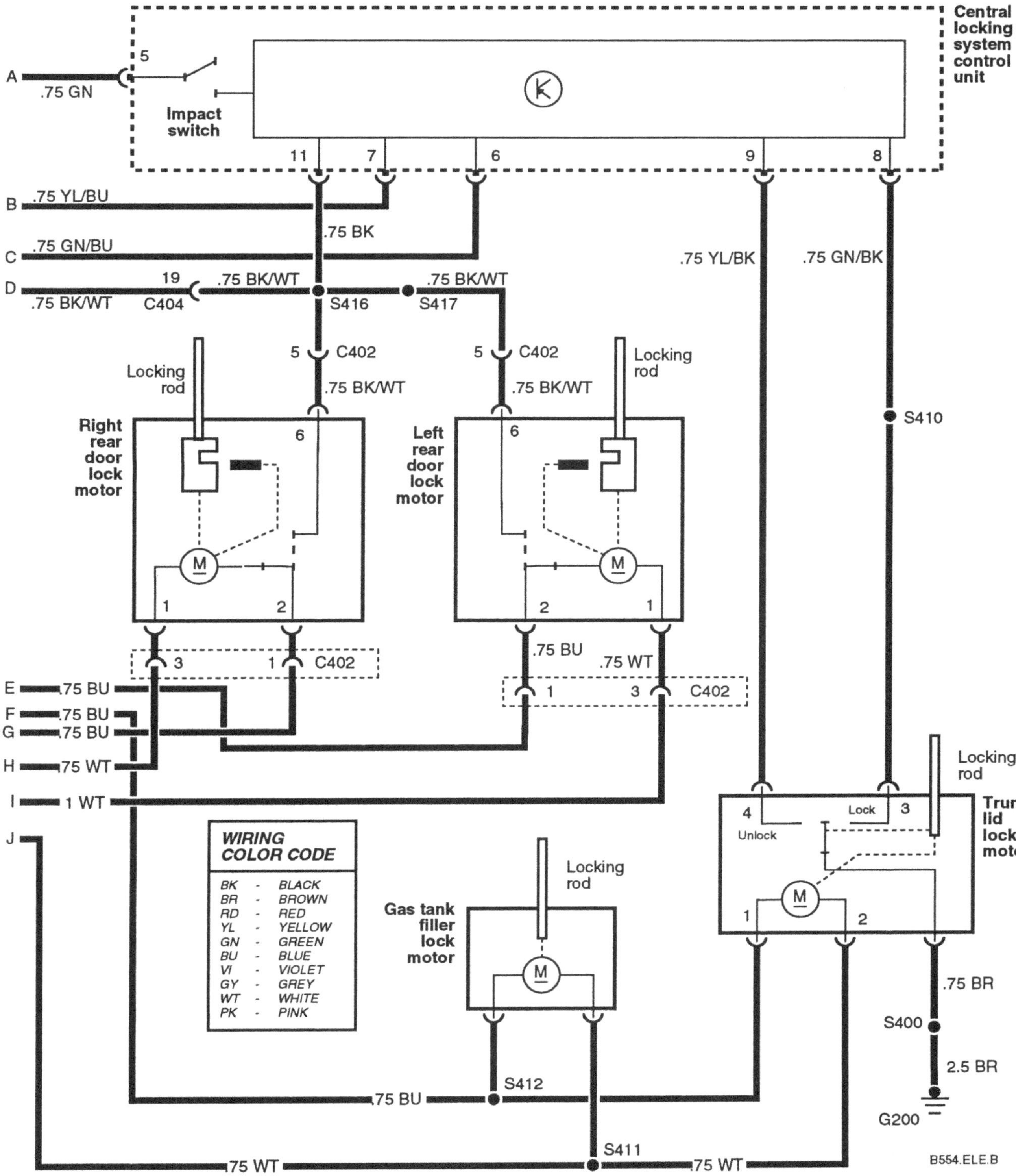
Central locking system control unit
A
.75 GN
5
Impact switch
11
7
6
9
8
B
.75 YL/BU
.75 BK
C
.75 GN/BU
D
19
.75 BK/WT
C404
S416
S417
.75 YL/BK
.75 GN/BK
5
C402
.75 BK/WT
Locking rod
Right rear door lock motor
Left rear door lock motor
6
M
1
2
3
1
.75 BU
.75 WT
E
F
G
H
I
J
1 WT
S410
Trunk lid lock motor
4
Unlock
Lock
Gas tank filler lock motor
.75 BR
S400
2.5 BR
G200
S412
S411
WIRING COLOR CODE
BK - BLACK
BR - BROWN
RD - RED
YL - YELLOW
GN - GREEN
BU - BLUE
VI - VIOLET
GY - GREY
WT - WHITE
PK - PINK
B554.ELE.B

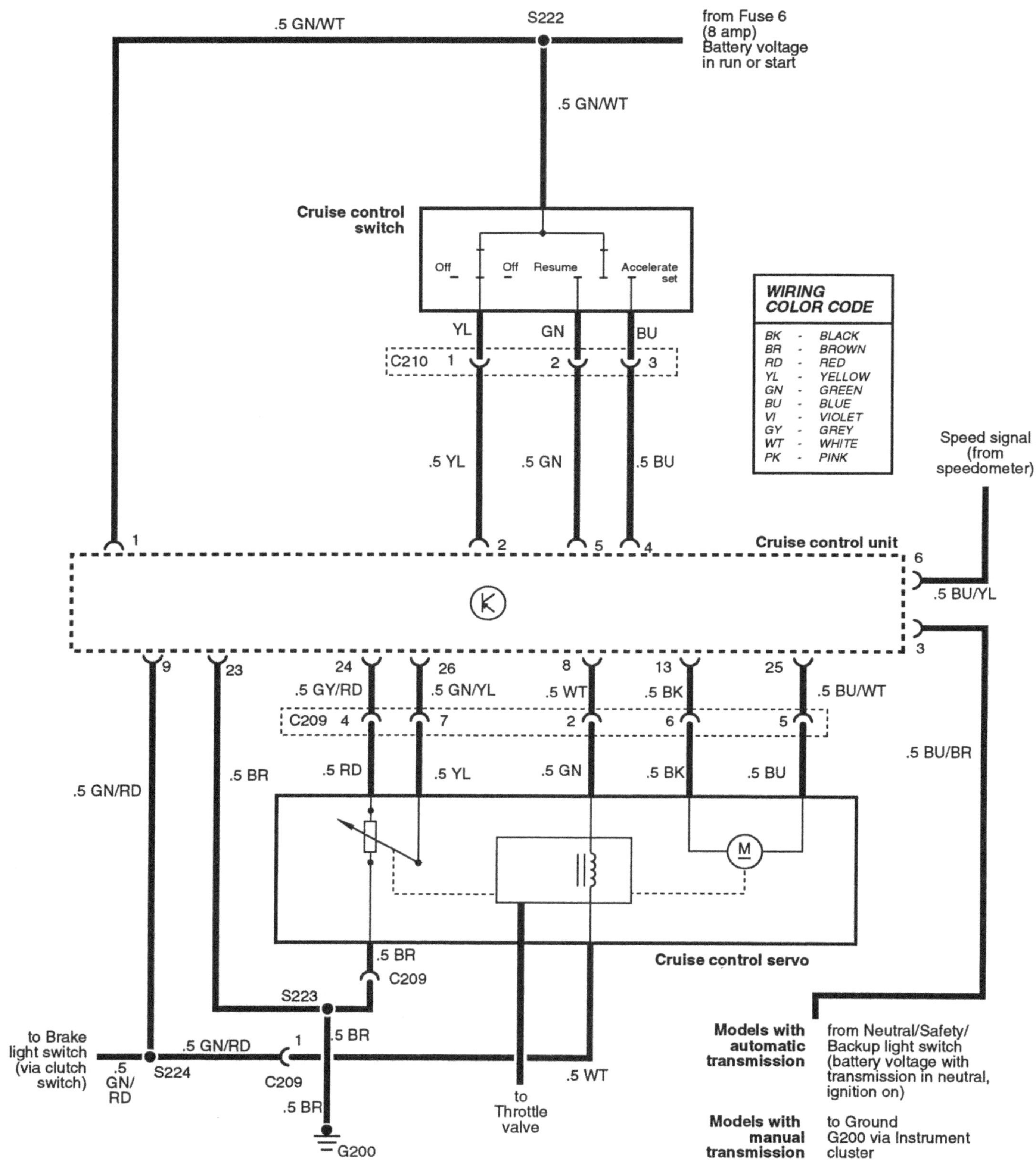

Fig. 10-7. Cruise control circuit.

B555.ELE.B

11. HARNESS CONNECTOR, GROUND, SPLICE, AND COMPONENT LOCATIONS

Table o, **Table p**, **Table q**, **Table r**, and **Table s** list the locations of the various harness connectors, grounds, splices (welded connections in the wiring harness), and various relays and components used throughout the circuit diagrams contained in this section. For the exact positions of the fuses and relays, see **12. Fuse/Relay Panel** below.

Table o. Wiring Harness Connector Locations

Connector	Location
C1	Rear of instrument cluster (blue, 26-pin)
C2	Rear of instrument cluster (white, 26-pin)
C3	Rear of instrument cluster
C4	Rear of instrument cluster
C5	Rear of instrument cluster (yellow, 26-pin)
C101	On side of fuse/relay panel (17pin)
C102	In fuse/relay panel (7-pin)
C103	Right side under dash (6-pin)
C104	Right side under dash (2-pin)
C105	Next to windshield washer bottle (1-pin)
C106	In fuse/relay panel (2-pin)
C107	In engine compartment, left front (2-pin)
C108	In engine compartment, right front (2-pin)
C110	Right front of engine compartment (2-pin)
C113	In front of radiator (3-pin)
C115	Left side of engine compartment near diagnostic connector (3-pin)
C116	Left side of engine compartment near diagnostic connector (3-pin)
C131	Right side of dashboard near Motronic control unit (1-pin)
C132	Right side of dashboard near Motronic control unit (1-pin)
C140	Right side of engine compartment at rear (4-pin)
C150	Left side of engine compartment near strut tower (2-pin)
C151	Right side of engine compartment near strut tower (2-pin)
C200	To left of steering column (10-pin)

continued

Table o. Wiring Harness Connector Locations (continued)

Connector	Location
C201	To left of steering column (6-pin)
C202	To left of steering column (13-pin)
C203	At the end of steering column (2-pin)
C204	Left side of heater/evaporator unit (13-pin)
C206	Below left side of dash on connector bracket (29-pin)
C208	Near clutch pedal (2-pin)
C209	Under left side of dash (7-pin)
C210	On left side of steering column (4-pin)
C212	Under left side of dash (1-pin)
C213	Under left side of dash (1-pin)
C214	In center of dash near heater blower (6-pin)
C215	Behind radio (2-pin)
C216	Behind radio (2-pin)
C217	Behind radio (2-pin)
C219	Near left shock tower (3-pin)
C220	Under left side of dash (1-pin)
C223	Behind radio (1-pin)
C235	Under left side of dash (1-pin)
C241	Right side of trunk (2-pin)
C245	Left side of steering column (1-pin)
C250	Below rear seat— center (13-pin)
C300	Near trunk light (2-pin)
C301	Center console near shifter (2-pin)
C302	Accessory connector
C303	Above headliner near dome light (2-pin)
C351	Below rear seat— right side (2-pin)
C352	Below rear seat— left side (2-pin)
C400	Behind left front speaker (13-pin)
C401	Behind right front speaker (13-pin)
C402	In left side B pillar (7-pin)
C403	In right side B pillar (7-pin)
C404	In left front door (5-pin)

continued on next page

Table o. Wiring Harness Connector Locations (continued)

Connector	Location
C405	In trunk near lock (2-pin)
C500	Near left front speaker (6-pin)
C501	Near left front speaker (4-pin)
C502	Near right front speaker (4-pin)
C503	In left front door (8-pin)
C550	Below center console (2-pin)
C551	Below center console (2-pin)
C560	In left front door (2-pin)

Table p. Wiring Harness Ground Locations

Ground point	Location
G102	Main ground above battery post
G103 (1982-1985 528e)	Main engine ground, on front of engine under diagnostic connector
G103 (1986-1988 528e)	Main engine ground, on rear side of engine above starter
G103 (533i, 535i)	Main engine ground, on rear top of engine
G104	Lower right side of engine
G200	Above brake pedal cluster
G201	At top of steering column
G301	Under rear seat— left side
G302	Right side of trunk
G600	Near sunroof switch

Table q. Wiring Harness Splice Locations

Splice (welded connection in wiring harness)	Harness or approximate location
S100	Main harness, front left corner in engine compartment
S101	Main harness, front right in engine compartment
S102	Main harness, front of engine compartment
S103	Right front corner, engine compartment
S104	Front of engine compartment compartment

continued

Table q. Wiring Harness Splice Locations (continued)

Splice (welded connection in wiring harness)	Harness or approximate location
S105	Engine harness
S106	Engine compartment, near firewall
S107	Engine compartment, near firewall
S109	Engine harness
S110	Engine harness
S111	Engine compartment, near firewall
S112	Engine compartment, near firewall
S113	Engine compartment, center left side
S114	Engine compartment, left side
S115	Engine compartment, near firewall
S116	Engine compartment, near firewall
S117	Engine compartment, left side
S118	Heated washer jets
S119	Heated washer jets
S120	Front of engine compartment
S121	Near transmission
S130	Engine compartment, near firewall
S131	Engine compartment, near firewall
S200	Near fuse/relay panel
S201	Near fuse/relay panel
S202	Engine compartment, near firewall
S203	Engine compartment, near firewall
S204	Engine compartment, near firewall
S205	Engine compartment, near firewall
S206	Under left side of dash
S207	Under left side of dash
S208	Under left side of dash
S209	Under left side of dash
S210	Under left side of dash
S211	Under left side of dash
S212	Under left side of dash
S213	Under left side of dash
S214	Under left side of dash
S215	Under left side of dash
S216	Under left side of dash
S217	Under left side of dash
S218	Under left side of dash

continued on next page

Table q. Wiring Harness Splice Locations (continued)

Splice (welded connection in wiring harness)	Harness or approximate location
S219	Under left side of dash
S220	Under left center of dash
S221	Under center side of dash
S222	Cruise control, left side of dash
S223	Cruise control, left side of dash
S224	Cruise control, left side of dash
S225	Center of dash
S226	Center of dash
S227	A/C harness, right side of dash
S228	A/C harness, right side of dash
S229	A/C harness, right side of dash
S231	Under left side of dash
S232	Power seats
S233	Power seats
S234	Power seats
S235	Center, near console
S240	Under left side of dash
S300	Rear harness, under left dash
S303	Rear harness, near center console
S304	Rear harness, under left dash
S305	Rear harness, under left dash
S306	Rear harness, under left dash
S307	Rear harness, under left dash
S308	Rear harness, under left dash
S309	Rear harness, under left dash
S310	Rear harness, near left A pillar
S311	Rear harness, near left A pillar
S312	Rear harness, near left A pillar
S313	Rear harness, near left A pillar
S314	Rear harness, near left A pillar
S315	Rear harness, left side of trunk
S316	Rear harness, left side of trunk
S317	Rear harness, left side of trunk
S318	Rear harness, under rear seat
S319	Rear harness, under rear seat
S320	Rear harness, under rear seat

continued

Table q. Wiring Harness Splice Locations (continued)

Splice (welded connection in wiring harness)	Harness or approximate location
S350	Rear harness, near left A pillar
S400	Central locking, near center console
S401	Central locking, near center console
S402	Central locking, near center console
S403	Central locking, under left dash
S404	Central locking, under left dash
S405	Central locking, near right A pillar
S406	Central locking, near right A pillar
S407	Central locking, near right A pillar
S408	Central locking, near right A pillar
S409	Central locking
S410	Central locking, near left A pillar
S411	Central locking, in trunk
S412	Central locking, in trunk
S414	Central locking, near center console
S415	Central locking, near center console
S416	Central locking, near right A pillar
S417	Central locking, near left A pillar
S418	Door harness, right front door
S420	Radio
S500	Door, driver's side
S501	Door, driver's side
S502	Door, driver's side
S503	Door, driver's side
S504	Door, driver's side
S540	Heated seats
S541	Heated seats
S552	Heated seats
S553	Heated seats
S600	Sunroof
S601	Sunroof
S700 (except 1988 528e)	ABS, near right A pillar
S700 (1988 528e)	Under right side of dash
S701	ABS, near right A pillar
S703 (except 1988 528e)	ABS
S703 (1988 528e)	Under right side of dash

22

Table r. Relay Locations

Relay	Location
Fuel pump	Right side of fuse/relay panel See **12. Fuse/Relay Panel**
Fog light relay	In fuse/relay panel
High beam relay	In fuse/relay panel
Horn relay	In fuse/relay panel
Load reduction relay	In fuse/relay panel
Low beam relay	In fuse/relay panel
Low beam check relay	Left front side of fuse/relay panel
On-board computer relay	To left side of steering column
Rear lights check relay	Luggage compartment, near trunk lock motor
Start relay (automatic transmission only)	Upper left corner of driver's footwell
Sunroof motor relay	In windshield header
Wiper control unit relay	In fuse/relay panel

Table s. Other Electrical Component Locations

Other Components	Location
ABS control unit	Beneath passenger's side of instrument panel
Active check control unit	In windshield header
Back-up light switch	On transmission side
Brake light switch	Above brake pedal
Central locking system control unit	In passenger's footwell, above speaker
Chime module	Beneath driver's side of instrument panel, next to cruise control unit

continued

Table s. Other Electrical Component Locations (continued)

Other Components	Location
Clutch switch	Above clutch pedal
Coolant level switch	In coolant expansion tank
Cruise control unit	Beneath instrument panel, to left of steering column
Cruise control servo	In engine compartment, in front of left shock tower
Flasher	In steering column, above lower steering column trim
Fuel tank sender	Beneath access panel in trunk
Heater blower motor	Behind firewall trim panel
Interior light timer control	In driver's footwell
Neutral/park/back up light switch	In center console, at base of shift lever
Oil level sensor	In oil pan, left side of engine
On-board computer horn and diode	Behind battery
On-board computer module	Below instrument panel
Seat belt warning timer	Beneath driver's side of instrument panel, near hood release
Starter	Left side of engine, rear
Sunroof motor	In windshield header
Windshield washer fluid level switch	In washer fluid reservoir in engine compartment
Windshield washer pump	In washer fluid reservoir in engine compartment
Wiper motor	Rear of engine compartment, behind firewall panel
Horns	Behind bumper
Horn brush/slip ring assembly	Beneath steering wheel on steering column

12. FUSE/RELAY PANEL

The fuses and relays are arranged together in one unit located in the engine compartment on the driver's side of the car.

Relays

Fig. 12-1 shows the fuse/relay panel location and identifies the relays.

1. Main relay (ex. 1985—1986 528e)
2. 1982—1984 528e/533i Purge valve relay
   1985—1986 528e Main relay
   1985—1986 535i Fuel pump relay
   1987—1988 528e/535i Fuel pump relay
3. 1982—1986 528e/533i Fuel pump relay
   1987 528e Accessory fuse box
   1988 528e Oxygen sensor heater relay
   1985—1988 535i Oxygen sensor relay
4. High beam relay
5. Low beam relay
6. Fog light relay
7. Wiper control unit
8. Horn relay
9. Load reduction relay
10. Low beam check relay
11. Low speed radiator cooling fan relay
12. High speed radiator cooling fan relay
13. 1988 528e Accessory fuse box

Fig. 12-1. Fuse/relay panel models covered by this manual.

NOTE —
- Relay locations are subject to change, and may vary from car to car, depending on options. If questions arise, please remember that an authorized BMW dealer is the best source for the most accurate and up-to-date information.

- Some relays incorporate several electrical functions. Although the relay will plug into the panel, do not substitute one relay for another.

Fuses

The fuses come in different colors that correspond to different current ratings. Each fuse is specifically chosen to protect its circuit against excess current flow that might damage the circuit components. When replacing fuses, it is never appropriate to substitute a fuse of a higher rating. Fig. 12-2 shows fuse locations. **Table t** identifies the fuse circuits and the correct fuse rating.

Ceramic type fuse rating and color	
• 8 amp	white
• 16 amp	red
• 25 amp	blue

CAUTION —
- *Only replace fuses with those of the same rating. Installing a fuse with higher rating will cause severe damage to the car's wiring and may also start a fire.*

- *A fuse with a lower rating may consistently blow, especially when the circuit is turned on. If the fuse in one particular location fails repeatedly, that is an indication of a problem in the circuit or in a component that should be repaired.*

NOTE —
Fuse designations and locations are subject to change, and may vary from car to car, depending on options. If questions arise, please remember that an authorized BMW dealer is the best source for the most accurate and up-to-date information.

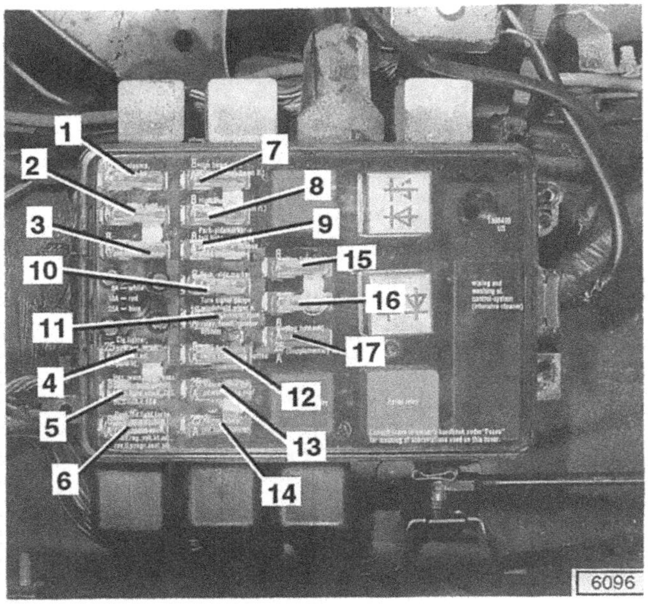

Fig. 12-1. Fuse/relay panel showing fuse locations.

Table t. Fuse Location and Designation

Fuse	Rating and color	Description
1	16 amp (red)	Fuel pumps
2	8 amp (white)	Headlight, right dual beam
3	8 amp (white)	Headlight, left dual beam
4	25 amp (blue)	Cigar lighter, power seats, power antenna (also see 5, 12)
5	8 amp (white)	Active check (also see fuses 6, 9, 10, 11), central locking, cruise control (also see fuse 6) (also see fuse 6), electronic transmission control, glove box, heated locks, seatbelt warning, interior lights (also see fuse 6), turn/warning lights (also see fuse 11), trunk, on board computer (also see fuse 6, 12), radio power antenna (also see fuse 4), service interval (also see fuse 6), instrument cluster (also see fuse 6)

continued

Table t. Fuse Location and Designation (continued)

Fuse	Rating and color	Description
6	8 amp (white)	Active check (also see fuses 5, 9, 10, 11) cruise control (also see fuse 5) back-up lights, transmission indicator lights instrument gauges, heater and air conditioning (also see fuse 14) fuel control, ignition, interior lights (also see fuse 5) on board computer (also see fuse 5, 12), power mirrors, power windows (also see circuit breaker) seatbelt warning (also see fuse 5), service interval, instrument cluster (also see fuse 5)
7	8 amp (white)	Headlight, right high beam
8	8 amp (white)	Headlight, left high beam
9	8 amp (white)	Active check (also see fuse 10, 11, 12), instrument lights (also see fuse 6, 14) front park, front side marker, taillights (also see fuse 10) rear marker and license lights
10	8 amp (white)	Front park, front side marker, taillights (also see fuse 9)
11	16 amp (red)	Active check (also see fuse 5, 6, 12), turn/warning lights, horn wiper/washer and heated jets
12	8 amp (white)	Active check (also see fuse 5, 6, 11), ABS, radio and power antenna (also see fuse 4, 5) brake lights, on board computer (also see fuse 5, 6),
13	16 amp (red)	Rear defogger, sunroof
14	25 amp (blue)	Auxiliary cooling fan (also see fuse 17), dash lights, heating and air conditioning (also see fuse 6)
15	8 amp (white)	Right fog light
16	8 amp (white)	Left fog light
17	25 amp (blue)	Auxiliary cooling fan (also see fuse 14)
Power window circuit breaker	25 amp	Power windows

INDEX

Subjects are indexed by section number in bold, followed by the page number(s) within the section where the subject can be found.
For example **1**:4 refers to section **1** FUNDAMENTALS, page 4.
Sections are as follows:

1	FUNDAMENTALS	**12**	MANUAL TRANSMISSION AND CLUTCH
2	LUBRICATION AND MAINTENANCE	**13**	MANUAL TRANSMISSION OVERHAUL
3	ENGINE MANAGEMENT—DRIVEABILITY	**14**	AUTOMATIC TRANSMISSION
4	ENGINE—SERVICE AND REPAIR	**15**	DRIVESHAFT AND FINAL DRIVE
5	ENGINE—RECONDITIONING	**16**	SUSPENSION—FRONT
6	FUEL SUPPLY	**17**	SUSPENSION—REAR
7	FUEL INJECTION	**18**	STEERING AND WHEEL ALIGNMENT
8	IGNITION	**19**	BRAKES
9	BATTERY, STARTER, ALTERNATOR	**20**	BODY AND INTERIOR
10	COOLING SYSTEM	**21**	HEATING AND AIR CONDITIONING
11	EXHAUST SYSTEM	**22**	ELECTRICAL SYSTEM

WARNING —
- *Automotive service and repair is serious business. You must be alert, use common sense, and exercise good judgement to prevent personal injury and complete the work safely.*
- *Before beginning any work on your vehicle, thoroughly read all the Cautions and Warnings listed near the front of this manual.*
- *Always read the complete procedure before you begin the work. Pay special attention to any Cautions and Warnings that accompany that procedure, or other information on a specific topic.*

INDEX

Subjects are indexed by section number in bold, followed by the page number(s) within the section where the subject can be found.
For example **1**:4 refers to section **1** FUNDAMENTALS, page 4.
Sections are as follows:

1 FUNDAMENTALS	**12** MANUAL TRANSMISSION AND CLUTCH
2 LUBRICATION AND MAINTENANCE	**13** MANUAL TRANSMISSION OVERHAUL
3 ENGINE MANAGEMENT—DRIVEABILITY	**14** AUTOMATIC TRANSMISSION
4 ENGINE—SERVICE AND REPAIR	**15** DRIVESHAFT AND FINAL DRIVE
5 ENGINE—RECONDITIONING	**16** SUSPENSION—FRONT
6 FUEL SUPPLY	**17** SUSPENSION—REAR
7 FUEL INJECTION	**18** STEERING AND WHEEL ALIGNMENT
8 IGNITION	**19** BRAKES
9 BATTERY, STARTER, ALTERNATOR	**20** BODY AND INTERIOR
10 COOLING SYSTEM	**21** HEATING AND AIR CONDITIONING
11 EXHAUST SYSTEM	**22** ELECTRICAL SYSTEM

WARNING —
- *Automotive service and repair is serious business. You must be alert, use common sense, and exercise good judgement to prevent personal injury and complete the work safely.*
- *Before beginning any work on your vehicle, thoroughly read all the Cautions and Warnings listed near the front of this manual.*
- *Always read the complete procedure before you begin the work. Pay special attention to any Cautions and Warnings that accompany that procedure, or other information on a specific topic.*

INDEX

Subjects are indexed by section number in bold, followed by the page number(s) within the section where the subject can be found.
For example **1**:4 refers to section **1** FUNDAMENTALS, page 4.
Sections are as follows:

1	FUNDAMENTALS	12	MANUAL TRANSMISSION AND CLUTCH
2	LUBRICATION AND MAINTENANCE	13	MANUAL TRANSMISSION OVERHAUL
3	ENGINE MANAGEMENT—DRIVEABILITY	14	AUTOMATIC TRANSMISSION
4	ENGINE—SERVICE AND REPAIR	15	DRIVESHAFT AND FINAL DRIVE
5	ENGINE—RECONDITIONING	16	SUSPENSION—FRONT
6	FUEL SUPPLY	17	SUSPENSION—REAR
7	FUEL INJECTION	18	STEERING AND WHEEL ALIGNMENT
8	IGNITION	19	BRAKES
9	BATTERY, STARTER, ALTERNATOR	20	BODY AND INTERIOR
10	COOLING SYSTEM	21	HEATING AND AIR CONDITIONING
11	EXHAUST SYSTEM	22	ELECTRICAL SYSTEM

WARNING —
- *Automotive service and repair is serious business. You must be alert, use common sense, and exercise good judgement to prevent personal injury and complete the work safely.*
- *Before beginning any work on your vehicle, thoroughly read all the Cautions and Warnings listed near the front of this manual.*
- *Always read the complete procedure before you begin the work. Pay special attention to any Cautions and Warnings that accompany that procedure, or other information on a specific topic.*

M

O

P

INDEX

Subjects are indexed by section number in bold, followed by the page number(s) within the section where the subject can be found.
For example **1**:4 refers to section **1** FUNDAMENTALS, page 4.
Sections are as follows:

1	FUNDAMENTALS	**12**	MANUAL TRANSMISSION AND CLUTCH
2	LUBRICATION AND MAINTENANCE	**13**	MANUAL TRANSMISSION OVERHAUL
3	ENGINE MANAGEMENT—DRIVEABILITY	**14**	AUTOMATIC TRANSMISSION
4	ENGINE—SERVICE AND REPAIR	**15**	DRIVESHAFT AND FINAL DRIVE
5	ENGINE—RECONDITIONING	**16**	SUSPENSION—FRONT
6	FUEL SUPPLY	**17**	SUSPENSION—REAR
7	FUEL INJECTION	**18**	STEERING AND WHEEL ALIGNMENT
8	IGNITION	**19**	BRAKES
9	BATTERY, STARTER, ALTERNATOR	**20**	BODY AND INTERIOR
10	COOLING SYSTEM	**21**	HEATING AND AIR CONDITIONING
11	EXHAUST SYSTEM	**22**	ELECTRICAL SYSTEM

WARNING —
- *Automotive service and repair is serious business. You must be alert, use common sense, and exercise good judgement to prevent personal injury and complete the work safely.*
- *Before beginning any work on your vehicle, thoroughly read all the Cautions and Warnings listed near the front of this manual.*
- *Always read the complete procedure before you begin the work. Pay special attention to any Cautions and Warnings that accompany that procedure, or other information on a specific topic.*

W, X, Y, Z